Brief Contents

Part I   Newton's Laws

Chapter 1   Concepts of Motion   2
Chapter 2   Kinematics in One Dimension   33
Chapter 3   Vectors and Coordinate Systems   69
Chapter 4   Kinematics in Two Dimensions   85
Chapter 5   Force and Motion   116
Chapter 6   Dynamics I: Motion Along a Line   138
Chapter 7   Newton's Third Law   167
Chapter 8   Dynamics II: Motion in a Plane   191

Part II   Conservation Laws

Chapter 9   Impulse and Momentum   220
Chapter 10   Energy   245
Chapter 11   Work   278

Part III   Applications of Newtonian Mechanics

Chapter 12   Rotation of a Rigid Body   312
Chapter 13   Newton's Theory of Gravity   354
Chapter 14   Oscillations   377
Chapter 15   Fluids and Elasticity   407

Part IV   Thermodynamics

Chapter 16   A Macroscopic Description of Matter   444
Chapter 17   Work, Heat, and the First Law of Thermodynamics   469
Chapter 18   The Micro/Macro Connection   502
Chapter 19   Heat Engines and Refrigerators   526

Part V   Waves and Optics

Chapter 20   Traveling Waves   560
Chapter 21   Superposition   591
Chapter 22   Wave Optics   627
Chapter 23   Ray Optics   655
Chapter 24   Optical Instruments   694

Part VI   Electricity and Magnetism

Chapter 25   Electric Charges and Forces   720
Chapter 26   The Electric Field   750
Chapter 27   Gauss's Law   780
Chapter 28   The Electric Potential   810
Chapter 29   Potential and Field   839
Chapter 30   Current and Resistance   867
Chapter 31   Fundamentals of Circuits   891
Chapter 32   The Magnetic Field   921
Chapter 33   Electromagnetic Induction   962
Chapter 34   Electromagnetic Fields and Waves   1003
Chapter 35   AC Circuits   1033

Part VII   Relativity and Quantum Physics

Chapter 36   Relativity   1060
Chapter 37   The Foundations of Modern Physics   1102
Chapter 38   Quantization   1125
Chapter 39   Wave Functions and Uncertainty   1156
Chapter 40   One-Dimensional Quantum Mechanics   1179
Chapter 41   Atomic Physics   1216
Chapter 42   Nuclear Physics   1248

Appendix A   Mathematics Review   A-1
Appendix B   Periodic Table of Elements   A-4
Appendix C   ActivPhysics OnLine Activities and PhET Simulations   A-5
Answers to Odd-Numbered Problems   A-7

Table of Problem-Solving Strategies

Note for users of the five-volume edition:
Volume 1 (pp. 1–443) includes chapters 1–15.
Volume 2 (pp. 444–559) includes chapters 16–19.
Volume 3 (pp. 560–719) includes chapters 20–24.
Volume 4 (pp. 720–1101) includes chapters 25–36.
Volume 5 (pp. 1102–1279) includes chapters 36–42.

Chapters 37–42 are not in the Standard Edition.

CHAPTER	PROBLEM-SOLVING STRATEGY		PAGE
Chapter 1	1.1	Motion diagrams	14
Chapter 1	1.2	General problem-solving strategy	22
Chapter 2	2.1	Kinematics with constant acceleration	49
Chapter 4	4.1	Projectile motion problems	94
Chapter 6	6.1	Equilibrium problems	139
Chapter 6	6.2	Dynamics problems	142
Chapter 7	7.1	Interacting-objects prcblems	175
Chapter 8	8.1	Circular-motion problems	207
Chapter 9	9.1	Conservation of momentum	230
Chapter 10	10.1	Conservation of mechanical energy	255
Chapter 11	11.1	Solving energy problems	297
Chapter 12	12.1	Rotational dynamics problems	327
Chapter 12	12.2	Static equilibrium problems	330
Chapter 17	17.1	Work in ideal-gas processes	474
Chapter 17	17.2	Calorimetry problems	484
Chapter 19	19.1	Heat-engine problems	535
Chapter 21	21.1	Interference of two waves	613
Chapter 25	25.1	Electrostatic forces and Coulomb's law	733
Chapter 26	26.1	The electric field of multiple point charges	752
Chapter 26	26.2	The electric field of a continuous distribution of charge	758
Chapter 27	27.1	Gauss's law	795
Chapter 28	28.1	Conservation of energy in charge interactions	820
Chapter 28	28.2	The electric potential of a continuous distribution of charge	829
Chapter 31	31.1	Resistor circuits	906
Chapter 32	32.1	The magnetic field of a current	928
Chapter 33	33.1	Electromagnetic induction	976
Chapter 36	36.1	Relativity	1083
Chapter 40	40.1	Quantum-mechanics problems	1184

THIRD EDITION

physics

FOR SCIENTISTS AND ENGINEERS
a strategic approach

VOLUME 4

randall d. knight

California Polytechnic State University
San Luis Obispo

PEARSON

Boston   Columbus   Indianapolis   New York   San Francisco   Upper Saddle River
Amsterdam   Cape Town   Dubai   London   Madrid   Milan   Munich   Paris   Montreal   Toronto
Delhi   Mexico City   Sao Paulo   Sydney   Hong Kong   Seoul   Singapore   Taipei   Tokyo

Publisher:	James Smith
Senior Development Editor:	Alice Houston, Ph.D.
Senior Project Editor:	Martha Steele
Assistant Editor:	Peter Alston
Media Producer:	Kelly Reed
Senior Administrative Assistant:	Cathy Glenn
Director of Marketing:	Christy Lesko
Executive Marketing Manager:	Kerry McGinnis
Managing Editor:	Corinne Benson
Production Project Manager:	Beth Collins
Production Management, Composition, and Interior Design:	Cenveo Publisher Services/Nesbitt Graphics, Inc.
Illustrations:	Rolin Graphics
Cover Design:	Yvo Riezebos Design
Manufacturing Buyer:	Jeff Sargent
Photo Research:	Eric Schrader
Image Lead:	Maya Melenchuk
Cover Printer:	Lehigh-Phoenix
Text Printer and Binder:	R.R. Donnelley/Willard
Cover Image:	Composite illustration by Yvo Riezebos Design
Photo Credits:	See page C-1

Library of Congress Cataloging-in-Publication Data

Knight, Randall Dewey.
Physics for scientists and engineers : a strategic approach / randall d. knight. -- 3rd ed.
  p. cm.
Includes bibliographical references and index.
ISBN 978-0-321-74090-8
1. Physics--Textbooks. I. Title.
QC23.2.K654 2012
530--dc23
2011033849
ISBN-13: 978-0-321-75316-8 ISBN-10: 0-321-75316-X (Volume 4)
ISBN-13: 978-0-321-74090-8 ISBN-10: 0-321-74090-4 (Student Edition)
ISBN-13: 978-0-321-76519-2 ISBN-10: 0-321-76519-2 (Instructor's Review Copy)
ISBN-13: 978-0-132-83212-0 ISBN-10: 0-132-83212-7 (NASTA Edition)

1 2 3 4 5 6 7 8 9 10—DOW—15 14 13 12 11

www.pearsonhighered.com

About the Author

Randy Knight has taught introductory physics for over 30 years at Ohio State University and California Polytechnic University, where he is currently Professor of Physics. Professor Knight received a bachelor's degree in physics from Washington University in St. Louis and a Ph.D. in physics from the University of California, Berkeley. He was a post-doctoral fellow at the Harvard-Smithsonian Center for Astrophysics before joining the faculty at Ohio State University. It was at Ohio State that he began to learn about the research in physics education that, many years later, led to this book.

Professor Knight's research interests are in the field of lasers and spectroscopy, and he has published over 25 research papers. He also directs the environmental studies program at Cal Poly, where, in addition to introductory physics, he teaches classes on energy, oceanography, and environmental issues. When he's not in the classroom or in front of a computer, you can find Randy hiking, sea kayaking, playing the piano, or spending time with his wife Sally and their seven cats.

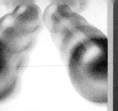

Builds problem-solving skills and confidence...

... through a carefully structured and research-proven program of problem-solving techniques and practice materials.

At the heart of the problem-solving instruction is the consistent 4-step MODEL/ VISUALIZE/ SOLVE/ ASSESS approach, used throughout the book and all supplements. ***Problem-Solving Strategies*** provide detailed guidance for particular topics and categories of problems, often drawing on key skills outlined in the step-by-step procedures of ***Tactics Boxes***. Problem-Solving Strategies and Tactics Boxes are also illustrated in dedicated MasteringPhysics ***Skill-Builder Tutorials***.

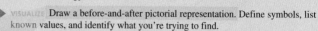

PROBLEM-SOLVING
STRATEGY 10.1 **Conservation of mechanical energy**

1 MODEL Choose a system that is isolated and has no friction or other losses of mechanical energy.

2 VISUALIZE Draw a before-and-after pictorial representation. Define symbols, list known values, and identify what you're trying to find.

3 SOLVE The mathematical representation is based on the law of conservation of mechanical energy:

$$K_f + U_f = K_i + U_i$$

4 ASSESS Check that your result has the correct units, is reasonable, and answers the question.

Exercise 8

TACTICS
BOX 9.1 **Drawing a before-and-after pictorial representation**

EXAMPLE 4.15 **Analyzing rotational data**

You've been assigned the task of measuring the start-up characteristics of a large industrial motor. After several seconds, when the motor has reached full speed, you know that the angular acceleration will be zero, but you hypothesize that the angular acceleration may be constant during the first couple of seconds as the motor speed increases. To find out, you attach a shaft encoder to the 3.0-cm-diameter axle. A shaft encoder is a device that converts the angular position of a shaft or axle to a signal that can be read by a computer. After setting the computer program to read four values a second, you start the motor and acquire the following data:

Time (s)	Angle(°)
0.00	0
0.25	16
0.50	69
0.75	161
1.00	267
1.25	428
1.50	620

a. Do the data support your hypothesis of a constant angular acceleration? If so, what is the angular acceleration? If not, is the angular acceleration increasing or decreasing with time?
b. A 76-cm-diameter blade is attached to the motor shaft. At what time does the acceleration of the tip of the blade reach 10 m/s²?

1 MODEL The axle is rotating with nonuniform circular motion. Model the tip of the blade as a particle.

2 VISUALIZE FIGURE 4.38 shows that the blade tip has both a tangential and a radial acceleration.

$\alpha = 2m$. If the graph is not a straight line, our observation of whether it curves upward or downward will tell us whether the angular acceleration us increasing or decreasing.

FIGURE 4.39 is the graph of θ versus t^2, and it confirms our hypothesis that the motor starts up with constant angular acceleration. The best-fit line, found using a spreadsheet, gives a slope of 274.6°/s². The units come not from the spreadsheet but by looking at the units of rise (°) over run (s² because we're graphing t^2 on the x-axis). Thus the angular acceleration is

$$\alpha = 2m = 549.2°/s^2 \times \frac{\pi \text{ rad}}{180°} = 9.6 \text{ rad/s}^2$$

where we used 180° = π rad to convert to SI units of rad/s².

FIGURE 4.39 Graph of θ versus t^2 for the motor shaft.

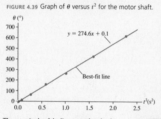

b. The magnitude of the linear acceleration is

$$a = \sqrt{a_r^2 + a_t^2}$$

Worked Examples walk the student carefully through detailed solutions, focusing on underlying reasoning and common pitfalls to avoid.

NEW! ***Data-based Examples*** (shown here) help students with the skill of drawing conclusions from laboratory data.

CHALLENGE EXAMPLE 10.10 **A rebounding pendulum**

A 200 g steel ball hangs on a 1.0-m-long string. The ball is pulled sideways so that the string is at a 45° angle, then released. At the very bottom of its swing the ball strikes a 500 g steel paperweight that is resting on a frictionless table. To what angle does the ball rebound?

NEW! ***Challenge Examples*** illustrate how to integrate multiple concepts and use more sophisticated reasoning.

 NEW! The Mastering Study Area also has ***Video Tutor Solutions***, created by Randy Knight's College Physics co-author Brian Jones. These engaging and helpful videos walk students through a representative problem for each main topic, often starting with a qualitative overview in the context of a lab- or real-world demo.

Promotes deeper understanding…

… using powerful techniques from multimedia learning theory that focus and structure student learning, and improve engagement and retention.

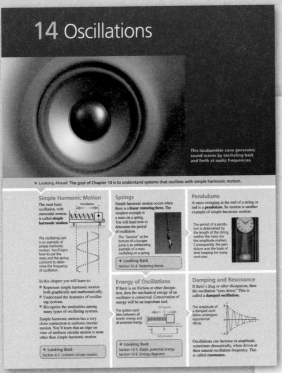

14 Oscillations

This loudspeaker cone generates sound waves by oscillating back and forth at audio frequencies.

▶ Looking Ahead The goal of Chapter 14 is to understand systems that oscillate with simple harmonic motion.

NEW! *Illustrated Chapter Previews* give an overview of the upcoming ideas for each chapter, setting them in context, explaining their utility, and tying them to existing knowledge (through *Looking Back* references).

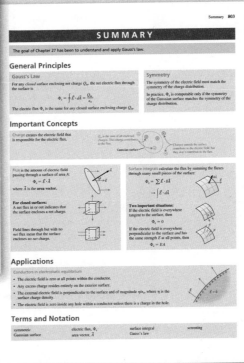

Critically acclaimed *Visual Chapter Summaries* and *Part Knowledge Structures* consolidate understanding by providing key concepts and principles in words, math, and figures and organizing these into a hierarchy.

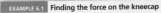

EXAMPLE 6.1 Finding the force on the kneecap

Your kneecap (patella) is attached by a tendon to your quadriceps muscle. This tendon pulls at a 10° angle relative to the femur, the bone of your upper leg. The patella is also attached to your lower leg (tibia) by a tendon that pulls parallel to the leg. To balance these forces, the lower end of your femur pushes outward on the patella. Bending your knee increases the tension in the tendons, and both have a tension of 60 N when the knee is bent to make a 70° angle between the upper and lower leg. What force does the femur exert on the kneecap in this position?

MODEL Model the kneecap as a particle in static equilibrium.

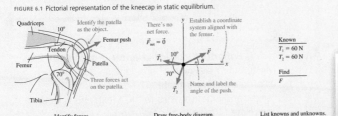

FIGURE 6.1 Pictorial representation of the kneecap in static equilibrium.

NEW! *Life-science and bioengineering examples* provide general interest, and specific context for biosciences students.

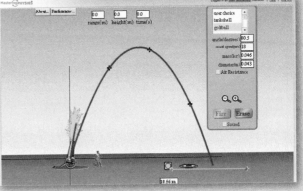

NEW! *PhET Simulations and Tutorials* allow students to explore real-life phenomena and discover the underlying physics. Sixteen tutorials are provided in the MasteringPhysics item library, and 76 PhET simulations are available in the Study Area and Pearson eText, along with the comprehensive library of ActivPhysics applets and applet-based tutorials.

NEW! *Video Tutor Demonstrations* feature "pause-and-predict" demonstrations of key physics concepts and incorporate assessment as the student progresses to actively engage them in understanding the key conceptual ideas underlying the physics principles.

Provides research-enhanced problems...

... extensively class-tested and calibrated using MasteringPhysics data.

Data captured by MasteringPhysics® has been thoroughly analyzed by the author to ensure an optimal range of difficulty (indicated in the textbook using a three-bar rating), problem types, and topic coverage are being met.

56. ‖ A uniform rod of mass M and length L swings as a pendulum on a pivot at distance $L/4$ from one end of the rod. Find an expression for the frequency f of small-angle oscillations.

57. ‖ A solid sphere of mass M and radius R is suspended from a thin rod, as shown in FIGURE P14.57. The sphere can swing back and forth at the bottom of the rod. Find an expression for the frequency f of small-angle oscillations.

FIGURE P14.57

An *increased emphasis on symbolic answers* encourages students to work algebraically.

58. ‖ A geologist needs to determine the local value of g. Unfortunately, his only tools are a meter stick, a saw, and a stopwatch. He starts by hanging the meter stick from one end and measuring its frequency as it swings. He then saws off 20 cm—using the centimeter markings—and measures the frequency again. After two more cuts, these are his data:

Length (cm)	Frequency (Hz)
100	0.61
80	0.67
60	0.79
40	0.96

Use the best-fit line of an appropriate graph to determine the local value of g.

NEW! *Data-based end-of-chapter problems* allow students to practice drawing conclusions from data (as demonstrated in the new data-based examples in the text).

59. ‖ Interestingly, there have been several studies using cadavers
BIO to determine the moments of inertia of human body parts, information that is important in biomechanics. In one study, the center of mass of a 5.0 kg lower leg was found to be 18 cm from the knee. When the leg was allowed to pivot at the knee and swing freely as a pendulum, the oscillation frequency was 1.6 Hz. What

NEW! *BIO problems* are set in life-science, bioengineering, or biomedical contexts.

Electromagnetic Induction and Electromagnetic Waves · CHAPTER 25   **25-7**

15. The graph shows how the magnetic field changes through a rectangular loop of wire with resistance R. Draw a graph of the current in the loop as a function of time. Let a counterclockwise current be positive, a clockwise current be negative.

a. What is the magnetic flux through the loop at $t = 0$?
b. Does this flux *change* between $t = 0$ and $t = t_1$?
c. Is there an induced current in the loop between $t = 0$ and $t = t_1$?
d. What is the magnetic flux through the loop at $t = t_2$?
e. What is the *change* in flux through the loop between t_1 and t_2?
f. What is the time interval between t_1 and t_2?
g. What is the magnitude of the induced emf between t_1 and t_2?
h. What is the magnitude of the induced current between t_1 and t_2?
i. Does the magnetic field point out of or into the loop?
f. Between t_1 and t_2, is the magnetic flux increasing or decreasing?
g. To oppose the *change* in the flux between t_1 and t_2, should the magnetic field of the induced current point out of or into the loop?
h. Is the induced current between t_1 and t_2 positive or negative?
i. Does the flux through the loop change after t_2?
j. Is there an induced current in the loop after t_2?
k. Use all this information to draw a graph of the induced current. Add appropriate labels on the vertical axis.

© 2010 Pearson Education, Inc.

NEW! *Student Workbook exercises* help students work through a full solution symbolically, structured around the relevant textbook Problem-Solving Strategy.

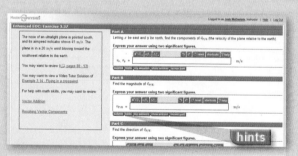

NEW! *Enhanced end-of-chapter problems* in MasteringPhysics now offer additional support such as problem-solving strategy hints, relevant math review and practice, links to the eText, and links to the related *Video Tutor Solution*.

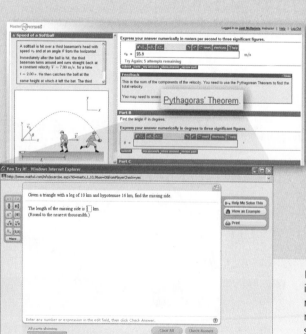

NEW! *Math Remediation* found within selected tutorials provide just-in-time math help and allow students to brush up on the most important mathematical concepts needed to successfully complete assignments. This new feature links students directly to math review and practice helping students make the connection between math and physics.

Make a difference with MasteringPhysics...

... the most effective and widely used online science tutorial, homework, and assessment system available.

MasteringPHYSICS www.masteringphysics.com

Pre-Built Assignments. For every chapter in the book, MasteringPhysics provides pre-built assignments that cover the material with a tested mix of tutorials and end-of-chapter problems of graded difficulty. Professors may use these assignments as-is or take them as a starting point for modification.

NEW! *Quizzing and Testing Enhancements.*
These include options to:
- Hide item titles.
- Add password protection.
- Limit access to completed assignments.
- Randomize question order in an assignment.

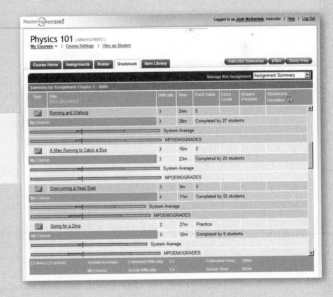

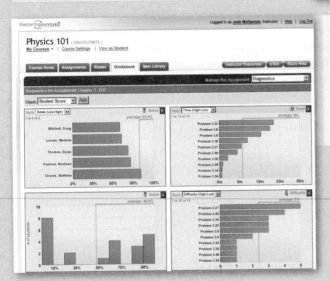

Gradebook

- Every assignment is graded automatically.
- Shades of red highlight vulnerable students and challenging assignments.
- The **Gradebook Diagnostics** screen provides your favorite weekly diagnostics, summarizing grade distribution, improvement in scores over the course, and much more.

Class Performance on Assignment. Click on a problem to see which step your students struggled with most, and even their most common wrong answers. Compare results at every stage with the national average or with your previous class.

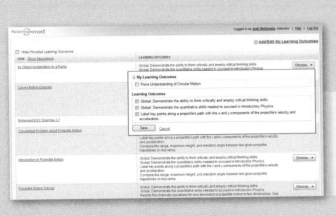

NEW! *Learning Outcomes.* In addition to being able to create your own learning outcomes to associate with questions in an assignment, you can now select content that is tagged to a large number of publisher-provided learning outcomes. You can also print or export student results based on learning outcomes for your own use or to incorporate into reports for your administration.

Preface to the Instructor

In 2003 we published *Physics for Scientists and Engineers: A Strategic Approach*. This was the first comprehensive introductory textbook built from the ground up on research into how students can more effectively learn physics. The development and testing that led to this book had been partially funded by the National Science Foundation. This first edition quickly became the most widely adopted new physics textbook in more than 30 years, meeting widespread critical acclaim from professors and students. For the second edition, and now the third, we have built on the research-proven instructional techniques introduced in the first edition and the extensive feedback from thousands of users to take student learning even further.

Objectives

My primary goals in writing *Physics for Scientists and Engineers: A Strategic Approach* have been:

- To produce a textbook that is more focused and coherent, less encyclopedic.
- To move key results from physics education research into the classroom in a way that allows instructors to use a range of teaching styles.
- To provide a balance of quantitative reasoning and conceptual understanding, with special attention to concepts known to cause student difficulties.
- To develop students' problem-solving skills in a systematic manner.
- To support an active-learning environment.

These goals and the rationale behind them are discussed at length in the *Instructor Guide* and in my small paperback book, *Five Easy Lessons: Strategies for Successful Physics Teaching*. Please request a copy from your local Pearson sales representative if it is of interest to you (ISBN 978-0-8053-8702-5).

FIVE EASY LESSONS

Strategies for Successful Physics Teaching

RANDALL D. KNIGHT

What's New to This Edition

For this third edition, we continue to apply the best results from educational research, and to refine and tailor them for this course and its students. At the same time, the extensive feedback we've received has led to many changes and improvements to the text, the figures, and the end-of-chapter problems. These include:

- New illustrated **Chapter Previews** give a visual overview of the upcoming ideas, set them in context, explain their utility, and tie them to existing knowledge (through **Looking Back** references). These previews build on the cognitive psychology concept of an "advance organizer."
- New **Challenge Examples** illustrate how to integrate multiple concepts and use more sophisticated reasoning in problem-solving, ensuring an optimal range of worked examples for students to study in preparation for homework problems.
- New **Data-based Examples** help students with the skill of drawing conclusions from laboratory data. Designed to supplement lab-based instruction, these examples also help students in general with mathematical reasoning, graphical interpretation, and assessment of results.

End-of-chapter problem enhancements include the following:

- **Data from Mastering Physics® have been thoroughly analyzed** to ensure an optimal range of difficulty, problem types, and topic coverage. In addition, the wording of every problem has been reviewed for clarity. Roughly 20% of the end-of-chapter problems are new or significantly revised.

- **Data-based problems** allow students to practice drawing conclusions from data (as demonstrated in the new data-based examples in the text).
- **An increased emphasis on symbolic answers** encourages students to work algebraically. The *Student Workbook* also contains new exercises to help students work through symbolic solutions.
- **Bio problems** are set in life-science, bioengineering, or biomedical contexts.

Targeted content changes have been carefully implemented throughout the book. These include:

- **Life-science and bioengineering worked examples and applications** focus on the physics of life-science situations in order to serve the needs of life-science students taking a calculus-based physics class.
- **Descriptive text throughout has been streamlined** to focus the presentation and generate a shorter text.
- The chapter on **Modern Optics and Matter Waves** has been re-worked into Chapters 38 and 39 to streamline the coverage of this material.

At the front of the book, you'll find an illustrated walkthrough of the new pedagogical features in this third edition. The *Preface to the Student* demonstrates how all the book's features are designed to help your students.

Textbook Organization

The 42-chapter extended edition (ISBN 978-0-321-73608-6/0-321-73608-7) of *Physics for Scientists and Engineers* is intended for a three-semester course. Most of the 36-chapter standard edition (ISBN 978-0-321-75294-9/0-321-75294-5), ending with relativity, can be covered in two semesters, although the judicious omission of a few chapters will avoid rushing through the material and give students more time to develop their knowledge and skills.

There's a growing sentiment that quantum physics is quickly becoming the province of engineers, not just scientists, and that even a two-semester course should include a reasonable introduction to quantum ideas. The *Instructor Guide* outlines a couple of routes through the book that allow most of the quantum physics chapters to be included in a two-semester course. I've written the book with the hope that an increasing number of instructors will choose one of these routes.

The full textbook is divided into seven parts: Part I: *Newton's Laws*, Part II: *Conservation Laws*, Part III: *Applications of Newtonian Mechanics*, Part IV: *Thermodynamics*, Part V: *Waves and Optics*, Part VI: *Electricity and Magnetism*, and Part VII: *Relativity and Quantum Physics*. Although I recommend covering the parts in this order (see below), doing so is by no means essential. Each topic is self-contained, and Parts III–VI can be rearranged to suit an instructor's needs. To facilitate a reordering of topics, the full text is available in the five individual volumes listed in the margin.

Organization Rationale: Thermodynamics is placed before waves because it is a continuation of ideas from mechanics. The key idea in thermodynamics is energy, and moving from mechanics into thermodynamics allows the uninterrupted development of this important idea. Further, waves introduce students to functions of two variables, and the mathematics of waves is more akin to electricity and magnetism than to mechanics. Thus moving from waves to fields to quantum physics provides a gradual transition of ideas and skills.

The purpose of placing optics with waves is to provide a coherent presentation of wave physics, one of the two pillars of classical physics. Optics as it is presented in introductory physics makes no use of the properties of electromagnetic fields. There's little reason other than historical tradition to delay optics until after E&M.

- **Extended edition,** with modern physics (ISBN 978-0-321-73608-6 / 0-321-73608-7): Chapters 1–42.
- **Standard edition** (ISBN 978-0-321-75294-9 / 0-321-75294-5): Chapters 1–36.
- **Volume 1** (ISBN 978-0-321-75291-8 / 0-321-75291-0) covers mechanics: Chapters 1–15.
- **Volume 2** (ISBN 978-0-321-75318-2 / 0-321-75318-6) covers thermodynamics: Chapters 16–19.
- **Volume 3** (ISBN 978-0-321-75317-5 / 0-321-75317-8) covers waves and optics: Chapters 20–24.
- **Volume 4** (ISBN 978-0-321-75316-8 / 0-321-75316-X) covers electricity and magnetism, plus relativity: Chapters 25–36.
- **Volume 5** (ISBN 978-0-321-75315-1 / 0-321-75315-1) covers relativity and quantum physics: Chapters 36–42.
- **Volumes 1–5** boxed set (ISBN 978-0-321-77265-7 / 0-321-77265-2).

The documented difficulties that students have with optics are difficulties with waves, not difficulties with electricity and magnetism. However, the optics chapters are easily deferred until the end of Part VI for instructors who prefer that ordering of topics.

The Student Workbook

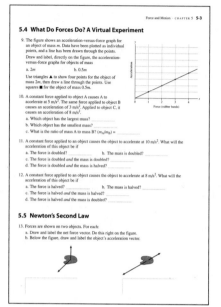

A key component of *Physics for Scientists and Engineers: A Strategic Approach* is the accompanying *Student Workbook*. The workbook bridges the gap between textbook and homework problems by providing students the opportunity to learn and practice skills prior to using those skills in quantitative end-of-chapter problems, much as a musician practices technique separately from performance pieces. The workbook exercises, which are keyed to each section of the textbook, focus on developing specific skills, ranging from identifying forces and drawing free-body diagrams to interpreting wave functions.

The workbook exercises, which are generally qualitative and/or graphical, draw heavily upon the physics education research literature. The exercises deal with issues known to cause student difficulties and employ techniques that have proven to be effective at overcoming those difficulties. The workbook exercises can be used in class as part of an active-learning teaching strategy, in recitation sections, or as assigned homework. More information about effective use of the *Student Workbook* can be found in the *Instructor Guide*.

Available versions: Extended (ISBN 978-0-321-75308-3/0-321-75308-9), Standard (ISBN 978-0-321-75309-0/0-321-75309-7), Volume 1 (ISBN 978-0-321-75314-4/0-321-75314-3), Volume 2 (ISBN 978-0-321-75313-7/0-321-75313-5), Volume 3 (ISBN 978-0-321-75312-0/0-321-75310-0), Volume 4 (ISBN 978-0-321-75311-3/0-321-75311-9), and Volume 5 (ISBN 978-0-321-75310-6/0-321-75310-0).

Instructor Supplements

- The **Instructor Guide for *Physics for Scientists and Engineers*** (ISBN 978-0-321-74765-5/0-321-74765-8) offers detailed comments and suggested teaching ideas for every chapter, an extensive review of what has been learned from physics education research, and guidelines for using active-learning techniques in your classroom. This invaluable guide is available on the Instructor Resource DVD, and via download, either from the MasteringPhysics Instructor Area or from the Instructor Resource Center (www.pearsonhighered.com/educator).

- The **Instructor Solutions** (ISBN 978-0-321-76940-4/0-321-76940-6), written by the author, Professor Larry Smith (Snow College), and Brett Kraabel (Ph.D., University of California, Santa Barbara), provide *complete* solutions to all the end-of-chapter problems. The solutions follow the four-step Model/Visualize/Solve/Assess procedure used in the Problem-Solving Strategies and in all worked examples. The solutions are available by chapter as editable Word® documents and as PDFs for your own use or for posting on your password-protected course website. Also provided are PDFs of handwritten solutions to all of the exercises in the *Student Workbook*, written by Professor James Andrews and Brian Garcar (Youngstown State University). All solutions are available

only via download, either from the MasteringPhysics Instructor Area or from the Instructor Resource Center (www.pearsonhighered.com/educator).

- The cross-platform **Instructor Resource DVD** (ISBN 978-0-321-75456-1/0-321-75456-5) provides a comprehensive library of more than 220 applets from **ActivPhysics OnLine** and 76 **PhET simulations**, as well as all figures, photos, tables, summaries, and key equations from the textbook in JPEG format. In addition, all the Problem-Solving Strategies, Tactics Boxes, and Key Equations are provided in editable Word format. PowerPoint® **Lecture Outlines** with embedded **Classroom Response System "Clicker" Questions** (including reading quizzes) are also provided.

- **MasteringPhysics®** (www.masteringphysics.com) is the most advanced, educationally effective, and widely used physics homework and tutorial system in the world. Eight years in development, it provides instructors with a library of extensively pre-tested end-of-chapter problems and rich, multipart, multistep tutorials that incorporate a wide variety of answer types, wrong answer feedback, individualized help (comprising hints or simpler sub-problems upon request), all driven by the largest metadatabase of student problem-solving in the world. NSF-sponsored published research (and subsequent studies) show that MasteringPhysics has dramatic educational results. MasteringPhysics allows instructors to build wide-ranging homework assignments of just the right difficulty and length and provides them with efficient tools to analyze in unprecedented detail both class trends and the work of any student.

MasteringPhysics routinely provides instant and individualized feedback and guidance to more than 100,000 students every day. A wide range of tools and support make MasteringPhysics fast and easy for instructors and students to learn to use. Extensive class tests show that by the end of their course, an unprecedented nine of ten students recommend MasteringPhysics as their preferred way to study physics and do homework.

For the third edition of *Physics for Scientists and Engineers,* MasteringPhysics now has the following functionalities:

- **Learning Outcomes:** In addition to being able to create their own learning outcomes to associate with questions in an assignment, professors can now select content that is tagged to a large number of publisher-provided learning outcomes. They can also print or export student results based on learning outcomes for their own use or to incorporate into reports for their administration.
- **Quizzing and Testing Enhancements:** These include options to hide item titles, add password protection, limit access to completed assignments, and to randomize question order in an assignment.
- **Math Remediation:** Found within selected tutorials, special links provide just-in-time math help and allow students to brush up on the most important mathematical concepts needed to successfully complete assignments. This new feature links students directly to math review and practice helping students make the connection between math and physics.
- **Enhanced End-of-Chapter Problems:** A subset of homework problems now offer additional support such as problem-solving strategy hints, relevant math review and practice, links to the eText, and links to the related Video Tutor Solution.
- **ActivPhysics OnLine™** (accessed through the Self Study area within www.masteringphysics.com) provides a comprehensive library of more than 220 tried and tested ActivPhysics core applets updated for web delivery using the latest online technologies. In addition, it provides a suite of highly regarded applet-based tutorials developed by education pioneers Alan Van Heuvelen and Paul D'Alessandris.

  The online exercises are designed to encourage students to confront misconceptions, reason qualitatively about physical processes, experiment quantitatively, and learn to think critically. The highly acclaimed ActivPhysics OnLine companion workbooks help students work through complex concepts and understand them more clearly. The applets from the ActivPhysics OnLine library are also available on the Instructor Resource DVD for this text.
- The **Test Bank** (ISBN 978-0-321-74766-2/0-321-74766-6) contains more than 2,000 high-quality problems, with a range of multiple-choice, true/false, short-answer, and regular homework-type questions. Test files are provided both in TestGen (an easy-to-use, fully networkable program for creating and editing quizzes and exams) and Word format. They are available only via download, either from the MasteringPhysics Instructor Area or from the Instructor Resource Center (www.pearsonhighered.com/educator).

Student Supplements

- The **Student Solutions Manuals Chapters 1–19** (ISBN 978-0-321-74767-9/0-321-74767-4) and **Chapters 20–42** (ISBN 978-0-321-77269-5/0-321-77269-5), written by the author, Professor Larry Smith (Snow College), and Brett Kraabel (Ph.D., University of California, Santa Barbara), provide *detailed* solutions to more than half of the odd-numbered end-of-chapter problems. The solutions follow the four-step Model/Visualize/Solve/Assess procedure used in the Problem-Solving Strategies and in all worked examples.
- **MasteringPhysics®** (www.masteringphysics.com) is a homework, tutorial, and assessment system based on years of research into how students work physics problems and precisely where they need help. Studies show that students who use MasteringPhysics significantly increase their scores compared to handwritten homework. MasteringPhysics achieves this improvement by providing students with instantaneous feedback specific to their wrong answers, simpler subproblems upon request when they get stuck, and partial credit for their method(s). This individualized, 24/7 Socratic tutoring is recommended by 9 out of 10 students to their peers as the most effective and time-efficient way to study.
- **Pearson eText** is available through MasteringPhysics, either automatically when MasteringPhysics is packaged with new books, or available as a purchased upgrade online. Allowing students access to the text wherever they have access to the Internet, Pearson eText comprises the full text, including figures that can be enlarged for better viewing. With eText, students are also able to pop up definitions and terms to help with vocabulary and the reading of the material. Students can also take notes in eText using the annotation feature at the top of each page.

- **Pearson Tutor Services** (www.pearsontutorservices.com) Each student's subscription to MasteringPhysics also contains complimentary access to Pearson Tutor Services, powered by Smarthinking, Inc. By logging in with their MasteringPhysics ID and password, they will be connected to highly qualified e-instructors who provide additional interactive online tutoring on the major concepts of physics. Some restrictions apply; offer subject to change.
- **(MP) ActivPhysics OnLine™** (accessed through the Self Study area within www.masteringphysics.com)

provides students with a suite of highly regarded applet-based tutorials (see above). The following workbooks help students work through complex concepts and understand them more clearly:
- **ActivPhysics OnLine Workbook, Volume 1: Mechanics • Thermal Physics • Oscillations & Waves** (ISBN 978-0-8053-9060-5/0-8053-9060-X)
- **ActivPhysics OnLine Workbook, Volume 2: Electricity & Magnetism • Optics • Modern Physics** (ISBN 978-0-8053-9061-2/0-8053-9061-8)

Acknowledgments

I have relied upon conversations with and, especially, the written publications of many members of the physics education research community. Those who may recognize their influence include Arnold Arons, Uri Ganiel, Ibrahim Halloun, Richard Hake, Ken Heller, Paula Heron, David Hestenes, Leonard Jossem, Jill Larkin, Priscilla Laws, John Mallinckrodt, Kandiah Manivannan, Lillian McDermott and members of the Physics Education Research Group at the University of Washington, David Meltzer, Edward "Joe" Redish, Fred Reif, Jeffery Saul, Rachel Scherr, Bruce Sherwood, Josip Slisko, David Sokoloff, Richard Steinberg, Ronald Thornton, Sheila Tobias, Alan Van Heuleven, and Michael Wittmann. John Rigden, founder and director of the Introductory University Physics Project, provided the impetus that got me started down this path. Early development of the materials was supported by the National Science Foundation as the *Physics for the Year 2000* project; their support is gratefully acknowledged.

I especially want to thank my editor Jim Smith, development editor Alice Houston, project editor Martha Steele, and all the other staff at Pearson for their enthusiasm and hard work on this project. Production project manager Beth Collins, Rose Kernan and the team at Nesbitt Graphics, Inc., and photo researcher Eric Schrader get a good deal of the credit for making this complex project all come together. Larry Smith and Brett Kraabel have done an outstanding job of checking the solutions to every end-of-chapter problem and updating the *Instructor Solutions Manual*. Jim Andrews and Brian Garcar must be thanked for so carefully writing out the solutions to *The Student Workbook* exercises, and Jason Harlow for putting together the Lecture Outlines. In addition to the reviewers and classroom testers listed below, who gave invaluable feedback, I am particularly grateful to Charlie Hibbard for his close scrutiny of every word and figure.

Finally, I am endlessly grateful to my wife Sally for her love, encouragement, and patience, and to our many cats, past and present, who understand clearly that their priority is not deadlines but "Pet me, pet me, pet me."

Randy Knight, September 2011
rknight@calpoly.edu

Reviewers and Classroom Testers

Special thanks go to our third edition review panel: Kyle Altman, Taner Edis, Kent Fisher, Marty Gelfand, Elizabeth George, Jason Harlow, Bob Jacobsen, David Lee, Gary Morris, Eric Murray, and Bruce Schumm.

Gary B. Adams, *Arizona State University*
Ed Adelson, *Ohio State University*
Kyle Altmann, *Elon University*
Wayne R. Anderson, *Sacramento City College*
James H. Andrews, *Youngstown State University*
Kevin Ankoviak, *Las Positas College*
David Balogh, *Fresno City College*
Dewayne Beery, *Buffalo State College*
Joseph Bellina, *Saint Mary's College*
James R. Benbrook, *University of Houston*
David Besson, *University of Kansas*

Randy Bohn, *University of Toledo*
Richard A. Bone, *Florida International University*
Gregory Boutis, *York College*
Art Braundmeier, *University of Southern Illinois, Edwardsville*
Carl Bromberg, *Michigan State University*
Meade Brooks, *Collin College*
Douglas Brown, *Cabrillo College*
Ronald Brown, *California Polytechnic State University, San Luis Obispo*
Mike Broyles, *Collin County Community College*
Debra Burris, *University of Central Arkansas*
James Carolan, *University of British Columbia*
Michael Chapman, *Georgia Tech University*
Norbert Chencinski, *College of Staten Island*
Kristi Concannon, *King's College*

Sean Cordry, *Northwestern College of Iowa*
Robert L. Corey, *South Dakota School of Mines*
Michael Crescimanno, *Youngstown State University*
Dennis Crossley, *University of Wisconsin–Sheboygan*
Wei Cui, *Purdue University*
Robert J. Culbertson, *Arizona State University*
Danielle Dalafave, *The College of New Jersey*
Purna C. Das, *Purdue University North Central*
Chad Davies, *Gordon College*
William DeGraffenreid, *California State University–Sacramento*
Dwain Desbien, *Estrella Mountain Community College*
John F. Devlin, *University of Michigan, Dearborn*
John DiBartolo, *Polytechnic University*
Alex Dickison, *Seminole Community College*
Chaden Djalali, *University of South Carolina*
Margaret Dobrowolska, *University of Notre Dame*
Sandra Doty, *Denison University*
Miles J. Dresser, *Washington State University*
Charlotte Elster, *Ohio University*
Robert J. Endorf, *University of Cincinnati*
Tilahun Eneyew, *Embry-Riddle Aeronautical University*
F. Paul Esposito, *University of Cincinnati*
John Evans, *Lee University*
Harold T. Evensen, *University of Wisconsin–Platteville*
Michael R. Falvo, *University of North Carolina*
Abbas Faridi, *Orange Coast College*
Nail Fazleev, *University of Texas–Arlington*
Stuart Field, *Colorado State University*
Daniel Finley, *University of New Mexico*
Jane D. Flood, *Muhlenberg College*
Michael Franklin, *Northwestern Michigan College*
Jonathan Friedman, *Amherst College*
Thomas Furtak, *Colorado School of Mines*
Alina Gabryszewska-Kukawa, *Delta State University*
Lev Gasparov, *University of North Florida*
Richard Gass, *University of Cincinnati*
J. David Gavenda, *University of Texas, Austin*
Stuart Gazes, *University of Chicago*
Katherine M. Gietzen, *Southwest Missouri State University*
Robert Glosser, *University of Texas, Dallas*
William Golightly, *University of California, Berkeley*
Paul Gresser, *University of Maryland*
C. Frank Griffin, *University of Akron*
John B. Gruber, *San Jose State University*
Stephen Haas, *University of Southern California*
John Hamilton, *University of Hawaii at Hilo*
Jason Harlow, *University of Toronto*
Randy Harris, *University of California, Davis*
Nathan Harshman, *American University*
J. E. Hasbun, *University of West Georgia*
Nicole Herbots, *Arizona State University*
Jim Hetrick, *University of Michigan–Dearborn*
Scott Hildreth, *Chabot College*
David Hobbs, *South Plains College*
Laurent Hodges, *Iowa State University*

Mark Hollabaugh, *Normandale Community College*
John L. Hubisz, *North Carolina State University*
Shane Hutson, *Vanderbilt University*
George Igo, *University of California, Los Angeles*
David C. Ingram, *Ohio University*
Bob Jacobsen, *University of California, Berkeley*
Rong-Sheng Jin, *Florida Institute of Technology*
Marty Johnston, *University of St. Thomas*
Stanley T. Jones, *University of Alabama*
Darrell Judge, *University of Southern California*
Pawan Kahol, *Missouri State University*
Teruki Kamon, *Texas A&M University*
Richard Karas, *California State University, San Marcos*
Deborah Katz, *U.S. Naval Academy*
Miron Kaufman, *Cleveland State University*
Katherine Keilty, *Kingwood College*
Roman Kezerashvili, *New York City College of Technology*
Peter Kjeer, *Bethany Lutheran College*
M. Kotlarchyk, *Rochester Institute of Technology*
Fred Krauss, *Delta College*
Cagliyan Kurdak, *University of Michigan*
Fred Kuttner, *University of California, Santa Cruz*
H. Sarma Lakkaraju, *San Jose State University*
Darrell R. Lamm, *Georgia Institute of Technology*
Robert LaMontagne, *Providence College*
Eric T. Lane, *University of Tennessee–Chattanooga*
Alessandra Lanzara, *University of California, Berkeley*
Lee H. LaRue, *Paris Junior College*
Sen-Ben Liao, *Massachusetts Institute of Technology*
Dean Livelybrooks, *University of Oregon*
Chun-Min Lo, *University of South Florida*
Olga Lobban, *Saint Mary's University*
Ramon Lopez, *Florida Institute of Technology*
Vaman M. Naik, *University of Michigan, Dearborn*
Kevin Mackay, *Grove City College*
Carl Maes, *University of Arizona*
Rizwan Mahmood, *Slippery Rock University*
Mani Manivannan, *Missouri State University*
Richard McCorkle, *University of Rhode Island*
James McDonald, *University of Hartford*
James McGuire, *Tulane University*
Stephen R. McNeil, *Brigham Young University–Idaho*
Theresa Moreau, *Amherst College*
Gary Morris, *Rice University*
Michael A. Morrison, *University of Oklahoma*
Richard Mowat, *North Carolina State University*
Eric Murray, *Georgia Institute of Technology*
Taha Mzoughi, *Mississippi State University*
Scott Nutter, *Northern Kentucky University*
Craig Ogilvie, *Iowa State University*
Benedict Y. Oh, *University of Wisconsin*
Martin Okafor, *Georgia Perimeter College*
Halina Opyrchal, *New Jersey Institute of Technology*
Yibin Pan, *University of Wisconsin–Madison*
Georgia Papaefthymiou, *Villanova University*
Peggy Perozzo, *Mary Baldwin College*

Preface to the Student

From Me to You

The most incomprehensible thing about the universe is that it is comprehensible.
— Albert Einstein

The day I went into physics class it was death.
— Sylvia Plath, *The Bell Jar*

Let's have a little chat before we start. A rather one-sided chat, admittedly, because you can't respond, but that's OK. I've talked with many of your fellow students over the years, so I have a pretty good idea of what's on your mind.

What's your reaction to taking physics? Fear and loathing? Uncertainty? Excitement? All of the above? Let's face it, physics has a bit of an image problem on campus. You've probably heard that it's difficult, maybe downright impossible unless you're an Einstein. Things that you've heard, your experiences in other science courses, and many other factors all color your *expectations* about what this course is going to be like.

It's true that there are many new ideas to be learned in physics and that the course, like college courses in general, is going to be much faster paced than science courses you had in high school. I think it's fair to say that it will be an *intense* course. But we can avoid many potential problems and difficulties if we can establish, here at the beginning, what this course is about and what is expected of you—and of me!

Just what is physics, anyway? Physics is a way of thinking about the physical aspects of nature. Physics is not better than art or biology or poetry or religion, which are also ways to think about nature; it's simply different. One of the things this course will emphasize is that physics is a human endeavor. The ideas presented in this book were not found in a cave or conveyed to us by aliens; they were discovered and developed by real people engaged in a struggle with real issues. I hope to convey to you something of the history and the process by which we have come to accept the principles that form the foundation of today's science and engineering.

You might be surprised to hear that physics is not about "facts." Oh, not that facts are unimportant, but physics is far more focused on discovering *relationships* that exist between facts and *patterns* that exist in nature than on learning facts for their own sake. As a consequence, there's not a lot of memorization when you study physics. Some—there are still definitions and equations to learn—but less than in many other courses. Our emphasis, instead, will be on thinking and reasoning. This is important to factor into your expectations for the course.

Perhaps most important of all, *physics is not math!* Physics is much broader. We're going to look for patterns and relationships in nature, develop the logic that relates different ideas, and search for the reasons *why* things happen as they do. In doing so, we're going to stress qualitative reasoning, pictorial and graphical reasoning, and reasoning by analogy. And yes, we will use math, but it's just one tool among many.

It will save you much frustration if you're aware of this physics–math distinction up front. Many of you, I know, want to find a formula and plug numbers into it—that is,

(a) X-ray diffraction pattern

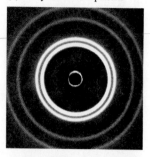

(b) Electron diffraction pattern

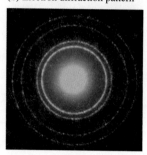

to do a math problem. Maybe that worked in high school science courses, but it is *not* what this course expects of you. We'll certainly do many calculations, but the specific numbers are usually the last and least important step in the analysis.

Physics is about recognizing patterns. For example, the top photograph is an x-ray diffraction pattern showing how a focused beam of x rays spreads out after passing through a crystal. The bottom photograph shows what happens when a focused beam of electrons is shot through the same crystal. What does the obvious similarity in these two photographs tell us about the nature of light and the nature of matter?

As you study, you'll sometimes be baffled, puzzled, and confused. That's perfectly normal and to be expected. Making mistakes is OK too *if* you're willing to learn from the experience. No one is born knowing how to do physics any more than he or she is born knowing how to play the piano or shoot basketballs. The ability to do physics comes from practice, repetition, and struggling with the ideas until you "own" them and can apply them yourself in new situations. There's no way to make learning effortless, at least for anything worth learning, so expect to have some difficult moments ahead. But also expect to have some moments of excitement at the joy of discovery. There will be instants at which the pieces suddenly click into place and you *know* that you understand a powerful idea. There will be times when you'll surprise yourself by successfully working a difficult problem that you didn't think you could solve. My hope, as an author, is that the excitement and sense of adventure will far outweigh the difficulties and frustrations.

Getting the Most Out of Your Course

Many of you, I suspect, would like to know the "best" way to study for this course. There is no best way. People are different, and what works for one student is less effective for another. But I do want to stress that *reading the text* is vitally important. Class time will be used to clarify difficulties and to develop tools for using the knowledge, but your instructor will *not* use class time simply to repeat information in the text. The basic knowledge for this course is written down on these pages, and the *number-one expectation* is that you will read carefully and thoroughly to find and learn that knowledge.

Despite there being no best way to study, I will suggest *one* way that is successful for many students. It consists of the following four steps:

1. **Read each chapter *before* it is discussed in class.** I cannot stress too strongly how important this step is. Class attendance is much more effective if you are prepared. When you first read a chapter, focus on learning new vocabulary, definitions, and notation. There's a list of terms and notations at the end of each chapter. Learn them! You won't understand what's being discussed or how the ideas are being used if you don't know what the terms and symbols mean.

2. **Participate actively in class.** Take notes, ask and answer questions, and participate in discussion groups. There is ample scientific evidence that *active participation* is much more effective for learning science than passive listening.

3. **After class, go back for a careful re-reading of the chapter.** In your second reading, pay closer attention to the details and the worked examples. Look for the *logic* behind each example (I've highlighted this to make it clear), not just at what formula is being used. Do the *Student Workbook* exercises for each section as you finish your reading of it.

4. **Finally, apply what you have learned to the homework problems at the end of each chapter.** I strongly encourage you to form a study group with two or three classmates. There's good evidence that students who study regularly with a group do better than the rugged individualists who try to go it alone.

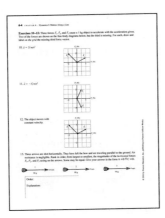

Did someone mention a workbook? The companion *Student Workbook* is a vital part of the course. Its questions and exercises ask you to reason *qualitatively,* to use graphical information, and to give explanations. It is through these exercises that you will learn what the concepts mean and will practice the reasoning skills appropriate to the chapter. You will then have acquired the baseline knowledge and confidence you need *before* turning to the end-of-chapter homework problems. In sports or in music, you would never think of performing before you practice, so why would you want to do so in physics? The workbook is where you practice and work on basic skills.

Many of you, I know, will be tempted to go straight to the homework problems and then thumb through the text looking for a formula that seems like it will work. That approach will not succeed in this course, and it's guaranteed to make you frustrated and discouraged. Very few homework problems are of the "plug and chug" variety where you simply put numbers into a formula. To work the homework problems successfully, you need a better study strategy—either the one outlined above or your own—that helps you learn the concepts and the relationships between the ideas.

A traditional guideline in college is to study two hours outside of class for every hour spent in class, and this text is designed with that expectation. Of course, two hours is an average. Some chapters are fairly straightforward and will go quickly. Others likely will require much more than two study hours per class hour.

Getting the Most Out of Your Textbook

Your textbook provides many features designed to help you learn the concepts of physics and solve problems more effectively.

- **TACTICS BOXES** give step-by-step procedures for particular skills, such as interpreting graphs or drawing special diagrams. Tactics Box steps are explicitly illustrated in subsequent worked examples, and these are often the starting point of a full *Problem-Solving Strategy*.

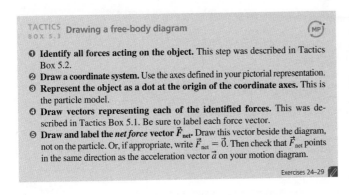

TACTICS BOX 5.3 Drawing a free-body diagram

❶ **Identify all forces acting on the object.** This step was described in Tactics Box 5.2.
❷ **Draw a coordinate system.** Use the axes defined in your pictorial representation.
❸ **Represent the object as a dot at the origin of the coordinate axes.** This is the particle model.
❹ **Draw vectors representing each of the identified forces.** This was described in Tactics Box 5.1. Be sure to label each force vector.
❺ **Draw and label the *net force* vector $\vec{F}_{net}$.** Draw this vector beside the diagram, not on the particle. Or, if appropriate, write $\vec{F}_{net} = \vec{0}$. Then check that $\vec{F}_{net}$ points in the same direction as the acceleration vector $\vec{a}$ on your motion diagram.

Exercises 24–29

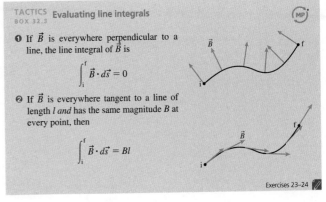

TACTICS BOX 32.3 Evaluating line integrals

❶ If $\vec{B}$ is everywhere perpendicular to a line, the line integral of $\vec{B}$ is

$$\int_i^f \vec{B} \cdot d\vec{s} = 0$$

❷ If $\vec{B}$ is everywhere tangent to a line of length *l* and has the same magnitude *B* at every point, then

$$\int_i^f \vec{B} \cdot d\vec{s} = Bl$$

Exercises 23–24

- **PROBLEM-SOLVING STRATEGIES** are provided for each broad class of problems— problems characteristic of a chapter or group of chapters. The strategies follow a consistent four-step approach to help you develop confidence and proficient problem-solving skills: MODEL, VISUALIZE, SOLVE, ASSESS.

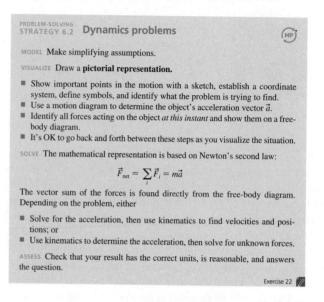

PROBLEM-SOLVING
STRATEGY 6.2   **Dynamics problems**   (MP)

MODEL Make simplifying assumptions.

VISUALIZE Draw a **pictorial representation.**

- Show important points in the motion with a sketch, establish a coordinate system, define symbols, and identify what the problem is trying to find.
- Use a motion diagram to determine the object's acceleration vector $\vec{a}$.
- Identify all forces acting on the object *at this instant* and show them on a free-body diagram.
- It's OK to go back and forth between these steps as you visualize the situation.

SOLVE The mathematical representation is based on Newton's second law:

$$\vec{F}_{net} = \sum_i \vec{F}_i = m\vec{a}$$

The vector sum of the forces is found directly from the free-body diagram. Depending on the problem, either

- Solve for the acceleration, then use kinematics to find velocities and positions; or
- Use kinematics to determine the acceleration, then solve for unknown forces.

ASSESS Check that your result has the correct units, is reasonable, and answers the question.

Exercise 22

- Worked **EXAMPLES** illustrate good problem-solving practices through the consistent use of the four-step problem-solving approach and, where appropriate, the Tactics Box steps. The worked examples are often very detailed and carefully lead you through the *reasoning* behind the solution as well as the numerical calculations. A careful study of the reasoning will help you apply the concepts and techniques to the new and novel problems you will encounter in homework assignments and on exams.
- **NOTE ▶** paragraphs alert you to common mistakes and point out useful tips for tackling problems.
- **STOP TO THINK** questions embedded in the chapter allow you to quickly assess whether you've understood the main idea of a section. A correct answer will give you confidence to move on to the next section. An incorrect answer will alert you to re-read the previous section.
- Blue annotations on figures help you better understand what the figure is showing. They will help you to interpret graphs; translate between graphs, math, and pictures; grasp difficult concepts through a visual analogy; and develop many other important skills.
- *Pencil sketches* provide practical examples of the figures you should draw yourself when solving a problem.

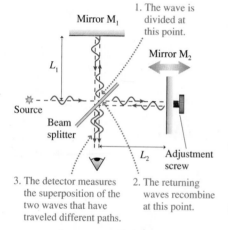

Annotated FIGURE showing the operation of the Michelson interferometer.

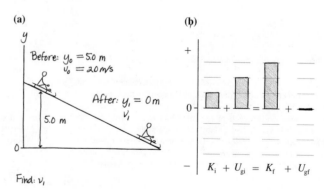

Pencil-sketch FIGURE showing a toboggan going down a hill and its energy bar chart.

- Each chapter begins with a *Chapter Preview*, a visual outline of the chapter ahead with recommendations of important topics you should review from previous chapters. A few minutes spent with the Preview will help you organize your thoughts so as to get the most out of reading the chapter.

- Schematic *Chapter Summaries* help you organize what you have learned into a hierarchy, from general principles (top) to applications (bottom). Side-by-side pictorial, graphical, textual, and mathematical representations are used to help you translate between these key representations.

- *Part Overviews* and *Summaries* provide a global framework for what you are learning. Each part begins with an overview of the chapters ahead and concludes with a broad summary to help you to connect the concepts presented in that set of chapters. KNOWLEDGE STRUCTURE tables in the Part Summaries, similar to the Chapter Summaries, help you to see the forest rather than just the trees.

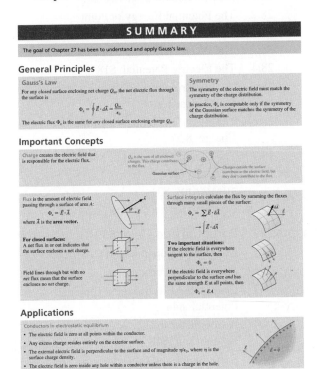

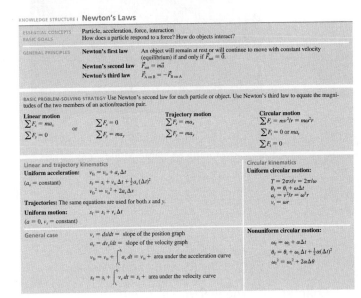

Now that you know more about what is expected of you, what can you expect of me? That's a little trickier because the book is already written! Nonetheless, the book was prepared on the basis of what I think my students throughout the years have expected—and wanted—from their physics textbook. Further, I've listened to the extensive feedback I have received from thousands of students like you, and their instructors, who used the first and second editions of this book.

You should know that these course materials—the text and the workbook—are based on extensive research about how students learn physics and the challenges they face. The effectiveness of many of the exercises has been demonstrated through extensive class testing. I've written the book in an informal style that I hope you will find appealing and that will encourage you to do the reading. And, finally, I have endeavored to make clear not only that physics, as a technical body of knowledge, is relevant to your profession but also that physics is an exciting adventure of the human mind.

I hope you'll enjoy the time we're going to spend together.

Detailed Contents

Volume 1 contains chapters 1-15; Volume 2 contains chapters 16-19; Volume 3 contains chapters 20-24; Volume 4 contains chapters 25-36; Volume 5 contains chapters 36-42.

Part VI Electricity and Magnetism

OVERVIEW   Phenomena and Theories 719

Chapter 25   Electric Charges and Forces 720
25.1   Developing a Charge Model 721
25.2   Charge 725
25.3   Insulators and Conductors 727
25.4   Coulomb's Law 731
25.5   The Field Model 736
         SUMMARY 743
         QUESTIONS AND PROBLEMS 744

Chapter 26   The Electric Field 750
26.1   Electric Field Models 751
26.2   The Electric Field of Multiple Point Charges 752
26.3   The Electric Field of a Continuous Charge Distribution 756
26.4   The Electric Fields of Rings, Planes, and Spheres 760
26.5   The Parallel-Plate Capacitor 764
26.6   Motion of a Charged Particle in an Electric Field 767
26.7   Motion of a Dipole in an Electric Field 770
         SUMMARY 773
         QUESTIONS AND PROBLEMS 774

Chapter 27   Gauss's Law 780
27.1   Symmetry 781
27.2   The Concept of Flux 783
27.3   Calculating Electric Flux 785
27.4   Gauss's Law 791
27.5   Using Gauss's Law 795

27.6   Conductors in Electrostatic Equilibrium 799
         SUMMARY 803
         QUESTIONS AND PROBLEMS 804

Chapter 28   The Electric Potential 810
28.1   Electric Potential Energy 811
28.2   The Potential Energy of Point Charges 814
28.3   The Potential Energy of a Dipole 817
28.4   The Electric Potential 818
28.5   The Electric Potential Inside a Parallel-Plate Capacitor 821
28.6   The Electric Potential of a Point Charge 826
28.7   The Electric Potential of Many Charges 828
         SUMMARY 831
         QUESTIONS AND PROBLEMS 832

Chapter 29   Potential and Field 839
29.1   Connecting Potential and Field 840
29.2   Sources of Electric Potential 842
29.3   Finding the Electric Field from the Potential 844
29.4   A Conductor in Electrostatic Equilibrium 848
29.5   Capacitance and Capacitors 849
29.6   The Energy Stored in a Capacitor 854
29.7   Dielectrics 855
         SUMMARY 860
         QUESTIONS AND PROBLEMS 861

Chapter 30   Current and Resistance 867
30.1   The Electron Current 868
30.2   Creating a Current 870
30.3   Current and Current Density 874
30.4   Conductivity and Resistivity 878
30.5   Resistance and Ohm's Law 880
         SUMMARY 885
         QUESTIONS AND PROBLEMS 886

Chapter 31   Fundamentals of Circuits 891
31.1   Circuit Elements and Diagrams 892
31.2   Kirchhoff's Laws and the Basic Circuit 892
31.3   Energy and Power 896

31.4 Series Resistors 898
31.5 Real Batteries 901
31.6 Parallel Resistors 903
31.7 Resistor Circuits 906
31.8 Getting Grounded 908
31.9 *RC* Circuits 909
SUMMARY 913
QUESTIONS AND PROBLEMS 914

Chapter 32 The Magnetic Field 921
32.1 Magnetism 922
32.2 The Discovery of the Magnetic Field 923
32.3 The Source of the Magnetic Field: Moving Charges 925
32.4 The Magnetic Field of a Current 927
32.5 Magnetic Dipoles 931
32.6 Ampère's Law and Solenoids 934
32.7 The Magnetic Force on a Moving Charge 940
32.8 Magnetic Forces on Current-Carrying Wires 946
32.9 Forces and Torques on Current Loops 948
32.10 Magnetic Properties of Matter 950
SUMMARY 954
QUESTIONS AND PROBLEMS 955

Chapter 33 Electromagnetic Induction 962
33.1 Induced Currents 963
33.2 Motional emf 964
33.3 Magnetic Flux 968
33.4 Lenz's Law 971
33.5 Faraday's Law 975
33.6 Induced Fields 978
33.7 Induced Currents: Three Applications 982
33.8 Inductors 984
33.9 *LC* Circuits 988
33.10 *LR* Circuits 991
SUMMARY 994
QUESTIONS AND PROBLEMS 995

Chapter 34 Electromagnetic Fields and Waves 1003
34.1 *E* or *B*? It Depends on Your Perspective 1004
34.2 The Field Laws Thus Far 1010
34.3 The Displacement Current 1011
34.4 Maxwell's Equations 1014
34.5 Electromagnetic Waves 1016

34.6 Properties of Electromagnetic Waves 1020
34.7 Polarization 1024
SUMMARY 1027
QUESTIONS AND PROBLEMS 1028

Chapter 35 AC Circuits 1033
35.1 AC Sources and Phasors 1034
35.2 Capacitor Circuits 1036
35.3 *RC* Filter Circuits 1038
35.4 Inductor Circuits 1041
35.5 The Series *RLC* Circuit 1042
35.6 Power in AC Circuits 1046
SUMMARY 1050
QUESTIONS AND PROBLEMS 1051
PART SUMMARY Electricity and Magnetism 1056

Part VII Relativity and Quantum Physics

OVERVIEW Contemporary Physics 1059

Chapter 36 Relativity 1060
36.1 Relativity: What's It All About? 1061
36.2 Galilean Relativity 1061
36.3 Einstein's Principle of Relativity 1066
36.4 Events and Measurements 1068
36.5 The Relativity of Simultaneity 1071
36.6 Time Dilation 1074
36.7 Length Contraction 1078
36.8 The Lorentz Transformations 1082
36.9 Relativistic Momentum 1087
36.10 Relativistic Energy 1090
SUMMARY 1096
QUESTIONS AND PROBLEMS 1097

Appendix A Mathematics Review A-1
Appendix B Periodic Table of Elements A-4
Appendix C ActivPhysics OnLine Activities and PhET Simulations A-5
Answers to Odd-Numbered Problems A-7
Credits C-1
Index I-1

VI Electricity and Magnetism

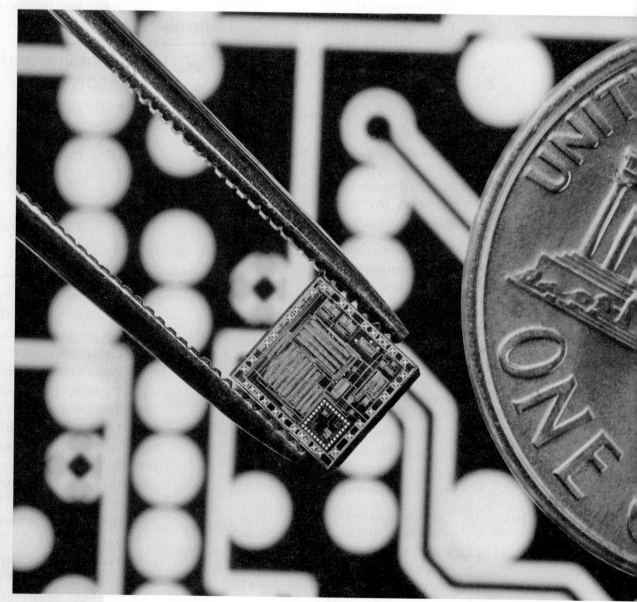

This integrated circuit contains millions of circuit elements. The density of circuit elements in integrated circuits has doubled about every 18 months for the past 30 years. Whether this trend continues depends on whether scientists and engineers can understand the physics of nanoscale electric circuits.

OVERVIEW

Phenomena and Theories

Amber, or fossilized tree resin, has long been prized for its beauty. Amber is of scientific interest today because biologists have learned how to recover DNA strands from million-year-old insects trapped in the resin. But amber has an ancient scientific connection as well. The Greek word for amber is *elektron.*

It has been known since antiquity that a piece of amber rubbed with fur can attract feathers or straw—seemingly magical powers to a pre-scientific society. It was also known to the ancient Greeks that certain stones from the region they called *Magnesia* could pick up pieces of iron. It is from these humble beginnings that we today have high-speed computers, lasers, and magnetic resonance imaging as well as such mundane modern-day miracles as the lightbulb.

The basic phenomena of electricity and magnetism are not as familiar as those of mechanics. You have spent your entire life exerting forces on objects and watching them move, but your experience with electricity and magnetism is probably much more limited. We will deal with this lack of experience by placing a large emphasis on the *phenomena* of electricity and magnetism.

We will begin by looking in detail at *electric charge* and the process of *charging* an object. It is easy to make systematic observations of how charges behave, and we will consider the forces between charges and how charges behave in different materials. Similarly, we will begin our study of magnetism by observing how magnets stick to some metals but not others and how magnets affect compass needles. But our most important observation will be that an electric current affects a compass needle in exactly the same way as a magnet. This observation, suggesting a close connection between electricity and magnetism, will eventually lead us to the discovery of electromagnetic waves.

Our goal in Part VI is to develop a theory to explain the phenomena of electricity and magnetism. The linchpin of our theory will be the entirely new concept of a *field.* Electricity and magnetism are about the long-range interactions of charges, both static charges and moving charges, and the field concept will help us understand how these interactions take place. We will want to know how fields are created by charges and how charges, in return, respond to the fields. Bit by bit, we will assemble a theory—based on the new concepts of electric and magnetic fields—that will allow us to understand, explain, and predict a wide range of electromagnetic behavior.

The story of electricity and magnetism is vast. The 19th-century formulation of the theory of electromagnetism, which led to sweeping revolutions in science and technology, has been called by no less than Einstein "the most important event in physics since Newton's time." Not surprisingly, all we can do in this text is develop some of the basic ideas and concepts, leaving many details and applications to later courses. Even so, our study of electricity and magnetism will explore some of the most exciting and important topics in physics.

25 Electric Charges and Forces

Electricity is one of the fundamental forces of nature. Lightning is a vivid manifestation of electric charges and forces.

▶ **Looking Ahead** The goal of Chapter 25 is to describe electric phenomena in terms of charges, forces, and fields.

Charge Model

Electric phenomena seem mysterious at first, but we'll find that we can understand them in terms of a **charge model:**

- There are two kinds of charge, called *positive* and *negative*.
- Two charges of the same kind repel; two opposite charges attract.
- Small neutral objects are attracted to a charge of either sign.

You'll learn how a comb rubbed through your hair picks up small pieces of paper.

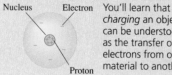

Coulomb's Law

The law governing the electric force is called **Coulomb's law.** It tells us how the force between charged particles depends on their charge and on the distance between them.

You'll find that Coulomb's law, like Newton's law of gravity, is an *inverse-square law*.

◀ **Looking Back**
Sections 3.2–3.4 Vector addition
Sections 13.3–13.4 Newton's theory of gravity

Field Model

How is a long-range force transmitted from one charge to another? We'll develop the idea that every charge alters the space around it by creating an **electric field.** It is the electric field that then exerts forces on other charges.

The liquid crystal displays (LCD) of your calculator, your digital watch, and your computer screen use electric fields to turn the pixels on and off.

Charges and Atoms

Electrons and protons—the constituents of atoms—are the basic charges of ordinary matter.

Nucleus   Electron

Proton

You'll learn that *charging* an object can be understood as the transfer of electrons from one material to another.

An object that is negative has an excess of electrons; a positively charged object is missing electrons.

Conductors and Insulators

There are two types of materials with very different electrical properties:

- **Conductors** are materials through or along which charge easily moves.
- **Insulators** are materials on or in which charge is immobile.

The metal wire—a conductor—carries a current of moving charges. It is separated from the support by a ceramic insulator.

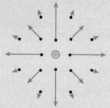

Point Charges

A charged particle, with no physical size, is called a **point charge.** You'll learn that real objects can be modeled as point charges if they are very small compared to the distances between them.

The electric field of a point charge will be important throughout our study of electricity.

25.1 Developing a Charge Model

You can receive a mildly unpleasant shock and produce a little spark if you touch a metal doorknob after walking across a carpet. Vigorously brushing your freshly washed hair makes all the hairs fly apart. A plastic comb that you've run through your hair will pick up bits of paper and other small objects, but a metal comb won't.

The common factor in these observations is that two objects are *rubbed* together. Why should rubbing an object cause forces and sparks? What kind of forces are these? Why do metallic objects behave differently from nonmetallic? These are the questions with which we begin our study of electricity.

Our first goal is to develop a model for understanding electric phenomena in terms of *charges* and *forces*. We will later use our contemporary knowledge of atoms to understand electricity on a microscopic level, but the basic concepts of electricity make *no* reference to atoms or electrons. The theory of electricity was well established long before the electron was discovered.

Experimenting with Charges

Let us enter a laboratory where we can make observations of electric phenomena. The major tools in the lab are:

- A variety of plastic and glass rods, each several centimeters long.
- A few metal rods with wood handles.
- Pieces of wool and silk.
- Small metal spheres, an inch or two in diameter, on wood stands.

Let's see what we can learn with these tools.

Discovering electricity I

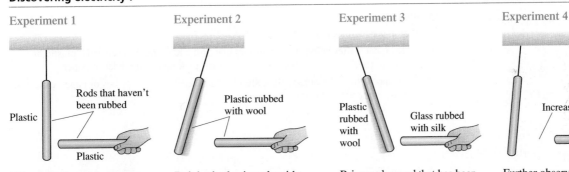

Experiment 1

Take a plastic rod that has been undisturbed for a long period of time and hang it by a thread. Pick up another undisturbed plastic rod and bring it close to the hanging rod. Nothing happens to either rod.

Plastic, *Rods that haven't been rubbed*, *Plastic*

Experiment 2

Rub both plastic rods with wool. Now the hanging rod tries to move away from the handheld rod when you bring the two close together. Two glass rods rubbed with silk also repel each other.

Plastic rubbed with wool

Experiment 3

Bring a glass rod that has been rubbed with silk close to a hanging plastic rod that has been rubbed with wool. These two rods *attract* each other.

Plastic rubbed with wool, *Glass rubbed with silk*

Experiment 4

Further observations show that:

- These forces are greater for rods that have been rubbed more vigorously.
- The strength of the forces decreases as the separation between the rods increases.

Increased distance

No forces were observed in Experiment 1. We will say that the original objects are **neutral.** Rubbing the rods (Experiments 2 and 3) somehow causes forces to be exerted between them. We will call the rubbing process **charging** and say that a rubbed rod is *charged.* For now, these are simply descriptive terms. The terms don't tell us anything about the process itself.

Experiment 2 shows that there is a *long-range repulsive force*, requiring no contact, between two identical objects that have been charged in the *same* way. Furthermore, Experiment 4 shows that the force between two charged objects depends on the distance between them. This is the first long-range force we've encountered since gravity was introduced in Chapter 5. It is also the first time we've observed a repulsive force, so right away we see that new ideas will be needed to understand electricity.

Experiment 3 is a puzzle. Two rods *seem* to have been charged in the same way, by rubbing, but these two rods *attract* each other rather than repel. Why does the outcome of Experiment 3 differ from that of Experiment 2? Back to the lab.

Discovering electricity II

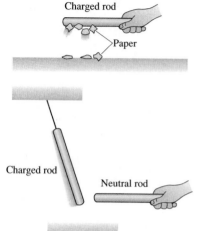

Charged rod

Paper

Experiment 5

Hold a charged (i.e., rubbed) plastic rod over small pieces of paper on the table. The pieces of paper leap up and stick to the rod. A charged glass rod does the same. However, a neutral rod has no effect on the pieces of paper.

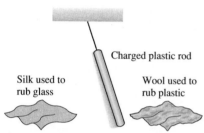

Charged rod

Neutral rod

Experiment 6

Rub a plastic rod with wool and a glass rod with silk. Hang both by threads, some distance apart. Both rods are attracted to a *neutral* (i.e., unrubbed) plastic rod that is held close. Interestingly, both are also attracted to a *neutral* glass rod. In fact, the charged rods are attracted to *any* neutral object, such as a finger, a piece of paper, or a metal rod.

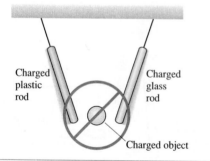

Charged plastic rod

Silk used to rub glass

Wool used to rub plastic

Experiment 7

Rub a hanging plastic rod with wool and then hold the *wool* close to the rod. The rod is weakly *attracted* to the wool. The plastic rod is *repelled* by a piece of silk that has been used to rub glass.

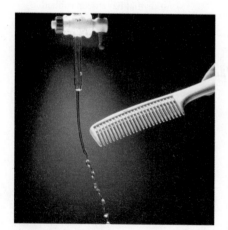

Charged plastic rod

Charged glass rod

Charged object

Experiment 8

Further experiments show that:

■ Other objects, after being rubbed, attract one of the hanging charged rods (plastic or glass) and repel the other. These objects always pick up small pieces of paper.

■ There appear to be *no* objects that, after being rubbed, pick up pieces of paper and attract *both* the charged plastic and glass rods.

A plastic comb that has been charged by running it through your hair attracts neutral objects—here drops of water.

Our first set of experiments found that charged objects exert forces on each other. The forces are sometimes attractive, sometimes repulsive. Experiments 5 and 6 show that there is an attractive force between a charged object and a *neutral* (uncharged) object. This discovery presents us with a problem: How can we tell if an object is charged or neutral? Because of the attractive force between a charged and a neutral object, simply observing an electric force does *not* imply that an object is charged.

However, an important characteristic of any *charged* object appears to be that **a charged object picks up small pieces of paper.** This behavior provides a straightforward test to answer the question, Is this object charged? An object that passes the test by picking up paper is charged; an object that fails the test is neutral.

These observations let us tentatively advance the first stages of a **charge model.**

Charge model, part I The basic postulates of our model are:

1. Frictional forces, such as rubbing, add something called **charge** to an object or remove it from the object. The process itself is called *charging*. More vigorous rubbing produces a larger quantity of charge.

2. There are two and only two kinds of charge. For now we will call these "plastic charge" and "glass charge." Other objects can sometimes be charged by rubbing, but the charge they receive is either "plastic charge" or "glass charge."
3. Two **like charges** (plastic/plastic or glass/glass) exert repulsive forces on each other. Two **opposite charges** (plastic/glass) attract each other.
4. The force between two charges is a long-range force. The size of the force increases as the quantity of charge increases and decreases as the distance between the charges increases.
5. *Neutral* objects have an *equal mixture* of both "plastic charge" and "glass charge." The rubbing process somehow manages to separate the two.

Postulate 2 is based on Experiment 8. If an object is charged (i.e., picks up paper), it always attracts one charged rod and repels the other. That is, it acts either "like plastic" or "like glass." If there were a third kind of charge, different from the first two, an object with that charge should pick up paper and attract *both* the charged plastic and glass rods. No such objects have ever been found.

The basis for postulate 5 is the observation in Experiment 7 that a charged plastic rod is attracted to the wool used to rub it but repelled by silk that has rubbed glass. It appears that rubbing glass causes the silk to acquire "plastic charge." The easiest way to explain this is to hypothesize that the silk starts out with equal amounts of "glass charge" and "plastic charge" and that the rubbing somehow transfers "glass charge" from the silk to the rod. This leaves an excess of "glass charge" on the rod and an excess of "plastic charge" on the silk.

While the charge model is *consistent* with the observations, it is by no means proved. One could easily imagine other hypotheses that are just as consistent with the limited observations we have made so far. We still have some large unexplained puzzles, such as why charged objects exert attractive forces on neutral objects.

Electric Properties of Materials

We still need to clarify how different types of materials respond to charges.

Discovering electricity III

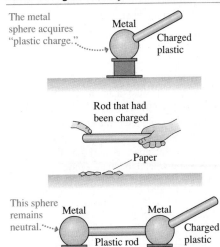

The metal sphere acquires "plastic charge."

Metal

Charged plastic

Rod that had been charged

Paper

This sphere remains neutral.

Metal   Metal

Plastic rod   Charged plastic

This sphere acquires "plastic charge."

Metal   Metal

Metal rod   Charged plastic

Experiment 9

Charge a plastic rod by rubbing it with wool. Touch a neutral metal sphere with the rubbed area of the rod. The metal sphere then picks up small pieces of paper and repels a charged, hanging plastic rod. The metal sphere appears to have acquired "plastic charge."

Experiment 10

Charge a plastic rod, then run your finger along it. After you've done so, the rod no longer picks up small pieces of paper or repels a charged, hanging plastic rod. Similarly, the metal sphere of Experiment 9 no longer repels the plastic rod after you touch it with your finger.

Experiment 11

Place two metal spheres close together with a plastic rod connecting them. Charge a second plastic rod, by rubbing, and touch it to one of the metal spheres. Afterward, the metal sphere that was touched picks up small pieces of paper and repels a charged, hanging plastic rod. The other metal sphere does neither.

Experiment 12

Repeat Experiment 11 with a metal rod connecting the two metal spheres. Touch one metal sphere with a charged plastic rod. Afterward, *both* metal spheres pick up small pieces of paper and repel a charged, hanging plastic rod.

Our final set of experiments has shown that

- Charge can be *transferred* from one object to another, but only when the objects *touch.* Contact is required. Removing charge from an object, which you can do by touching it, is called **discharging.**
- There are two types or classes of materials with very different electric properties. We call these *conductors* and *insulators.*

Experiment 12, in which a metal rod is used, is in sharp contrast to Experiment 11. Charge somehow *moves through* or along a metal rod, from one sphere to the other, but remains *fixed in place* on a plastic or glass rod. Let us define **conductors** as those materials through or along which charge easily moves and **insulators** as those materials on or in which charges remain immobile. Glass and plastic are insulators; metal is a conductor.

This information lets us add two more postulates to our charge model:

Charge model, part II

6. There are two types of materials. Conductors are materials through or along which charge easily moves. Insulators are materials on or in which charges remain fixed in place.
7. Charge can be transferred from one object to another by contact.

NOTE ▶ Both insulators and conductors can be charged. They differ in the *mobility* of the charge. ◀

We have by no means exhausted the number of experiments and observations we might try. Early scientific investigators were faced with all of these results, plus many others. Moreover, many of these experiments are hard to reproduce with much accuracy. How should we make sense of it all? The charge model seems promising, but certainly not proven. We have not yet explained how charged objects exert attractive forces on *neutral* objects, nor have we explained what charge is, how it is transferred, or *why* it moves through some objects but not others. Nonetheless, we will take advantage of our historical hindsight and continue to pursue this model. Homework problems will let you practice using the model to explain other observations.

EXAMPLE 25.1 **Transferring charge**

In Experiment 12, touching one metal sphere with a charged plastic rod caused a second metal sphere to become charged with the same type of charge as the rod. Use the postulates of the charge model to explain this.

SOLVE We need the following ideas from the charge model:

1. Charge is transferred upon contact.
2. Metal is a conductor.
3. Like charges repel.

The plastic rod was charged by rubbing with wool. The charge doesn't move around on the rod, because it is an insulator, but some of the "plastic charge" is transferred to the metal upon contact. Once in the metal, which is a conductor, the charges are free to move around. Furthermore, because like charges repel, these plastic charges quickly move as far apart as they possibly can. Some move through the connecting metal rod to the second sphere. Consequently, the second sphere acquires "plastic charge."

STOP TO THINK 25.1   To determine if an object has "glass charge," you need to

a. See if the object attracts a charged plastic rod.
b. See if the object repels a charged glass rod.
c. Do both a and b.
d. Do either a or b.

25.2 Charge

As you probably know, the modern names for the two types of charge are *positive charge* and *negative charge.* You may be surprised to learn that the names were coined by Benjamin Franklin. Franklin found that charge behaves like positive and negative numbers. If a plastic rod is charged twice, by rubbing, and twice transfers charge to a metal sphere, the electric forces exerted by the sphere are doubled. That is, $2 + 2 = 4$. But the sphere is found to be neutral after receiving equal amounts of "plastic charge" and "glass charge." This is like $2 + (-2) = 0$.

So what is positive and what is negative? It's entirely up to us! Franklin established the convention that **a glass rod that has been rubbed with silk is *positively* charged.** That's it. Any other object that repels a charged glass rod is also positively charged. Any charged object that attracts a charged glass rod is negatively charged. Thus **a plastic rod rubbed with wool is negative.** It was only long afterward, with the discovery of electrons and protons, that electrons were found to be attracted to a charged glass rod while protons were repelled. Thus *by convention* electrons have a negative charge and protons a positive charge.

Atoms and Electricity

Now let's fast forward to the 21st century. The theory of electricity was developed without knowledge of atoms, but there is no reason for us to continue to overlook this important part of our contemporary perspective. FIGURE 25.1 shows that an atom consists of a very small and dense *nucleus* (diameter $\sim 10^{-14}$ m) surrounded by much less massive orbiting *electrons.* The electron orbital frequencies are so enormous ($\sim 10^{15}$ revolutions per second) that the electrons seem to form an **electron cloud** of diameter $\sim 10^{-10}$ m, a factor 10^4 larger than the nucleus. In fact, the wave–particle duality of quantum physics destroys any notion of a well-defined electron trajectory, and *all* we know about the electrons is the size and shape of the electron cloud.

Experiments at the end of the 19th century revealed that electrons are particles with both mass and a negative charge. The nucleus is a composite structure consisting of *protons,* positively charged particles, and neutral *neutrons.* The atom is held together by the attractive electric force between the positive nucleus and the negative electrons.

One of the most important discoveries is that **charge, like mass, is an inherent property of electrons and protons.** It's no more possible to have an electron without charge than it is to have an electron without mass. As far as we know today, electrons and protons have charges of opposite sign but *exactly* equal magnitude. (Very careful experiments have never found any difference.) This atomic-level unit of charge, called the **fundamental unit of charge,** is represented by the symbol e. Table 25.1 shows the masses and charges of protons and electrons. We need to define a unit of charge, which we will do in Section 25.4, before we can specify how much charge e is.

The Micro/Macro Connection

Electrons and protons are the basic charges of ordinary matter. Consequently, the various observations we made in Section 25.1 need to be explained in terms of electrons and protons.

NOTE ▶ Electrons and protons are particles of matter. Their motion is governed by Newton's laws. Electrons can move from one object to another when the objects are in contact, but neither electrons nor protons can leap through the air from one object to another. An object does not become charged simply from being close to a charged object. ◀

FIGURE 25.1 An atom.

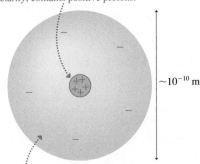

The nucleus, exaggerated for clarity, contains positive protons.

$\sim 10^{-10}$ m

The electron cloud is negatively charged.

TABLE 25.1 Protons and electrons

Particle	Mass (kg)	Charge
Proton	1.67×10^{-27}	$+e$
Electron	9.11×10^{-31}	$-e$

Charge is represented by the symbol q (or sometimes Q). A macroscopic object, such as a plastic rod, has charge

$$q = N_p e - N_e e = (N_p - N_e)e \qquad (25.1)$$

where N_p and N_e are the number of protons and electrons contained in the object. Most macroscopic objects have an *equal number* of protons and electrons and therefore have $q = 0$. An object with no *net* charge (i.e., $q = 0$) is said to be *electrically neutral*.

NOTE ▶ *Neutral* does *not* mean "no charges" but, instead, means that there is no *net* charge. ◀

A charged object has an unequal number of protons and electrons. An object is positively charged if $N_p > N_e$. It is negatively charged if $N_p < N_e$. Notice that an object's charge is always an integer multiple of e. That is, the amount of charge on an object varies by small but discrete steps, not continuously. This is called **charge quantization.**

In practice, objects acquire a positive charge not by gaining protons, as you might expect, but by losing electrons. Protons are *extremely* tightly bound within the nucleus and cannot be added to or removed from atoms. Electrons, on the other hand, are bound rather loosely and can be removed without great difficulty. The process of removing an electron from the electron cloud of an atom is called **ionization.** An atom that is missing an electron is called a *positive ion*. Its *net* charge is $q = +e$.

Some atoms can accommodate an *extra* electron and thus become a *negative ion* with net charge $q = -e$. A saltwater solution is a good example. When table salt (the chemical sodium chloride, NaCl) dissolves, it separates into positive sodium ions Na⁺ and negative chlorine ions Cl⁻. FIGURE 25.2 shows positive and negative ions.

All the charging processes we observed in Section 25.1 involved rubbing and friction. The forces of friction cause molecular bonds at the surface to break as the two materials slide past each other. Molecules are electrically neutral, but FIGURE 25.3 shows that *molecular ions* can be created when one of the bonds in a large molecule is broken. The positive molecular ions remain on one material and the negative ions on the other, so one of the objects being rubbed ends up with a net positive charge and the other with a net negative charge. This is the way in which a plastic rod is charged by rubbing with wool or a comb is charged by passing through your hair.

Charge Conservation and Charge Diagrams

One of the important discoveries about charge is the **law of conservation of charge:** Charge is neither created nor destroyed. Charge can be transferred from one object to another as electrons and ions move about, but the *total* amount of charge remains constant. For example, charging a plastic rod by rubbing it with wool transfers electrons from the wool to the plastic as the molecular bonds break. The wool is left with a positive charge equal in magnitude but opposite in sign to the negative charge of the rod: $q_{wool} = -q_{plastic}$. The *net* charge remains zero.

Diagrams are going to be an important tool for understanding and explaining charges and the forces on charged objects. As you begin to use diagrams, it will be important to make explicit use of charge conservation. The net number of plusses and minuses drawn on your diagrams should *not* change as you show them moving around.

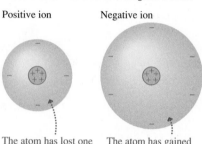

FIGURE 25.2 Positive and negative ions.

Positive ion   Negative ion

The atom has lost one electron, giving it a net positive charge.

The atom has gained one electron, giving it a net negative charge.

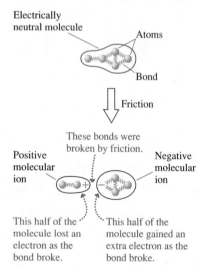

FIGURE 25.3 Charging by friction usually creates molecular ions as bonds are broken.

Electrically neutral molecule

Atoms

Bond

Friction

These bonds were broken by friction.

Positive molecular ion

Negative molecular ion

This half of the molecule lost an electron as the bond broke.

This half of the molecule gained an extra electron as the bond broke.

STOP TO THINK 25.2 Rank in order, from most positive to most negative, the charges q_a to q_e of these five systems.

Proton	Electron	17 protons 19 electrons	1,000,000 protons 1,000,000 electrons	Glass ball missing 3 electrons
•	•			◦
(a)	(b)	(c)	(d)	(e)

25.3 Insulators and Conductors

You have seen that there are two classes of materials as defined by their electrical properties: insulators and conductors. It's time for a closer look at these materials.

FIGURE 25.4 looks inside an insulator and a metallic conductor. The electrons in the insulator are all tightly bound to the positive nuclei and not free to move around. Charging an insulator by friction leaves patches of molecular ions on the surface, but these patches are immobile.

In metals, the outer atomic electrons (called the *valence electrons* in chemistry) are only weakly bound to the nuclei. As the atoms come together to form a solid, these outer electrons become detached from their parent nuclei and are free to wander about through the entire solid. The solid *as a whole* remains electrically neutral, because we have not added or removed any electrons, but the electrons are now rather like a negatively charged gas or liquid—what physicists like to call a **sea of electrons**—permeating an array of positively charged **ion cores.**

The primary consequence of this structure is that electrons in a metal are highly mobile. They can quickly and easily move through the metal in response to electric forces. The motion of charges through a material is what we will later call a **current,** and the charges that physically move are called the **charge carriers.** The charge carriers in metals are electrons.

Metals aren't the only conductors. Ionic solutions, such as salt water, are also good conductors. But the charge carriers in an ionic solution are the ions, not electrons. We'll focus on metallic conductors because of their importance in applications of electricity.

Charging

Insulators are often charged by rubbing. The charge diagrams of FIGURE 25.5 show that the charges on the rod are on the surface and that charge is conserved. The charge can be transferred to another object upon contact, but it doesn't move around on the rod.

FIGURE 25.5 An insulating rod is charged by rubbing.

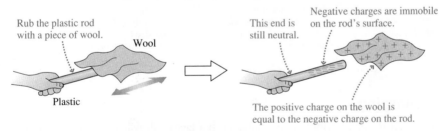

Metals usually cannot be charged by rubbing, but Experiment 9 showed that a metal sphere can be charged by contact with a charged plastic rod. FIGURE 25.6 gives a pictorial explanation. An essential idea is that **the electrons in a conductor are free to move.** Once charge is transferred to the metal, repulsive forces between the negative charges cause the electrons to move apart from each other.

Note that the newly added electrons do not themselves need to move to the far corners of the metal. Because of the repulsive forces, the newcomers simply "shove" the entire electron sea a little to the side. The electron sea takes an extremely short time to adjust itself to the presence of the added charge, typically less than 10^{-9} s. For all practical purposes, a conductor responds *instantaneously* to the addition or removal of charge.

Other than this very brief interval during which the electron sea is adjusting, the charges in an *isolated* conductor are in static equilibrium. That is, the charges are at rest and there is no net force on any charge. This condition is called **electrostatic equilibrium.** If there *were* a net force on one of the charges, it would quickly move to an equilibrium point at which the force was zero.

FIGURE 25.4 A microscopic look at insulators and conductors.

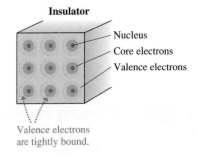

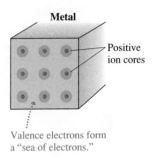

FIGURE 25.6 A conductor is charged by contact with a charged plastic rod.

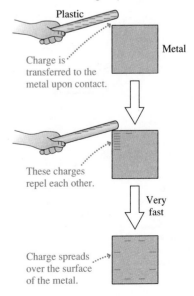

Electrostatic equilibrium has an important consequence:

In an isolated conductor, any excess charge is located on the surface of the conductor.

To see this, suppose there *were* an excess electron in the interior of an isolated conductor. The extra electron would upset the electrical neutrality of the interior and exert forces on nearby electrons, causing them to move. But their motion would violate the assumption of static equilibrium, so we're forced to conclude that there cannot be any excess electrons in the interior. Any excess electrons push each other apart until they're all on the surface.

EXAMPLE 25.2 Charging an electroscope

Many electricity demonstrations are carried out with the help of an *electroscope* like the one shown in FIGURE 25.7. Touching the sphere at the top of an electroscope with a charged plastic rod causes the leaves to fly apart and remain hanging at an angle. Use charge diagrams to explain why.

MODEL We'll use the charge model and the model of a conductor as a material through which electrons move.

VISUALIZE FIGURE 25.8 uses a series of charge diagrams to show the charging of an electroscope.

FIGURE 25.7 A charged electroscope.

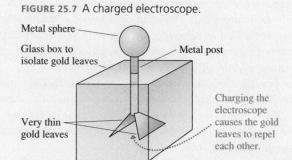

FIGURE 25.8 The process by which an electroscope is charged.

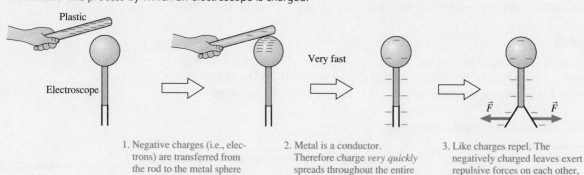

1. Negative charges (i.e., electrons) are transferred from the rod to the metal sphere upon contact.

2. Metal is a conductor. Therefore charge *very quickly* spreads throughout the entire electroscope.

3. Like charges repel. The negatively charged leaves exert repulsive forces on each other, causing them to spread apart.

FIGURE 25.9 Touching a charged metal discharges it.

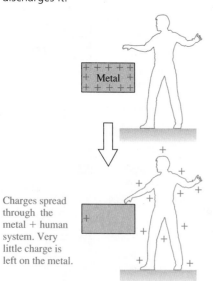

Charges spread through the metal + human system. Very little charge is left on the metal.

Discharging

Pure water is not a terribly good conductor, but nearly all water contains a variety of dissolved minerals that float around as ions. Dissolved table salt, as we noted previously, separates into Na^+ and Cl^- ions. These ions are the charge carriers, allowing salt water to be a fairly good conductor.

The human body consists largely of salt water. Consequently, and occasionally tragically, humans are reasonably good conductors. This fact allows us to understand how it is that *touching* a charged object discharges it, as we observed in Experiment 10. As FIGURE 25.9 shows, the net effect of touching a charged metal is that it and the conducting human together become a much larger conductor than the metal alone. Any excess charge that was initially confined to the metal can now spread over the larger metal + human conductor. This may not entirely discharge the metal, but in typical circumstances, where the human is much larger than the metal, the residual charge remaining on the metal is much reduced from the original charge. The metal, for most practical purposes, is discharged. In essence, two conductors in contact "share" the charge that was originally on just one of them.

Moist air is a conductor, although a rather poor one. Charged objects in air slowly lose their charge as the object shares its charge with the air. The earth itself is a giant conductor because of its water, moist soil, and a variety of ions. Any object that is physically connected to the earth through a conductor is said to be **grounded.** The effect of being grounded is that the object shares any excess charge it has with the entire earth! But the earth is so enormous that any conductor attached to the earth will be completely discharged.

The purpose of *grounding* objects, such as circuits and appliances, is to prevent the buildup of any charge on the objects. The third prong on appliances and electronics that have a three-prong plug is the ground connection. The building wiring physically connects that third wire deep into the ground somewhere just outside the building, often by attaching it to a metal water pipe that goes underground.

Charge Polarization

One observation from Section 25.1 still needs an explanation. How do charged objects of either sign exert an attractive force on a *neutral* object? To begin answering this question, FIGURE 25.10 shows a positively charged rod held close to—but not touching—a *neutral* electroscope. The leaves move apart and stay apart as long as you hold the rod near, but they quickly collapse when it is removed. Can we understand this behavior?

We can, and FIGURE 25.11a shows how. Although the metal as a whole is still electrically neutral, we say that the object has been *polarized*. **Charge polarization** is a slight separation of the positive and negative charges in a neutral object. Charge polarization produces an excess positive charge on the leaves of the electroscope shown in FIGURE 25.11b, so they repel each other. But because the electroscope has no *net* charge, the electron sea quickly readjusts once the rod is removed.

FIGURE 25.10 A charged rod held close to an electroscope causes the leaves to repel each other.

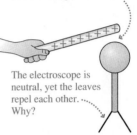

Bring a positively charged glass rod close to an electroscope without touching the sphere.

The electroscope is neutral, yet the leaves repel each other. Why?

FIGURE 25.11 A charged rod polarizes a metal.

(a) The sea of electrons is attracted to the rod and shifts so that there is excess negative charge on the near surface.

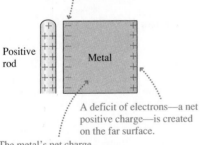

Positive rod

Metal

A deficit of electrons—a net positive charge—is created on the far surface.

The metal's net charge is still zero, but it has been *polarized* by the charged rod.

(b) The electroscope is polarized by the charged rod. The sea of electrons shifts toward the positive rod.

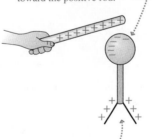

Although the net charge on the electroscope is still zero, the leaves have excess positive charge and repel each other.

Why don't *all* the electrons in Figure 25.11a rush to the side near the positive charge? Once the electron sea shifts slightly, the stationary positive ions begin to exert a force, a restoring force, pulling the electrons back to the right. The equilibrium position for the sea of electrons is just far enough to the left that the forces due to the external charge and the positive ions are in balance. In practice, the displacement of the electron sea is usually *less than* 10^{-15} *m*!

Charge polarization explains not only why the electroscope leaves deflect but also how a charged object exerts an attractive force on a neutral object. FIGURE 25.12 on the next page shows a positively charged rod near a neutral piece of metal. Because the electric force decreases with distance, the attractive force on the electrons at the top surface is *slightly greater* than the repulsive force on the ions at the bottom. The net force toward the charged rod is called a **polarization force.** The polarization force arises because the charges in the metal are separated, *not* because the rod and metal are oppositely charged.

FIGURE 25.12 The polarization force on a neutral piece of metal is due to the slight charge separation.

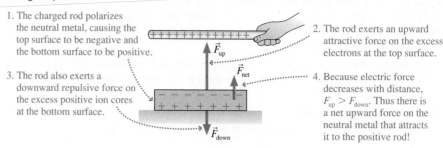

1. The charged rod polarizes the neutral metal, causing the top surface to be negative and the bottom surface to be positive.

2. The rod exerts an upward attractive force on the excess electrons at the top surface.

3. The rod also exerts a downward repulsive force on the excess positive ion cores at the bottom surface.

4. Because electric force decreases with distance, $F_{up} > F_{down}$. Thus there is a net upward force on the neutral metal that attracts it to the positive rod!

A negatively charged rod would push the electron sea slightly away, polarizing the metal to have a positive upper surface charge and a negative lower surface charge. Once again, these are the conditions for the charge to exert a *net attractive force* on the metal. Thus our charge model explains how a charged object of *either* sign attracts neutral pieces of metal.

The Electric Dipole

Now let's consider a slightly trickier situation. Why does a charged rod pick up paper, which is an insulator rather than a metal? First consider what happens when we bring a positive charge near an atom. As FIGURE 25.13a shows, the charge polarizes the atom. The electron cloud doesn't move far, because the force from the positive nucleus pulls it back, but the center of positive charge and the center of negative charge are now slightly separated.

FIGURE 25.13 A neutral atom is polarized by an external charge, forming an *electric dipole*.

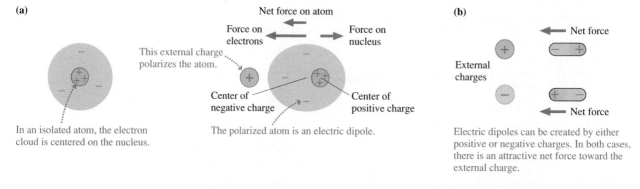

(a)

In an isolated atom, the electron cloud is centered on the nucleus.

This external charge polarizes the atom.

Net force on atom

Force on electrons

Force on nucleus

Center of negative charge

Center of positive charge

The polarized atom is an electric dipole.

(b)

External charges

Net force

Net force

Electric dipoles can be created by either positive or negative charges. In both cases, there is an attractive net force toward the external charge.

FIGURE 25.14 The atoms in an insulator are polarized by an external charge.

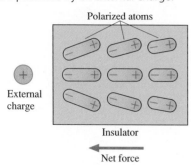

Polarized atoms

External charge

Insulator

Net force

Two opposite charges with a slight separation between them form what is called an **electric dipole.** FIGURE 25.13b shows that an external charge of either sign polarizes the atom to produce an electric dipole with the near end opposite in sign to the charge. (The actual distortion from a perfect sphere is minuscule, nothing like the distortion shown in the figure.) The attractive force on the dipole's near end *slightly* exceeds the repulsive force on its far end because the near end is closer to the charge. The net force, an *attractive* force between the charge and the atom, is another example of a polarization force.

An insulator has no sea of electrons to shift if an external charge is brought close. Instead, as FIGURE 25.14 shows, all the individual atoms inside the insulator become polarized. The polarization force acting *on each atom* produces a net polarization force toward the external charge. This solves the puzzle. A charged rod picks up pieces of paper by

■ Polarizing the atoms in the paper,
■ Then exerting an attractive polarization force on each atom.

This is important. Make sure you understand all the steps in the reasoning.

STOP TO THINK 25.3   An electroscope is positively charged by *touching* it with a positive glass rod. The electroscope leaves spread apart and the glass rod is removed. Then a negatively charged plastic rod is brought close to the top of the electroscope, but it doesn't touch. What happens to the leaves?

    a. The leaves get closer together.
    b. The leaves spread farther apart.
    c. One leaf moves higher, the other lower.
    d. The leaves don't move.

Charging by Induction

Charge polarization is responsible for an interesting and counterintuitive way of charging an electroscope. **FIGURE 25.15** shows a positively charged glass rod held near an electroscope but not touching it, while a person touches the electroscope with a finger. Unlike what happens in Figure 25.10, the electroscope leaves hardly move.

FIGURE 25.15 Charging by induction.

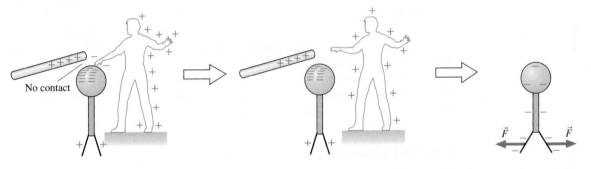

1. The charged rod polarizes the electroscope + person conductor. The leaves repel slightly due to polarization, but overall the electroscope has an excess of electrons and the person has a deficit of electrons.

2. The negative charge on the electroscope is isolated when contact is broken.

3. When the rod is removed, the leaves first collapse as the polarization vanishes, then repel as the excess negative charge spreads out. The electroscope has been *negatively* charged.

Charge polarization occurs, as it did in Figure 25.10, but this time in the much larger electroscope + person conductor. If the person removes his or her finger while the system is polarized, the electroscope is left with a *net* negative charge and the person has a net positive charge. The electroscope has been charged *opposite to the rod* in a process called **charging by induction.**

25.4 Coulomb's Law

The first three sections have established a *model* of charges and electric forces. This model has successfully provided a qualitative explanation of electric phenomena; now it's time to become quantitative. Experiment 4 in Section 25.1 found that the electric force decreases with distance. The force law that describes this behavior is known as *Coulomb's law*.

Charles Coulomb was one of many scientists investigating electricity in the late 18th century. Coulomb had the idea of studying electric forces using the torsion balance scheme by which Cavendish had measured the value of the gravitational constant G (see Section 13.4). This was a difficult experiment. Despite many obstacles, Coulomb announced in 1785 that the electric force obeys an *inverse-square law* analogous to Newton's law of gravity. Today we know it as **Coulomb's law.**

> **Coulomb's law:**
>
> 1. If two charged particles having charges q_1 and q_2 are a distance r apart, the particles exert forces on each other of magnitude
>
> $$F_{1 \text{ on } 2} = F_{2 \text{ on } 1} = \frac{K|q_1||q_2|}{r^2} \qquad (25.2)$$
>
> where K is called the **electrostatic constant**. These forces are an action/reaction pair, equal in magnitude and opposite in direction.
> 2. The forces are directed along the line joining the two particles. The forces are *repulsive* for two like charges and *attractive* for two opposite charges.

FIGURE 25.16 Attractive and repulsive forces between charged particles.

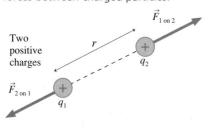

Two positive charges

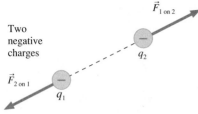

Two negative charges

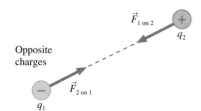

Opposite charges

We sometimes speak of the "force between charge q_1 and charge q_2," but keep in mind that we are really dealing with charged *objects* that also have a mass, a size, and other properties. Charge is not some disembodied entity that exists apart from matter. Coulomb's law describes the force between charged *particles,* which are also called **point charges.** A charged particle, which is an extension of the particle model we used in Part I, has a mass and a charge but has no size.

Coulomb's law looks much like Newton's law of gravity, but there is one important difference: The charge q can be either positive or negative. Consequently, the absolute value signs in Equation 25.2 are especially important. The first part of Coulomb's law gives only the *magnitude* of the force, which is always positive. The direction must be determined from the second part of the law. FIGURE 25.16 shows the forces between different combinations of positive and negative charges.

Units of Charge

Coulomb had no *unit* of charge, so he was unable to determine a value for K, whose numerical value depends on the units of both charge and distance. The SI unit of charge, the **coulomb** (C), is derived from the SI unit of *current,* so we'll have to await the study of current in Chapter 30 before giving a precise definition. For now we'll note that the fundamental unit of charge e has been measured to have the value

$$e = 1.60 \times 10^{-19} \text{ C}$$

This is a very small amount of charge. Stated another way, 1 C is the net charge of roughly 6.25×10^{18} protons.

NOTE ▶ The amount of charge produced by rubbing plastic or glass rods is typically in the range 1 nC (10^{-9} C) to 100 nC (10^{-7} C). This corresponds to an excess or deficit of 10^{10} to 10^{12} electrons. ◀

Once the unit of charge is established, torsion balance experiments such as Coulomb's can be used to measure the electrostatic constant K. In SI units

$$K = 8.99 \times 10^9 \text{ N m}^2/\text{C}^2$$

It is customary to round this to $K = 9.0 \times 10^9 \text{ N m}^2/\text{C}^2$ for all but extremely precise calculations, and we will do so.

Surprisingly, we will find that Coulomb's law is not explicitly used in much of the theory of electricity. While it *is* the basic force law, most of our future discussion and calculations will be of things called *fields* and *potentials.* It turns out that we can make many future equations easier to use if we rewrite Coulomb's law in a somewhat more complicated way. Let's define a new constant, called the **permittivity constant** ϵ_0 (pronounced "epsilon zero" or "epsilon naught"), as

$$\epsilon_0 = \frac{1}{4\pi K} = 8.85 \times 10^{-12} \text{ C}^2/\text{N m}^2$$

Rewriting Coulomb's law in terms of ϵ_0 gives us

$$F = \frac{1}{4\pi\epsilon_0} \frac{|q_1||q_2|}{r^2} \qquad (25.3)$$

It will be easiest when using Coulomb's law directly to use the electrostatic constant K. However, in later chapters we will switch to the second version with ϵ_0.

Using Coulomb's Law

Coulomb's law is a force law, and forces are vectors. It has been many chapters since we made much use of vectors and vector addition, but these mathematical techniques will be essential in our study of electricity and magnetism.

There are two important observations regarding Coulomb's law:

1. **Coulomb's law applies only to point charges.** A point charge is an idealized material object with charge and mass but with no size or extension. For practical purposes, two charged objects can be modeled as point charges if they are much smaller than the separation between them.
2. **Electric forces, like other forces, can be superimposed.** If multiple charges 1, 2, 3, . . . are present, the *net* electric force on charge j due to all other charges is

$$\vec{F}_{\text{net}} = \vec{F}_{1 \text{ on } j} + \vec{F}_{2 \text{ on } j} + \vec{F}_{3 \text{ on } j} + \cdots \qquad (25.4)$$

where each of the $\vec{F}_{i \text{ on } j}$ is given by Equation 25.2 or 25.3.

These conditions are the basis of a strategy for using Coulomb's law to solve electrostatic force problems.

PROBLEM-SOLVING
STRATEGY 25.1   **Electrostatic forces and Coulomb's law**   (MP)

MODEL Identify point charges or objects that can be modeled as point charges.

VISUALIZE Use a *pictorial representation* to establish a coordinate system, show the positions of the charges, show the force vectors on the charges, define distances and angles, and identify what the problem is trying to find. This is the process of translating words to symbols.

SOLVE The mathematical representation is based on Coulomb's law:

$$F_{1 \text{ on } 2} = F_{2 \text{ on } 1} = \frac{K|q_1||q_2|}{r^2}$$

- Show the directions of the forces—repulsive for like charges, attractive for opposite charges—on the pictorial representation.
- When possible, do graphical vector addition on the pictorial representation. While not exact, it tells you the type of answer you should expect.
- Write each force vector in terms of its x- and y-components, then add the components to find the net force. Use the pictorial representation to determine which components are positive and which are negative.

ASSESS Check that your result has the correct units, is reasonable, and answers the question.

Exercise 26

EXAMPLE 25.3 **The point of zero force**

Two positively charged particles q_1 and $q_2 = 3q_1$ are 10.0 cm apart on the x-axis. Where (other than at infinity) could a third charge q_3 be placed so as to experience no net force?

MODEL Model the charged particles as point charges.

VISUALIZE FIGURE 25.17 establishes a coordinate system with q_1 at the origin. We first need to identify the region of space in which q_3 must be located. We have no information about the sign of q_3, so apparently the position for which we are looking will work for either sign. You can see from the figure that the forces at point A, above the axis, and at point B, outside the charges, cannot possibly add to zero. However, at point C on the x-axis *between* the charges, the two forces are oppositely directed.

FIGURE 25.17 A pictorial representation of the charges and forces.

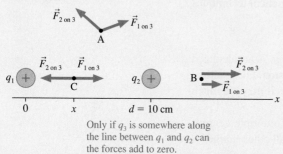

Only if q_3 is somewhere along the line between q_1 and q_2 can the forces add to zero.

SOLVE The mathematical problem is to find the position for which the forces $\vec{F}_{1\,on\,3}$ and $\vec{F}_{2\,on\,3}$ are equal in magnitude. If q_3 is distance x from q_1, it is distance $d - x$ from q_2. The *magnitudes* of the forces are

$$F_{1\,on\,3} = \frac{Kq_1|q_3|}{r_{13}^2} = \frac{Kq_1|q_3|}{x^2}$$

$$F_{2\,on\,3} = \frac{Kq_2|q_3|}{r_{23}^2} = \frac{K(3q_1)|q_3|}{(d-x)^2}$$

Charges q_1 and q_2 are positive and do not need absolute value signs. Equating the two forces gives

$$\frac{Kq_1|q_3|}{x^2} = \frac{3Kq_1|q_3|}{(d-x)^2}$$

The term $Kq_1|q_3|$ cancels. Multiplying by $x^2(d-x)^2$ gives

$$(d-x)^2 = 3x^2$$

which can be rearranged into the quadratic equation

$$2x^2 + 2dx - d^2 = 2x^2 + 20x - 100 = 0$$

where we used $d = 10$ cm and x is in cm. The solutions to this equation are

$$x = +3.66 \text{ cm and } -13.66 \text{ cm}$$

Both are points where the *magnitudes* of the two forces are equal, but $x = -13.66$ cm is a point where the magnitudes are equal while the directions are the same. The solution we want, which is between the charges, is $x = 3.66$ cm. Thus the point to place q_3 is 3.66 cm from q_1 along the line joining q_1 and q_2.

ASSESS q_1 is smaller than q_2, so we expect the point at which the forces balance to be closer to q_1 than to q_2. The solution seems reasonable. Note that the problem statement has no coordinates, so "$x = 3.66$ cm" is *not* an acceptable answer. You need to describe the position relative to q_1 and q_2.

EXAMPLE 25.4 **Three charges**

Three charged particles with $q_1 = -50$ nC, $q_2 = +50$ nC, and $q_3 = +30$ nC are placed on the corners of the 5.0 cm × 10.0 cm rectangle shown in FIGURE 25.18. What is the net force on charge q_3 due to the other two charges? Give your answer both in component form and as a magnitude and direction.

FIGURE 25.18 The three charges.

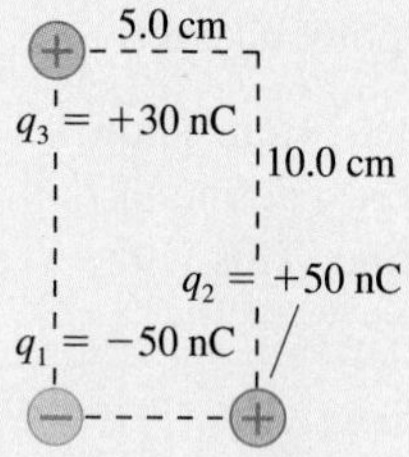

MODEL Model the charged particles as point charges.

VISUALIZE The pictorial representation of FIGURE 25.19 establishes a coordinate system. q_1 and q_3 are opposite charges, so force vector $\vec{F}_{1\,on\,3}$ is an attractive force toward q_1. q_2 and q_3 are like charges, so force vector $\vec{F}_{2\,on\,3}$ is a repulsive force away from q_2. q_1 and q_2 have equal magnitudes, but $\vec{F}_{2\,on\,3}$ has been drawn shorter than $\vec{F}_{1\,on\,3}$ because q_2 is farther from q_3. Vector addition has been used to draw the net force vector and to define the angle ϕ.

SOLVE The question asks for a *force*, so our answer will be the *vector* sum $\vec{F}_3 = \vec{F}_{1\,on\,3} + \vec{F}_{2\,on\,3}$. We need to write $\vec{F}_{1\,on\,3}$ and $\vec{F}_{2\,on\,3}$ in component form. The magnitude of force $\vec{F}_{1\,on\,3}$ can be found using Coulomb's law:

FIGURE 25.19 A pictorial representation of the charges and forces.

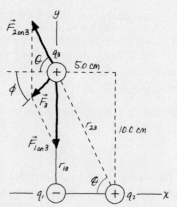

$$F_{1\,on\,3} = \frac{K|q_1||q_3|}{r_{13}^2}$$

$$= \frac{(9.0 \times 10^9 \text{ N m}^2/\text{C}^2)(50 \times 10^{-9} \text{ C})(30 \times 10^{-9} \text{ C})}{(0.100 \text{ m})^2}$$

$$= 1.35 \times 10^{-3} \text{ N}$$

where we used $r_{13} = 10.0$ cm.

The pictorial representation shows that $\vec{F}_{1\,on\,3}$ points downward, in the negative y-direction, so

$$\vec{F}_{1\,on\,3} = -1.35 \times 10^{-3}\hat{j} \text{ N}$$

To calculate $\vec{F}_{2\,on\,3}$ we first need the distance r_{23} between the charges:

$$r_{23} = \sqrt{(5.0 \text{ cm})^2 + (10.0 \text{ cm})^2} = 11.2 \text{ cm}$$

The magnitude of $\vec{F}_{2\,on\,3}$ is thus

$$F_{2\,on\,3} = \frac{K|q_2||q_3|}{r_{23}^2}$$

$$= \frac{(9.0 \times 10^9 \text{ N m}^2/\text{C}^2)(50 \times 10^{-9} \text{ C})(30 \times 10^{-9} \text{ C})}{(0.112 \text{ m})^2}$$

$$= 1.08 \times 10^{-3} \text{ N}$$

This is only a magnitude. The *vector* $\vec{F}_{2\,on\,3}$ is

$$\vec{F}_{2\,on\,3} = -F_{2\,on\,3}\cos\theta\,\hat{i} + F_{2\,on\,3}\sin\theta\,\hat{j}$$

where angle θ is defined in the figure and the signs (negative x-component, positive y-component) were determined from the pictorial representation. From the geometry of the rectangle,

$$\theta = \tan^{-1}\left(\frac{10.0 \text{ cm}}{5.0 \text{ cm}}\right) = \tan^{-1}(2.0) = 63.4°$$

Thus $\vec{F}_{2\,on\,3} = (-4.83\hat{i} + 9.66\hat{j}) \times 10^{-4}$ N. Now we can add $\vec{F}_{1\,on\,3}$ and $\vec{F}_{2\,on\,3}$ to find

$$\vec{F}_3 = \vec{F}_{1\,on\,3} + \vec{F}_{2\,on\,3} = (-4.83\hat{i} - 3.84\hat{j}) \times 10^{-4} \text{ N}$$

This would be an acceptable answer for many problems, but sometimes we need the net force as a magnitude and direction. With angle ϕ as defined in the figure, these are

$$F_3 = \sqrt{F_{3x}^2 + F_{3y}^2} = 6.2 \times 10^{-4} \text{ N}$$

$$\phi = \tan^{-1}\left|\frac{F_{3y}}{F_{3x}}\right| = 38°$$

Thus $\vec{F}_3 = (6.2 \times 10^{-4} \text{ N}, 38°$ below the negative x-axis).

ASSESS The forces are not large, but they are typical of electrostatic forces. Even so, you'll soon see that these forces can produce very large accelerations because the masses of the charged objects are usually very small.

EXAMPLE 25.5 **Lifting a glass bead**

A small plastic sphere charged to -10 nC is held 1.0 cm above a small glass bead at rest on a table. The bead has a mass of 15 mg and a charge of $+10$ nC. Will the glass bead "leap up" to the plastic sphere?

MODEL Model the plastic sphere and glass bead as point charges.

VISUALIZE FIGURE 25.20 establishes a y-axis, identifies the plastic sphere as q_1 and the glass bead as q_2, and shows a free-body diagram.

SOLVE If $F_{1\,on\,2}$ is less than the gravitational force $F_G = m_{bead}g$, then the bead will remain at rest on the table with $\vec{F}_{1\,on\,2} + \vec{F}_G + \vec{n} = \vec{0}$. But if $F_{1\,on\,2}$ is greater than $m_{bead}g$, the glass bead will accelerate upward from the table. Using the values provided, we have

$$F_{1\,on\,2} = \frac{K|q_1||q_2|}{r^2} = 9.0 \times 10^{-3} \text{ N}$$

$$F_G = m_{bead}g = 1.5 \times 10^{-4} \text{ N}$$

$F_{1\,on\,2}$ exceeds $m_{bead}g$ by a factor of 60, so the glass bead will leap upward.

FIGURE 25.20 A pictorial representation of the charges and forces.

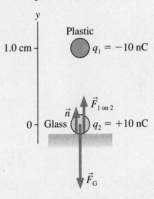

ASSESS The values used in this example are realistic for spheres ≈ 2 mm in diameter. In general, as in this example, electric forces are *significantly* larger than gravitational forces. Consequently, we can neglect gravity when working electric-force problems unless the particles are fairly massive.

EXAMPLE 25.6 **A point charge and a charged wire**

Coulomb's inverse-square law applies only to the electric force between two point charges. Your lab assignment for the week is to discover the law describing the force between a point charge and a long, straight, charged metal wire. It is postulated that the force on a point charge q is characterized by a *power law* $F \propto qr^n$, where r is the distance from the wire. To test this hypothesis and, if it is correct, to determine the exponent n, you first set up a long, straight metal wire and charge it by connecting it to a high-voltage

Continued

power supply. You then charge a small plastic ball and, using a sensitive force probe, measure the force on the ball at different distances from the wire. Your data are as follows:

Distance (cm)	Force (μN)
2.0	895
4.0	455
6.0	310
8.0	215
10.0	185

Is the force described by a power law? And if so, what is the exponent?

MODEL Model the small plastic ball as a point charge.

SOLVE A power law is represented by a linear log-log graph. To see why, we can write the postulated force law as

$$F = cqr^n$$

where c is an unknown proportionality constant. If we take the logarithm of both sides, applying the rules $\log(ab) = \log a + \log b$ and $\log a^n = n \log a$, we get

$$\log F = \log(cqr^n) = \log(cq) + \log r^n = \log(cq) + n \log r$$

If we plot $\log F$ on the y-axis against $\log r$ on the x-axis—a log-log graph—it should be a straight line with slope n. A nonlinear log-log graph would disprove the hypothesis that the force is characterized by a power law.

FIGURE 25.21 is a graph of $\log F$ versus $\log r$. It is clearly linear, which validates the postulated power-law force. And because distances were measured to only two significant figures, the experimental slope of -0.997 is consistent with the simpler $n = -1$. Thus our data show that the force between a point charge and a long, charged wire can be characterized as $F \propto q/r$.

FIGURE 25.21 A log-log graph of force versus distance.

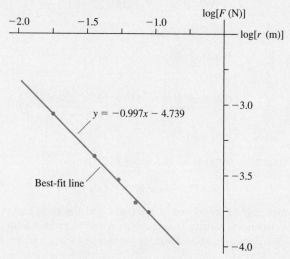

$y = -0.997x - 4.739$

Best-fit line

ASSESS The force depends inversely on the distance. The inverse-square dependence of Coulomb's law describes only the force between two point charges.

STOP TO THINK 25.4 Charged spheres A and B exert repulsive forces on each other. $q_A = 4q_B$. Which statement is true?

a. $F_{A \text{ on } B} > F_{B \text{ on } A}$ b. $F_{A \text{ on } B} = F_{B \text{ on } A}$ c. $F_{A \text{ on } B} < F_{B \text{ on } A}$

25.5 The Field Model

FIGURE 25.22 If charge A moves, how long does it take the force vector on B to respond?

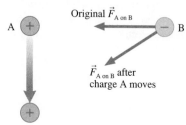

Original $\vec{F}_{A \text{ on } B}$

$\vec{F}_{A \text{ on } B}$ after charge A moves

Electric and magnetic forces, like gravity, are *long-range* forces; no contact is required for one charged particle to exert a force on another. But this raises some troubling issues. For example, consider the charged particles A and B in **FIGURE 25.22**. If A suddenly starts moving, as shown by the arrow, the force vector on B must pivot to follow A. Does this happen *instantly*? Or is there some *delay* between when A moves and when the force $\vec{F}_{A \text{ on } B}$ responds?

Neither Coulomb's law nor Newton's law of gravity is dependent on time, so the answer from the perspective of Newtonian physics has to be "instantly." Yet most scientists found this troubling. What if A is 100,000 light years from B? Will B respond *instantly* to an event 100,000 light years away? The idea of instantaneous transmission of forces had become unbelievable to most scientists by the beginning of the 19th century. But if there is a delay, how long is it? How does the information to "change force" get sent from A to B? These were the issues when a young Michael Faraday appeared on the scene.

Michael Faraday is one of the most interesting figures in the history of science. Because of the late age at which he started his education—he was a teenager—he never became fluent in mathematics. In place of equations, Faraday's brilliant and insightful mind developed many ingenious *pictorial* methods for thinking about and describing physical phenomena. By far the most important of these was the field.

The Concept of a Field

Faraday was particularly impressed with the pattern that iron filings make when sprinkled around a magnet, as seen in FIGURE 25.23. The pattern's regularity and the curved lines suggested to Faraday that the *space itself* around the magnet is filled with some kind of magnetic influence. Perhaps the magnet in some way alters the space around it. In this view, a piece of iron near the magnet responds not directly to the magnet but, instead, to the alteration of space caused by the magnet. This space alteration, whatever it is, is the *mechanism* by which the long-range force is exerted.

FIGURE 25.24 illustrates Faraday's idea. The Newtonian view was that A and B interact directly. In Faraday's view, A first alters or modifies the space around it, and particle B then comes along and interacts with this altered space. The alteration of space becomes the *agent* by which A and B interact. Furthermore, this alteration could easily be imagined to take a finite time to propagate outward from A, perhaps in a wave-like fashion. If A changes, B responds only when the new alteration of space reaches it. The interaction between B and this alteration of space is a *local* interaction, rather like a contact force.

Faraday's idea came to be called a **field.** The term "field," which comes from mathematics, describes a function that assigns a vector to every point in space. When used in physics, a field conveys the idea that the physical entity exists at every point in space. That is, indeed, what Faraday was suggesting about how long-range forces operate. The charge makes an alteration *everywhere* in space. Other charges then respond to the alteration at their position. The alteration of the space around a mass is called the *gravitational field*. Similarly, the space around a charge is altered to create the **electric field.**

NOTE ▶ The concept of a field is in sharp contrast to the concept of a particle. A particle exists at *one* point in space. The purpose of Newton's laws of motion is to determine how the particle moves from point to point along a trajectory. A field exists simultaneously at *all* points in space. ◀

Faraday's idea was not taken seriously at first; it seemed too vague and nonmathematical to scientists steeped in the Newtonian tradition of particles and forces. But the significance of the concept of field grew as electromagnetic theory developed during the first half of the 19th century. What seemed at first a pictorial "gimmick" came to be seen as more and more essential for understanding electric and magnetic forces.

Faraday's field ideas were finally placed on a mathematical foundation in 1865 by James Clerk Maxwell. Maxwell was able to describe completely all the known behaviors of electric and magnetic fields in four equations, known today as Maxwell's equations. We will explore aspects of Maxwell's theory as we go along, then look at the full implications of Maxwell's equations in Chapter 34.

The Electric Field

We begin our investigation of electric fields by postulating a **field model** that describes how charges interact:

1. Some charges, which we will call the **source charges,** alter the space around them by creating an *electric field* $\vec{E}$.
2. A separate charge *in* the electric field experiences a force $\vec{F}$ exerted *by the field*.

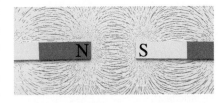

FIGURE 25.23 Iron filings sprinkled around the ends of a magnet suggest that the influence of the magnet extends into the space around it.

FIGURE 25.24 Newton's and Faraday's ideas about long-range forces.

In the Newtonian view, A exerts a force directly on B.

In Faraday's view, A alters the space around it. (The wavy lines are poetic license. We don't know what the alteration looks like.)

Particle B then responds to the altered space. The altered space is the agent that exerts the force on B.

Suppose some set of charges—the source charges—have created an electric field. We can learn about the field by using a different charge q as a *probe charge*. As we move the probe charge around from point to point in space, it experiences a changing electric force $\vec{F}_{\text{on }q}$ due to the other charges. This suggests that "something" is present at each point in space to cause the force that charge q experiences. We can use the force on the probe charge to define the electric field $\vec{E}$ at the point (x, y, z) as

$$\vec{E}(x, y, z) \equiv \frac{\vec{F}_{\text{on }q} \text{ at } (x, y, z)}{q} \tag{25.5}$$

We're *defining* the electric field as a force-to-charge ratio; hence the units of the electric field are newtons per coulomb, or N/C. The magnitude E of the electric field is called the **electric field strength.**

If a probe charge q experiences an electric force at a point in space, as FIGURE 25.25a shows, we say that there is an electric field at that point causing the force. Further, we *define* the electric field at that point to be the vector given by Equation 25.5. FIGURE 25.25b shows the electric field only at two points, but you can imagine "mapping out" the electric field by moving charge q all through space.

NOTE ▶ Probe charge q also creates an electric field. But charges don't exert forces on themselves, so q is measuring only the electric field of *other* charges. ◀

The basic idea of the field model is that **the field is the agent that exerts an electric force on a charged particle.** Notice three important ideas about the field:

1. Equation 25.5 assigns a *vector* to *every point* in space. That is, the electric field is a *vector field.* Electric field diagrams will show a sample of the vectors, but there is an electric field vector at every point whether one is shown or not.
2. If q is positive, the electric field vector points in the same direction as the force on the charge.
3. Because q appears in Equation 25.5, it may seem that the electric field depends on the size of the charge used to probe the field. It doesn't. We know from Coulomb's law that the force $\vec{F}_{\text{on }q}$ is proportional to q. Thus the electric field defined in Equation 25.5 is *independent* of the charge q that probes the field. The electric field depends only on the source charges that create the field.

In practice we often want to turn Equation 25.5 around and find the force exerted by a known field. That is, a charged particle with charge q at a point in space where the electric field is $\vec{E}$ experiences an electric force

$$\vec{F}_{\text{on }q} = q\vec{E} \tag{25.6}$$

If q is positive, the force on the particle is in the direction of $\vec{E}$. The force on a negative charge is *opposite* the direction of $\vec{E}$.

FIGURE 25.25 Charge q is a probe of the electric field.

(a)

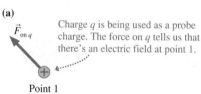

Charge q is being used as a probe charge. The force on q tells us that there's an electric field at point 1.

Point 1

Point 2

Now charge q is placed at point 2. There's also an electric field here that differs from the field at point 1.

(b)

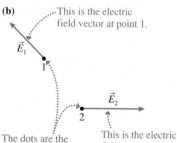

This is the electric field vector at point 1.

This is the electric field vector at point 2.

The dots are the points at which the field is known.

EXAMPLE 25.7 **Electric forces in a cell**

Every cell in your body is electrically active in various ways. For example, nerve propagation occurs when large electric fields in the cell membranes of neurons cause ions to move through the cell walls. The field strength in a typical cell membrane is 1.0×10^7 N/C. What is the magnitude of the electric force on a singly charged calcium ion?

MODEL The ion is a point charge in an electric field. A singly charged ion is missing one electron and has net charge $q = +e$.

SOLVE A charged particle in an electric field experiences an electric force $\vec{F}_{\text{on }q} = q\vec{E}$. In this case, the magnitude of the force is

$$F = eE = (1.6 \times 10^{-19}\text{ C})(1.0 \times 10^7\text{ N/C}) = 1.6 \times 10^{-12}\text{ N}$$

ASSESS This may seem like an incredibly tiny force, but it is applied to a particle with mass $m \sim 10^{-26}$ kg. The ion would have an unimaginable acceleration ($F/m \sim 10^{14}$ m/s^2) were it not for resistive forces as it moves through the membrane. Even so, an ion can cross the cell wall in less than 1 μs.

STOP TO THINK 25.5 An electron is placed at the position marked by the dot. The force on the electron is

$\xrightarrow{\vec{E}}$ $\xrightarrow{\vec{E}}$

•

$\xrightarrow{\vec{E}}$ $\xrightarrow{\vec{E}}$

a. Zero. b. To the right. c. To the left.
d. There's not enough information to tell.

The Electric Field of a Point Charge

We will begin to put the definition of the electric field to full use in the next chapter. For now, to develop the ideas, we will determine the electric field of a single point charge q. FIGURE 25.26a shows charge q and a point in space at which we would like to know the electric field. We need a second charge, shown as q' in FIGURE 25.26b, to serve as a probe of the electric field.

For the moment, assume both charges are positive. The force on q', which is repulsive and directed straight away from q, is given by Coulomb's law:

$$\vec{F}_{\text{on }q'} = \left(\frac{1}{4\pi\epsilon_0} \frac{qq'}{r^2}, \text{ away from } q \right) \quad (25.7)$$

It's customary to use $1/4\pi\epsilon_0$ rather than K for field calculations. Equation 25.5 defined the electric field in terms of the force on a probe charge, thus the electric field at this point is

$$\vec{E} = \frac{\vec{F}_{\text{on }q'}}{q'} = \left(\frac{1}{4\pi\epsilon_0} \frac{q}{r^2}, \text{ away from } q \right) \quad (25.8)$$

The electric field is shown in FIGURE 25.26c.

NOTE ▶ The expression for the electric field is similar to Coulomb's law. To distinguish the two, remember that Coulomb's law has a product of two charges in the numerator. It describes the force between *two* charges. The electric field has a single charge in the numerator. It is the field of *a* charge. ◀

If we calculate the field at a sufficient number of points in space, we can draw a **field diagram** such as the one shown in FIGURE 25.27. Notice that the field vectors all point straight away from charge q. Also notice how quickly the arrows decrease in length due to the inverse-square dependence on r.

Keep these three important points in mind when using field diagrams:

1. The diagram is just a representative sample of electric field vectors. The field exists at all the other points. A well-drawn diagram can tell you fairly well what the field would be like at a neighboring point.
2. The arrow indicates the direction and the strength of the electric field *at the point to which it is attached*—that is, at the point where the *tail* of the vector is placed. In this chapter, we indicate the point at which the electric field is measured with

FIGURE 25.26 Charge q' is used to probe the electric field of point charge q.

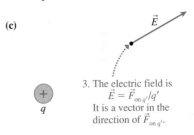

(a) What is the electric field of q at this point?

Point charge

q

(b) 1. Place q' at the point to probe the field.

$\vec{F}_{\text{on }q'}$

q'

r

2. Measure the force on q'.

q

(c)

$\vec{E}$

3. The electric field is $\vec{E} = \vec{F}_{\text{on }q'}/q'$ It is a vector in the direction of $\vec{F}_{\text{on }q'}$.

q

FIGURE 25.27 The electric field of a positive point charge.

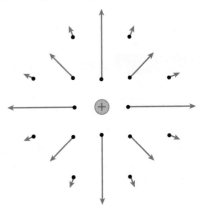

a dot. The length of any vector is significant only relative to the lengths of other vectors.

3. Although we have to draw a vector across the page, from one point to another, an electric field vector is *not* a spatial quantity. It does not "stretch" from one point to another. Each vector represents the electric field at *one point* in space.

Unit Vector Notation

FIGURE 25.28 Using the unit vector $\hat{r}$.

(a)

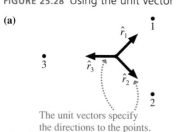

The unit vectors specify the directions to the points.

(b)

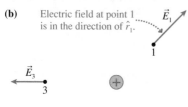

Electric field at point 1 is in the direction of $\hat{r}_1$.

$\vec{E}_2$ is in the direction of $\hat{r}_2$.

FIGURE 25.29 The electric field of a negative point charge.

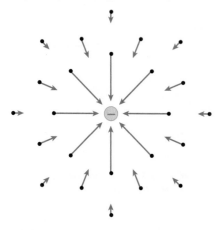

Equation 25.8 is precise, but it's not terribly convenient. Furthermore, what happens if the source charge q is negative? We need a more concise notation to write the electric field, a notation that will allow q to be either positive or negative.

The basic need is to express "away from q" in mathematical notation. "Away from q" is a *direction* in space. To guide us, recall that we already have a notation for expressing certain directions—namely, the unit vectors $\hat{i}$, $\hat{j}$, and $\hat{k}$. For example, unit vector $\hat{i}$ means "in the direction of the positive x-axis." With a minus sign, $-\hat{i}$ means "in the direction of the negative x-axis." Unit vectors, with a magnitude of 1 and no units, provide purely directional information.

With this in mind, let's define the unit vector $\hat{r}$ to be a vector of length 1 that points from the origin to a point of interest. Unit vector $\hat{r}$ provides no information at all about the distance to the point; it merely specifies the direction.

FIGURE 25.28a shows unit vectors $\hat{r}_1$, $\hat{r}_2$, and $\hat{r}_3$ pointing toward points 1, 2, and 3. Unlike $\hat{i}$ and $\hat{j}$, unit vector $\hat{r}$ does not have a fixed direction. Instead, unit vector $\hat{r}$ specifies the direction "straight outward from this point." But that's just what we need to describe the electric field vector. FIGURE 25.28b shows the electric fields at points 1, 2, and 3 due to a positive charge at the origin. No matter which point you choose, the electric field at that point is "straight outward" from the charge. In other words, the electric field $\vec{E}$ points in the direction of the unit vector $\hat{r}$.

With this notation, the electric field at distance r from a point charge q is

$$\vec{E} = \frac{1}{4\pi\epsilon_0}\frac{q}{r^2}\hat{r} \qquad \text{(electric field of a point charge)} \qquad (25.9)$$

where $\hat{r}$ is the unit vector from the charge toward the point at which we want to know the field. Equation 25.9 is identical to Equation 25.8, but written in a notation in which the unit vector $\hat{r}$ expresses the idea "away from q."

Equation 25.9 works equally well if q is negative. A negative sign in front of a vector simply reverses its direction, so the unit vector $-\hat{r}$ points *toward* charge q. FIGURE 25.29 shows the electric field of a negative point charge. It looks like the electric field of a positive point charge except that the vectors point inward, toward the charge, instead of outward.

We'll end this chapter with three examples of the electric field of a point charge. Chapter 26 will expand these ideas to the electric fields of multiple charges and of extended objects.

EXAMPLE 25.8 **Calculating the electric field**

A -1.0 nC charged particle is located at the origin. Points 1, 2, and 3 have (x, y) coordinates $(1\text{ cm}, 0\text{ cm})$, $(0\text{ cm}, 1\text{ cm})$, and $(1\text{ cm}, 1\text{ cm})$, respectively. Determine the electric field $\vec{E}$ at these points, then show the vectors on an electric field diagram.

MODEL The electric field is that of a negative point charge.

VISUALIZE The electric field points straight *toward* the origin. It will be weaker at $(1\text{ cm}, 1\text{ cm})$, which is farther from the charge.

SOLVE The electric field is

$$\vec{E} = \frac{1}{4\pi\epsilon_0}\frac{q}{r^2}\hat{r}$$

where $q = -1.0$ nC $= -1.0 \times 10^{-9}$ C. The distance r is 1.0 cm $= 0.010$ m for points 1 and 2 and $\left(\sqrt{2} \times 1.0\text{ cm}\right) = 0.0141$ m for point 3. The *magnitude* of $\vec{E}$ at the three points is

$$E_1 = E_2 = \frac{1}{4\pi\epsilon_0}\frac{|q|}{r_1^2}$$

$$= \frac{(9.0\times10^9\,\mathrm{N\,m^2/C^2})(1.0\times10^{-9}\,\mathrm{C})}{(0.010\,\mathrm{m})^2} = 90{,}000\,\mathrm{N/C}$$

$$E_3 = \frac{1}{4\pi\epsilon_0}\frac{|q|}{r_3^2}$$

$$= \frac{(9.0\times10^9\,\mathrm{N\,m^2/C^2})(1.0\times10^{-9}\,\mathrm{C})}{(0.0141\,\mathrm{m})^2} = 45{,}000\,\mathrm{N/C}$$

Because q is negative, the field at each of these positions points directly at charge q. The electric field vectors, in component form, are

$$\vec{E}_1 = -90{,}000\,\hat{\imath}\ \mathrm{N/C}$$

$$\vec{E}_2 = -90{,}000\,\hat{\jmath}\ \mathrm{N/C}$$

$$\vec{E}_3 = -E_3\cos45°\,\hat{\imath} - E_3\sin45°\,\hat{\jmath}$$

$$= (-31{,}800\,\hat{\imath} - 31{,}800\,\hat{\jmath})\ \mathrm{N/C}$$

These vectors are shown on the electric field diagram of FIGURE 25.30.

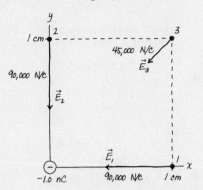

FIGURE 25.30 The electric field diagram of a −1.0 nC charged particle.

EXAMPLE 25.9 **The electric field of a proton**

The electron in a hydrogen atom orbits the proton at a radius of 0.053 nm.

a. What is the proton's electric field strength at the position of the electron?
b. What is the magnitude of the electric force on the electron?

SOLVE a. The proton's charge is $q = e$. Its electric field strength at the distance of the electron is

$$E = \frac{1}{4\pi\epsilon_0}\frac{e}{r^2} = \frac{1}{4\pi\epsilon_0}\frac{1.6\times10^{-19}\,\mathrm{C}}{(5.3\times10^{-11}\,\mathrm{m})^2} = 5.1\times10^{11}\,\mathrm{N/C}$$

Notice how large this field is in comparison to the field of Example 25.8.
b. We could use Coulomb's law to find the force on the electron, but the whole point of knowing the electric field is that we can use it directly to find the force on a charge in the field. The magnitude of the force on the electron is

$$F_\mathrm{on\,elec} = |q_e|E_\mathrm{of\,proton}$$

$$= (1.60\times10^{-19}\,\mathrm{C})(5.1\times10^{11}\,\mathrm{N/C})$$

$$= 8.2\times10^{-8}\,\mathrm{N}$$

STOP TO THINK 25.6 Rank in order, from largest to smallest, the electric field strengths E_a to E_d at points a to d.

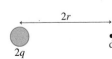

A charge in static equilibrium

A horizontal electric field causes the charged ball in FIGURE 25.31 to hang at a 15° angle, as shown. The spring is plastic, so it doesn't discharge the ball, and in its equilibrium position the spring extends only to the vertical dashed line. What is the electric field strength?

FIGURE 25.31 A charged ball hanging in static equilibrium.

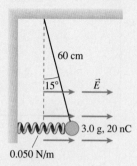

60 cm

15° $\vec{E}$

3.0 g, 20 nC

0.050 N/m

MODEL Model the ball as a point charge in static equilibrium. The electric force on the ball is $\vec{F}_E = q\vec{E}$. The charge is positive, so the force is in the same direction as the field.

VISUALIZE FIGURE 25.32 is a free-body diagram for the ball.

SOLVE The ball is in static equilibrium, so the net force on the ball must be zero. With the field applied, the spring is stretched by $\Delta x = L\sin\theta = (0.60 \text{ m})(\sin 15°) = 0.16 \text{ m}$, where L is the string length, and exerts a pulling force $F_{sp} = k\,\Delta x$ to the left.

Newton's first law, which we've not used in quite some time, is

$$\sum F_x = F_E - F_{sp} - T\sin\theta = 0$$

$$\sum F_y = T\cos\theta - F_G = T\cos\theta - mg = 0$$

FIGURE 25.32 The free-body diagram.

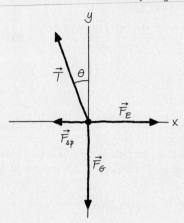

From the y-equation,

$$T = \frac{mg}{\cos\theta}$$

The x-equation is then

$$qE - k\,\Delta x - mg\tan\theta = 0$$

We can solve this for the electric field strength:

$$E = \frac{mg\tan\theta + k\,\Delta x}{q}$$

$$= \frac{(0.0030 \text{ kg})(9.8 \text{ m/s}^2)\tan 15° + (0.050 \text{ N/m})(0.16 \text{ m})}{20 \times 10^{-9}\text{ C}}$$

$$= 7.9 \times 10^5 \text{ N/C}$$

ASSESS We don't yet have a way of judging whether this is a reasonable field strength, but we'll see in the next chapter that this is typical of the electric field strength near an object that has been charged by rubbing.

SUMMARY

The goal of Chapter 25 has been to describe electric phenomena in terms of charges, forces, and fields.

General Principles

Coulomb's Law

The forces between two charged particles q_1 and q_2 separated by distance r are

$$F_{1 \text{ on } 2} = F_{2 \text{ on } 1} = \frac{K|q_1||q_2|}{r^2}$$

These forces are an action/reaction pair directed along the line joining the particles.

- The forces are repulsive for two like charges, attractive for two opposite charges.
- The net force on a charge is the sum of the forces from all other charges.
- The unit of charge is the coulomb (C).
- The electrostatic constant is $K = 9.0 \times 10^9 \text{ N m}^2/\text{C}^2$.

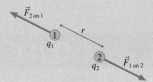

Important Concepts

The Charge Model

There are two kinds of charge, positive and negative.

- Fundamental charges are protons and electrons, with charge $\pm e$ where $e = 1.60 \times 10^{-19}$ C.
- Objects are charged by adding or removing electrons.
- The amount of charge is $q = (N_p - N_e)e$.
- An object with an equal number of protons and electrons is neutral, meaning no *net* charge.

Charged objects exert electric forces on each other.

- Like charges repel, opposite charges attract.
- The force increases as the charge increases.
- The force decreases as the distance increases.

There are two types of material, insulators and conductors.

- Charge remains fixed in or on an insulator.
- Charge moves easily through or along conductors.
- Charge is transferred by contact between objects.

Charged objects attract neutral objects.

- Charge polarizes metal by shifting the electron sea.
- Charge polarizes atoms, creating electric dipoles.
- The polarization force is always an attractive force.

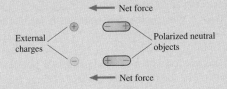

The Field Model

Charges interact with each other via the electric field $\vec{E}$.

- Charge A alters the space around it by creating an electric field.

- The field is the agent that exerts a force. The force on charge q_B is $\vec{F}_{\text{on B}} = q_B \vec{E}$.

An electric field is identified and measured in terms of the force on a **probe charge** q:

$$\vec{E} = \vec{F}_{\text{on } q}/q$$

- The electric field exists at all points in space.
- An electric field vector shows the field only at one point, the point at the tail of the vector.

The electric field of a **point charge** is

$$\vec{E} = \frac{1}{4\pi\epsilon_0} \frac{q}{r^2} \hat{r}$$

Terms and Notation

neutral	electron cloud	electrostatic equilibrium	coulomb, C
charging	fundamental unit of charge, e	grounded	permittivity constant, ϵ_0
charge model	charge quantization	charge polarization	field
charge, q or Q	ionization	polarization force	electric field, $\vec{E}$
like charges	law of conservation of charge	electric dipole	field model
opposite charges	sea of electrons	charging by induction	source charge
discharging	ion core	Coulomb's law	electric field strength, E
conductor	current	electrostatic constant, K	field diagram
insulator	charge carriers	point charge	

CONCEPTUAL QUESTIONS

1. Can an insulator be charged? If so, how would you charge an insulator? If not, why not?

2. Can a conductor be charged? If so, how would you charge a conductor? If not, why not?

3. Four lightweight balls A, B, C, and D are suspended by threads. Ball A has been touched by a plastic rod that was rubbed with wool. When the balls are brought close together, without touching, the following observations are made:
   • Balls B, C, and D are attracted to ball A.
   • Balls B and D have no effect on each other.
   • Ball B is attracted to ball C.
   What are the charge states (glass, plastic, or neutral) of balls A, B, C, and D? Explain.

4. Charged plastic and glass rods hang by threads.
   a. An object repels the plastic rod. Can you predict what it will do to the glass rod? If so, what? If not, why not?
   b. A different object attracts the plastic rod. Can you predict what it will do to the glass rod? If so, what? If not, why not?

5. A lightweight metal ball hangs by a thread. When a charged rod is held near, the ball moves toward the rod, touches the rod, then quickly "flies away" from the rod. Explain this behavior.

6. Suppose there exists a third type of charge in addition to the two types we've called glass and plastic. Call this third type X charge. What experiment or series of experiments would you use to test whether an object has X charge? State clearly how each possible outcome of the experiments is to be interpreted.

7. A negatively charged electroscope has separated leaves.
   a. Suppose you bring a negatively charged rod close to the top of the electroscope, but not touching. How will the leaves respond? Use both charge diagrams and words to explain.
   b. How will the leaves respond if you bring a positively charged rod close to the top of the electroscope, but not touching? Use both charge diagrams and words to explain.

8. The two oppositely charged metal spheres in FIGURE Q25.8 have equal quantities of charge. They are brought into contact with a neutral metal rod. What is the final charge state of each sphere and of the rod? Use both charge diagrams and words to explain.

FIGURE Q25.8

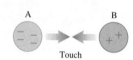

FIGURE Q25.9

9. Metal sphere A in FIGURE Q25.9 has 4 units of negative charge and metal sphere B has 2 units of positive charge. The two spheres are brought into contact. What is the final charge state of each sphere? Explain.

10. Metal spheres A and B in FIGURE Q25.10 are initially neutral and are touching. A positively charged rod is brought near A, but not touching. Is A now positive, negative, or neutral? Use both charge diagrams and words to explain.

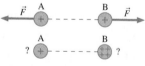

FIGURE Q25.10   **FIGURE Q25.11**

11. If you bring your finger near a lightweight, negatively charged hanging ball, the ball swings over toward your finger as shown in FIGURE Q25.11. Use charge diagrams and words to explain this observation.

12. Reproduce FIGURE Q25.12 on your paper. Then draw a dot (or dots) on the figure to show the position (or positions) where an electron would experience no net force.

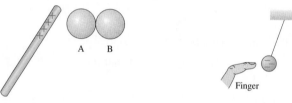

FIGURE Q25.12

13. Charges A and B in FIGURE Q25.13 are equal. Each charge exerts a force on the other of magnitude F. Suppose the charge of B is increased by a factor of 4, but everything else is unchanged. In terms of F, (a) what is the magnitude of the force on A, and (b) what is the magnitude of the force on B?

FIGURE Q25.13

14. The electric field strength at one point near a point charge is 1000 N/C. What is the field strength if (a) the distance from the point charge is doubled, and (b) the distance from the point charge is halved?

15. The electric force on a charged particle in an electric field is F. What will be the force if the particle's charge is tripled and the electric field strength is halved?

EXERCISES AND PROBLEMS

Problems labeled [ ] integrate material from earlier chapters.

Exercises

Section 25.1 Developing a Charge Model

Section 25.2 Charge

1. | A plastic rod is charged to -12 nC by rubbing.
   a. Have electrons been added to the rod or protons removed? Explain.
   b. How many electrons have been added or protons removed?
2. | A glass rod is charged to $+8.0$ nC by rubbing.
   a. Have electrons been removed from the rod or protons added? Explain.
   b. How many electrons have been removed or protons added?
3. | A glass rod that has been charged to $+12$ nC touches a metal sphere. Afterward, the rod's charge is $+8.0$ nC.
   a. What kind of charged particle was transferred between the rod and the sphere, and in which direction? That is, did it move from the rod to the sphere or from the sphere to the rod?
   b. How many charged particles were transferred?
4. | A plastic rod that has been charged to -15 nC touches a metal sphere. Afterward, the rod's charge is -10 nC.
   a. What kind of charged particle was transferred between the rod and the sphere, and in which direction? That is, did it move from the rod to the sphere or from the sphere to the rod?
   b. How many charged particles were transferred?
5. ‖ What is the total charge of all the protons in 1.0 mol of He gas?
6. ‖‖ What is the total charge of all the electrons in 1.0 L of liquid water?

Section 25.3 Insulators and Conductors

7. | Figure 25.8 showed how an electroscope becomes negatively charged. The leaves will also repel each other if you touch the electroscope with a positively charged glass rod. Use a series of charge diagrams to explain what happens and why the leaves repel each other.
8. | A plastic balloon that has been rubbed with wool will stick to a wall.
   a. Can you conclude that the wall is charged? If not, why not? If so, where does the charge come from?
   b. Draw a series of charge diagrams showing how the balloon is held to the wall.
9. | Two neutral metal spheres on wood stands are touching. A negatively charged rod is held directly above the top of the left sphere, not quite touching it. While the rod is there, the right sphere is moved so that the spheres no longer touch. Then the rod is withdrawn. Afterward, what is the charge state of each sphere? Use charge diagrams to explain your answer.
10. ‖ You have two neutral metal spheres on wood stands. Devise a procedure for charging the spheres so that they will have like charges of *exactly* equal magnitude. Use charge diagrams to explain your procedure.
11. ‖ You have two neutral metal spheres on wood stands. Devise a procedure for charging the spheres so that they will have opposite charges of *exactly* equal magnitude. Use charge diagrams to explain your procedure.

Section 25.4 Coulomb's Law

12. ‖ Two 1.0 kg masses are 1.0 m apart (center to center) on a frictionless table. Each has $+10$ μC of charge.
   a. What is the magnitude of the electric force on one of the masses?
   b. What is the initial acceleration of this mass if it is released and allowed to move?
13. ‖ Two small plastic spheres each have a mass of 2.0 g and a charge of -50.0 nC. They are placed 2.0 cm apart (center to center).
   a. What is the magnitude of the electric force on each sphere?
   b. By what factor is the electric force on a sphere larger than its weight?
14. ‖ A small glass bead has been charged to $+20$ nC. A metal ball bearing 1.0 cm above the bead feels a 0.018 N downward electric force. What is the charge on the ball bearing?
15. | Two protons are 2.0 fm apart.
   a. What is the magnitude of the electric force on one proton due to the other proton?
   b. What is the magnitude of the gravitational force on one proton due to the other proton?
   c. What is the ratio of the electric force to the gravitational force?
16. | What is the net electric force on charge A in FIGURE EX25.16?

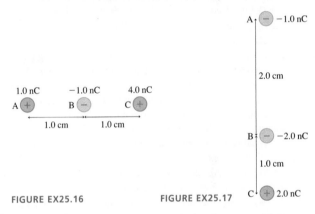

FIGURE EX25.16   FIGURE EX25.17

17. | What is the net electric force on charge B in FIGURE EX25.17?
18. | Object A, which has been charged to $+4.0$ nC, is at the origin. Object B, which has been charged to -8.0 nC, is at $(x, y) = (0.0 \text{ cm}, 2.0 \text{ cm})$. Determine the electric force on each object. Write each force vector in component form.
19. | A small plastic bead has been charged to -15 nC. What are the magnitude and direction of the acceleration of (a) a proton and (b) an electron that is 1.0 cm from the center of the bead?

Section 25.5 The Field Model

20. | What are the strength and direction of the electric field 1.0 mm from (a) a proton and (b) an electron?
21. | The electric field at a point in space is $\vec{E} = (400\,\hat{i} + 100\,\hat{j})$ N/C.
   a. What is the electric force on a proton at this point? Give your answer in component form.
   b. What is the electric force on an electron at this point? Give your answer in component form.
   c. What is the magnitude of the proton's acceleration?
   d. What is the magnitude of the electron's acceleration?

22. ‖ What magnitude charge creates a 1.0 N/C electric field at a point 1.0 m away?

23. ⎮ What are the strength and direction of the electric field 4.0 cm from a small plastic bead that has been charged to −8.0 nC?

24. ‖ The electric field 2.0 cm from a small object points away from the object with a strength of 270,000 N/C. What is the object's charge?

25. ‖ What are the strength and direction of an electric field that will balance the weight of a 1.0 g plastic sphere that has been charged to −3.0 nC?

26. ‖ A +12 nC charge is located at the origin.
    a. What are the electric fields at the positions $(x, y) =$ (5.0 cm, 0 cm), (−5.0 cm, 5.0 cm), and (−5.0 cm, −5.0 cm)? Write each electric field vector in component form.
    b. Draw a field diagram showing the electric field vectors at these points.

27. ‖ A −12 nC charge is located at $(x, y) = $ (1.0 cm, 0 cm). What are the electric fields at the positions $(x, y) = $ (5.0 cm, 0 cm), (−5.0 cm, 0 cm), and (0 cm, 5.0 cm)? Write each electric field vector in component form.

Problems

28. ‖‖ Pennies today are copper-covered zinc, but older pennies are 3.1 g of solid copper. What are the total positive charge and total negative charge in a solid copper penny that is electrically neutral?

29. ⎮ A 2.0 g plastic bead charged to −4.0 nC and a 4.0 g glass bead charged to +8.0 nC are 2.0 cm apart (center to center). What are the accelerations of (a) the plastic bead and (b) the glass bead?

30. ‖ The nucleus of a ^{125}Xe atom (an isotope of the element xenon with mass 125 u) is 6.0 fm in diameter. It has 54 protons and charge $q = +54e$.
    a. What is the electric force on a proton 2.0 fm from the surface of the nucleus?
    b. What is the proton's acceleration?
    Hint: Treat the spherical nucleus as a point charge.

31. ‖ Two 1.0 g spheres are charged equally and placed 2.0 cm apart. When released, they begin to accelerate at 150 m/s². What is the magnitude of the charge on each sphere?

32. ‖ Objects A and B are both positively charged. Both have a mass of 100 g, but A has twice the charge of B. When A and B are placed 10 cm apart, B experiences an electric force of 0.45 N.
    a. What is the charge on A?
    b. If the objects are released, what is the initial acceleration of A?

33. ‖ What is the force $\vec{F}$ on the 1.0 nC charge in FIGURE P25.33? Give your answer as a magnitude and a direction.

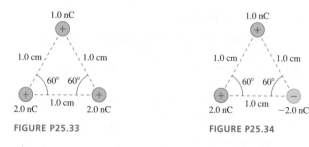

FIGURE P25.33   FIGURE P25.34

34. ‖ What is the force $\vec{F}$ on the 1.0 nC charge in FIGURE P25.34? Give your answer as a magnitude and a direction.

35. ‖ What is the force $\vec{F}$ on the −10 nC charge in FIGURE P25.35? Give your answer as a magnitude and an angle measured cw or ccw (specify which) from the +x-axis.

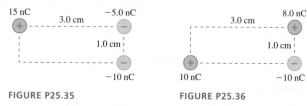

FIGURE P25.35   FIGURE P25.36

36. ‖ What is the force $\vec{F}$ on the −10 nC charge in FIGURE P25.36? Give your answer as a magnitude and an angle measured cw or ccw (specify which) from the +x-axis.

37. ‖ What is the force $\vec{F}$ on the 5.0 nC charge in FIGURE P25.37? Give your answer as a magnitude and an angle measured cw or ccw (specify which) from the +x-axis.

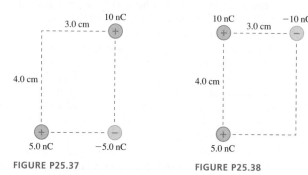

FIGURE P25.37   FIGURE P25.38

38. ‖ What is the force $\vec{F}$ on the 5.0 nC charge in FIGURE P25.38? Give your answer as a magnitude and an angle measured cw or ccw (specify which) from the +x-axis.

39. ‖‖ What is the force $\vec{F}$ on the 1.0 nC charge in the middle of FIGURE P25.39 due to the four other charges? Give your answer in component form.

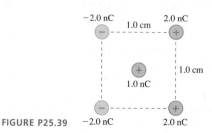

FIGURE P25.39

40. ‖ What is the force $\vec{F}$ on the 1.0 nC charge at the bottom in FIGURE P25.40? Give your answer in component form.

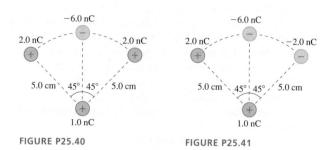

FIGURE P25.40   FIGURE P25.41

41. ‖ What is the force $\vec{F}$ on the 1.0 nC charge at the bottom in FIGURE P25.41? Give your answer in component form.

42. ⫼ A +2.0 nC charge is at the origin and a −4.0 nC charge is at $x = 1.0$ cm.
    a. At what x-coordinate could you place a proton so that it would experience no net force?
    b. Would the net force be zero for an electron placed at the same position? Explain.

43. ⫼ The net force on the 1.0 nC charge in FIGURE P25.43 is zero. What is q?

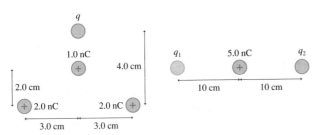

FIGURE P25.43   FIGURE P25.44

44. ⫼ Charge q_2 in FIGURE P25.44 is in static equilibrium. What is q_1?

45. ⫼ A positive point charge Q is located at $x = a$ and a negative point charge $−Q$ is at $x = −a$. A positive charge q can be placed anywhere on the y-axis. Find an expression for $(F_{net})_x$, the x-component of the net force on q.

46. ⫼ A positive point charge Q is located at $x = a$ and a negative point charge $−Q$ is at $x = −a$. A positive charge q can be placed anywhere on the x-axis. Find an expression for $(F_{net})_x$, the x-component of the net force on q, when (a) $|x| < a$ and (b) $|x| > a$.

47. ⫼ FIGURE P25.47 shows four charges at the corners of a square of side L. What is the magnitude of the net force on q?

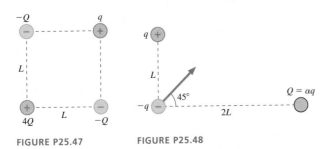

FIGURE P25.47   FIGURE P25.48

48. ⫼ FIGURE P25.48 shows three charges and the net force on charge $−q$. Charge Q is some multiple α of q. What is α?

49. ⫼ Two positive point charges q and $4q$ are at $x = 0$ and $x = L$, respectively, and free to move. A third charge is placed so that the entire three-charge system is in static equilibrium. What are the magnitude, sign, and x-coordinate of the third charge?

50. ⫼ Suppose the magnitude of the proton charge differs from the magnitude of the electron charge by a mere 1 part in 10^9.
    a. What would be the force between two 2.0-mm-diameter copper spheres 1.0 cm apart? Assume that each copper atom has an equal number of electrons and protons.
    b. Would this amount of force be detectable? What can you conclude from the fact that no such forces are observed?

51. ⫼ In a simple model of the hydrogen atom, the electron moves in a circular orbit of radius 0.053 nm around a stationary proton. How many revolutions per second does the electron make?

52. ⫼ You have two small, 2.0 g balls that have been given equal but opposite charges, but you don't know the magnitude of the charge. To find out, you place the balls distance d apart on a slippery horizontal surface, release them, and use a motion detector to measure the initial acceleration of one of the balls toward the other. After repeating this for several different separation distances, your data are as follows:

Distance (cm)	Acceleration (m/s²)
2.0	0.74
3.0	0.30
4.0	0.19
5.0	0.10

Use an appropriate graph of the data to determine the magnitude of the charge.

53. ⫼ A 0.10 g honeybee acquires a charge of +23 pC while flying.
BIO a. The earth's electric field near the surface is typically (100 N/C, downward). What is the ratio of the electric force on the bee to the bee's weight?
    b. What electric field (strength and direction) would allow the bee to hang suspended in the air?

54. ⫼ As a science project, you've invented an "electron pump" that moves electrons from one object to another. To demonstrate your invention, you bolt a small metal plate to the ceiling, connect the pump between the metal plate and yourself, and start pumping electrons from the metal plate to you. How many electrons must be moved from the metal plate to you in order for you to hang suspended in the air 2.0 m below the ceiling? Your mass is 60 kg. **Hint:** Assume that both you and the plate can be modeled as point charges.

55. ⫼ You have a lightweight spring whose unstretched length is 4.0 cm. First, you attach one end of the spring to the ceiling and hang a 1.0 g mass from it. This stretches the spring to a length of 5.0 cm. You then attach two small plastic beads to the opposite ends of the spring, lay the spring on a frictionless table, and give each plastic bead the same charge. This stretches the spring to a length of 4.5 cm. What is the magnitude of the charge (in nC) on each bead?

56. ⫼ An electric dipole consists of two opposite charges $\pm q$ separated by a small distance s. The product $p = qs$ is called the *dipole moment*. FIGURE P25.56 shows an electric dipole perpendicular to an electric field $\vec{E}$. Find an expression in terms of p and E for the magnitude of the torque that the electric field exerts on the dipole.

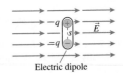

FIGURE P25.56   Electric dipole

57. ⫼ You sometimes create a spark when you touch a doorknob after shuffling your feet on a carpet. Why? The air always has a few free electrons that have been kicked out of atoms by cosmic rays. If an electric field is present, a free electron is accelerated until it collides with an air molecule. It will transfer its kinetic energy to the molecule, then accelerate, then collide, then accelerate, collide, and so on. If the electron's kinetic energy just before a collision is 2.0×10^{-18} J or more, it has sufficient energy to kick an electron out of the molecule it hits. Where there was one free electron, now there are two! Each of these can then

accelerate, hit a molecule, and kick out another electron. Then there will be four free electrons. In other words, as FIGURE P25.57 shows, a sufficiently strong electric field causes a "chain reaction" of electron production. This is called a *breakdown* of the air. The current of moving electrons is what gives you the shock, and a spark is generated when the electrons recombine with the positive ions and give off excess energy as a burst of light.

a. The average distance an electron travels between collisions is 2.0 μm. What acceleration must an electron have to gain 2.0×10^{-18} J of kinetic energy in this distance?

b. What force must act on an electron to give it the acceleration found in part a?

c. What strength electric field will exert this much force on an electron? This is the *breakdown field strength*. **Note:** The measured breakdown field strength is a little less than your calculated value because our model of the process is a bit too simple. Even so, your calculated value is close.

d. Suppose a free electron in air is 1.0 cm away from a point charge. What minimum charge must this point charge have to cause a breakdown of the air and create a spark?

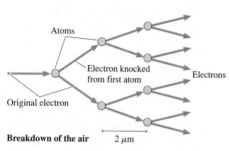

FIGURE P25.57

58. ‖ Two 5.0 g point charges on 1.0-m-long threads repel each other after being charged to $+100$ nC, as shown in FIGURE P25.58. What is the angle θ? You can assume that θ is a small angle.

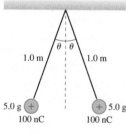

FIGURE P25.58   FIGURE P25.59

59. ‖ Two 3.0 g point charges on 1.0-m-long threads repel each other after being equally charged, as shown in FIGURE P25.59. What is the charge q?

60. ‖ What are the electric fields at points 1, 2, and 3 in FIGURE P25.60? Give your answer in component form.

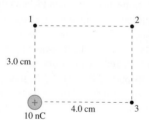

FIGURE P25.60

61. ‖ What are the electric fields at points 1 and 2 in FIGURE P25.61? Give your answer as a magnitude and direction.

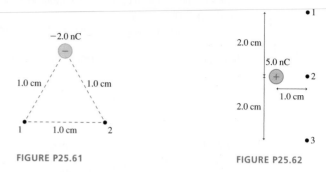

FIGURE P25.61   FIGURE P25.62

62. ‖ What are the electric fields at points 1, 2, and 3 in FIGURE P25.62? Give your answer in component form.

63. ‖ A -10.0 nC charge is located at position $(x, y) = (2.0$ cm, 1.0 cm). At what (x, y) position(s) is the electric field
a. $-225{,}000\,\hat{\imath}$ N/C?
b. $(161{,}000\,\hat{\imath} - 80{,}500\,\hat{\jmath})$ N/C?
c. $(28{,}800\,\hat{\imath} + 21{,}600\,\hat{\jmath})$ N/C?

64. ‖ A 10.0 nC charge is located at position $(x, y) = (1.0$ cm, 2.0 cm). At what (x, y) position(s) is the electric field
a. $-225{,}000\,\hat{\imath}$ N/C?
b. $(161{,}000\,\hat{\imath} + 80{,}500\,\hat{\jmath})$ N/C?
c. $(21{,}600\,\hat{\imath} - 28{,}800\,\hat{\jmath})$ N/C?

65. ‖ Three 1.0 nC charges are placed as shown in FIGURE P25.65. Each of these charges creates an electric field $\vec{E}$ at a point 3.0 cm in front of the middle charge.

$q_1 = 1.0$ nC
$q_2 = 1.0$ nC
$q_3 = 1.0$ nC

FIGURE P25.65

a. What are the three fields $\vec{E}_1$, $\vec{E}_2$, and $\vec{E}_3$ created by the three charges? Write your answer for each as a vector in component form.

b. Do you think that electric fields obey a principle of superposition? That is, is there a "net field" at this point given by $\vec{E}_{net} = \vec{E}_1 + \vec{E}_2 + \vec{E}_3$? Use what you learned in this chapter and previously in our study of forces to argue why this is or is not true.

c. If it is true, what is $\vec{E}_{net}$?

66. ‖ An electric field $\vec{E} = 100{,}000\,\hat{\imath}$ N/C causes the 5.0 g point charge in FIGURE P25.66 to hang at a 20° angle. What is the charge on the ball?

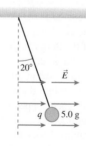

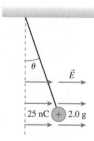

FIGURE P25.66   FIGURE P25.67

67. ‖ An electric field $\vec{E} = 200{,}000\,\hat{\imath}$ N/C causes the point charge in FIGURE P25.67 to hang at an angle. What is θ?

In Problems 68 through 71 you are given the equation(s) used to solve a problem. For each of these,

  a. Write a realistic problem for which this is the correct equation(s).
  b. Finish the solution of the problem.

68. $\dfrac{(9.0 \times 10^9 \,\text{N}\,\text{m}^2/\text{C}^2) \times N \times (1.60 \times 10^{-19}\,\text{C})}{(1.0 \times 10^{-6}\,\text{m})^2}$

$= 1.5 \times 10^6\,\text{N/C}$

69. $\dfrac{(9.0 \times 10^9\,\text{N}\,\text{m}^2/\text{C}^2)q^2}{(0.0150\,\text{m})^2} = 0.020\,\text{N}$

70. $\dfrac{(9.0 \times 10^9\,\text{N}\,\text{m}^2/\text{C}^2)(15 \times 10^{-9}\,\text{C})}{r^2} = 54{,}000\,\text{N/C}$

71. $\sum F_x = 2 \times \dfrac{(9.0 \times 10^9\,\text{N}\,\text{m}^2/\text{C}^2)(1.0 \times 10^{-9}\,\text{C})q}{\left((0.020\,\text{m})/\sin 30°\right)^2} \times \cos 30°$

$= 5.0 \times 10^{-5}\,\text{N}$

$\sum F_y = 0\,\text{N}$

Challenge Problems

72. A 2.0-mm-diameter copper ball is charged to $+50$ nC. What fraction of its electrons have been removed?

73. Three 3.0 g balls are tied to 80-cm-long threads and hung from a *single* fixed point. Each of the balls is given the same charge q. At equilibrium, the three balls form an equilateral triangle in a horizontal plane with 20 cm sides. What is q?

74. The identical small spheres shown in FIGURE CP25.74 are charged to $+100$ nC and -100 nC. They hang as shown in a 100,000 N/C electric field. What is the mass of each sphere?

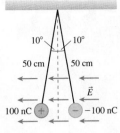

FIGURE CP25.74

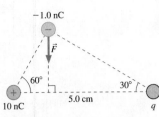

FIGURE CP25.75

75. The force on the -1.0 nC charge is as shown in FIGURE CP25.75. What is the magnitude of this force?

76. In Section 25.3 we claimed that a charged object exerts a net attractive force on an electric dipole. Let's investigate this. FIGURE CP25.76 shows a *permanent* electric dipole consisting of charges $+q$ and $-q$ separated by the fixed distance s. Charge $+Q$ is distance r from the center of the dipole. We'll assume, as is usually the case in practice, that $s \ll r$.

  a. Write an expression for the net force exerted on the dipole by charge $+Q$.
  b. Is this force toward $+Q$ or away from $+Q$? Explain.
  c. Use the *binomial approximation* $(1 + x)^{-n} \approx 1 - nx$ if $x \ll 1$ to show that your expression from part a can be written $F_{\text{net}} = 2KqQs/r^3$.
  d. How can an electric force have an inverse-cube dependence? Doesn't Coulomb's law say that the electric force depends on the inverse square of the distance? Explain.

FIGURE CP25.76

Stop to Think 25.1: b. Charged objects are attracted to neutral objects, so an attractive force is inconclusive. Repulsion is the only sure test.

Stop to Think 25.2: $q_e(+3e) > q_a(+1e) > q_d(0) > q_b(-1e) > q_c(-2e)$.

Stop to Think 25.3: a. The negative plastic rod will polarize the electroscope by pushing electrons down toward the leaves. This will partially neutralize the positive charge the leaves had acquired from the glass rod.

Stop to Think 25.4: b. The two forces are an action/reaction pair, opposite in direction but *equal* in magnitude.

Stop to Think 25.5: c. There's an electric field at *all* points, whether an $\vec{E}$ vector is shown or not. The electric field at the dot is to the right. But an electron is a negative charge, so the force of the electric field on the electron is to the left.

Stop to Think 25.6: $E_b > E_a > E_d > E_c$.

26 The Electric Field

In a plasma ball, electrons follow the electric field lines outward from the center electrode. The streamers appear where gas atoms emit light after the high-speed electrons collide with them.

▶ **Looking Ahead** The goal of Chapter 26 is to learn how to calculate and use the electric field.

Fields of Multiple Charges

You'll learn that the electric field due to several point charges is the vector sum of the individual fields.

You'll also learn to use **electric field lines.** This figure shows the electric field lines of a *dipole,* two equal but opposite point charges.

◀ **Looking Back**
Section 25.5 The electric field of a point charge

The Field of a Continuous Distribution of Charge

You'll learn a strategy for computing the electric field of a macroscopic charged object, such as a charged rod or a disk of charge.

- A charged object can be described by its **charge density,** the charge per unit length, area, or volume.
- The vector sum of electric fields will become an integral. We'll develop a step-by-step approach to setting up and evaluating these integrals.

We'll calculate the electric field of charged wires, charged disks, planes of charge, and spheres of charge.

The electric field of a plane of charge is perpendicular to the plane. Many practical devices can be modeled as planes or lines of charge.

Uniform Electric Fields

Two parallel conducting plates with equal but opposite charges are called a **parallel-plate capacitor.**

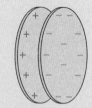

You'll learn that parallel-plate capacitors are important for creating a **uniform electric field.**

Charges in Electric Fields

Electric fields exert forces on charged particles. You'll learn to calculate the trajectories of charged particles moving in electric fields.

Older televisions and computer monitors use a *cathode-ray tube.* The picture is formed as a changing electric field sweeps an electron beam back and forth across the screen.

◀ **Looking Back**
Section 4.3 Projectile motion

Dipoles in Electric Fields

You learned in Chapter 25 that charged objects of either sign attract a neutral object. We'll understand better why this happens.

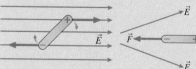

An electric field exerts a *torque* on a dipole, causing it to align with the field.

A nonuniform field exerts a force on a dipole, drawing it toward the stronger field.

26.1 Electric Field Models

Chapter 25 made a distinction between those charged particles that are the *sources* of an electric field and other charged particles that *experience* and move in the electric field. This is a very important distinction. Most of this chapter will be concerned with the *sources* of the electric field. Only at the end, once we know how to calculate the electric field, will we look at what happens to charges that find themselves *in* an electric field.

The electric fields used in science and engineering are often caused by fairly complicated distributions of charge. Sometimes these fields require exact calculations, but much of the time we can understand the essential physics on the basis of simplified *models* of the electric field.

FIGURE 26.1 Four basic electric field models.

A point charge
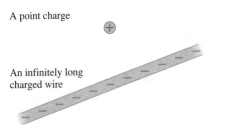

An infinitely long
charged wire

An infinitely wide
charged plane
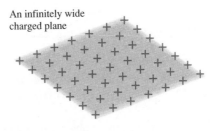

A charged sphere
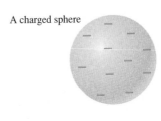

Four widely used electric field models, illustrated in **FIGURE 26.1**, are:

- The electric field of a point charge.
- The electric field of an infinitely long charged wire.
- The electric field of an infinitely wide charged plane.
- The electric field of a charged sphere.

Small charged objects can often be modeled as point charges or charged spheres. Real wires aren't infinitely long, but in many practical situations this approximation is perfectly reasonable. As we derive and use these electric fields, we'll consider the conditions under which they are appropriate models.

Our starting point is the electric field of a point charge q:

$$\vec{E} = \frac{1}{4\pi\epsilon_0} \frac{q}{r^2} \hat{r} \qquad \text{(electric field of a point charge)} \qquad (26.1)$$

where $\hat{r}$ is a unit vector pointing away from q and $\epsilon_0 = 8.85 \times 10^{-12} \text{ C}^2/\text{N m}^2$ is the permittivity constant. **FIGURE 26.2** reminds you of the electric fields of point charges. Although we have to give each vector we draw a length, keep in mind that each arrow represents the electric field *at a point*. The electric field is not a spatial quantity that "stretches" from one end of the arrow to the other.

The electric field was defined as $\vec{E} = \vec{F}_{\text{on }q}/q$, where $\vec{F}_{\text{on }q}$ is the electric force on charge q. Forces add as vectors, so the net force on q due to a group of point charges is the vector sum

$$\vec{F}_{\text{on }q} = \vec{F}_{1 \text{ on }q} + \vec{F}_{2 \text{ on }q} + \cdots$$

Consequently, the net electric field due to a group of point charges is

$$\vec{E}_{\text{net}} = \frac{\vec{F}_{\text{on }q}}{q} = \frac{\vec{F}_{1 \text{ on }q}}{q} + \frac{\vec{F}_{2 \text{ on }q}}{q} + \cdots = \vec{E}_1 + \vec{E}_2 + \cdots = \sum_i \vec{E}_i \qquad (26.2)$$

where $\vec{E}_i$ is the field from point charge i.

Equation 26.2, which is the primary tool for calculating electric fields, tells us that **the net electric field is the *vector sum* of the electric fields due to each charge.** In other words, electric fields obey the *principle of superposition*.

Knowing typical electric field strengths will also be helpful. The values in Table 26.1 on the next page will help you judge the reasonableness of your solutions to problems.

FIGURE 26.2 The electric field of a positive and a negative point charge.

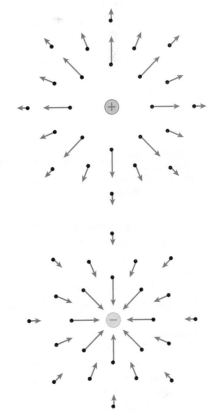

TABLE 26.1 Typical electric field strengths

Field location	Field strength (N/C)
Inside a current-carrying wire	10^{-3}–10^{-1}
Near the earth's surface	10^{2}–10^{4}
Near objects charged by rubbing	10^{3}–10^{6}
Electric breakdown in air, causing a spark	3×10^{6}
Inside an atom	10^{11}

26.2 The Electric Field of Multiple Point Charges

Suppose the source of an electric field is a group of point charges $q_1, q_2, \ldots$. According to Equation 26.2, the net electric field $\vec{E}_{net}$ at each point in space is a superposition of the electric fields due to each individual charge:

$$(E_{net})_x = (E_1)_x + (E_2)_x + \cdots = \sum (E_i)_x$$
$$(E_{net})_y = (E_1)_y + (E_2)_y + \cdots = \sum (E_i)_y \qquad (26.3)$$
$$(E_{net})_z = (E_1)_z + (E_2)_z + \cdots = \sum (E_i)_z$$

Sometimes you'll want to write $\vec{E}_{net}$ in component form:

$$\vec{E}_{net} = (E_{net})_x \hat{i} + (E_{net})_y \hat{j} + (E_{net})_z \hat{k}$$

At other times you will give $\vec{E}_{net}$ as a magnitude and a direction.

PROBLEM-SOLVING STRATEGY 26.1 **The electric field of multiple point charges**

MODEL Model charged objects as point charges.

VISUALIZE For the pictorial representation:

- Establish a coordinate system and show the locations of the charges.
- Identify the point P at which you want to calculate the electric field.
- Draw the electric field of each charge at P.
- Use symmetry to determine if any components of $\vec{E}_{net}$ are zero.

SOLVE The mathematical representation is $\vec{E}_{net} = \sum \vec{E}_i$.

- For each charge, determine its distance from P and the angle of $\vec{E}_i$ from the axes.
- Calculate the field strength of each charge's electric field.
- Write each vector $\vec{E}_i$ in component form.
- Sum the vector components to determine $\vec{E}_{net}$.
- If needed, determine the magnitude and direction of $\vec{E}_{net}$.

ASSESS Check that your result has the correct units, is reasonable, and agrees with any known limiting cases.

Exercise 16

EXAMPLE 26.1 **The electric field of three equal point charges**

Three equal point charges q are located on the y-axis at $y = 0$ and at $y = \pm d$. What is the electric field at a point on the x-axis?

MODEL This problem is a step along the way to understanding the electric field of a charged wire. We'll assume that q is positive when drawing pictures, but the solution should allow for the possibility that q is negative. The question does not ask about any specific point, so we will be looking for a symbolic expression in terms of the unspecified position x.

VISUALIZE FIGURE 26.3 shows the charges, the coordinate system, and the three electric field vectors $\vec{E}_1$, $\vec{E}_2$, and $\vec{E}_3$. Each of these fields points *away from* its source charge because of the assumption that q is positive. We need to find the vector sum $\vec{E}_{net} = \vec{E}_1 + \vec{E}_2 + \vec{E}_3$.

Before rushing into a calculation, we can make our task *much* easier by first thinking qualitatively about the situation. For example, the fields $\vec{E}_1$, $\vec{E}_2$, and $\vec{E}_3$ all lie in the xy-plane, hence we can conclude without any calculations that $(E_{net})_z = 0$. Next, we

FIGURE 26.3 Calculating the electric field of three equal point charges.

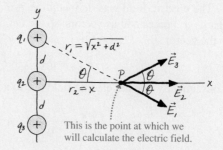

This is the point at which we will calculate the electric field.

look at the y-components of the fields. The fields $\vec{E}_1$ and $\vec{E}_3$ have equal magnitudes and are tilted away from the x-axis by the same angle θ. Consequently, the y-components of $\vec{E}_1$ and $\vec{E}_3$ will *cancel* when added. $\vec{E}_2$ has no y-component, so we can conclude that $(E_{net})_y = 0$. The only component we need to calculate is $(E_{net})_x$.

SOLVE We're ready to calculate. The x-component of the field is

$$(E_{net})_x = (E_1)_x + (E_2)_x + (E_3)_x = 2(E_1)_x + (E_2)_x$$

where we used the fact that fields $\vec{E}_1$ and $\vec{E}_3$ have *equal* x-components. Vector $\vec{E}_2$ has *only* the x-component

$$(E_2)_x = E_2 = \frac{1}{4\pi\epsilon_0}\frac{q_2}{r_2^2} = \frac{1}{4\pi\epsilon_0}\frac{q}{x^2}$$

where $r_2 = x$ is the distance from q_2 to the point at which we are calculating the field. Vector $\vec{E}_1$ is at angle θ from the x-axis, so its x-component is

$$(E_1)_x = E_1\cos\theta = \frac{1}{4\pi\epsilon_0}\frac{q_1}{r_1^2}\cos\theta$$

where r_1 is the distance from q_1. This expression for $(E_1)_x$ is correct, but it is not yet sufficient. Both the distance r_1 and the angle θ vary with the position x and need to be expressed as functions of x. From the Pythagorean theorem, $r_1 = (x^2 + d^2)^{1/2}$. Then from trigonometry,

$$\cos\theta = \frac{x}{r_1} = \frac{x}{(x^2 + d^2)^{1/2}}$$

By combining these pieces, we see that $(E_1)_x$ is

$$(E_1)_x = \frac{1}{4\pi\epsilon_0}\frac{q}{x^2+d^2}\frac{x}{(x^2+d^2)^{1/2}} = \frac{1}{4\pi\epsilon_0}\frac{xq}{(x^2+d^2)^{3/2}}$$

This expression is a bit complex, but notice that the dimensions of $x/(x^2+d^2)^{3/2}$ are $1/m^2$, as they *must* be for the field of a point charge. Checking dimensions is a good way to verify that you haven't made algebra errors.

We can now combine $(E_1)_x$ and $(E_2)_x$ to write the x-component of $\vec{E}_{net}$ as

$$(E_{net})_x = 2(E_1)_x + (E_2)_x = \frac{q}{4\pi\epsilon_0}\left[\frac{1}{x^2} + \frac{2x}{(x^2+d^2)^{3/2}}\right]$$

The other two components of $\vec{E}_{net}$ are zero, hence the electric field of the three charges at a point on the x-axis is

$$\vec{E}_{net} = \frac{q}{4\pi\epsilon_0}\left[\frac{1}{x^2} + \frac{2x}{(x^2+d^2)^{3/2}}\right]\hat{i}$$

ASSESS This is the electric field only at points *on the x-axis.* Furthermore, this expression is valid only for $x > 0$. The electric field to the left of the charges points in the opposite direction, but our expression doesn't change sign for negative x. (This is a consequence of how we wrote $(E_2)_x$.) We would need to modify this expression to use it for negative values of x. The good news, though, is that our expression *is* valid for both positive and negative q. A negative value of q makes $(E_{net})_x$ negative, which would be an electric field pointing to the left, toward the negative charges.

Let's explore this example a bit more. There are two limiting cases for which we know what the result should be. First, let x become really, really small. As the point in Figure 26.3 approaches the origin, the fields $\vec{E}_1$ and $\vec{E}_3$ become opposite to each other and cancel. Thus as $x \to 0$, the field *should* be that of the single point charge q at the origin, a field we already know. Is it? Notice that

$$\lim_{x\to 0}\frac{2x}{(x^2+d^2)^{3/2}} = 0 \tag{26.4}$$

Thus $E_{net} \to q/4\pi\epsilon_0 x^2$ as $x \to 0$, the expected field of a single point charge.

Now consider the opposite situation, where x becomes extremely large. From very far away, the three source charges will seem to merge into a single charge of size $3q$, just as three very distant lightbulbs appear to be a single light. Thus the field for $x \gg d$ *should* be that of a point charge $3q$. Is it?

The field is zero in the limit $x \to \infty$. That doesn't tell us much, so we don't want to go *that* far away. We simply want x to be very large in comparison to the spacing d between the source charges. If $x \gg d$, then the denominator of the second term of $\vec{E}_{net}$ is well approximated by $(x^2+d^2)^{3/2} \approx (x^2)^{3/2} = x^3$. Thus

$$\lim_{x\gg d}\left[\frac{1}{x^2} + \frac{2x}{(x^2+d^2)^{3/2}}\right] = \frac{1}{x^2} + \frac{2x}{x^3} = \frac{3}{x^2} \tag{26.5}$$

Consequently, the net electric field far from the source charges is

$$\vec{E}_{net}(x\gg d) = \frac{1}{4\pi\epsilon_0}\frac{(3q)}{x^2}\hat{i} \tag{26.6}$$

As expected, this is the field of a point charge $3q$. These checks of limiting cases provide confidence in the result of the calculation.

FIGURE 26.4 is a graph of the field strength E_{net} for the three charges of Example 26.1. Although we don't have any numerical values, we can specify x as a multiple of the charge separation d. Notice how the graph matches the field of a single point charge when $x \ll d$ and matches the field of a point charge $3q$ when $x \gg d$.

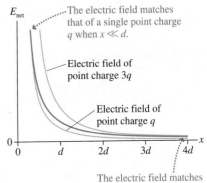

FIGURE 26.4 The electric field strength along a line perpendicular to three equal point charges.

FIGURE 26.5 Permanent and induced electric dipoles.

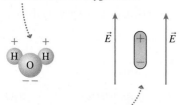

A water molecule is a *permanent* dipole because the negative electrons spend more time with the oxygen atom.

This dipole is *induced*, or stretched, by the electric field acting on the + and − charges.

The Electric Field of a Dipole

Two equal but opposite charges separated by a small distance form an *electric dipole*. FIGURE 26.5 shows two examples. In a *permanent electric dipole*, such as the water molecule, the oppositely charged particles maintain a small permanent separation. We can also create an electric dipole, as you learned in Chapter 25, by polarizing a neutral atom with an external electric field. This is an *induced electric dipole*.

FIGURE 26.6 shows that we can represent an electric dipole, whether permanent or induced, by two opposite charges $\pm q$ separated by the small distance s. The dipole has zero net charge, but it *does* have an electric field. Consider a point on the positive y-axis. This point is slightly closer to $+q$ than to $-q$, so the fields of the two charges do not cancel. You can see in the figure that $\vec{E}_{\text{dipole}}$ points in the positive y-direction. Similarly, vector addition shows that $\vec{E}_{\text{dipole}}$ points in the negative y-direction at points along the x-axis.

Let's calculate the electric field of a dipole at a point on the axis of the dipole. This is the y-axis in Figure 26.6. The point is distance $r_+ = y - s/2$ from the positive charge and $r_- = y + s/2$ from the negative charge. The net electric field at this point has only a y-component, and the sum of the fields of the two point charges gives

$$(E_{\text{dipole}})_y = (E_+)_y + (E_-)_y = \frac{1}{4\pi\epsilon_0}\frac{q}{(y - \frac{1}{2}s)^2} + \frac{1}{4\pi\epsilon_0}\frac{(-q)}{(y + \frac{1}{2}s)^2}$$
$$= \frac{q}{4\pi\epsilon_0}\left[\frac{1}{(y - \frac{1}{2}s)^2} - \frac{1}{(y + \frac{1}{2}s)^2}\right] \qquad (26.7)$$

Combining the two terms over a common denominator, we find

$$(E_{\text{dipole}})_y = \frac{q}{4\pi\epsilon_0}\left[\frac{2ys}{(y - \frac{1}{2}s)^2(y + \frac{1}{2}s)^2}\right] \qquad (26.8)$$

We omitted some of the algebraic steps, but be sure you can do this yourself. Some of the homework problems will require similar algebra.

In practice, we almost always observe the electric field of a dipole only for distances $y \gg s$—that is, for distances much larger than the charge separation. In such cases, the denominator can be approximated $(y - \frac{1}{2}s)^2(y + \frac{1}{2}s)^2 \approx y^4$. With this approximation, Equation 26.8 becomes

$$(E_{\text{dipole}})_y \approx \frac{1}{4\pi\epsilon_0}\frac{2qs}{y^3} \qquad (26.9)$$

It is useful to define the **dipole moment** $\vec{p}$, shown in FIGURE 26.7, as the vector

$$\vec{p} = (qs, \text{ from the negative to the positive charge}) \qquad (26.10)$$

The direction of $\vec{p}$ identifies the orientation of the dipole, and the dipole-moment magnitude $p = qs$ determines the electric field strength. The SI units of the dipole moment are $C\,m$.

We can use the dipole moment to write a succinct expression for the electric field at a point on the axis of a dipole:

$$\vec{E}_{\text{dipole}} \approx \frac{1}{4\pi\epsilon_0}\frac{2\vec{p}}{r^3} \qquad \text{(on the axis of an electric dipole)} \qquad (26.11)$$

where r is the distance measured from the *center* of the dipole. We've switched from y to r because we've now specified that Equation 26.11 is valid only along the axis of the dipole. Notice that the electric field along the axis points in the direction of the dipole moment $\vec{p}$.

A homework problem will let you calculate the electric field in the plane that bisects the dipole. This is the field shown on the x-axis in Figure 26.6, but it could

FIGURE 26.6 The dipole electric field at two points.

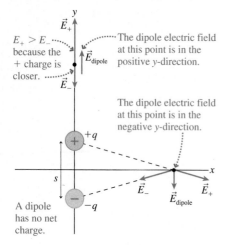

$E_+ > E_-$ because the + charge is closer.

The dipole electric field at this point is in the positive y-direction.

The dipole electric field at this point is in the negative y-direction.

A dipole has no net charge.

FIGURE 26.7 The dipole moment.

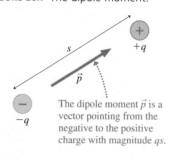

The dipole moment $\vec{p}$ is a vector pointing from the negative to the positive charge with magnitude qs.

equally well be the field on the z-axis as it comes out of the page. The field, for $r \gg s$, is

$$\vec{E}_{\text{dipole}} \approx -\frac{1}{4\pi\epsilon_0}\frac{\vec{p}}{r^3} \quad \text{(bisecting plane)} \qquad (26.12)$$

This field is *opposite* to $\vec{p}$, and it is only half the strength of the on-axis field at the same distance.

NOTE ▶ Do these inverse-cube equations violate Coulomb's law? Not at all. Coulomb's law describes the force between two *point charges,* and from Coulomb's law we found that the electric field of a *point charge* varies with the inverse square of the distance. But a dipole is not a point charge. The field of a dipole decreases more rapidly than that of a point charge, which is to be expected because the dipole is, after all, electrically neutral. ◀

EXAMPLE 26.2 **The electric field of a water molecule**

The water molecule H_2O has a permanent dipole moment of magnitude 6.2×10^{-30} C m. What is the electric field strength 1.0 nm from a water molecule at a point on the dipole's axis?

MODEL The size of a molecule is ≈ 0.1 nm. Thus $r \gg s$, and we can use Equation 26.11 for the on-axis electric field of the molecule's dipole moment.

SOLVE The on-axis electric field strength at $r = 1.0$ nm is

$$E \approx \frac{1}{4\pi\epsilon_0}\frac{2p}{r^3} = (9.0 \times 10^9 \text{ N m}^2/\text{C}^2)\frac{2(6.2 \times 10^{-30} \text{ C m})}{(1.0 \times 10^{-9} \text{ m})^3}$$
$$= 1.1 \times 10^8 \text{ N/C}$$

ASSESS By referring to Table 26.1 you can see that the field strength is "strong" compared to our everyday experience with charged objects but "weak" compared to the electric field inside the atoms themselves. This seems reasonable.

Picturing the Electric Field

We can't see the electric field. Consequently, we need pictorial tools to help us visualize it in a region of space. One method, introduced in Chapter 25, is to picture the electric field by drawing electric field vectors at various points in space. Another way to picture the field is to draw **electric field lines.**

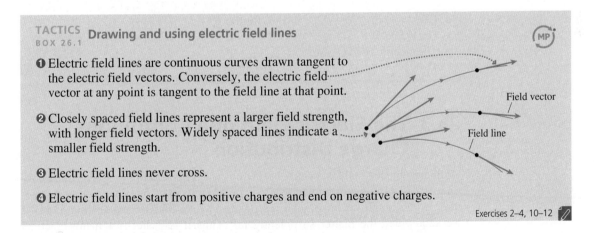

TACTICS
BOX 26.1 **Drawing and using electric field lines**

❶ Electric field lines are continuous curves drawn tangent to the electric field vectors. Conversely, the electric field vector at any point is tangent to the field line at that point.

❷ Closely spaced field lines represent a larger field strength, with longer field vectors. Widely spaced lines indicate a smaller field strength.

❸ Electric field lines never cross.

❹ Electric field lines start from positive charges and end on negative charges.

Field vector

Field line

Exercises 2–4, 10–12

Step 3 is required to make sure that $\vec{E}$ has a unique direction at every point in space. Step 4 follows from the fact that electric fields are created by charges. However, we will have to modify step 4 in Chapter 33 when we find another way to create an electric field.

FIGURE 26.8a on the next page represents the electric field of a dipole as a field-vector diagram. FIGURE 26.8b shows the same field using electric field lines. Notice how the on-axis field points in the direction of $\vec{p}$, both above and below the dipole, while the field in the bisecting plane points opposite to $\vec{p}$. At most points, however, $\vec{E}$ has components both parallel to $\vec{p}$ and perpendicular to $\vec{p}$.

FIGURE 26.8 The electric field of a dipole: (a) field vectors, (b) field lines.

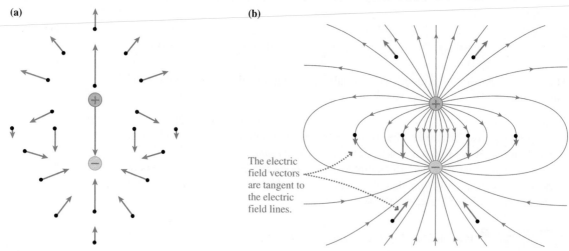

The electric field vectors are tangent to the electric field lines.

FIGURE 26.9 The electric field of two equal positive charges.

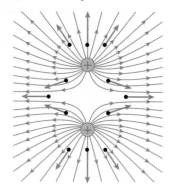

FIGURE 26.9 shows the electric field of two same-sign charges. A careful comparison of Figures 26.8b and 26.9 is worthwhile. Make sure you can explain the similarities and differences.

Neither field-vector diagrams nor field-line diagrams are perfect pictorial representations of an electric field. The field vectors are somewhat harder to draw, and they show the field at only a few points, but they do clearly indicate the direction and strength of the electric field at those points. Field-line diagrams perhaps look more elegant, and they're sometimes easier to sketch, but there's no formula for knowing where to draw the lines. We'll use both field-vector diagrams and field-line diagrams, depending on the circumstances.

STOP TO THINK 26.1   At the dot, the electric field points

a. Left.
c. Up.
e. The electric field is zero.

b. Right.
d. Down.

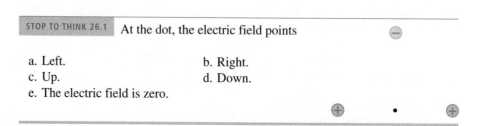

26.3 The Electric Field of a Continuous Charge Distribution

Ordinary objects—tables, chairs, beakers of water—seem to our senses to be continuous distributions of matter. There is no obvious evidence for an atomic structure, even though we have good reasons to believe that we would find atoms if we subdivided the matter sufficiently far. Thus it is easier, for many practical purposes, to consider matter to be continuous and to talk about the *density* of matter. Density—the number of kilograms of matter per cubic meter—allows us to describe the distribution of matter *as if* the matter were continuous rather than atomic.

Much the same situation occurs with charge. If a charged object contains a large number of excess electrons—for example, 10^{12} extra electrons on a metal rod—it is not practical to track every electron. It makes more sense to consider the charge to be *continuous* and to describe how it is distributed over the object.

FIGURE 26.10a shows an object of length L, such as a plastic rod or a metal wire, with charge Q spread uniformly along it. (We will use an uppercase Q for the total charge of an object, reserving lowercase q for individual point charges.) The **linear charge density** λ is defined to be

$$\lambda = \frac{Q}{L} \qquad (26.13)$$

Linear charge density, which has units of C/m, is the amount of charge *per meter* of length. The linear charge density of a 20-cm-long wire with 40 nC of charge is 2.0 nC/cm or 2.0×10^{-7} C/m.

NOTE ▶ The linear charge density λ is analogous to the linear mass density μ that you used in Chapter 20 to find the speed of a wave on a string. ◀

We'll also be interested in charged surfaces. FIGURE 26.10b shows a two-dimensional distribution of charge across a surface of area A. We define the **surface charge density** η (lowercase Greek eta) to be

$$\eta = \frac{Q}{A} \qquad (26.14)$$

Surface charge density, with units of C/m^2, is the amount of charge *per square meter*. A 1.0 mm $\times$ 1.0 mm square on a surface with $\eta = 2.0 \times 10^{-4}$ C/m^2 contains 2.0×10^{-10} C or 0.20 nC of charge. (The volume charge density $\rho = Q/V$, measured in C/m^3, will be used in Chapter 27.)

Figure 26.10 and the definitions of Equations 26.13 and 26.14 assume that the object is **uniformly charged,** meaning that the charges are evenly spread over the object. We will assume objects are uniformly charged unless noted otherwise.

NOTE ▶ Some textbooks represent the surface charge density with the symbol σ. Because σ is also used to represent *conductivity,* an idea we'll introduce in Chapter 30, we've selected a different symbol for surface charge density. ◀

FIGURE 26.10 One-dimensional and two-dimensional continuous charge distributions.

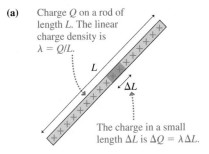

(a) Charge Q on a rod of length L. The linear charge density is $\lambda = Q/L$.

The charge in a small length ΔL is $\Delta Q = \lambda \Delta L$.

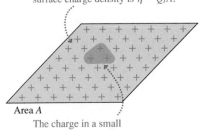

(b) Charge Q on a surface of area A. The surface charge density is $\eta = Q/A$.

Area A

The charge in a small area ΔA is $\Delta Q = \eta \Delta A$.

STOP TO THINK 26.2 A piece of plastic is uniformly charged with surface charge density η_a. The plastic is then broken into a large piece with surface charge density η_b and a small piece with surface charge density η_c. Rank in order, from largest to smallest, the surface charge densities η_a to η_c.

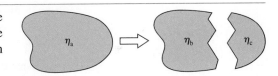

A Problem-Solving Strategy

Our goal is to find the electric field of a continuous distribution of charge, such as a charged rod or a charged disk. We have two basic tools to work with:

■ The electric field of a point charge, and
■ The principle of superposition.

We can apply these tools to a continuous distribution of charge if we follow a three-step strategy:

1. Divide the total charge Q into many small point-like charges ΔQ.
2. Use our knowledge of the electric field of a point charge to find the electric field of each ΔQ.
3. Calculate the net field $\vec{E}_{net}$ by summing the fields of all the ΔQ.

In practice, as you may have guessed, we'll let the sum become an integral.

The difficulty with electric field calculations is not the summation or integration itself, which is the last step, but setting up the calculation and knowing *what* to integrate. We will go step by step through several examples to illustrate the procedures. However, we first need to flesh out the steps of the problem-solving strategy. The aim of this problem-solving strategy is to break a difficult problem down into small steps that are individually manageable.

PROBLEM-SOLVING
STRATEGY 26.2

The electric field of a continuous distribution of charge

MODEL Model the distribution as a simple shape, such as a line of charge or a disk of charge. Assume the charge is uniformly distributed.

VISUALIZE For the pictorial representation:

❶ Draw a picture and establish a coordinate system.
❷ Identify the point P at which you want to calculate the electric field.
❸ Divide the total charge Q into small pieces of charge ΔQ, using shapes for which you *already know* how to determine $\vec{E}$. This is often, but not always, a division into point charges.
❹ Draw the electric field vector at P for one or two small pieces of charge. This will help you identify distances and angles that need to be calculated.
❺ Look for symmetries of the charge distribution that simplify the field. You may conclude that some components of $\vec{E}$ are zero.

SOLVE The mathematical representation is $\vec{E}_{net} = \sum \vec{E}_i$.

- Use superposition to form an algebraic expression for *each* of the three components of $\vec{E}$ (unless you are sure one or more is zero) at point P.
- Let the (x, y, z) coordinates of the point remain variables.
- Replace the small charge ΔQ with an equivalent expression involving a charge density and a coordinate, such as dx, that describes the shape of charge ΔQ. **This is the critical step in making the transition from a sum to an integral** because you need a coordinate to serve as the integration variable.
- Express all angles and distances in terms of the coordinates.
- Let the sum become an integral. The integration will be over the *one* coordinate variable that is related to ΔQ. The integration limits for this variable must "cover" the entire charged object.

ASSESS Check that your result is consistent with any limits for which you know what the field should be.

EXAMPLE 26.3 **The electric field of a line of charge**

FIGURE 26.11 shows a thin, uniformly charged rod of length L with total charge Q. Find the electric field strength at radial distance r in the plane that bisects the rod.

FIGURE 26.11 A thin, uniformly charged rod.

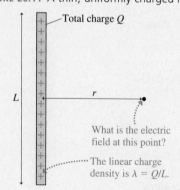

MODEL The rod is thin, so we'll assume the charge lies along a line and forms what we call a *line of charge*. This is an important

charge distribution that models the electric field of a charged rod or a charged metal wire. The rod's linear charge density is $\lambda = Q/L$.

VISUALIZE FIGURE 26.12 illustrates the five steps of the problem-solving strategy. We've chosen a coordinate system in which the rod lies along the y-axis and point P, in the bisecting plane, is on the x-axis. We've then divided the rod into N small segments of charge ΔQ, each of which can be modeled as a point charge. For every ΔQ in the bottom half of the wire with a field that points to the right and up, there's a matching ΔQ in the top half whose field points to the right and down. The y-components of these two fields cancel, hence the net electric field on the x-axis points straight away from the rod. The only component we need to calculate is E_x. (This is the same reasoning on the basis of symmetry that we used in Example 26.1.)

SOLVE Each of the little segments of charge can be modeled as a point charge. We know the electric field of a point charge, so we can write the x-component of $\vec{E}_i$, the electric field of segment i, as

$$(E_i)_x = E_i \cos\theta_i = \frac{1}{4\pi\epsilon_0} \frac{\Delta Q}{r_i^2} \cos\theta_i$$

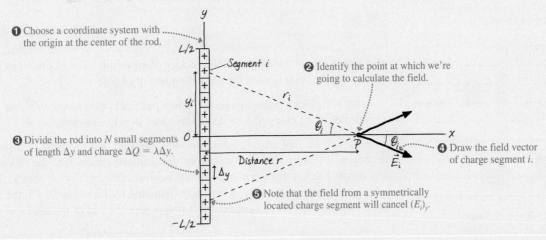

FIGURE 26.12 Calculating the electric field of a line of charge.

❶ Choose a coordinate system with the origin at the center of the rod.

❷ Identify the point at which we're going to calculate the field.

❸ Divide the rod into N small segments of length Δy and charge $\Delta Q = \lambda \Delta y$.

❹ Draw the field vector of charge segment i.

❺ Note that the field from a symmetrically located charge segment will cancel $(E_i)_y$.

where r_i is the distance from charge i to point P. You can see from the figure that $r_i = (y_i^2 + r^2)^{1/2}$ and $\cos\theta_i = r/r_i = r/(y_i^2 + r^2)^{1/2}$. With these, $(E_i)_x$ is

$$(E_i)_x = \frac{1}{4\pi\epsilon_0}\frac{\Delta Q}{y_i^2 + r^2}\frac{r}{\sqrt{y_i^2 + r^2}}$$

$$= \frac{1}{4\pi\epsilon_0}\frac{r\,\Delta Q}{(y_i^2 + r^2)^{3/2}}$$

Compare this result to the very similar calculation we did in Example 26.1. If we now sum this expression over all the charge segments, the net x-component of the electric field is

$$E_x = \sum_{i=1}^{N}(E_i)_x = \frac{1}{4\pi\epsilon_0}\sum_{i=1}^{N}\frac{r\,\Delta Q}{(y_i^2 + r^2)^{3/2}}$$

This is the same superposition we did for the $N = 3$ case in Example 26.1. The only difference is that we have now written the result as an explicit summation so that N can have any value. We want to let $N \to \infty$ and to replace the sum with an integral, but we can't integrate over Q; it's not a geometric quantity. This is where the linear charge density enters. The quantity of charge in each segment is related to its length Δy by $\Delta Q = \lambda\,\Delta y = (Q/L)\Delta y$. In terms of the linear charge density, the electric field is

$$E_x = \frac{Q/L}{4\pi\epsilon_0}\sum_{i=1}^{N}\frac{r\,\Delta y}{(y_i^2 + r^2)^{3/2}}$$

Now we're ready to let the sum become an integral. If we let $N \to \infty$, then each segment becomes an infinitesimal length $\Delta y \to dy$ while the discrete position variable y_i becomes the continuous integration variable y. The sum from $i = 1$ at the bottom end of the line of charge to $i = N$ at the top end will be replaced with an integral from $y = -L/2$ to $y = +L/2$. Thus in the limit $N \to \infty$,

$$E_x = \frac{Q/L}{4\pi\epsilon_0}\int_{-L/2}^{L/2}\frac{r\,dy}{(y^2 + r^2)^{3/2}}$$

This is a standard integral that you have learned to do in calculus and that can be found in Appendix A. Note that r is a *constant* as far as this integral is concerned. Integrating gives

$$E_x = \frac{Q/L}{4\pi\epsilon_0}\frac{y}{r\sqrt{y^2 + r^2}}\bigg|_{-L/2}^{L/2}$$

$$= \frac{Q/L}{4\pi\epsilon_0}\left[\frac{L/2}{r\sqrt{(L/2)^2 + r^2}} - \frac{-L/2}{r\sqrt{(-L/2)^2 + r^2}}\right]$$

$$= \frac{1}{4\pi\epsilon_0}\frac{Q}{r\sqrt{r^2 + (L/2)^2}}$$

Because E_x is the *only* component of the field, the electric field strength E_{rod} at distance r from the center of a charged rod is

$$E_{rod} = \frac{1}{4\pi\epsilon_0}\frac{|Q|}{r\sqrt{r^2 + (L/2)^2}}$$

The field strength must be positive, so we added absolute value signs to Q to allow for the possibility that the charge could be negative. The only restriction is to remember that this is the electric field at a point in the plane that bisects the rod.

ASSESS Suppose we are at a point *very* far from the rod. If $r \gg L$, the length of the rod is not relevant and the rod appears to be a point charge Q in the distance. Thus in the *limiting case* $r \gg L$, we expect the rod's electric field to be that of a point charge. If $r \gg L$, the square root becomes $(r^2 + (L/2)^2)^{1/2} \approx (r^2)^{1/2} = r$ and the electric field strength at distance r becomes $E_{rod} \approx Q/4\pi\epsilon_0 r^2$, the field of a point charge. The fact that our expression of E_{rod} has the correct limiting behavior gives us confidence that we haven't made any mistakes in its derivation.

An Infinite Line of Charge

What if the rod or wire becomes very long, becoming a **line of charge,** while the linear charge density λ remains constant? To answer this question, we can rewrite the expression for E_{rod} by factoring $(L/2)^2$ out of the denominator:

$$E_{rod} = \frac{1}{4\pi\epsilon_0}\frac{|Q|}{r\cdot L/2}\frac{1}{\sqrt{1 + 4r^2/L^2}} = \frac{1}{4\pi\epsilon_0}\frac{2|\lambda|}{r}\frac{1}{\sqrt{1 + 4r^2/L^2}}$$

FIGURE 26.13 The electric field of an infinite line of charge.

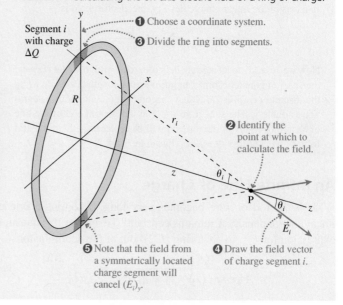

The field points straight away from the line at all points.

Infinite line of charge

The field strength decreases with distance.

where $|\lambda| = |Q|/L$ is the magnitude of the linear charge density. If we now let $L \rightarrow \infty$, the last term becomes simply 1 and we're left with

$$E_{\text{line}} = \frac{1}{4\pi\epsilon_0} \frac{2|\lambda|}{r} \qquad (26.15)$$

FIGURE 26.13 shows the electric field vectors of an infinite line of positive charge. The vectors would point inward for a negative line of charge.

NOTE ▶ Unlike a point charge, for which the field decreases as $1/r^2$, the field of an infinitely long charged wire decreases more slowly—as only $1/r$. ◀

Although no real wire is infinitely long, the fact that the field of a point charge decreases inversely with the square of the distance means that the electric field at any point is determined primarily by the nearest charges. Consequently, the field of a realistic finite-length wire is well approximated by Equation 26.15, the field of an infinitely long line of charge, except at points near the end of the wire.

STOP TO THINK 26.3   Which of the following actions will increase the electric field strength at the position of the dot?

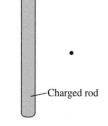

a. Make the rod longer without changing the charge.
b. Make the rod shorter without changing the charge.
c. Make the rod wider without changing the charge.
d. Make the rod narrower without changing the charge.
e. Add charge to the rod.
f. Remove charge from the rod.
g. Move the dot farther from the rod.
h. Move the dot closer to the rod.

Charged rod

26.4 The Electric Fields of Rings, Disks, Planes, and Spheres

In this section we'll derive the electric fields for several important charge distributions.

EXAMPLE 26.4   The electric field of a ring of charge

A thin ring of radius R is uniformly charged with total charge Q. Find the electric field at a point on the axis of the ring (perpendicular to the ring).

MODEL Because the ring is thin, we'll assume the charge lies along a circle of radius R. You can think of this as a line of charge of length $2\pi R$ wrapped into a circle. The linear charge density along the ring is $\lambda = Q/2\pi R$.

VISUALIZE FIGURE 26.14 shows the ring and illustrates the five steps of the problem-solving strategy. We've chosen a coordinate system in which the ring lies in the xy-plane and point P is on the z-axis. We've then divided the ring into N small segments of charge ΔQ, each of which can be modeled as a point charge. As you can see from the figure, the component of the field perpendicular to the axis cancels for two diametrically opposite segments. Thus we need to calculate only the z-component E_z.

SOLVE The z-component of the electric field due to segment i is

$$(E_i)_z = E_i \cos\theta_i = \frac{1}{4\pi\epsilon_0} \frac{\Delta Q}{r_i^2} \cos\theta_i$$

FIGURE 26.14 Calculating the on-axis electric field of a ring of charge.

Segment i with charge ΔQ

❶ Choose a coordinate system.

❸ Divide the ring into segments.

R

r_i

z

θ_i

❷ Identify the point at which to calculate the field.

P

θ_i

z

$\vec{E}_i$

❺ Note that the field from a symmetrically located charge segment will cancel $(E_i)_y$.

❹ Draw the field vector of charge segment i.

You can see from the figure that *every* segment of the ring, independent of i, has

$$r_i = \sqrt{z^2 + R^2}$$

$$\cos\theta_i = \frac{z}{r_i} = \frac{z}{\sqrt{z^2 + R^2}}$$

Consequently, the field of segment i is

$$(E_i)_z = \frac{1}{4\pi\epsilon_0}\frac{\Delta Q}{z^2 + R^2}\frac{z}{\sqrt{z^2 + R^2}} = \frac{1}{4\pi\epsilon_0}\frac{z}{(z^2 + R^2)^{3/2}}\Delta Q$$

The net electric field is found by summing $(E_i)_z$ due to all N segments:

$$E_z = \sum_{i=1}^{N}(E_i)_z = \frac{1}{4\pi\epsilon_0}\frac{z}{(z^2 + R^2)^{3/2}}\sum_{i=1}^{N}\Delta Q$$

We were able to bring all terms involving z to the front because z is a constant as far as the summation is concerned. Surprisingly, we don't need to convert the sum to an integral to complete this calculation. The sum of all the ΔQ around the ring is simply the ring's total charge, $\sum\Delta Q = Q$, hence the field on the axis is

$$(E_{\text{ring}})_z = \frac{1}{4\pi\epsilon_0}\frac{zQ}{(z^2 + R^2)^{3/2}}$$

This expression is valid for both positive and negative z (i.e., on either side of the ring) and for both positive and negative charge.

ASSESS It will be left as a homework problem to show that this result gives the expected limit when $z \gg R$.

FIGURE 26.15 shows two representations of the on-axis electric field of a positively charged ring. **FIGURE 26.15a** shows that the electric field vectors point away from the ring, increasing in length until reaching a maximum when $|z| \approx R$, then decreasing. The graph of $(E_{\text{ring}})_z$ in **FIGURE 26.15b** confirms that the field strength has a maximum on either side of the ring. Notice that the electric field at the center of the ring is zero, even though this point is surrounded by charge. You might want to spend a minute thinking about why this has to be the case.

A Disk of Charge

FIGURE 26.16 shows a disk of radius R that is uniformly charged with charge Q. This is a mathematical disk, with no thickness, and its surface charge density is

$$\eta = \frac{Q}{A} = \frac{Q}{\pi R^2} \quad (26.16)$$

We would like to calculate the on-axis electric field of this disk. Our problem-solving strategy tells us to divide a continuous charge into segments for which we already know how to find $\vec{E}$. Because we now know the on-axis electric field of a ring of charge, let's divide the disk into N very narrow rings of radius r and width Δr. One such ring, with radius r_i and charge ΔQ_i, is shown.

We need to be careful with notation. The R in Example 26.4 was the radius of the ring. Now we have many rings, and the radius of ring i is r_i. Similarly, Q was the charge on the ring. Now the charge on ring i is ΔQ_i, a small fraction of the total charge on the disk. With these changes, the electric field of ring i, with radius r_i, is

$$(E_i)_z = \frac{1}{4\pi\epsilon_0}\frac{z\,\Delta Q_i}{(z^2 + r_i^2)^{3/2}} \quad (26.17)$$

The on-axis electric field of the charged disk is the sum of the electric fields of all of the rings:

$$(E_{\text{disk}})_z = \sum_{i=1}^{N}(E_i)_z = \frac{z}{4\pi\epsilon_0}\sum_{i=1}^{N}\frac{\Delta Q_i}{(z^2 + r_i^2)^{3/2}} \quad (26.18)$$

The critical step, as always, is to relate ΔQ to a coordinate. Because we now have a surface, rather than a line, the charge in ring i is $\Delta Q = \eta\,\Delta A_i$, where ΔA_i is the area of ring i. We can find ΔA_i, as you've learned to do in calculus, by "unrolling" the ring to form a narrow rectangle of length $2\pi r_i$ and height Δr. Thus the area of ring i is $\Delta A_i = 2\pi r_i\,\Delta r$ and the charge is $\Delta Q_i = 2\pi\eta r_i\,\Delta r$. With this substitution, Equation 26.18 becomes

$$(E_{\text{disk}})_z = \frac{\eta z}{2\epsilon_0}\sum_{i=1}^{N}\frac{r_i\,\Delta r}{(z^2 + r_i^2)^{3/2}} \quad (26.19)$$

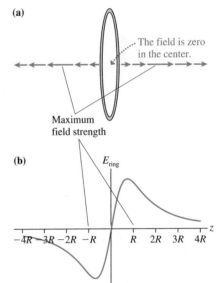

FIGURE 26.15 The on-axis electric field of a ring of charge.

(a)

The field is zero in the center.

Maximum field strength

(b)

E_{ring}

$-4R$ $-3R$ $-2R$ $-R$ R $2R$ $3R$ $4R$ z

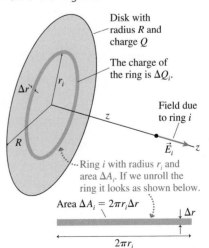

FIGURE 26.16 Calculating the on-axis field of a charged disk.

Disk with radius R and charge Q

The charge of the ring is ΔQ_i.

Field due to ring i

$\vec{E}_i$

Ring i with radius r_i and area ΔA_i. If we unroll the ring it looks as shown below.

Area $\Delta A_i = 2\pi r_i\,\Delta r$

$\downarrow \Delta r$

$2\pi r_i$

As $N \rightarrow \infty$, $\Delta r \rightarrow dr$ and the sum becomes an integral. Adding all the rings means integrating from $r = 0$ to $r = R$; thus

$$(E_{\text{disk}})_z = \frac{\eta z}{2\epsilon_0} \int_0^R \frac{r\,dr}{(z^2 + r^2)^{3/2}} \tag{26.20}$$

All that remains is to carry out the integration. This is straightforward if we make the variable change $u = z^2 + r^2$. Then $du = 2r\,dr$ or, equivalently, $r\,dr = \frac{1}{2}\,du$. At the lower integration limit $r = 0$, our new variable is $u = z^2$. At the upper limit $r = R$, the new variable is $u = z^2 + R^2$.

NOTE ▶ When changing variables in a definite integral, you *must* also change the limits of integration. ◀

With this variable change the integral becomes

$$(E_{\text{disk}})_z = \frac{\eta z}{2\epsilon_0} \frac{1}{2} \int_{z^2}^{z^2+R^2} \frac{du}{u^{3/2}} = \frac{\eta z}{4\epsilon_0} \frac{-2}{u^{1/2}} \bigg|_{z^2}^{z^2+R^2} = \frac{\eta z}{2\epsilon_0} \left[\frac{1}{z} - \frac{1}{\sqrt{z^2 + R^2}} \right] \tag{26.21}$$

If we multiply through by z, the on-axis electric field of a charged disk with surface charge density $\eta = Q/\pi R^2$ is

$$(E_{\text{disk}})_z = \frac{\eta}{2\epsilon_0} \left[1 - \frac{z}{\sqrt{z^2 + R^2}} \right] \tag{26.22}$$

NOTE ▶ This expression is valid only for $z > 0$. The field for $z < 0$ has the same magnitude but points in the opposite direction. ◀

It's a bit difficult see what Equation 26.22 is telling us, so let's compare it to what we already know. First, you can see that the quantity in square brackets is dimensionless. The surface charge density $\eta = Q/A$ has the same units as q/r^2, so $\eta/2\epsilon_0$ has the same units as $q/4\pi\epsilon_0 r^2$. This tells us that $\eta/2\epsilon_0$ really is an electric field.

Next, let's move very far away from the disk. At distance $z \gg R$, the disk appears to be a point charge Q in the distance and the field of the disk should approach that of a point charge. If we let $z \rightarrow \infty$ in Equation 26.22, so that $z^2 + R^2 \approx z^2$, we find $(E_{\text{disk}})_z \rightarrow 0$. This is true, but not quite what we wanted. We need to let z be very large in comparison to R, but not so large as to make E_{disk} vanish. That requires a little more care in taking the limit.

We can cast Equation 26.22 into a somewhat more useful form by factoring the z^2 out of the square root to give

$$(E_{\text{disk}})_z = \frac{\eta}{2\epsilon_0} \left[1 - \frac{1}{\sqrt{1 + R^2/z^2}} \right] \tag{26.23}$$

Now $R^2/z^2 \ll 1$ if $z \gg R$, so the second term in the square brackets is of the form $(1 + x)^{-1/2}$ where $x \ll 1$. We can then use the *binomial approximation*

$$(1 + x)^n \approx 1 + nx \quad \text{if} \quad x \ll 1 \qquad \text{(binomial approximation)}$$

to simplify the expression in square brackets:

$$1 - \frac{1}{\sqrt{1 + R^2/z^2}} = 1 - (1 + R^2/z^2)^{-1/2} \approx 1 - \left(1 + \left(-\frac{1}{2} \right) \frac{R^2}{z^2} \right) = \frac{R^2}{2z^2} \tag{26.24}$$

This is a good approximation when $z \gg R$. Substituting this approximation into Equation 26.23, we find that the electric field of the disk for $z \gg R$ is

$$(E_{\text{disk}})_z \approx \frac{\eta}{2\epsilon_0} \frac{R^2}{2z^2} = \frac{Q/\pi R^2}{4\epsilon_0} \frac{R^2}{z^2} = \frac{1}{4\pi\epsilon_0} \frac{Q}{z^2} \quad \text{if} \quad z \gg R \tag{26.25}$$

This is, indeed, the field of a point charge Q, giving us confidence in Equation 26.22 for the on-axis electric field of a disk of charge.

NOTE ▶ The binomial approximation is an important tool for looking at the limiting cases of electric fields. ◀

EXAMPLE 26.5 **The electric field of a charged disk**

A 10-cm-diameter plastic disk is charged uniformly with an extra 10^{11} electrons. What is the electric field 1.0 mm above the surface at a point near the center?

MODEL Model the plastic disk as a uniformly charged disk. We are seeking the on-axis electric field. Because the charge is negative, the field will point *toward* the disk.

SOLVE The total charge on the plastic square is $Q = N(-e) = -1.60 \times 10^{-8}$ C. The surface charge density is

$$\eta = \frac{Q}{A} = \frac{Q}{\pi R^2} = \frac{-1.60 \times 10^{-8}\,\text{C}}{\pi (0.050\,\text{m})^2} = -2.04 \times 10^{-6}\,\text{C/m}^2$$

The electric field at $z = 0.0010$ m, given by Equation 26.23, is

$$E_z = \frac{\eta}{2\epsilon_0}\left[1 - \frac{1}{\sqrt{1 + R^2/z^2}}\right] = -1.1 \times 10^5\ \text{N/C}$$

The minus sign indicates that the field points *toward*, rather than away from, the disk. As a vector,

$$\vec{E} = (1.1 \times 10^5\ \text{N/C, toward the disk})$$

ASSESS The total charge, -16 nC, is typical of the amount of charge produced on a small plastic object by rubbing or friction. Thus 10^5 N/C is a typical electric field strength near an object that has been charged by rubbing.

A Plane of Charge

Many electronic devices use charged, flat surfaces—disks, squares, rectangles, and so on—to steer electrons along the proper paths. These charged surfaces are called **electrodes.** Although any real electrode is finite in extent, we can often model an electrode as an infinite **plane of charge.** As long as the distance z to the electrode is small in comparison to the distance to the edges, we can reasonably treat the edges *as if* they are infinitely far away.

The electric field of a plane of charge is found from the on-axis field of a charged disk by letting the radius $R \to \infty$. That is, a disk with infinite radius is an infinite plane. From Equation 26.22, we see that the electric field of a plane of charge with surface charge density η is:

$$E_{\text{plane}} = \frac{\eta}{2\epsilon_0} = \text{constant} \qquad (26.26)$$

This is a simple result, but what does it tell us? First, the field strength is directly proportional to the charge density η: More charge, bigger field. Second, and more interesting, the field strength is the same at *all* points in space, independent of the distance z. The field strength 1000 m from the plane is the same as the field strength 1 mm from the plane.

How can this be? It seems that the field should get weaker as you move away from the plane of charge. But remember that we are dealing with an *infinite* plane of charge. What does it mean to be "close to" or "far from" an infinite object? For a disk of finite radius R, whether a point at distance z is "close to" or "far from" the disk is a comparison of z to R. If $z \ll R$, the point is close to the disk. If $z \gg R$, the point is far from the disk. But as $R \to \infty$, we have no *scale* for distinguishing near and far. In essence, *every* point in space is "close to" a disk of infinite radius.

No real plane is infinite in extent, but we can interpret Equation 26.26 as saying that the field of a surface of charge, regardless of its shape, is a constant $\eta/2\epsilon_0$ for those points whose distance z to the surface is much smaller than their distance to the edge. Eventually, when $z \gg R$, the charged surface will begin to look like a point charge Q and the field will have to decrease as $1/z^2$.

We do need to note that the derivation leading to Equation 26.26 considered only $z > 0$. For a positively charged plane, with $\eta > 0$, the electric field points *away from* the plane on both sides of the plane. This requires $E_z < 0$ ($\vec{E}$ pointing in the negative

FIGURE 26.17 Two views of the electric field of a plane of charge.

Perspective view

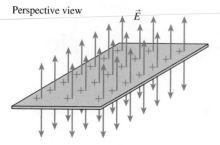

Edge view

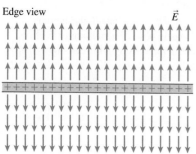

z-direction) on the side with $z < 0$. Thus a complete description of the electric field, valid for both sides of the plane and for either sign of η, is

$$(E_{\text{plane}})_z = \begin{cases} +\dfrac{\eta}{2\epsilon_0} & z > 0 \\[2ex] -\dfrac{\eta}{2\epsilon_0} & z < 0 \end{cases} \qquad (26.27)$$

FIGURE 26.17 shows two views of the electric field of a positively charged plane. All the arrows would be reversed for a negatively charged plane. It would have been very difficult to anticipate this result from Coulomb's law or from the electric field of a single point charge, but step by step we have been able to use the concept of the electric field to look at increasingly complex distributions of charge.

A Sphere of Charge

The one last charge distribution for which we need to know the electric field is a **sphere of charge.** This problem is analogous to wanting to know the gravitational field of a spherical planet or star. The procedure for calculating the field of a sphere of charge is the same as we used for lines and planes, but the integrations are significantly more difficult. We will skip the details of the calculations and, for now, simply assert the result without proof. In Chapter 27 we'll use an alternative procedure to find the field of a sphere of charge.

A sphere of charge Q and radius R, be it a uniformly charged sphere or just a spherical shell, has an electric field *outside* the sphere ($r \geq R$) that is exactly the same as that of a point charge Q located at the center of the sphere:

FIGURE 26.18 The electric field of a sphere of positive charge.

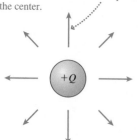

The electric field outside a sphere or spherical shell is the same as the field of a point charge Q at the center.

$$\vec{E}_{\text{sphere}} = \frac{Q}{4\pi\epsilon_0 r^2}\hat{r} \qquad \text{for } r \geq R \qquad (26.28)$$

This assertion is analogous to our earlier assertion that the gravitational force between stars and planets can be computed as if all the mass is at the center.

FIGURE 26.18 shows the electric field of a sphere of positive charge. The field of a negative sphere would point inward.

STOP TO THINK 26.4 Rank in order, from largest to smallest, the electric field strengths E_a to E_e at these five points near a plane of charge.

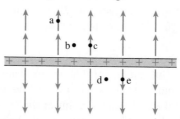

FIGURE 26.19 A parallel-plate capacitor.

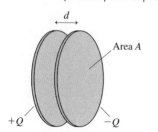

26.5 The Parallel-Plate Capacitor

FIGURE 26.19 shows two electrodes, one with charge $+Q$ and the other with $-Q$, placed face-to-face a distance d apart. This arrangement of two electrodes, charged equally but oppositely, is called a **parallel-plate capacitor.** Capacitors play important roles in many electric circuits. Our goal is to find the electric field both inside the capacitor (i.e., between the plates) and outside the capacitor.

NOTE ▶ The *net* charge of a capacitor is zero. Capacitors are charged by transferring electrons from one plate to the other. The plate that gains N electrons has charge $-Q = N(-e)$. The plate that loses electrons has charge $+Q$. ◀

Let's begin with a qualitative investigation. FIGURE 26.20 is an enlarged view of the capacitor plates, seen from the side. Because opposite charges attract, all of the charge is on the *inner* surfaces of the two plates. Thus the inner surfaces of the plates can be modeled as two planes of charge with equal but opposite surface charge densities. As you can see from the figure, at all points in space the electric field $\vec{E}_+$ of the positive plate points *away from* the plane of positive charges. Similarly, the field $\vec{E}_-$ of the negative plate everywhere points *toward* the plane of negative charges.

NOTE ▶ You might think the right capacitor plate would somehow "block" the electric field created by the positive plate and prevent the presence of an $\vec{E}_+$ field to the right of the capacitor. To see that it doesn't, consider an analogous situation with gravity. The strength of gravity above a table is the same as its strength below it. Just as the table doesn't block the earth's gravitational field, intervening matter or charges do not alter or block an object's electric field. ◀

Inside the capacitor, $\vec{E}_+$ and $\vec{E}_-$ are parallel and of equal strength. Their superposition creates a net electric field inside the capacitor that points from the positive plate to the negative plate. Outside the capacitor, $\vec{E}_+$ and $\vec{E}_-$ point in opposite directions and, because the field of a plane of charge is independent of the distance from the plane, have equal magnitudes. Consequently, the fields $\vec{E}_+$ and $\vec{E}_-$ add to zero outside the capacitor plates.

We can calculate the fields between the capacitor plates from the field of an infinite charged plane. Between the electrodes, $\vec{E}_+$ is of magnitude $\eta/2\epsilon_0$ and points from the positive toward the negative side. The field $\vec{E}_-$ is *also* of magnitude $\eta/2\epsilon_0$ and *also* points from positive to negative. Thus the electric field inside the capacitor is

$$\vec{E}_{capacitor} = \vec{E}_+ + \vec{E}_- = \left(\frac{\eta}{\epsilon_0}, \text{from positive to negative}\right)$$

$$= \left(\frac{Q}{\epsilon_0 A}, \text{from positive to negative}\right) \quad (26.29)$$

where A is the surface area of each electrode. Outside the capacitor plates, where $\vec{E}_+$ and $\vec{E}_-$ have equal magnitudes but *opposite* directions, $\vec{E} = \vec{0}$.

FIGURE 26.21a shows the electric field of an ideal parallel-plate capacitor constructed from two infinite charged planes. Now, it's true that no real capacitor is infinite in extent, but the ideal parallel-plate capacitor is a very good approximation for all but the most precise calculations as long as the electrode separation d is much smaller than the electrodes' size. FIGURE 26.21b shows that the interior field of a real capacitor is virtually identical to that of an ideal capacitor but that the exterior field isn't quite zero. This weak field outside the capacitor is called the **fringe field**. We will keep things simple by always assuming the plates are very close together and using Equation 26.29 for the field inside a parallel-plate capacitor.

NOTE ▶ The shape of the electrodes—circular or square or any other shape—is not relevant as long as the electrodes are very close together. ◀

FIGURE 26.20 The electric fields inside and outside a parallel-plate capacitor.

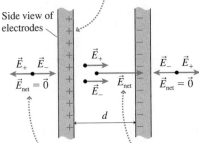

The capacitor's charge resides on the inner surfaces as planes of charge.

Outside the capacitor, $\vec{E}_+$ and $\vec{E}_-$ are opposite, so the net field is zero.

Inside the capacitor, $\vec{E}_+$ and $\vec{E}_-$ are parallel, so the net field is large.

FIGURE 26.21 The electric field of a capacitor.

(a) Ideal capacitor

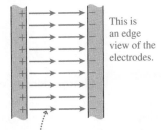

This is an edge view of the electrodes.

The field is uniform, pointing from the positive to the negative electrode.

(b) Real capacitor

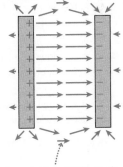

A weak fringe field extends outside the electrodes.

EXAMPLE 26.6 **The electric field inside a capacitor**

Two 1.0 cm × 2.0 cm rectangular electrodes are 1.0 mm apart. What charge must be placed on each electrode to create a uniform electric field of strength 2.0×10^6 N/C? How many electrons must be moved from one electrode to the other to accomplish this?

MODEL The electrodes can be modeled as a parallel-plate capacitor because the spacing between them is much smaller than their lateral dimensions.

SOLVE The electric field strength inside the capacitor is $E = Q/\epsilon_0 A$. Thus the charge to produce a field of strength E is

$$Q = (8.85 \times 10^{-12} \text{ C}^2/\text{N m}^2)(2.0 \times 10^{-4} \text{ m}^2)(2.0 \times 10^6 \text{ N/C})$$

$$= 3.5 \times 10^{-9} \text{ C} = 3.5 \text{ nC}$$

The positive plate must be charged to +3.5 nC and the negative plate to −3.5 nC. In practice, the plates are charged by using a

Continued

battery to move electrons from one plate to the other. The number of electrons in 3.5 nC is

$$N = \frac{Q}{e} = \frac{3.5 \times 10^{-9}\,\text{C}}{1.60 \times 10^{-19}\,\text{C/electron}} = 2.2 \times 10^{10}\ \text{electrons}$$

Thus 2.2×10^{10} electrons are moved from one electrode to the other. Note that the capacitor *as a whole* has no net charge.

ASSESS The plate spacing does not enter the result. As long as the spacing is much smaller than the plate dimensions, as is true in this example, the field is independent of the spacing.

Uniform Electric Fields

FIGURE 26.22 A uniform electric field.

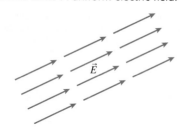

FIGURE 26.22 shows an electric field that is the *same*—in strength and direction—at every point in a region of space. This is called a **uniform electric field.** A uniform electric field is analogous to the uniform gravitational field near the surface of the earth. Uniform fields are of great practical significance because, as you will see in the next section, computing the trajectory of a charged particle moving in a uniform electric field is a straightforward process.

The easiest way to produce a uniform electric field is with a parallel-plate capacitor, as you can see in Figure 26.21a. Indeed, our interest in capacitors is due in large measure to the fact that the electric field is uniform. Many electric field problems refer to a uniform electric field. Such problems carry an implicit assumption that the action is taking place *inside* a parallel-plate capacitor.

EXAMPLE 26.7 **Charge density on a cell wall**

Example 25.7 noted that the electric field strength in the cell wall of a neuron is typically 1.0×10^7 N/C. This electric field is established because the outer surface of the cell wall is positive and the inner surface negative. What is a typical surface charge density on the surface of a cell wall?

MODEL Although cells are roughly spherical, the wall thickness is much less than the radius of the cell. Locally, at a point inside the cell wall, the curvature is negligible, so we can model the cell wall as a parallel-plate capacitor.

VISUALIZE FIGURE 26.23 shows a section of the cell wall. The charges are due to ions, not electrons, but that doesn't affect our analysis.

SOLVE The electric field strength inside a capacitor is $E = \eta/\epsilon_0$. The surface charge density needed to produce a known field is

$$\eta = \epsilon_0 E = (8.85 \times 10^{-12}\ \text{C}^2/\text{N m}^2)(1.0 \times 10^7\ \text{N/C})$$
$$= 8.9 \times 10^{-5}\ \text{C/m}^2$$

FIGURE 26.23 The electric field inside the cell wall is due to charges on the surfaces.

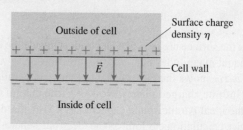

ASSESS The charge density may seem rather large, but cells are very small. A typical cell is $\approx 10\ \mu\text{m}$ in diameter, with a surface area of $\approx 3 \times 10^{-10}\ \text{m}^2$. At a surface charge density of $9 \times 10^{-5}\ \text{C/m}^2$, the total charge on the outer surface of the cell is $\approx 3 \times 10^{-14}\ \text{C}$, or $\approx 200{,}000$ ions.

STOP TO THINK 26.5 Rank in order, from largest to smallest, the forces F_a to F_e a proton would experience if placed at points a to e in this parallel-plate capacitor.

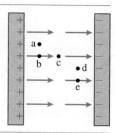

26.6 Motion of a Charged Particle in an Electric Field

Our motivation for introducing the concept of the electric field was to understand the long-range electric interaction of charges. We said that some charges, the *source charges,* create an electric field. Other charges then respond to that electric field. The first five sections of this chapter have focused on the electric field of the source charges. Now we turn our attention to the second half of the interaction.

FIGURE 26.24 shows a particle of charge q and mass m at a point where an electric field $\vec{E}$ has been produced by *other* charges, the source charges. The electric field exerts a force

$$\vec{F}_{\text{on } q} = q\vec{E}$$

on the charged particle. Notice that the force on a negatively charged particle is *opposite* in direction to the electric field vector. Signs are important!

FIGURE 26.24 The electric field exerts a force on a charged particle.

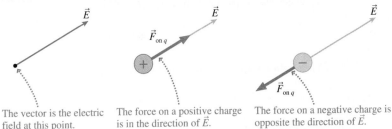

The vector is the electric field at this point.

The force on a positive charge is in the direction of $\vec{E}$.

The force on a negative charge is opposite the direction of $\vec{E}$.

If $\vec{F}_{\text{on } q}$ is the only force acting on q, it causes the charged particle to accelerate with

$$\vec{a} = \frac{\vec{F}_{\text{on } q}}{m} = \frac{q}{m}\vec{E} \qquad (26.30)$$

This acceleration is the *response* of the charged particle to the source charges that created the electric field. The ratio q/m is especially important for the dynamics of charged-particle motion. It is called the **charge-to-mass ratio.** Two *equal* charges, say a proton and a Na^+ ion, will experience *equal* forces $\vec{F} = q\vec{E}$ if placed at the same point in an electric field, but their accelerations will be *different* because they have different masses and thus different charge-to-mass ratios. Two particles with different charges and masses *but* with the same charge-to-mass ratio will undergo the same acceleration and follow the same trajectory.

Motion in a Uniform Field

The motion of a charged particle in a *uniform* electric field is especially important for its basic simplicity and because of its many valuable applications. A uniform field is *constant* at all points—constant in both magnitude and direction—within the region of space where the charged particle is moving. It follows, from Equation 26.30, that **a charged particle in a uniform electric field will move with constant acceleration.** The magnitude of the acceleration is

$$a = \frac{qE}{m} = \text{constant} \qquad (26.31)$$

where E is the electric field strength, and the direction of $\vec{a}$ is parallel or antiparallel to $\vec{E}$, depending on the sign of q.

"DNA fingerprints" are measured with the technique of *gel electrophoresis.* A solution of DNA fragments is placed in a well at one end of a plate covered with gel. The fragments are negatively charged when in solution, and they begin to migrate through the gel when a uniform electric field is established parallel to the surface of the plate. Because the gel exerts a drag force, the fragments move at a terminal speed inversely proportional to their size. Thus gel electrophoresis sorts the DNA fragments by size, and fluorescent markers allow the results to be seen.

Identifying the motion of a charged particle in a uniform field as being one of constant acceleration brings into play all the kinematic machinery that we developed in Chapters 2 and 4 for constant-acceleration motion. The basic trajectory of a charged particle in a uniform field is a *parabola,* analogous to the projectile motion of a mass in the near-earth uniform gravitational field. In the special case of a charged particle moving parallel to the electric field vectors, the motion is one-dimensional, analogous to the one-dimensional vertical motion of a mass tossed straight up or falling straight down.

NOTE ▶ The gravitational acceleration $\vec{a}_{\text{grav}}$ always points straight down. The electric field acceleration $\vec{a}_{\text{elec}}$ can point in *any* direction. You must determine the electric field $\vec{E}$ in order to learn the direction of $\vec{a}$. ◀

EXAMPLE 26.8 An electron moving across a capacitor

Two 6.0-cm-diameter electrodes are spaced 5.0 mm apart. They are charged by transferring 1.0×10^{11} electrons from one electrode to the other. An electron is released from rest at the surface of the negative electrode. How long does it take the electron to cross to the positive electrode? What is its speed as it collides with the positive electrode? Assume the space between the electrodes is a vacuum.

MODEL The electrodes form a parallel-plate capacitor. The electric field inside a parallel-plate capacitor is a uniform field, so the electron will have constant acceleration.

VISUALIZE FIGURE 26.25 shows an edge view of the capacitor and the electron. The force on the negative electron is *opposite* the electric field, so the electron is repelled by the negative electrode as it accelerates across the gap of width d.

FIGURE 26.25 An electron accelerates across a capacitor (plate separation exaggerated).

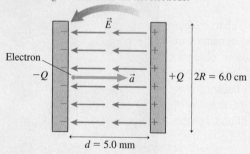

The capacitor was charged by transferring 10^{11} electrons from the right electrode to the left electrode.

$d = 5.0$ mm

SOLVE The electrodes are not point charges, so we cannot use Coulomb's law to find the force on the electron. Instead, we must analyze the electron's motion in terms of the electric field inside the capacitor. The field is the agent that exerts the force on the electron, causing it to accelerate. The electric field strength inside a parallel-plate capacitor with charge $Q = Ne$ is

$$E = \frac{\eta}{\epsilon_0} = \frac{Q}{\epsilon_0 A} = \frac{Ne}{\epsilon_0 \pi R^2} = 639{,}000 \text{ N/C}$$

The electron's acceleration in this field is

$$a = \frac{eE}{m} = 1.1 \times 10^{17} \text{ m/s}^2$$

where we used the electron mass $m = 9.11 \times 10^{-31}$ kg. This is an enormous acceleration compared to accelerations we're familiar with for macroscopic objects. We can use one-dimensional kinematics, with $x_i = 0$ and $v_i = 0$, to find the time required for the electron to cross the capacitor:

$$x_f = d = \frac{1}{2} a (\Delta t)^2$$

$$\Delta t = \sqrt{\frac{2d}{a}} = 3.0 \times 10^{-10} \text{ s} = 0.30 \text{ ns}$$

The electron's speed as it reaches the positive electrode is

$$v = a \, \Delta t = 3.3 \times 10^7 \text{ m/s}$$

ASSESS We used e rather than $-e$ to find the acceleration because we already knew the direction; we needed only the magnitude. The electron's speed, after traveling a mere 5 mm, is approximately 10% the speed of light.

Parallel electrodes such as those in Example 26.8 are often used to accelerate charged particles. If the positive plate has a small hole in the center, a *beam* of electrons will pass through the hole, after accelerating across the capacitor gap, and emerge with a speed of 3.3×10^7 m/s. This is the basic idea of the *electron gun* used until quite recently in *cathode-ray tube* (CRT) devices such as televisions and computer display terminals. (A negatively charged electrode is called a *cathode,* so the physicists who first learned to produce electron beams in the late 19th century called them *cathode rays.*) The following example shows that parallel electrodes can also be used to deflect charged particles sideways.

EXAMPLE 26.9 **Deflecting an electron beam**

An electron gun creates a beam of electrons moving horizontally with a speed of 3.3×10^7 m/s. The electrons enter a 2.0-cm-long gap between two parallel electrodes where the electric field is $\vec{E} = (5.0 \times 10^4$ N/C, down). In which direction, and by what angle, is the electron beam deflected by these electrodes?

MODEL The electric field between the electrodes is uniform. Assume that the electric field outside the electrodes is zero.

VISUALIZE FIGURE 26.26 shows an electron moving through the electric field. The electric field points down, so the force on the (negative) electrons is upward. The electrons will follow a parabolic trajectory, analogous to that of a ball thrown horizontally, except that the electrons "fall up" rather than down.

FIGURE 26.26 The deflection of an electron beam in a uniform electric field.

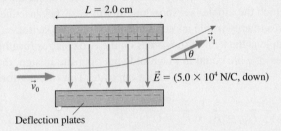

Deflection plates

SOLVE This is a two-dimensional motion problem. The electron enters the capacitor with velocity *vector* $\vec{v}_0 = v_{0x}\hat{\imath} = 3.3 \times 10^7 \hat{\imath}$ m/s and leaves with velocity $\vec{v}_1 = v_{1x}\hat{\imath} + v_{1y}\hat{\jmath}$. The electron's angle of travel upon leaving the electric field is

$$\theta = \tan^{-1}\left(\frac{v_{1y}}{v_{1x}}\right)$$

This is the *deflection angle*. To find θ we must compute the final velocity vector $\vec{v}_1$.

There is no horizontal force on the electron, so $v_{1x} = v_{0x} = 3.3 \times 10^7$ m/s. The electron's upward acceleration has magnitude

$$a = \frac{eE}{m} = \frac{(1.60 \times 10^{-19}\ \text{C})(5.0 \times 10^4\ \text{N/C})}{9.11 \times 10^{-31}\ \text{kg}}$$

$$= 8.78 \times 10^{15}\ \text{m/s}^2$$

We can use the fact that the horizontal velocity is constant to determine the time interval Δt needed to travel length 2.0 cm:

$$\Delta t = \frac{L}{v_{0x}} = \frac{0.020\ \text{m}}{3.3 \times 10^7\ \text{m/s}} = 6.06 \times 10^{-10}\ \text{s}$$

Vertical acceleration will occur during this time interval, resulting in a final vertical velocity

$$v_{1y} = v_{0y} + a\,\Delta t = 5.3 \times 10^6\ \text{m/s}$$

The electron's velocity as it leaves the capacitor is thus

$$\vec{v}_1 = (3.3 \times 10^7\,\hat{\imath} + 5.3 \times 10^6\,\hat{\jmath})\ \text{m/s}$$

and the deflection angle θ is

$$\theta = \tan^{-1}\left(\frac{v_{1y}}{v_{1x}}\right) = 9.1°$$

ASSESS We know that the electron beam in a cathode-ray tube can be deflected enough to cover the screen, so a deflection angle of 9° seems reasonable. Our neglect of the gravitational force is seen to be justified because the acceleration of the electrons is enormous in comparison to the free-fall acceleration g.

Example 26.9 demonstrates how an electron beam is steered to a point on the screen of a cathode-ray tube. First, a high-speed electron beam is created by an electron gun like that of Example 26.8. The beam then passes first through a set of *vertical deflection plates,* as in Example 26.9, then through a second set of *horizontal deflection plates.* After leaving the deflection plates, it travels in a straight line (through vacuum, to eliminate collisions with air molecules) to the screen of the CRT, where it strikes a phosphor coating on the inside surface and makes a dot of light. Properly choosing the electric fields within the deflection plates steers the electron beam to any point on the screen.

STOP TO THINK 26.6 Which electric field is responsible for the proton's trajectory?

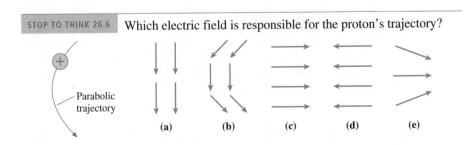

26.7 Motion of a Dipole in an Electric Field

Let us conclude this chapter by returning to one of the more striking puzzles we faced when making the observations at the beginning of Chapter 25. There you found that charged objects of *either* sign exert forces on neutral objects, such as when a comb used to brush your hair picks up pieces of paper. Our qualitative understanding of the *polarization force* was that it required two steps:

- The charge polarizes the neutral object, creating an induced electric dipole.
- The charge then exerts an attractive force on the near end of the dipole that is slightly stronger than the repulsive force on the far end.

We are now in a position to make that understanding more quantitative.

Dipoles in a Uniform Field

FIGURE 26.27a shows an electric dipole in a *uniform* external electric field $\vec{E}$ that has been created by source charges we do not see. That is, $\vec{E}$ is *not* the field of the dipole but, instead, is a field to which the dipole is responding. In this case, because the field is uniform, the dipole is presumably inside an unseen parallel-plate capacitor.

The net force on the dipole is the sum of the forces on the two charges forming the dipole. Because the charges $\pm q$ are equal in magnitude but opposite in sign, the two forces $\vec{F}_+ = +q\vec{E}$ and $\vec{F}_- = -q\vec{E}$, are also equal but opposite. Thus the net force on the dipole is

$$\vec{F}_{net} = \vec{F}_+ + \vec{F}_- = \vec{0} \tag{26.32}$$

There is no net force on a dipole in a uniform electric field.

There may be no net force, but the electric field *does* affect the dipole. Because the two forces in Figure 26.27a are in opposite directions but not aligned with each other, the electric field exerts a *torque* on the dipole and causes the dipole to *rotate*.

The torque causes the dipole to rotate until it is aligned with the electric field, as shown in FIGURE 26.27b. In this position, the dipole experiences not only no net force but also no torque. Thus Figure 26.27b represents the *equilibrium position* for a dipole in a uniform electric field. Notice that the positive end of the dipole is in the direction in which $\vec{E}$ points.

FIGURE 26.28 shows a sample of permanent dipoles, such as water molecules, in an external electric field. All the dipoles rotate until they are aligned with the electric field. This is the mechanism by which the sample becomes *polarized*. Once the dipoles are aligned, there is an excess of positive charge at one end of the sample and an excess of negative charge at the other end. The excess charges at the ends of the sample are the basis of the polarization forces we discussed in Section 25.3.

It's not hard to calculate the torque. Recall from Chapter 12 that the magnitude of a torque is the product of the force and the moment arm. FIGURE 26.29 shows that there are two forces of the same magnitude ($F_+ = F_- = qE$), each with the same moment arm ($d = \frac{1}{2}s\sin\theta$). Thus the torque on the dipole is

$$\tau = 2 \times dF_+ = 2(\tfrac{1}{2}s\sin\theta)(qE) = pE\sin\theta \tag{26.33}$$

where $p = qs$ was our definition of the dipole moment. The torque is zero when the dipole is aligned with the field, making $\theta = 0$.

Recall from Chapter 12 that the torque can be written in a compact mathematical form as the cross product between two vectors. The terms p and E in Equation 26.33 are the magnitudes of vectors, and θ is the angle between them. Thus in vector notation, the torque exerted on a dipole moment $\vec{p}$ by an electric field $\vec{E}$ is

$$\vec{\tau} = \vec{p} \times \vec{E} \tag{26.34}$$

The torque is greatest when $\vec{p}$ is perpendicular to $\vec{E}$, zero when $\vec{p}$ is aligned with or opposite to $\vec{E}$.

FIGURE 26.27 A dipole in a uniform electric field.

(a) The electric field exerts a torque on this dipole.

(b)
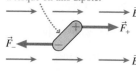
This dipole is in equilibrium.

FIGURE 26.28 A sample of permanent dipoles is *polarized* in an electric field.

The dipoles align with the electric field.
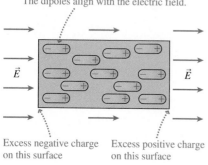
Excess negative charge on this surface   Excess positive charge on this surface

FIGURE 26.29 The torque on a dipole.
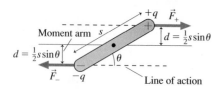
In terms of vectors, $\vec{\tau} = \vec{p} \times \vec{E}$.

EXAMPLE 26.10 **The angular acceleration of a dipole dumbbell**

Two 1.0 g balls are connected by a 2.0-cm-long insulating rod of negligible mass. One ball has a charge of +10 nC, the other a charge of −10 nC. The rod is held in a 1.0×10^4 N/C uniform electric field at an angle of 30° with respect to the field, then released. What is its initial angular acceleration?

MODEL The two oppositely charged balls form an electric dipole. The electric field exerts a torque on the dipole, causing an angular acceleration.

VISUALIZE **FIGURE 26.30** shows the dipole in the electric field.

FIGURE 26.30 The dipole of Example 26.10.

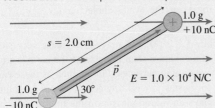

SOLVE The dipole moment is $p = qs = (1.0 \times 10^{-8}\, \text{C}) \times (0.020\, \text{m}) = 2.0 \times 10^{-10}\, \text{C m}$. The torque exerted on the dipole moment by the electric field is

$$\tau = pE \sin\theta = (2.0 \times 10^{-10}\, \text{C m})(1.0 \times 10^4\, \text{N/C}) \sin 30°$$
$$= 1.0 \times 10^{-6}\, \text{N m}$$

You learned in Chapter 12 that a torque causes an angular acceleration $\alpha = \tau/I$, where I is the moment of inertia. The dipole rotates about its center of mass, which is at the center of the rod, so the moment of inertia is

$$I = m_1 r_1^2 + m_2 r_2^2 = 2m\left(\frac{1}{2}s\right)^2 = \frac{1}{2}ms^2 = 2.0 \times 10^{-7}\, \text{kg m}^2$$

Thus the rod's angular acceleration is

$$\alpha = \frac{\tau}{I} = \frac{1.0 \times 10^{-6}\, \text{N m}}{2.0 \times 10^{-7}\, \text{kg m}^2} = 5.0\, \text{rad/s}^2$$

ASSESS This value of α is the initial angular acceleration, when the rod is first released. The torque and the angular acceleration will decrease as the rod rotates toward alignment with $\vec{E}$.

Dipoles in a Nonuniform Field

Suppose that a dipole is placed in a nonuniform electric field, one in which the field strength changes with position. For example, **FIGURE 26.31** shows a dipole in the nonuniform field of a point charge. The first response of the dipole is to rotate until it is aligned with the field, with the dipole's positive end pointing in the same direction as the field. Now, however, there is a slight difference between the forces acting on the two ends of the dipole. This difference occurs because the electric field, which depends on the distance from the point charge, is stronger at the end of the dipole nearest the charge. This causes a net force to be exerted on the dipole.

Which way does the force point? **FIGURE 26.31a** shows a positive point charge. Once the dipole is aligned, the leftward attractive force on its negative end is slightly stronger than the rightward repulsive force on its positive end. This causes a net force to the *left,* toward the point charge. The dipole in **FIGURE 26.31b** aligns in the opposite orientation in the field of a negative point charge, but the net force is still to the left.

As you can see, **the net force on a dipole is toward the direction of the strongest field.** Because any finite-size charged object, such as a charged rod or a charged disk, has a field strength that increases as you get closer to the object, we can conclude that **a dipole will experience a net force toward any charged object.**

FIGURE 26.31 An aligned dipole is drawn toward a point charge.

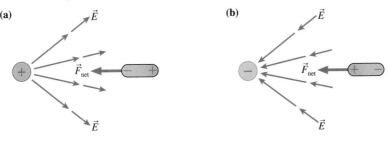

EXAMPLE 26.11 **The force on a water molecule**

The water molecule H_2O has a permanent dipole moment of magnitude 6.2×10^{-30} C m. A water molecule is located 10 nm from a Na^+ ion in a saltwater solution. What force does the ion exert on the water molecule?

VISUALIZE **FIGURE 26.32** shows the ion and the dipole. The forces are an action/reaction pair.

FIGURE 26.32 The interaction between an ion and a permanent dipole.

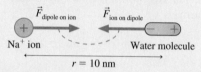

Na^+ ion   Water molecule

$r = 10$ nm

SOLVE A Na^+ ion has charge $q = +e$. The electric field of the ion aligns the water's dipole moment and exerts a net force on it. We could calculate the net force on the dipole as the small difference between the attractive force on its negative end and the repulsive force on its positive end. Alternatively, we know from Newton's third law that the force $\vec{F}_{\text{dipole on ion}}$ has the same magnitude as the force $\vec{F}_{\text{ion on dipole}}$ that we are seeking. We calculated the on-axis field of a dipole in Section 26.2. An ion of charge $q = e$ will experience a force of magnitude $F = qE_{\text{dipole}} = eE_{\text{dipole}}$ when placed in that field. The dipole's electric field, which we found in Equation 26.11, is

$$E_{\text{dipole}} = \frac{1}{4\pi\epsilon_0} \frac{2p}{r^3}$$

The force on the ion at distance $r = 1.0 \times 10^{-8}$ m is

$$F_{\text{dipole on ion}} = eE_{\text{dipole}} = \frac{1}{4\pi\epsilon_0} \frac{2ep}{r^3} = 1.8 \times 10^{-14} \text{ N}$$

Thus the force on the water molecule is $F_{\text{ion on dipole}} = 1.8 \times 10^{-14}$ N.

ASSESS While 1.8×10^{-14} N may seem like a very small force, it is $\approx 10^{11}$ times larger than the size of the earth's gravitational force on these atomic particles. Forces such as these cause water molecules to cluster around any ions that are in solution. This clustering plays an important role in the microscopic physics of solutions studied in chemistry and biochemistry.

CHALLENGE EXAMPLE 26.12 **An orbiting proton**

In a vacuum chamber, a proton orbits a 1.0-cm-diameter metal ball 1.0 mm above the surface with a period of 1.0 μs. What is the charge on the ball?

MODEL Model the ball as a charged sphere. The electric field of a charged sphere is the same as that of a point charge at the center, so the radius of the ball is irrelevant. Assume that the gravitational force on the proton is extremely small compared to the electric force and can be neglected.

VISUALIZE **FIGURE 26.33** shows the orbit and the force on the proton.

FIGURE 26.33 An orbiting proton.

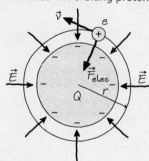

SOLVE The ball must be negative, with an inward electric field exerting an inward electric force on the positive proton. This is exactly the necessary condition for uniform circular motion. Recall from Chapter 8 that Newton's second law for uniform circular motion is $(F_{\text{net}})_r = mv^2/r$. Here the only radial force has magnitude $F_{\text{elec}} = eE$, so the proton will move in a circular orbit if

$$eE = \frac{mv^2}{r}$$

The electric field strength of a sphere of charge Q at distance r is $E = Q/4\pi\epsilon_0 r^2$. From Chapter 4, orbital speed and period are related by $v = $ circumference/period $= 2\pi r/T$. With these substitutions, Newton's second law becomes

$$\frac{eQ}{4\pi\epsilon_0 r^2} = \frac{4\pi^2 m}{T^2} r$$

Solving for Q, we find

$$Q = \frac{16\pi^3 \epsilon_0 m r^3}{eT^2} = 9.9 \times 10^{-12} \text{ C}$$

where we used $r = 6.0$ mm as the radius of the proton's orbit. Q is the *magnitude* of the charge on the ball. Including the sign, we have

$$Q_{\text{ball}} = -9.9 \times 10^{-12} \text{ C}$$

ASSESS This is not a lot of charge, but it shouldn't take much charge to affect the motion of something as light as a proton.

SUMMARY

The goal of Chapter 26 has been to learn how to calculate and use the electric field.

General Principles

Sources of $\vec{E}$

Electric fields are created by charges.

Two major tools for calculating $\vec{E}$ are

- The field of a point charge:

$$\vec{E} = \frac{1}{4\pi\epsilon_0}\frac{q}{r^2}\hat{r}$$

- The principle of superposition

Multiple point charges

Use superposition: $\vec{E} = \vec{E}_1 + \vec{E}_2 + \vec{E}_3 + \cdots$

Continuous distribution of charge

- Divide the charge into segments ΔQ for which you already know the field.
- Find the field of each ΔQ.
- Find $\vec{E}$ by summing the fields of all ΔQ.

The summation usually becomes an integral. A critical step is replacing ΔQ with an expression involving a **charge density** (λ or η) and an integration coordinate.

Consequences of $\vec{E}$

The electric field exerts a force on a charged particle:

$$\vec{F} = q\vec{E}$$

The force causes acceleration:

$$\vec{a} = (q/m)\vec{E}$$

Trajectories of charged particles are calculated with kinematics.

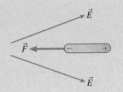

The electric field exerts a torque on a dipole:

$$\tau = pE\sin\theta$$

The torque tends to align the dipoles with the field.

In a nonuniform electric field, a dipole has a net force in the direction of increasing field strength.

Applications

Electric dipole

$$-q \boxed{\quad\longrightarrow\quad} +q$$

$$\underleftrightarrow{s}$$

The electric dipole moment is

$\vec{p} = (qs$, from negative to positive)

Field on axis: $\vec{E} = \frac{1}{4\pi\epsilon_0}\frac{2\vec{p}}{r^3}$

Field in bisecting plane: $\vec{E} = -\frac{1}{4\pi\epsilon_0}\frac{\vec{p}}{r^3}$

Infinite line of charge with linear charge density λ

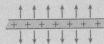

$\vec{E} = \left(\frac{1}{4\pi\epsilon_0}\frac{2\lambda}{r}\text{, perpendicular to line}\right)$

Infinite plane of charge with surface charge density η

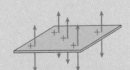

$\vec{E} = \left(\frac{\eta}{2\epsilon_0}\text{, perpendicular to plane}\right)$

Sphere of charge

Same as a point charge Q for $r > R$

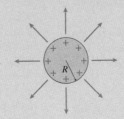

Parallel-plate capacitor

The electric field inside an ideal capacitor is a **uniform electric field:**

$\vec{E} = \left(\frac{\eta}{\epsilon_0}\text{, from positive to negative}\right)$

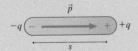

A real capacitor has a weak **fringe field** around it.

Terms and Notation

dipole moment, $\vec{p}$	uniformly charged	plane of charge	fringe field
electric field line	line of charge	sphere of charge	uniform electric field
linear charge density, λ	electrode	parallel-plate capacitor	charge-to-mass ratio, q/m
surface charge density, η			

CONCEPTUAL QUESTIONS

1. You've been assigned the task of determining the magnitude and direction of the electric field at a point in space. Give a step-by-step procedure of how you will do so. List any objects you will use, any measurements you will make, and any calculations you will need to perform. Make sure that your measurements do not disturb the charges that are creating the field.

2. Reproduce FIGURE Q26.2 on your paper. For each part, draw a dot or dots on the figure to show any position or positions (other than infinity) where $\vec{E} = \vec{0}$.

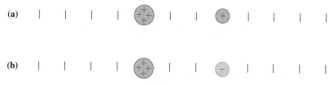

FIGURE Q26.2

3. Rank in order, from largest to smallest, the electric field strengths E_1 to E_4 at points 1 to 4 in FIGURE Q26.3. Explain.

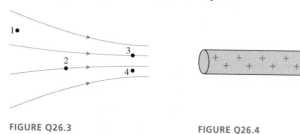

FIGURE Q26.3   FIGURE Q26.4

4. A small segment of wire in FIGURE Q26.4 contains 10 nC of charge.
   a. The segment is shrunk to one-third of its original length. What is the ratio λ_f/λ_i, where λ_i and λ_f are the initial and final linear charge densities?
   b. A proton is very far from the wire. What is the ratio F_f/F_i of the electric force on the proton after the segment is shrunk to the force before the segment was shrunk?
   c. Suppose the original segment of wire is stretched to 10 times its original length. How much charge must be *added* to the wire to keep the linear charge density unchanged?

5. An electron experiences a force of magnitude F when it is 1 cm from a very long, charged wire with linear charge density λ. If the charge density is doubled, at what distance from the wire will a proton experience a force of the same magnitude F?

6. FIGURE Q26.6 shows a hollow soda straw that has been uniformly charged with positive charge. What is the electric field at the center (inside) of the straw? Explain.

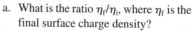
Inside straw

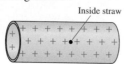

FIGURE Q26.6

7. The irregularly shaped area of charge in FIGURE Q26.7 has surface charge density η_i. Each dimension (x and y) of the area is reduced by a factor of 3.163.

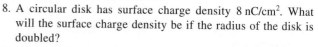

FIGURE Q26.7

   a. What is the ratio η_f/η_i, where η_f is the final surface charge density?
   b. An electron is very far from the area. What is the ratio F_f/F_i of the electric force on the electron after the area is reduced to the force before the area was reduced?

8. A circular disk has surface charge density 8 nC/cm². What will the surface charge density be if the radius of the disk is doubled?

9. A sphere of radius R has charge Q. The electric field strength at distance $r > R$ is E_i. What is the ratio E_f/E_i of the final to initial electric field strengths if (a) Q is halved, (b) R is halved, and (c) r is halved (but is still $> R$)? Each part changes only one quantity; the other quantities have their initial values.

10. The ball in FIGURE Q26.10 is suspended from a large, uniformly charged positive plate. It swings with period T. If the ball is discharged, will the period increase, decrease, or stay the same? Explain.

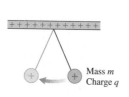

FIGURE Q26.10   FIGURE Q26.11

11. Rank in order, from largest to smallest, the electric field strengths E_1 to E_5 at the five points in FIGURE Q26.11. Explain.

12. A parallel-plate capacitor consists of two square plates, size $L \times L$, separated by distance d. The plates are given charge $\pm Q$. What is the ratio E_f/E_i of the final to initial electric field strengths if (a) Q is doubled, (b) L is doubled, and (c) d is doubled? Each part changes only one quantity; the other quantities have their initial values.

13. A small object is released at point 3 in the center of the capacitor in FIGURE Q26.11. For each situation, does the object move to the right, to the left, or remain in place? If it moves, does it accelerate or move at constant speed?
   a. A positive object is released from rest.
   b. A neutral but polarizable object is released from rest.
   c. A negative object is released from rest.

14. A proton and an electron are released from rest in the center of a capacitor.
    a. Is the force ratio F_p/F_e greater than 1, less than 1, or equal to 1? Explain.
    b. Is the acceleration ratio a_p/a_e greater than 1, less than 1, or equal to 1? Explain.

15. Three charges are placed at the corners of the triangle in FIGURE Q26.15. The ++ charge has twice the quantity of charge of the two − charges; the net charge is zero. Is the triangle in equilibrium? If so, explain why. If not, draw the equilibrium orientation.

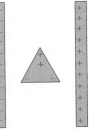

FIGURE Q26.15

EXERCISES AND PROBLEMS

Problems labeled ▓ integrate material from earlier chapters.

Exercises

Section 26.2 The Electric Field of Multiple Point Charges

1. ‖ What are the strength and direction of the electric field at the position indicated by the dot in FIGURE EX26.1? Specify the direction as an angle above or below horizontal.

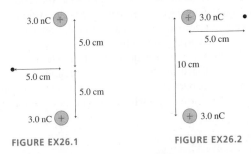

FIGURE EX26.1   FIGURE EX26.2

2. ‖ What are the strength and direction of the electric field at the position indicated by the dot in FIGURE EX26.2? Specify the direction as an angle above or below horizontal.

3. ‖ What are the strength and direction of the electric field at the position indicated by the dot in FIGURE EX26.3? Specify the direction as an angle above or below horizontal.

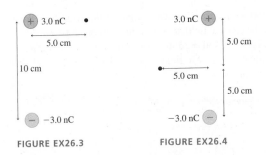

FIGURE EX26.3   FIGURE EX26.4

4. ‖ What are the strength and direction of the electric field at the position indicated by the dot in FIGURE EX26.4? Specify the direction as an angle above or below horizontal.

5. ‖ An electric dipole is formed from ± 1.0 nC charges spaced 2.0 mm apart. The dipole is at the origin, oriented along the x-axis. What is the electric field strength at the points (a) $(x, y) = (10$ cm, 0 cm$)$ and (b) $(x, y) = (0$ cm, 10 cm$)$?

6. ‖ An electric dipole is formed from two charges, $\pm q$, spaced 1.0 cm apart. The dipole is at the origin, oriented along the y-axis.

The electric field strength at the point $(x, y) = (0$ cm, 10 cm$)$ is 360 N/C.
   a. What is the charge q? Give your answer in nC.
   b. What is the electric field strength at the point $(x, y) = (10$ cm, 0 cm$)$?

Section 26.3 The Electric Field of a Continuous Charge Distribution

7. ‖ The electric field strength 10.0 cm from a very long charged wire is 2000 N/C. What is the electric field strength 5.0 cm from the wire?

8. ‖ A 10-cm-long thin glass rod uniformly charged to $+10$ nC and a 10-cm-long thin plastic rod uniformly charged to -10 nC are placed side by side, 4.0 cm apart. What are the electric field strengths E_1 to E_3 at distances 1.0 cm, 2.0 cm, and 3.0 cm from the glass rod along the line connecting the midpoints of the two rods?

9. ‖ Two 10-cm-long thin glass rods uniformly charged to $+10$ nC are placed side by side, 4.0 cm apart. What are the electric field strengths E_1 to E_3 at distances 1.0 cm, 2.0 cm, and 3.0 cm to the right of the rod on the left along the line connecting the midpoints of the two rods?

10. ‖ A small glass bead charged to $+6.0$ nC is 4.0 cm from a thin, uniformly charged, 10-cm-long glass rod. The bead is repelled from the rod with a force of 840 μN. What is the total charge on the rod?

Section 26.4 The Electric Fields of Rings, Disks, Planes, and Spheres

11. ‖ Two 10-cm-diameter charged rings face each other, 20 cm apart. The left ring is charged to -20 nC and the right ring is charged to $+20$ nC.
    a. What is the electric field $\vec{E}$, both magnitude and direction, at the midpoint between the two rings?
    b. What is the force $\vec{F}$ on a -1.0 nC charge placed at the midpoint?

12. ‖ Two 10-cm-diameter charged rings face each other, 20 cm apart. Both rings are charged to $+20$ nC. What is the electric field strength at (a) the midpoint between the two rings and (b) the center of the left ring?

13. ‖ Two 10-cm-diameter charged disks face each other, 20 cm apart. The left disk is charged to -50 nC and the right disk is charged to $+50$ nC.
    a. What is the electric field $\vec{E}$, both magnitude and direction, at the midpoint between the two disks?
    b. What is the force $\vec{F}$ on a -1.0 nC charge placed at the midpoint?

14. ‖ Two 10-cm-diameter charged disks face each other, 20 cm apart. Both disks are charged to +50 nC. What is the electric field strength at (a) the midpoint between the two disks and (b) a point on the axis 5.0 cm from one disk?

15. ‖ The electric field strength 2.0 cm from a 10-cm-diameter metal ball is 50,000 N/C. What is the charge (in nC) on the ball?

16. ‖ A 20 cm × 20 cm horizontal metal electrode is uniformly charged to +80 nC. What is the electric field strength 2.0 mm above the center of the electrode?

Section 26.5 The Parallel-Plate Capacitor

17. ‖ Two circular disks spaced 0.50 mm apart form a parallel-plate capacitor. Transferring 3.0×10^9 electrons from one disk to the other causes the electric field strength to be 2.0×10^5 N/C. What are the diameters of the disks?

18. ‖ A parallel-plate capacitor is formed from two 6.0-cm-diameter electrodes spaced 2.0 mm apart. The electric field strength inside the capacitor is 1.0×10^6 N/C. What is the charge (in nC) on each electrode?

19. ‖ Air "breaks down" when the electric field strength reaches 3.0×10^6 N/C, causing a spark. A parallel-plate capacitor is made from two 4.0 cm × 4.0 cm disks. How many electrons must be transferred from one disk to the other to create a spark between the disks?

Section 26.6 Motion of a Charged Particle in an Electric Field

20. ‖ A 0.10 g glass bead is charged by the removal of 1.0×10^{10} electrons. What electric field $\vec{E}$ (strength and direction) will cause the bead to hang suspended in the air?

21. ‖ Two 2.0-cm-diameter disks face each other, 1.0 mm apart. They are charged to ±10 nC.
   a. What is the electric field strength between the disks?
   b. A proton is shot from the negative disk toward the positive disk. What launch speed must the proton have to just barely reach the positive disk?

22. ‖ An electron in a uniform electric field increases its speed from 2.0×10^7 m/s to 4.0×10^7 m/s over a distance of 1.2 cm. What is the electric field strength?

23. ‖ The surface charge density on an infinite charged plane is -2.0×10^{-6} C/m². A proton is shot straight away from the plane at 2.0×10^6 m/s. How far does the proton travel before reaching its turning point?

24. ‖ A 1.0-μm-diameter oil droplet (density 900 kg/m³) is negatively charged with the addition of 25 extra electrons. It is released from rest 2.0 mm from a very wide plane of positive charge, after which it accelerates toward the plane and collides with a speed of 3.5 m/s. What is the surface charge density of the plane?

Section 26.7 Motion of a Dipole in an Electric Field

25. ‖ The permanent electric dipole moment of the water molecule (H_2O) is 6.2×10^{-30} C m. What is the maximum possible torque on a water molecule in a 5.0×10^8 N/C electric field?

26. ‖ A point charge Q is distance r from the center of a dipole consisting of charges ±q separated by distance s. The charge is located in the plane that bisects the dipole. At this instant, what are (a) the force (magnitude and direction) and (b) the magnitude of the torque on the dipole? You can assume $r \gg s$.

27. ‖ An ammonia molecule (NH_3) has a permanent electric dipole moment 5.0×10^{-30} C m. A proton is 2.0 nm from the molecule in the plane that bisects the dipole. What is the electric force of the molecule on the proton?

Problems

28. ‖ What are the strength and direction of the electric field at the position indicated by the dot in FIGURE P26.28? Give your answer (a) in component form and (b) as a magnitude and angle measured cw or ccw (specify which) from the positive x-axis.

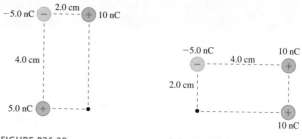

FIGURE P26.28   FIGURE P26.29

29. ‖ What are the strength and direction of the electric field at the position indicated by the dot in FIGURE P26.29? Give your answer (a) in component form and (b) as a magnitude and angle measured cw or ccw (specify which) from the positive x-axis.

30. ‖ What are the strength and direction of the electric field at the position indicated by the dot in FIGURE P26.30? Give your answer (a) in component form and (b) as a magnitude and angle measured cw or ccw (specify which) from the positive x-axis.

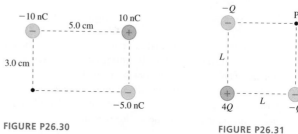

FIGURE P26.30   FIGURE P26.31

31. ‖ FIGURE P26.31 shows three charges at the corners of a square. Write the electric field at point P in component form.

32. ‖ Charges $-q$ and $+2q$ in FIGURE P26.32 are located at $x = \pm a$. Determine the electric field at points 1 to 4. Write each field in component form.

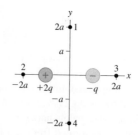

FIGURE P26.32

33. ‖ Two positive charges q are on the y-axis at $y = \pm \frac{1}{2}s$.
   a. Find an expression for the electric field strength at distance x on the axis that bisects the two charges.
   b. For $q = 1.0$ nC and $s = 6.0$ mm, evaluate E at $x = 0, 2, 4, 6,$ and 10 mm.

34. ‖ Derive Equation 26.12 for the field $\vec{E}_{dipole}$ in the plane that bisects an electric dipole.

35. ‖ Three charges are on the y-axis. Charges $-q$ are at $y = \pm d$ and charge $+2q$ is at $y = 0$.
    a. Find an expression for the electric field $\vec{E}$ at a point on the x-axis.
    b. Verify that your answer to part a has the expected behavior as x becomes very small and very large.

36. ‖ FIGURE P26.36 is a cross section of two infinite lines of charge that extend out of the page. Both have linear charge density λ. Find an expression for the electric field strength E at height y above the midpoint between the lines.

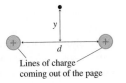

FIGURE P26.36

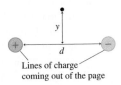

FIGURE P26.37

37. ‖‖‖ FIGURE P26.37 is a cross section of two infinite lines of charge that extend out of the page. The linear charge densities are $\pm \lambda$. Find an expression for the electric field strength E at height y above the midpoint between the lines.

38. ‖ Two infinite lines of charge, each with linear charge density λ, lie along the x- and y-axes, crossing at the origin. What is the electric field strength at position (x, y)?

39. ‖ The electric field 5.0 cm from a very long charged wire is (2000 N/C, toward the wire). What is the charge (in nC) on a 1.0-cm-long segment of the wire?

40. ‖ FIGURE P26.40 shows a thin rod of length L with total charge Q.
    a. Find an expression for the electric field strength at point P on the axis of the rod at distance r from the center.
    b. Verify that your expression has the expected behavior if $r \gg L$.
    c. Evaluate E at $r = 3.0$ cm if $L = 5.0$ cm and $Q = 3.0$ nC.

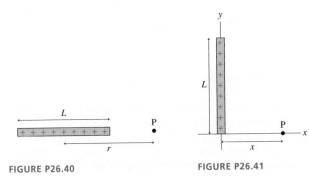

FIGURE P26.40

FIGURE P26.41

41. ‖‖‖ FIGURE P26.41 shows a thin rod of length L with total charge Q. Find an expression for the electric field $\vec{E}$ at point P. Give your answer in component form.

42. ‖ Show that the on-axis electric field of a ring of charge has the expected behavior when $z \ll R$ and when $z \gg R$.

43. ‖ A ring of radius R has total charge Q.
    a. At what distance along the z-axis is the electric field strength a maximum?
    b. What is the electric field strength at this point?

44. ‖ Charge Q is uniformly distributed along a thin, flexible rod of length L. The rod is then bent into the semicircle shown in FIGURE P26.44.

    a. Find an expression for the electric field $\vec{E}$ at the center of the semicircle.
    Hint: A small piece of arc length Δs spans a small angle $\Delta \theta = \Delta s/R$, where R is the radius.
    b. Evaluate the field strength if $L = 10$ cm and $Q = 30$ nC.

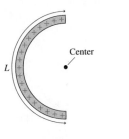
FIGURE P26.44

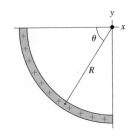
FIGURE P26.45

45. ‖ A plastic rod with linear charge density λ is bent into the quarter circle shown in FIGURE P26.45. We want to find the electric field at the origin.
    a. Write expressions for the x- and y-components of the electric field at the origin due to a small piece of charge at angle θ.
    b. Write, but do not evaluate, definite integrals for the x- and y-components of the net electric field at the origin.
    c. Evaluate the integrals and write $\vec{E}_{net}$ in component form.

46. ‖ You've hung two very large sheets of plastic facing each other with distance d between them, as shown in FIGURE P26.46. By rubbing them with wool and silk, you've managed to give one sheet a uniform surface charge density $\eta_1 = -\eta_0$ and the other a uniform surface charge density $\eta_2 = +3\eta_0$. What are the electric field vectors at points 1, 2, and 3?

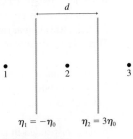
FIGURE P26.46

47. ‖ Two 2.0-cm-diameter insulating spheres have a 6.0 cm space between them. One sphere is charged to $+10$ nC, the other to -15 nC. What is the electric field strength at the midpoint between the two spheres?

48. ‖ Two parallel plates 1.0 cm apart are equally and oppositely charged. An electron is released from rest at the surface of the negative plate and simultaneously a proton is released from rest at the surface of the positive plate. How far from the negative plate is the point at which the electron and proton pass each other?

49. ‖ A parallel-plate capacitor has 2.0 cm × 2.0 cm electrodes with surface charge densities $\pm 1.0 \times 10^{-6}$ C/m². A proton traveling parallel to the electrodes at 1.0×10^6 m/s enters the center of the gap between them. By what distance has the proton been deflected sideways when it reaches the far edge of the capacitor? Assume the field is uniform inside the capacitor and zero outside the capacitor.

50. ‖ An electron is launched at a 45° angle and a speed of 5.0×10^6 m/s from the positive plate of the parallel-plate capacitor shown in FIGURE P26.50. The electron lands 4.0 cm away.
    a. What is the electric field strength inside the capacitor?
    b. What is the smallest possible spacing between the plates?

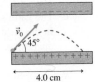
FIGURE P26.50

51. ‖ The two parallel plates in FIGURE P26.51 are 2.0 cm apart and the electric field strength between them is 1.0×10^4 N/C. An electron is launched at a 45° angle from the positive plate. What is the maximum initial speed v_0 the electron can have without hitting the negative plate?

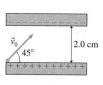

FIGURE P26.51   FIGURE P26.52

52. ‖ A problem of practical interest is to make a beam of electrons turn a 90° corner. This can be done with the parallel-plate capacitor shown in FIGURE P26.52. An electron with kinetic energy 3.0×10^{-17} J enters through a small hole in the bottom plate of the capacitor.
   a. Should the bottom plate be charged positive or negative relative to the top plate if you want the electron to turn to the right? Explain.
   b. What strength electric field is needed if the electron is to emerge from an exit hole 1.0 cm away from the entrance hole, traveling at right angles to its original direction?
   Hint: The difficulty of this problem depends on how you choose your coordinate system.
   c. What minimum separation d_{min} must the capacitor plates have?

53. ‖ The combustion of fossil fuels produces micron-sized particles of soot, one of the major components of air pollution. The terminal speeds of these particles are extremely small, so they remain suspended in air for very long periods of time. Furthermore, very small particles almost always acquire small amounts of charge from cosmic rays and various atmospheric effects, so their motion is influenced not only by gravity but also by the earth's weak electric field. Consider a small spherical particle of radius r, density ρ, and charge q. A small sphere moving with speed v experiences a drag force $F_{drag} = 6\pi\eta r v$, where η is the viscosity of the air. (This differs from the drag force you learned in Chapter 6 because there we considered macroscopic rather than microscopic objects.)
   a. A particle falling at its terminal speed v_{term} is in dynamic equilibrium with no net force. Write Newton's first law for this particle falling in the presence of a *downward* electric field of strength E, then solve to find an expression for v_{term}.
   b. Soot is primarily carbon, and carbon in the form of graphite has a density of 2200 kg/m³. In the absence of an electric field, what is the terminal speed in mm/s of a 1.0-μm-diameter graphite particle? The viscosity of air at 20°C is 1.8×10^{-5} kg/m s.
   c. The earth's electric field is typically (150 N/C, downward). In this field, what is the terminal speed in mm/s of a 1.0-μm-diameter graphite particle that has acquired 250 extra electrons?

54. ‖ A 2.0-mm-diameter glass sphere has a charge of +1.0 nC. What speed does an electron need to orbit the sphere 1.0 mm above the surface?

55. ‖ In a classical model of the hydrogen atom, the electron orbits the proton in a circular orbit of radius 0.053 nm. What is the orbital frequency? The proton is so much more massive than the electron that you can assume the proton is at rest.

56. ‖ In a classical model of the hydrogen atom, the electron orbits a stationary proton in a circular orbit. What is the radius of the orbit for which the orbital frequency is 1.0×10^{12} s⁻¹?

57. ‖ An electric field can *induce* an electric dipole in a neutral atom or molecule by pushing the positive and negative charges in opposite directions. The dipole moment of an induced dipole is directly proportional to the electric field. That is, $\vec{p} = \alpha\vec{E}$, where α is called the *polarizability* of the molecule. A bigger field stretches the molecule farther and causes a larger dipole moment.
   a. What are the units of α?
   b. An ion with charge q is distance r from a molecule with polarizability α. Find an expression for the force $\vec{F}_{ion\ on\ dipole}$.

58. ‖ Show that an infinite line of charge with linear charge density λ exerts an attractive force on an electric dipole with magnitude $F = 2\lambda p/4\pi\epsilon_0 r^2$. Assume that r is much larger than the charge separation in the dipole.

In Problems 59 through 62 you are given the equation(s) used to solve a problem. For each of these
   a. Write a realistic problem for which this is the correct equation(s).
   b. Finish the solution of the problem.

59. $(9.0 \times 10^9 \text{ N m}^2/\text{C}^2)\dfrac{(2.0 \times 10^{-9} \text{ C})s}{(0.025 \text{ m})^3} = 1150$ N/C

60. $(9.0 \times 10^9 \text{ N m}^2/\text{C}^2)\dfrac{2(2.0 \times 10^{-7} \text{ C/m})}{r} = 25{,}000$ N/C

61. $\dfrac{\eta}{2\epsilon_0}\left[1 - \dfrac{z}{\sqrt{z^2 + R^2}}\right] = \dfrac{1}{2}\dfrac{\eta}{2\epsilon_0}$

62. $2.0 \times 10^{12} \text{ m/s}^2 = \dfrac{(1.60 \times 10^{-19} \text{ C})E}{(1.67 \times 10^{-27} \text{ kg})}$

$E = \dfrac{Q}{(8.85 \times 10^{-12} \text{ C}^2/\text{N m}^2)(0.020 \text{ m})^2}$

Challenge Problems

63. Your physics assignment is to figure out a way to use electricity to launch a small 6.0-cm-long plastic drink stirrer. You decide that you'll charge the little plastic rod by rubbing it with fur, then hold it near a long, charged wire, as shown in FIGURE CP26.63. When you let go, the electric force of the wire on the plastic rod will shoot it away. Suppose you can uniformly charge the plastic stirrer to 10 nC and that the linear charge density of the long wire is 1.0×10^{-7} C/m. What is the net electric force on the plastic stirrer if the end closest to the wire is 2.0 cm away?

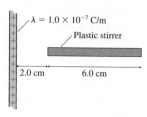

FIGURE CP26.63

64. Three 10-cm-long rods form an equilateral triangle in a plane. Two of the rods are charged to +10 nC, the third to −10 nC. What is the electric field strength at the center of the triangle?

65. A rod of length L lies along the y-axis with its center at the origin. The rod has a nonuniform linear charge density $\lambda = a|y|$, where a is a constant with the units C/m^2.
    a. Draw a graph of λ versus y over the length of the rod.
    b. Determine the constant a in terms of L and the rod's total charge Q.
    Hint: This requires an integration. Think about how to handle the absolute value sign.
    c. Find the electric field strength of the rod at distance x on the x-axis.

66. a. An infinitely long *sheet* of charge of width L lies in the xy-plane between $x = -L/2$ and $x = L/2$. The surface charge density is η. Derive an expression for the electric field $\vec{E}$ at height z above the centerline of the sheet.
    b. Verify that your expression has the expected behavior if $z \ll L$ and if $z \gg L$.
    c. Draw a graph of field strength E versus z.

67. a. An infinitely long *sheet* of charge of width L lies in the xy-plane between $x = -L/2$ and $x = L/2$. The surface charge density is η. Derive an expression for the electric field $\vec{E}$ along the x-axis for points outside the sheet $(x > L/2)$.
    b. Verify that your expression has the expected behavior if $x \gg L$.
    Hint: $\ln(1 + u) \approx u$ if $u \ll 1$.
    c. Draw a graph of field strength E versus x for $x > L/2$.

68. One type of ink-jet printer, called an electrostatic ink-jet printer, forms the letters by using deflecting electrodes to steer charged ink drops up and down vertically as the ink jet sweeps horizontally across the page. The ink jet forms 30-μm-diameter drops of ink, charges them by spraying 800,000 electrons on the surface, and shoots them toward the page at a speed of 20 m/s. Along the way, the drops pass through two horizontal, parallel electrodes that are 6.0 mm long, 4.0 mm wide, and spaced 1.0 mm apart. The distance from the center of the electrodes to the paper is 2.0 cm. To form the tallest letters, which have a height of 6.0 mm, the drops need to be deflected upward (or downward) by 3.0 mm. What electric field strength is needed between the electrodes to achieve this deflection? Ink, which consists of dye particles suspended in alcohol, has a density of 800 kg/m^3.

69. A proton orbits a long charged wire, making 1.0×10^6 revolutions per second. The radius of the orbit is 1.0 cm. What is the wire's linear charge density?

70. A *positron* is an elementary particle identical to an electron except that its charge is $+e$. An electron and a positron can rotate about their center of mass as if they were a dumbbell connected by a massless rod. What is the orbital frequency for an electron and a positron 1.0 nm apart?

71. You have a summer intern position with a company that designs and builds nanomachines. An engineer with the company is designing a microscopic oscillator to help keep time, and you've been assigned to help him analyze the design. He wants to place a negative charge at the center of a very small, positively charged metal ring. His claim is that the negative charge will undergo simple harmonic motion at a frequency determined by the amount of charge on the ring.
    a. Consider a negative charge near the center of a positively charged ring centered on the z-axis. Show that there is a restoring force on the charge if it moves along the z-axis but stays close to the center of the ring. That is, show there's a force that tries to keep the charge at $z = 0$.
    b. Show that for *small* oscillations, with amplitude $\ll R$, a particle of mass m with charge $-q$ undergoes simple harmonic motion with frequency

$$f = \frac{1}{2\pi} \sqrt{\frac{qQ}{4\pi\epsilon_0 mR^3}}$$

    R and Q are the radius and charge of the ring.
    c. Evaluate the oscillation frequency for an electron at the center of a 2.0-μm-diameter ring charged to 1.0×10^{-13} C.

STOP TO THINK ANSWERS

Stop to Think 26.1: c. From symmetry, the fields of the positive charges cancel. The net field is that of the negative charge, which is toward the charge.

Stop to Think 26.2: $\eta_c = \eta_b = \eta_a$. All pieces of a uniformly charged surface have the same surface charge density.

Stop to Think 26.3: b, e, and h. b and e both increase the linear charge density λ.

Stop to Think 26.4: $E_a = E_b = E_c = E_d = E_e$. The field strength of a charged plane is the same at all distances from the plane. An electric field diagram shows the electric field vectors at only a few points; the field exists at all points.

Stop to Think 26.5: $F_a = F_b = F_c = F_d = F_e$. The field strength inside a capacitor is the same at all points, hence the force on a charge is the same at all points. The electric field exists at all points whether or not a vector is shown at that point.

Stop to Think 26.6: c. Parabolic trajectories require *constant* acceleration and thus a *uniform* electric field. The proton has an initial velocity component to the left, but it's being pushed back to the right.

27 Gauss's Law

An electric field image of blood plasma from healthy blood. The wire in the center creates the electric field. Variations in the shape and color of the pattern can give early warning of cancer.

▶ **Looking Ahead** The goal of Chapter 27 is to understand and apply Gauss's law.

Symmetry

You'll learn how the shape of some important electric fields, those with a high degree of **symmetry,** can be deduced from the shape of the charge distribution. The idea of symmetry plays an important role in science and mathematics.

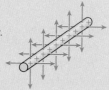

An infinitely long charged wire has *cylindrical symmetry*. The electric field of the wire must have the same symmetry.

Electric Flux

The amount of electric field passing through a surface is called the **electric flux.** You'll learn how to calculate the flux through open and closed surfaces.

The electric flux is analogous to the amount of air or water flowing through a loop.

◀ **Looking Back**
Section 11.3 The vector dot product

Gauss's Law

In Chapter 26, you learned to calculate electric fields based on the superposition of the fields of point charges. In this chapter, you'll learn a different way to calculate electric fields based on the idea of electric flux.

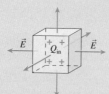

Gauss's law says that the electric flux through a *closed surface* is proportional to the charge Q_{in} enclosed within the surface. This will be the basis of a powerful problem-solving strategy for finding the electric fields of highly symmetric charge distributions.

Gauss's law is a more general statement about the nature of electric fields than is Coulomb's law. It is the first of the four equations that we'll later call *Maxwell's equations*, the governing equations of electricity and magnetism.

◀ **Looking Back**
Section 25.5 The field of a point charge
Section 26.2 Electric field lines

Using Gauss's Law

You'll learn how Gauss's law can be used to find the electric field both inside and outside of charged spheres, cylinders, and planes. In these highly symmetric situations, Gauss's law is much easier to use than superposition.

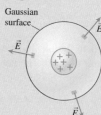

To find the field of a sphere of charge, you'll draw a *Gaussian surface* around the sphere and then calculate the electric flux through the surface.

Conductors

Gauss's law can be used to establish several important properties of conductors in **electrostatic equilibrium.**

- Excess charge is on the surface.
- The interior electric field is zero.

The metal grid in the door of a microwave oven shields the room because the electric field inside the metal must be zero. It turns out that the holes don't matter because they are very small compared to the wavelength of the microwaves.

27.1 Symmetry

Suppose we knew only two things about electric fields:

1. The field points away from positive charges, toward negative charges, and
2. An electric field exerts a force on a charged particle.

From this information alone, what can we deduce about the electric field of the infinitely long charged cylinder shown in FIGURE 27.1?

We don't know if the cylinder's diameter is large or small. We don't know if the charge density is the same at the outer edge as along the axis. All we know is that the charge is positive and the charge distribution has *cylindrical symmetry*. We say that a charge distribution is **symmetric** if there is a group of *geometric transformations* that don't cause any *physical* change.

To make this idea concrete, suppose you close your eyes while a friend transforms a charge distribution in one of the following three ways. He or she can

- *Translate* (that is, displace) the charge parallel to an axis,
- *Rotate* the charge about an axis, or
- *Reflect* the charge in a mirror.

When you open your eyes, will you be able to tell if the charge distribution has been changed? You might tell by observing a visual difference in the distribution. Or the results of an experiment with charged particles could reveal that the distribution has changed. If nothing you can see or do reveals any change, then we say that the charge distribution is symmetric under that particular transformation.

FIGURE 27.2 shows that the charge distribution of Figure 27.1 is symmetric with respect to

- Translation parallel to the cylinder axis. Shifting an infinitely long cylinder by 1 mm or 1000 m makes no noticeable or measurable change.
- Rotation by any angle about the cylinder axis. Turning a cylinder about its axis by 1° or 100° makes no detectable change.
- Reflections in any plane containing or perpendicular to the cylinder axis. Exchanging top and bottom, front and back, or left and right makes no detectable change.

A charge distribution that is symmetric under these three groups of geometric transformations is said to be *cylindrically symmetric*. Other charge distributions have other types of symmetries. Some charge distributions have no symmetry at all.

Our interest in symmetry can be summed up in a single statement:

The symmetry of the electric field must match the symmetry of the charge distribution.

If this were not true, you could use the electric field to test whether the charge distribution had undergone a transformation.

Now we're ready to see what we can learn about the electric field in Figure 27.1. Could the field look like FIGURE 27.3a? (Imagine this picture rotated about the axis.) That is, is this a *possible* field? This field looks the same if it's translated parallel to the

FIGURE 27.1 A charge distribution with cylindrical symmetry.

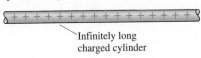
Infinitely long charged cylinder

FIGURE 27.2 Transformations that don't change an infinite cylinder of charge.

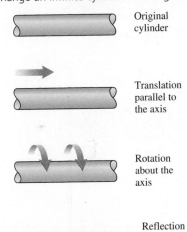
Original cylinder

Translation parallel to the axis

Rotation about the axis

Reflection in plane containing the axis

Reflection perpendicular to the axis

FIGURE 27.3 Could the field of a cylindrical charge distribution look like this?

(a) Is this a possible electric field of an infinitely long charged cylinder? Suppose the charge and the field are reflected in a plane perpendicular to the axis.

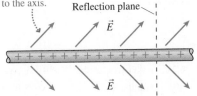
Reflection plane

$\vec{E}$

$\vec{E}$

Reflect

(b) The charge distribution is not changed by the reflection, but the field is. This field doesn't match the symmetry of the cylinder, so the cylinder's field can't look like this.

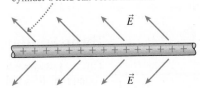
$\vec{E}$

$\vec{E}$

FIGURE 27.4 Or might the field of a cylindrical charge distribution look like this?

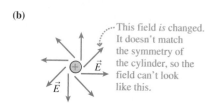

(a)

End view of cylinder — Reflection plane

The charge distribution is not changed by reflecting it in a plane containing the axis.

$\vec{E}$

Reflect

(b)

This field *is* changed. It doesn't match the symmetry of the cylinder, so the field can't look like this.

$\vec{E}$

cylinder axis, if up and down are exchanged by reflecting the field in a plane coming out of the page, or if you rotate the cylinder about its axis.

But the proposed field fails one test: reflection in a plane perpendicular to the axis, a reflection that exchanges left and right. This reflection, which would *not* make any change in the charge distribution itself, produces the field shown in FIGURE 27.3b. This change in the field is detectable because a positively charged particle would now have a component of motion to the left instead of to the right.

The field of Figure 27.3a, which makes a distinction between left and right, is not cylindrically symmetric and thus is *not* a possible field. In general, **the electric field of a cylindrically symmetric charge distribution cannot have a component parallel to the cylinder axis.**

Well then, what about the electric field shown in FIGURE 27.4a? Here we're looking down the axis of the cylinder. The electric field vectors are restricted to planes perpendicular to the cylinder and thus do not have any component parallel to the cylinder axis. This field is symmetric for rotations about the axis, but it's *not* symmetric for a reflection in a plane containing the axis.

The field of FIGURE 27.4b, after this reflection, is easily distinguishable from the field of Figure 27.4a. Thus **the electric field of a cylindrically symmetric charge distribution cannot have a component tangent to the circular cross section.**

FIGURE 27.5 shows the only remaining possible field shape. The electric field is radial, pointing straight out from the cylinder like the bristles on a bottle brush. This is the one electric field shape matching the symmetry of the charge distribution.

FIGURE 27.5 This is the only shape for the electric field that matches the symmetry of the charge distribution.

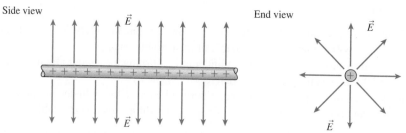

Side view $\vec{E}$ End view $\vec{E}$

What Good Is Symmetry?

Given how little we assumed about Figure 27.1—that the charge distribution is cylindrically symmetric and that electric fields point away from positive charges—we've been able to deduce a great deal about the electric field. In particular, we've deduced the *shape* of the electric field.

Now, shape is not everything. We've learned nothing about the strength of the field or how strength changes with distance. Is E constant? Does it decrease like $1/r$ or $1/r^2$? We don't yet have a complete description of the field, but knowing what shape the field *has* to have will make finding the field strength a much easier task.

That's the good of symmetry. Symmetry arguments allow us to *rule out* many conceivable field shapes as simply being incompatible with the symmetry of the charge distribution. Knowing what doesn't happen, or can't happen, is often as useful as knowing what does happen. By the process of elimination, we're led to the one and only shape the field can possibly have. Reasoning on the basis of symmetry is a sometimes subtle but always powerful means of reasoning.

Three Fundamental Symmetries

Three fundamental symmetries appear frequently in electrostatics. The first row of FIGURE 27.6 shows the simplest form of each symmetry. The second row shows a more complex, but more realistic, situation with the same symmetry.

FIGURE 27.6 Three fundamental symmetries.

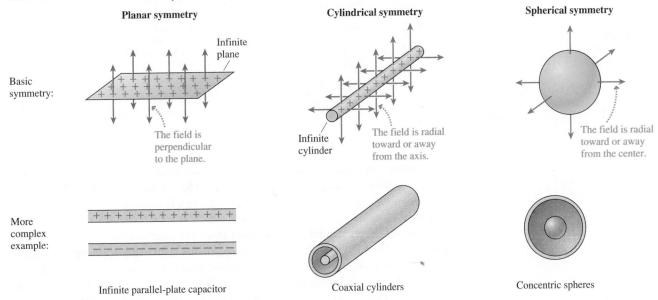

NOTE ▶ Figures must be finite in extent, but the planes and cylinders in Figure 27.6 are assumed to be infinite. ◀

Objects do exist that are extremely close to being perfect spheres, but no real cylinder or plane can be infinite in extent. Even so, the fields of infinite planes and cylinders are good models for the fields of finite planes and cylinders at points not too close to an edge or an end. The fields that we'll study in this chapter, even if idealized, have many important applications.

STOP TO THINK 27.1   A uniformly charged rod has a *finite* length L. The rod is symmetric under rotations about the axis and under reflection in any plane containing the axis. It is *not* symmetric under translations or under reflections in a plane perpendicular to the axis unless that plane bisects the rod. Which field shape or shapes match the symmetry of the rod?

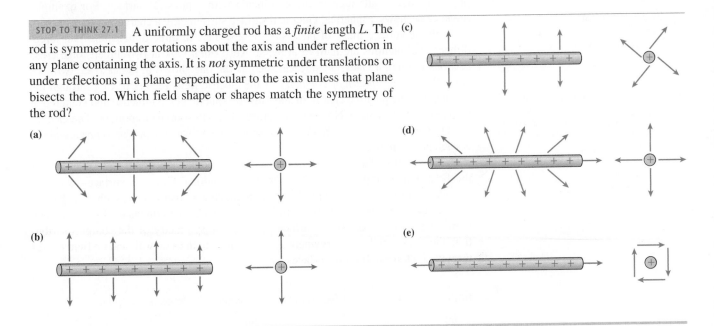

27.2 The Concept of Flux

FIGURE 27.7a on the next page shows an opaque box surrounding a region of space. We can't see what's in the box, but there's an electric field vector coming out of each face of the box. Can you figure out what's in the box?

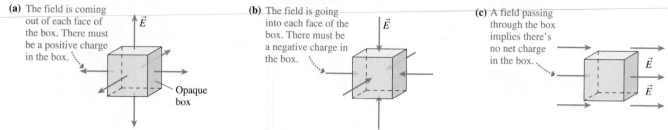

FIGURE 27.7 Although we can't see into the boxes, the electric fields passing through the faces tell us something about what's in them.

(a) The field is coming out of each face of the box. There must be a positive charge in the box.

Opaque box

$\vec{E}$

(b) The field is going into each face of the box. There must be a negative charge in the box.

$\vec{E}$

(c) A field passing through the box implies there's no net charge in the box.

$\vec{E}$
$\vec{E}$

Of course you can. Because electric fields point away from positive charges, and the electric field is coming out of every face of the box, it seems clear that the box contains a positive charge or charges. Similarly, the box in FIGURE 27.7b certainly contains a negative charge.

What can we tell about the box in FIGURE 27.7c? The electric field points into the box on the left. An equal electric field points out on the right. This might be the electric field between a large positive electrode somewhere out of sight on the left and a large negative electrode off to the right. An electric field passes through the box, but we see no evidence there's any charge (or at least any net charge) inside the box.

These examples suggest that the electric field as it passes into, out of, or through the box is in some way connected to the charge within the box. However, these simple pictures don't tell us how much charge there is or where within the box the charge is located. Perhaps a better box would be more informative.

Suppose we surround a region of space with a *closed surface*, a surface that divides space into distinct inside and outside regions. Within the context of electrostatics, a closed surface through which an electric field passes is called a **Gaussian surface,** named after the 19th-century mathematician Karl Gauss who developed the mathematical foundations of geometry. This is an imaginary, mathematical surface, not a physical surface, although it might coincide with a physical surface. For example, FIGURE 27.8a shows a spherical Gaussian surface surrounding a charge.

A closed surface must, of necessity, be a surface in three dimensions. But three-dimensional pictures are hard to draw, so we'll often look at two-dimensional cross sections through a Gaussian surface, such as the one shown in FIGURE 27.8b. Now we can tell from the *spherical symmetry* of the electric field vectors poking through the surface that the positive charge inside must be spherically symmetric and centered at the *center* of the sphere. Notice two features that will soon be important: The electric field is everywhere *perpendicular* to the spherical surface and has the *same magnitude* at each point on the surface.

A Gaussian surface is most useful when it matches the shape and symmetry of the field. For example, FIGURE 27.9a shows a cylindrical Gaussian surface—a *closed* cylinder—surrounding some kind of cylindrical charge distribution, such as a charged wire. FIGURE 27.9b simplifies the drawing by showing two-dimensional end and side views. Because the Gaussian surface matches the symmetry of the charge distribution, the electric field is everywhere *perpendicular* to the side wall and *no* field passes through the top and bottom surfaces.

FIGURE 27.8 Gaussian surface surrounding a charge. A two-dimensional cross section is usually easier to draw.

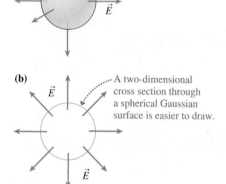

(a)
A Gaussian surface is a closed surface around a charge.
$\vec{E}$
$\vec{E}$

(b)
A two-dimensional cross section through a spherical Gaussian surface is easier to draw.
$\vec{E}$
$\vec{E}$

FIGURE 27.9 A Gaussian surface is most useful when it matches the shape of the field.

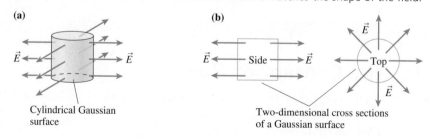

(a)
$\vec{E}$
$\vec{E}$
Cylindrical Gaussian surface

(b)
$\vec{E}$ ← Side → $\vec{E}$
$\vec{E}$
Top
$\vec{E}$
Two-dimensional cross sections of a Gaussian surface

For contrast, consider the spherical surface in FIGURE 27.10a. This is also a Gaussian surface, and the protruding electric field tells us there's a positive charge inside. It might be a point charge located on the left side, but we can't really say. A Gaussian surface that doesn't match the symmetry of the charge distribution isn't terribly useful.

The nonclosed surface of FIGURE 27.10b doesn't provide much help either. What appears to be a uniform electric field to the right could be due to a large positive plate on the left, a large negative plate on the right, or both. A nonclosed surface doesn't provide enough information.

These examples lead us to two conclusions:

1. The electric field, in some sense, "flows" *out of* a closed surface surrounding a region of space containing a net positive charge and *into* a closed surface surrounding a net negative charge. The electric field may flow *through* a closed surface surrounding a region of space in which there is no net charge, but the *net flow* is zero.
2. The electric field pattern through the surface is particularly simple if the closed surface matches the symmetry of the charge distribution inside.

The electric field doesn't really flow like a fluid, but the metaphor is a useful one. The Latin word for flow is *flux,* and the amount of electric field passing through a surface is called the **electric flux.** Our first conclusions, stated in terms of electric flux, are

- There is an outward flux through a closed surface around a net positive charge.
- There is an inward flux through a closed surface around a net negative charge.
- There is no net flux through a closed surface around a region of space in which there is no net charge.

This chapter has been entirely qualitative thus far as we've established pictorially what we mean by symmetry, the idea of flux, and the fact that the electric flux through a closed surface has something to do with the charge inside. Understanding these qualitative ideas is essential, but to go further we need to make these ideas quantitative and precise. In the next section, you'll learn how to calculate the electric flux through a surface. Then, in the section following that, we'll establish a precise relationship between the net flux through a Gaussian surface and the enclosed charge. That relationship, Gauss's law, will allow us to determine the electric fields of some interesting and useful charge distributions.

FIGURE 27.10 Not every surface is useful for learning about charge.

(a)

A Gaussian surface that doesn't match the symmetry of the electric field isn't very useful.

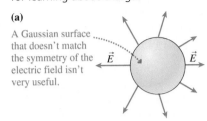

(b)

A nonclosed surface doesn't provide enough information about the charges.

STOP TO THINK 27.2   This box contains

a. A positive charge.   b. A negative charge.
c. No charge.   d. A net positive charge.
e. A net negative charge.   f. No net charge.

27.3 Calculating Electric Flux

Let's start with a brief overview of where this section will take us. We'll begin with a definition of flux that is easy to understand, then we'll turn that simple definition into a formidable-looking integral. We need the integral because the simple definition applies only to uniform electric fields and flat surfaces. Those are good starting points, but we'll soon need to calculate the flux of nonuniform fields through curved surfaces.

Mathematically, the flux of a nonuniform field through a curved surface is described by a special kind of integral called a *surface integral.* It's quite possible that you have not yet encountered surface integrals in your calculus course, and the "novelty factor" contributes to making this integral look worse than it really is. We will emphasize over and over the idea that an integral is just a fancy way of doing a sum, in this case the sum of the small amounts of flux through many small pieces of a surface.

The good news is that *every* surface integral we need to evaluate in this chapter, or that you will need to evaluate for the homework problems, is either zero or is so easy that you will be able to do it in your head. This seems like an astounding claim, but you will soon see it is true. The key will be to make effective use of the *symmetry* of the electric field.

Now that you've been warned, you needn't panic at the sight of the mathematical notation that will be introduced. We'll go step by step, and you'll see that, at least as far as electrostatics is concerned, calculating the electric flux is not difficult.

The Basic Definition of Flux

Imagine holding a rectangular wire loop of area A in front of a fan. As FIGURE 27.11 shows, the volume of air flowing through the loop each second depends on the angle between the loop and the direction of flow. The flow is maximum through a loop that is perpendicular to the airflow; no air goes through the same loop if it lies parallel to the flow.

FIGURE 27.11 The amount of air flowing through a loop depends on the angle between $\vec{v}$ and $\hat{n}$.

(a)

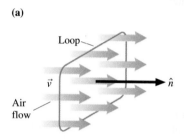

The air flowing through the loop is maximum when $\theta = 0°$.

(b)

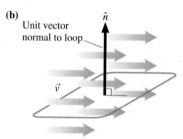

No air flows through the loop when $\theta = 90°$.

(c) The loop is tilted by angle θ.

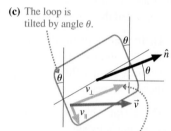

$v_\perp = v\cos\theta$ is the component of the air velocity perpendicular to the loop.

The flow direction is identified by the velocity vector $\vec{v}$. We can identify the loop's orientation by defining a unit vector $\hat{n}$ normal to the plane of the loop. Angle θ is then the angle between $\vec{v}$ and $\hat{n}$. The loop perpendicular to the flow in FIGURE 27.11a has $\theta = 0°$; the loop parallel to the flow in FIGURE 27.11b has $\theta = 90°$. You can think of θ as the angle by which a loop has been tilted away from perpendicular.

NOTE ▶ A surface has two sides, so $\hat{n}$ could point either way. We'll choose the side that makes $\theta \le 90°$. ◀

You can see from FIGURE 27.11c that the velocity vector $\vec{v}$ can be decomposed into components $v_\perp = v\cos\theta$ perpendicular to the loop and $v_\parallel = v\sin\theta$ parallel to the loop. Only the perpendicular component $v_\perp$ carries air *through* the loop. Consequently, the volume of air flowing through the loop each second is

$$\text{volume of air per second } (\text{m}^3/\text{s}) = v_\perp A = vA\cos\theta \qquad (27.1)$$

$\theta = 0°$ is the orientation for maximum flow through the loop, as expected, and no air flows through the loop if it is tilted $90°$.

An electric field doesn't flow in a literal sense, but we can apply the same idea to an electric field passing through a surface. FIGURE 27.12 shows a surface of area A in a uniform electric field $\vec{E}$. Unit vector $\hat{n}$ is normal to the surface, and θ is the angle between $\hat{n}$ and $\vec{E}$. Only the component $E_\perp = E\cos\theta$ passes *through* the surface.

With this in mind, and using Equation 27.1 as an analog, let's define the *electric flux* Φ_e (uppercase Greek phi) as

$$\Phi_e = E_\perp A = EA\cos\theta \qquad (27.2)$$

The electric flux measures the amount of electric field passing through a surface of area A if the normal to the surface is tilted at angle θ from the field.

FIGURE 27.12 An electric field passing through a surface.

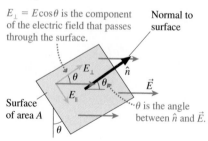

$E_\perp = E\cos\theta$ is the component of the electric field that passes through the surface.

Normal to surface

Surface of area A

θ is the angle between $\hat{n}$ and $\vec{E}$.

Equation 27.2 looks very much like a vector dot product: $\vec{E} \cdot \vec{A} = EA\cos\theta$. For this idea to work, let's define an **area vector** $\vec{A} = A\hat{n}$ to be a vector in the direction of $\hat{n}$—that is, *perpendicular* to the surface—with a magnitude A equal to the area of the surface. Vector $\vec{A}$ has units of m^2. FIGURE 27.13a shows two area vectors.

FIGURE 27.13 The electric flux can be defined in terms of the area vector $\vec{A}$.

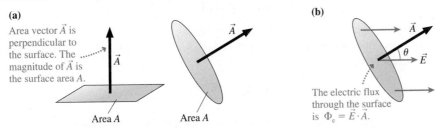

FIGURE 27.13b shows an electric field passing through a surface of area A. The angle between vectors $\vec{A}$ and $\vec{E}$ is the same angle used in Equation 27.2 to define the electric flux, so Equation 27.2 really is a dot product. We can define the electric flux more concisely as

$$\Phi_e = \vec{E} \cdot \vec{A} \quad \text{(electric flux of a constant electric field)} \quad (27.3)$$

Writing the flux as a dot product helps make clear how angle θ is defined: θ is the angle between the electric field and a line *perpendicular* to the plane of the surface.

NOTE ▶ Figure 27.13b shows a circular area, but the shape of the surface is not relevant. However, Equation 27.3 is restricted to a *constant* electric field passing through a *planar* surface. ◀

EXAMPLE 27.1 **The electric flux inside a parallel-plate capacitor**

Two 100 cm^2 parallel electrodes are spaced 2.0 cm apart. One is charged to +5.0 nC, the other to −5.0 nC. A 1.0 cm × 1.0 cm surface between the electrodes is tilted to where its normal makes a 45° angle with the electric field. What is the electric flux through this surface?

MODEL Assume the surface is located near the center of the capacitor where the electric field is uniform. The electric flux doesn't depend on the shape of the surface.

VISUALIZE The surface is square, rather than circular, but otherwise the situation looks like Figure 27.13b.

SOLVE In Chapter 26, we found the electric field inside a parallel-plate capacitor to be

$$E = \frac{Q}{\epsilon_0 A_{plates}} = \frac{5.0 \times 10^{-9} \text{ C}}{(8.85 \times 10^{-12} \text{ C}^2/\text{Nm}^2)(1.0 \times 10^{-2} \text{ m}^2)}$$
$$= 5.65 \times 10^4 \text{ N/C}$$

A 1.0 cm × 1.0 cm surface has $A = 1.0 \times 10^{-4} \text{ m}^2$. The electric flux through this surface is

$$\Phi_e = \vec{E} \cdot \vec{A} = EA\cos\theta$$
$$= (5.65 \times 10^4 \text{ N/C})(1.0 \times 10^{-4} \text{ m}^2)\cos 45°$$
$$= 4.0 \text{ Nm}^2/\text{C}$$

ASSESS The units of electric flux are the product of electric field and area units: $N\,m^2/C$.

The Electric Flux of a Nonuniform Electric Field

Our initial definition of the electric flux assumed that the electric field $\vec{E}$ was constant over the surface. How should we calculate the electric flux if $\vec{E}$ varies from point to point on the surface? We can answer this question by returning to the analogy of air flowing through a loop. Suppose the airflow varies from point to point. We can still find the total volume of air passing through the loop each second by dividing the loop into many small areas, finding the flow through each small area, then adding them. Similarly, **the electric flux through a surface can be calculated as the sum of the fluxes through smaller pieces of the surface.** Because flux is a scalar, adding fluxes is easier than adding electric fields.

FIGURE 27.14 A surface in a nonuniform electric field.

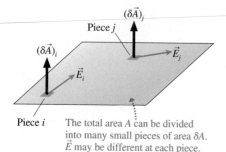

Piece j

Piece i The total area A can be divided into many small pieces of area δA. $\vec{E}$ may be different at each piece.

FIGURE 27.14 shows a surface in a nonuniform electric field. Imagine dividing the surface into many small pieces of area δA. Each little area has an area vector $\delta \vec{A}$ perpendicular to the surface. Two of the little pieces are shown in the figure. The electric fluxes through these two pieces differ because the electric fields are different.

Consider the small piece i where the electric field is $\vec{E}_i$. The small electric flux $\delta \Phi_i$ through area $(\delta \vec{A})_i$ is

$$\delta \Phi_i = \vec{E}_i \cdot (\delta \vec{A})_i \tag{27.4}$$

The flux through every other little piece of the surface is found the same way. The total electric flux through the entire surface is then the sum of the fluxes through each of the small areas:

$$\Phi_e = \sum_i \delta \Phi_i = \sum_i \vec{E}_i \cdot (\delta \vec{A})_i \tag{27.5}$$

Now let's go to the limit $\delta \vec{A} \to d\vec{A}$. That is, the little areas become infinitesimally small, and there are infinitely many of them. Then the sum becomes an integral, and the electric flux through the surface is

$$\Phi_e = \int_{\text{surface}} \vec{E} \cdot d\vec{A} \tag{27.6}$$

The integral in Equation 27.6 is called a **surface integral.**

Equation 27.6 may look rather frightening if you haven't seen surface integrals before. Despite its appearance, a surface integral is no more complicated than integrals you know from calculus. After all, what does $\int f(x)\,dx$ really mean? This expression is a shorthand way to say "Divide the x-axis into many little segments of length δx, evaluate the function $f(x)$ in each of them, then add up $f(x)\,\delta x$ for all the segments along the line." The integral in Equation 27.6 differs only in that we're dividing a surface into little pieces instead of a line into little segments. In particular, we're summing the fluxes through a vast number of very tiny pieces.

You may be thinking, "OK, I understand the idea, but I don't know what to *do*. In calculus, I learned formulas for evaluating integrals such as $\int x^2\,dx$. How do I evaluate a surface integral?" This is a good question. We'll deal with evaluation shortly, and it will turn out that the surface integrals in electrostatics are quite easy to evaluate. But don't confuse *evaluating* the integral with understanding what the integral *means*. The surface integral in Equation 27.6 is simply a shorthand notation for the summation of the electric fluxes through a vast number of very tiny pieces of a surface.

The electric field might be different at every point on the surface, but suppose it isn't. That is, suppose the surface is in a uniform electric field $\vec{E}$. A field that is the same at every single point on a surface is a constant as far as the integration of Equation 27.6 is concerned, so we can take it outside the integral. In that case,

$$\Phi_e = \int_{\text{surface}} \vec{E} \cdot d\vec{A} = \int_{\text{surface}} E \cos\theta \, dA = E \cos\theta \int_{\text{surface}} dA \tag{27.7}$$

The integral that remains in Equation 27.7 tells us to add up all the little areas into which the full surface was subdivided. But the sum of all the little areas is simply the area of the surface:

$$\int_{\text{surface}} dA = A \tag{27.8}$$

This idea—that the surface integral of dA is the area of the surface—is one we'll use to evaluate most of the surface integrals of electrostatics. If we substitute Equation 27.8 into Equation 27.7, we find that the electric flux in a uniform electric field is $\Phi_e = EA \cos\theta$. We already knew this, from Equation 27.2, but it was important to see that the surface integral of Equation 27.6 gives the correct result for the case of a uniform electric field.

The Flux Through a Curved Surface

Most of the Gaussian surfaces we considered in the last section were curved surfaces. FIGURE 27.15 shows an electric field passing through a curved surface. How do we find the electric flux through this surface? Just as we did for a flat surface!

Divide the surface into many small pieces of area δA. For each, define the area vector $\delta \vec{A}$ perpendicular to the surface *at that point*. Compared to Figure 27.14, the only difference that the curvature of the surface makes is that the $\delta \vec{A}$ are no longer parallel to each other. Find the small electric flux $\delta \Phi_i = \vec{E}_i \cdot (\delta \vec{A})_i$ through each little area, then add them all up. The result, once again, is

$$\Phi_e = \int_{surface} \vec{E} \cdot d\vec{A} \qquad (27.9)$$

We *assumed*, in deriving this expression the first time, that the surface was flat and that all the $\delta \vec{A}$ were parallel to each other. But that assumption wasn't necessary. The *meaning* of Equation 27.9—a summation of the fluxes through a vast number of very tiny pieces—is unchanged if the pieces lie on a curved surface.

We seem to be getting more and more complex, using surface integrals first for nonuniform fields and now for curved surfaces. But consider the two situations shown in FIGURE 27.16. The electric field $\vec{E}$ in FIGURE 27.16a is everywhere tangent, or parallel, to the curved surface. We don't need to know the magnitude of $\vec{E}$ to recognize that $\vec{E} \cdot d\vec{A}$ is *zero at every point* on the surface because $\vec{E}$ is perpendicular to $d\vec{A}$ at every point. Thus $\Phi_e = 0$. A tangent electric field never pokes through the surface, so it has no flux through the surface.

The electric field in FIGURE 27.16b is everywhere perpendicular to the surface *and* has the same magnitude E at every point. $\vec{E}$ differs in direction at different points on a curved surface, but at any particular point $\vec{E}$ is parallel to $d\vec{A}$ and $\vec{E} \cdot d\vec{A}$ is simply EdA. In this case,

$$\Phi_e = \int_{surface} \vec{E} \cdot d\vec{A} = \int_{surface} E\,dA = E \int_{surface} dA = EA \qquad (27.10)$$

As we evaluated the integral, the fact that E has the same magnitude at every point on the surface allowed us to bring the constant value outside the integral. We then used the fact that the integral of dA over the surface is the surface area A.

We can summarize these two situations with a Tactics Box.

TACTICS BOX 27.1 Evaluating surface integrals

❶ If the electric field is everywhere tangent to a surface, the electric flux through the surface is $\Phi_e = 0$.
❷ If the electric field is everywhere perpendicular to a surface *and* has the same magnitude E at every point, the electric flux through the surface is $\Phi_e = EA$.

These two results will be of immeasurable value for using Gauss's law because *every* flux we'll need to calculate will be one of these situations. This is the basis for our earlier claim that the evaluation of surface integrals is not going to be difficult.

The Electric Flux Through a Closed Surface

Our final step, to calculate the electric flux through a closed surface such as a box, a cylinder, or a sphere, requires nothing new. We've already learned how to calculate the electric flux through flat and curved surfaces, and a closed surface is nothing more than a surface that happens to be closed.

However, the mathematical notation for the surface integral over a closed surface differs slightly from what we've been using. It is customary to use a little circle on

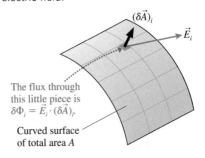
FIGURE 27.15 A curved surface in an electric field.

The flux through this little piece is $\delta \Phi_i = \vec{E}_i \cdot (\delta \vec{A})_i$.
Curved surface of total area A

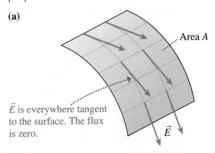
FIGURE 27.16 Electric fields that are everywhere tangent to or everywhere perpendicular to a curved surface.

(a)

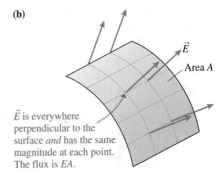

Area A

$\vec{E}$ is everywhere tangent to the surface. The flux is zero.

(b)

$\vec{E}$
Area A

$\vec{E}$ is everywhere perpendicular to the surface *and* has the same magnitude at each point. The flux is EA.

the integral sign to indicate that the surface integral is to be performed over a closed surface. With this notation, the electric flux through a closed surface is

$$\Phi_e = \oint \vec{E} \cdot d\vec{A} \qquad (27.11)$$

Only the notation has changed. The electric flux is still the summation of the fluxes through a vast number of tiny pieces, pieces that now cover a closed surface.

NOTE ▶ A closed surface has a distinct inside and outside. The area vector $d\vec{A}$ is defined to always point *toward the outside*. This removes an ambiguity that was present for a single surface, where $d\vec{A}$ could point to either side. ◀

EXAMPLE 27.2 Calculating the electric flux through a closed cylinder

A charge distribution with cylindrical symmetry has created the electric field $\vec{E} = E_0(r^2/r_0^2)\hat{r}$, where E_0 and r_0 are constants and where unit vector $\hat{r}$ lies in the xy-plane. Calculate the electric flux through a closed cylinder of length L and radius R that is centered along the z-axis.

MODEL The electric field extends radially outward from the z-axis with cylindrical symmetry. The z-component is $E_z = 0$. The cylinder is a Gaussian surface.

VISUALIZE FIGURE 27.17a is a view of the electric field looking along the z-axis. The field strength increases with increasing radial distance, and it's symmetric about the z-axis. **FIGURE 27.17b** is the closed Gaussian surface for which we need to calculate the electric flux. We can place the cylinder anywhere along the z-axis because the electric field extends forever in that direction.

SOLVE To calculate the flux, we divide the closed cylinder into three surfaces: the top, the bottom, and the cylindrical wall. The electric field is tangent to the surface at every point on the top and bottom surfaces. Hence, according to step 1 in Tactics Box 27.1, the flux through those two surfaces is zero. For the cylindrical wall, the electric field is perpendicular to the surface at every point *and* has the constant magnitude $E = E_0(R^2/r_0^2)$ at every point on the surface. Thus, from step 2 in Tactics Box 27.1,

$$\Phi_{wall} = EA_{wall}$$

If we add the three pieces, the net flux through the closed surface is

$$\Phi_e = \oint \vec{E} \cdot d\vec{A} = \Phi_{top} + \Phi_{bottom} + \Phi_{wall} = 0 + 0 + EA_{wall}$$

$$= EA_{wall}$$

We've evaluated the surface integral, using the two steps in Tactics Box 27.1, and there was nothing to it! To finish, all we need to recall is that the surface area of a cylindrical wall is circumference × height, or $A_{wall} = 2\pi RL$. Thus

$$\Phi_e = \left(E_0\frac{R^2}{r_0^2}\right)(2\pi RL) = \frac{2\pi LR^3}{r_0^2}E_0$$

ASSESS LR^3/r_0^2 has units of m², an area, so this expression for Φ_e has units of N m²/C. These are the correct units for electric flux, giving us confidence in our answer. Notice the important role played by symmetry. The electric field was perpendicular to the wall and of constant value at every point on the wall *because* the Gaussian surface had the same symmetry as the charge distribution. We would not have been able to evaluate the surface integral in such an easy way for a surface of any other shape. Symmetry is the key.

FIGURE 27.17 The electric field and the closed surface through which we will calculate the electric flux.

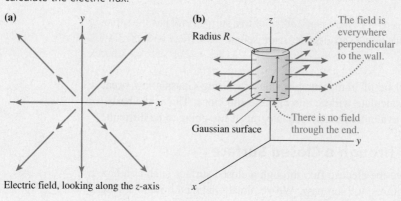

Example 27.2 illustrated a two-step approach to performing a flux integral over a closed surface. In summary:

TACTICS
BOX 27.2 Finding the flux through a closed surface

❶ Divide the closed surface into pieces that are everywhere tangent to the electric field and everywhere perpendicular to the electric field.

❷ Use Tactics Box 27.1 to evaluate the surface integrals over these surfaces, then add the results.

Exercise 11

STOP TO THINK 27.3 The total electric flux through this box is

a. $0 \text{ N m}^2/\text{C}$
b. $1 \text{ N m}^2/\text{C}$
c. $2 \text{ N m}^2/\text{C}$
d. $4 \text{ N m}^2/\text{C}$
e. $6 \text{ N m}^2/\text{C}$
f. $8 \text{ N m}^2/\text{C}$

$\vec{E} = (1 \text{ N/C, up})$

Plane of charge

Cross section of a
$1 \text{ m} \times 1 \text{ m} \times 1 \text{ m}$ box

$\vec{E} = (1 \text{ N/C, down})$

27.4 Gauss's Law

The last section was long, but knowing how to calculate the electric flux through a closed surface is essential for the main topic of this chapter: Gauss's law. Gauss's law is equivalent to Coulomb's law for static charges, although Gauss's law will look very different.

The purpose of learning Gauss's law is twofold:

■ Gauss's law allows the electric fields of some continuous distributions of charge to be found much more easily than does Coulomb's law.

■ Gauss's law is valid for *moving* charges, but Coulomb's law is not (although it's a very good approximation for velocities that are much less than the speed of light). Thus Gauss's law is ultimately a more fundamental statement about electric fields than is Coulomb's law.

Let's start with Coulomb's law for the electric field of a point charge. FIGURE 27.18 shows a spherical Gaussian surface of radius r centered on a positive charge q. Keep in mind that this is an imaginary, mathematical surface, not a physical surface. There is a net flux through this surface because the electric field points outward at every point on the surface. To evaluate the flux, given formally by the surface integral of Equation 27.11, notice that the electric field is perpendicular to the surface at every point on the surface *and*, from Coulomb's law, it has the same magnitude $E = q/4\pi\epsilon_0 r^2$ at every point on the surface. This simple situation arises because **the Gaussian surface has the same symmetry as the electric field.**

Thus we know, without having to do any hard work, that the flux integral is

$$\Phi_e = \oint \vec{E} \cdot d\vec{A} = EA_{\text{sphere}} \qquad (27.12)$$

The surface area of a sphere of radius r is $A_{\text{sphere}} = 4\pi r^2$. If we use A_{sphere} and the Coulomb-law expression for E in Equation 27.12, we find that the electric flux through the spherical surface is

$$\Phi_e = \frac{q}{4\pi\epsilon_0 r^2} 4\pi r^2 = \frac{q}{\epsilon_0} \qquad (27.13)$$

FIGURE 27.18 A spherical Gaussian surface surrounding a point charge.

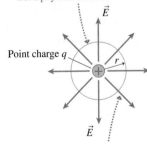

Cross section of a Gaussian sphere of radius r. This is a mathematical surface, not a physical surface.

$\vec{E}$

Point charge q

r

$\vec{E}$

The electric field is everywhere perpendicular to the surface *and* has the same magnitude at every point.

You should examine the logic of this calculation closely. We really did evaluate the surface integral of Equation 27.11, although it may appear, at first, as if we didn't do much. The integral was easily evaluated, we reiterate for emphasis, because the closed surface on which we performed the integration matched the *symmetry* of the charge distribution.

NOTE ▶ We found Equation 27.13 for a positive charge, but it applies equally to negative charges. According to Equation 27.13, Φ_e is negative if q is negative. And that's what we would expect from the basic definition of flux, $\vec{E} \cdot \vec{A}$. The electric field of a negative charge points inward, while the area vector of a closed surface points outward, making the dot product negative. ◀

FIGURE 27.19 The electric flux is the same through *every* sphere centered on a point charge.

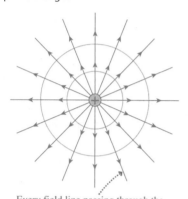

Every field line passing through the smaller sphere also passes through the larger sphere. Hence the flux through the two spheres is the same.

FIGURE 27.20 An arbitrary Gaussian surface can be approximated with spherical and radial pieces.

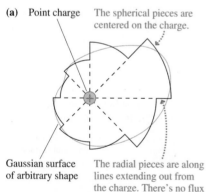

(a) Point charge   The spherical pieces are centered on the charge.

Gaussian surface of arbitrary shape

The radial pieces are along lines extending out from the charge. There's no flux through these.

(b)

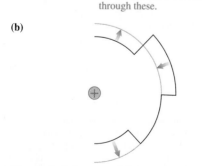

The spherical pieces can slide in or out to form a complete sphere. Hence the flux through the pieces is the same as the flux through a sphere.

Electric Flux Is Independent of Surface Shape and Radius

Notice something interesting about Equation 27.13. The electric flux depends on the amount of charge but *not* on the radius of the sphere. Although this may seem a bit surprising, it's really a direct consequence of what we *mean* by flux. Think of the fluid analogy with which we introduced the term "flux." If fluid flows outward from a central point, all the fluid crossing a small-radius spherical surface will, at some later time, cross a large-radius spherical surface. No fluid is lost along the way, and no new fluid is created. Similarly, the point charge in FIGURE 27.19 is the only source of electric field. Every electric field line passing through a small-radius spherical surface also passes through a large-radius spherical surface. Hence the electric flux is independent of r.

NOTE ▶ This argument hinges on the fact that Coulomb's law is an inverse-square force law. The electric field strength, which is proportional to $1/r^2$, decreases with distance. But the surface area, which increases in proportion to r^2, exactly compensates for this decrease. Consequently, the electric flux of a point charge through a spherical surface is independent of the radius of the sphere. ◀

This conclusion about the flux has an extremely important generalization. FIGURE 27.20a shows a point charge and a closed Gaussian surface of arbitrary shape and dimensions. All we know is that the charge is *inside* the surface. What is the electric flux through this arbitrary surface?

One way to answer the question is to approximate the surface as a patchwork of spherical and radial pieces. The spherical pieces are centered on the charge and the radial pieces lie along lines extending outward from the charge. (Figure 27.20 is a two-dimensional drawing so you need to imagine these arcs as actually being pieces of a spherical shell.) The figure, of necessity, shows fairly large pieces that don't match the actual surface all that well. However, we can make this approximation as good as we want by letting the pieces become sufficiently small.

The electric field is everywhere tangent to the radial pieces. Hence the electric flux through the radial pieces is zero. The spherical pieces, although at varying distances from the charge, form a *complete sphere*. That is, any line drawn radially outward from the charge will pass through exactly one spherical piece, and no radial lines can avoid passing through a spherical piece. You could even imagine, as FIGURE 27.20b shows, sliding the spherical pieces in and out *without changing the angle they subtend* until they come together to form a complete sphere.

Consequently, the electric flux through these spherical pieces that, when assembled, form a complete sphere must be exactly the same as the flux q/ϵ_0 through a spherical Gaussian surface. In other words, **the flux through *any* closed surface surrounding a point charge q is**

$$\Phi_e = \oint \vec{E} \cdot d\vec{A} = \frac{q}{\epsilon_0} \tag{27.14}$$

This surprisingly simple result is a consequence of the fact that Coulomb's law is an inverse-square force law. Even so, the reasoning that got us to Equation 27.14 is rather subtle and well worth reviewing.

Charge Outside the Surface

The closed surface shown in FIGURE 27.21a has a point charge q outside the surface but no charges inside. Now what can we say about the flux? By approximating this surface with spherical and radial pieces *centered on the charge,* as we did in Figure 27.20, we can reassemble the surface into the equivalent surface of FIGURE 27.21b. This closed surface consists of sections of two spherical shells, and it is equivalent in the sense that the electric flux through this surface is the same as the electric flux through the original surface of Figure 27.21a.

FIGURE 27.21 A point charge outside a Gaussian surface.

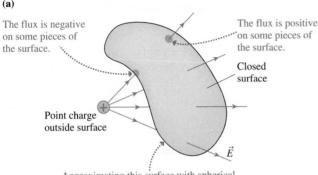

(a)

The flux is negative on some pieces of the surface.

The flux is positive on some pieces of the surface.

Closed surface

Point charge outside surface

$\vec{E}$

Approximating this surface with spherical and radial pieces allows it to be reassembled as the surface in part (b) that has the same flux.

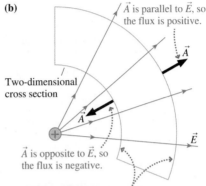

(b)

$\vec{A}$ is parallel to $\vec{E}$, so the flux is positive.

$\vec{A}$

Two-dimensional cross section

$\vec{A}$

$\vec{E}$

$\vec{A}$ is opposite to $\vec{E}$, so the flux is negative.

The fluxes through these surfaces are equal but opposite. The net flux is zero.

If the electric field were a fluid flowing outward from the charge, all the fluid *entering* the closed region through the first spherical surface would later *exit* the region through the second spherical surface. There is no *net* flow into or out of the closed region. Similarly, every electric field line entering this closed volume through one spherical surface exits through the other spherical surface.

Mathematically, the electric fluxes through the two spherical surfaces have the same magnitude because Φ_e is independent of r. But they have *opposite signs* because the outward-pointing area vector $\vec{A}$ is parallel to $\vec{E}$ on one surface but opposite to $\vec{E}$ on the other. The sum of the fluxes through the two surfaces is zero, and we are led to the conclusion that **the net electric flux is zero through a closed surface that does not contain any net charge.** Charges outside the surface do not produce a net flux through the surface.

This isn't to say that the flux through a small piece of the surface is zero. In fact, as Figure 27.21a shows, nearly every piece of the surface has an electric field either entering or leaving and thus has a nonzero flux. But some of these are positive and some are negative. When summed over the *entire* surface, the positive and negative contributions exactly cancel to give no *net* flux.

Multiple Charges

Finally, consider an arbitrary Gaussian surface and a group of charges $q_1, q_2, q_3, \cdots$ such as those shown in FIGURE 27.22. What is the net electric flux through the surface?

By definition, the net flux is

$$\Phi_e = \oint \vec{E} \cdot d\vec{A}$$

From the principle of superposition, the electric field is $\vec{E} = \vec{E}_1 + \vec{E}_2 + \vec{E}_3 + \cdots$, where $\vec{E}_1, \vec{E}_2, \vec{E}_3, \cdots$ are the fields of the individual charges. Thus the flux is

$$\Phi_e = \oint \vec{E}_1 \cdot d\vec{A} + \oint \vec{E}_2 \cdot d\vec{A} + \oint \vec{E}_3 \cdot d\vec{A} + \cdots \qquad (27.15)$$

$$= \Phi_1 + \Phi_2 + \Phi_3 + \cdots$$

FIGURE 27.22 Charges both inside and outside a Gaussian surface.

The fluxes due to charges outside the surface are all zero.

q_3

q_2

q_1

Two-dimensional cross section of a Gaussian surface

Total charge inside is Q_{in}.

The fluxes due to charges inside the surface add.

where Φ_1, Φ_2, Φ_3, ... are the fluxes through the Gaussian surface due to the individual charges. That is, the net flux is the sum of the fluxes due to individual charges. But we know what those are: q/ϵ_0 for the charges inside the surface and zero for the charges outside. Thus

$$\Phi_e = \left(\frac{q_1}{\epsilon_0} + \frac{q_2}{\epsilon_0} + \cdots + \frac{q_i}{\epsilon_0} \text{ for all charges inside the surface} \right)$$
$$+(0 + 0 + \cdots + 0 \text{ for all charges outside the surface)} \qquad (27.16)$$

We define

$$Q_{in} = q_1 + q_2 + \cdots + q_i \text{ for all charges inside the surface} \qquad (27.17)$$

as the total charge enclosed *within* the surface. With this definition, we can write our result for the net electric flux in a very neat and compact fashion. For any *closed* surface enclosing total charge Q_{in}, the net electric flux through the surface is

$$\Phi_e = \oint \vec{E} \cdot d\vec{A} = \frac{Q_{in}}{\epsilon_0} \qquad (27.18)$$

This result for the electric flux is known as **Gauss's law.**

What Does Gauss's Law Tell Us?

In one sense, Gauss's law doesn't say anything new or anything that we didn't already know from Coulomb's law. After all, we derived Gauss's law from Coulomb's law. But in another sense, Gauss's law is more important than Coulomb's law. Gauss's law states a very general property of electric fields—namely, that charges create electric fields in just such a way that the net flux of the field is the same through *any* surface surrounding the charges, no matter what its size and shape may be. This fact may have been implied by Coulomb's law, but it was by no means obvious. And Gauss's law will turn out to be particularly useful later when we combine it with other electric and magnetic field equations.

Gauss's law is the mathematical statement of our observations in Section 27.2. There we noticed a net "flow" of electric field out of a closed surface containing charges. Gauss's law quantifies this idea by making a specific connection between the "flow," now measured as electric flux, and the amount of charge.

But is it useful? Although to some extent Gauss's law is a formal statement about electric fields, not a tool for solving practical problems, there are exceptions: Gauss's law will allow us to find the electric fields of some very important and very practical charge distributions much more easily than if we had to rely on Coulomb's law. We'll consider some examples in the next section.

STOP TO THINK 27.4 These are two-dimensional cross sections through three-dimensional closed spheres and a cube. Rank in order, from largest to smallest, the electric fluxes Φ_a to Φ_e through surfaces a to e.

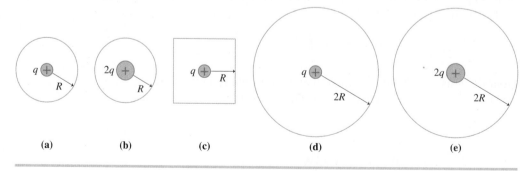

(a) (b) (c) (d) (e)

27.5 Using Gauss's Law

In this section, we'll use Gauss's law to determine the electric fields of several important charge distributions. Some of these you already know, from Chapter 26; others will be new. Three important observations can be made about using Gauss's law:

1. Gauss's law applies only to a *closed* surface, called a Gaussian surface.
2. A Gaussian surface is not a physical surface. It need not coincide with the boundary of any physical object (although it could if we wished). It is an imaginary, mathematical surface in the space surrounding one or more charges.
3. We can't find the electric field from Gauss's law alone. We need to apply Gauss's law in situations where, from symmetry and superposition, we already can guess the *shape* of the field.

These observations and our previous discussion of symmetry and flux lead to the following strategy for solving electric field problems with Gauss's law.

PROBLEM-SOLVING
STRATEGY 27.1   **Gauss's law**

MODEL   Model the charge distribution as a distribution with symmetry.

VISUALIZE   Draw a picture of the charge distribution.

■ Determine the symmetry of its electric field.
■ Choose and draw a Gaussian surface with the *same symmetry*.
■ You need not enclose all the charge within the Gaussian surface.
■ Be sure every part of the Gaussian surface is either tangent to or perpendicular to the electric field.

SOLVE   The mathematical representation is based on Gauss's law

$$\Phi_e = \oint \vec{E} \cdot d\vec{A} = \frac{Q_{in}}{\epsilon_0}$$

■ Use Tactics Boxes 27.1 and 27.2 to evaluate the surface integral.

ASSESS   Check that your result has the correct units, is reasonable, and answers the question.

Exercise 19

EXAMPLE 27.3   **Outside a sphere of charge**

In Chapter 26 we asserted, without proof, that the electric field outside a sphere of total charge Q is the same as the field of a point charge Q at the center. Use Gauss's law to prove this result.

MODEL   The charge distribution within the sphere need not be uniform (i.e., the charge density might increase or decrease with r), but it must have spherical symmetry in order for us to use Gauss's law. We will assume that it does.

VISUALIZE   FIGURE 27.23 shows a sphere of charge Q and radius R. We want to find $\vec{E}$ outside this sphere, for distances $r > R$. The spherical symmetry of the charge distribution tells us that the electric field must point *radially outward* from the sphere. Although Gauss's law is true for any surface surrounding the charged sphere, it is useful only if we choose a Gaussian surface to match the spherical symmetry of the charge distribution and the field. Thus a spherical surface of radius $r > R$ concentric with

FIGURE 27.23 A spherical Gaussian surface surrounding a sphere of charge.

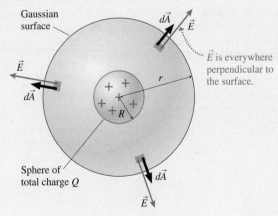

Continued

the charged sphere will be our Gaussian surface. Because this surface surrounds the entire sphere of charge, the enclosed charge is simply $Q_{in} = Q$.

SOLVE Gauss's law is

$$\Phi_e = \oint \vec{E} \cdot d\vec{A} = \frac{Q_{in}}{\epsilon_0} = \frac{Q}{\epsilon_0}$$

To calculate the flux, notice that the electric field is everywhere perpendicular to the spherical surface. And although we don't know the electric field magnitude E, spherical symmetry dictates that E must have the same value at all points equally distant from the center of the sphere. Thus we have the simple result that the net flux through the Gaussian surface is

$$\Phi_e = EA_{sphere} = 4\pi r^2 E$$

where we used the fact that the surface area of a sphere is $A_{sphere} = 4\pi r^2$. With this result for the flux, Gauss's law is

$$4\pi r^2 E = \frac{Q}{\epsilon_0}$$

Thus the electric field at distance r outside a sphere of charge is

$$E_{outside} = \frac{1}{4\pi\epsilon_0} \frac{Q}{r^2}$$

Or in vector form, making use of the fact that $\vec{E}$ is radially outward,

$$\vec{E}_{outside} = \frac{1}{4\pi\epsilon_0} \frac{Q}{r^2} \hat{r}$$

where $\hat{r}$ is a radial unit vector.

ASSESS The field is exactly that of a point charge Q, which is what we wanted to show.

The derivation of the electric field of a sphere of charge depended crucially on a proper choice of the Gaussian surface. We would not have been able to evaluate the flux integral so simply for any other choice of surface. It's worth noting that the result of Example 27.3 can also be proven by the superposition of point-charge fields, but it requires a difficult three-dimensional integral and about a page of algebra. We obtained the answer using Gauss's law in just a few lines. Where Gauss's law works, it works *extremely* well! However, it works only in situations, such as this, with a very high degree of symmetry.

EXAMPLE 27.4 **Inside a sphere of charge**

What is the electric field *inside* a uniformly charged sphere?

MODEL We haven't considered a situation like this before. To begin, we don't know if the field strength is increasing or decreasing as we move outward from the center of the sphere. But the field inside must have spherical symmetry. That is, the field must point radially inward or outward, and the field strength can depend only on r. This is sufficient information to solve the problem because it allows us to choose a Gaussian surface.

VISUALIZE FIGURE 27.24 shows a spherical Gaussian surface with radius $r \leq R$ inside, and *concentric with,* the sphere of charge. This surface matches the symmetry of the charge distribution, hence $\vec{E}$ is perpendicular to this surface and the field strength E has the same value at all points on the surface.

FIGURE 27.24 A spherical Gaussian surface inside a uniform sphere of charge.

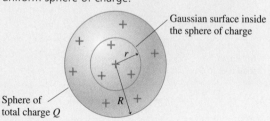

Gaussian surface inside the sphere of charge

Sphere of total charge Q

SOLVE The flux integral is identical to that of Example 27.3:

$$\Phi_e = EA_{sphere} = 4\pi r^2 E$$

Consequently, Gauss's law is

$$\Phi_e = 4\pi r^2 E = \frac{Q_{in}}{\epsilon_0}$$

The difference between this example and Example 27.3 is that Q_{in} is no longer the total charge of the sphere. Instead, Q_{in} is the amount of charge *inside* the Gaussian sphere of radius r. Because the charge distribution is *uniform,* the volume charge density is

$$\rho = \frac{Q}{V_R} = \frac{Q}{\frac{4}{3}\pi R^3}$$

The charge enclosed in a sphere of radius r is thus

$$Q_{in} = \rho V_r = \left(\frac{Q}{\frac{4}{3}\pi R^3} \right) \left(\frac{4}{3}\pi r^3 \right) = \frac{r^3}{R^3} Q$$

The amount of enclosed charge increases with the cube of the distance r from the center and, as expected, $Q_{in} = Q$ if $r = R$. With this expression for Q_{in}, Gauss's law is

$$4\pi r^2 E = \frac{(r^3/R^3)Q}{\epsilon_0}$$

Thus the electric field at radius r inside a uniformly charged sphere is

$$E_{\text{inside}} = \frac{1}{4\pi\epsilon_0}\frac{Q}{R^3}r$$

The electric field strength inside the sphere increases *linearly* with the distance r from the center.

ASSESS The field inside and the field outside a sphere of charge match at the boundary of the sphere, $r = R$, where both give $E = Q/4\pi\epsilon_0 R^2$. In other words, the field strength is *continuous* as we cross the boundary of the sphere. These results are shown graphically in **FIGURE 27.25**.

FIGURE 27.25 The electric field strength of a uniform sphere of charge of radius R.

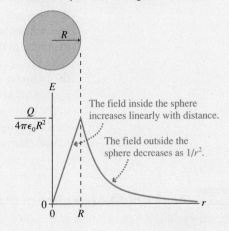

The field inside the sphere increases linearly with distance.

The field outside the sphere decreases as $1/r^2$.

EXAMPLE 27.5 The electric field of a long, charged wire

In Chapter 26, we used superposition to find the electric field of an infinitely long line of charge with linear charge density (C/m) λ. It was not an easy derivation. Find the electric field using Gauss's law.

MODEL A long, charged wire can be modeled as an infinitely long line of charge.

VISUALIZE FIGURE 27.26 shows an infinitely long line of charge. We can use the symmetry of the situation to see that the only possible shape of the electric field is to point straight into or out from the wire, rather like the bristles on a bottle brush. The shape of the field suggests that we choose our Gaussian surface to be a cylinder of radius r and length L, centered on the wire. Because Gauss's law refers to *closed* surfaces, we must include the ends of the cylinder as part of the surface.

FIGURE 27.26 A Gaussian surface around a charged wire.

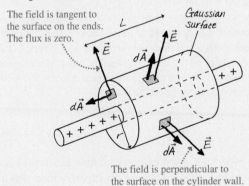

The field is tangent to the surface on the ends. The flux is zero.

The field is perpendicular to the surface on the cylinder wall.

SOLVE Gauss's law is

$$\Phi_e = \oint \vec{E}\cdot d\vec{A} = \frac{Q_{\text{in}}}{\epsilon_0}$$

where Q_{in} is the charge *inside* the closed cylinder. We have two tasks: to evaluate the flux integral, and to determine how much charge is inside the closed surface. The wire has linear charge density λ, so the amount of charge inside a cylinder of length L is simply

$$Q_{\text{in}} = \lambda L$$

Finding the net flux is just as straightforward. We can divide the flux through the entire closed surface into the flux through each end plus the flux through the cylindrical wall. The electric field $\vec{E}$, pointing straight out from the wire, is tangent to the end surfaces at every point. Thus the flux through these two surfaces is zero. On the wall, $\vec{E}$ is perpendicular to the surface and has the same strength E at every point. Thus

$$\Phi_e = \Phi_{\text{top}} + \Phi_{\text{bottom}} + \Phi_{\text{wall}} = 0 + 0 + EA_{\text{cyl}} = 2\pi rLE$$

where we used $A_{\text{cyl}} = 2\pi rL$ as the surface area of a cylindrical wall of radius r and length L. Once again, the proper choice of the Gaussian surface reduces the flux integral merely to finding a surface area. With these expressions for Q_{in} and Φ_e, Gauss's law is

$$\Phi_e = 2\pi rLE = \frac{Q_{\text{in}}}{\epsilon_0} = \frac{\lambda L}{\epsilon_0}$$

Thus the electric field at distance r from a long, charged wire is

$$E_{\text{wire}} = \frac{\lambda}{2\pi\epsilon_0 r}$$

ASSESS This agrees exactly with the result of the more complex derivation in Chapter 26. Notice that the result does not depend on our choice of L. A Gaussian surface is an imaginary device, not a physical object. We needed a finite-length cylinder to do the flux calculation, but the electric field of an *infinitely* long wire can't depend on the length of an imaginary cylinder.

Example 27.5, for the electric field of a long, charged wire, contains a subtle but important idea, one that often occurs when using Gauss's law. The Gaussian cylinder of length L encloses only some of the wire's charge. The pieces of the charged wire outside the cylinder are not enclosed by the Gaussian surface and consequently do not contribute anything to the net flux. Even so, *they are essential* to the use of Gauss's law because it takes the *entire* charged wire to produce an electric field with cylindrical symmetry. In other words, the wire outside the cylinder may not contribute to the flux, but it affects the *shape* of the electric field. Our ability to write $\Phi_e = EA_{cyl}$ depended on knowing that E is the same at every point on the wall of the cylinder. That would not be true for a charged wire of finite length, so we cannot use Gauss's law to find the electric field of a finite-length charged wire.

EXAMPLE 27.6 **The electric field of a plane of charge**

Use Gauss's law to find the electric field of an infinite plane of charge with surface charge density (C/m²) η.

MODEL A uniformly charged flat electrode can be modeled as an infinite plane of charge.

VISUALIZE **FIGURE 27.27** shows a uniformly charged plane with surface charge density η. We will assume that the plane extends infinitely far in all directions, although we obviously have to show "edges" in our drawing. The planar symmetry allows the electric field to point only straight toward or away from the plane. With this in mind, choose as a Gaussian surface a cylinder with length L and faces of area A centered on the plane of charge. Although we've drawn them as circular, the shape of the faces is not relevant.

FIGURE 27.27 The Gaussian surface extends to both sides of a plane of charge.

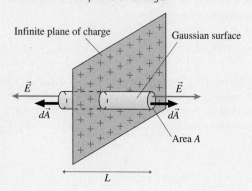

Infinite plane of charge   Gaussian surface

$\vec{E}$ $\vec{E}$
$d\vec{A}$ $d\vec{A}$

Area A

L

SOLVE The electric field is perpendicular to both faces of the cylinder, so the total flux through both faces is $\Phi_{faces} = 2EA$. (The fluxes add rather than cancel because the area vector $\vec{A}$ points *outward* on each face.) There's *no* flux through the wall of the cylinder because the field vectors are tangent to the wall. Thus the net flux is simply

$$\Phi_e = 2EA$$

The charge inside the cylinder is the charge contained in area A of the plane. This is

$$Q_{in} = \eta A$$

With these expressions for Q_{in} and Φ_e, Gauss's law is

$$\Phi_e = 2EA = \frac{Q_{in}}{\epsilon_0} = \frac{\eta A}{\epsilon_0}$$

Thus the electric field of an infinite charged plane is

$$E_{plane} = \frac{\eta}{2\epsilon_0}$$

This agrees with the result in Chapter 26.

ASSESS This is another example of a Gaussian surface enclosing only some of the charge. Most of the plane's charge is outside the Gaussian surface and does not contribute to the flux, but it does affect the shape of the field. We wouldn't have planar symmetry, with the electric field exactly perpendicular to the plane, without all the rest of the charge on the plane.

The plane of charge is an especially good example of how powerful Gauss's law can be. Finding the electric field of a plane of charge via superposition was a difficult and tedious derivation. With Gauss's law, once you see how to apply it, the problem is simple enough to solve in your head!

You might wonder, then, why we bothered with superposition at all. The reason is that Gauss's law, powerful though it may be, is effective only in a limited number of situations where the field is highly symmetric. Superposition always works, even if the derivation is messy, because superposition goes directly back to the fields of individual point charges. It's good to use Gauss's law when you can, but superposition is often the only way to attack real-world charge distributions.

STOP TO THINK 27.5 Which Gaussian surface would allow you to use Gauss's law to determine the electric field outside a uniformly charged cube?

a. A sphere whose center coincides with the center of the charged cube
b. A cube whose center coincides with the center of the charged cube and that has parallel faces
c. Either a or b
d. Neither a nor b

27.6 Conductors in Electrostatic Equilibrium

Consider a charged conductor, such as a charged metal electrode, in electrostatic equilibrium. That is, there is no current through the conductor and the charges are all stationary. One very important conclusion is that **the electric field is zero at all points within a conductor in electrostatic equilibrium.** That is, $\vec{E}_{in} = \vec{0}$. If this weren't true, the electric field would cause the charge carriers to move and thus violate the assumption that all the charges are at rest. Let's use Gauss's law to see what else we can learn.

At the Surface of a Conductor

FIGURE 27.28 shows a Gaussian surface just barely inside the physical surface of a conductor that's in electrostatic equilibrium. The electric field is zero at all points within the conductor, hence the electric flux Φ_e through this Gaussian surface must be zero. But if $\Phi_e = 0$, Gauss's law tells us that $Q_{in} = 0$. That is, there's no net charge within this surface. There are charges—electrons and positive ions—but no *net* charge.

If there's no net charge in the interior of a conductor in electrostatic equilibrium, then **all the excess charge on a charged conductor resides on the exterior surface of the conductor.** Any charges added to a conductor quickly spread across the surface until reaching positions of electrostatic equilibrium, but there is no net charge *within* the conductor.

There may be no electric field within a charged conductor, but the presence of net charge requires an exterior electric field in the space outside the conductor. **FIGURE 27.29** shows that **the electric field right at the surface of the conductor has to be perpendicular to the surface.** To see that this is so, suppose $\vec{E}_{surface}$ had a component tangent to the surface. This component of $\vec{E}_{surface}$ would exert a force on the surface charges and cause a surface current, thus violating the assumption that all charges are at rest. The only exterior electric field consistent with electrostatic equilibrium is one that is perpendicular to the surface.

We can use Gauss's law to relate the field strength at the surface to the charge density on the surface. **FIGURE 27.30** shows a small Gaussian cylinder with faces very slightly above and below the surface of a charged conductor. The charge inside this Gaussian cylinder is ηA, where η is the surface charge density at this point on the conductor. There's a flux $\Phi = AE_{surface}$ through the outside face of this cylinder but, unlike Example 27.6 for the plane of charge, *no* flux through the inside face because $\vec{E}_{in} = \vec{0}$ within the conductor. Furthermore, there's no flux through the wall of the cylinder because $\vec{E}_{surface}$ is perpendicular to the surface. Thus the net flux is $\Phi_e = AE_{surface}$. Gauss's law is

$$\Phi_e = AE_{surface} = \frac{Q_{in}}{\epsilon_0} = \frac{\eta A}{\epsilon_0} \tag{27.19}$$

from which we can conclude that the electric field at the surface of a charged conductor is

$$\vec{E}_{surface} = \left(\frac{\eta}{\epsilon_0}, \text{ perpendicular to surface}\right) \tag{27.20}$$

FIGURE 27.28 A Gaussian surface just inside a conductor that's in electrostatic equilibrium.

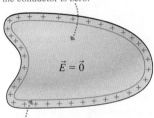

The electric field inside the conductor is zero.

The flux through the Gaussian surface is zero. There's no net charge inside the conductor. Hence all the excess charge is on the surface.

FIGURE 27.29 The electric field at the surface of a charged conductor.

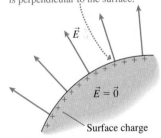

The electric field at the surface is perpendicular to the surface.

Surface charge

FIGURE 27.30 A Gaussian surface extending through the surface of the conductor has a flux only through the outer face.

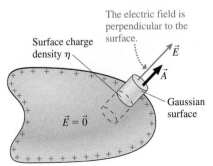

The electric field is perpendicular to the surface.

Surface charge density η

Gaussian surface

In general, the surface charge density η is *not* constant on the surface of a conductor but depends on the shape of the conductor. If we can determine η, by either calculating it or measuring it, then Equation 27.20 tells us the electric field at that point on the surface. Alternatively, we can use Equation 27.20 to deduce the charge density on the conductor's surface if we know the electric field just outside the conductor.

Charges and Fields Within a Conductor

FIGURE 27.31 shows a charged conductor with a hole inside. Can there be charge on this interior surface? To find out, we place a Gaussian surface around the hole, infinitesimally close but entirely within the conductor. The electric flux Φ_e through this Gaussian surface is zero because the electric field is zero everywhere inside the conductor. Thus we must conclude that $Q_{in} = 0$. There's no net charge inside this Gaussian surface and thus no charge on the surface of the hole. Any excess charge resides on the *exterior* surface of the conductor, not on any interior surfaces.

Furthermore, because there's no electric field inside the conductor and no charge inside the hole, the electric field inside the hole must also be zero. This conclusion has an important practical application. For example, suppose we need to exclude the electric field from the region in FIGURE 27.32a enclosed within dashed lines. We can do so by surrounding this region with the neutral conducting box of FIGURE 27.32b.

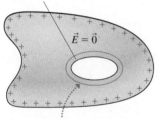

FIGURE 27.31 A Gaussian surface surrounding a hole inside a conductor in electrostatic equilibrium.

A hollow completely enclosed by the conductor

$\vec{E} = \vec{0}$

The flux through the Gaussian surface is zero. There's no net charge inside, hence no charge on this interior surface.

FIGURE 27.32 The electric field can be excluded from a region of space by surrounding it with a conducting box.

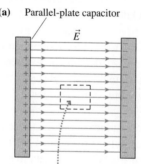

(a) Parallel-plate capacitor

$\vec{E}$

We want to exclude the electric field from this region.

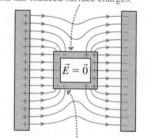

(b) The conducting box has been polarized and has induced surface charges.

$\vec{E} = \vec{0}$

The electric field is perpendicular to all conducting surfaces.

This region of space is now a hole inside a conductor, thus the interior electric field is zero. The use of a conducting box to exclude electric fields from a region of space is called **screening**. Solid metal walls are ideal, but in practice wire screen or wire mesh—sometimes called a *Faraday cage*—provides sufficient screening for all but the most sensitive applications. The price we pay is that the exterior field is now very complicated.

Finally, FIGURE 27.33 shows a charge q inside a hole within a neutral conductor. The electric field *within* the conductor is still zero, hence the electric flux through the Gaussian surface is zero. But $\Phi_e = 0$ requires $Q_{in} = 0$. Consequently, the charge inside the hole attracts an equal charge of opposite sign, and charge $-q$ now lines the inner surface of the hole.

The conductor as a whole is neutral, so moving $-q$ to the surface of the hole must leave $+q$ behind somewhere else. Where is it? It can't be in the interior of the conductor, as we've seen, and that leaves only the exterior surface. In essence, an internal charge polarizes the conductor just as an external charge would. Net charge $-q$ moves to the inner surface and net charge $+q$ is left behind on the exterior surface.

In summary, conductors in electrostatic equilibrium have the properties described in Tactics Box 27.3.

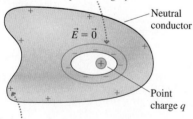

FIGURE 27.33 A charge in the hole causes a net charge on the interior and exterior surfaces.

The flux through the Gaussian surface is zero, hence there's no *net* charge inside this surface. There must be charge $-q$ on the inside surface to balance point charge q.

$\vec{E} = \vec{0}$

Neutral conductor

Point charge q

The outer surface must have charge $+q$ so that the conductor remains neutral.

TACTICS
BOX 27.3
Finding the electric field of a conductor in electrostatic equilibrium

❶ The electric field is zero at all points within the volume of the conductor.
❷ Any excess charge resides entirely on the *exterior* surface.
❸ The external electric field at the surface of a charged conductor is perpendicular to the surface and of magnitude η/ϵ_0, where η is the surface charge density at that point.
❹ The electric field is zero inside any hole within a conductor unless there is a charge in the hole.

Exercises 20–24

EXAMPLE 27.7 **The electric field at the surface of a charged metal sphere**

A 2.0-cm-diameter brass sphere has been given a charge of 2.0 nC. What is the electric field strength at the surface?

MODEL Brass is a conductor. The excess charge resides on the surface.

VISUALIZE The charge distribution has spherical symmetry. The electric field points radially outward from the surface.

SOLVE We can solve this problem in two ways. One uses the fact that a sphere is the one shape for which any excess charge will spread out to a *uniform* surface charge density. Thus

$$\eta = \frac{q}{A_{\text{sphere}}} = \frac{q}{4\pi R^2} = \frac{2.0 \times 10^{-9}\ \text{C}}{4\pi(0.010\ \text{m})^2} = 1.59 \times 10^{-6}\ \text{C/m}^2$$

From Equation 27.20, we know the electric field at the surface has strength

$$E_{\text{surface}} = \frac{\eta}{\epsilon_0} = \frac{1.59 \times 10^{-6}\ \text{C/m}^2}{8.85 \times 10^{-12}\ \text{C}^2/\text{N m}^2} = 1.8 \times 10^5\ \text{N/C}$$

Alternatively, we could have used the result, obtained earlier in the chapter, that the electric field strength outside a sphere of charge Q is $E_{\text{outside}} = Q_{\text{in}}/(4\pi\epsilon_0 r^2)$. But $Q_{\text{in}} = q$ and, at the surface, $r = R$. Thus

$$E_{\text{surface}} = \frac{1}{4\pi\epsilon_0}\frac{q}{R^2} = (9.0 \times 10^9\ \text{N m}^2/\text{C}^2)\frac{2.0 \times 10^{-9}\ \text{C}}{(0.010\ \text{m})^2}$$

$$= 1.8 \times 10^5\ \text{N/C}$$

As we can see, both methods lead to the same result.

CHALLENGE EXAMPLE 27.8 **The electric field of a slab of charge**

An infinite slab of charge of thickness $2a$ is centered in the xy-plane. The charge density is $\rho = \rho_0(1 - |z|/a)$. Find the electric field strengths inside and outside this slab of charge.

MODEL The charge density is not uniform. Starting at ρ_0 in the xy-plane, it decreases linearly with distance above and below the xy-plane until reaching zero at $z = \pm a$, the edges of the slab.

VISUALIZE FIGURE 27.34 shows an edge view of the slab of charge and, as Gaussian surfaces, side views of two cylinders with cross-section area A. By symmetry, the electric field must point away from the xy-plane; the field cannot have an x- or y-component.

FIGURE 27.34 Two cylindrical Gaussian surfaces for an infinite slab of charge.

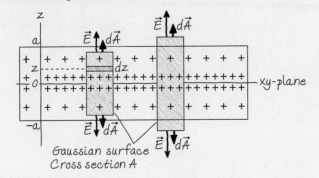

Gaussian surface Cross section A

SOLVE Gauss's law is

$$\Phi_e = \oint \vec{E} \cdot d\vec{A} = \frac{Q_{\text{in}}}{\epsilon_0}$$

With symmetry, finding the net flux is straightforward. The electric field is perpendicular to the faces of the cylinders and pointing outward, so the total flux through the faces is $\Phi_{\text{faces}} = 2EA$, where E may depend on distance z. The field is parallel to the walls of the cylinders, so $\Phi_{\text{wall}} = 0$. Thus the net flux is simply

$$\Phi_e = 2EA$$

Because the charge density is not uniform, we need to integrate to find Q_{in}, the charge *inside* the cylinder. We can slice the cylinder into small slabs of infinitesimal thickness dz and volume $dV = A\,dz$. Figure 27.34 shows one such little slab at distance z from the xy-plane. The charge in this little slab is

$$dq = \rho\,dV = \rho_0\left(1 - \frac{z}{a}\right)A\,dz$$

where we assumed that z is positive. Because the charge is symmetric about $z = 0$, we can avoid difficulties with the absolute value sign in the charge density by integrating from 0 and

Continued

multiplying by 2. For the Gaussian cylinder that ends inside the slab of charge, at distance z, the total charge inside is

$$Q_{in} = \int dq = 2\int_0^z \rho_0\left(1 - \frac{z}{a}\right)A\,dz$$

$$= 2\rho_0 A\left[z\Big|_0^z - \frac{1}{2a}z^2\Big|_0^z\right]$$

$$= 2\rho_0 Az\left(1 - \frac{z}{2a}\right)$$

Gauss's law inside the slab is then

$$\Phi_e = 2E_{inside}A = \frac{Q_{in}}{\epsilon_0} = \frac{2\rho_0 Az}{\epsilon_0}\left(1 - \frac{z}{2a}\right)$$

The area A cancels, as it must because it was an arbitrary choice, leaving

$$E_{inside} = \frac{\rho_0 z}{\epsilon_0}\left(1 - \frac{z}{2a}\right)$$

The field strength is zero at $z = 0$, then increases as z increases. This expression is valid only above the xy-plane, for $z > 0$, but the field strength is symmetric on the other side.

For the Gaussian cylinder that extends outside the slab of charge, the integral for Q has to end at $z = a$. Thus

$$Q_{in} = 2\rho_0 Aa\left(1 - \frac{a}{2a}\right) = \rho_0 Aa$$

independent of distance z. With this, Gauss's law gives

$$E_{outside} = \frac{Q_{in}}{2\epsilon_0 A} = \frac{\rho_0 a}{2\epsilon_0}$$

This matches E_{inside} at the surface, $z = a$, so the field is continuous as it crosses the boundary.

ASSESS Outside a sphere of charge, the field is the same as that of a point charge at the center. Similarly, the field outside an infinite slab of charge should be the same as that of an infinite charged plane. We found, by integration, that the total charge in an area A of the slab is $Q = \rho_0 Aa$. If we squished this charge into a plane, the surface charge density would be $\eta = Q/A = \rho_0 a$. Thus our expression for $E_{outside}$ could be written $\eta/2\epsilon_0$, which matches the field we found in Example 27.6 for a plane of charge.

SUMMARY

The goal of Chapter 27 has been to understand and apply Gauss's law.

General Principles

Gauss's Law

For any *closed* surface enclosing net charge Q_{in}, the net electric flux through the surface is

$$\Phi_e = \oint \vec{E} \cdot d\vec{A} = \frac{Q_{in}}{\epsilon_0}$$

The electric flux Φ_e is the same for *any* closed surface enclosing charge Q_{in}.

Symmetry

The symmetry of the electric field must match the symmetry of the charge distribution.

In practice, Φ_e is computable only if the symmetry of the Gaussian surface matches the symmetry of the charge distribution.

Important Concepts

Charge creates the electric field that is responsible for the electric flux.

Q_{in} is the sum of all enclosed charges. This charge contributes to the flux.

Gaussian surface

Charges outside the surface contribute to the electric field, but they don't contribute to the flux.

Flux is the amount of electric field passing through a surface of area A:

$$\Phi_e = \vec{E} \cdot \vec{A}$$

where $\vec{A}$ is the **area vector.**

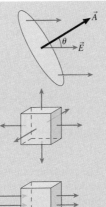

For closed surfaces:
A net flux in or out indicates that the surface encloses a net charge.

Field lines through but with no *net* flux mean that the surface encloses no *net* charge.

Surface integrals calculate the flux by summing the fluxes through many small pieces of the surface:

$$\Phi_e = \sum \vec{E} \cdot \delta\vec{A}$$

$$\rightarrow \int \vec{E} \cdot d\vec{A}$$

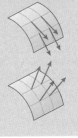

Two important situations:
If the electric field is everywhere tangent to the surface, then

$$\Phi_e = 0$$

If the electric field is everywhere perpendicular to the surface *and* has the same strength E at all points, then

$$\Phi_e = EA$$

Applications

Conductors in electrostatic equilibrium

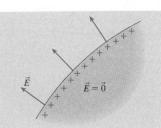

- The electric field is zero at all points within the conductor.
- Any excess charge resides entirely on the exterior surface.
- The external electric field is perpendicular to the surface and of magnitude η/ϵ_0, where η is the surface charge density.
- The electric field is zero inside any hole within a conductor unless there is a charge in the hole.

Terms and Notation

symmetric	electric flux, Φ_e	surface integral	screening
Gaussian surface	area vector, $\vec{A}$	Gauss's law	

CONCEPTUAL QUESTIONS

1. Suppose you have the uniformly charged cube in FIGURE Q27.1. Can you use symmetry alone to deduce the *shape* of the cube's electric field? If so, sketch and describe the field shape. If not, why not?

FIGURE Q27.1

2. FIGURE Q27.2 shows cross sections of three-dimensional closed surfaces. They have a flat top and bottom surface above and below the plane of the page. However, the electric field is everywhere parallel to the page, so there is no flux through the top or bottom surface. The electric field is uniform over each face of the surface. For each, does the surface enclose a net positive charge, a net negative charge, or no net charge? Explain.

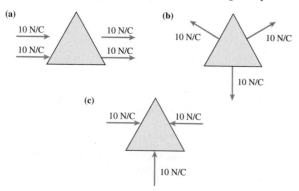

FIGURE Q27.2

3. The square and circle in FIGURE Q27.3 are in the same uniform field. The diameter of the circle equals the edge length of the square. Is Φ_{square} larger than, smaller than, or equal to Φ_{circle}? Explain.

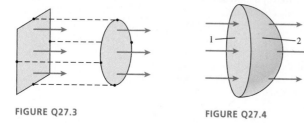

FIGURE Q27.3   FIGURE Q27.4

4. In FIGURE Q27.4, where the field is uniform, is Φ_1 larger than, smaller than, or equal to Φ_2? Explain.

5. What is the electric flux through each of the surfaces in FIGURE Q27.5? Give each answer as a multiple of q/ϵ_0.

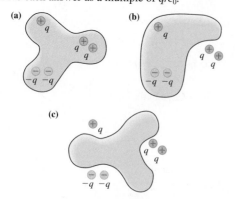

FIGURE Q27.5

6. What is the electric flux through each of the surfaces A to E in FIGURE Q27.6? Give each answer as a multiple of q/ϵ_0.

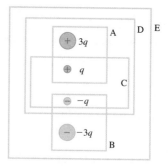

FIGURE Q27.6

7. The charged balloon in FIGURE Q27.7 expands as it is blown up, increasing in size from the initial to final diameters shown. Do the electric field strengths at points 1, 2, and 3 increase, decrease, or stay the same? Explain your reasoning for each.

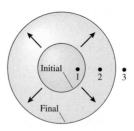

FIGURE Q27.7

8. The two spheres in FIGURE Q27.8 surround equal charges. Three students are discussing the situation.
Student 1: The fluxes through spheres A and B are equal because they enclose equal charges.
Student 2: But the electric field on sphere B is weaker than the electric field on sphere A. The flux depends on the electric field strength, so the flux through A is larger than the flux through B.
Student 3: I thought we learned that flux was about surface area. Sphere B is larger than sphere A, so I think the flux through B is larger than the flux through A.
Which of these students, if any, do you agree with? Explain.

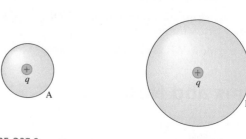

FIGURE Q27.8

9. The sphere and ellipsoid in FIGURE Q27.9 surround equal charges. Four students are discussing the situation.

Student 1: The fluxes through A and B are equal because the average radius is the same.

Student 2: I agree that the fluxes are equal, but that's because they enclose equal charges.

Student 3: The electric field is not perpendicular to the surface for B, and that makes the flux through B less than the flux through A.

Student 4: I don't think that Gauss's law even applies to a situation like B, so we can't compare the fluxes through A and B.

Which of these students, if any, do you agree with? Explain.

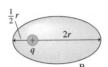

FIGURE Q27.9

10. A small, metal sphere hangs by an insulating thread within the larger, hollow conducting sphere of FIGURE Q27.10. A conducting wire extends from the small sphere through, but not touching, a small hole in the hollow sphere. A charged rod is used to transfer positive charge to the protruding wire. After the charged rod has touched the wire and been removed, are the following surfaces positive, negative, or not charged? Explain.

a. The small sphere.
b. The inner surface of the hollow sphere.
c. The outer surface of the hollow sphere.

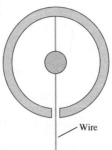

FIGURE Q27.10

EXERCISES AND PROBLEMS

Exercises

Section 27.1 Symmetry

1. | FIGURE EX27.1 shows two cross sections of two infinitely long coaxial cylinders. The inner cylinder has a positive charge, the outer cylinder has an equal negative charge. Draw this figure on your paper, then draw electric field vectors showing the shape of the electric field.

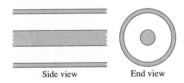

Side view   End view

FIGURE EX27.1

FIGURE EX27.2

2. | FIGURE EX27.2 shows a cross section of two concentric spheres. The inner sphere has a negative charge. The outer sphere has a positive charge larger in magnitude than the charge on the inner sphere. Draw this figure on your paper, then draw electric field vectors showing the shape of the electric field.

3. | FIGURE EX27.3 shows a cross section of two infinite parallel planes of charge. Draw this figure on your paper, then draw electric field vectors showing the shape of the electric field.

+ + + + + + + + + + + + + + + + + +

FIGURE EX27.3   + + + + + + + + + + + + + + + + + +

Section 27.2 The Concept of Flux

4. | The electric field is constant over each face of the cube shown in FIGURE EX27.4. Does the box contain positive charge, negative charge, or no charge? Explain.

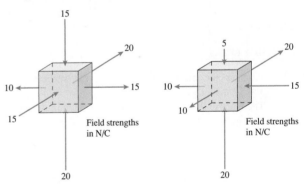

FIGURE EX27.4   FIGURE EX27.5

5. | The electric field is constant over each face of the cube shown in FIGURE EX27.5. Does the box contain positive charge, negative charge, or no charge? Explain.

6. | The cube in FIGURE EX27.6 contains negative charge. The electric field is constant over each face of the cube. Does the missing electric field vector on the front face point in or out? What strength must this field exceed?

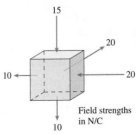

FIGURE EX27.6

7. | The cube in FIGURE EX27.7 contains negative charge. The electric field is constant over each face of the cube. Does the missing electric field vector on the front face point in or out? What strength must this field exceed?

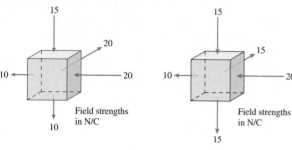

FIGURE EX27.7   FIGURE EX27.8

8. | The cube in FIGURE EX27.8 contains no net charge. The electric field is constant over each face of the cube. Does the missing electric field vector on the front face point in or out? What is the field strength?

Section 27.3 Calculating Electric Flux

9. ‖ What is the electric flux through the surface shown in FIGURE EX27.9?

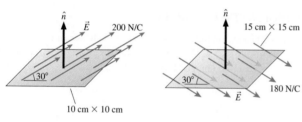

FIGURE EX27.9   FIGURE EX27.10

10. ‖ What is the electric flux through the surface shown in FIGURE EX27.10?

11. ‖ The electric flux through the surface shown in FIGURE EX27.11 is 25 N m²/C. What is the electric field strength?

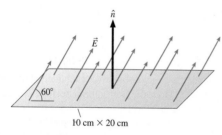

FIGURE EX27.11

12. ‖ A 2.0 cm × 3.0 cm rectangle lies in the xy-plane. What is the electric flux through the rectangle if
   a. $\vec{E} = (100\hat{i} + 50\hat{k})$ N/C?
   b. $\vec{E} = (100\hat{i} + 50\hat{j})$ N/C?

13. ‖ A 2.0 cm × 3.0 cm rectangle lies in the xz-plane. What is the electric flux through the rectangle if
   a. $\vec{E} = (100\hat{i} + 50\hat{k})$ N/C?
   b. $\vec{E} = (100\hat{i} + 50\hat{j})$ N/C?

14. ‖ A 3.0-cm-diameter circle lies in the xz-plane in a region where the electric field is $\vec{E} = (1500\hat{i} + 1500\hat{j} - 1500\hat{k})$ N/C. What is the electric flux through the circle?

15. ‖ A 1.0 cm × 1.0 cm × 1.0 cm box with its edges aligned with the xyz-axes is in the electric field $\vec{E} = (350x + 150)\hat{i}$ N/C, where x is in meters. What is the net electric flux through the box?

16. | What is the net electric flux through the two cylinders shown in FIGURE EX27.16? Give your answer in terms of R and E.

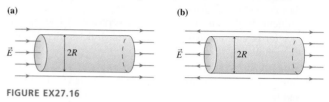

FIGURE EX27.16

Section 27.4 Gauss's Law

Section 27.5 Using Gauss's Law

17. | FIGURE EX27.17 shows three charges. Draw these charges on your paper four times. Then draw two-dimensional cross sections of three-dimensional closed surfaces through which the electric flux is (a) $2q/\epsilon_0$, (b) q/ϵ_0, (c) 0, and (d) $5q/\epsilon_0$.

FIGURE EX27.17   FIGURE EX27.18

18. | FIGURE EX27.18 shows three charges. Draw these charges on your paper four times. Then draw two-dimensional cross sections of three-dimensional closed surfaces through which the electric flux is (a) $-q/\epsilon_0$, (b) q/ϵ_0, (c) $3q/\epsilon_0$, and (d) $4q/\epsilon_0$.

19. | FIGURE EX27.19 shows three Gaussian surfaces and the electric flux through each. What are the three charges q_1, q_2, and q_3?

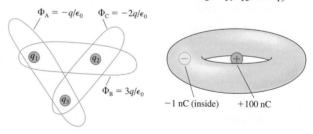

FIGURE EX27.19   FIGURE EX27.20

20. ‖ What is the net electric flux through the torus (i.e., doughnut shape) of FIGURE EX27.20?

21. | What is the net electric flux through the cylinder of FIGURE EX27.21?

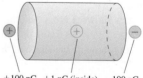

FIGURE EX27.21   +100 nC   +1 nC (inside)   −100 nC

22. ‖ The net electric flux through an octahedron is −1000 N m²/C. How much charge is enclosed within the octahedron?

23. ‖ 55.3 million excess electrons are inside a closed surface. What is the net electric flux through the surface?

Section 27.6 Conductors in Electrostatic Equilibrium

24. ‖ The electric field strength just above one face of a copper penny is 2000 N/C. What is the surface charge density on this face of the penny?

25. | A spark occurs at the tip of a metal needle if the electric field strength exceeds 3.0×10^6 N/C, the field strength at which air breaks down. What is the minimum surface charge density for producing a spark?

26. | The conducting box in FIGURE EX27.26 has been given an excess negative charge. The surface density of excess electrons at the center of the top surface is 5.0×10^{10} electrons/m². What are the electric field strengths E_1 to E_3 at points 1 to 3?

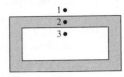

FIGURE EX27.26

27. | A thin, horizontal, 10-cm-diameter copper plate is charged to 3.5 nC. If the electrons are uniformly distributed on the surface, what are the strength and direction of the electric field
   a. 0.1 mm above the center of the top surface of the plate?
   b. at the plate's center of mass?
   c. 0.1 mm below the center of the bottom surface of the plate?

28. ‖ FIGURE EX27.28 shows a hollow cavity within a neutral conductor. A point charge Q is inside the cavity. What is the net electric flux through the closed surface that surrounds the conductor?

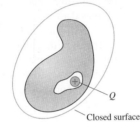

FIGURE EX27.28

Problems

29. | FIGURE P27.29 shows four sides of a 3.0 cm × 3.0 cm × 3.0 cm cube.
   a. What are the electric fluxes Φ_1 to Φ_4 through sides 1 to 4?
   b. What is the net flux through these four sides?

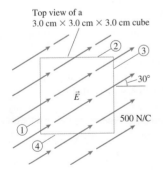

FIGURE P27.29

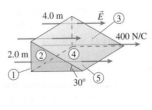

FIGURE P27.30

30. ‖ Find the electric fluxes Φ_1 to Φ_5 through surfaces 1 to 5 in FIGURE P27.30.

31. ‖ A tetrahedron has an equilateral triangle base with 20-cm-long edges and three equilateral triangle sides. The base is parallel to the ground, and a vertical uniform electric field of strength 200 N/C passes upward through the tetrahedron.
   a. What is the electric flux through the base?
   b. What is the electric flux through each of the three sides?

32. | Charges $q_1 = -4Q$ and $q_2 = +2Q$ are located at $x = -a$ and $x = +a$, respectively. What is the net electric flux through a sphere of radius $2a$ centered (a) at the origin and (b) at $x = 2a$?

33. ‖ A 10 nC point charge is at the center of a 2.0 m × 2.0 m × 2.0 m cube. What is the electric flux through the top surface of the cube?

34. ‖ The electric flux is 300 N m²/C through two opposing faces of a 2.0 cm × 2.0 cm × 2.0 cm box. The flux through each of the other faces is 100 N m²/C. How much charge is inside the box?

35. ‖ A spherically symmetric charge distribution produces the electric field $\vec{E} = (200/r)\hat{r}$ N/C, where r is in m.
   a. What is the electric field strength at $r = 10$ cm?
   b. What is the electric flux through a 20-cm-diameter spherical surface that is concentric with the charge distribution?
   c. How much charge is inside this 20-cm-diameter spherical surface?

36. ‖ A spherically symmetric charge distribution produces the electric field $\vec{E} = (5000r^2)\hat{r}$ N/C, where r is in m.
   a. What is the electric field strength at $r = 20$ cm?
   b. What is the electric flux through a 40-cm-diameter spherical surface that is concentric with the charge distribution?
   c. How much charge is inside this 40-cm-diameter spherical surface?

37. ‖ A neutral conductor contains a hollow cavity in which there is a +100 nC point charge. A charged rod then transfers −50 nC to the conductor. Afterward, what is the charge (a) on the inner wall of the cavity wall, and (b) on the exterior surface of the conductor?

38. ‖ A hollow metal sphere has inner radius a and outer radius b. The hollow sphere has charge $+2Q$. A point charge $+Q$ sits at the center of the hollow sphere.
   a. Determine the electric fields in the three regions $r \leq a$, $a < r < b$, and $r \geq b$.
   b. How much charge is on the inside surface of the hollow sphere? On the exterior surface?

39. ‖ A 20-cm-radius ball is uniformly charged to 80 nC.
   a. What is the ball's volume charge density (C/m³)?
   b. How much charge is enclosed by spheres of radii 5, 10, and 20 cm?
   c. What is the electric field strength at points 5, 10, and 20 cm from the center?

40. ‖ FIGURE P27.40 shows a solid metal sphere at the center of a hollow metal sphere. What is the total charge on (a) the exterior of the inner sphere, (b) the inside surface of the hollow sphere, and (c) the exterior surface of the hollow sphere?

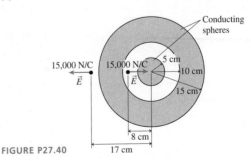

FIGURE P27.40

41. ‖ The earth has a vertical electric field at the surface, pointing down, that averages 100 N/C. This field is maintained by various atmospheric processes, including lightning. What is the excess charge on the surface of the earth?

42. ‖ Figure 27.32b showed a conducting box inside a parallel-plate capacitor. The electric field inside the box is $\vec{E} = \vec{0}$. Suppose the surface charge on the exterior of the box could be frozen. Draw a picture of the electric field inside the box after the box, with its frozen charge, is removed from the capacitor.
Hint: Superposition.

43. ‖ A hollow metal sphere has 6 cm and 10 cm inner and outer radii, respectively. The surface charge density on the inside surface is -100 nC/m². The surface charge density on the exterior surface is $+100$ nC/m². What are the strength and direction of the electric field at points 4, 8, and 12 cm from the center?

44. ‖ A positive point charge q sits at the center of a hollow spherical shell. The shell, with radius R and negligible thickness, has net charge $-2q$. Find an expression for the electric field strength (a) inside the sphere, $r < R$, and (b) outside the sphere, $r > R$. In what direction does the electric field point in each case?

45. ‖ Find the electric field inside and outside a hollow plastic ball of radius R that has charge Q uniformly distributed on its outer surface.

46. ‖ A uniformly charged ball of radius a and charge $-Q$ is at the center of a hollow metal shell with inner radius b and outer radius c. The hollow sphere has net charge $+2Q$. Determine the electric field strength in the four regions $r \leq a$, $a < r < b$, $b \leq r \leq c$, and $r > c$.

47. ‖ The three parallel planes of charge shown in FIGURE P27.47 have surface charge densities $-\frac{1}{2}\eta$, η, and $-\frac{1}{2}\eta$. Find the electric fields $\vec{E}_1$ to $\vec{E}_4$ in regions 1 to 4.

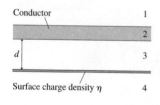

FIGURE P27.47

48. ‖ An infinite slab of charge of thickness $2z_0$ lies in the xy-plane between $z = -z_0$ and $z = +z_0$. The volume charge density ρ (C/m³) is a constant.
   a. Use Gauss's law to find an expression for the electric field strength inside the slab ($-z_0 \leq z \leq z_0$).
   b. Find an expression for the electric field strength above the slab ($z \geq z_0$).
   c. Draw a graph of E from $z = 0$ to $z = 3z_0$.

49. ‖ FIGURE P27.49 shows an infinitely wide conductor parallel to and distance d from an infinitely wide plane of charge with surface charge density η. What are the electric fields $\vec{E}_1$ to $\vec{E}_4$ in regions 1 to 4?

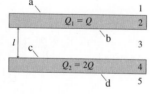

FIGURE P27.49   FIGURE P27.50

50. ‖ FIGURE P27.50 shows two very large slabs of metal that are parallel and distance l apart. Each slab has a total surface area (top + bottom) A. The thickness of each slab is so small in comparison to its lateral dimensions that the surface area around the sides is negligible. Metal 1 has total charge $Q_1 = Q$ and metal 2

has total charge $Q_2 = 2Q$. Assume Q is positive. In terms of Q and A, determine
   a. The electric field strengths E_1 to E_5 in regions 1 to 5.
   b. The surface charge densities η_a to η_d on the four surfaces a to d.

51. ‖ A long, thin straight wire with linear charge density λ runs down the center of a thin, hollow metal cylinder of radius R. The cylinder has a net linear charge density 2λ. Assume λ is positive. Find expressions for the electric field strength (a) inside the cylinder, $r < R$, and (b) outside the cylinder, $r > R$. In what direction does the electric field point in each of the cases?

52. ‖ A very long, uniformly charged cylinder has radius R and linear charge density λ. Find the cylinder's electric field (a) outside the cylinder, $r \geq R$, and (b) inside the cylinder, $r \leq R$. (c) Show that your answers to parts a and b match at the boundary, $r = R$.

53. ‖ A spherical shell has inner radius R_{in} and outer radius R_{out}. The shell contains total charge Q, uniformly distributed. The interior of the shell is empty of charge and matter.
   a. Find the electric field outside the shell, $r \geq R_{out}$.
   b. Find the electric field in the interior of the shell, $r \leq R_{in}$.
   c. Find the electric field within the shell, $R_{in} \leq r \leq R_{out}$.
   d. Show that your solutions match at both the inner and outer boundaries.

54. ‖ An early model of the atom, proposed by Rutherford after his discovery of the atomic nucleus, had a positive point charge $+Ze$ (the nucleus) at the center of a sphere of radius R with uniformly distributed negative charge $-Ze$. Z is the atomic number, the number of protons in the nucleus and the number of electrons in the negative sphere.
   a. Show that the electric field inside this atom is

$$E_{in} = \frac{Ze}{4\pi\epsilon_0}\left(\frac{1}{r^2} - \frac{r}{R^3}\right)$$

   b. What is E at the surface of the atom? Is this the expected value? Explain.
   c. A uranium atom has $Z = 92$ and $R = 0.10$ nm. What is the electric field strength at $r = \frac{1}{2}R$?

Challenge Problems

55. All examples of Gauss's law have used highly symmetric surfaces where the flux integral is either zero or EA. Yet we've claimed that the net $\Phi_e = Q_{in}/\epsilon_0$ is independent of the surface. This is worth checking. FIGURE CP27.55 shows a cube of edge length L centered on a long thin wire with linear charge density λ. The flux through one face of the cube is *not* simply EA because, in this case, the electric field varies in both strength and direction. But you can calculate the flux by actually doing the flux integral.

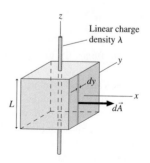

FIGURE CP27.55

a. Consider the face parallel to the yz-plane. Define area $d\vec{A}$ as a strip of width dy and height L with the vector pointing in the x-direction. One such strip is located at position y. Use the known electric field of a wire to calculate the electric flux $d\Phi$ through this little area. Your expression should be written in terms of y, which is a variable, and various constants. It should not explicitly contain any angles.

b. Now integrate $d\Phi$ to find the total flux through this face.

c. Finally, show that the net flux through the cube is $\Phi_e = Q_{in}/\epsilon_0$.

56. An infinite cylinder of radius R has a linear charge density λ. The volume charge density (C/m^3) within the cylinder ($r \le R$) is $\rho(r) = r\rho_0/R$, where ρ_0 is a constant to be determined.

a. Draw a graph of ρ versus x for an x-axis that crosses the cylinder perpendicular to the cylinder axis. Let x range from $-2R$ to $2R$.

b. The charge within a small volume dV is $dq = \rho \, dV$. The integral of $\rho \, dV$ over a cylinder of length L is the total charge $Q = \lambda L$ within the cylinder. Use this fact to show that $\rho_0 = 3\lambda/2\pi R^2$.

Hint: Let dV be a cylindrical shell of length L, radius r, and thickness dr. What is the volume of such a shell?

c. Use Gauss's law to find an expression for the electric field E inside the cylinder, $r \le R$.

d. Does your expression have the expected value at the surface, $r = R$? Explain.

57. A sphere of radius R has total charge Q. The volume charge density (C/m^3) within the sphere is $\rho(r) = C/r^2$, where C is a constant to be determined.

a. The charge within a small volume dV is $dq = \rho \, dV$. The integral of $\rho \, dV$ over the entire volume of the sphere is the total charge Q. Use this fact to determine the constant C in terms of Q and R.

Hint: Let dV be a spherical shell of radius r and thickness dr. What is the volume of such a shell?

b. Use Gauss's law to find an expression for the electric field E inside the sphere, $r \le R$.

c. Does your expression have the expected value at the surface, $r = R$? Explain.

58. A sphere of radius R has total charge Q. The volume charge density (C/m^3) within the sphere is

$$\rho = \rho_0\left(1 - \frac{r}{R}\right)$$

This charge density decreases linearly from ρ_0 at the center to zero at the edge of the sphere.

a. Show that $\rho_0 = 3Q/\pi R^3$.

b. Show that the electric field inside the sphere points radially outward with magnitude

$$E = \frac{Qr}{4\pi\epsilon_0 R^3}\left(4 - 3\frac{r}{R}\right)$$

c. Show that your result of part b has the expected value at $r = R$.

59. A spherical ball of charge has radius R and total charge Q. The electric field strength inside the ball ($r \le R$) is $E(r) = E_{max}(r^4/R^4)$.

a. What is E_{max} in terms of Q and R?

b. Find an expression for the volume charge density $\rho(r)$ inside the ball as a function of r.

c. Verify that your charge density gives the total charge Q when integrated over the volume of the ball.

STOP TO THINK ANSWERS

Stop to Think 27.1: a and d. Symmetry requires the electric field to be unchanged if front and back are reversed, if left and right are reversed, or if the field is rotated about the wire's axis. Fields a and d both have the proper symmetry. Other factors would now need to be considered to determine the correct field.

Stop to Think 27.2: e. The net flux is into the box.

Stop to Think 27.3: c. There's no flux through the four sides. The flux is positive 1 N m^2/C through both the top and bottom because $\vec{E}$ and $\vec{A}$ both point outward.

Stop to Think 27.4: $\Phi_b = \Phi_e > \Phi_a = \Phi_c = \Phi_d$. The flux through a closed surface depends only on the amount of enclosed charge, not the size or shape of the surface.

Stop to Think 27.5: d. A cube doesn't have enough symmetry to use Gauss's law. The electric field of a charged cube is *not* constant over the face of a cubic Gaussian surface, so we can't evaluate the surface integral for the flux.

28 The Electric Potential

City lights seen from space show where millions of lightbulbs are transforming electric energy into light and thermal energy.

▶ **Looking Ahead** The goals of Chapter 28 are to calculate and use the electric potential and electric potential energy.

Electric Energy

Energy allows things to happen. You want your lights to light, your computer to compute, and your stereo to keep your neighbors awake. All these require energy—*electric* energy.

This is the first of two chapters that explore electric energy and its connection to electric forces and fields.

Lightning is a dramatic example of the transformation of electric energy into light, sound, and thermal energy.

You'll learn to calculate the electric potential energy of charged particles and to solve problems using conservation of mechanical energy.

There's a close connection between electric potential energy and gravitational potential energy because both forces obey inverse-square laws.

◀ **Looking Back**
Sections 10.2–10.5 Kinetic energy, potential energy, and conservation

◀ **Looking Back**
Sections 11.2–11.5 Work and potential energy

The Electric Potential

Just as source charges create an electric field, they also create an **electric potential.** A charge moving in an electric potential has an electric potential energy.

The unit of electric potential is the **volt,** perhaps the most well known of all electrical units. A voltmeter reads the *potential difference* between two points.

Using Electric Potential

Charged particles *accelerate* as they move through a potential difference.

You'll learn to use the electric potential and a conservation of energy problem-solving strategy to solve problems about the motion of charged particles.

◀ **Looking Back**
Section 10.6 Energy diagrams

Calculating Electric Potential

You'll learn how to calculate the electric potential for several important charge distributions.

Elevation graph   **Equipotential surfaces**

You'll also learn to use several different representations of the electric potential.

◀ **Looking Back**
Section 26.3 Calculating electric fields

Sources of Electric Potential

In practice, electric potential is created by separating positive and negative charges—an idea we'll explore more thoroughly in Chapter 29.

A battery is the most common source of electric potential. As you'll learn, its *voltage* is the potential difference between separated charges—the plus and minus terminals.

28.1 Electric Potential Energy

In electricity, just as in mechanics, it takes energy to make things happen. It's been many chapters since we dealt much with work and energy, but these ideas will now be *essential* to our story. Consequently, the Looking Back recommendations in the chapter preview are especially important. You will recall that a system's mechanical energy $E_{mech} = K + U$ is conserved for particles that interact with each other via *conservative forces*, where K and U are the kinetic and potential energy. That is,

$$\Delta E_{mech} = \Delta K + \Delta U = 0 \qquad (28.1)$$

We need to be careful with notation because we are now using E to represent the electric field strength. To avoid confusion, we will represent mechanical energy either as the explicit sum $K + U$ or as E_{mech}, with an explicit subscript.

NOTE ▶ Recall that for any X, the *change* in X is $\Delta X = X_{final} - X_{initial}$. ◀

The kinetic energy $K = \sum K_i$, where $K_i = \frac{1}{2}m_i v_i^2$, is the sum of the kinetic energies of all the particles in the system. The potential energy U is the *interaction energy* of the system. In particular, we defined the *change* in potential energy in terms of the work W done by the forces of interaction as the system moves from an initial position or configuration i to a final position or configuration f:

$$\Delta U = U_f - U_i = -W_{interaction forces} \qquad \text{(position i} \rightarrow \text{position f)} \quad (28.2)$$

This formal definition of ΔU is rather abstract and will make more sense when we see specific applications.

A *constant* force does work

$$W = \vec{F} \cdot \Delta \vec{r} = F \Delta r \cos \theta \qquad (28.3)$$

on a particle that undergoes a linear displacement $\Delta \vec{r}$, where θ is the angle between the force $\vec{F}$ and $\Delta \vec{r}$. FIGURE 28.1 reminds you of the three special cases $\theta = 0°$, $90°$, and $180°$. It also shows that, in general, the work is done by the force component F_r in the direction of motion.

NOTE ▶ Work is *not* the oft-remembered "force times distance." Work is force times distance only in the one very special case in which the force is both constant *and* parallel to the displacement. ◀

If the force is *not* constant or the displacement is *not* along a linear path, we can calculate the work by dividing the path into many small segments. FIGURE 28.2 shows how this is done. The work done as the particle moves distance ds is $F_s ds$, where F_s is the force component parallel to ds (i.e., the component in the direction of motion). The total work done on the particle is

$$W = \sum_j (F_s)_j \Delta s_j \rightarrow \int_{s_i}^{s_f} F_s \, ds = \int_i^f \vec{F} \cdot d\vec{s} \qquad (28.4)$$

The second integral recognizes that $F_s ds = F \cos \theta \, ds$ is equivalent to the dot product $\vec{F} \cdot d\vec{s}$, allowing us to write the work in vector notation. As with Gauss's law, this integral looks more formidable than it really is. We'll look at examples shortly.

Finally, recall that a *conservative force* is one for which the work done as a particle moves from position i to position f is *independent of the path followed*. In other words, the integral in Equation 28.4 gives the same value for *any* path between points i and f. We'll assert for now, and prove later, that **the electric force is a conservative force.**

Uniform Fields

Gravity, like electricity, is a long-range force. Much as we defined the electric field $\vec{E} = \vec{F}_{on\,q}/q$, we can also define a gravitational field—the agent that exerts gravitational forces on masses—as $\vec{F}_{on\,m}/m$. But $\vec{F}_{on\,m} = m\vec{g}$ near the earth's surface; thus

FIGURE 28.1 The work done by a constant force.

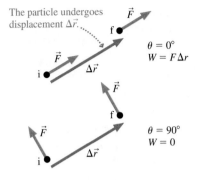

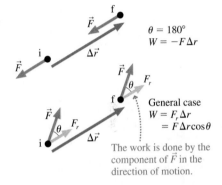

The work is done by the component of $\vec{F}$ in the direction of motion.

FIGURE 28.2 The work done along a curved path or by a variable force.

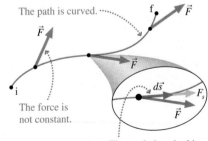

The path is curved.

The force is not constant.

The work done in this small segment of the motion is $F_s ds = \vec{F} \cdot d\vec{s}$.

the familiar $\vec{g} = (9.80 \text{ N/kg, down})$ is really the gravitational field! Notice how we've written the units of $\vec{g}$ as N/kg, as is appropriate for a field, but you can easily show that N/kg = m/s². The gravitational field near the earth's surface is a *uniform* field in the downward direction.

FIGURE 28.3 shows a particle of mass m falling in the gravitational field. The gravitational force is in the same direction as the particle's displacement, so the gravitational field does a *positive* amount of work on the particle. The gravitational force is constant, hence the work done by gravity is

$$W_{\text{grav}} = F_G \, \Delta r \cos 0° = mg|y_f - y_i| = mgy_i - mgy_f \qquad (28.5)$$

We have to be careful with signs because Δr, the magnitude of the displacement vector, must be a positive number.

Now we can see how the definition of ΔU in Equation 28.2 makes sense. The *change* in gravitational potential energy is

$$\Delta U_{\text{grav}} = U_f - U_i = -W_{\text{grav}}(\text{i} \rightarrow \text{f}) = mgy_f - mgy_i \qquad (28.6)$$

Comparing the initial and final terms on the two sides of the equation, we see that the gravitational potential energy near the earth is the familiar quantity

$$U_{\text{grav}} = U_0 + mgy \qquad (28.7)$$

where U_0 is the value of U_{grav} at $y = 0$. We usually choose $U_0 = 0$, in which case $U_{\text{grav}} = mgy$, but such a choice is not necessary. The zero point of potential energy is an arbitrary choice because we have defined ΔU rather than U.

The uniform electric field between the plates of the parallel-plate capacitor of FIGURE 28.4 looks very much like the uniform gravitational field near the earth's surface. The one difference is that $\vec{g}$ always points down whereas the positive-to-negative electric field can point in any direction. To deal with this, let's define a coordinate axis s that points *from* the negative plate, which we define to be $s = 0$, *toward* the positive plate. The electric field $\vec{E}$ then points in the negative s-direction, just as the gravitational field $\vec{g}$ points in the negative y-direction. This s-axis, which is valid no matter how the capacitor is oriented, is analogous to the y-axis used for gravitational potential energy.

A positive charge q inside the capacitor speeds up and gains kinetic energy as it "falls" toward the negative plate. Is the charge losing potential energy as it gains kinetic energy? Indeed it is, and the calculation of the potential energy is just like the calculation of gravitational potential energy. The electric field exerts a *constant* force $F = qE$ on the charge in the direction of motion; thus the work done on the charge by the electric field is

$$W_{\text{elec}} = F \, \Delta r \cos 0° = qE|s_f - s_i| = qEs_i - qEs_f \qquad (28.8)$$

where we again have to be careful with the signs because $s_f < s_i$.

The work done by the electric field causes the charge to experience a change in *electric* potential energy given by

$$\Delta U_{\text{elec}} = U_f - U_i = -W_{\text{elec}}(\text{i} \rightarrow \text{f}) = qEs_f - qEs_i \qquad (28.9)$$

Comparing the initial and final terms on the two sides of the equation, we see that the **electric potential energy** of charge q in a uniform electric field is

$$U_{\text{elec}} = U_0 + qEs \qquad (28.10)$$

where s is measured from the negative plate and U_0 is the potential energy at the negative plate ($s = 0$). It will often be convenient to choose $U_0 = 0$, but the choice has no physical consequences because it doesn't affect ΔU_{elec}, the *change* in the electric potential energy. Only the *change* is significant.

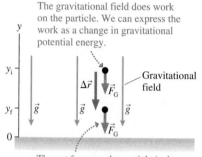

FIGURE 28.3 Potential energy is transformed into kinetic energy as a particle moves in a gravitational field.

The gravitational field does work on the particle. We can express the work as a change in gravitational potential energy.

The net force on the particle is down. It gains kinetic energy (i.e., speeds up) as it loses potential energy.

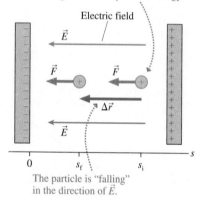

FIGURE 28.4 The electric field does work on the charged particle.

The electric field does work on the particle. We can express the work as a change in electric potential energy.

The particle is "falling" in the direction of $\vec{E}$.

Equation 28.10 was derived with the assumption that q is positive, but it is valid for either sign of q. A negative value for q in Equation 28.10 causes the potential energy U_{elec} to become *more negative* as s increases. As FIGURE 28.5 shows, a negative charge gains kinetic energy as it moves *away from* the negative plate of the capacitor.

FIGURE 28.5 A charged particle of either sign gains kinetic energy as it moves in the direction of decreasing potential energy.

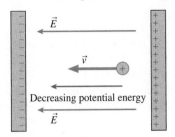

The potential energy of a positive charge decreases in the direction of $\vec{E}$. The charge gains kinetic energy as it moves toward the negative plate.

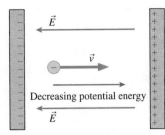

The potential energy of a negative charge decreases in the direction opposite to $\vec{E}$. The charge gains kinetic energy as it moves away from the negative plate.

NOTE ▶ Although Equation 28.10 is often called "the potential energy of charge q," it is really the potential energy of the charge + capacitor system. To the extent that the charges on the capacitor plate stay fixed, we're justified in thinking of this as the potential energy of just the charge q. ◀

FIGURE 28.6 is the *energy diagram* for a positively charged particle in a uniform electric field. Recall that an energy diagram is a graphical representation of how the kinetic and potential energy are transformed as a particle moves. The potential energy, given by Equation 28.10, increases linearly with distance, but the particle's total mechanical energy E_{mech} is fixed. If a positively charged particle is projected against a uniform field, it gradually slows (transforming kinetic to potential energy) until reaching the *turning point* where $U_{\text{elec}} = E_{\text{mech}}$.

FIGURE 28.6 The energy diagram for a positively charged particle in a uniform electric field.

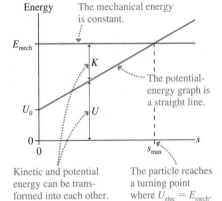

EXAMPLE 28.1   **Conservation of energy**

A 2.0 cm × 2.0 cm parallel-plate capacitor with a 2.0 mm spacing is charged to ±1.0 nC. First a proton, then an electron are released from rest at the midpoint of the capacitor.

a. What is each particle's change in electric potential energy from its release until it collides with one of the plates?
b. What is each particle's speed as it reaches the plate?

MODEL The mechanical energy of each particle is conserved. A parallel-plate capacitor has a uniform electric field.

VISUALIZE FIGURE 28.7 is a before-and-after pictorial representation, as you learned to draw in Part II. On the energy diagram of Figure 28.6, each particle is released at the turning point ($K = 0$) and moves toward lower potential energy. Thus the proton moves toward the negative plate, the electron toward the positive plate.

SOLVE a. The s-axis was defined to point from the negative toward the positive plate of the capacitor. Both charged particles have $s_i = \frac{1}{2}d$, where $d = 2.0$ mm is the plate separation. The positive proton loses potential energy and gains kinetic energy

FIGURE 28.7 A proton and an electron in a capacitor.

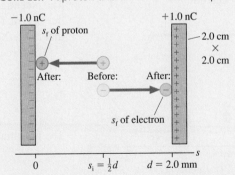

as it moves toward the negative plate. For the proton, with $q = +e$ and $s_f = 0$, the change in potential energy is

$$\Delta U_p = U_f - U_i = (U_0 + 0) - \left(U_0 + eE\frac{d}{2}\right) = -\frac{1}{2}eEd$$

Continued

where we used the electric potential energy for a charge in a uniform electric field. ΔU_p is negative, as expected. Notice that U_0 cancels when ΔU is calculated.

The electron moves toward the positive plate, which is the direction of decreasing potential energy for a negative charge. The electron has $q = -e$ and ends at $s_f = d$. Thus

$$\Delta U_e = U_f - U_i = (U_0 + (-e)Ed) - \left(U_0 + (-e)E\frac{d}{2}\right)$$
$$= -\frac{1}{2}eEd$$

Both particles have the *same* change in potential energy. The capacitor's electric field is

$$E = \frac{\eta}{\epsilon_0} = \frac{Q}{\epsilon_0 A} = 2.82 \times 10^5 \text{ N/C}$$

Using $d = 0.0020$ m, we find

$$\Delta U_p = \Delta U_e = -4.5 \times 10^{-17} \text{ J}$$

b. The law of conservation of energy is $\Delta K + \Delta U = 0$. Both particles are released from rest; hence $\Delta K = K_f - 0 = \frac{1}{2}mv_f^2$. Thus $\frac{1}{2}mv_f^2 = -\Delta U$, or

$$v_f = \sqrt{\frac{-2\,\Delta U}{m}} = \begin{cases} 2.3 \times 10^5 \text{ m/s for the proton} \\ 1.0 \times 10^7 \text{ m/s for the electron} \end{cases}$$

where we used the masses of the proton and the electron.

ASSESS Even though both particles have the same ΔU, the electron reaches a much faster final speed due to its much smaller mass.

STOP TO THINK 28.1 A glass rod is positively charged. The figure shows an end view of the rod. A negatively charged particle moves in a circular arc around the glass rod. Is the work done on the charged particle by the rod's electric field positive, negative, or zero?

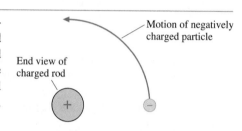

28.2 The Potential Energy of Point Charges

Now that we've introduced the idea of electric potential energy, let's look at *the* fundamental interaction of electricity—the force between two point charges. This force, given by Coulomb's law, varies with the distance between the two charges; hence we need to use the integral expression of Equation 28.4 to calculate the work done.

FIGURE 28.8a shows two charges q_1 and q_2, which we will assume to be like charges. The potential energy of their interaction can be found by calculating the work done by the electric field of q_1 on q_2 as q_2 moves from position x_i to position x_f. We'll assume that q_1 has been glued down and is unable to move, as shown in FIGURE 28.8b.

The force is entirely in the direction of motion, so $F_s\,ds = F_{1\text{ on }2}\,dx$. Thus

$$W_{elec} = \int_{x_i}^{x_f} F_{1\text{ on }2}\,dx = \int_{x_i}^{x_f} \frac{Kq_1q_2}{x^2}\,dx = Kq_1q_2\left.\frac{-1}{x}\right|_{x_i}^{x_f} = -\frac{Kq_1q_2}{x_f} + \frac{Kq_1q_2}{x_i} \quad (28.11)$$

The potential energy of the two charges is related to the work done by

$$\Delta U_{elec} = U_f - U_i = -W_{elec}(i \to f) = \frac{Kq_1q_2}{x_f} - \frac{Kq_1q_2}{x_i} \quad (28.12)$$

By comparing the left and right sides of the equation we see that the potential energy of the two-point-charge system is

$$U_{elec} = \frac{Kq_1q_2}{x} \quad (28.13)$$

We could include a constant U_0, as we did in Equation 28.10, for the potential energy of a charge in a uniform electric field, but it is customary to set $U_0 = 0$.

FIGURE 28.8 The interaction between two point charges.

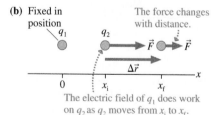

(a) $\vec{F}_{2\text{ on }1}$ q_1 q_2 $\vec{F}_{1\text{ on }2}$

Like charges exert repulsive forces.

(b) Fixed in position
q_1 q_2
The force changes with distance.
$\vec{F}$ $\vec{F}$
$\Delta\vec{r}$
0 x_i x_f

The electric field of q_1 does work on q_2 as q_2 moves from x_i to x_f.

We chose to integrate along the *x*-axis for convenience, but what is really important is the *distance* between the charges. Thus a more general expression for the electric potential energy is

$$U_{elec} = \frac{Kq_1q_2}{r} = \frac{1}{4\pi\epsilon_0}\frac{q_1q_2}{r} \qquad \text{(two point charges)} \qquad (28.14)$$

This is explicitly the energy *of the system*, not the energy of just q_1 or q_2.

NOTE ▶ The electric potential energy of two point charges looks *almost* the same as the force between the charges. The difference is the *r* in the denominator of the potential energy compared to the r^2 in Coulomb's law. ◀

Three important points need to be noted:

■ The choice $U_0 = 0$ is equivalent to saying that the potential energy of two charged particles is zero only when they are infinitely far apart. This makes sense because two charged particles cease interacting only when they are infinitely far apart.
■ We derived Equation 28.14 for two like charges, but it is equally valid for two opposite charges. The potential energy of two like charges is *positive* and of two opposite charges is *negative*.
■ Because the electric field outside a *sphere of charge* is the same as that of a point charge at the center, Equation 28.14 is also the electric potential energy of two charged spheres. Distance *r* is the distance between their centers.

FIGURE 28.9a shows the potential-energy curve—a hyperbola—for two like charges as a function of the distance *r* between them. Distances must be positive numbers, so the graph shows only $r > 0$. Also shown is the total energy line for two charged particles shot toward each other with equal but opposite momenta. Recall, from Chapter 10, that the total energy line is horizontal because the mechanical energy is conserved.

You can see that the total energy line crosses the potential-energy curve at r_{min}. This is a turning point. The two charges gradually slow down, because of the repulsive force between them, until the distance separating them is r_{min}. At this point, the kinetic energy is zero and both particles are instantaneously at rest. Both then reverse direction and move apart, speeding up as they go. r_{min} is the *distance of closest approach*.

Two opposite charges are a little trickier because of the negative energies. Negative total energies seem troubling at first, but they characterize *bound systems*. FIGURE 28.9b shows two oppositely charged particles shot apart from each other with equal but opposite momenta. If $E_{mech} < 0$, as shown, then their total energy line crosses the potential-energy curve at r_{max}. That is, the particles slow down, lose kinetic energy, reverse directions at *maximum separation* r_{max}, and then "fall" back together. They cannot escape from each other. Although moving in three dimensions rather than one, the electron and proton of a hydrogen atom are a realistic example of a bound system, and their mechanical energy is negative.

Two oppositely charged particles *can* escape from each other if $E_{mech} > 0$. They'll slow down, but eventually the potential energy vanishes and the particles still have kinetic energy. The threshold condition for escape is $E_{mech} = 0$, which will allow the particles to reach infinite separation ($U \rightarrow 0$) at infinitesimally slow speed ($K \rightarrow 0$). The initial speed that gives $E_{mech} = 0$ is called the *escape speed*.

NOTE ▶ Real particles can't be infinitely far apart, but because U_{elec} decreases with distance, there comes a point when $U_{elec} = 0$ is an excellent approximation. Two charged particles for which $U_{elec} \approx 0$ are sometimes described as "far apart" or "far away." ◀

FIGURE 28.9 The potential-energy diagrams for two like charges and two opposite charges.

(a) Like charges

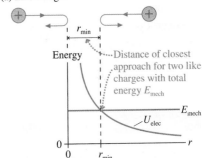

(b) Opposite charges

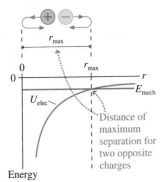

The Electric Force Is a Conservative Force

Potential energy can be defined only if the force is *conservative*, meaning that the work done on the particle as it moves from position i to position f is independent of the path followed between i and f. FIGURE 28.10 demonstrates that electric force is indeed conservative.

FIGURE 28.10 The work done on q_2 is independent of the path from i to f.

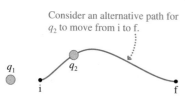

Consider an alternative path for q_2 to move from i to f.

Approximate the path using circular arcs and radial lines centered on q_1.

The electric force is a *central force*. As a result, zero work is done as q_2 moves along a circular arc because the force is perpendicular to the displacement.

All the work is done along the radial line segments, which are equivalent to a straight line from i to f. This is the work that was calculated in Equation 28.11.

EXAMPLE 28.2 **Approaching a charged sphere**

A proton is fired from far away at a 1.0-mm-diameter glass sphere that has been charged to $+100$ nC. What initial speed must the proton have to just reach the surface of the glass?

MODEL Energy is conserved. The glass sphere can be treated as a charged particle, so the potential energy is that of two point charges. The proton starts "far away," which we interpret as sufficiently far to make $U_i \approx 0$.

VISUALIZE FIGURE 28.11 shows the before-and-after pictorial representation. To "just reach" the glass sphere means that the proton comes to rest, $v_f = 0$, as it reaches $r_f = 0.50$ mm, the *radius* of the sphere.

SOLVE Conservation of energy $K_f + U_f = K_i + U_i$ is

$$0 + \frac{Kq_p q_{sphere}}{r_f} = \frac{1}{2}mv_i^2 + 0$$

FIGURE 28.11 A proton approaching a glass sphere.

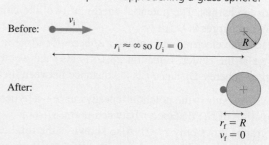

Before: v_i $r_i \approx \infty$ so $U_i = 0$ R

After: $r_f = R$ $v_f = 0$

The proton charge is $q_p = e$. With this, we can solve for the proton's initial speed:

$$v_i = \sqrt{\frac{2Keq_{sphere}}{mr_f}} = 1.86 \times 10^7 \text{ m/s}$$

EXAMPLE 28.3 **Escape velocity**

An interaction between two elementary particles causes an electron and a positron (a positive electron) to be shot out back to back with equal speeds. What minimum speed must each have when they are 100 fm apart in order to escape each other?

MODEL Energy is conserved. The particles end "far apart," which we interpret as sufficiently far to make $U_f \approx 0$.

VISUALIZE FIGURE 28.12 shows the before-and-after pictorial representation. The minimum speed to escape is the speed that allows the particles to reach $r_f = \infty$ with $v_f = 0$.

FIGURE 28.12 An electron and a positron flying apart.

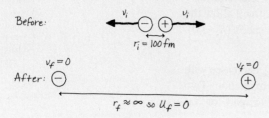

Before: v_i (−)(+) v_i
$r_i = 100$ fm

$v_f = 0$
After: (−)   (+) $v_f = 0$

$r_f \approx \infty$ so $U_f = 0$

SOLVE Here it is essential to interpret U_{elec} as the potential energy of the electron + positron system. Similarly, K is the *total* kinetic energy of the system. The electron and the positron, with equal masses and equal speeds, have equal kinetic energies. Conservation of energy $K_f + U_f = K_i + U_i$ is

$$0 + 0 + 0 = \frac{1}{2}mv_i^2 + \frac{1}{2}mv_i^2 + \frac{Kq_e q_p}{r_i} = mv_i^2 - \frac{Ke^2}{r_i}$$

Using $r_i = 100$ fm $= 1.0 \times 10^{-13}$ m, we can calculate the minimum initial speed to be

$$v_i = \sqrt{\frac{Ke^2}{mr_i}} = 5.0 \times 10^7 \text{ m/s}$$

ASSESS v_i is a little more than 10% the speed of light, just about the limit of what a "classical" calculation can predict. We would need to use the theory of relativity if v_i were any larger.

Multiple Point Charges

If more than two charges are present, the potential energy is the sum of the potential energies due to all pairs of charges:

$$U_{\text{elec}} = \sum_{i<j} \frac{Kq_iq_j}{r_{ij}} \qquad (28.15)$$

where r_{ij} is the distance between q_i and q_j. The summation contains the $i < j$ restriction to ensure that each pair of charges is counted only once.

NOTE ▶ For energy conservation problems, it's necessary to calculate only the potential energy for those pairs of charges for which the distance r_{ij} changes. The potential energy of charges that don't move is an additive constant with no physical consequences. ◀

EXAMPLE 28.4 **Launching an electron**

Three electrons are spaced 1.0 mm apart along a vertical line. The outer two electrons are fixed in position.

a. Is the center electron at a point of stable or unstable equilibrium?
b. If the center electron is displaced horizontally by a small distance, what will its speed be when it is very far away?

MODEL Energy is conserved. The outer two electrons don't move, so we don't need to include the potential energy of their interaction.

VISUALIZE FIGURE 28.13 shows the situation.

FIGURE 28.13 Three electrons.

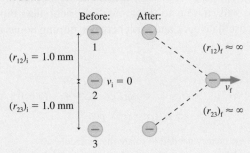

SOLVE a. The center electron is in equilibrium *exactly* in the center because the two electric forces on it balance. But if it moves a little to the right or left, no matter how little, then the horizontal components of the forces from both outer electrons will push the center electron farther away. This is an unstable equilibrium for horizontal displacements, like being on the top of a hill.

b. A small displacement will cause the electron to move away. If the displacement is only infinitesimal, the initial conditions are $(r_{12})_i = (r_{23})_i = 1.0$ mm and $v_i = 0$. "Far away" is interpreted as $r_f \to \infty$, where $U_f \approx 0$. There are now *two* terms in the potential energy, so conservation of energy $K_f + U_f = K_i + U_i$ gives

$$\frac{1}{2}mv_f^2 + 0 + 0 = 0 + \left[\frac{Kq_1q_2}{(r_{12})_i} + \frac{Kq_2q_3}{(r_{23})_i}\right]$$

$$= \left[\frac{Ke^2}{(r_{12})_i} + \frac{Ke^2}{(r_{23})_i}\right]$$

This is easily solved to give

$$v_f = \sqrt{\frac{2}{m}\left[\frac{Ke^2}{(r_{12})_i} + \frac{Ke^2}{(r_{23})_i}\right]} = 1000 \text{ m/s}$$

STOP TO THINK 28.2 Rank in order, from largest to smallest, the potential energies U_a to U_d of these four pairs of charges. Each + symbol represents the same amount of charge.

(a)   (b)   (c)   (d)

28.3 The Potential Energy of a Dipole

The electric dipole has been our model for understanding how charged objects interact with neutral objects. In Chapter 26 we found that an electric field exerts a *torque* on a dipole. We can complete the picture by calculating the potential energy of an electric dipole in a uniform electric field.

FIGURE 28.14 shows a dipole in an electric field $\vec{E}$. Recall that the dipole moment $\vec{p}$ is a vector that points from $-q$ to q with magnitude $p = qs$. The forces $\vec{F}_+$ and $\vec{F}_-$ exert a torque on the dipole, but now we're interested in calculating the *work* done by these forces as the dipole rotates from angle ϕ_i to angle ϕ_f.

FIGURE 28.14 The electric field does work as a dipole rotates.

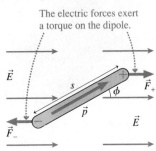

The electric forces exert a torque on the dipole.

When a force component F_s acts through a small displacement ds, the force does work $dW = F_s\,ds$. If we exploit the rotational-linear motion analogy from Chapter 12, where torque τ is the analog of force and angular displacement $\Delta\phi$ is the analog of linear displacement, then a torque acting through a small angular displacement $d\phi$ does work $dW = \tau\,d\phi$. From Chapter 26, the torque on the dipole in Figure 28.14 is $\tau = -pE\sin\phi$, where the minus sign is due to the torque trying to cause a clockwise rotation. Thus the work done by the electric field on the dipole as it rotates through the small angle $d\phi$ is

$$dW_{\text{elec}} = -pE\sin\phi\,d\phi \qquad (28.16)$$

The total work done by the electric field as the dipole turns from ϕ_i to ϕ_f is

$$W_{\text{elec}} = -pE\int_{\phi_i}^{\phi_f}\sin\phi\,d\phi = pE\cos\phi_f - pE\cos\phi_i \qquad (28.17)$$

The potential energy associated with the work done on the dipole is

$$\Delta U_{\text{dipole}} = U_f - U_i = -W_{\text{elec}}(i \rightarrow f) = -pE\cos\phi_f + pE\cos\phi_i \quad (28.18)$$

By comparing the left and right sides of Equation 28.18, we see that the potential energy of an electric dipole $\vec{p}$ in a uniform electric field $\vec{E}$ is

$$U_{\text{dipole}} = -pE\cos\phi = -\vec{p}\cdot\vec{E} \qquad (28.19)$$

FIGURE 28.15 shows the energy diagram of a dipole. The potential energy is minimum at $\phi = 0°$ where the dipole is aligned with the electric field. This is a point of stable equilibrium. A dipole exactly opposite $\vec{E}$, at $\phi = \pm 180°$, is at a point of unstable equilibrium. Any disturbance will cause it to flip around. A frictionless dipole with mechanical energy E_{mech} will oscillate back and forth between turning points on either side of $\phi = 0°$.

FIGURE 28.15 The energy of a dipole in an electric field.

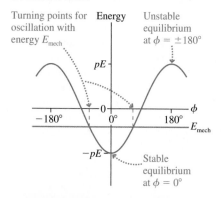

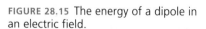

EXAMPLE 28.5 **Rotating a molecule**

The water molecule is a permanent electric dipole with dipole moment 6.2×10^{-30} C m. A water molecule is aligned in an electric field with field strength 1.0×10^7 N/C. How much energy is needed to rotate the molecule 90°?

MODEL The molecule is at the point of minimum energy. It won't spontaneously rotate 90°. However, an external force that supplies energy, such as a collision with another molecule, can cause the water molecule to rotate.

SOLVE The molecule starts at $\phi_i = 0°$ and ends at $\phi_f = 90°$. The increase in potential energy is

$$\Delta U_{\text{dipole}} = U_f - U_i = -pE\cos 90° - (-pE\cos 0°)$$
$$= pE = 6.2 \times 10^{-23}\,\text{J}$$

This is the energy needed to rotate the molecule 90°.

ASSESS ΔU_{dipole} is significantly less than $k_B T$ at room temperature. Thus collisions with other molecules can easily supply the energy to rotate the water molecules and keep them from staying aligned with the electric field.

This battery is labeled 1.5 Volts. As we'll soon see, a battery is a source of electric potential.

28.4 The Electric Potential

We introduced the concept of the *electric field* in Chapter 25 because action at a distance raised concerns and difficulties. The field provides an intermediary through which two charges exert forces on each other. Charge q_1 somehow alters the space around it by creating an electric field $\vec{E}_1$. Charge q_2 then responds to the field, experiencing force $\vec{F} = q_2\vec{E}_1$.

We face the same kinds of difficulties when we try to understand electric potential energy. For a mass on a spring, we can *see* how the energy is stored in the stretched or compressed spring. But when we say two charged particles have a potential energy, an energy that can be converted to a tangible kinetic energy of motion, *where is the energy?* It's indisputable that two positive charges fly apart when you release them, gaining kinetic energy, but there's no obvious place that the energy had been stored.

In defining the electric field, we chose to separate the charges that are the *source* of the field from the charge *in* the field. The force on charge q is related to the electric field of the source charges by

force on q by sources = [charge q] × [alteration of space by the source charges]

Let's try a similar procedure for the potential energy. The electric potential energy is due to the interaction of charge q with other charges, so let's write

potential energy of q + sources

= [charge q] × [*potential* for interaction of the source charges]

FIGURE 28.16 shows this idea schematically.

In analogy with the electric field, we will define the **electric potential** V (or, for brevity, just *the potential*) as

$$V \equiv \frac{U_{q+\text{sources}}}{q} \qquad (28.20)$$

Charge q is used as a probe to determine the electric potential, but the value of V is *independent of q.* **The electric potential, like the electric field, is a property of the source charges.**

In practice, we're usually more interested in knowing the potential energy if a charge q happens to be at a point in space where the electric potential of the source charges is V. Turning Equation 28.20 around, we see that the electric potential energy is

$$U_{q+\text{sources}} = qV \qquad (28.21)$$

Once the potential has been determined, it's very easy to find the potential energy.

The unit of electric potential is the joule per coulomb, which is called the **volt** V:

$$1 \text{ volt} = 1 \text{ V} \equiv 1 \text{ J/C}$$

This unit is named for Alessandro Volta, who invented the electric battery in the year 1800. Microvolts (μV), millivolts (mV), and kilovolts (kV) are commonly used units.

NOTE ▶ Once again, commonly used symbols are in conflict. The symbol V is widely used to represent *volume,* and now we're introducing the same symbol to represent *potential.* To make matters more confusing, V is the abbreviation for *volts.* In printed text, V for potential is italicized and V for volts is not, but you can't make such a distinction in handwritten work. This is not a pleasant state of affairs, but these are the commonly accepted symbols. It's incumbent upon you to be especially alert to the *context* in which a symbol is used. ◀

Using the Electric Potential

The electric potential is an abstract idea, and it will take some practice to see just what it means and how it is useful. We'll use multiple representations—words, pictures, graphs, and analogies—to explain and describe the electric potential.

NOTE ▶ It is unfortunate that the terms *potential* and *potential energy* are so much alike. Despite the similar names, they are very different concepts and are not interchangeable. Table 28.1 will help you to distinguish between the two. ◀

Basically, knowing the electric potential in a region of space allows us to determine whether a charged particle speeds up or slows down as it moves through that region. FIGURE 28.17 on the next page illustrates this idea. Here a group of source charges, which remains hidden offstage, has created an electric potential V that increases from left to right. A charged particle q, which for now we'll assume to be positive, has electric

FIGURE 28.16 Source charges alter the space around them by creating an electric potential.

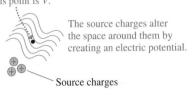

The potential at this point is V.

The source charges alter the space around them by creating an electric potential.

Source charges

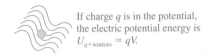

If charge q is in the potential, the electric potential energy is $U_{q+\text{sources}} = qV$.

TABLE 28.1 Distinguishing electric potential and potential energy

The *electric potential* is a property of the source charges and, as you'll soon see, is related to the electric field. The electric potential is present whether or not a charged particle is there to experience it. Potential is measured in J/C, or V.

The *electric potential energy* is the interaction energy of a charged particle with the source charges. Potential energy is measured in J.

FIGURE 28.17 A charged particle speeds up or slows down as it moves through a potential difference.

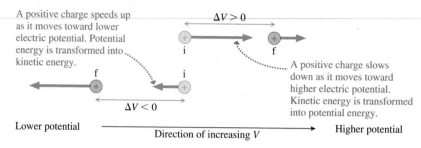

A positive charge speeds up as it moves toward lower electric potential. Potential energy is transformed into kinetic energy.

$\Delta V > 0$

A positive charge slows down as it moves toward higher electric potential. Kinetic energy is transformed into potential energy.

$\Delta V < 0$

Lower potential   Direction of increasing V Higher potential

potential energy $U = qV$. If the particle moves to the right, its potential energy increases and so, by energy conservation, its kinetic energy must decrease. **A positive charge slows down as it moves into a region of higher electric potential.**

It is customary to say that the particle moves through a **potential difference** $\Delta V = V_f - V_i$. The potential difference between two points is often called the **voltage.** The particle moving to the right moves through a positive potential difference ($\Delta V > 0$ because $V_f > V_i$), so we can say that a positively charged particle slows down as it moves through a positive potential difference.

The particle moving to the left in Figure 28.17 travels in the direction of decreasing electric potential—through a negative potential difference—and is losing potential energy. It speeds up as it transforms potential energy into kinetic energy. A negatively charged particle would slow down because its potential energy qV would increase as V decreases. Table 28.2 summarizes these ideas.

If a particle moves through a potential difference ΔV, its electric potential energy changes by $\Delta U = q\,\Delta V$. We can write the conservation of energy equation in terms of the electric potential as $\Delta K + \Delta U = \Delta K + q\,\Delta V = 0$ or, as is often more practical,

$$K_f + qV_f = K_i + qV_i \qquad (28.22)$$

Conservation of energy is the basis of a powerful problem-solving strategy.

TABLE 28.2 Charged particles moving in an electric potential

| | Electric potential | |
	Increasing ($\Delta V > 0$)	Decreasing ($\Delta V < 0$)
+ charge	Slows down	Speeds up
− charge	Speeds up	Slows down

PROBLEM-SOLVING STRATEGY 28.1   **Conservation of energy in charge interactions**

MODEL Check whether there are any dissipative forces that would keep the mechanical energy from being conserved.

VISUALIZE Draw a before-and-after pictorial representation. Define symbols that will be used in the problem, list known values, and identify what you're trying to find.

SOLVE The mathematical representation is based on the law of conservation of mechanical energy:

$$K_f + qV_f = K_i + qV_i$$

■ Is the electric potential given in the problem statement? If not, you'll need to use a known potential, such as that of a point charge, or calculate the potential using the procedure given later, in Problem-Solving Strategy 28.2.
■ K_i and K_f are the sums of the kinetic energies of all moving particles.
■ Some problems may need additional conservation laws, such as conservation of charge or conservation of momentum.

ASSESS Check that your result has the correct units, is reasonable, and answers the question.

Exercise 22

EXAMPLE 28.6   **Moving through a potential difference**

A proton with a speed of 2.0×10^5 m/s enters a region of space in which source charges have created an electric potential. What is the proton's speed after it moves through a potential difference of 100 V? What will be the final speed if the proton is replaced by an electron?

MODEL Energy is conserved. The electric potential determines the potential energy.

VISUALIZE FIGURE 28.18 is a before-and-after pictorial representation of a charged particle moving through a potential difference. A positive charge *slows down* as it moves into a region of higher potential ($K \rightarrow U$). A negative charge *speeds up* ($U \rightarrow K$).

FIGURE 28.18 A charged particle moving through a potential difference.

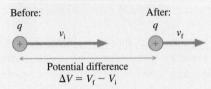

Before:

After:

Potential difference
$\Delta V = V_f - V_i$

SOLVE The potential energy of charge q is $U = qV$. Conservation of energy, now expressed in terms of the electric potential V, is $K_f + qV_f = K_i + qV_i$, or

$$K_f = K_i - q\,\Delta V$$

where $\Delta V = V_f - V_i$ is the potential difference through which the particle moves. In terms of the speeds, energy conservation is

$$\frac{1}{2}mv_f^2 = \frac{1}{2}mv_i^2 - q\,\Delta V$$

We can solve this for the final speed:

$$v_f = \sqrt{v_i^2 - \frac{2q}{m}\Delta V}$$

For a proton, with $q = e$, the final speed is

$$(v_f)_p = \sqrt{(2.0 \times 10^5 \text{ m/s})^2 - \frac{2(1.60 \times 10^{-19} \text{ C})(100 \text{ V})}{1.67 \times 10^{-27} \text{ kg}}}$$

$$= 1.4 \times 10^5 \text{ m/s}$$

An electron, though, with $q = -e$ and a different mass, speeds up to $(v_f)_e = 5.9 \times 10^6$ m/s.

ASSESS The electric potential *already existed* in space due to other charges that are not explicitly seen in the problem. The electron and proton have nothing to do with creating the potential. Instead, they *respond* to the potential by having potential energy $U = qV$.

STOP TO THINK 28.3   A proton is released from rest at point B, where the potential is 0 V. Afterward, the proton

a. Remains at rest at B.
b. Moves toward A with a steady speed.
c. Moves toward A with an increasing speed.
d. Moves toward C with a steady speed.
e. Moves toward C with an increasing speed.

-100 V   0 V $+100$ V

A•   B•   C•

28.5 The Electric Potential Inside a Parallel-Plate Capacitor

We began this chapter with the potential energy of a charge inside a parallel-plate capacitor. Now let's investigate the electric potential. FIGURE 28.19 shows two parallel electrodes, separated by distance d, with surface charge density $\pm \eta$. As a specific example, we'll let $d = 3.00$ mm and $\eta = 4.42 \times 10^{-9}$ C/m^2. The electric field inside the capacitor, as you learned in Chapter 26, is

$$\vec{E} = \left(\frac{\eta}{\epsilon_0}, \text{ from positive toward negative} \right)$$

$$= (500 \text{ N/C, from right to left})$$

(28.23)

This electric field is due to the *source charges* on the capacitor plates.

FIGURE 28.19 A parallel-plate capacitor.

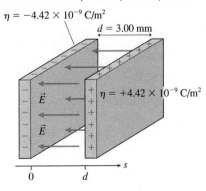

$\eta = -4.42 \times 10^{-9}$ C/m^2

$d = 3.00$ mm

$\eta = +4.42 \times 10^{-9}$ C/m^2

In Section 28.1, we found that the electric potential energy of a charge q in the uniform electric field of a parallel-plate capacitor is

$$U_{elec} = U_{q+sources} = qEs \qquad (28.24)$$

We've set the constant term U_0 to zero. U_{elec} is the energy of q interacting with the source charges on the capacitor plates.

Our new view of the interaction is to separate the role of charge q from the role of the source charges by defining the electric potential $V = U_{q+sources}/q$. Thus the electric potential inside a parallel-plate capacitor is

$$V = Es \qquad \text{(electric potential inside a parallel-plate capacitor)} \qquad (28.25)$$

where s **is the distance from the *negative* electrode.** The electric potential, like the electric field, exists at *all points* inside the capacitor. The electric potential is created by the source charges on the capacitor plates and exists whether or not charge q is inside the capacitor.

FIGURE 28.20 illustrates the important point that the electric potential increases linearly from the negative plate, where $V_- = 0$, to the positive plate, where $V_+ = Ed$. Let's define the *potential difference* ΔV_C between the two capacitor plates to be

$$\Delta V_C = V_+ - V_- = Ed \qquad (28.26)$$

In our specific example, $\Delta V_C = (500 \text{ N/C})(0.0030 \text{ m}) = 1.5 \text{ V}$. The units work out because $1.5 \text{ (N m)/C} = 1.5 \text{ J/C} = 1.5 \text{ V}$.

NOTE ▶ People who work with circuits would call ΔV_C "the voltage across the capacitor" or simply "the capacitor voltage." ◀

Equation 28.26 has an interesting implication. Thus far, we've determined the electric field inside a capacitor by specifying the surface charge density η on the plates. Alternatively, we could specify the capacitor voltage ΔV_C (i.e., the potential difference between the capacitor plates) and then determine the electric field strength as

$$E = \frac{\Delta V_C}{d} \qquad (28.27)$$

In fact, this is how E is determined in practical applications because it's easy to measure ΔV_C with a voltmeter but difficult, in practice, to know the value of η.

Equation 28.27 implies that the units of electric field are volts per meter, or V/m. We have been using electric field units of newtons per coulomb. In fact, as you can show as a homework problem, these units are equivalent to each other. That is,

$$1 \text{ N/C} = 1 \text{ V/m}$$

NOTE ▶ Volts per meter are the electric field units used by scientists and engineers in practice. We will now adopt them as our standard electric field units. ◀

Returning to the electric potential, we can substitute Equation 28.27 for E into Equation 28.25 for V. Thus the electric potential inside the capacitor is

$$V = Es = \frac{s}{d} \Delta V_C \qquad (28.28)$$

The potential increases linearly from $V_- = 0 \text{ V}$ at the negative plate ($s = 0$) to $V_+ = \Delta V_C$ at the positive plate ($s = d$).

Let's explore the electric potential inside the capacitor by looking at several different, but related, ways that the potential can be represented graphically.

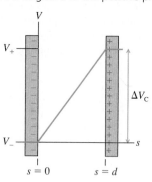

FIGURE 28.20 The electric potential of a parallel-plate capacitor increases linearly from the negative to the positive plate.

Graphical representations of the electric potential inside a capacitor

A graph of potential versus *s*. You can see the potential increasing from 0.0 V at the negative plate to 1.5 V at the positive plate.	A three-dimensional view showing **equipotential surfaces.** These are mathematical surfaces, not physical surfaces, with the same value of *V* at every point. The equipotential surfaces of a capacitor are planes parallel to the capacitor plates. The capacitor plates are also equipotential surfaces.	A two-dimensional **contour map.** The capacitor plates and the equipotential surfaces are seen edge-on, so you need to imagine them extending above and below the plane of the page.	A three-dimensional **elevation graph.** The potential is graphed vertically versus the *s*-coordinate on one axis and a generalized "*yz*-coordinate" on the other axis. Viewing the right face of the elevation graph gives you the potential graph.

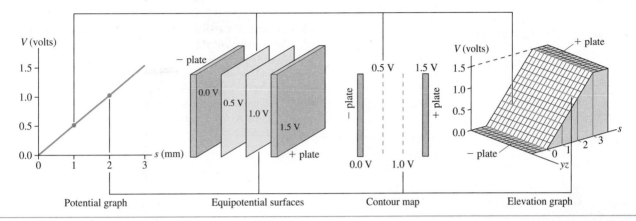

Potential graph   Equipotential surfaces   Contour map   Elevation graph

These four graphical representations show the same information from different perspectives, and the connecting lines help you see how they are related. If you think of the elevation graph as a "mountain," then the contour lines on the contour map are like the lines of a topographic map.

The potential graph and the contour map are the two representations most widely used in practice because they are easy to draw. Their limitation is that they are trying to convey three-dimensional information in a two-dimensional presentation. When you see graphs or contour maps, you need to imagine the three-dimensional equipotential surfaces or the three-dimensional elevation graph.

There's nothing special about showing equipotential surfaces or contour lines every 0.5 V. We chose these intervals because they were convenient. As an alternative, FIGURE 28.21 shows how the contour map looks if the contour lines are spaced every 0.3 V. Contour lines and equipotential surfaces are *imaginary* lines and surfaces drawn to help us visualize how the potential changes in space. Drawing the map more than one way reinforces the idea that there is an electric potential at *every* point inside the capacitor, not just at the points where we happened to draw a contour line or an equipotential surface.

Figure 28.21 also shows the electric field vectors. Notice that

- The electric field vectors are perpendicular to the equipotential surfaces.
- The electric field points in the direction of decreasing potential. In other words, the electric field points "downhill" on a graph or map of the electric potential.

Chapter 29 will present a more in-depth exploration of the connection between the electric field and the electric potential. There you will find that these observations are always true. They are not unique to the parallel-plate capacitor.

Finally, you might wonder how we can arrange a capacitor to have a surface charge density of precisely 4.42×10^{-9} C/m^2. Simple! As FIGURE 28.22 shows, we use wires to attach the capacitor plates to a 1.5 V battery. This is another topic that we'll explore in Chapter 29, but it's worth noting now that **a battery is a source of potential.** That's why batteries are labeled in volts, and it's a major reason we need to thoroughly understand the concept of potential.

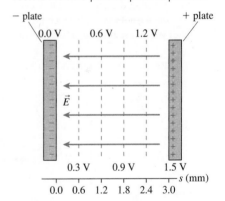

FIGURE 28.21 The contour lines of the electric potential and the electric field vectors inside a parallel-plate capacitor.

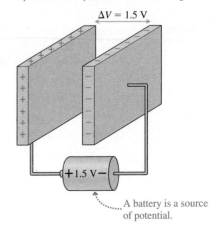

FIGURE 28.22 Using a battery to charge a capacitor to a precise value of ΔV_C.

EXAMPLE 28.7 **Measuring the speed of a proton**

The lab in which you work has a small proton accelerator. You've been assigned the task of measuring the speed of the protons as they emerge from the accelerator. To do so, you decide to measure how much voltage is needed across a parallel-plate capacitor to stop the protons. The capacitor you choose has a 2.0 mm plate separation and a small hole in one plate that you shoot the protons through. By filling the space between the plates with a low-density gas, you can see (with a microscope) a slight glow from the region where the protons collide with and excite the gas molecules. The width of the glow tells you how far the protons travel before being stopped and reversing direction. Varying the voltage across the capacitor gives the following data:

Capacitor voltage (V)	Glow width (mm)
1000	1.7
1250	1.3
1500	1.1
1750	1.0
2000	0.8

What value will you report for the speed of the protons?

MODEL Energy is conserved. The proton's potential energy can be found from the capacitor's electric potential.

VISUALIZE FIGURE 28.23 shows a before-and-after pictorial representation of the proton entering the capacitor with speed v_i, which we want to find, and later reaching a turning point with $v_f = 0$ m/s after traveling distance s_f = glow width. For the protons to slow and stop, the hole through which they pass has to be in the negative plate. We've established an s-axis with $s = 0$ at this point.

FIGURE 28.23 A proton being stopped in a capacitor.

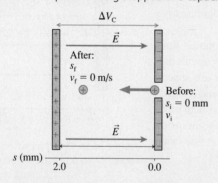

SOLVE The conservation of energy equation, with the proton having charge $q = e$, is $K_f + eV_f = K_i + eV_i$. The initial potential energy is zero, because the capacitor's electric potential is zero at

$s_i = 0$, and the final kinetic energy is zero. Using Equation 28.28 for the potential inside the capacitor, we have

$$eV_f = e\left(\frac{s_f}{d}\Delta V_C\right) = K_i = \frac{1}{2}mv_i^2$$

Solving for the distance traveled, we find

$$s_f = \frac{dmv_i^2}{2e}\frac{1}{\Delta V_C}$$

Thus a graph of the distance traveled versus the *inverse* of the capacitor voltage should be a straight line with zero y-intercept and slope $dmv_i^2/2e$. We can use the experimentally determined slope to find the proton speed.

FIGURE 28.24 is a graph of s_f versus $1/\Delta V_C$. It has the expected shape, and the slope of the best-fit line is seen to be 1.72 V m. The units are those of the rise-over-run. Using the slope, we calculate the proton speed:

$$v_i = \sqrt{\frac{2e}{dm} \times \text{slope}} = \sqrt{\frac{2(1.60 \times 10^{-19}\,\text{C})(1.72\,\text{V m})}{(0.0020\,\text{m})(1.67 \times 10^{-27}\,\text{kg})}}$$

$$= 4.1 \times 10^5\,\text{m/s}$$

FIGURE 28.24 A graph of the data.

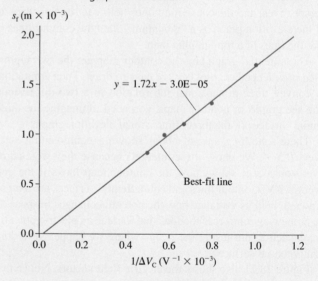

ASSESS This would be a very high speed for a macroscopic object but quite typical of the speeds of charged particles.

In writing the electric potential inside a parallel-plate capacitor, we made the choice that $V_- = 0$ V at the negative plate. But that is not the only possible choice. FIGURE 28.25 shows three parallel-plate capacitors, each having the same capacitor voltage $\Delta V_C = V_+ - V_- = 100$ V, but each with a different choice for the location of the zero point of the electric potential. Notice the *terminal symbols* (lines with small circles at the end) showing how the potential, from a battery or a power supply, is applied to each plate; these symbols are common in electronics.

FIGURE 28.25 These three choices for $V = 0$ represent the same physical situation. These are contour maps, showing the edges of the equipotential surfaces.

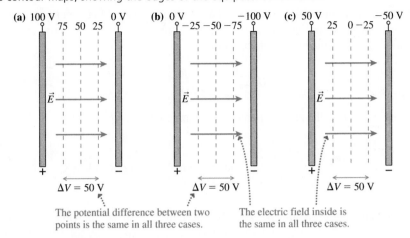

The potential difference between two points is the same in all three cases.

The electric field inside is the same in all three cases.

The important thing to notice is that the three contour maps in Figure 28.25 represent the *same physical situation*. The potential difference between any two points is the same in all three maps. The electric field is the same in all three. We may *prefer* one of these figures over the others, but there is no measurable physical difference between them.

EXAMPLE 28.8   The force on an ion

Example 26.7 noted that a cell wall can be modeled as a parallel-plate capacitor, with the outer surface of the cell wall being positive while the inner surface is negative. The potential difference between the inside of the cell and the outside is called the *membrane potential*. Suppose a molecular ion with charge $5e$ is embedded within the 5.0-nm-thick wall of a cell with a membrane potential of -70 mV, typical for a nerve cell in its resting state. What is the force on the molecular ion?

MODEL Model the cell wall as a parallel-plate capacitor with the inner surface being the negative plate. Although the walls are actually curved, and not large flat planes, the parallel-plate approximation is valid if the wall thickness is much less than the radius of the cell. The capacitor voltage is $\Delta V_C = 70$ mV $= 0.070$ V. The membrane potential is negative because the potential inside the cell is less than the potential outside, but ΔV_C, the capacitor voltage, is the *magnitude* of the potential difference and thus always positive.

SOLVE The force on a charged particle is $\vec{F} = q\vec{E}$. The electric field strength inside the parallel-plate capacitor of the cell wall is

$$E = \frac{\Delta V_C}{d} = \frac{0.070 \text{ V}}{5.0 \times 10^{-9} \text{ m}} = 1.4 \times 10^7 \text{ V/m}$$

Notice that we're now using V/m rather than N/C as the units of electric field. Because the field points from positive to negative, the field vector is $\vec{E} = (1.4 \times 10^7$ V/m, toward inside). Thus the force on an ion with $q = 5e = 8.0 \times 10^{-19}$ C is

$$\vec{F} = q\vec{E} = (1.1 \times 10^{-11} \text{ N, toward inside})$$

ASSESS For cells to function, a steady flow of molecules must pass back and forth through the cell wall. Although the details of how this happens are very complex, a key idea is that a potential difference between the inside and outside of the cell creates an electric field that pushes positive ions toward the inside, negative ions toward the outside.

STOP TO THINK 28.4 Rank in order, from largest to smallest, the potentials V_a to V_e at the points a to e.

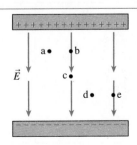

28.6 The Electric Potential of a Point Charge

FIGURE 28.26 Measuring the electric potential of charge q.

To determine the potential of q at this point . . .

q

q'

r

. . . place charge q' at the point as a probe and measure the potential energy $U_{q'+q}$.

q

Another important electric potential is that of a point charge. Let q in FIGURE 28.26 be the source charge, and let a second charge q' probe the electric potential of q. The potential energy of the two point charges is

$$U_{q'+q} = \frac{1}{4\pi\epsilon_0}\frac{qq'}{r} \qquad (28.29)$$

Thus, by definition, the electric potential of charge q is

$$V = \frac{U_{q'+q}}{q'} = \frac{1}{4\pi\epsilon_0}\frac{q}{r} \qquad \text{(electric potential of a point charge)} \quad (28.30)$$

The potential of Equation 28.30 extends through all of space, showing the influence of charge q, but it weakens with distance as $1/r$. This expression for V assumes that we have chosen $V = 0$ V to be at $r = \infty$. This is the most logical choice for a point charge because the influence of charge q ends at infinity.

The expression for the electric potential of charge q is similar to that for the electric field of charge q. The difference most quickly seen is that V depends on $1/r$ whereas $\vec{E}$ depends on $1/r^2$. But it is also important to notice that **the potential is a scalar** whereas the field is a vector. Thus the mathematics of using the potential are much easier than the vector mathematics using the electric field requires.

EXAMPLE 28.9 **Calculating the potential of a point charge**

What is the electric potential 1.0 cm from a +1.0 nC charge? What is the potential difference between a point 1.0 cm away and a second point 3.0 cm away?

SOLVE The potential at $r = 1.0$ cm is

$$V_{1\,cm} = \frac{1}{4\pi\epsilon_0}\frac{q}{r} = (9.0\times10^9\,\text{N}\,\text{m}^2/\text{C}^2)\frac{1.0\times10^{-9}\,\text{C}}{0.010\,\text{m}}$$

$$= 900\,\text{V}$$

We can similarly calculate $V_{3\,cm} = 300$ V. Thus the potential difference between these two points is $\Delta V = V_{1\,cm} - V_{3\,cm} = 600$ V.

ASSESS 1 nC is typical of the electrostatic charge produced by rubbing, and you can see that such a charge creates a fairly large potential nearby. Why are we not shocked and injured when working with the "high voltages" of such charges? The sensation of being shocked is a result of current, not potential. Some high-potential sources simply do not have the ability to generate much current. We will look at this issue in Chapter 31.

Visualizing the Potential of a Point Charge

FIGURE 28.27 shows four graphical representations of the electric potential of a point charge. These match the four representations of the electric potential inside a capacitor, and a comparison of the two is worthwhile. This figure assumes that q is positive; you may want to think about how the representations would change if q were negative.

FIGURE 28.27 Four graphical representations of the electric potential of a point charge.

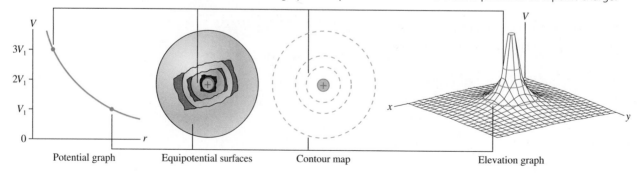

Potential graph   Equipotential surfaces   Contour map   Elevation graph

STOP TO THINK 28.5   Rank in order, from largest to smallest, the potential differences ΔV_{ab}, ΔV_{ac}, and ΔV_{bc} between points a and b, points a and c, and points b and c.

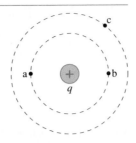

The Electric Potential of a Charged Sphere

In practice, you are more likely to work with a charged sphere, of radius R and total charge Q, than with a point charge. Outside a uniformly charged sphere, the electric potential is identical to that of a point charge Q at the center. That is,

$$V = \frac{1}{4\pi\epsilon_0}\frac{Q}{r} \qquad \text{(sphere of charge, } r \geq R\text{)} \qquad (28.31)$$

We can cast this result in a more useful form. It is customary to speak of charging an electrode, such as a sphere, "to" a certain potential, as in "Bob charged the sphere to a potential of 3000 volts." This potential, which we will call V_0, is the potential right on the surface of the sphere. We can see from Equation 28.31 that

$$V_0 = V(\text{at } r = R) = \frac{Q}{4\pi\epsilon_0 R} \qquad (28.32)$$

Consequently, a sphere of radius R that is charged to potential V_0 has total charge

$$Q = 4\pi\epsilon_0 R V_0 \qquad (28.33)$$

If we substitute this expression for Q into Equation 28.31, we can write the potential outside a sphere that is charged to potential V_0 as

$$V = \frac{R}{r}V_0 \qquad \text{(sphere charged to potential } V_0\text{)} \qquad (28.34)$$

Equation 28.34 tells us that the potential of a sphere is V_0 on the surface and decreases inversely with the distance. The potential at $r = 3R$ is $\frac{1}{3}V_0$.

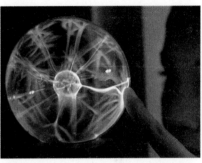

A *plasma ball* consists of a small metal ball charged to a potential of about 2000 V inside a hollow glass sphere. The glass sphere is filled with gas—typically neon or argon because of the colors they produce—at a pressure of about 0.01 atm. The electric field of the high-voltage ball is sufficient to cause a gas breakdown at this pressure, creating "lightning bolts" between the ball and the glass sphere.

EXAMPLE 28.10   **A proton and a charged sphere**

A proton is released from rest at the surface of a 1.0-cm-diameter sphere that has been charged to $+1000$ V.

a. What is the charge of the sphere?
b. What is the proton's speed at 1.0 cm from the sphere?

MODEL Energy is conserved. The potential outside the charged sphere is the same as the potential of a point charge at the center.

VISUALIZE **FIGURE 28.28** shows the situation.

FIGURE 28.28 A sphere and a proton.

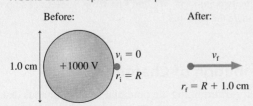

SOLVE a. The charge of the sphere is

$$Q = 4\pi\epsilon_0 R V_0 = 0.56 \times 10^{-9}\,\text{C} = 0.56\,\text{nC}$$

b. A sphere charged to $V_0 = +1000$ V is positively charged. The proton will be repelled by this charge and move away from the sphere. The conservation of energy equation $K_f + eV_f = K_i + eV_i$, with Equation 28.34 for the potential of a sphere, is

$$\frac{1}{2}mv_f^2 + \frac{eR}{r_f}V_0 = \frac{1}{2}mv_i^2 + \frac{eR}{r_i}V_0$$

The proton starts from the surface of the sphere, $r_i = R$, with $v_i = 0$. When the proton is 1.0 cm from the *surface* of the sphere, it has $r_f = 1.0$ cm $+ R = 1.5$ cm. Using these, we can solve for v_f:

$$v_f = \sqrt{\frac{2eV_0}{m}\left(1 - \frac{R}{r_f}\right)} = 3.6 \times 10^5\,\text{m/s}$$

ASSESS This example illustrates how the ideas of electric potential and potential energy work together, yet they are *not* the same thing.

28.7 The Electric Potential of Many Charges

Suppose there are many source charges q_1, q_2, ... The electric potential V at a point in space is the sum of the potentials due to each charge:

$$V = \sum_i \frac{1}{4\pi\epsilon_0} \frac{q_i}{r_i} \qquad (28.35)$$

where r_i is the distance from charge q_i to the point in space where the potential is being calculated. In other words, **the electric potential, like the electric field, obeys the principle of superposition.**

As an example, the contour map and elevation graph in FIGURE 28.29 show that the potential of an electric dipole is the sum of the potentials of the positive and negative charges. Potentials such as these have many practical applications. For example, electrical activity within the body can be monitored by measuring equipotential lines on the skin. Figure 28.29c shows that the equipotentials near the heart are a slightly distorted but recognizable electric dipole.

FIGURE 28.29 The electric potential of an electric dipole.

(a) Contour map

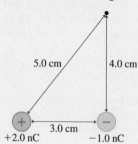

Equipotential surfaces

(b) Elevation graph

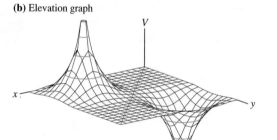

(c)

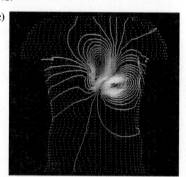

Equipotentials on the chest of a human are a slightly distorted electric dipole.

EXAMPLE 28.11 **The potential of two charges**

What is the electric potential at the point indicated in FIGURE 28.30?

FIGURE 28.30 Finding the potential of two charges.

5.0 cm   4.0 cm

+2.0 nC   3.0 cm   −1.0 nC

MODEL The potential is the sum of the potentials due to each charge.

SOLVE The potential at the indicated point is

$$V = \frac{1}{4\pi\epsilon_0} \frac{q_1}{r_1} + \frac{1}{4\pi\epsilon_0} \frac{q_2}{r_2}$$

$$= (9.0 \times 10^9 \text{ N m}^2/\text{C}^2)\left(\frac{2.0 \times 10^{-9} \text{ C}}{0.050 \text{ m}} + \frac{-1.0 \times 10^{-9} \text{ C}}{0.040 \text{ m}}\right)$$

$$= 135 \text{ V}$$

ASSESS The potential is a *scalar*, so we found the net potential by adding two numbers. We don't need any angles or components to calculate the potential.

A Continuous Distribution of Charge

Equation 28.35 is the basis for determining the potential of a continuous distribution of charge, such as a charged rod or a charged disk. The procedure is much like the one you learned in Chapter 26 for calculating the electric field of a continuous distribution of charge, but *easier* because the potential is a scalar. We will continue to assume that the object is *uniformly charged,* meaning that the charges are evenly spaced over the object.

The electric potential of a continuous distribution of charge

MODEL Model the charges as a simple shape, such as a line or a disk. Assume the charge is uniformly distributed.

VISUALIZE For the pictorial representation:

❶ Draw a picture and establish a coordinate system.
❷ Identify the point P at which you want to calculate the electric potential.
❸ Divide the total charge Q into small pieces of charge ΔQ, using shapes for which you *already know* how to determine V. This division is often, but not always, into point charges.
❹ Identify distances that need to be calculated.

SOLVE The mathematical representation is $V = \sum V_i$.

■ Use superposition to form an algebraic expression for the potential at P.
■ Let the (x, y, z) coordinates of the point remain as variables.
■ Replace the small charge ΔQ with an equivalent expression involving a *charge density* and a *coordinate,* such as dx, that describes the shape of charge ΔQ. **This is the critical step in making the transition from a sum to an integral** because you need a coordinate to serve as the integration variable.
■ All distances must be expressed in terms of the coordinates.
■ Let the sum become an integral. The integration will be over the coordinate variable that is related to ΔQ. The integration limits for this variable will depend on the coordinate system you have chosen. Carry out the integration and simplify the result.

ASSESS Check that your result is consistent with any limits for which you know what the potential should be.

Exercise 29

EXAMPLE 28.12 **The potential of a ring of charge**

A thin, uniformly charged ring of radius R has total charge Q. Find the potential at distance z on the axis of the ring.

MODEL Because the ring is thin, we'll assume the charge lies along a circle of radius R.

VISUALIZE FIGURE 28.31 illustrates the four steps of the problem-solving strategy. We've chosen a coordinate system in which the ring lies in the xy-plane and point P is on the z-axis. We've then divided the ring into N small segments of charge ΔQ, each of which can be modeled as a point charge. The distance r_i between segment i and point P is

$$r_i = \sqrt{R^2 + z^2}$$

Note that r_i is a constant distance, the same for every charge segment.

SOLVE The potential V at P is the sum of the potentials due to each segment of charge:

$$V = \sum_{i=1}^{N} V_i = \sum_{i=1}^{N} \frac{1}{4\pi\epsilon_0} \frac{\Delta Q}{r_i} = \frac{1}{4\pi\epsilon_0} \frac{1}{\sqrt{R^2 + z^2}} \sum_{i=1}^{N} \Delta Q$$

FIGURE 28.31 Finding the potential of a ring of charge.

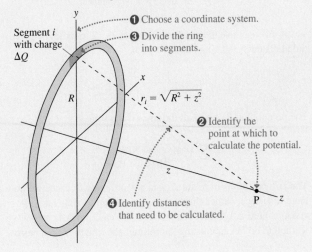

❶ Choose a coordinate system.
❸ Divide the ring into segments.
Segment i with charge ΔQ
$r_i = \sqrt{R^2 + z^2}$
❷ Identify the point at which to calculate the potential.
❹ Identify distances that need to be calculated.

Continued

We were able to bring all terms involving z to the front because z is a constant as far as the summation is concerned. Surprisingly, we don't need to convert the sum to an integral to complete this calculation. The sum of all the ΔQ charge segments around the ring is simply the ring's total charge, $\sum(\Delta Q) = Q$; hence the electric potential on the axis of a charged ring is

$$V_{\text{ring on axis}} = \frac{1}{4\pi\epsilon_0}\frac{Q}{\sqrt{R^2 + z^2}}$$

ASSESS From far away, the ring appears as a point charge Q in the distance. Thus we expect the potential of the ring to be that of a point charge when $z \gg R$. You can see that $V_{\text{ring}} \approx Q/4\pi\epsilon_0 z$ when $z \gg R$, which is, indeed, the potential of a point charge Q.

CHALLENGE EXAMPLE 28.13 | **The potential of a charged dime**

A 17.5-mm-diameter dime is charged to +5.00 nC.

a. What is the potential of the dime?

b. What is the potential energy of an electron 1.00 cm above the dime?

MODEL Model the dime as a thin, uniformly charged disk of radius R and charge Q. The disk has uniform surface charge density $\eta = Q/A = Q/\pi R^2$. We can take advantage of now knowing the on-axis potential of a ring of charge.

VISUALIZE Orient the disk in the xy-plane, as shown in FIGURE 28.32, with point P at distance z. Then divide the disk into *rings* of equal width Δr. Ring i has radius r_i and charge ΔQ_i.

FIGURE 28.32 Finding the potential of a disk of charge.

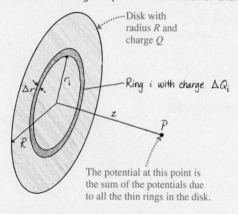

Disk with radius R and charge Q

Ring i with charge ΔQ_i

The potential at this point is the sum of the potentials due to all the thin rings in the disk.

SOLVE We can use the result of Example 28.12 to write the potential at distance z of ring i as

$$V_i = \frac{1}{4\pi\epsilon_0}\frac{\Delta Q_i}{\sqrt{r_i^2 + z^2}}$$

The potential at P due to all the rings is the sum

$$V = \sum_i V_i = \frac{1}{4\pi\epsilon_0}\sum_{i=1}^{N}\frac{\Delta Q_i}{\sqrt{r_i^2 + z^2}}$$

The critical step is to relate ΔQ_i to a coordinate. Because we now have a surface, rather than a line, the charge in ring i is $\Delta Q_i = \eta \Delta A_i$, where ΔA_i is the area of ring i. We can find ΔA_i, as you've learned to do in calculus, by "unrolling" the ring to form a narrow

rectangle of length $2\pi r_i$ and height Δr. Thus the area of ring i is $\Delta A_i = 2\pi r_i \Delta r$ and the charge is

$$\Delta Q_i = \eta \Delta A_i = \frac{Q}{\pi R^2}2\pi r_i \Delta r = \frac{2Q}{R^2}r_i \Delta r$$

With this substitution, the potential at P is

$$V = \frac{1}{4\pi\epsilon_0}\sum_{i=1}^{N}\frac{2Q}{R^2}\frac{r_i \Delta r_i}{\sqrt{r_i^2 + z^2}} \to \frac{Q}{2\pi\epsilon_0 R^2}\int_0^R \frac{r\,dr}{\sqrt{r^2 + z^2}}$$

where, in the last step, we let $N \to \infty$ and the sum become an integral. This integral can be found in Appendix A, but it's not hard to evaluate with a change of variables. Let $u = r^2 + z^2$, in which case $r\,dr = \frac{1}{2}du$. Changing variables requires that we also change the integration limits. You can see that $u = z^2$ when $r = 0$, and $u = R^2 + z^2$ when $r = R$. With these changes, the on-axis potential of a charged disk is

$$V_{\text{disk on axis}} = \frac{Q}{2\pi\epsilon_0 R^2}\int_{z^2}^{R^2+z^2}\frac{\frac{1}{2}du}{u^{1/2}} = \frac{Q}{2\pi\epsilon_0 R^2}u^{1/2}\Big|_{z^2}^{R^2+z^2}$$

$$= \frac{Q}{2\pi\epsilon_0 R^2}\left(\sqrt{R^2 + z^2} - z\right)$$

We can find the potential V_0 of the disk itself by setting $z = 0$, giving $V_0 = Q/2\pi\epsilon_0 R$. In other words, placing charge Q on a disk of radius R charges it to potential V_0. The on-axis potential of the disk can be written in terms of V_0 as

$$V_{\text{disk on axis}} = V_0\left[\sqrt{1 + (z/R)^2} - (z/R)\right]$$

Now we can evaluate the case of the charged dime.

a. The potential of the dime is the potential of a disk at $z = 0$:

$$V_0 = \frac{Q}{2\pi\epsilon_0 R} = 10{,}300 \text{ V}$$

b. To calculate the potential energy $U = qV$ of charge q, we first need to determine the potential of the disk at $z = 1.0$ cm. This is

$$V = V_0\left[\sqrt{1 + (z/R)^2} - (z/R)\right] = 3870 \text{ V}$$

The electron's charge is $q = -e = -1.60 \times 10^{-19}$ C, so its potential energy at $z = 1.00$ cm is $U = qV = -6.19 \times 10^{-16}$ J.

ASSESS Although we had to go through a number of steps, this procedure is easier than evaluating the electric field because we do not have to worry about vector components.

SUMMARY

The goals of Chapter 28 have been to calculate and use the electric potential and electric potential energy.

General Principles

Sources of V

The electric potential, like the electric field, is created by charges.

Two major tools for calculating V are

- The potential of a point charge $V = \dfrac{1}{4\pi\epsilon_0}\dfrac{q}{r}$

- The principle of superposition

Multiple point charges

Use superposition: $V = V_1 + V_2 + V_3 + \cdots$

Continuous distribution of charge

- Divide the charge into point-like ΔQ.

- Find the potential of each ΔQ.

- Find V by summing the potentials of all ΔQ.

The summation usually becomes an integral. A critical step is replacing ΔQ with an expression involving a charge density and an integration coordinate. Calculating V is usually easier than calculating $\vec{E}$ because the potential is a scalar.

Consequences of V

A charged particle has potential energy

$$U = qV$$

at a point where source charges have created an electric potential V.

The electric force is a conservative force, so the mechanical energy is conserved for a charged particle in an electric potential:

$$K_{\mathrm{f}} + qV_{\mathrm{f}} = K_{\mathrm{i}} + qV_{\mathrm{i}}$$

The potential energy of **two point charges** separated by distance r is

$$U_{q_1 + q_2} = \frac{Kq_1q_2}{r} = \frac{1}{4\pi\epsilon_0}\frac{q_1q_2}{r}$$

The **zero point** of potential and potential energy is chosen to be convenient. For point charges, we let $U = 0$ when $r \to \infty$.

The potential energy in an electric field of an **electric dipole** with dipole moment $\vec{p}$ is

$$U_{\mathrm{dipole}} = -pE\cos\theta = -\vec{p}\cdot\vec{E}$$

Applications

Graphical representations of the potential:

Potential graph

Equipotential surfaces

Contour map

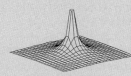
Elevation graph

Sphere of charge Q

Same as a point charge if $r \geq R$

Parallel-plate capacitor

$V = Es$, where s is measured from the negative plate. The electric field inside is

$$E = \frac{\Delta V_{\mathrm{C}}}{d}$$

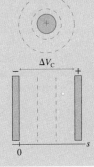

Units

Electric potential: $1\ \mathrm{V} = 1\ \mathrm{J/C}$

Electric field: $1\ \mathrm{V/m} = 1\ \mathrm{N/C}$

Terms and Notation

electric potential energy, U
electric potential, V
volt, V

potential difference, ΔV
voltage, ΔV
equipotential surface

contour map
elevation graph

CONCEPTUAL QUESTIONS

1. a. Charge q_1 is distance r from a positive point charge Q. Charge $q_2 = q_1/3$ is distance $2r$ from Q. What is the ratio U_1/U_2 of their potential energies due to their interactions with Q?
    b. Charge q_1 is distance s from the negative plate of a parallel-plate capacitor. Charge $q_2 = q_1/3$ is distance $2s$ from the negative plate. What is the ratio U_1/U_2 of their potential energies?

2. FIGURE Q28.2 shows the potential energy of a proton ($q = +e$) and a lead nucleus ($q = +82e$). The horizontal scale is in units of *femtometers*, where 1 fm $= 10^{-15}$ m.
    a. A proton is fired toward a lead nucleus from very far away. How much initial kinetic energy does the proton need to reach a turning point 10 fm from the nucleus? Explain.
    b. How much kinetic energy does the proton of part a have when it is 20 fm from the nucleus and moving toward it, before the collision?

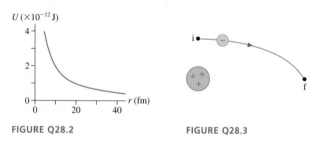

FIGURE Q28.2   FIGURE Q28.3

3. An electron moves along the trajectory of FIGURE Q28.3 from i to f.
    a. Does the electric potential energy increase, decrease, or stay the same? Explain.
    b. Is the electron's speed at f greater than, less than, or equal to its speed at i? Explain.

4. Two protons are launched with the same speed from point 1 inside the parallel-plate capacitor of FIGURE Q28.4. Points 2 and 3 are the same distance from the negative plate.
    a. Is $\Delta U_{1\rightarrow 2}$, the change in potential energy along the path $1 \rightarrow 2$, larger than, smaller than, or equal to $\Delta U_{1\rightarrow 3}$?
    b. Is the proton's speed v_2 at point 2 larger than, smaller than, or equal to v_3? Explain.

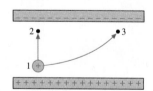

FIGURE Q28.4

5. Rank in order, from most positive to most negative, the potential energies U_a to U_f of the six electric dipoles in the uniform electric field of FIGURE Q28.5. Explain.

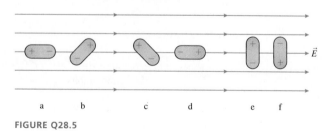

FIGURE Q28.5

6. FIGURE Q28.6 shows the electric potential along the x-axis.
    a. Draw a graph of the potential energy of a 0.1 C charged particle. Provide a numerical scale for both axes.
    b. If the charged particle is shot toward the right from $x = 1$ m with 1.0 J of kinetic energy, where is its turning point? Use your graph to explain.

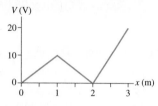

FIGURE Q28.6

7. A capacitor with plates separated by distance d is charged to a potential difference ΔV_C. All wires and batteries are disconnected, then the two plates are pulled apart (with insulated handles) to a new separation of distance $2d$.
    a. Does the capacitor charge Q change as the separation increases? If so, by what factor? If not, why not?
    b. Does the electric field strength E change as the separation increases? If so, by what factor? If not, why not?
    c. Does the potential difference ΔV_C change as the separation increases? If so, by what factor? If not, why not?

8. Rank in order, from largest to smallest, the electric potentials V_a to V_e at points a to e in FIGURE Q28.8. Explain.

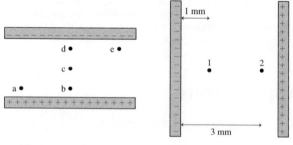

FIGURE Q28.8   FIGURE Q28.9

9. FIGURE Q28.9 shows two points inside a capacitor. Let $V = 0$ V at the negative plate.
    a. What is the ratio V_2/V_1 of the electric potentials? Explain.
    b. What is the ratio E_2/E_1 of the electric field strengths?

10. FIGURE Q28.10 shows two points near a positive point charge.
    a. What is the ratio V_2/V_1 of the electric potentials? Explain.
    b. What is the ratio E_2/E_1 of the electric field strengths?

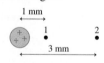

FIGURE Q28.10

11. FIGURE Q28.11 shows three points in the vicinity of two point charges. The charges have equal magnitudes. Rank in order, from most positive to most negative, the potentials V_a to V_c.

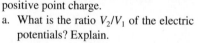

FIGURE Q28.11

12. Reproduce FIGURE Q28.12 on your paper. Then draw a dot (or dots) on the figure to show the position (or positions) at which the electric potential is zero.

FIGURE Q28.12

EXERCISES AND PROBLEMS

Problems labeled [ ] integrate material from earlier chapters.

Exercises

Section 28.1 Electric Potential Energy

1. ‖ The electric field strength is 50,000 N/C inside a parallel-plate capacitor with a 2.0 mm spacing. A proton is released from rest at the positive plate. What is the proton's speed when it reaches the negative plate?

2. ‖ The electric field strength is 20,000 N/C inside a parallel-plate capacitor with a 1.0 mm spacing. An electron is released from rest at the negative plate. What is the electron's speed when it reaches the positive plate?

3. ‖ A proton is released from rest at the positive plate of a parallel-plate capacitor. It crosses the capacitor and reaches the negative plate with a speed of 50,000 m/s. What will be the final speed of an electron released from rest at the negative plate?

4. | A proton is released from rest at the positive plate of a parallel-plate capacitor. It crosses the capacitor and reaches the negative plate with a speed of 50,000 m/s. The experiment is repeated with a He$^+$ ion (charge e, mass 4 u). What is the ion's speed at the negative plate?

Section 28.2 The Potential Energy of Point Charges

5. ‖ What is the electric potential energy of the proton in FIGURE EX28.5? The electrons are fixed and cannot move.

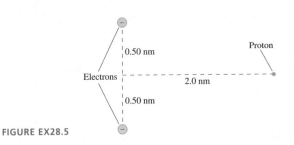

FIGURE EX28.5

6. ‖ What is the electric potential energy of the group of charges in FIGURE EX28.6?

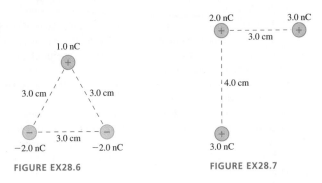

FIGURE EX28.6   FIGURE EX28.7

7. ‖ What is the electric potential energy of the group of charges in FIGURE EX28.7?

Section 28.3 The Potential Energy of a Dipole

8. | A water molecule perpendicular to an electric field has 1.0×10^{-21} J more potential energy than a water molecule aligned with the field. The dipole moment of a water molecule is 6.2×10^{-30} C m. What is the strength of the electric field?

9. | FIGURE EX28.9 shows the potential energy of an electric dipole. Consider a dipole that oscillates between ±60°.
   a. What is the dipole's mechanical energy?
   b. What is the dipole's kinetic energy when it is aligned with the electric field?

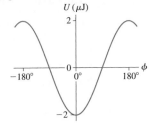

FIGURE EX28.9

Section 28.4 The Electric Potential

10. | What is the speed of a proton that has been accelerated from rest through a potential difference of -1000 V?

11. | What is the speed of an electron that has been accelerated from rest through a potential difference of 1000 V?

12. ‖ What potential difference is needed to accelerate an electron from rest to a speed of 2.0×10^6 m/s?

13. ‖ What potential difference is needed to accelerate a He$^+$ ion (charge $+e$, mass 4 u) from rest to a speed of 2.0×10^6 m/s?

14. | A proton with an initial speed of 800,000 m/s is brought to rest by an electric field.
   a. Did the proton move into a region of higher potential or lower potential?
   b. What was the potential difference that stopped the proton?

15. ‖ An electron with an initial speed of 500,000 m/s is brought to rest by an electric field.
   a. Did the electron move into a region of higher potential or lower potential?
   b. What was the potential difference that stopped the electron?

Section 28.5 The Electric Potential Inside a Parallel-Plate Capacitor

16. | Show that 1 V/m = 1 N/C.

17. | a. What is the potential of an ordinary AA or AAA battery? (If you're not sure, find one and look at the label.)
   b. An AA battery is connected to a parallel-plate capacitor having 4.0 cm × 4.0 cm plates spaced 1.0 mm apart. How much charge does the battery supply to each plate?

18. | Two 2.00 cm × 2.00 cm plates that form a parallel-plate capacitor are charged to ±0.708 nC. What are the electric field strength inside and the potential difference across the capacitor if the spacing between the plates is (a) 1.00 mm and (b) 2.00 mm?

19. | A 3.0-cm-diameter parallel-plate capacitor has a 2.0 mm spacing. The electric field strength inside the capacitor is 1.0×10^5 V/m.
   a. What is the potential difference across the capacitor?
   b. How much charge is on each plate?

20. ‖ Two 2.0-cm-diameter disks spaced 2.0 mm apart form a parallel-plate capacitor. The electric field between the disks is 5.0×10^5 V/m.
   a. What is the voltage across the capacitor?
   b. An electron is launched from the negative plate. It strikes the positive plate at a speed of 2.0×10^7 m/s. What was the electron's speed as it left the negative plate?

Section 28.6 The Electric Potential of a Point Charge

21. | a. What is the electric potential at points A, B, and C in FIGURE EX28.21?
   b. What are the potential differences ΔV_{AB} and ΔV_{BC}?

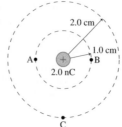

FIGURE EX28.21

22. ‖ A 1.0-mm-diameter ball bearing has 2.0×10^9 excess electrons. What is the ball bearing's potential?

23. | In a semiclassical model of the hydrogen atom, the electron orbits the proton at a distance of 0.053 nm.
   a. What is the electric potential of the proton at the position of the electron?
   b. What is the electron's potential energy?

Section 28.7 The Electric Potential of Many Charges

24. | What is the electric potential at the point indicated with the dot in FIGURE EX28.24?

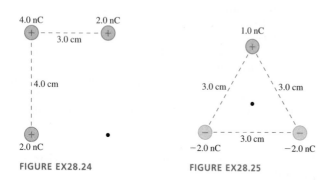

FIGURE EX28.24   FIGURE EX28.25

25. | What is the electric potential at the point indicated with the dot in FIGURE EX28.25?

26. ‖ The electric potential at the dot in FIGURE EX28.26 is 3140 V. What is charge q?

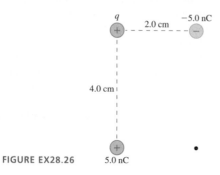

FIGURE EX28.26

27. ‖ A -2.0 nC charge and a $+2.0$ nC charge are located on the x-axis at $x = -1.0$ cm and $x = +1.0$ cm, respectively.
   a. Other than at infinity, is there a position or positions on the x-axis where the electric field is zero? If so, where?
   b. Other than at infinity, at what position or positions on the x-axis is the electric potential zero?
   c. Sketch graphs of the electric field strength and the electric potential along the x-axis.

28. ‖ Two point charges q_a and q_b are located on the x-axis at $x = a$ and $x = b$. FIGURE EX28.28 is a graph of E_x, the x-component of the electric field.
   a. What are the signs of q_a and q_b?
   b. What is the ratio $|q_a/q_b|$?
   c. Draw a graph of V, the electric potential, as a function of x.

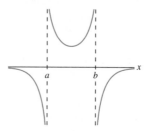

FIGURE EX28.28

29. ‖ Two point charges q_a and q_b are located on the x-axis at $x = a$ and $x = b$. FIGURE EX28.29 is a graph of V, the electric potential.
   a. What are the signs of q_a and q_b?
   b. What is the ratio $|q_a/q_b|$?
   c. Draw a graph of E_x, the x-component of the electric field, as a function of x.

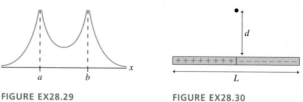

FIGURE EX28.29   FIGURE EX28.30

30. | The two halves of the rod in FIGURE EX28.30 are uniformly charged to $\pm Q$. What is the electric potential at the point indicated by the dot?

Problems

31. | Two positive point charges are 5.0 cm apart. If the electric potential energy is 72 μJ, what is the magnitude of the force between the two charges?

32. ‖ Two point charges 2.0 cm apart have an electric potential energy -180 μJ. The total charge is 30 nC. What are the two charges?

33. ‖ A -10.0 nC point charge and a $+20.0$ nC point charge are 15.0 cm apart on the x-axis.
   a. What is the electric potential at the point on the x-axis where the electric field is zero?
   b. What is the magnitude of the electric field at the point on the x-axis, between the charges, where the electric potential is zero?

34. ‖ A $+3.0$ nC charge is at $x = 0$ cm and a -1.0 nC charge is at $x = 4$ cm. At what point or points on the x-axis is the electric potential zero?

35. ‖ A -3.0 nC charge is on the x-axis at $x = -9$ cm and a $+4.0$ nC charge is on the x-axis at $x = 16$ cm. At what point or points on the y-axis is the electric potential zero?

36. ‖ Two small metal cubes with masses 2.0 g and 4.0 g are tied together by a 5.0-cm-long massless string and are at rest on a frictionless surface. Each is charged to $+2.0\ \mu C$.
   a. What is the energy of this system?
   b. What is the tension in the string?
   c. The string is cut. What is the speed of each cube when they are far apart?
   Hint: There are *two* conserved quantities. Make use of both.

37. ‖ The four 1.0 g spheres shown in **FIGURE P28.37** are released simultaneously and allowed to move away from each other. What is the speed of each sphere when they are very far apart?

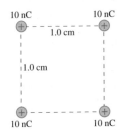

FIGURE P28.37

38. ‖ A proton's speed as it passes point A is 50,000 m/s. It follows the trajectory shown in **FIGURE P28.38**. What is the proton's speed at point B?

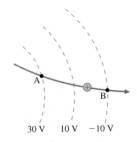

FIGURE P28.38   30 V   10 V   −10 V

39. ‖ Living cells "pump" singly ionized sodium ions, Na^+, from
BIO the inside of the cell to the outside to maintain a membrane potential $\Delta V_{membrane} = V_{in} - V_{out} = -70\ mV$. It is called *pumping* because work must be done to move a positive ion from the negative inside of the cell to the positive outside, and it must go on continuously because sodium ions "leak" back through the cell wall by diffusion.
   a. How much work must be done to move one sodium ion from the inside of the cell to the outside?
   b. At rest, the human body uses energy at the rate of approximately 100 W to maintain basic metabolic functions. It has been estimated that 20% of this energy is used to operate the sodium pumps of the body. Estimate—to one significant figure—the number of sodium ions pumped per second.

40. ‖ An arrangement of source charges produces the electric potential $V = 5000x^2$ along the x-axis, where V is in volts and x is in meters. What is the maximum speed of a 1.0 g, 10 nC charged particle that moves in this potential with turning points at $\pm 8.0\ cm$?

41. ‖ A proton moves along the x-axis, where an arrangement of source charges has created the electric potential $V = 6000x^2$, where V is in volts and x is in meters. By exploiting the analogy with the potential energy of a mass on a spring, determine the proton's oscillation frequency.

42. ‖ In **FIGURE P28.42**, a proton is fired with a speed of 200,000 m/s from the midpoint of the capacitor toward the positive plate.
   a. Show that this is insufficient speed to reach the positive plate.
   b. What is the proton's speed as it collides with the negative plate?

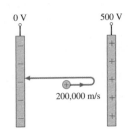

FIGURE P28.42

43. ‖ The electron gun in an old TV picture tube accelerates electrons between two parallel plates 1.2 cm apart with a 25 kV potential difference between them. The electrons enter through a small hole in the negative plate, accelerate, then exit through a small hole in the positive plate. Assume that the holes are small enough not to affect the electric field or potential.
   a. What is the electric field strength between the plates?
   b. With what speed does an electron exit the electron gun if its entry speed is close to zero?
   NOTE ▶ The exit speed is so fast that we really need to use the theory of relativity to compute an accurate value. Your answer to part b is in the right range but a little too big. ◀

44. ‖ An uncharged parallel-plate capacitor with spacing d is horizontal. A small bead with mass m and positive charge q is shot straight up from the bottom plate with speed v_0. It reaches maximum height y_{max} before falling back. Then the capacitor is charged with the bottom plate negative. Find an expression for the capacitor voltage ΔV_C for which the bead's maximum height is reduced to $\frac{1}{2}y_{max}$. Ignore air resistance.

45. ‖ A room with 3.0-m-high ceilings has a metal plate on the floor with V = 0 V and a separate metal plate on the ceiling. A 1.0 g glass ball charged to +4.9 nC is shot straight up at 5.0 m/s. How high does the ball go if the ceiling voltage is (a) $+3.0 \times 10^6$ V and (b) -3.0×10^6 V?

46. ‖ In *proton-beam therapy*, a high-energy beam of protons is
BIO fired at a tumor. As the protons stop in the tumor, their kinetic energy breaks apart the tumor's DNA, thus killing the tumor cells. For one patient, it is desired to deposit 0.10 J of proton energy in the tumor. To create the proton beam, protons are accelerated from rest through a 10,000 kV potential difference. What is the total charge of the protons that must be fired at the tumor?

47. ‖ What is the escape speed of an electron launched from the surface of a 1.0-cm-diameter glass sphere that has been charged to 10 nC?

48. ‖ An electric dipole consists of 1.0 g spheres charged to ± 2.0 nC at the ends of a 10-cm-long massless rod. The dipole rotates on a frictionless pivot at its center. The dipole is held perpendicular to a uniform electric field with field strength 1000 V/m, then released. What is the dipole's angular velocity at the instant it is aligned with the electric field?

49. ‖‖ Three electrons form an equilateral triangle 1.0 nm on each side. A proton is at the center of the triangle. What is the potential energy of this group of charges?

50. ‖‖ A 2.0-mm-diameter glass bead is positively charged. The potential difference between a point 2.0 mm from the bead and a point 4.0 mm from the bead is 500 V. What is the charge on the bead?

51. ‖ Your lab assignment for the week is to measure the amount of charge on the 6.0-cm-diameter metal sphere of a Van de Graaff generator. To do so, you're going to use a spring with spring constant 0.65 N/m to launch a small, 1.5 g bead horizontally toward the sphere. You can reliably charge the bead to 2.5 nC, and your plan is to use a video camera to measure the bead's closest approach to the sphere as you change the compression of the spring. Your data are as follows:

Compression (cm)	Closest approach (cm)
1.6	5.5
1.9	2.6
2.2	1.6
2.5	0.4

Use an appropriate graph of the data to determine the sphere's charge in nC. You can assume that the bead's motion is entirely horizontal and that the spring is so far away that the bead has no interaction with the sphere as it's launched.

52. ‖ A proton is fired from far away toward the nucleus of an iron atom. Iron is element number 26, and the diameter of the nucleus is 9.0 fm. What initial speed does the proton need to just reach the surface of the nucleus? Assume the nucleus remains at rest.

53. ‖ A proton is fired from far away toward the nucleus of a mercury atom. Mercury is element number 80, and the diameter of the nucleus is 14.0 fm. If the proton is fired at a speed of 4.0×10^7 m/s, what is its closest approach to the surface of the nucleus? Assume the nucleus remains at rest.

54. ‖ In the form of radioactive decay known as *alpha decay,* an unstable nucleus emits a helium-atom nucleus, which is called an *alpha particle.* An alpha particle contains two protons and two neutrons, thus having mass $m = 4$ u and charge $q = 2e$. Suppose a uranium nucleus with 92 protons decays into thorium, with 90 protons, and an alpha particle. The alpha particle is initially at rest at the surface of the thorium nucleus, which is 15 fm in diameter. What is the speed of the alpha particle when it is detected in the laboratory? Assume the thorium nucleus remains at rest.

55. ‖ One form of nuclear radiation, *beta decay,* occurs when a neutron changes into a proton, an electron, and a neutral particle called a *neutrino:* $n \rightarrow p^+ + e^- + \nu$ where ν is the symbol for a neutrino. When this change happens to a neutron within the nucleus of an atom, the proton remains behind in the nucleus while the electron and neutrino are ejected from the nucleus. The ejected electron is called a *beta particle.* One nucleus that exhibits beta decay is the isotope of hydrogen ^{3}H, called *tritium,* whose nucleus consists of one proton (making it hydrogen) and two neutrons (giving tritium an atomic mass $m = 3$ u). Tritium is radioactive, and it decays to helium: ^{3}H $\rightarrow$ ^{3}He $+ e^- + \nu$.
   a. Is charge conserved in the beta decay process? Explain.
   b. Why is the final product a helium atom? Explain.
   c. The nuclei of both ^{3}H and ^{3}He have radii of 1.5×10^{-15} m. With what minimum speed must the electron be ejected if it is to escape from the nucleus and not fall back?

56. ‖ The sun is powered by *fusion,* with four protons fusing together to form a helium nucleus (two of the protons turn into neutrons) and, in the process, releasing a large amount of thermal energy. The process happens in several steps, not all at once. In one step, two protons fuse together, with one proton then becoming a neutron, to form the "heavy hydrogen" isotope *deuterium* (^{2}H).

A proton is essentially a 2.4-fm-diameter sphere of charge, and fusion occurs only if two protons come into contact with each other. This requires extraordinarily high temperatures due to the strong repulsion between the protons. Recall that the average kinetic energy of a gas particle is $\frac{3}{2}k_BT$.
   a. Suppose two protons, each with exactly the average kinetic energy, have a head-on collision. What is the minimum temperature for fusion to occur?
   b. Your answer to part a is much hotter than the 15 million K in the core of the sun. If the temperature were as high as you calculated, every proton in the sun would fuse almost instantly and the sun would explode. For the sun to last for billions of years, fusion can occur only in collisions between two protons with kinetic energies much higher than average. Only a very tiny fraction of the protons have enough kinetic energy to fuse when they collide, but that fraction is enough to keep the sun going. Suppose two protons with the same kinetic energy collide head-on and just barely manage to fuse. By what factor does each proton's energy exceed the average kinetic energy at 15 million K?

57. ‖ Two 10-cm-diameter electrodes 0.50 cm apart form a parallel-plate capacitor. The electrodes are attached by metal wires to the terminals of a 15 V battery. After a long time, the capacitor is disconnected from the battery but is not discharged. What are the charge on each electrode, the electric field strength inside the capacitor, and the potential difference between the electrodes
   a. Right after the battery is disconnected?
   b. After insulating handles are used to pull the electrodes away from each other until they are 1.0 cm apart?
   c. After the original electrodes (not the modified electrodes of part b) are expanded until they are 20 cm in diameter?

58. ‖ Two 10-cm-diameter electrodes 0.50 cm apart form a parallel-plate capacitor. The electrodes are attached by metal wires to the terminals of a 15 V battery. What are the charge on each electrode, the electric field strength inside the capacitor, and the potential difference between the electrodes
   a. While the capacitor is attached to the battery?
   b. After insulating handles are used to pull the electrodes away from each other until they are 1.0 cm apart? The electrodes remain connected to the battery during this process.
   c. After the original electrodes (not the modified electrodes of part b) are expanded until they are 20 cm in diameter while remaining connected to the battery?

59. ‖ a. Find an algebraic expression for the electric field strength E_0 at the surface of a charged sphere in terms of the sphere's potential V_0 and radius R.
   b. What is the electric field strength at the surface of a 1.0-cm-diameter marble charged to 500 V?

60. ‖ Two spherical drops of mercury each have a charge of 0.10 nC and a potential of 300 V at the surface. The two drops merge to form a single drop. What is the potential at the surface of the new drop?

61. ‖ A Van de Graaff generator is a device for generating a large electric potential by building up charge on a hollow metal sphere. A typical classroom-demonstration model has a diameter of 30 cm.
   a. How much charge is needed on the sphere for its potential to be 500,000 V?
   b. What is the electric field strength just outside the surface of the sphere when it is charged to 500,000 V?

62. ‖ A thin spherical shell of radius R has total charge Q. What is the electric potential at the center of the shell?

63. | **FIGURE P28.63** shows two uniformly charged spheres. What is the potential difference between points a and b? Which point is at the higher potential?

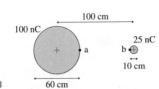

FIGURE P28.63

Hint: The potential at any point is the superposition of the potentials due to *all* charges.

64. ‖ An electric dipole with dipole moment p is oriented along the y-axis.
   a. Find an expression for the electric potential on the y-axis at a point where y is much larger than the charge spacing s. Write your expression in terms of the dipole moment p.
   b. The dipole moment of a water molecule is 6.2×10^{-30} C m. What is the electric potential 1.0 nm from a water molecule along the axis of the dipole?

65. ‖ Two positive point charges q are located on the y-axis at $y = \pm \frac{1}{2}s$.
   a. Find an expression for the potential along the x-axis.
   b. Draw a graph of V versus x for $-\infty < x < \infty$. For comparison, use a dotted line to show the potential of a point charge $2q$ located at the origin.

66. ‖ The arrangement of charges shown in **FIGURE P28.66** is called a *linear electric quadrupole*. The positive charges are located at $y = \pm s$. Notice that the net charge is zero. Find an expression for the electric potential on the y-axis at distances $y \gg s$.

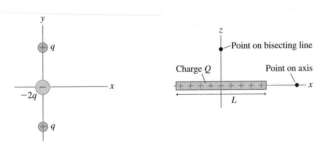

FIGURE P28.66   FIGURE P28.67

67. ‖ **FIGURE P28.67** shows a thin rod of length L and charge Q. Find an expression for the electric potential a distance x away from the center of the rod on the axis of the rod.

68. ‖‖ **FIGURE P28.67** showed a thin rod of length L and charge Q. Find an expression for the electric potential a distance z away from the center of rod on the line that bisects the rod.

69. | **FIGURE P28.69** shows a thin rod with charge Q that has been bent into a semi-circle of radius R. Find an expression for the electric potential at the center.

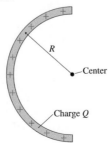

FIGURE P28.69

70. ‖ A disk with a hole has inner radius R_{in} and outer radius R_{out}. The disk is uniformly charged with total charge Q. Find an expression for the on-axis electric potential at distance z from the center of the disk. Verify that your expression has the correct behavior when $R_{in} \to 0$.

In Problems 71 through 73 you are given the equation(s) used to solve a problem. For each of these,
   a. Write a realistic problem for which this is the correct equation(s).
   b. Finish the solution of the problem.

71. $\dfrac{(9.0 \times 10^9 \text{ N m}^2/\text{C}^2)q_1 q_2}{0.030 \text{ m}} = 90 \times 10^{-6} \text{ J}$

$q_1 + q_2 = 40 \text{ nC}$

72. $\frac{1}{2}(1.67 \times 10^{-27} \text{ kg})(2.5 \times 10^6 \text{ m/s})^2 + 0 =$

$\frac{1}{2}(1.67 \times 10^{-27} \text{ kg})v_i^2 +$

$\dfrac{(9.0 \times 10^9 \text{ N m}^2/\text{C}^2)(2.0 \times 10^{-9} \text{ C})(1.60 \times 10^{-19} \text{ C})}{0.0010 \text{ m}}$

73. $\dfrac{(9.0 \times 10^9 \text{ N m}^2/\text{C}^2)(3.0 \times 10^{-9} \text{ C})}{0.030 \text{ m}} +$

$\dfrac{(9.0 \times 10^9 \text{ N m}^2/\text{C}^2)(3.0 \times 10^{-9} \text{ C})}{(0.030 \text{ m}) + d} = 1200 \text{ V}$

Challenge Problems

74. A proton and an alpha particle ($q = +2e$, $m = 4$ u) are fired directly toward each other from far away, each with an initial speed of $0.010c$. What is their distance of closest approach, as measured between their centers?

75. Bead A has a mass of 15 g and a charge of -5.0 nC. Bead B has a mass of 25 g and a charge of -10.0 nC. The beads are held 12 cm apart (measured between their centers) and released. What maximum speed is achieved by each bead?

76. Two 2.0-mm-diameter beads, C and D, are 10 mm apart, measured between their centers. Bead C has mass 1.0 g and charge 2.0 nC. Bead D has mass 2.0 g and charge -1.0 nC. If the beads are released from rest, what are the speeds v_C and v_D at the instant the beads collide?

77. An electric dipole has dipole moment p. If $r \gg s$, where s is the separation between the charges, show that the electric potential of the dipole can be written

$$V = \frac{1}{4\pi\epsilon_0} \frac{p \cos\theta}{r^2}$$

where r is the distance from the center of the dipole and θ is the angle from the dipole axis.

78. Electrodes of area A are spaced distance d apart to form a parallel-plate capacitor. The electrodes are charged to $\pm q$.
   a. What is the infinitesimal increase in electric potential energy dU if an infinitesimal amount of charge dq is moved from the negative electrode to the positive electrode?
   b. An uncharged capacitor can be charged to $\pm Q$ by transferring charge dq over and over and over. Use your answer to part a to show that the potential energy of a capacitor charged to $\pm Q$ is $U_{cap} = \frac{1}{2}Q \Delta V_C$.

79. A sphere of radius R has charge q.
    a. What is the infinitesimal increase in electric potential energy dU if an infinitesimal amount of charge dq is brought from infinity to the surface of the sphere?
    b. An uncharged sphere can acquire total charge Q by the transfer of charge dq over and over and over. Use your answer to part a to find an expression for the potential energy of a sphere of radius R with total charge Q.
    c. Your answer to part b is the amount of energy needed to assemble a charged sphere. It is often called the *self-energy* of the sphere. What is the self-energy of a proton, assuming it to be a charged sphere with a diameter of 1.0×10^{-15} m?

80. The wire in FIGURE CP28.80 has linear charge density λ. What is the electric potential at the center of the semicircle?

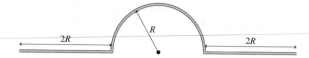

FIGURE CP28.80

81. A circular disk of radius R and total charge Q has the charge distributed with surface charge density $\eta = cr$, where c is a constant. Find an expression for the electric potential at distance z on the axis of the disk. Your expression should include R and Q, but not c.

82. A hollow cylindrical shell of length L and radius R has charge Q uniformly distributed along its length. What is the electric potential at the center of the cylinder?

STOP TO THINK ANSWERS

Stop to Think 28.1: Zero. The motion is always perpendicular to the electric force.

Stop to Think 28.2: $U_b = U_d > U_a = U_c$. The potential energy depends inversely on r. The effects of doubling the charge and doubling the distance cancel each other.

Stop to Think 28.3: c. The proton gains speed by losing potential energy. It loses potential energy by moving in the direction of decreasing electric potential.

Stop to Think 28.4: $V_a = V_b > V_c > V_d = V_e$. The potential decreases steadily from the positive to the negative plate. It depends only on the distance from the positive plate.

Stop to Think 28.5: $\Delta V_{ac} = \Delta V_{bc} > \Delta V_{ab}$. The potential depends only on the *distance* from the charge, not the direction. $\Delta V_{ab} = 0$ because these points are at the same distance.

29 Potential and Field

These solar cells are *photovoltaic* cells, meaning that light creates a voltage—a potential difference.

▶ **Looking Ahead** The goal of Chapter 29 is to understand how the electric potential is related to the electric field.

Field and Potential

The electric potential and the electric field are intimately connected. They are two different perspectives of how source charges alter the space around them.

You'll learn to:

- Use the electric potential to find the electric field.
- Use the electric field to find the electric potential.

The mathematical connection is analogous to that between force and potential energy.

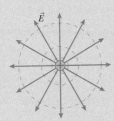

The electric field and the electric potential can be related to each other geometrically.

You'll also learn that:

- The electric field is always perpendicular to equipotential surfaces.
- The electric field points "downhill" in the direction of decreasing potential.
- The electric field is stronger where equipotential lines are closer together.

◀ **Looking Back**
Sections 28.4–28.6 Electric potential and its graphical representations

Sources of Potential

A potential difference—a *voltage*—is created by separating positive and negative charges.

We'll develop a *charge escalator* model of a battery in which chemical reactions separate charge to create a potential difference.

You'll learn that any nonelectrical means of separating charge—in batteries, photocells, and generators—does work and develops what we'll call an **emf.**

Conductors

You'll learn several important characteristics of conductors in **electrostatic equilibrium,** with stationary charges.

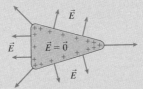

- Any excess charge is on the surface.
- The interior electric field is zero.
- The exterior electric field is perpendicular to the surface.
- The entire conductor is an equipotential.

Capacitors

Capacitors are circuit elements that store charge and energy. They are used in devices ranging from high-speed computers to heart defibrillators.

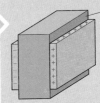

The flash on your camera uses energy stored in a capacitor. The capacitor can discharge in a few microseconds, much faster than a battery can provide energy.

You'll learn to:

- Work with combinations of capacitors called *in series* and *in parallel*.
- Calculate the energy stored in a capacitor's electric field.
- Understand capacitors with dielectrics.

Dielectric

An insulator between the capacitor plates is called a **dielectric.** It changes the capacitor properties in many useful ways.

◀ **Looking Back**
Section 26.5 Parallel-plate capacitors
Section 26.7 Dipoles in electric fields

29.1 Connecting Potential and Field

FIGURE 29.1 The four key ideas.

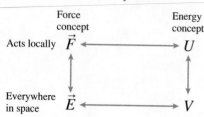

FIGURE 29.1 shows the four key ideas of force, field, potential energy, and potential. The electric field and the electric potential were based on force and potential energy. We know, from Chapters 10 and 11, that force and potential energy are closely related. The focus of this chapter is to establish a similar relationship between the electric field and the electric potential. **The electric potential and electric field are not two distinct entities but, instead, two different perspectives or two different mathematical representations of how source charges alter the space around them.**

If this is true, we should be able to find the electric potential from the electric field. Chapter 28 introduced all the pieces we need to do so. We used the potential energy of charge q and the source charges to define the electric potential as

$$V \equiv \frac{U_{q+\text{sources}}}{q} \tag{29.1}$$

Potential energy is defined in terms of the work done by force $\vec{F}$ on charge q as it moves from position i to position f:

$$\Delta U = -W(\text{i} \rightarrow \text{f}) = -\int_{s_i}^{s_f} F_s \, ds = -\int_i^f \vec{F} \cdot d\vec{s} \tag{29.2}$$

But the force exerted on charge q by the electric field is $\vec{F} = q\vec{E}$. Putting these three pieces together, you can see that the charge q cancels out and the potential difference between two points in space is

$$\Delta V = V_f - V_i = -\int_{s_i}^{s_f} E_s \, ds = -\int_i^f \vec{E} \cdot d\vec{s} \tag{29.3}$$

where s is the position along a line from point i to point f. That is, we can find the potential difference between two points if we know the electric field.

We can think of an integral as an area under a curve. Thus a graphical interpretation of Equation 29.3 is

$$V_f = V_i - (\text{area under the } E_s\text{-versus-}s \text{ curve between } s_i \text{ and } s_f) \tag{29.4}$$

Notice, because of the minus sign in Equation 29.3, that the area is *subtracted* from V_i.

EXAMPLE 29.1 | **Finding the potential**

FIGURE 29.2 is a graph of E_x, the x-component of the electric field, versus position along the x-axis. Find and graph $V(x)$. Assume $V = 0$ V at $x = 0$ m.

FIGURE 29.2 Graph of E_x versus x.

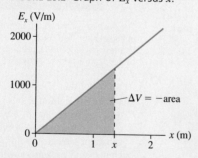

MODEL The potential difference is the *negative* of the area under the curve.

VISUALIZE E_x is positive throughout this region of space, meaning that $\vec{E}$ points in the positive x-direction.

SOLVE If we integrate from $x = 0$, then $V_i = V(x = 0) = 0$. The potential for $x > 0$ is the negative of the triangular area under the

E_x curve. We can see that $E_x = 1000x$ V/m, where x is in m. Thus

$$V_f = V(x) = 0 - (\text{area under the } E_x \text{ curve})$$

$$= -\tfrac{1}{2} \times \text{base} \times \text{height} = -\tfrac{1}{2}(x)(1000x) = -500x^2 \text{ V}$$

FIGURE 29.3 shows that the electric potential in this region of space is parabolic, decreasing from 0 V at $x = 0$ m to -2000 V at $x = 2$ m.

FIGURE 29.3 Graph of V versus x.

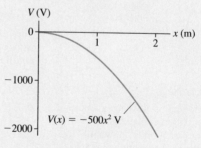

ASSESS The electric field points in the direction in which V is *decreasing*. We'll soon see that this is a general rule.

Finding the potential from the electric field

❶ Draw a picture and identify the point at which you wish to find the potential.
   Call this position f.
❷ Choose the zero point of the potential, often at infinity. Call this position i.
❸ Establish a coordinate axis from i to f along which you already know or can
   easily determine the electric field component E_s.
❹ Carry out the integration of Equation 29.3 to find the potential.

To see how this works, let's use the electric field of a point charge to find its electric potential. FIGURE 29.4 identifies a point P at $s_f = r$ at which we want to know the potential and calls this position f. We've chosen position i to be at $s_i = \infty$ and identified that as the zero point of the potential. The integration of Equation 29.3 is straight inward along the radial line from i to f:

$$\Delta V = V(r) - V(\infty) = -\int_\infty^r E_s \, ds = \int_r^\infty E_s \, ds \qquad (29.5)$$

The electric field is radially outward. Its s-component is

$$E_s = \frac{1}{4\pi\epsilon_0}\frac{q}{s^2}$$

Thus the potential at distance r from a point charge q is

$$V(r) = V(\infty) + \frac{q}{4\pi\epsilon_0}\int_r^\infty \frac{ds}{s^2} = V(\infty) + \frac{q}{4\pi\epsilon_0}\frac{-1}{s}\Big|_r^\infty = 0 + \frac{1}{4\pi\epsilon_0}\frac{q}{r} \qquad (29.6)$$

We've rediscovered the potential of a point charge that you learned in Chapter 28:

$$V_{\text{point charge}} = \frac{1}{4\pi\epsilon_0}\frac{q}{r} \qquad (29.7)$$

FIGURE 29.4 Finding the potential of a point charge.

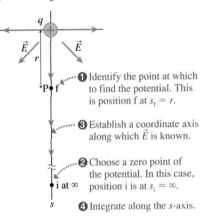

❶ Identify the point at which to find the potential. This is position f at $s_f = r$.

❸ Establish a coordinate axis along which $\vec{E}$ is known.

❷ Choose a zero point of the potential. In this case, position i is at $s_i = \infty$.

❹ Integrate along the s-axis.

EXAMPLE 29.2 | The potential of a parallel-plate capacitor

In Chapter 26, the electric field inside a capacitor was found to be

$$\vec{E} = \left(\frac{Q}{\epsilon_0 A}, \text{from positive to negative}\right)$$

Find the electric potential inside the capacitor. Let $V = 0$ V at the negative plate.

MODEL The electric field inside a capacitor is a uniform field.

VISUALIZE FIGURE 29.5 shows the capacitor and establishes a point P where we want to find the potential. We've chosen an s-axis measured from the negative plate, which is the zero point of the potential.

SOLVE We'll integrate along the s-axis from $s_i = 0$ (where $V_f = 0$ V) to $s_f = s$. Notice that $\vec{E}$ points in the negative s-direction, so $E_s = -Q/\epsilon_0 A$. $Q/\epsilon_0 A$ is a constant, so

$$V(s) = V_f = V_i - \int_0^s E_s \, ds = -\left(-\frac{Q}{\epsilon_0 A}\right)\int_0^s ds = \frac{Q}{\epsilon_0 A}s = Es$$

ASSESS $V = Es$ is the capacitor potential we deduced in Chapter 28 by working directly with the potential energy. The

FIGURE 29.5 Finding the potential inside a capacitor.

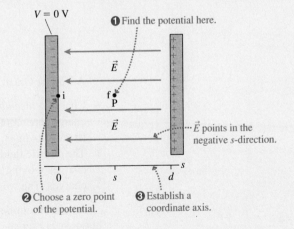

$V = 0$ V

❶ Find the potential here.

$\vec{E}$

$\vec{E}$

$\vec{E}$ points in the negative s-direction.

❷ Choose a zero point of the potential.

❸ Establish a coordinate axis.

potential increases linearly from $V = 0$ at the negative plate to $V = Ed$ at the positive plate. Here we found the potential by explicitly recognizing the connection between the potential and the field.

29.2 Sources of Electric Potential

FIGURE 29.6 A charge separation creates a potential difference.

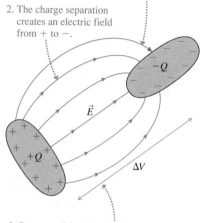

FIGURE 29.6 A charge separation creates a potential difference.

1. Charge is separated by moving electrons from one electrode to the other.

2. The charge separation creates an electric field from + to −.

$\vec{E}$

$+Q$

ΔV

3. Because of the electric field, there's a potential difference between the electrodes.

A *separation of charge* creates an electric potential difference. Shuffling your feet on the carpet transfers electrons from the carpet to you, creating a potential difference between you and a doorknob that causes a spark and a shock as you touch it. Charging a capacitor by moving electrons from one plate to the other creates a potential difference across the capacitor.

In fact, as FIGURE 29.6 shows, *any* separation of charge causes a potential difference. The charge separation between the two electrodes creates an electric field $\vec{E}$ pointing from the positive toward the negative electrode. As a consequence, there is a potential difference between the electrodes that is given by

$$\Delta V = V_{pos} - V_{neg} = -\int_{neg}^{pos} E_s \, ds$$

where the integral runs from any point on the negative electrode to any point on the positive. The key idea is that **we can create a potential difference by creating a charge separation.**

The **Van de Graaff generator** shown in FIGURE 29.7a is a mechanical charge separator—essentially a fancy foot shuffler. A moving plastic or leather belt is charged, then the charge is mechanically transported via the conveyor belt to the spherical electrode at the top of the insulating column. The charging of the belt could be done by friction, but in practice a *corona discharge* created by the strong electric field at the tip of a needle is more efficient and reliable.

FIGURE 29.7 A Van de Graaff generator.

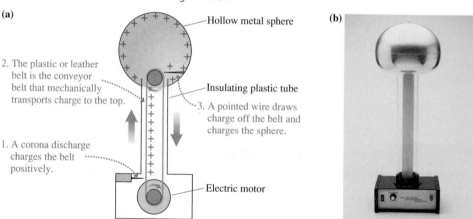

(a)

2. The plastic or leather belt is the conveyor belt that mechanically transports charge to the top.

1. A corona discharge charges the belt positively.

Hollow metal sphere

Insulating plastic tube

3. A pointed wire draws charge off the belt and charges the sphere.

Electric motor

(b)

A Van de Graaff generator has two noteworthy features:

- Charge is *mechanically* transported from the negative side to the positive side. This charge separation creates a potential difference between the spherical electrode and its surroundings.
- The electric field of the spherical electrode exerts a downward force on the positive charges moving up the belt. Consequently, *work must be done* to "lift" the positive charges. The work is done by the electric motor that runs the belt.

A classroom-demonstration Van de Graaff generator like the one shown in FIGURE 29.7b creates a potential difference of several hundred thousand volts between the upper sphere and its surroundings. The maximum potential is reached when the electric field near the sphere becomes large enough to cause a breakdown of the air. This produces a spark and temporarily discharges the sphere. A large Van de Graaff generator surrounded by vacuum can reach a potential of 20 MV or more. These generators are used to accelerate protons for nuclear physics experiments.

Batteries and emf

The most common source of electric potential is a **battery.** A battery consists of chemicals, called *electrolytes,* sandwiched between two electrodes made of different metals. Chemical reactions in the electrolytes separate charge by moving positive ions to one electrode and negative ions to the other. In other words, chemical reactions, rather than a mechanical conveyor belt, transport charge from one electrode to the other. The procedure is different, but the outcome is the same: a potential difference.

We can sidestep the chemistry details by introducing the **charge escalator model** of a battery shown in FIGURE 29.8. The escalator separates charge by "lifting" positive charges from the negative terminal to the positive terminal. Lifting positive charges to a positive terminal requires that work be done, and the chemical reactions within the battery provide the energy to do this work. When the chemicals are used up, the reactions cease, and the battery is dead.

By separating the charge, the charge escalator establishes a potential difference ΔV_{bat} between the terminals. The value of ΔV_{bat} is determined by the specific chemical reactions employed by the battery. To see how, suppose the chemical reactions do work W_{chem} to move charge q from the negative to the positive terminal. In an **ideal battery,** in which there are no internal energy losses, the charge gains electric potential energy $\Delta U = W_{chem}$. This is analogous to a book gaining gravitational potential energy as you do work to lift it from the floor to a shelf.

The quantity W_{chem}/q, which is the work done *per charge* by the charge escalator, is called the **emf** of the battery, pronounced as the sequence of three letters "e-m-f." The symbol for emf is $\mathcal{E}$, a script E, and the units are those of the electric potential: joules per coulomb, or volts. The *rating* of a battery, such as 1.5 V or 9 V, is the battery's emf. Originally the term emf was an abbreviation of "electromotive force." That is an outdated term (work per charge is not a force!), so today we just call it emf and it is not an abbreviation of anything.

By definition, the electric potential is related to the electric potential energy of charge q by $\Delta V = \Delta U/q$. But $\Delta U = W_{chem}$ for the charges in a battery, hence the potential difference between the terminals of an ideal battery is

$$\Delta V_{bat} = \frac{W_{chem}}{q} = \mathcal{E} \qquad \text{(ideal battery)} \qquad (29.8)$$

In other words, a battery constructed to have an emf of 1.5 V (i.e., the chemical reactions do 1.5 J of work to separate 1 C of charge) creates a 1.5 V potential difference between its positive and negative terminals. In practice, the measured potential difference ΔV_{bat} between the terminals of a real battery, called the **terminal voltage,** is usually slightly less than $\mathcal{E}$. You will learn the reason for this in Chapter 31.

Many consumer goods, from flashlights to digital cameras, use more than one battery. Why? A particular type of battery, such as an AA or AAA battery, produces a fixed emf determined by the chemical reactions inside. The emf of one battery, often 1.5 V, is not sufficient to light a lightbulb or power a camera. But just as you can reach the third floor of a building by taking three escalators in succession, we can produce a larger potential difference by placing two or more batteries *in series*. FIGURE 29.9 shows two batteries with the positive terminal of one literally touching the negative terminal of the next. Flashlight batteries usually are arranged like this. Other devices, such as cameras, achieve the same effect by using conducting metal wires between one battery and the next. Either way, the total potential difference of batteries in series is simply the sum of their individual terminal voltages:

$$\Delta V_{series} = \Delta V_1 + \Delta V_2 + \cdots \qquad \text{(batteries in series)} \qquad (29.9)$$

Electric generators, photocells, and other sources of potential difference use different means to separate charges, but otherwise they function exactly the same as a battery. The common feature of all such devices is that they use a *nonelectrical* means to separate charge and, thus, to create a potential difference. The emf $\mathcal{E}$ of any device is the work done per charge to separate the charge.

FIGURE 29.8 The charge escalator model of a battery.

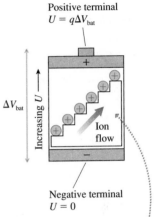

The charge escalator "lifts" charge from the negative side to the positive side. Charge q gains energy $\Delta U = q\Delta V_{bat}$.

Flashlight batteries are placed in series to create twice the potential difference of one battery.

FIGURE 29.9 Batteries in series.

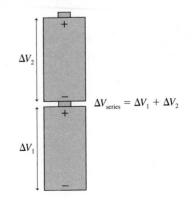

What total potential difference is created by these three batteries?

29.3 Finding the Electric Field from the Potential

FIGURE 29.10 The electric field does work on charge q.

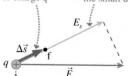

A very small displacement of charge q

E_s, the component of $\vec{E}$ in the direction of motion, is essentially constant over the small distance Δs.

FIGURE 29.10 shows two points i and f separated by a very small distance Δs, so small that the electric field is essentially constant over this very short distance. The work done by the electric field as a charge q moves through this small distance is $W = F_s \Delta s = qE_s \Delta s$. Consequently, the potential difference between these two points is

$$\Delta V = \frac{\Delta U_{q + \text{sources}}}{q} = \frac{-W}{q} = -E_s \Delta s \qquad (29.10)$$

In terms of the potential, the component of the electric field in the s-direction is $E_s = -\Delta V/\Delta s$. In the limit $\Delta s \to 0$,

$$E_s = -\frac{dV}{ds} \qquad (29.11)$$

Now we have reversed Equation 29.3 and can find the electric field from the potential. We'll begin with examples where the field is parallel to a coordinate axis, then we'll look at what Equation 29.11 tells us about the geometry of the field and the potential.

Field Parallel to a Coordinate Axis

The derivative in Equation 29.11 gives E_s, the component of the electric field parallel to the displacement $\Delta \vec{s}$. It doesn't tell us about the electric field component perpendicular to $\Delta \vec{s}$. Thus Equation 29.11 is most useful if we can use symmetry to select a coordinate axis that is parallel to $\vec{E}$ and along which the perpendicular component of $\vec{E}$ is known to be zero.

For example, suppose we knew the potential of a point charge to be $V = q/4\pi\epsilon_0 r$ but didn't remember the electric field. Symmetry requires that the field point straight outward from the charge, with only a radial component E_r. If we choose the s-axis to be in the radial direction, parallel to $\vec{E}$, we can use Equation 29.11 to find

$$E_r = -\frac{dV}{dr} = -\frac{d}{dr}\left(\frac{q}{4\pi\epsilon_0 r}\right) = \frac{1}{4\pi\epsilon_0}\frac{q}{r^2} \qquad (29.12)$$

This is, indeed, the well-known electric field of a point charge.

Equation 29.11 is especially useful for a continuous distribution of charge because calculating V, which is a scalar, is usually much easier than calculating the vector $\vec{E}$ directly from the charge. Once V is known, $\vec{E}$ is found simply by taking a derivative.

EXAMPLE 29.3 **The electric field of a ring of charge**

In Chapter 28, we found the on-axis potential of a ring of radius R and charge Q to be

$$V_{\text{ring}} = \frac{1}{4\pi\epsilon_0}\frac{Q}{\sqrt{z^2 + R^2}}$$

Find the on-axis electric field of a ring of charge.

SOLVE Symmetry requires the electric field along the axis to point straight outward from the ring with only a z-component E_z. The electric field at position z is

$$E_z = -\frac{dV}{dz} = -\frac{d}{dz}\left(\frac{1}{4\pi\epsilon_0}\frac{Q}{\sqrt{z^2 + R^2}}\right)$$

$$= \frac{1}{4\pi\epsilon_0}\frac{zQ}{(z^2 + R^2)^{3/2}}$$

ASSESS This result is in perfect agreement with the electric field we found in Chapter 26, but this calculation was easier because we didn't have to deal with angles.

A geometric interpretation of Equation 29.11 is that the electric field is the negative of the *slope* of the *V*-versus-*s* graph. This interpretation should be familiar. You learned in Chapter 11 that the force on a particle is the negative of the slope of the potential-energy graph: $F = -dU/ds$. In fact, Equation 29.11 is simply $F = -dU/ds$ with both sides divided by q to yield E and V. This geometric interpretation is an important step in developing an understanding of potential.

EXAMPLE 29.4 **Finding *E* from the slope of *V***

FIGURE 29.11 is a graph of the electric potential in a region of space where $\vec{E}$ is parallel to the *x*-axis. Draw a graph of E_x versus *x*.

FIGURE 29.11 Graph of *V* versus position *x*.

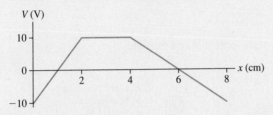

MODEL The electric field is the *negative* of the slope of the potential graph.

SOLVE There are three regions of different slope:

$$0 < x < 2 \text{ cm} \quad \begin{cases} \Delta V/\Delta x = (20 \text{ V})/(0.020 \text{ m}) = 1000 \text{ V/m} \\ E_x = -1000 \text{ V/m} \end{cases}$$

$$2 < x < 4 \text{ cm} \quad \begin{cases} \Delta V/\Delta x = 0 \text{ V/m} \\ E_x = 0 \text{ V/m} \end{cases}$$

$$4 < x < 8 \text{ cm} \quad \begin{cases} \Delta V/\Delta x = (-20 \text{ V})/(0.040 \text{ m}) = -500 \text{ V/m} \\ E_x = 500 \text{ V/m} \end{cases}$$

The results are shown in **FIGURE 29.12**.

FIGURE 29.12 Graph of E_x versus position *x*.

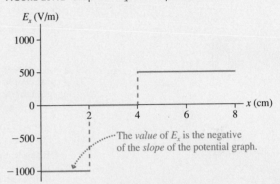

The *value* of E_x is the negative of the *slope* of the potential graph.

ASSESS The electric field $\vec{E}$ points to the left (E_x is negative) for $0 < x < 2$ cm and to the right (E_x is positive) for $4 < x < 8$ cm. Notice that **the electric field is zero in a region of space where the potential is not changing.**

STOP TO THINK 29.2 Which potential graph describes the electric field at the left?

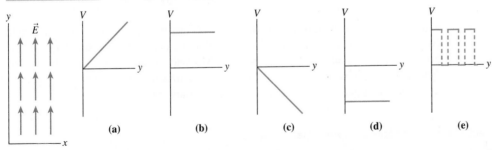

The Geometry of Potential and Field

Equations 29.3 for *V* in terms of E_s and 29.11 for E_s in terms of *V* have profound implications for the geometry of the potential and the field. **FIGURE 29.13** shows two equipotential surfaces, with V_+ positive relative to V_-. To learn about the electric field $\vec{E}$ at point P, allow a charge to move through the two displacements $\Delta \vec{s}_1$ and $\Delta \vec{s}_2$. Displacement $\Delta \vec{s}_1$ is *tangent* to the equipotential surface, hence a charge moving in this direction experiences *no* potential difference. According to Equation 29.11, the electric field component along a direction of *constant* potential is $E_s = -dV/ds = 0$. In other words, the electric field component tangent to the equipotential is $E_{\parallel} = 0$.

FIGURE 29.13 The electric field at P is related to the shape of the equipotential surfaces.

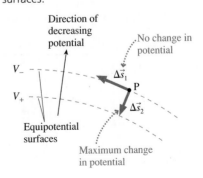

Direction of decreasing potential

No change in potential

Equipotential surfaces

Maximum change in potential

Displacement $\Delta \vec{s}_2$ is *perpendicular* to the equipotential surface. There is a potential difference along $\Delta \vec{s}_2$, hence the electric field component is

$$E_\perp = -\frac{dV}{ds} \approx -\frac{\Delta V}{\Delta s} = -\frac{V_+ - V_-}{\Delta s_2}$$

You can see that the electric field is inversely proportional to Δs_2, the spacing between the equipotential surfaces. Furthermore, because $(V_+ - V_-) > 0$, the minus sign tells us that the electric field is *opposite* in direction to $\Delta \vec{s}_2$. In other words, $\vec{E}$ is **perpendicular to the equipotential surfaces and points "downhill" in the direction** of *decreasing* potential.

These important ideas about the geometry of the potential and the field are summarized in FIGURE 29.14.

FIGURE 29.14 The geometry of the potential and the field.

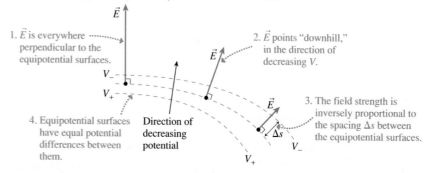

1. $\vec{E}$ is everywhere perpendicular to the equipotential surfaces.

2. $\vec{E}$ points "downhill," in the direction of decreasing V.

3. The field strength is inversely proportional to the spacing Δs between the equipotential surfaces.

4. Equipotential surfaces have equal potential differences between them.

Direction of decreasing potential

Mathematically, we can calculate the individual components of $\vec{E}$ at any point by extending Equation 29.11 to three dimensions:

$$\vec{E} = E_x \hat{\imath} + E_y \hat{\jmath} + E_z \hat{k} = -\left(\frac{\partial V}{\partial x} \hat{\imath} + \frac{\partial V}{\partial y} \hat{\jmath} + \frac{\partial V}{\partial z} \hat{k} \right) \qquad (29.13)$$

where $\partial V/\partial x$ is the partial derivative of V with respect to x while y and z are held constant. You may recognize from calculus that the expression in parentheses is the *gradient* of V, written ∇V. Thus, $\vec{E} = -\nabla V$. More advanced treatments of the electric field make extensive use of this mathematical relationship, but for the most part we'll limit our investigations to those we can analyze graphically.

EXAMPLE 29.5 **Finding the electric field from the equipotential surfaces**

In FIGURE 29.15 a 1 cm × 1 cm grid is superimposed on a contour map of the potential. Estimate the strength and direction of the electric field at points 1, 2, and 3. Show your results graphically by drawing the electric field vectors on the contour map.

FIGURE 29.15 Equipotential lines.

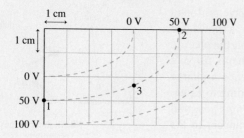

MODEL The electric field is perpendicular to the equipotential lines, points "downhill," and depends on the slope of the potential hill.

VISUALIZE The potential is highest on the bottom and the right. An elevation graph of the potential would look like the lower-right quarter of a bowl or a football stadium.

SOLVE Some distant but unseen source charges have created an electric field and potential. We do not need to see the source charges to relate the field to the potential. Because $E \approx -\Delta V/\Delta s$, the electric field is stronger where the equipotential lines are closer together and weaker where they are farther apart. If Figure 29.15 were a topographic map, you would interpret the closely spaced contour lines at the bottom of the figure as a steep slope.

FIGURE 29.16 shows how measurements of Δs from the grid are combined with values of ΔV to determine $\vec{E}$. Point 3 requires an estimate of the spacing between the 0 V and the 100 V surfaces. Notice that we're using the 0 V and 100 V equipotential surfaces to determine $\vec{E}$ at a point on the 50 V equipotential.

ASSESS The *directions* of $\vec{E}$ are found by drawing downhill vectors perpendicular to the equipotentials. The distances between the equipotential surfaces are needed to determine the field strengths.

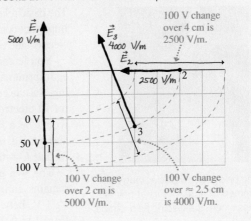

FIGURE 29.16 The electric field at points 1 to 3.

Kirchhoff's Loop Law

FIGURE 29.17 shows two points, 1 and 2, in a region of electric field and potential. You learned in Chapter 28 that the work done in moving a charge between points 1 and 2 is *independent of the path*. Consequently, the potential difference between points 1 and 2 along any two paths that join them is $\Delta V = 20$ V. This must be true in order for the idea of an equipotential surface to make sense.

Now consider the path 1–a–b–c–2–d–1 that ends where it started. What is the potential difference "around" this closed path? The potential increases by 20 V in moving from 1 to 2, but then decreases by 20 V in moving from 2 back to 1. Thus $\Delta V = 0$ V around the closed path.

The numbers are specific to this example, but the idea applies to any loop (i.e., a closed path) through an electric field. The situation is analogous to hiking on the side of a mountain. You may walk uphill during parts of your hike and downhill during other parts, but if you return to your starting point your *net* change of elevation is zero. So for any path that starts and ends at the same point, we can conclude that

$$\Delta V_{\text{loop}} = \sum_i (\Delta V)_i = 0 \qquad (29.14)$$

Stated in words, **the sum of all the potential differences encountered while moving around a loop or closed path is zero.** This statement is known as **Kirchhoff's loop law.**

Kirchhoff's loop law is a statement of energy conservation because a charge that moves around a loop and returns to its starting point has $\Delta U = q\,\Delta V = 0$. Kirchhoff's loop law and a second Kirchhoff's law you'll meet in the next chapter will turn out to be the two fundamental principles of circuit analysis.

FIGURE 29.17 The potential difference between points 1 and 2 is the same along either path.

The potential difference along path 1-a-b-c-2 is $\Delta V = 0$ V + 10 V + 0 V + 10 V = 20 V.

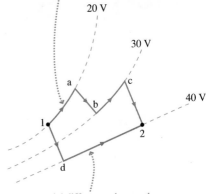

The potential difference along path 1-d-2 is $\Delta V = 20$ V + 0 V = 20 V.

STOP TO THINK 29.3   Which set of equipotential surfaces matches this electric field?

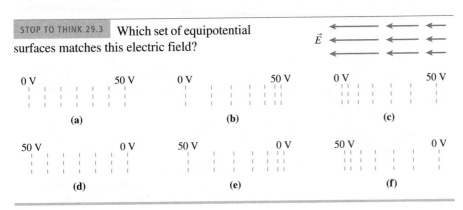

A corona discharge, with crackling noises and glimmers of light, occurs at pointed metal tips where the electric field can be very strong.

FIGURE 29.18 All points inside a conductor in electrostatic equilibrium are at the same potential.

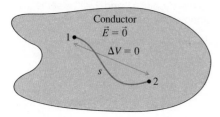

29.4 A Conductor in Electrostatic Equilibrium

The basic relationships between potential and field allow us to draw some interesting and important conclusions about conductors. Consider a conductor, such as a metal, that is in electrostatic equilibrium. The conductor may be charged, but all the charges are at rest.

You learned in Chapter 25 that any excess charges on a conductor in electrostatic equilibrium are always located on the *surface* of the conductor. Using similar reasoning, we can conclude that **the electric field is zero at any interior point of a conductor in electrostatic equilibrium.** Why? If the field were other than zero, then there would be a force $\vec{F} = q\vec{E}$ on the charge carriers and they would move, creating a current. But there are no currents in a conductor in electrostatic equilibrium, so it must be that $\vec{E} = \vec{0}$ at all interior points.

The two points inside the conductor in FIGURE 29.18 are connected by a line that remains entirely inside the conductor. We can find the potential difference $\Delta V = V_2 - V_1$ between these points by using Equation 29.3 to integrate E_s along the line from 1 to 2. But $E_s = 0$ at all points along the line, because $\vec{E} = \vec{0}$; thus the value of the integral is zero and $\Delta V = 0$. In other words, **any two points inside a conductor in electrostatic equilibrium are at the same potential.**

When a conductor is in electrostatic equilibrium, the *entire conductor* is at the same potential. If we charge a metal sphere, then the entire sphere is at a single potential. Similarly, a charged metal rod or wire is at a single potential *if* it is in electrostatic equilibrium.

If $\vec{E} = \vec{0}$ inside a charged conductor but $\vec{E} \neq \vec{0}$ outside, what happens right at the surface? If the entire conductor is at the same potential, then the surface is an equipotential surface. You have seen that the electric field is always perpendicular to an equipotential surface, hence **the exterior electric field $\vec{E}$ of a charged conductor is perpendicular to the surface.**

We can also conclude that the electric field, and thus the surface charge density, is largest at sharp points. This follows from our earlier discovery that the field at the surface of a sphere of radius R can be written $E = V_0/R$. If we approximate the rounded corners of a conductor with sections of spheres, all of which are at the same potential V_0, the field strength will be largest at the corners with the smallest radii of curvature—the sharpest points.

FIGURE 29.19 Electric properties of a conductor in electrostatic equilibrium.

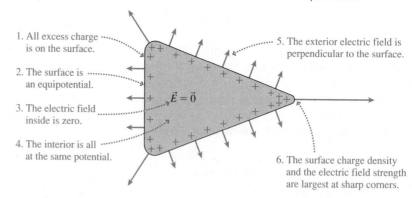

1. All excess charge is on the surface.

2. The surface is an equipotential.

3. The electric field inside is zero.

4. The interior is all at the same potential.

5. The exterior electric field is perpendicular to the surface.

6. The surface charge density and the electric field strength are largest at sharp corners.

$\vec{E} = \vec{0}$

FIGURE 29.19 summarizes what we know about conductors in electrostatic equilibrium. These are important and practical conclusions because conductors are the primary components of electrical devices.

We can use similar reasoning to estimate the electric field and potential between two charged conductors. As an example, FIGURE 29.20 shows a negatively charged metal sphere near a flat metal plate. The surfaces of the sphere and the flat plate are equipotentials, hence the electric field must be perpendicular to both. Close to a surface, the electric field is still *nearly* perpendicular to the surface. Consequently, **an equipotential surface close to an electrode must roughly match the shape of the electrode.**

FIGURE 29.20 Estimating the field and potential between two charged conductors.

The field lines are perpendicular to the equipotential surfaces.

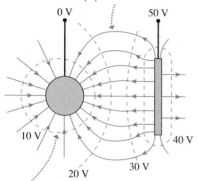

0 V   50 V

10 V   40 V

20 V   30 V

The equipotential surfaces gradually change from the shape of one electrode to that of the other.

In between, the equipotential surfaces *gradually* change as they "morph" from one electrode shape to the other. It's not hard to sketch a contour map showing a plausible set of equipotential surfaces. You can then draw electric field lines (field lines are easier to draw than field vectors) that are perpendicular to the equipotentials, point "downhill," and are closer together where the contour line spacing is smaller.

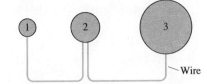

STOP TO THINK 29.4 Three charged metal spheres of different radii are connected by a thin metal wire. The potential and electric field at the surface of each sphere are V and E. Which of the following is true?

a. $V_1 = V_2 = V_3$ and $E_1 = E_2 = E_3$
b. $V_1 = V_2 = V_3$ and $E_1 > E_2 > E_3$
c. $V_1 > V_2 > V_3$ and $E_1 = E_2 = E_3$
d. $V_1 > V_2 > V_3$ and $E_1 > E_2 > E_3$
e. $V_3 > V_2 > V_1$ and $E_3 = E_2 = E_1$
f. $V_3 > V_2 > V_1$ and $E_3 > E_2 > E_1$

29.5 Capacitance and Capacitors

We introduced the parallel-plate capacitor in Chapter 26 and have made frequent use of it since. We've assumed that the capacitor is charged, but we haven't really addressed the issue of *how* it gets charged. FIGURE 29.21 shows the two plates of a capacitor connected with conducting wires to the two terminals of a battery. What happens? And how is the potential difference ΔV_C across the capacitor related to the battery's potential difference ΔV_{bat}?

FIGURE 29.21a shows the situation shortly after the capacitor is connected to the battery and before it is fully charged. The battery's charge escalator is moving charge from one capacitor plate to the other, and it is this work done by the battery that charges the capacitor. (The connecting wires are conductors, and you learned in Chapter 25 that charges can move through conductors as a *current*.) The capacitor voltage ΔV_C steadily increases as the charge separation continues.

Capacitors are important elements in electric circuits. They come in a variety of sizes and shapes.

FIGURE 29.21 A parallel-plate capacitor is charged by a battery.

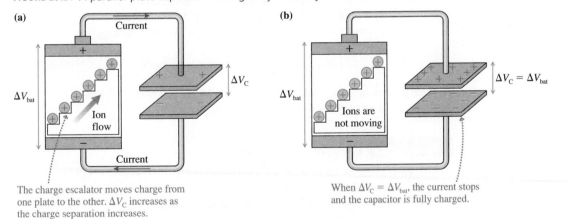

(a) The charge escalator moves charge from one plate to the other. ΔV_C increases as the charge separation increases.

(b) When $\Delta V_C = \Delta V_{bat}$, the current stops and the capacitor is fully charged.

But this process cannot continue forever. The growing positive charge on the upper capacitor plate exerts a repulsive force on new charges coming up the escalator, and eventually the capacitor charge gets so large that no new charges can arrive. The capacitor in FIGURE 29.21b is now *fully charged*. In Chapter 31 we'll analyze how long the charging process takes, but it is typically less than a nanosecond for a capacitor connected directly to a battery with copper wires.

Once the capacitor is fully charged, with charges no longer in motion, the positive capacitor plate, the upper wire, and the positive terminal of the battery form a single

The keys on most computer keyboards are capacitor switches. Pressing the key pushes two capacitor plates closer together, increasing their capacitance. A larger capacitor can hold more charge, so a momentary current carries charge from the battery (or power supply) to the capacitor. This current is sensed, and the keystroke is then recorded. Capacitor switches are much more reliable than make-and-break contact switches.

conductor in electrostatic equilibrium. This is an important idea, and it wasn't true while the capacitor was charging. As you just learned, any two points in a conductor in electrostatic equilibrium are at the same potential. Thus the positive plate of a fully charged capacitor is at the same potential as the positive terminal of the battery.

Similarly, the negative plate of a fully charged capacitor is at the same potential as the negative terminal of the battery. Consequently, the potential difference ΔV_C between the capacitor plates exactly matches the potential difference ΔV_{bat} between the battery terminals. **A capacitor attached to a battery charges until $\Delta V_C = \Delta V_{bat}$.** Once the capacitor is charged, you can disconnect it from the battery; it will maintain this charge and potential difference until and unless something—a current—allows positive charge to move back to the negative plate. An ideal capacitor in vacuum would stay charged forever.

You learned in Chapter 28 that a parallel-plate capacitor's potential difference is related to the electric field inside by $\Delta V_C = Ed$, where d is the separation between the plates. And you know from Chapter 26 that a capacitor's electric field is

$$E = \frac{Q}{\epsilon_0 A} \tag{29.15}$$

where A is the surface area of the plates. Combining these gives

$$Q = \frac{\epsilon_0 A}{d} \Delta V_C \tag{29.16}$$

In other words, **the charge on the capacitor plates is directly proportional to the potential difference between the plates.**

The ratio of the charge Q to the potential difference ΔV_C is called the **capacitance** C:

$$C \equiv \frac{Q}{\Delta V_C} = \frac{\epsilon_0 A}{d} \quad \text{(parallel-plate capacitor)} \tag{29.17}$$

Capacitance is a purely *geometric* property of two electrodes because it depends only on their surface area and spacing. The SI unit of capacitance is the **farad,** named in honor of Michael Faraday. One farad is defined as

$$1 \text{ farad} = 1 \text{ F} \equiv 1 \text{ C/V}$$

One farad turns out to be an enormous amount of capacitance. Practical capacitors are usually measured in units of microfarads (μF) or picofarads (1 pF = 10^{-12} F).

With this definition of capacitance, Equation 29.17 can be written

$$Q = C \Delta V_C \quad \text{(charge on a capacitor)} \tag{29.18}$$

The charge on a capacitor is determined jointly by the potential difference supplied by a battery *and* a property of the electrodes called capacitance.

EXAMPLE 29.6 **Charging a capacitor**

The spacing between the plates of a 1.0 μF capacitor is 0.050 mm.

a. What is the surface area of the plates?

b. How much charge is on the plates if this capacitor is attached to a 1.5 V battery?

MODEL Assume the battery is ideal and the capacitor is a parallel-plate capacitor.

SOLVE a. From the definition of capacitance,

$$A = \frac{dC}{\epsilon_0} = 5.65 \text{ m}^2$$

b. The charge is $Q = C \Delta V_C = 1.5 \times 10^{-6} \text{ C} = 1.5 \ \mu\text{C}$.

ASSESS The surface area needed to construct a 1.0 μF capacitor (a fairly typical value) is enormous. We'll see in Section 29.7 how the area can be reduced by inserting an insulator between the capacitor plates.

Forming a Capacitor

The parallel-plate capacitor is important because it is straightforward to analyze and it produces a uniform electric field. But capacitors and capacitance are not limited to flat, parallel electrodes. *Any* two electrodes, regardless of their shape, form a capacitor.

FIGURE 29.22 shows two arbitrary electrodes charged to $\pm Q$. The net charge, as was the case with a parallel-plate capacitor, is zero. By definition, the capacitance of the two electrodes is

$$C = \frac{Q}{\Delta V_C} \tag{29.19}$$

where ΔV_C is the potential difference between the positive and negative electrodes. It might appear that the capacitance depends on the amount of charge, but the potential difference is proportional to Q. Consequently, **the capacitance depends only on the geometry of the electrodes.**

To make use of Equation 29.19, you must be able to determine the potential difference between the electrodes when they are charged to $\pm Q$. You can do so if you know the electric field—say from Gauss's law—by carrying out the integration of Equation 29.3. Several homework problems will let you try this to calculate the capacitance of electrodes with other geometries.

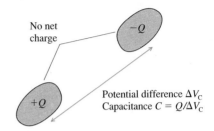

FIGURE 29.22 Any two electrodes form a capacitor.

Combinations of Capacitors

In practice, two or more capacitors are sometimes joined together. FIGURE 29.23 illustrates two basic combinations: **parallel capacitors** and **series capacitors**. Notice that a capacitor, no matter what its actual geometric shape, is represented in *circuit diagrams* by two parallel lines.

FIGURE 29.23 Parallel and series capacitors.

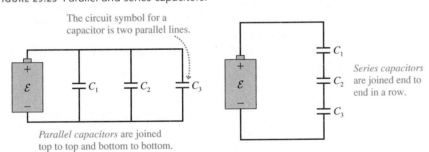

NOTE ▶ The terms "parallel capacitors" and "parallel-plate capacitor" do not describe the same thing. The former term describes how two or more capacitors are connected to each other, the latter describes how a particular capacitor is constructed. ◀

As we'll show, parallel or series capacitors (or, as is sometimes said, capacitors "in parallel" or "in series") can be represented by a single **equivalent capacitance.** We'll demonstrate this first with the two parallel capacitors C_1 and C_2 of FIGURE 29.24a. Because the two top electrodes are connected by a conducting wire, they form a single conductor in electrostatic equilibrium. Thus the two top electrodes are at the same potential. Similarly, the two connected bottom electrodes are at the same potential. Consequently, two (or more) capacitors in parallel each have the *same* potential difference ΔV_C between the two electrodes.

The charges on the two capacitors are $Q_1 = C_1 \Delta V_C$ and $Q_2 = C_2 \Delta V_C$. Altogether, the battery's charge escalator moved total charge $Q = Q_1 + Q_2$ from the negative

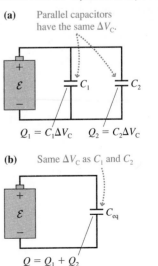

FIGURE 29.24 Replacing two parallel capacitors with an equivalent capacitor.

electrodes to the positive electrodes. Suppose, as in FIGURE 29.24b, we replaced the two capacitors with a single capacitor having charge $Q = Q_1 + Q_2$ and potential difference ΔV_C. This capacitor is equivalent to the original two in the sense that the battery can't tell the difference. In either case, the battery has to establish the same potential difference and move the same amount of charge.

By definition, the capacitance of this equivalent capacitor is

$$C_{eq} = \frac{Q}{\Delta V_C} = \frac{Q_1 + Q_2}{\Delta V_C} = \frac{Q_1}{\Delta V_C} + \frac{Q_2}{\Delta V_C} = C_1 + C_2 \qquad (29.20)$$

This analysis hinges on the fact that **parallel capacitors each have the same potential difference ΔV_C.** We could easily extend this analysis to more than two capacitors. If capacitors $C_1, C_2, C_3, \ldots$ are in parallel, their equivalent capacitance is

$$C_{eq} = C_1 + C_2 + C_3 + \cdots \qquad \text{(parallel capacitors)} \qquad (29.21)$$

Neither the battery nor any other part of a circuit can tell if the parallel capacitors are replaced by a single capacitor having capacitance C_{eq}.

Now consider the two series capacitors in FIGURE 29.25a. The center section, consisting of the bottom plate of C_1, the top plate of C_2, and the connecting wire, is electrically isolated. The battery cannot remove charge from or add charge to this section. If it starts out with no net charge, it must end up with no net charge. As a consequence, the two capacitors in series have equal charges $\pm Q$. The battery transfers Q from the bottom of C_2 to the top of C_1. This transfer polarizes the center section, as shown, but it still has $Q_{net} = 0$.

The potential differences across the two capacitors are $\Delta V_1 = Q/C_1$ and $\Delta V_2 = Q/C_2$. The total potential difference across both capacitors is $\Delta V_C = \Delta V_1 + \Delta V_2$. Suppose, as in FIGURE 29.25b, we replaced the two capacitors with a single capacitor having charge Q and potential difference $\Delta V_C = \Delta V_1 + \Delta V_2$. This capacitor is equivalent to the original two because the battery has to establish the same potential difference and move the same amount of charge in either case.

By definition, the capacitance of this equivalent capacitor is $C_{eq} = Q/\Delta V_C$. The inverse of the equivalent capacitance is thus

$$\frac{1}{C_{eq}} = \frac{\Delta V_C}{Q} = \frac{\Delta V_1 + \Delta V_2}{Q} = \frac{\Delta V_1}{Q} + \frac{\Delta V_2}{Q} = \frac{1}{C_1} + \frac{1}{C_2} \qquad (29.22)$$

This analysis hinges on the fact that **series capacitors each have the same charge Q.** We could easily extend this analysis to more than two capacitors. If capacitors $C_1, C_2, C_3, \ldots$ are in series, their equivalent capacitance is

$$C_{eq} = \left(\frac{1}{C_1} + \frac{1}{C_2} + \frac{1}{C_3} + \cdots \right)^{-1} \qquad \text{(series capacitors)} \qquad (29.23)$$

NOTE ▶ Be careful to avoid the common error of adding the inverses but forgetting to invert the sum. ◀

Let's summarize the key facts before looking at a numerical example:

- Parallel capacitors all have the same potential difference ΔV_C. Series capacitors all have the same amount of charge $\pm Q$.
- The equivalent capacitance of a parallel combination of capacitors is *larger* than any single capacitor in the group. The equivalent capacitance of a series combination of capacitors is *smaller* than any single capacitor in the group.

FIGURE 29.25 Replacing two series capacitors with an equivalent capacitor.

(a) Series capacitors have the same Q.

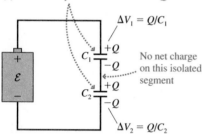

$\Delta V_1 = Q/C_1$

No net charge on this isolated segment

$\Delta V_2 = Q/C_2$

(b) Same Q as C_1 and C_2

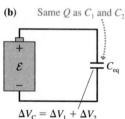

$\Delta V_C = \Delta V_1 + \Delta V_2$
Same total potential difference as C_1 and C_2

EXAMPLE 29.7 **A capacitor circuit**

Find the charge on and the potential difference across each of the three capacitors in FIGURE 29.26.

FIGURE 29.26 A capacitor circuit.

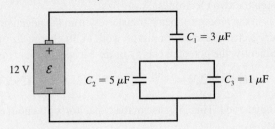

MODEL Assume the battery is ideal, with $\Delta V_{bat} = \mathcal{E} = 12$ V. Use the results for parallel and series capacitors.

SOLVE The three capacitors are neither in parallel nor in series, but we can break them down into smaller groups that are. A useful method of *circuit analysis* is first to combine elements until reaching a single equivalent element, then to reverse the process and calculate values for each element. FIGURE 29.27 shows the analysis of this circuit. Notice that we redraw the circuit after every step. The equivalent capacitance of the 3 μF and 6 μF capacitors in series is found from

$$C_{eq} = \left(\frac{1}{3\,\mu F} + \frac{1}{6\,\mu F}\right)^{-1} = \left(\frac{2}{6} + \frac{1}{6}\right)^{-1} \mu F = 2\,\mu F$$

FIGURE 29.27 Analyzing the capacitor circuit.

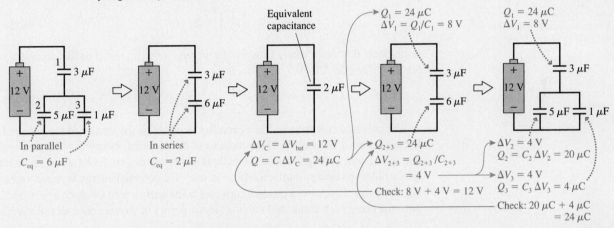

Once we get to the single equivalent capacitance, we find that $\Delta V_C = \Delta V_{bat} = 12$ V and $Q = C\Delta V_C = 24\,\mu$C. Now we can reverse direction. Capacitors in series all have the same charge, so the charge on C_1 and on C_{2+3} is $\pm 24\,\mu$C. This is enough to determine that $\Delta V_1 = 8$ V and $\Delta V_{2+3} = 4$ V. Capacitors in parallel all have the same potential difference, so $\Delta V_2 = \Delta V_3 = 4$ V. This is enough to find that $Q_2 = 20\,\mu$C and $Q_3 = 4\,\mu$C. The charge on and the potential difference across each of the three capacitors are shown in the final step of Figure 29.27.

ASSESS Notice that we had two important checks of internal consistency. $\Delta V_1 + \Delta V_{2+3} = 8$ V + 4 V add up to the 12 V we had found for the 2 μF equivalent capacitor. Then $Q_2 + Q_3 = 20\,\mu$C + 4 μC add up to the 24 μC we had found for the 6 μF equivalent capacitor. We'll do much more circuit analysis of this type in the next chapter, but it's worth noting now that circuit analysis becomes nearly foolproof *if* you make use of these checks of internal consistency.

STOP TO THINK 29.5 Rank in order, from largest to smallest, the equivalent capacitance $(C_{eq})_a$ to $(C_{eq})_d$ of circuits a to d.

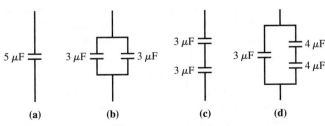

29.6 The Energy Stored in a Capacitor

FIGURE 29.28 The charge escalator does work on charge dq as the capacitor is being charged.

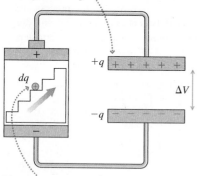

The instantaneous charge on the plates is $\pm q$.

The instantaneous charge on the plates is $\pm q$.

$+q$

dq

ΔV

$-q$

The charge escalator does work $dq\,\Delta V$ to move charge dq from the negative plate to the positive plate.

Capacitors are important elements in electric circuits because of their ability to store energy. FIGURE 29.28 shows a capacitor being charged. The instantaneous value of the charge on the two plates is $\pm q$, and this charge separation has established a potential difference $\Delta V = q/C$ between the two electrodes.

An additional charge dq is in the process of being transferred from the negative to the positive electrode. The battery's charge escalator must do work to lift charge dq "uphill" to a higher potential. Consequently, the potential energy of dq + capacitor increases by

$$dU = dq\,\Delta V = \frac{q\,dq}{C} \qquad (29.24)$$

NOTE ▶ Energy must be conserved. This increase in the capacitor's potential energy is provided by the battery. ◀

The total energy transferred from the battery to the capacitor is found by integrating Equation 29.24 from the start of charging, when $q = 0$, until the end, when $q = Q$. Thus we find that the energy stored in a charged capacitor is

$$U_C = \frac{1}{C} \int_0^Q q\,dq = \frac{Q^2}{2C} \qquad (29.25)$$

In practice, it is often easier to write the stored energy in terms of the capacitor's potential difference $\Delta V_C = Q/C$. This is

$$U_C = \frac{Q^2}{2C} = \frac{1}{2} C (\Delta V_C)^2 \qquad (29.26)$$

The potential energy stored in a capacitor depends on the *square* of the potential difference across it. This result is reminiscent of the potential energy $U = \frac{1}{2}k(\Delta x)^2$ stored in a spring, and a charged capacitor really is analogous to a stretched spring. A stretched spring holds the energy until we release it, then that potential energy is transformed into kinetic energy. Likewise, a charged capacitor holds energy until we discharge it. Then the potential energy is transformed into the kinetic energy of moving charges (the current).

EXAMPLE 29.8 **Storing energy in a capacitor**

How much energy is stored in a 2.0 μF capacitor that has been charged to 5000 V? What is the average power dissipation if this capacitor is discharged in 10 μs?

SOLVE The energy stored in the charged capacitor is

$$U_C = \frac{1}{2} C (\Delta V_C)^2 = \frac{1}{2} (2.0 \times 10^{-6}\ \text{F})(5000\ \text{V})^2 = 25\ \text{J}$$

If this energy is released in 10 μs, the average power dissipation is

$$P = \frac{\Delta E}{\Delta t} = \frac{25\ \text{J}}{1.0 \times 10^{-5}\ \text{s}} = 2.5 \times 10^6\ \text{W} = 2.5\ \text{MW}$$

ASSESS The stored energy is equivalent to raising a 1 kg mass 2.5 m. This is a rather large amount of energy, which you can see by imagining the damage a 1 kg mass could do after falling 2.5 m. When this energy is released very quickly, which is possible in an electric circuit, it provides an *enormous* amount of power.

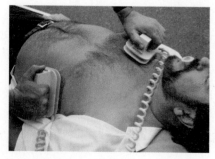

A defibrillator, which can restore a normal heartbeat, discharges a capacitor through the patient's chest.

The usefulness of a capacitor stems from the fact that it can be charged slowly (the charging rate is usually limited by the battery's ability to transfer charge) but then can release the energy very quickly. A mechanical analogy would be using a crank to slowly stretch the spring of a catapult, then quickly releasing the energy to launch a massive rock.

The capacitor described in Example 29.8 is typical of the capacitors used in high-power pulsed lasers. The capacitor is charged relatively slowly, in about 0.1 s, then quickly discharged into the laser tube to generate a high-power laser pulse. Exactly the same thing occurs, only on a smaller scale, in the flash unit of a camera. The camera batteries charge a capacitor, then the energy stored in the capacitor is quickly discharged into a *flashlamp*. The charging process in a camera takes several seconds, which is why you can't fire a camera flash twice in quick succession.

An important medical application of capacitors is the *defibrillator*. A heart attack or a serious injury can cause the heart to enter a state known as *fibrillation* in which the heart muscles twitch randomly and cannot pump blood. A strong electric shock through the chest completely stops the heart, giving the cells that control the heart's rhythm a chance to restore the proper heartbeat. A defibrillator has a large capacitor that can store up to 360 J of energy. This energy is released in about 2 ms through two "paddles" pressed against the patient's chest. It takes several seconds to charge the capacitor, which is why, on television medical shows, you hear an emergency room doctor or nurse shout "Charging!"

The Energy in the Electric Field

We can "see" the potential energy of a stretched spring in the tension of the coils. If a charged capacitor is analogous to a stretched spring, where is the stored energy? It's in the electric field!

FIGURE 29.29 shows a parallel-plate capacitor in which the plates have area A and are separated by distance d. The potential difference across the capacitor is related to the electric field inside the capacitor by $\Delta V_C = Ed$. The capacitance, which we found in Equation 29.17, is $C = \epsilon_0 A/d$. Substituting these into Equation 29.26, we find that the energy stored in the capacitor is

$$U_C = \frac{1}{2}C(\Delta V_C)^2 = \frac{1}{2}\frac{\epsilon_0 A}{d}(Ed)^2 = \frac{\epsilon_0}{2}(Ad)E^2 \qquad (29.27)$$

The quantity Ad is the volume *inside* the capacitor, the region in which the capacitor's electric field exists. (Recall that an ideal capacitor has $\vec{E} = \vec{0}$ everywhere except between the plates.) Although we talk about "the energy stored in the capacitor," Equation 29.27 suggests that, strictly speaking, **the energy is stored in the capacitor's electric field.**

Because Ad is the volume in which the energy is stored, we can define an **energy density** u_E of the electric field:

$$u_E = \frac{\text{energy stored}}{\text{volume in which it is stored}} = \frac{U_C}{Ad} = \frac{\epsilon_0}{2}E^2 \qquad (29.28)$$

The energy density has units J/m^3. We've derived Equation 29.28 for a parallel-plate capacitor, but it turns out to be the correct expression for any electric field.

From this perspective, charging a capacitor stores energy in the capacitor's electric field as the field grows in strength. Later, when the capacitor is discharged, the energy is released as the field collapses.

We first introduced the electric field as a way to visualize how a long-range force operates. But if the field can store energy, the field must be real, not merely a pictorial device. We'll explore this idea further in Chapter 34, where we'll find that the energy transported by a light wave—the very real energy of warm sunshine—is the energy of electric and magnetic fields.

FIGURE 29.29 A capacitor's energy is stored in the electric field.

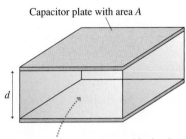

Capacitor plate with area A

The capacitor's energy is stored in the electric field in volume Ad between the plates.

29.7 Dielectrics

FIGURE 29.30a shows a parallel-plate capacitor with the plates separated by vacuum, the perfect insulator. Suppose the capacitor is charged to voltage $(\Delta V_C)_0$, then disconnected from the battery. The charge on the plates will be $\pm Q_0$, where $Q_0 = C_0(\Delta V_C)_0$. We'll use a subscript 0 in this section to refer to a vacuum-insulated capacitor.

FIGURE 29.30 Vacuum-insulated and dielectric-filled capacitors.

(a)

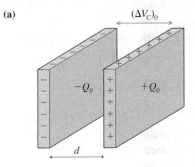

Capacitance C_0 in vacuum

(b)

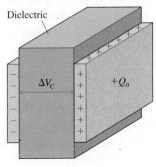

Capacitance $C > C_0$

Now suppose, as in FIGURE 29.30b, an insulating material, such as oil or glass or plastic, is slipped between the capacitor plates. We'll assume for now that the insulator is of thickness d and completely fills the space. An insulator in an electric field is called a **dielectric,** for reasons that will soon become clear, so we call this a *dielectric-filled capacitor*. How does a dielectric-filled capacitor differ from the vacuum-insulated capacitor?

The charge on the capacitor plates does not change. The insulator doesn't allow charge to move through it, and the capacitor has been disconnected from the battery, so no charge can be added to or removed from either plate. That is, $Q = Q_0$. Nonetheless, measurements of the capacitor voltage with a voltmeter would find that the voltage has decreased: $\Delta V_C < (\Delta V_C)_0$. Consequently, based on the definition of capacitance, the capacitance has increased:

$$C = \frac{Q}{\Delta V_C} > \frac{Q_0}{(\Delta V_C)_0} = C_0$$

Example 29.6 found that the plate size needed to make a 1 μF capacitor is unreasonably large. It appears that we can get more capacitance *with the same plates* by filling the capacitor with an insulator.

We can utilize two tools you learned in Chapter 26, superposition and polarization, to understand the properties of dielectric-filled capacitors. Figure 26.28 showed how an insulating material becomes *polarized* in an external electric field. FIGURE 29.31a reproduces the basic ideas from that earlier figure. The electric dipoles in Figure 29.31a could be permanent dipoles, such as water molecules, or simply induced dipoles due to a slight charge separation in the atoms. However the dipoles originate, their alignment in the electric field—the *polarization* of the material—produces an excess positive charge on one surface, an excess negative charge on the other. The insulator as a whole is still neutral, but the external electric field separates positive and negative charge.

FIGURE 29.31b represents the polarized insulator as simply two sheets of charge with surface charge densities $\pm \eta_{induced}$. The size of $\eta_{induced}$ depends both on the strength of the electric field and on the properties of the insulator. These two sheets of charge create an electric field—a situation we analyzed in Chapter 26. In essence, the two sheets of induced charge act just like the two charged plates of a parallel-plate capacitor. The **induced electric field** (keep in mind that this field is due to the insulator responding to the external electric field) is

$$\vec{E}_{induced} = \begin{cases} \left(\dfrac{\eta_{induced}}{\epsilon_0}, \text{from positive to negative} \right) & \text{inside the insulator} \\ \vec{0} & \text{outside the insulator} \end{cases} \quad (29.29)$$

It is because an insulator in an electric field has *two* sheets of induced *electric* charge that we call it a *dielectric*, with the prefix *di*, meaning *two*, the same as in "diatomic" and "dipole."

FIGURE 29.32 shows what happens when you insert a dielectric into a capacitor. The capacitor plates have their own surface charge density $\eta_0 = Q_0/A$. This creates the electric field $\vec{E}_0 = (\eta_0/\epsilon_0$, from positive to negative) into which the dielectric is placed. The dielectric responds with induced surface charge density $\eta_{induced}$ and the induced electric field $\vec{E}_{induced}$. Notice that $\vec{E}_{induced}$ points *opposite* to $\vec{E}_0$. By the principle of superposition, another important lesson from Chapter 26, the net electric field between the capacitor plates is the *vector* sum of these two fields:

$$\vec{E} = \vec{E}_0 + \vec{E}_{induced} = (E_0 - E_{induced}, \text{from positive to negative}) \quad (29.30)$$

The presence of the dielectric weakens the electric field, from E_0 to $E_0 - E_{induced}$, but the field still points from the positive capacitor plate to the negative capacitor plate. The field is weakened because the induced surface charge in the dielectric acts to counter the electric field of the capacitor plates.

FIGURE 29.31 An insulator in an external electric field.

(a) The insulator is polarized.

Excess positive charge on this surface   Excess negative charge on this surface

$\vec{E}_0$ $\vec{E}_0$

(b) The polarized insulator—a dielectric—can be represented as two sheets of surface charge. This surface charge creates an electric field inside the insulator.

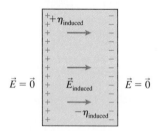

$+\eta_{induced}$

$\vec{E} = \vec{0}$ $\vec{E}_{induced}$ $\vec{E} = \vec{0}$

$-\eta_{induced}$

FIGURE 29.32 The consequences of filling a capacitor with a dielectric.

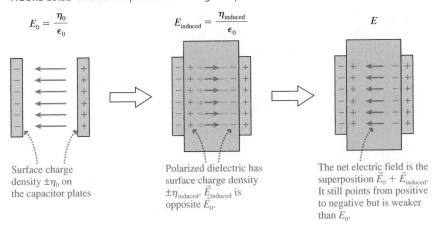

Surface charge density $\pm\eta_0$ on the capacitor plates

Polarized dielectric has surface charge density $\pm\eta_{induced}$. $\vec{E}_{induced}$ is opposite $\vec{E}_0$.

The net electric field is the superposition $\vec{E}_0 + \vec{E}_{induced}$. It still points from positive to negative but is weaker than E_0.

Let's define the **dielectric constant** κ (Greek *kappa*) as

$$\kappa \equiv \frac{E_0}{E} \qquad (29.31)$$

Equivalently, the field strength inside a dielectric in an external field is $E = E_0/\kappa$. The dielectric constant is the factor by which a dielectric *weakens* an electric field, so $\kappa \geq 1$. You can see from the definition that κ is a pure number with no units.

The dielectric constant, like density or specific heat, is a property of a material. Easily polarized materials have larger dielectric constants than materials not easily polarized. Vacuum has $\kappa = 1$ exactly, and low-pressure gases have $\kappa \approx 1$. (Air has $\kappa_{air} = 1.00$ to three significant figures, so we won't worry about the very slight effect air has on capacitors.) Table 29.1 lists the dielectric constants for different materials.

The electric field inside the capacitor, although weakened, is still uniform. Consequently, the potential difference across the capacitor is

$$\Delta V_C = Ed = \frac{E_0}{\kappa}d = \frac{(\Delta V_C)_0}{\kappa} \qquad (29.32)$$

where $(\Delta V_C)_0 = E_0 d$ was the voltage of the vacuum-insulated capacitor. The presence of a dielectric reduces the capacitor voltage, the observation with which we started this section. Now we see why; it is due to the polarization of the material. Further, the new capacitance is

$$C = \frac{Q}{\Delta V_C} = \frac{Q_0}{(\Delta V_C)_0/\kappa} = \kappa\frac{Q_0}{(\Delta V_C)_0} = \kappa C_0 \qquad (29.33)$$

Filling a capacitor with a dielectric increases the capacitance by a factor equal to the dielectric constant. This ranges from virtually no increase for an air-filled capacitor to a capacitance 300 times larger if the capacitor is filled with strontium titanate.

We'll leave it as a homework problem to show that the induced surface charge density is

$$\eta_{induced} = \eta_0\left(1 - \frac{1}{\kappa}\right) \qquad (29.34)$$

$\eta_{induced}$ ranges from nearly zero when $\kappa \approx 1$ to $\approx \eta_0$ when $\kappa \gg 1$.

NOTE ▶ We assumed that the capacitor was disconnected from the battery after being charged, so Q couldn't change. If you insert a dielectric while a capacitor is attached to a battery, then it will be ΔV_C, fixed at the battery voltage, that can't change. In this case, more charge will flow from the battery until $Q = \kappa Q_0$. In both cases, the capacitance increases to $C = \kappa C_0$. ◀

TABLE 29.1 Properties of dielectrics

Material	Dielectric constant κ	Dielectric strength $E_{max}(10^6 \text{ V/m})$
Vacuum	1	—
Air (1 atm)	1.0006	3
Teflon	2.1	60
Polystyrene plastic	2.6	24
Mylar	3.1	7
Paper	3.7	16
Pyrex glass	4.7	14
Pure water (20°C)	80	—
Titanium dioxide	110	6
Strontium titanate	300	8

EXAMPLE 29.9 **A water-filled capacitor**

A 5.0 nF parallel-plate capacitor is charged to 160 V. It is then disconnected from the battery and immersed in distilled water. What are (a) the capacitance and voltage of the water-filled capacitor and (b) the energy stored in the capacitor before and after its immersion?

MODEL Pure distilled water is a good insulator. (The conductivity of tap water is due to dissolved ions.) Thus the immersed capacitor has a dielectric between the electrodes.

SOLVE a. From Table 29.1, the dielectric constant of water is $\kappa = 80$. The presence of the dielectric increases the capacitance to

$$C = \kappa C_0 = 80 \times 5.0 \text{ nF} = 400 \text{ nF}$$

At the same time, the voltage decreases to

$$\Delta V_C = \frac{(\Delta V_C)_0}{\kappa} = \frac{160 \text{ V}}{80} = 2.0 \text{ V}$$

b. The presence of a dielectric does not alter the derivation leading to Equation 29.26 for the energy stored in a capacitor. Right after being disconnected from the battery, the stored energy was

$$(U_C)_0 = \frac{1}{2} C_0 (\Delta V_C)_0^2 = \frac{1}{2}(5.0 \times 10^{-9} \text{ F})(160 \text{ V})^2 = 6.4 \times 10^{-5} \text{ J}$$

After being immersed, the stored energy is

$$U_C = \frac{1}{2} C (\Delta V_C)^2 = \frac{1}{2}(400 \times 10^{-9} \text{ F})(2.0 \text{ V})^2 = 8.0 \times 10^{-7} \text{ J}$$

ASSESS Water, with its large dielectric constant, has a *big* effect on the capacitor. But where did the energy go? We learned in Chapter 26 that a dipole is drawn into a region of stronger electric field. The electric field inside the capacitor is much stronger than just outside the capacitor, so the polarized dielectric is actually *pulled* into the capacitor. The "lost" energy is the work the capacitor's electric field did pulling in the dielectric.

EXAMPLE 29.10 **Energy density of a defibrillator**

A defibrillator unit contains a 150 μF capacitor that is charged to 2100 V. The capacitor plates are separated by a 0.050-mm-thick insulator with dielectric constant 120.

a. What is the area of the capacitor plates?
b. What are the stored energy and the energy density in the electric field when the capacitor is charged?

MODEL Model the defibrillator as a parallel-plate capacitor with a dielectric.

SOLVE a. The capacitance of a parallel-plate capacitor in a vacuum is $C_0 = \epsilon_0 A / d$. A dielectric increases the capacitance by the factor κ, to $C = \kappa C_0$, so the area of the capacitor plates is

$$A = \frac{Cd}{\kappa \epsilon_0} = \frac{(150 \times 10^{-6} \text{ F})(5.0 \times 10^{-5} \text{ m})}{120 \,(8.85 \times 10^{-12} \text{ C}^2/\text{N m}^2)} = 7.1 \text{ m}^2$$

Although the surface area is very large, Figure 29.33 below shows how very large sheets of very thin metal can be rolled up into capacitors that you hold in your hand.

b. The energy stored in the capacitor is

$$U_c = \frac{1}{2} C (\Delta V_C)^2 = \frac{1}{2}(150 \times 10^{-6} \text{ F})(2100 \text{ V})^2 = 330 \text{ J}$$

Because the dielectric has increased C by a factor of κ, the energy density of Equation 29.28 is increased by a factor of κ to $u_E = \frac{1}{2} \kappa \epsilon_0 E^2$. The electric field strength in the capacitor is

$$E = \frac{\Delta V_C}{d} = \frac{2100 \text{ V}}{5.0 \times 10^{-5} \text{ m}} = 4.2 \times 10^7 \text{ V/m}$$

Consequently, the energy density is

$$u_E = \frac{1}{2}(120)(8.85 \times 10^{-12} \text{ C}^2/\text{N m}^2)(4.2 \times 10^7 \text{ V/m})^2$$
$$= 9.4 \times 10^5 \text{ J/m}^3$$

ASSESS 330 J is a substantial amount of energy—equivalent to that of a 1 kg mass traveling at 25 m/s. And it can be delivered very quickly as the capacitor is discharged through the patient's chest.

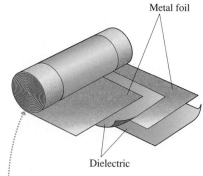

FIGURE 29.33 A practical capacitor.

Metal foil

Dielectric

Many real capacitors are a rolled-up sandwich of metal foils and thin, insulating dielectrics.

Solid or liquid dielectrics allow a set of electrodes to have more capacitance than they would if filled with air. Not surprisingly, as FIGURE 29.33 shows, this is important in the production of practical capacitors. In addition, dielectrics allow capacitors to be charged to higher voltages. All materials have a maximum electric field they can sustain without *breakdown*—the production of a spark. The breakdown electric field of air, as we've noted previously, is about 3×10^6 V/m. In general, a material's maximum sustainable electric field is called its **dielectric strength.** Table 29.1 includes dielectric strengths for air and the solid dielectrics. (The breakdown of water is extremely sensitive to ions and impurities in the water, so water doesn't have a well-defined dielectric strength.)

Many materials have dielectric strengths much larger than air. Teflon, for an example, has a dielectric strength 20 times that of air. Consequently, a Teflon-filled capacitor can be safely charged to a voltage 20 times larger than an air-filled capacitor with the same plate separation. An air-filled capacitor with a plate separation of 0.2 mm can be charged only to 600 V, but a capacitor with a 0.2-mm-thick Teflon sheet could be charged to 12,000 V.

A Geiger counter

The radiation detector known as a *Geiger counter* consists of a 25-mm-diameter cylindrical metal tube, sealed at the ends, with a 1.0-mm-diameter wire along its axis. The wire and cylinder are separated by a low-pressure gas whose dielectric strength is 1.0×10^6 V/m. What is the maximum potential difference between the wire and the tube?

MODEL Model the Geiger counter as two long, concentric, conducting cylinders. To avoid breakdown of the gas, the field strength at the surface of the wire—the point of maximum field strength—must not exceed the dielectric strength.

VISUALIZE FIGURE 29.34 shows a cross section of the Geiger counter tube. Applying a potential difference between the inner and outer cylinders charges it like a capacitor; indeed, it *is* a cylindrical capacitor. We've chosen to let the outer cylinder be positive, with an inward-pointing electric field, but a negative outer cylinder would lead to the same answer since it's only the field strength that we're interested in.

FIGURE 29.34 Cross section of a Geiger counter tube.

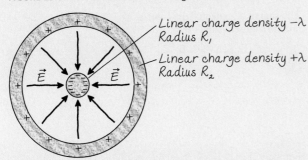

Linear charge density $-\lambda$
Radius R_1

Linear charge density $+\lambda$
Radius R_2

SOLVE Gauss's law tells us that the electric field between the cylinders is due only to the charge on the inner cylinder. Thus $\vec{E}$ is the field of a long, charged wire—a field we found in Chapter 26 using superposition and again in Chapter 27 using Gauss's law. It is

$$\vec{E} = \left(\frac{\lambda}{2\pi\epsilon_0 r}, \text{ inward} \right)$$

where λ is the magnitude of the linear charge density. We need to connect this field to the potential difference between the wire and the outer cylinder. For that, we need to use Equation 29.3:

$$\Delta V = V_f - V_i = -\int_{s_i}^{s_f} E_s\, ds$$

We'll integrate along a radial line from $s_i = R_1$ on the surface of the inner cylinder to $s_f = R_2$ at the outer cylinder. The field component E_s is negative because the field points inward. Thus the potential difference is

$$\Delta V = -\int_{R_1}^{R_2} \left(-\frac{\lambda}{2\pi\epsilon_0 s} \right) ds = \frac{\lambda}{2\pi\epsilon_0} \int_{R_1}^{R_2} \frac{ds}{s}$$

$$= \frac{\lambda}{2\pi\epsilon_0} \ln s \Big|_{R_1}^{R_2} = \frac{\lambda}{2\pi\epsilon_0} \ln\left(\frac{R_2}{R_1} \right)$$

We see that the applied potential difference and the linear charge density are related by

$$\frac{\lambda}{2\pi\epsilon_0} = \frac{\Delta V}{\ln(R_2/R_1)}$$

Using this in the expression for $\vec{E}$, we find the electric field strength at distance r is

$$E = \frac{\Delta V}{r \ln(R_2/R_1)}$$

The field strength is a maximum at the surface of the wire, where it reaches

$$E_{max} = \frac{\Delta V}{R_1 \ln(R_2/R_1)}$$

The maximum applied voltage will bring E_{max} to the dielectric strength, $E_{max} = 1.0 \times 10^6$ V/m. Thus the maximum potential difference between the wire and the tube is

$$\Delta V_{max} = R_1 E_{max} \ln\left(\frac{R_2}{R_1} \right)$$

$$= (5.0 \times 10^{-4}\text{ m})(1.0 \times 10^6\text{ V/m})\ln(25)$$

$$= 1600\text{ V}$$

ASSESS This is the *maximum* possible voltage, but it's not practical to operate right at the maximum. Real Geiger counters operate with typically a 1000 V potential difference to avoid an accidental breakdown of the gas. If a high-speed charged particle from a radioactive decay then happens to pass through the tube, it will collide with and ionize a number of the gas atoms. Because the tube is already very close to breakdown, the addition of these extra ions and electrons is enough to push it over the edge: A breakdown of the gas occurs, with a spark jumping across the tube. The "clicking" sounds of a Geiger counter are made by amplifying the current pulses associated with the sparks.

SUMMARY

The goal of Chapter 29 has been to understand how the electric potential is related to the electric field.

General Principles

Connecting V and $\vec{E}$

The electric potential and the electric field are two different perspectives of how source charges alter the space around them. V and $\vec{E}$ are related by

$$\Delta V = V_f - V_i = -\int_{s_i}^{s_f} E_s \, ds$$

where s is measured from point i to point f and E_s is the component of $\vec{E}$ parallel to the line of integration.

Graphically

ΔV = the negative of the area under the E_s graph

and

$$E_s = -\frac{dV}{ds}$$

= the negative of the slope of the potential graph

The Geometry of Potential and Field

The electric field

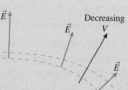

- Is perpendicular to the equipotential surfaces.
- Points "downhill" in the direction of decreasing V.
- Is inversely proportional to the spacing Δs between the equipotential surfaces.

Conservation of Energy

The sum of all potential differences around a closed path is zero.

$$\sum (\Delta V)_i = 0$$

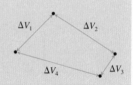

Important Concepts

A **battery** is a source of potential. The charge escalator in a battery uses chemical reactions to move charges from the negative terminal to the positive terminal:

$$\Delta V_{bat} = \mathcal{E}$$

where the emf $\mathcal{E}$ is the work per charge done by the charge escalator.

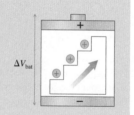

For a conductor in electrostatic equilibrium

- The interior electric field is zero.
- The exterior electric field is perpendicular to the surface.
- The surface is an equipotential.
- The interior is at the same potential as the surface.

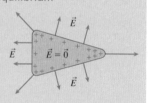

Applications

Capacitors

The **capacitance** of two conductors charged to $\pm Q$ is

$$C = \frac{Q}{\Delta V_C}$$

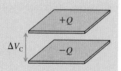

A parallel-plate capacitor has

$$C = \frac{\epsilon_0 A}{d}$$

Filling the space between the plates with a dielectric of dielectric constant κ increases the capacitance to $C = \kappa C_0$.

The energy stored in a capacitor is $u_C = \frac{1}{2} C (\Delta V_C)^2$.

This energy is stored in the electric field at density $u_E = \frac{1}{2} \kappa \epsilon_0 E^2$.

Combinations of capacitors

Series capacitors

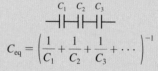

$$C_{eq} = \left(\frac{1}{C_1} + \frac{1}{C_2} + \frac{1}{C_3} + \cdots \right)^{-1}$$

Parallel capacitors

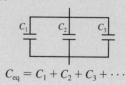

$$C_{eq} = C_1 + C_2 + C_3 + \cdots$$

Terms and Notation

Van de Graaff generator	terminal voltage, ΔV_{bat}	series capacitors	dielectric constant, κ
battery	Kirchhoff's loop law	equivalent capacitance, C_{eq}	dielectric strength
charge escalator model	capacitance, C	energy density, u_E	
ideal battery	farad, F	dielectric	
emf, $\mathcal{E}$	parallel capacitors	induced electric field	

CONCEPTUAL QUESTIONS

1. **FIGURE Q29.1** shows the x-component of $\vec{E}$ as a function of x. Draw a graph of V versus x in this same region of space. Let $V = 0$ V at $x = 0$ m and include an appropriate vertical scale.

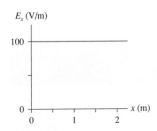

FIGURE Q29.1

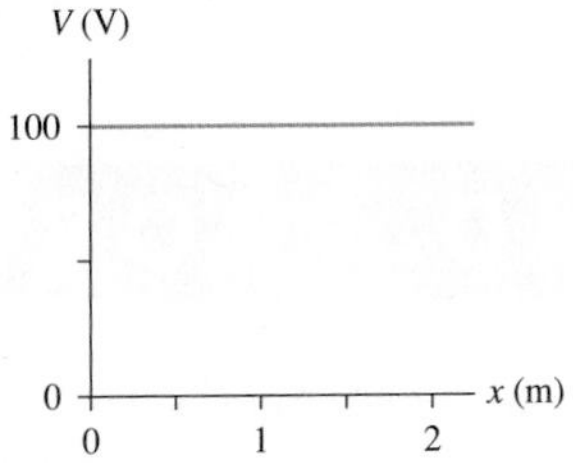

FIGURE Q29.2

2. **FIGURE Q29.2** shows the electric potential as a function of x. Draw a graph of E_x versus x in this same region of space.

3. a. Suppose that $\vec{E} = \vec{0}$ V/m throughout some region of space. Can you conclude that $V = 0$ V in this region? Explain.

   b. Suppose that $V = 0$ V throughout some region of space. Can you conclude that $\vec{E} = \vec{0}$ V/m in this region? Explain.

4. For each contour map in **FIGURE Q29.4**, estimate the electric fields $\vec{E}_1$ and $\vec{E}_2$ at points 1 and 2. Don't forget that $\vec{E}$ is a vector.

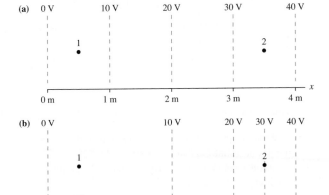

FIGURE Q29.4

5. An electron is released from rest at $x = 2$ m in the potential shown in **FIGURE Q29.5**. Does it move? If so, to the left or to the right? Explain.

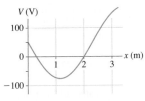

FIGURE Q29.5

6. **FIGURE Q29.6** shows an electric field diagram. Dashed lines 1 and 2 are two surfaces in space, not physical objects.

   a. Is the electric potential at point a higher than, lower than, or equal to the electric potential at point b? Explain.

   b. Rank in order, from largest to smallest, the magnitudes of the potential differences ΔV_{ab}, ΔV_{cd}, and ΔV_{ef}.

   c. Is surface 1 an equipotential surface? What about surface 2? Explain why or why not.

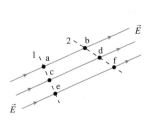

FIGURE Q29.6

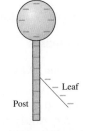

FIGURE Q29.7

7. **FIGURE Q29.7** shows a negatively charged electroscope. The gold leaf stands away from the rigid metal post. Is the electric potential of the leaf higher than, lower than, or equal to the potential of the post? Explain.

8. The two metal spheres in **FIGURE Q29.8** are connected by a metal wire with a switch in the middle. Initially the switch is open. Sphere 1, with the larger radius, is given a positive charge. Sphere 2, with the smaller radius, is neutral. Then the switch is closed. Afterward, sphere 1 has charge Q_1, is at potential V_1, and the electric field strength at its surface is E_1. The values for sphere 2 are Q_2, V_2, and E_2.

   a. Is V_1 larger than, smaller than, or equal to V_2? Explain.

   b. Is Q_1 larger than, smaller than, or equal to Q_2? Explain.

   c. Is E_1 larger than, smaller than, or equal to E_2? Explain.

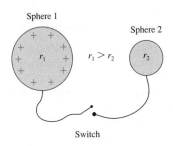

FIGURE Q29.8

9. FIGURE Q29.9 shows a 3 V battery with metal wires attached to each end. What are the potential differences ΔV_{12}, ΔV_{23}, ΔV_{34}, and ΔV_{14}?

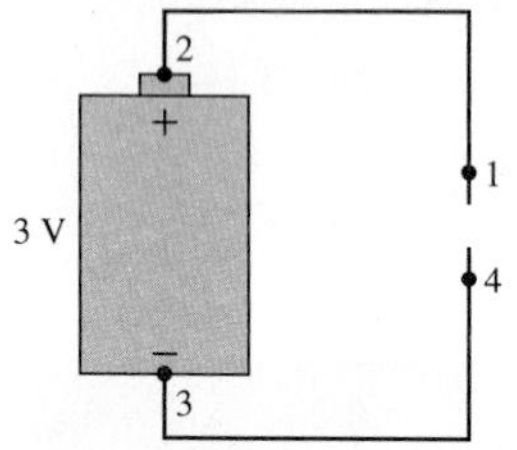

FIGURE Q29.9

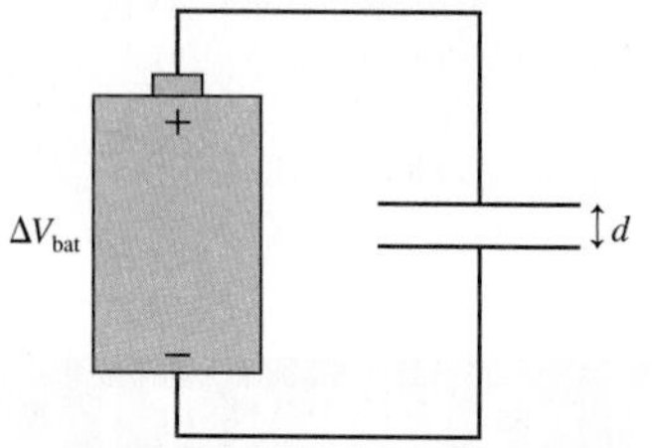

FIGURE Q29.10

10. The parallel-plate capacitor in FIGURE Q29.10 is connected to a battery having potential difference ΔV_{bat}. Without breaking any of the connections, insulating handles are used to increase the plate separation to $2d$.

a. Does the potential difference ΔV_C change as the separation increases? If so, by what factor? If not, why not?
b. Does the capacitance change? If so, by what factor? If not, why not?
c. Does the capacitor charge Q change? If so, by what factor? If not, why not?

11. Rank in order, from largest to smallest, the potential differences $(\Delta V_C)_1$ to $(\Delta V_C)_4$ of the four capacitors in FIGURE Q29.11. Explain.

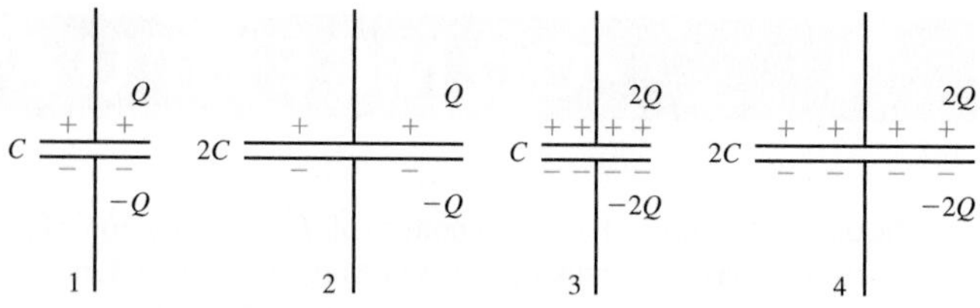

FIGURE Q29.11

EXERCISES AND PROBLEMS

Problems labeled ▭ integrate material from earlier chapters.

Exercises

Section 29.1 Connecting Potential and Field

1. | | What is the potential difference between $x_i = 10$ cm and $x_f = 30$ cm in the uniform electric field $E_x = 1000$ V/m?
2. | | What is the potential difference between $y_i = -5$ cm and $y_f = 5$ cm in the uniform electric field $\vec{E} = (20{,}000\hat{\imath} - 50{,}000\hat{\jmath})$ V/m?
3. | | FIGURE EX29.3 is a graph of E_x. What is the potential difference between $x_i = 1.0$ m and $x_f = 3.0$ m?

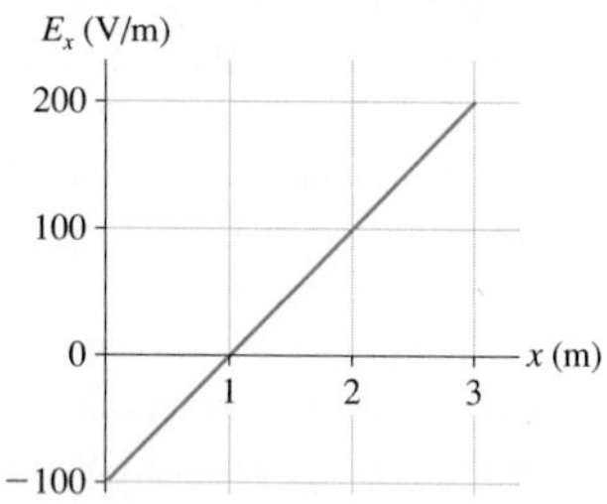

FIGURE EX29.3

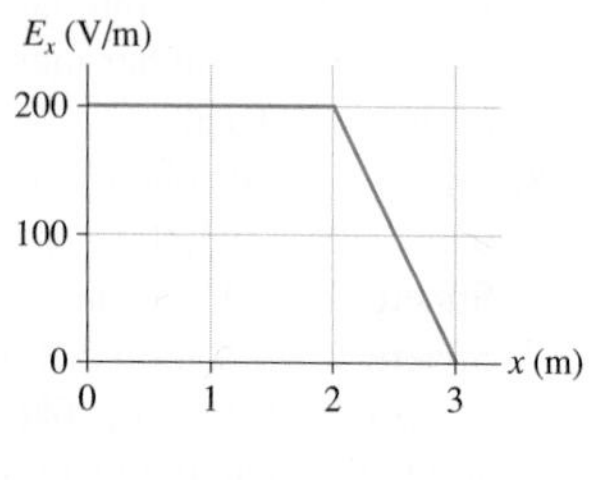

FIGURE EX29.4

4. | | FIGURE EX29.4 is a graph of E_x. The potential at the origin is -50 V. What is the potential at $x = 3.0$ m?

Section 29.2 Sources of Electric Potential

5. | How much work does the charge escalator do to move 1.0 μC of charge from the negative terminal to the positive terminal of a 1.5 V battery?
6. | | How much work does the electric motor of a Van de Graaff generator do to lift a positive ion ($q = e$) if the potential of the spherical electrode is 1.0 MV?
7. | | How much charge does a 9.0 V battery transfer from the negative to the positive terminal while doing 27 J of work?
8. | Light from the sun allows a solar cell to move electrons from the positive to the negative terminal, doing 2.4×10^{-19} J of work per electron. What is the emf of this solar cell?

Section 29.3 Finding the Electric Field from the Potential

9. | What are the magnitude and direction of the electric field at the dot in FIGURE EX29.9?

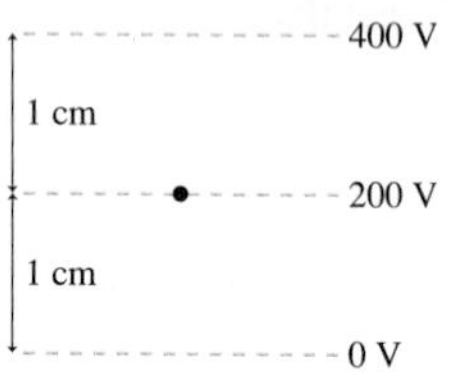

FIGURE EX29.9

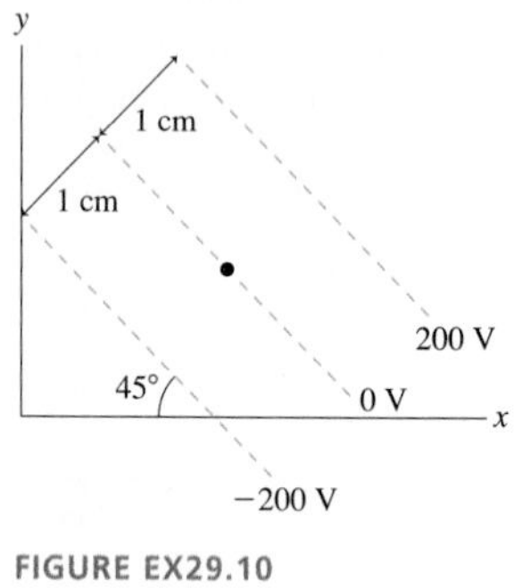

FIGURE EX29.10

10. | What are the magnitude and direction of the electric field at the dot in FIGURE EX29.10?
11. | | FIGURE EX29.11 is a graph of V versus x. Draw the corresponding graph of E_x versus x.

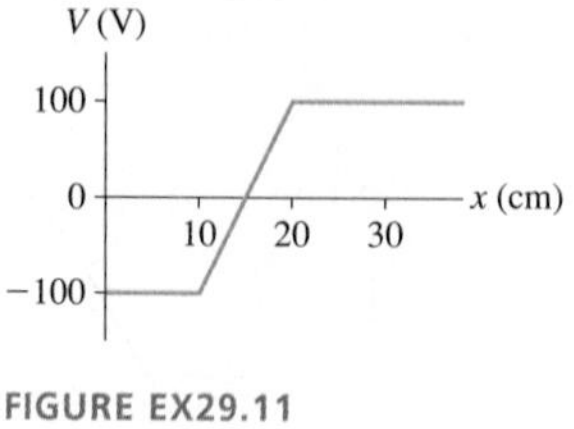

FIGURE EX29.11

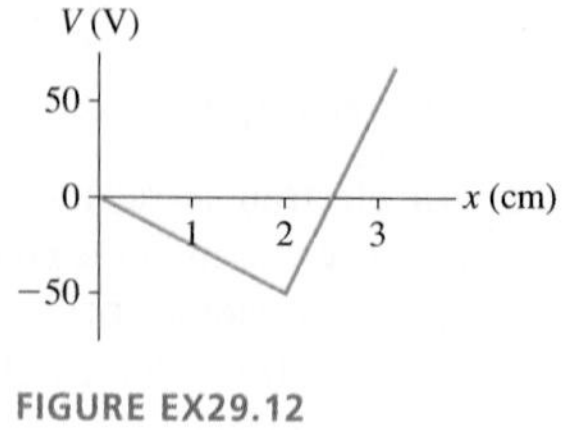

FIGURE EX29.12

12. | | FIGURE EX29.12 is a graph of V versus x. Draw the corresponding graph of E_x versus x.

13. ‖ The electric potential in a region of uniform electric field is -1000 V at $x = -1.0$ m and $+1000$ V at $x = +1.0$ m. What is E_x?

14. ‖ The electric potential along the x-axis is $V = 100x^2$ V, where x is in meters. What is E_x at (a) $x = 0$ m and (b) $x = 1$ m?

15. ‖ The electric potential along the x-axis is $V = 100e^{-2x}$ V, where x is in meters. What is E_x at (a) $x = 1.0$ m and (b) $x = 2.0$ m?

16. │ What is the potential difference ΔV_{34} in FIGURE EX29.16?

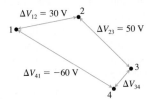

FIGURE EX29.16

Section 29.5 Capacitance and Capacitors

17. │ Two 3.0-cm-diameter aluminum electrodes are spaced 0.50 mm apart. The electrodes are connected to a 100 V battery.
    a. What is the capacitance?
    b. What is the magnitude of the charge on each electrode?

18. ‖ You need to construct a 100 pF capacitor for a science project. You plan to cut two $L \times L$ metal squares and insert small spacers between their corners. The thinnest spacers you have are 0.20 mm thick. What is the proper value of L?

19. │ A switch that connects a battery to a 10 μF capacitor is closed. Several seconds later you find that the capacitor plates are charged to ± 30 μC. What is the emf of the battery?

20. │ A 6 μF capacitor, a 10 μF capacitor, and a 16 μF capacitor are connected in series. What is their equivalent capacitance?

21. │ A 6 μF capacitor, a 10 μF capacitor, and a 16 μF capacitor are connected in parallel. What is their equivalent capacitance?

22. │ You need a capacitance of 50 μF, but you don't happen to have a 50 μF capacitor. You do have a 30 μF capacitor. What additional capacitor do you need to produce a total capacitance of 50 μF? Should you join the two capacitors in parallel or in series?

23. │ You need a capacitance of 50 μF, but you don't happen to have a 50 μF capacitor. You do have a 75 μF capacitor. What additional capacitor do you need to produce a total capacitance of 50 μF? Should you join the two capacitors in parallel or in series?

24. ‖ What is the capacitance of the two metal spheres shown in FIGURE EX29.24?

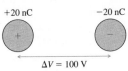

FIGURE EX29.24

Section 29.6 The Energy Stored in a Capacitor

25. ‖ To what potential should you charge a 1.0 μF capacitor to store 1.0 J of energy?

26. ‖ FIGURE EX29.26 shows Q versus t for a 2.0 μF capacitor. Draw a graph showing U_C versus t.

FIGURE EX29.26

27. │ Capacitor 2 has half the capacitance and twice the potential difference as capacitor 1. What is the ratio U_{C1}/U_{C2}?

28. ‖ 50 pJ of energy is stored in a 2.0 cm $\times$ 2.0 cm $\times$ 2.0 cm region of uniform electric field. What is the electric field strength?

29. ‖ A 2.0-cm-diameter parallel-plate capacitor with a spacing of 0.50 mm is charged to 200 V. What are (a) the total energy stored in the electric field and (b) the energy density?

Section 29.7 Dielectrics

30. ‖ Two 4.0 cm $\times$ 4.0 cm metal plates are separated by a 0.20-mm-thick piece of Teflon.
    a. What is the capacitance?
    b. What is the maximum potential difference between the plates?

31. ‖ Two 5.0 mm $\times$ 5.0 mm electrodes with a 0.10-mm-thick sheet of Mylar between them are attached to a 9.0 V battery. Without disconnecting the battery, the Mylar is withdrawn. (Very small spacers keep the electrode separation unchanged.) What are the charge, potential difference, and electric field (a) before and (b) after the Mylar is withdrawn?

32. ‖ A typical cell has a layer of negative charge on the inner
BIO surface of the cell wall and a layer of positive charge on the outside surface, thus making the cell wall a capacitor. What is the capacitance of a 50-μm-diameter cell with a 7.0-nm-thick cell wall whose dielectric constant is 9.0? Because the cell's diameter is much larger than the wall thickness, it is reasonable to ignore the curvature of the cell and think of it as a parallel-plate capacitor.

Problems

33. ‖ a. Which point in FIGURE P29.33, A or B, has a larger electric potential?
    b. What is the potential difference between A and B?

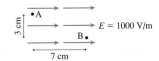

FIGURE P29.33

34. ‖‖ The electric field in a region of space is $E_x = -1000x$ V/m, where x is in meters.
    a. Graph E_x versus x over the region -1 m $\leq x \leq 1$ m.
    b. What is the potential difference between $x_i = -20$ cm and $x_f = 30$ cm?

35. ‖ The electric field in a region of space is $E_x = 5000x$ V/m, where x is in meters.
    a. Graph E_x versus x over the region -1 m $\leq x \leq 1$ m.
    b. Find an expression for the potential V at position x. As a reference, let $V = 0$ V at the origin.
    c. Graph V versus x over the region -1 m $\leq x \leq 1$ m.

36. ‖ An infinitely long cylinder of radius R has linear charge density λ. The potential on the surface of the cylinder is V_0, and the electric field outside the cylinder is $E_r = \lambda/2\pi\epsilon_0 r$. Find the potential relative to the surface at a point that is distance r from the axis, assuming $r > R$.

37. ‖ FIGURE P29.37 is an edge view of three charged metal electrodes. Let the left electrode be the zero point of the electric potential. What are V and $\vec{E}$ at (a) $x = 0.5$ cm, (b) $x = 1.5$ cm, and (c) $x = 2.5$ cm?

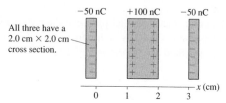

FIGURE P29.37

38. ‖ FIGURE P29.38 shows a graph of V versus x in a region of space. The potential is independent of y and z. What is E_x at (a) $x = -2$ cm, (b) $x = 0$ cm, and (c) $x = 2$ cm?

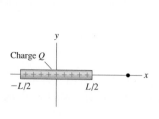

FIGURE P29.38

39. ‖ Use the on-axis potential of a charged disk from Chapter 28 to find the on-axis electric field of a charged disk.

40. ‖ a. Use the methods of Chapter 28 to find the potential at distance x on the axis of the charged rod shown in FIGURE P29.40.
    b. Use the result of part a to find the electric field at distance x on the axis of a rod.

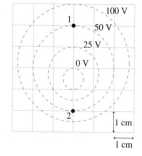

FIGURE P29.40   FIGURE P29.41

41. ‖ Determine the magnitude and direction of the electric field at points 1 and 2 in FIGURE P29.41.

42. ‖ It is postulated that the radial electric field of a group of charges falls off as $E_r = C/r^n$, where r is the distance from the center of the group and n is an unknown exponent. To test this hypothesis, you make a *field probe* consisting of two needle tips spaced 1.00 mm apart. You orient the needles so that a line between the tips points to the center of the charges, then use a voltmeter to read the potential difference between the tips. After you take measurements at several distances from the center of the group, your data are as follows:

Distance (cm)	Potential difference (mV)
2.0	34.7
4.0	6.6
6.0	2.1
8.0	1.2
10.0	0.6

Use an appropriate graph of the data to determine the constants C and n.

43. ‖ The electric potential in a region of space is $V = (150x^2 - 200y^2)$ V, where x and y are in meters. What are the strength and direction of the electric field at $(x, y) = (2.0$ m, 2.0 m)? Give the direction as an angle cw or ccw (specify which) from the positive x-axis.

44. ‖ The electric potential in a region of space is $V = 200/\sqrt{x^2 + y^2}$, where x and y are in meters. What are the strength and direction of the electric field at $(x, y) = (2.0$ m, 1.0 m)? Give the direction as an angle cw or ccw (specify which) from the positive x-axis.

45. ‖ Metal sphere 1 has a positive charge of 6.0 nC. Metal sphere 2, which is twice the diameter of sphere 1, is initially uncharged. The spheres are then connected together by a long, thin metal wire. What are the final charges on each sphere?

46. ‖ The metal spheres in FIGURE P29.46 are charged to ±300 V. Draw this figure on your paper, then draw a plausible contour map of the potential, showing and labeling the −300 V, −200 V, −100 V, ..., 300 V equipotential surfaces.

FIGURE P29.46

47. ‖ The potential at the center of a 4.0-cm-diameter copper sphere is 500 V, relative to $V = 0$ V at infinity. How much excess charge is on the sphere?

48. ‖ Two 2.0 cm × 2.0 cm metal electrodes are spaced 1.0 mm apart and connected by wires to the terminals of a 9.0 V battery.
    a. What are the charge on each electrode and the potential difference between them?
    The wires are disconnected, and insulated handles are used to pull the plates apart to a new spacing of 2.0 mm.
    b. What are the charge on each electrode and the potential difference between them?

49. ‖ Two 2.0 cm × 2.0 cm metal electrodes are spaced 1.0 mm apart and connected by wires to the terminals of a 9.0 V battery.
    a. What are the charge on each electrode and the potential difference between them?
    While the plates are still connected to the battery, insulated handles are used to pull them apart to a new spacing of 2.0 mm.
    b. What are the charge on each electrode and the potential difference between them?

50. ‖ Find expressions for the equivalent capacitance of (a) N identical capacitors C in parallel and (b) N identical capacitors C in series.

51. ‖ What is the equivalent capacitance of the three capacitors in FIGURE P29.51?

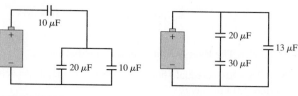

FIGURE P29.51   FIGURE P29.52

52. ‖ What is the equivalent capacitance of the three capacitors in FIGURE P29.52?

53. ‖ What are the charge on and the potential difference across each capacitor in FIGURE P29.53?

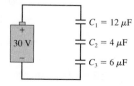

FIGURE P29.53

54. ‖ What are the charge on and the potential difference across each capacitor in FIGURE P29.54?

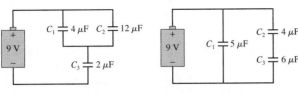

FIGURE P29.54   FIGURE P29.55

55. ‖ What are the charge on and the potential difference across each capacitor in FIGURE P29.55?

56. ‖ You have three 12 μF capacitors. Draw diagrams showing how you could arrange all three so that their equivalent capacitance is (a) 4.0 μF, (b) 8.0 μF, (c) 18 μF, and (d) 36 μF.

57. ‖ Six identical capacitors with capacitance C are connected as shown in FIGURE P29.57.
    a. What is the equivalent capacitance of these six capacitors?
    b. What is the potential difference between points a and b?

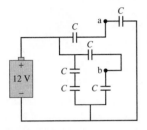

FIGURE P29.57   FIGURE P29.58

58. ‖ What is the capacitance of the two electrodes in FIGURE P29.58? **Hint:** Can you think of this as a combination of capacitors?

59. ‖ Initially, the switch in FIGURE P29.59 is in position A and capacitors C_2 and C_3 are uncharged. Then the switch is flipped to position B. Afterward, what are the charge on and the potential difference across each capacitor?

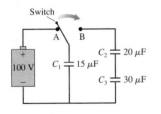
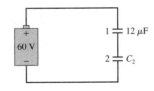

FIGURE P29.59   FIGURE P29.60

60. ‖ A battery with an emf of 60 V is connected to the two capacitors shown in FIGURE P29.60. Afterward, the charge on capacitor 2 is 450 μC. What is the capacitance of capacitor 2?

61. ‖ Capacitors $C_1 = 10$ μF and $C_2 = 20$ μF are each charged to 10 V, then disconnected from the battery without changing the charge on the capacitor plates. The two capacitors are then connected in parallel, with the positive plate of C_1 connected to the negative plate of C_2 and vice versa. Afterward, what are the charge on and the potential difference across each capacitor?

62. ‖ An isolated 5.0 μF parallel-plate capacitor has 4.0 mC of charge. An external force changes the distance between the electrodes until the capacitance is 2.0 μF. How much work is done by the external force?

63. ‖ A parallel-plate capacitor is constructed from two 10 cm × 10 cm electrodes spaced 1.0 mm apart. The capacitor plates are charged to ± 10 nC, then disconnected from the battery.
    a. How much energy is stored in the capacitor?
    b. Insulating handles are used to pull the capacitor plates apart until the spacing is 2.0 mm. Now how much energy is stored in the capacitor?
    c. Energy must be conserved. How do you account for the difference between a and b?

64. ‖ What is the energy density in the electric field at the surface of a 1.0-cm-diameter sphere charged to a potential of 1000 V?

65. ‖ BIO The 90 μF capacitor in a defibrillator unit supplies an average of 6500 W of power to the chest of the patient during a discharge lasting 5.0 ms. To what voltage is the capacitor charged?

66. ‖ The flash unit in a camera uses a 3.0 V battery to charge a capacitor. The capacitor is then discharged through a flashlamp. The discharge takes 10 μs, and the average power dissipated in the flashlamp is 10 W. What is the capacitance of the capacitor?

67. ‖ You need to use a motor and lightweight cable to lift a 2.0 kg copper weight to a height of 3.0 m. To do so, you've decided to use a 1000 V power supply to charge a capacitor, then run the motor by letting the capacitor discharge through it. If the motor is 90% efficient (that is, 10% of the energy supplied to the motor is dissipated as heat), what minimum capacitance do you need?

68. ‖ Two 5.0-cm-diameter metal disks separated by a 0.50-mm-thick piece of Pyrex glass are charged to a potential difference of 1000 V. What are (a) the surface charge density on the disks and (b) the surface charge density on the glass?

69. ‖ BIO A typical cell has a membrane potential of −70 mV, meaning that the potential inside the cell is 70 mV less than the potential outside due to a layer of negative charge on the inner surface of the cell wall and a layer of positive charge on the outer surface. This effectively makes the cell wall a charged capacitor. Because a cell's diameter is much larger than the wall thickness, it is reasonable to ignore the curvature of the cell and think of it as a parallel-plate capacitor. How much energy is stored in the electric field of a 50-μm-diameter cell with a 7.0-nm-thick cell wall whose dielectric constant is 9.0?

70. ‖ BIO A nerve cell in its resting state has a membrane potential of −70 mV, meaning that the potential inside the cell is 70 mV less than the potential outside due to a layer of negative charge on the inner surface of the cell wall and a layer of positive charge on the outer surface. This effectively makes the cell wall a charged capacitor. When the nerve cell fires, sodium ions, Na^+, flood through the cell wall to briefly switch the membrane potential to +40 mV. Model the central body of a nerve cell—the *soma*—as a 50-μm-diameter sphere with a 7.0-nm-thick cell wall whose dielectric constant is 9.0. Because a cell's diameter is much larger than the wall thickness, it is reasonable to ignore the curvature of the cell and think of it as a parallel-plate capacitor. How many sodium ions enter the cell as it fires?

71. ‖ Derive Equation 29.34 for the induced surface charge density on the dielectric in a capacitor.

72. ‖ A vacuum-insulated parallel-plate capacitor with plate separation d has capacitance C_0. What is the capacitance if an insulator with dielectric constant κ and thickness is $d/2$ slipped between the electrodes?

In Problems 73 through 75 you are given the equation(s) used to solve a problem. For each of these, you are to
   a. Write a realistic problem for which this is the correct equation(s).
   b. Finish the solution of the problem.

73. $2az$ V/m $= -\dfrac{dV}{dz}$, where a is a constant with units of V/m^2

  $V(z = 0) = 10$ V

74. 400 nC $= (100$ V$)\,C$

  $C = \dfrac{(8.85 \times 10^{-12}\ \text{C}^2/\text{N m}^2)(0.10\ \text{m} \times 0.10\ \text{m})}{d}$

75. $\left(\dfrac{1}{3\ \mu\text{F}} + \dfrac{1}{6\ \mu\text{F}}\right)^{-1} + C = 4\ \mu\text{F}$

Challenge Problems

76. The electric potential in a region of space is $V = 100(x^2 - y^2)$ V, where x and y are in meters.
   a. Draw a contour map of the potential, showing and labeling the -400 V, -100 V, 0 V, $+100$ V, and $+400$ V equipotential surfaces.
   b. Find an expression for the electric field $\vec{E}$ at position (x, y).
   c. Draw the electric field lines on your diagram of part a.
77. An electric dipole at the origin consists of two charges $\pm q$ spaced distance s apart along the y-axis.
   a. Find an expression for the potential $V(x, y)$ at an arbitrary point in the xy-plane. Your answer will be in terms of q, s, x, and y.
   b. Use the binomial approximation to simplify your result of part a when $s \ll x$ and $s \ll y$.
   c. Assuming $s \ll x$ and y, find expressions for E_x and E_y, the components of $\vec{E}$ for a dipole.
   d. What is the on-axis field $\vec{E}$? Does your result agree with Equation 26.11?
   e. What is the field $\vec{E}$ on the bisecting axis? Does your result agree with Equation 26.12?
78. Charge is uniformly distributed with charge density ρ inside a very long cylinder of radius R. Find the potential difference between the surface and the axis of the cylinder.
79. Consider a uniformly charged sphere of radius R and total charge Q. The electric field E_{out} *outside* the sphere ($r \geq R$) is simply that of a point charge Q. In Chapter 27, we used Gauss's law to find that the electric field E_{in} *inside* the sphere ($r \leq R$) is radially outward with field strength

$$E_{\text{in}} = \frac{1}{4\pi\epsilon_0} \frac{Q}{R^3} r$$

   a. The electric potential V_{out} *outside* the sphere is that of a point charge Q. Find an expression for the electric potential V_{in} at position r inside the sphere. As a reference, let $V_{\text{in}} = V_{\text{out}}$ at the surface of the sphere.
   b. What is the ratio $V_{\text{center}}/V_{\text{surface}}$?
   c. Graph V versus r for $0 \leq r \leq 3R$.
80. a. Find an expression for the capacitance of a *spherical capacitor*, consisting of concentric spherical shells of radii R_1 (inner shell) and R_2 (outer shell).
   b. A spherical capacitor with a 1.0 mm gap between the shells has a capacitance of 100 pF. What are the diameters of the two spheres?
81. High-frequency signals are often transmitted along a *coaxial cable*, such as the one shown in FIGURE CP29.81. For example, the cable TV hookup coming into your home is a coaxial cable. The signal is carried on a wire of radius R_1 while the outer conductor of radius R_2 is grounded (i.e., at $V = 0$ V). An insulating material fills the space between them, and an insulating plastic coating goes around the outside.

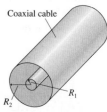

FIGURE CP29.81

   a. Find an expression for the capacitance per meter of a coaxial cable. Assume that the insulating material between the cylinders is air.
   b. Evaluate the capacitance per meter of a cable having $R_1 = 0.50$ mm and $R_2 = 3.0$ mm.
82. Each capacitor in FIGURE CP29.82 has capacitance C. What is the equivalent capacitance between points a and b?

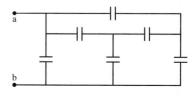

FIGURE CP29.82

STOP TO THINK ANSWERS

Stop to Think 29.1: 5.0 V. The potentials add, but $\Delta V_2 = -1.0$ V because the charge escalator goes *down* by 1.0 V.

Stop to Think 29.2: c. E_y is the negative of the slope of the V-versus-y graph. E_y is positive because $\vec{E}$ points up, so the graph has a negative slope. E_y has constant magnitude, so the slope has a constant value.

Stop to Think 29.3: c. $\vec{E}$ points "downhill," so V must decrease from right to left. E is larger on the left than on the right, so the contour lines must be closer together on the left.

Stop to Think 29.4: b. Because of the connecting wire, the three spheres form a single conductor in electrostatic equilibrium. Thus all points are at the same potential. The electric field of a sphere is related to the sphere's potential by $E = V/R$, so a smaller-radius sphere has a larger E.

Stop to Think 29.5: $(C_{\text{eq}})_b > (C_{\text{eq}})_a = (C_{\text{eq}})_d > (C_{\text{eq}})_c.$ $(C_{\text{eq}})_b = 3\ \mu\text{F} + 3\ \mu\text{F} = 6\ \mu\text{F}$. The equivalent capacitance of series capacitors is less than any capacitor in the group, so $(C_{\text{eq}})_c < 3\ \mu\text{F}$. Only d requires any real calculation. The two 4 μF capacitors are in series and are equivalent to a single 2 μF capacitor. The 2 μF equivalent capacitor is in parallel with 3 μF, so $(C_{\text{eq}})_d = 5\ \mu\text{F}$.

30 Current and Resistance

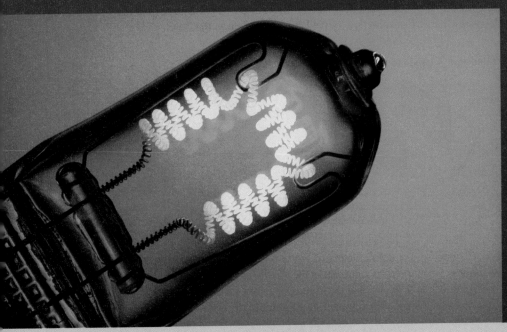

A lightbulb filament is a very thin tungsten wire—coiled repeatedly to increase its length—heated until it glows by passing a current through it.

▶ **Looking Ahead** The goal of Chapter 30 is to learn how and why charge moves through a conductor as what we call a current.

A Model of Conduction

You'll learn to use a model of conduction to understand many of the properties of current.

A nonuniform surface charge distribution, typically established when the ends of a wire are connected to the terminals of a battery, creates an electric field in the wire.

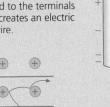

The electric field pushes the **sea of electrons** opposite the field direction, but the electrons undergo frequent collisions with the positive ions of the crystal lattice. The net result is a slow but sustained flow of charges at the **drift speed v_d**. This is the **electron current.**

For historical reasons, current is defined to be in the direction that positive charges would move. Current is measured in **amperes,** where one ampere (or one amp) is a charge flow rate of 1 coulomb per second.

Current

Current is the flow of charge through a conductor. But we can't see the charges moving, so how do we know they do?

You'll learn that the flow of charge can be recognized by its effects. These include heating wires and deflecting compass needles. These are *indicators* of a current.

◀ **Looking Back**
Section 26.6 The motion of charge in an electric field

Conservation of Current

Any charge entering one end of a wire must be balanced by an equal charge leaving the other end.

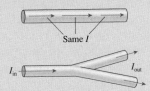

As a consequence, you'll learn that the current is the same from one end of a wire to the other. At a junction, the sum of the currents entering must equal the sum of the currents leaving.

Resistance

Collisions of electrons with the crystal lattice cause conductors to resist the flow of charges. You'll learn to use:
- **Resistivity,** an electric property of a material, such as copper.
- **Resistance,** a property of a specific wire based on its geometry and the material of which it is made.

Heater wires, such as those in toasters, are made of an alloy called *nichrome* because its resistivity is larger than that of ordinary metals.

Ohm's Law

You'll discover that the current I through a conductor is determined by the potential difference ΔV across the conductor and the conductor's resistance R.

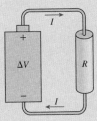

Ohm's law is $I = \dfrac{\Delta V}{R}$

◀ **Looking Back**
Section 29.2 Sources of potential

30.1 The Electron Current

We've focused thus far on situations in which charges are in static equilibrium. Now it's time to explore the *controlled* motion of charges—currents. Let's begin with a simple question: How does a capacitor get discharged? FIGURE 30.1a shows a charged capacitor. If, as in FIGURE 30.1b, we connect the two capacitor plates with a metal wire, a conductor, the plates quickly become neutral; that is, the capacitor has been *discharged.* Charge has somehow moved from one plate to the other.

FIGURE 30.1 A capacitor is discharged by a metal wire.

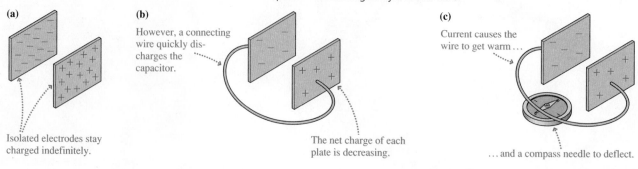

(a)

Isolated electrodes stay charged indefinitely.

(b)

However, a connecting wire quickly discharges the capacitor.

The net charge of each plate is decreasing.

(c)

Current causes the wire to get warm ...

... and a compass needle to deflect.

In Chapter 25, we defined **current** as the motion of charges. It would seem that the capacitor is discharged by a current in the connecting wire. Let's see what else we can observe. FIGURE 30.1c shows that the connecting wire gets warm. If the wire is very thin in places, such as the thin filament in a lightbulb, the wire gets hot enough to glow. The current-carrying wire also deflects a compass needle, an observation we'll explore further in Chapter 32. For now, we will use "makes the wire warm" and "deflects a compass needle" as *indicators* that a current is present in a wire.

Charge Carriers

FIGURE 30.2 The sea of electrons is a model of how conduction electrons behave in a metal.

Ions (the metal atoms minus valence electrons) occupy fixed positions in the lattice.

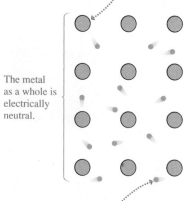

The metal as a whole is electrically neutral.

The conduction electrons are free to move around. They are bound to the solid as a whole, not to any particular atom.

The charges that move in a conductor are called the *charge carriers.* FIGURE 30.2 reminds you of the microscopic model of a metallic conductor that we introduced in Chapter 25. The outer electrons of metal atoms—the valence electrons—are only weakly bound to the nuclei. When the atoms come together to form a solid, the outer electrons become detached from their parent nuclei to form a fluid-like *sea of electrons* that can move through the solid. That is, **electrons are the charge carriers in metals.** Notice that the metal as a whole remains electrically neutral. This is not a perfect model because it overlooks some quantum effects, but it provides a reasonably good description of current in a metal.

NOTE ▶ Electrons are the charge carriers in *metals.* Other conductors, such as ionic solutions or semiconductors, have different charge carriers. We will focus on metals because of their importance to circuits, but don't think that electrons are *always* the charge carrier. ◀

The conduction electrons in a metal, like molecules in a gas, undergo random thermal motions, but there is no *net* motion. We can change that by pushing on the sea of electrons with an electric field, causing the entire sea of electrons to move in one direction like a gas or liquid flowing through a pipe. This net motion, which takes place at what we'll call the **drift speed** v_d, is superimposed on top of the random thermal motions of the individual electrons. The drift speed is quite small. As we'll establish later, 10^{-4} m/s is a fairly typical value for v_d.

As FIGURE 30.3 shows, the entire sea of electrons moves from left to right at the drift speed. Suppose an observer could count the electrons as they pass through this cross section of the wire. Let's define the **electron current** i_e to be the number of electrons *per second* that pass through a cross section of a wire or other conductor. The units of electron current are s^{-1}. Stated another way, the number N_e of electrons that pass through the cross section during the time interval Δt is

$$N_e = i_e \Delta t \qquad (30.1)$$

Increasing the drift speed will increase the number of electrons passing through a wire each second—that is, will increase the electron current. To quantify this idea, FIGURE 30.4 shows the sea of electrons moving through a wire at the drift speed v_d. The electrons passing through a particular cross section of the wire during the interval Δt are shaded. How many of them are there?

FIGURE 30.3 The electron current.

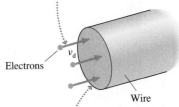

The sea of electrons flows through a wire at the drift speed v_d, much like a fluid flowing through a pipe.

Electrons

Wire

The electron current i_e is the number of electrons passing through this cross section of the wire per second.

FIGURE 30.4 The sea of electrons moves to the right with drift speed v_d.

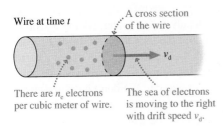

Wire at time t

A cross section of the wire

v_d

There are n_e electrons per cubic meter of wire.

The sea of electrons is moving to the right with drift speed v_d.

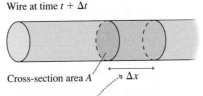

Wire at time $t + \Delta t$

Cross-section area A

Δx

The sea of electrons has moved forward distance $\Delta x = v_d \Delta t$. The shaded volume is $V = A \Delta x$.

The electrons travel distance $\Delta x = v_d \Delta t$ to the right during the interval Δt, forming a cylinder of charge with volume $V = A \Delta x$. If the *number density* of conduction electrons is n_e electrons per cubic meter, then the total number of electrons in the cylinder is

$$N_e = n_e V = n_e A \Delta x = n_e A v_d \Delta t \qquad (30.2)$$

Comparing Equations 30.2 and 30.1, you can see that the electron current in the wire is

$$i_e = n_e A v_d \qquad (30.3)$$

You can increase the electron current—the number of electrons per second moving through the wire—by making them move faster, by having more of them per cubic meter, or by increasing the size of the pipe they're flowing through. That all makes sense.

In most metals, each atom contributes one valence electron to the sea of electrons. Thus the number of conduction electrons per cubic meter is the same as the number of atoms per cubic meter, a quantity that can be determined from the metal's mass density. Table 30.1 gives values of the conduction-electron density n_e for several metals.

TABLE 30.1 Conduction-electron density in metals

Metal	Electron density (m^{-3})
Aluminum	6.0×10^{28}
Copper	8.5×10^{28}
Iron	8.5×10^{28}
Gold	5.9×10^{28}
Silver	5.8×10^{28}

EXAMPLE 30.1 **The size of the electron current**

What is the electron current in a 2.0-mm-diameter copper wire if the electron drift speed is 1.0×10^{-4} m/s?

SOLVE This is a straightforward calculation. The wire's cross-section area is $A = \pi r^2 = 3.14 \times 10^{-6} \, m^2$. Table 30.1 gives the electron density for copper as $8.5 \times 10^{28} \, m^{-3}$. Thus we find

$$i_e = n_e A v_d = 2.7 \times 10^{19} \, s^{-1}$$

ASSESS This is an incredible number of electrons to pass through a section of the wire every second. The number is high not because the sea of electrons moves fast—in fact, it moves at literally a snail's pace—but because the density of electrons is so enormous. This is a fairly typical electron current.

STOP TO THINK 30.1 These four wires are made of the same metal. Rank in order, from largest to smallest, the electron currents i_a to i_d.

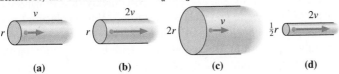

(a) (b) (c) (d)

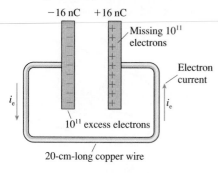

FIGURE 30.5 How long does it take to discharge this capacitor?

−16 nC   +16 nC

Missing 10^{11} electrons

Electron current

i_e i_e

10^{11} excess electrons

20-cm-long copper wire

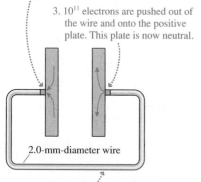

FIGURE 30.6 The sea of electrons needs only a minuscule rearrangement to discharge the capacitor.

1. The 10^{11} excess electrons on the negative plate move into the wire. The length of wire needed to accommodate these electrons is only 4×10^{-13} m.

3. 10^{11} electrons are pushed out of the wire and onto the positive plate. This plate is now neutral.

2.0-mm-diameter wire

2. The sea of 5×10^{22} electrons in the wire is pushed to the side. It moves only 4×10^{-13} m, taking almost no time.

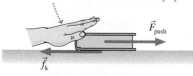

FIGURE 30.7 An electron current is sustained by pushing on the sea of electrons with an electric field.

Because of friction, a steady push is needed to move the book at steady speed.

$\vec{F}_{push}$

$\vec{f}_k$

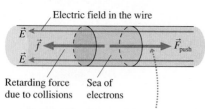

Electric field in the wire

$\vec{E}$

$\vec{f}$ $\vec{F}_{push}$

$\vec{E}$

Retarding force due to collisions   Sea of electrons

Because of collisions with atoms, a steady push is needed to move the sea of electrons at steady speed.

Discharging a Capacitor

FIGURE 30.5 shows a capacitor charged to ±16 nC as it is being discharged by a 2.0-mm-diameter, 20-cm-long copper wire. *How long does it take* to discharge the capacitor? We've noted that a fairly typical drift speed of the electron current through a wire is 10^{-4} m/s. At this rate, it would take 2000 s, or about a half hour, for an electron to travel 20 cm. We should have time to go for a cup of coffee while we wait for the discharge to occur!

But this isn't what happens. As far as our senses are concerned, the discharge of a capacitor by a copper wire is instantaneous. So what's wrong with our simple calculation?

The important point we overlooked is that the wire is *already full* of electrons. As an analogy, think of water in a hose. If the hose is already full of water, adding a drop to one end immediately (or very nearly so) pushes a drop out the other end. Likewise with the wire. As soon as the excess electrons move from the negative capacitor plate into the wire, they immediately (or very nearly so) push an equal number of electrons out the other end of the wire and onto the positive plate, thus neutralizing it. We don't have to wait for electrons to move all the way through the wire from one plate to the other. Instead, we just need to slightly rearrange the charges on the plates *and* in the wire.

Let's do a rough estimate of how much rearrangement is needed and how long the discharge takes. Using the conduction-electron density of copper in Table 30.1, we can calculate that there are 5×10^{22} conduction electrons in the wire. The negative plate in **FIGURE 30.6**, with $Q = -16$ nC, has 10^{11} excess electrons, far fewer than in the wire. In fact, the length of copper wire needed to hold 10^{11} electrons is a mere 4×10^{-13} m, only about 1% the diameter of an atom.

The instant the wire joins the capacitor plates together, the repulsive forces between the excess 10^{11} electrons on the negative plate cause them to push their way into the wire. As they do, 10^{11} electrons are squeezed out of the final 4×10^{-13} m of the wire and onto the positive plate. If the electrons all move together, and if they move at the typical drift speed of 10^{-4} m/s—both less than perfect assumptions but fine for making an estimate—it takes 4×10^{-9} s, or 4 ns, to move 4×10^{-13} m and discharge the capacitor. And, indeed, this is the right order of magnitude for how long the electrons take to rearrange themselves so that the capacitor plates are neutral.

STOP TO THINK 30.2 Why does the light in a room come on instantly when you flip a switch several meters away?

30.2 Creating a Current

Suppose you want to slide a book across the table to your friend. You give it a quick push to start it moving, but it begins slowing down because of friction as soon as you take your hand off. The book's kinetic energy is transformed into thermal energy, leaving the book and the table slightly warmer. The only way to keep the book moving at a constant speed is to *continue pushing it*.

As **FIGURE 30.7** shows, the sea of electrons is similar to the book. If you push the sea of electrons, you create a current of electrons moving through the conductor. But the electrons aren't moving in a vacuum. Collisions between the electrons and the atoms of the metal transform the electrons' kinetic energy into the thermal energy of the metal, making the metal warmer. (Recall that "makes the wire warm" is one of our indicators of a current.) Consequently, the sea of electrons will quickly slow down and stop *unless you continue pushing*. How do you push on electrons? With an electric field!

One of the important conclusions of Chapter 27 was that $\vec{E} = \vec{0}$ inside a conductor in electrostatic equilibrium. But a conductor with electrons moving through it is *not* in electrostatic equilibrium. **An electron current is a nonequilibrium motion of charges sustained by an internal electric field.**

Thus the quick answer to "What creates a current?" is "An electric field." But why is there an electric field in a current-carrying wire?

Establishing the Electric Field in a Wire

FIGURE 30.8a shows two metal wires attached to the plates of a charged capacitor. The wires are conductors, so some of the charges on the capacitor plates become spread out along the wires as a surface charge. (Remember that all excess charge on a conductor is located on the surface.)

This is an electrostatic situation, with no current and no charges in motion. Consequently—because this is always true in electrostatic equilibrium—the electric field inside the wire is zero. Symmetry requires there to be equal amounts of charge to either side of each point to make $\vec{E} = \vec{0}$ at that point; hence the surface charge density must be uniform along each wire except near the ends (where the details need not concern us). We implied this uniform density in Figure 30.8a by drawing equally spaced $+$ and $-$ symbols along the wire. Remember that a positively charged surface is a surface that is *missing* electrons.

FIGURE 30.8 The surface charge on the wires before and after they are connected.

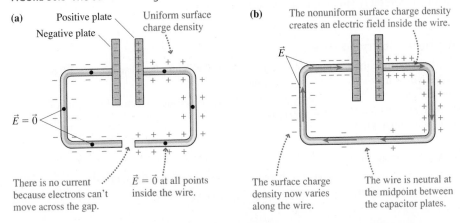

Now we connect the ends of the wires together. What happens? The excess electrons on the negative wire suddenly have an opportunity to move onto the positive wire that is missing electrons. Within a *very* brief interval of time ($\approx 10^{-9}$ s), the sea of electrons shifts slightly and the surface charge is rearranged into a *nonuniform* distribution like that shown in FIGURE 30.8b. The surface charge near the positive and negative plates remains strongly positive and negative because of the large amount of charge on the capacitor plates, but the midpoint of the wire, halfway between the positive and negative plates, is now electrically neutral. The new surface charge density on the wire varies from positive at the positive capacitor plate through zero at the midpoint to negative at the negative plate.

This nonuniform distribution of surface charge has an *extremely* important consequence. FIGURE 30.9 shows a section from a wire on which the surface charge density becomes more positive toward the left and more negative toward the right. Calculating

FIGURE 30.9 A varying surface charge distribution creates an internal electric field inside the wire.

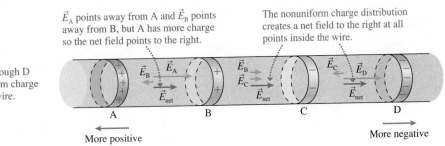

the exact electric field is complicated, but we can understand the basic idea if we *model* this section of wire with four circular rings of charge.

In Chapter 26, we found that the on-axis field of a ring of charge

- Points away from a positive ring, toward a negative ring;
- Is proportional to the amount of charge on the ring; and
- Decreases with distance away from the ring.

The field at the midpoint between rings A and B is well approximated as $\vec{E}_{net} \approx \vec{E}_A + \vec{E}_B$. Ring A has more charge than ring B, so $\vec{E}_{net}$ points away from A.

The analysis of Figure 30.9 leads to a very important conclusion:

> A *nonuniform* distribution of surface charges along a wire creates a net electric field *inside* the wire that points from the more positive end of the wire toward the more negative end of the wire. This is the internal electric field $\vec{E}$ that pushes the electron current through the wire.

Note that the surface charges are *not* the moving charges of the current. Further, the current—the moving charges—is *inside* the wire, not on the surface. In fact, as the next example shows, the electric field inside a current-carrying wire can be established with an extremely small amount of surface charge.

EXAMPLE 30.2 **The surface charge on a current-carrying wire**

Table 26.1 in Chapter 26 gave a typical electric field strength in a current-carrying wire as 0.01 N/C or, as we would now say, 0.01 V/m. (We'll verify this value later in this chapter.) Two 2.0-mm-diameter rings are 2.0 mm apart. They are charged to $\pm Q$. What value of Q causes the electric field at the midpoint to be 0.010 V/m?

MODEL Use the on-axis electric field of a ring of charge from Chapter 26.

VISUALIZE FIGURE 30.10 shows the two rings. Both contribute equally to the field strength, so the electric field strength of the

FIGURE 30.10 The electric field of two charged rings.

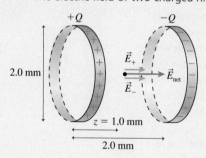

positive ring is $E_+ = 0.0050$ V/m. The distance $z = 1.0$ mm is half the ring spacing.

SOLVE Chapter 26 found the on-axis electric field of a ring of charge Q to be

$$E_+ = \frac{1}{4\pi\epsilon_0}\frac{zQ}{(z^2 + R^2)^{3/2}}$$

Thus the charge needed to produce the desired field is

$$\begin{aligned}Q &= \frac{4\pi\epsilon_0(z^2 + R^2)^{3/2}}{z}E_+ \\ &= \frac{((0.0010\ \text{m})^2 + (0.0010\ \text{m})^2)^{3/2}}{(9.0\times10^9\ \text{N}\,\text{m}^2/\text{C}^2)(0.0010\ \text{m})}(0.0050\ \text{V/m}) \\ &= 1.6\times10^{-18}\ \text{C}\end{aligned}$$

ASSESS The electric field of a ring of charge is largest at $z \approx R$, so these two rings are a simple but reasonable model for estimating the electric field inside a 2.0-mm-diameter wire. We find that the surface charge needed to establish the electric field is *very small*. A mere 10 electrons have to be moved from one ring to the other to charge them to $\pm 1.6\times10^{-18}$ C. The resulting electric field is sufficient to drive a sizable electron current through the wire.

STOP TO THINK 30.3 The two charged rings are a model of the surface charge distribution along a wire. Rank in order, from largest to smallest, the electron currents E_a to E_e at the midpoint between the rings.

(a)

(b)

(c)

(d)

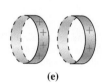

(e)

A Model of Conduction

Electrons don't just magically move through a wire as a current. They move because an electric field inside the wire—a field created by a nonuniform surface charge density on the wire—pushes on the sea of electrons to create the electron current. The field has to *keep* pushing because the electrons continuously lose energy in collisions with the positive ions that form the structure of the solid. These collisions provide a drag force, much like friction.

We will model the conduction electrons—those electrons that make up the sea of electrons—as free particles moving through the lattice of the metal. In the absence of an electric field, the electrons, like the molecules in a gas, move randomly in all directions with a distribution of speeds. If we assume that the average thermal energy of the electrons is given by the same $\frac{3}{2}k_BT$ that applies to an ideal gas, we can calculate that the average electron speed at room temperature is $\approx 10^5$ m/s. This estimate turns out, for quantum physics reasons, to be not quite right, but it correctly indicates that the conduction electrons are moving very fast.

However, an individual electron does not travel far before colliding with an ion and being scattered to a new direction. FIGURE 30.11a shows that an electron bounces back and forth between collisions, but its *average* velocity is zero, and it undergoes no *net* displacement. This is similar to molecules in a container of gas.

Suppose we now turn on an electric field. FIGURE 30.11b shows that the steady electric force causes the electrons to move along *parabolic trajectories* between collisions. Because of the curvature of the trajectories, the negatively charged electrons begin to drift slowly in the direction opposite the electric field. The motion is similar to a ball moving in a pinball machine with a slight downward tilt. An individual electron ricochets back and forth between the ions at a high rate of speed, but now there is a slow *net* motion in the "downhill" direction. Even so, this net displacement is a *very* small effect superimposed on top of the much larger thermal motion. Figure 30.11b has greatly exaggerated the rate at which the drift would occur.

Suppose an electron just had a collision with an ion and has rebounded with velocity $\vec{v}_0$. The acceleration of the electron between collisions is

$$a_x = \frac{F}{m} = \frac{eE}{m} \tag{30.4}$$

where E is the electric field strength inside the wire and m is the mass of the electron. (We'll assume that $\vec{E}$ points in the negative x-direction.) The field causes the x-component of the electron's velocity to increase linearly with time:

$$v_x = v_{0x} + a_x \Delta t = v_{0x} + \frac{eE}{m} \Delta t \tag{30.5}$$

The electron speeds up, with increasing kinetic energy, until its next collision with an ion. The collision transfers much of the electron's kinetic energy to the ion and thus to the thermal energy of the metal. **This energy transfer is the "friction" that raises the temperature of the wire.** The electron then rebounds, in a random direction, with a new initial velocity $\vec{v}_0$, and starts the process all over.

FIGURE 30.12a on the next page shows how the velocity abruptly changes due to a collision. Notice that the acceleration (the slope of the line) is the same before and after the collision. FIGURE 30.12b follows an electron through a series of collisions. You can see that each collision "resets" the velocity. The primary observation we can make from Figure 30.12b is that this repeated process of speeding up and colliding gives the electron a nonzero *average* velocity. **The magnitude of the electron's average velocity, due to the electric field, is the *drift speed* v_d of the electron.**

If we observe all the electrons in the metal at one instant of time, their average velocity is

$$v_d = \overline{v_x} = \overline{v_{0x}} + \frac{eE}{m}\overline{\Delta t} \tag{30.6}$$

FIGURE 30.11 A microscopic view of a conduction electron moving through a metal.

(a) No electric field

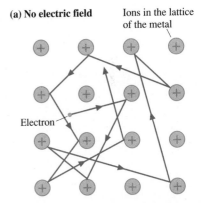

The electron has frequent collisions with ions, but it undergoes no net displacement.

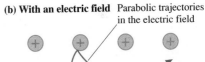

(b) With an electric field

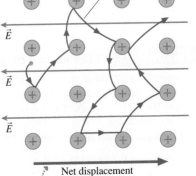

A net displacement in the direction opposite to $\vec{E}$ is superimposed on the random thermal motion.

FIGURE 30.12 The electron velocity as a function of time.

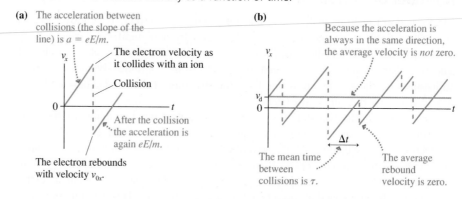

where a bar over a quantity indicates an average value. The average value of v_{0x}, the velocity with which an electron rebounds after a collision, is zero. We know this because, in the absence of an electric field, the sea of electrons moves neither right nor left.

The quantity Δt is the time between collisions, so the average value of Δt is the **mean time between collisions,** which we designate τ. The mean time between collisions, analogous to the mean free path between collisions in the kinetic theory of gases, depends on the metal's temperature but can be considered a constant in the equations below.

Thus the average speed at which the electrons are pushed along by the electric field is

$$v_{d} = \frac{e\tau}{m}E \qquad (30.7)$$

We can complete our model of conduction by using Equation 30.7 for v_d in the electron-current equation $i_e = n_e A v_d$. Upon doing so, we find that an electric field strength E in a wire of cross-section area A causes an electron current

$$i_{e} = \frac{n_e e\tau A}{m}E \qquad (30.8)$$

The electron density n_e and the mean time between collisions τ are properties of the metal.

Equation 30.8 is the main result of this model of conduction. We've found that **the electron current is directly proportional to the electric field strength.** A stronger electric field pushes the electrons faster and thus increases the electron current.

EXAMPLE 30.3 **Collisions in a copper wire**

Example 30.1 found the electron current to be $2.7 \times 10^{19} \ \text{s}^{-1}$ for a 2.0-mm-diameter copper wire in which the electron drift speed is 1.0×10^{-4} m/s. If an internal electric field of 0.020 V/m is needed to sustain this current, a typical value, how many collisions per second, on average, do electrons in copper undergo?

MODEL Use the model of conduction.

SOLVE From Equation 30.7, the mean time between collisions is

$$\tau = \frac{mv_d}{eE} = 2.8 \times 10^{-14} \ \text{s}$$

The average number of collisions per second is the inverse:

$$\text{Collision rate} = \frac{1}{\tau} = 3.5 \times 10^{13} \ \text{s}^{-1}$$

ASSESS This was another straightforward calculation simply to illustrate the incredibly large collision rate of conduction electrons.

30.3 Current and Current Density

We have developed the idea of a current as the motion of electrons through metals. But the properties of currents were known and used for a century before the discovery that electrons are the charge carriers in metals. We need to connect our ideas about the electron current to the conventional definition of current.

Because the coulomb is the unit of charge, and because currents are charges in motion, it seemed quite natural in the 19th century to define current as the *rate,* in coulombs per second, at which charge moves through a wire. If Q is the total amount of charge that has moved past a point in the wire, we define the current I in the wire to be the rate of charge flow:

$$I \equiv \frac{dQ}{dt} \qquad (30.9)$$

For a *steady current,* which will be our primary focus, the amount of charge delivered by current I during the time interval Δt is

$$Q = I\,\Delta t \qquad (30.10)$$

The SI unit for current is the coulomb per second, which is called the **ampere** A:

$$1 \text{ ampere} = 1\text{ A} \equiv 1 \text{ coulomb per second} = 1 \text{ C/s}$$

The current unit is named after the French scientist André Marie Ampère, who made major contributions to the study of electricity and magnetism in the early 19th century. The *amp* is an informal abbreviation of ampere. Household currents are typically ≈ 1 A. For example, the current through a 100 watt lightbulb is 0.85 A, meaning that 0.85 C of charge flow through the bulb every second. Currents in consumer electronics, such as stereos and computers, are much less. They are typically measured in milliamps ($1 \text{ mA} = 10^{-3}$ A) or microamps ($1 \text{ } \mu\text{A} = 10^{-6}$ A).

Equation 30.10 is closely related to Equation 30.1, which said that the number of electrons delivered during a time interval Δt is $N_e = i_e\,\Delta t$. Each electron has charge of magnitude e; hence the total charge of N_e electrons is $Q = eN_e$. Consequently, the conventional current I and the electron current i_e are related by

$$I = \frac{Q}{\Delta t} = \frac{eN_e}{\Delta t} = ei_e \qquad (30.11)$$

Because electrons are the charge carriers, the rate at which charge moves is e times the rate at which the electrons move.

EXAMPLE 30.4   **The current in a copper wire**

The electron current in the copper wire of Examples 30.1 and 30.3 was 2.7×10^{19} electrons/s. What is the current I? How much charge flows through a cross section of the wire each hour?

SOLVE The current in the wire is

$I = ei_e = (1.60 \times 10^{-19}\text{ C})(2.7 \times 10^{19}\text{ s}^{-1}) = 4.3$ A

The amount of charge passing through the wire in $1\text{ h} = 3600\text{ s}$ is

$Q = I\,\Delta t = (4.3\text{ A})(3600\text{ s}) = 16{,}000$ C

In one sense, the current I and the electron current i_e differ by only a scale factor. The electron current i_e, the rate at which electrons move through a wire, is more *fundamental* because it looks directly at the charge carriers. The current I, the rate at which the charge of the electrons moves through the wire, is more *practical* because we can measure charge more easily than we can count electrons.

Despite the close connection between i_e and I, there's one extremely important distinction. Because currents were known and studied before it was known what the charge carriers are, **the direction of current is *defined* to be the direction in which positive charges *seem* to move.** Thus the direction of the current I is the same as that of the internal electric field $\vec{E}$. But because the charge carriers turned out to be negative, at least for a metal, **the direction of the current I in a metal is opposite the direction of motion of the electrons.**

FIGURE 30.13 The current I is opposite the direction of motion of the electrons in a metal.

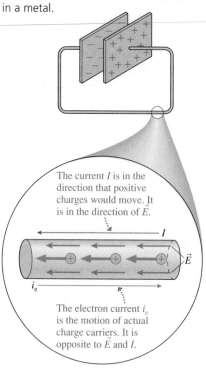

FIGURE 30.13 The current I is opposite the direction of motion of the electrons in a metal.

The current I is in the direction that positive charges would move. It is in the direction of $\vec{E}$.

The electron current i_e is the motion of actual charge carriers. It is opposite to $\vec{E}$ and I.

The situation shown in FIGURE 30.13 may seem disturbing, but it makes no real difference. A capacitor is discharged regardless of whether positive charges move toward the negative plate or negative charges move toward the positive plate. The primary application of current is the analysis of circuits, and in a circuit—a macroscopic device—we simply can't tell what is moving through the wires. All of our calculations will be correct and all of our circuits will work perfectly well if we choose to think of current as the flow of positive charge. The distinction is important only at the microscopic level.

The Current Density in a Wire

We found the electron current in a wire of cross-section area A to be $i_e = n_e A v_d$. Thus the current I is

$$I = e i_e = n_e e v_d A \qquad (30.12)$$

The quantity $n_e e v_d$ depends on the charge carriers and on the internal electric field that determines the drift speed, whereas A is simply a physical dimension of the wire. It will be useful to separate these quantities by defining the **current density** J in a wire as the current per square meter of cross section:

$$J = \text{current density} \equiv \frac{I}{A} = n_e e v_d \qquad (30.13)$$

The current density has units of A/m^2. A specific piece of metal, shaped into a wire with cross-section area A, carries current $I = JA$.

EXAMPLE 30.5 **Finding the electron drift speed**

A 1.0 A current passes through a 1.0-mm-diameter aluminum wire. What are the current density and the drift speed of the electrons in the wire?

SOLVE We can find the drift speed from the current density. The current density is

$$J = \frac{I}{A} = \frac{I}{\pi r^2} = \frac{1.0 \text{ A}}{\pi(0.00050 \text{ m})^2} = 1.3 \times 10^6 \text{ A/m}^2$$

The electron drift speed is thus

$$v_d = \frac{J}{n_e e} = 1.3 \times 10^{-4} \text{ m/s} = 0.13 \text{ mm/s}$$

where the conduction-electron density for aluminum was taken from Table 30.1.

ASSESS We earlier used 1.0×10^{-4} m/s as a typical electron drift speed. This example shows where that value comes from.

Conservation of Current

FIGURE 30.14 How does the brightness of bulb A compare to that of bulb B?

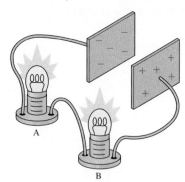

FIGURE 30.14 shows two lightbulbs in the wire connecting two charged capacitor plates. Both bulbs glow as the capacitor is discharged. How do you think the brightness of bulb A compares to that of bulb B? Is one brighter than the other? Or are they equally bright? Think about this before going on.

You might have predicted that B is brighter than A because the current I, which carries positive charges from plus to minus, reaches B first. In order to be glowing, B must use up some of the current, leaving less for A. Or perhaps you realized that the actual charge carriers are electrons, moving from minus to plus. The conventional current I may be mathematically equivalent, but physically it's the negative electrons rather than positive charge that actually move. Because the electron current gets to A first, you might have predicted that A is brighter than B.

In fact, both bulbs are equally bright. This is an important observation, one that demands an explanation. After all, "something" gets used up to make the bulb glow, so why don't we observe a decrease in the current? Current is the amount of charge moving through the wire per second. There are only two ways to decrease I: either decrease the amount of charge, or decrease the charge's drift speed through the wire. Electrons, the charge carriers, are charged particles. The lightbulb can't destroy electrons without violating both the law of conservation of mass and the law of conservation of charge. Thus the amount of charge (i.e., the *number* of electrons) cannot be changed by a lightbulb.

Do charges slow down after passing through the bulb? This is a little trickier, so consider the fluid analogy shown in FIGURE 30.15. Suppose the water flows into one end at a rate of 2.0 kg/s. Is it possible that the water, after turning a paddle wheel, flows out the other end at a rate of only 1.5 kg/s? That is, does turning the paddle wheel cause the water current to decrease?

We can't destroy water molecules any more than we can destroy electrons, we can't increase the density of water by pushing the molecules closer together, and there's nowhere to store extra water inside the pipe. Each drop of water entering the left end pushes a drop out the right end; hence water flows out at the exactly the same rate it flows in.

The same is true for electrons in a wire. **The rate of electrons leaving a lightbulb (or any other device) is exactly the same as the rate of electrons entering the lightbulb. The current does not change.** A lightbulb doesn't "use up" current, but it *does*—like the paddlewheel in the fluid analogy—use energy. The kinetic energy of the electrons is dissipated by their collisions with the ions in the lattice of the metal (the atomic-level friction) as the electrons move through the atoms, making the wire hotter until, in the case of the lightbulb filament, it glows. The lightbulb affects the amount of current *everywhere* in the wire, a process we'll examine later in the chapter, but the current doesn't change as it passes through the bulb.

There are many issues that we'll need to look at before we can say that we understand how currents work, and we'll take them one at a time. For now, we draw a first important conclusion:

Law of conservation of current The current is the same at all points in a current-carrying wire.

The law of conservation of current is really a practical application of the law of conservation of charge.

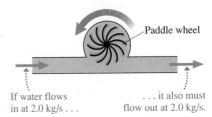

FIGURE 30.15 Water flowing through a pipe.

If water flows in at 2.0 kg/s . . . . . . it also must flow out at 2.0 kg/s.

FIGURE 30.16 The sum of the currents into a junction must equal the sum of the currents leaving the junction.

(a)

The current in a wire is the same at all points.

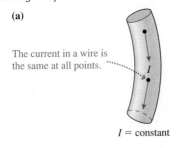

I = constant

(b)

Junction

Input currents

Output currents

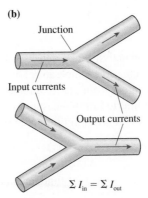

$\sum I_{\text{in}} = \sum I_{\text{out}}$

FIGURE 30.16a summarizes the law of conservation in a single wire. But what about FIGURE 30.16b, where two wires merge into one and another wire splits into two? A point where a wire branches is called a **junction.** The presence of a junction doesn't change our basic reasoning. We cannot create or destroy electrons in the wire, and neither can we store them in the junction. The rate at which electrons flow into one *or many* wires must be exactly balanced by the rate at which they flow out of others. For a *junction,* the law of conservation of charge requires that

$$\sum I_{\text{in}} = \sum I_{\text{out}} \qquad (30.14)$$

where, as usual, the Σ symbol means summation.

This basic conservation statement—that the sum of the currents into a junction equals the sum of the currents leaving—is called **Kirchhoff's junction law.** The junction law, together with *Kirchhoff's loop law* that you met in Chapter 29, will play an important role in circuit analysis in the next chapter.

> **STOP TO THINK 30.4** What are the magnitude and the direction of the current in the fifth wire?

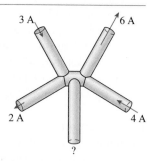

30.4 Conductivity and Resistivity

The current density $J = n_e e v_d$ is directly proportional to the electron drift speed v_d. We earlier used the microscopic model of conduction to find that the drift speed is $v_d = e\tau E/m$, where τ is the mean time between collisions and m is the mass of an electron. Combining these, we find the current density is

$$J = n_e e v_d = n_e e\left(\frac{e\tau E}{m}\right) = \frac{n_e e^2 \tau}{m} E \tag{30.15}$$

The quantity $n_e e^2 \tau/m$ depends *only* on the conducting material. According to Equation 30.15, a given electric field strength will generate a larger current density in a material with a larger electron density n_e or longer times τ between collisions than in materials with smaller values. In other words, such a material is a *better conductor* of current.

It makes sense, then, to define the **conductivity** σ of a material as

$$\sigma = \text{conductivity} = \frac{n_e e^2 \tau}{m} \tag{30.16}$$

Conductivity, like density, characterizes a material as a whole. All pieces of copper (at the same temperature) have the same value of σ, but the conductivity of copper is different from that of aluminum. Notice that the mean time between collisions τ can be inferred from measured values of the conductivity.

With this definition of conductivity, Equation 30.15 becomes

$$J = \sigma E \tag{30.17}$$

This is a result of fundamental importance. Equation 30.17 tells us three things:

1. Current is caused by an electric field exerting forces on the charge carriers.
2. The current density, and hence the current $I = JA$, depends linearly on the strength of the electric field. To double the current, you must double the strength of the electric field that pushes the charges along.
3. The current density also depends on the *conductivity* of the material. Different conducting materials have different conductivities because they have different values of the electron density and, especially, different values of the mean time between electron collisions with the lattice of atoms.

The value of the conductivity is affected by the structure of a metal, by any impurities, and by the temperature. As the temperature increases, so do the thermal vibrations of the lattice atoms. This makes them "bigger targets" and causes collisions to be more frequent, thus lowering τ and decreasing the conductivity. Metals conduct better at low temperatures than at high temperatures.

For many practical applications of current it will be convenient to use the inverse of the conductivity, called the **resistivity:**

$$\rho = \text{resistivity} = \frac{1}{\sigma} = \frac{m}{n_e e^2 \tau} \qquad (30.18)$$

The resistivity of a material tells us how reluctantly the electrons move in response to an electric field. Table 30.2 gives measured values of the resistivity and conductivity for several metals and for carbon. You can see that they vary quite a bit, with copper and silver being the best two conductors.

The units of conductivity, from Equation 30.17, are those of J/E, namely $A\,C/N\,m^2$. These are clearly awkward. In the next section we will introduce a new unit called the *ohm*, symbolized by Ω (uppercase Greek omega). It will then turn out that resistivity has units of Ω m and conductivity has units of $\Omega^{-1}\,m^{-1}$.

This woman is measuring her percentage body fat by gripping a device that sends a small electric current through her body. Because muscle and fat have different resistivities, the amount of current allows the fat-to-muscle ratio to be determined.

EXAMPLE 30.6 **The electric field in a wire**

A 2.0-mm-diameter aluminum wire carries a current of 800 mA. What is the electric field strength inside the wire?

SOLVE The electric field strength is

$$E = \frac{J}{\sigma} = \frac{I}{\sigma \pi r^2} = \frac{0.80\ A}{(3.5 \times 10^7\ \Omega^{-1}\,m^{-1})\pi(0.0010\ m)^2} = 0.0072\ V/m$$

where the conductivity of aluminum was taken from Table 30.2.

ASSESS This is a *very* small field in comparison with those we calculated in Chapters 25 and 26. This calculation justifies the claim in Table 26.1 that a typical electric field strength inside a current-carrying wire is ≈ 0.01 V/m. It takes *very few* surface charges on a wire to create the weak electric field necessary to push a considerable current through the wire. The reason, once again, is the enormous value of the charge-carrier density n_e. Even though the electric field is very tiny and the drift speed is agonizingly slow, a wire can carry a substantial current due to the vast number of charge carriers able to move.

TABLE 30.2 Resistivity and conductivity of conducting materials

Material	Resistivity (Ω m)	Conductivity ($\Omega^{-1}\,m^{-1}$)
Aluminum	2.8×10^{-8}	3.5×10^7
Copper	1.7×10^{-8}	6.0×10^7
Gold	2.4×10^{-8}	4.1×10^7
Iron	9.7×10^{-8}	1.0×10^7
Silver	1.6×10^{-8}	6.2×10^7
Tungsten	5.6×10^{-8}	1.8×10^7
Nichrome*	1.5×10^{-6}	6.7×10^5
Carbon	3.5×10^{-5}	2.9×10^4

*Nickel-chromium alloy used for heating wires.

Superconductivity

In 1911, the Dutch physicist Kamerlingh Onnes was studying the conductivity of metals at very low temperatures. Scientists had just recently discovered how to liquefy helium, and this opened a whole new field of *low-temperature physics*. As we noted above, metals become better conductors (i.e., they have higher conductivity and lower resistivity) at lower temperatures. But the effect is gradual. Onnes, however, found that mercury suddenly and dramatically loses *all* resistance to current when cooled below a temperature of 4.2 K. This complete loss of resistance at low temperatures is called **superconductivity.**

Later experiments established that the resistivity of a superconducting metal is not just small, it is truly zero. The electrons are moving in a frictionless environment, and charge will continue to move through a superconductor *without an electric field.* Superconductivity was not understood until the 1950s, when it was explained as being a specific quantum effect.

Superconducting wires can carry enormous currents because the wires are not heated by electrons colliding with the atoms. Very strong magnetic fields can be created with superconducting electromagnets, but applications remained limited for many decades because all known superconductors required temperatures less than 20 K. This situation changed dramatically in 1986 with the discovery of *high-temperature superconductors*. These ceramic-like materials are superconductors at temperatures as "high" as 125 K. Although $-150°C$ may not seem like a high tem-

Superconductors have unusual magnetic properties. Here a small permanent magnet levitates above a disk of the high-temperature superconductor $YBa_2Cu_3O_7$ that has been cooled to liquid-nitrogen temperature.

perature to you, the technology for producing such temperatures is simple and inexpensive. Thus many new superconductor applications are likely to appear in coming years.

STOP TO THINK 30.5 Rank in order, from largest to smallest, the current densities J_a to J_d in these four wires.

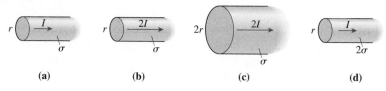

(a)　　　　(b)　　　　(c)　　　　(d)

30.5 Resistance and Ohm's Law

FIGURE 30.17 The current I is related to the potential difference ΔV.

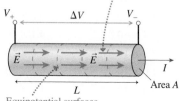

The potential difference creates an electric field inside the conductor and causes charges to flow through it.

Equipotential surfaces are perpendicular to the electric field.

FIGURE 30.17 shows a section of a conductor in which an electric field $\vec{E}$ is creating current I by pushing the charge carriers. We found in Chapter 29 that an electric field requires a potential difference. Further, the electric field points "downhill" and is perpendicular to the equipotential surfaces. Thus it should come as no surprise that current is related to potential difference.

Recall that the electric field component E_s is related to the potential by $E_s = -dV/ds$. We're interested in only the electric field strength $E = |E_s|$, so the minus sign isn't relevant. The field strength is constant inside a constant-diameter conductor (a consequence of conservation of current); thus

$$E = \frac{\Delta V}{\Delta s} = \frac{\Delta V}{L} \tag{30.19}$$

where $\Delta V = V_+ - V_-$ is the potential difference between the ends of a conductor of length L. Equation 30.19 is an important result: The electric field strength inside a constant-diameter conductor—the field that drives the current forward—is simply the potential difference between the ends of the conductor divided by its length.

Now we can use E to find the current I in the conductor. We found earlier that the current density is $J = \sigma E$, and the current in a wire of cross-section area A is related to the current density by $I = JA$. Thus

$$I = JA = A\sigma E = \frac{A}{\rho}E \tag{30.20}$$

where $\rho = 1/\sigma$ is the resistivity.

Combining Equations 30.19 and 30.20, we see that the current is

$$I = \frac{A}{\rho L}\Delta V \tag{30.21}$$

That is, **the current is proportional to the potential difference between the ends of a conductor.** We can cast Equation 30.21 into a more useful form if we define the **resistance** of a conductor to be

$$R = \frac{\rho L}{A} \tag{30.22}$$

The resistance is a property of a *specific* conductor because it depends on the conductor's length and diameter as well as on the resistivity of the material from which it is made.

The SI unit of resistance is the **ohm,** defined as

$$1 \text{ ohm} = 1\ \Omega \equiv 1\ \text{V/A}$$

The ohm is the basic unit of resistance, although kilohms (1 kΩ = 10^3 Ω) and megohms (1 MΩ = 10^6 Ω) are widely used. You can now see from Equation 30.22 why the resistivity ρ has units of Ω m while the units of conductivity σ are $\Omega^{-1}\text{m}^{-1}$.

The resistance of a wire or conductor increases as the length increases. This seems reasonable because it should be harder to push electrons through a longer wire than a shorter one. Decreasing the cross-section area also increases the resistance. This again seems reasonable because the same electric field can push more electrons through a fat wire than a skinny one.

NOTE ▶ It is important to distinguish between resistivity and resistance. *Resistivity* describes just the *material,* not any particular piece of it. *Resistance* characterizes a specific piece of the conductor with a specific geometry. The relationship between resistivity and resistance is analogous to that between mass density and mass. ◀

The definition of resistance allows us to write the current through a conductor as

$$I = \frac{\Delta V}{R} \quad \text{(Ohm's law)} \quad (30.23)$$

In other words, establishing a potential difference ΔV between the ends of a conductor of resistance R creates an electric field that, in turn, causes a current $I = \Delta V/R$ through the conductor. The smaller the resistance, the larger the current. This simple relationship between potential difference and current is known as **Ohm's law.**

EXAMPLE 30.7 The resistivity of a leaf

Resistivity measurements on the leaves of corn plants are a good way to assess stress and the plant's overall health. To determine resistivity, the current is measured when a voltage is applied between two electrodes placed 20 cm apart on a leaf that is 2.5 cm wide and 0.20 mm thick. The following data are obtained by using several different voltages:

Voltage (V)	Current (μA)
5.0	2.3
10.0	5.1
15.0	7.5
20.0	10.3
25.0	12.2

What is the resistivity of the leaf tissue?

MODEL Model the leaf as a bar of length $L = 0.20$ m with a rectangular cross-section area $A = (0.025 \text{ m})(2.0 \times 10^{-4} \text{ m}) = 5.0 \times 10^{-6} \text{ m}^2$. The potential difference creates an electric field inside the leaf and causes a current. The current and the potential difference are related by Ohm's law.

SOLVE We can find the leaf's resistivity ρ from its resistance R. Ohm's law

$$I = \frac{1}{R}\Delta V$$

tells us that a graph of current versus potential difference should be a straight line through the origin with slope $1/R$. The graph of

the data in FIGURE 30.18 is as expected. Using the slope of the best-fit line, 0.50 μA/V, we find the leaf's resistance to be

$$R = \frac{1}{0.50 \, \mu\text{A/V}} = 2.0 \times 10^6 \, \frac{\text{V}}{\text{A}} = 2.0 \times 10^6 \, \Omega$$

We can now use Equation 30.22 to find the resistivity:

$$\rho = \frac{AR}{L} = \frac{(5.0 \times 10^{-6} \text{ m}^2)(2.0 \times 10^6 \, \Omega)}{0.20 \text{ m}} = 50 \, \Omega \, \text{m}$$

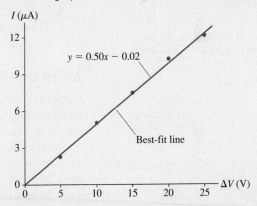

FIGURE 30.18 A graph of current versus potential difference.

ASSESS This is a huge resistivity compared to metals, but that's not surprising; the conductivity of the salty fluids in a leaf is certainly much less than that of a metal. In fact, this value is typical of the resistivities of plant and animal tissues.

Batteries and Current

Our study of current has focused on the discharge of a capacitor because we can understand where all the charges are and how they move. By contrast, we can't easily see what's happening to the charges inside a battery. Nonetheless, current in most "real" circuits is driven by a battery rather than by a capacitor. Just like the wire

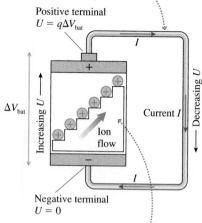

FIGURE 30.19 A battery's charge escalator causes a sustained current in a wire.

The charge "falls downhill" through the wire, but a current can be sustained because of the charge escalator.

Positive terminal
$U = q\Delta V_{bat}$

ΔV_{bat}

Increasing U

Ion flow

Current I

Decreasing U

Negative terminal
$U = 0$

The charge escalator "lifts" charge from the negative side to the positive side. Charge q gains energy $\Delta U = q\Delta V_{bat}$.

FIGURE 30.20 Current-versus-potential-difference graphs for ohmic and nonohmic materials.

(a) Ohmic material

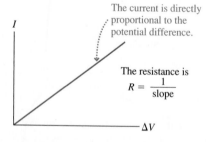

The current is directly proportional to the potential difference.

The resistance is
$R = \dfrac{1}{slope}$

I

ΔV

(b) Nonohmic materials

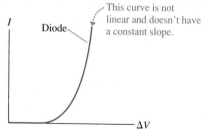

I

Diode

This curve is not linear and doesn't have a constant slope.

ΔV

discharging a capacitor, a wire connecting two battery terminals gets warm, deflects a compass needle, and makes a lightbulb glow brightly. These indicators tell us that charges flow through the wire from one terminal to the other.

The one major difference between a capacitor and a battery is the duration of the current. The current discharging a capacitor is transient, ceasing as soon as the excess charge on the capacitor plates is removed. In contrast, the current supplied by a battery is *sustained*.

We can use the charge escalator model of a battery to understand why. FIGURE 30.19 shows the charge escalator creating a potential difference ΔV_{bat} by lifting positive charge from the negative terminal to the positive terminal. Once at the positive terminal, positive charges can move *through the wire* as current I. In essence, the charges are "falling downhill" through the wire, losing the energy they gained on the escalator. This energy transfer to the wire warms the wire.

Eventually the charges find themselves back at the negative terminal of the battery, where they can ride the escalator back up and repeat the journey. A battery, unlike a charged capacitor, has an internal source of energy (the chemical reactions) that keeps the charge escalator running. It is the charge escalator that *sustains* the current in the wire by providing a continually renewed supply of charge at the battery terminals.

An important consequence of the charge escalator model, one you learned in the previous chapter, is that **a battery is a source of potential difference**. It is true that charges flow through a wire connecting the battery terminals, but current is a *consequence* of the battery's potential difference. The battery's emf is the *cause;* current, heat, light, sound, and so on are all *effects* that happen when the battery is used in certain ways.

Distinguishing cause and effect will be vitally important for understanding how a battery functions in a circuit. The reasoning is as follows:

1. A battery is a source of potential difference ΔV_{bat}. An ideal battery has $\Delta V_{bat} = \mathcal{E}$.
2. The battery creates a potential difference $\Delta V_{wire} = \Delta V_{bat}$ between the ends of a wire.
3. The potential difference ΔV_{wire} causes an electric field $E = \Delta V_{wire}/L$ in the wire.
4. The electric field establishes a current $I = JA = \sigma AE$ in the wire.
5. The magnitude of the current is determined *jointly* by the battery and the wire's resistance R to be $I = \Delta V_{wire}/R$.

More on Ohm's Law

Circuit textbooks often write Ohm's law as $V = IR$ rather than $I = \Delta V/R$. This can be misleading until you have sufficient experience with circuit analysis. First, Ohm's law relates the current to the potential *difference* between the ends of the conductor. Engineers and circuit designers *mean* "potential difference" when they use the symbol V, but the symbol is easily misinterpreted as simply "the potential." Second, $V = IR$ or even $\Delta V = IR$ suggests that a current I causes a potential difference ΔV. As you have seen, current is a *consequence* of a potential difference; hence $I = \Delta V/R$ is a better description of cause and effect.

Despite its name, Ohm's law is *not* a law of nature. It is limited to those materials whose resistance R remains constant—or very nearly so—during use. The materials to which Ohm's law applies are called *ohmic*. FIGURE 30.20a shows that the current through an ohmic material is directly proportional to the potential difference. Doubling the potential difference doubles the current. Metal and other conductors are ohmic devices.

Because the resistance of metals is small, a circuit made exclusively of metal wires would have enormous currents and would quickly deplete the battery. It is useful to limit the current in a circuit with ohmic devices, called **resistors**, whose resistance is significantly larger than the metal wires. Resistors are made with poorly conducting materials, such as carbon, or by depositing very thin metal films on an insulating substrate.

Some materials and devices are *nonohmic*, meaning that the current through the device is *not* directly proportional to the potential difference. For example, FIGURE 30.20b shows the *I*-versus-ΔV graph of a commonly used semiconductor device called a *diode*. Diodes do not have a well-defined resistance. Batteries, where $\Delta V = \mathcal{E}$ is determined by chemical reactions, and capacitors, where the relationship between *I* and ΔV differs from that of a resistor, are important nonohmic devices.

We can identify three important classes of ohmic circuit materials:

1. *Wires* are metals with very small resistivities ρ and thus very small resistances ($R \ll 1\,\Omega$). An **ideal wire** has $R = 0\,\Omega$; hence the potential difference between the ends of an ideal wire is $\Delta V = 0$ V *even if there is a current in it.* We will usually adopt the *ideal-wire model* of assuming that any connecting wires in a circuit are ideal.
2. *Resistors* are poor conductors with resistances usually in the range 10^1 to $10^6\,\Omega$. They are used to control the current in a circuit. Most resistors in a circuit have a specified value of *R*, such as 500 Ω. The filament in a lightbulb (a tungsten wire with a high resistance due to an extremely small cross-section area *A*) functions as a resistor as long as it is glowing, but the filament is slightly nonohmic because the value of its resistance when hot is larger than its room-temperature value.
3. *Insulators* are materials such as glass, plastic, or air. An **ideal insulator** has $R = \infty\,\Omega$; hence there is no current in an insulator even if there is a potential difference across it ($I = \Delta V/R = 0$ A). This is why insulators can be used to hold apart two conductors at different potentials. All practical insulators have $R \gg 10^9\,\Omega$ and can be treated, for our purposes, as ideal.

NOTE ▶ Ohm's law will be an important part of circuit analysis in the next chapter because resistors are essential components of almost any circuit. However, it is important that you apply Ohm's law *only* to the resistors and not to anything else. ◀

FIGURE 30.21a shows a resistor connected to a battery with current-carrying wires. Current must be conserved; hence the current *I* through the resistor is the same as the current in each wire. Because the wire's resistance is *much* less than that of the resistor, $R_{wire} \ll R_{resist}$, the potential difference $\Delta V_{wire} = IR_{wire}$ between the ends of each wire is *much* less than the potential difference $\Delta V_{resist} = IR_{resist}$ across the resistor. FIGURE 30.21b shows the potential along the wire-resistor-wire combination. You can see the large *voltage drop*, or potential difference, across the resistor. The voltage drops across the two wires are much smaller.

If we assume ideal wires with $R_{wire} = 0\,\Omega$, then $\Delta V_{wire} = 0$ V and *all* the voltage drop occurs across the resistor. In this *ideal-wire model,* shown in FIGURE 30.21c, the segments of the graph corresponding to the wires are horizontal. As we begin circuit analysis in the next chapter, we will assume that all wires are ideal unless stated otherwise. Thus our analysis will be focused on the resistors.

EXAMPLE 30.8 **A battery and a resistor**

What resistor would have a 15 mA current if connected across the terminals of a 9.0 V battery?

MODEL Assume the resistor is connected to the battery with ideal wires.

SOLVE Connecting the resistor to the battery with ideal wires makes $\Delta V_{resist} = \Delta V_{bat} = 9.0$ V. From Ohm's law, the resistance giving a 15 mA current is

$$R = \frac{\Delta V_{resist}}{I} = \frac{9.0\text{ V}}{0.015\text{ A}} = 600\,\Omega$$

The resistors used in circuits range from a few ohms to millions of ohms of resistance.

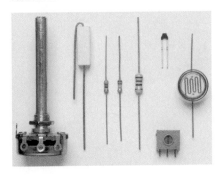

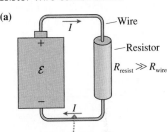

FIGURE 30.21 The potential along a wire-resistor-wire combination.

(a)

The current is constant along the wire-resistor-wire combination.

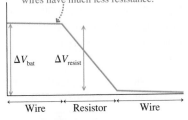

(b) The voltage drop along the wires is much less than across the resistor because the wires have much less resistance.

(c) In the ideal-wire model, with $R_{wire} = 0\,\Omega$, there is no voltage drop along the wires. All the voltage drop is across the resistor; thus $\Delta V_{resist} = \Delta V_{bat}$.

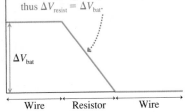

STOP TO THINK 30.6 A wire connects the positive and negative terminals of a battery. Two identical wires connect the positive and negative terminals of an identical battery. Rank in order, from largest to smallest, the currents I_a to I_d at points a to d.

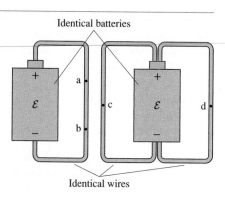

CHALLENGE EXAMPLE 30.9 **Measuring body composition**

The woman in the photo on page 879 is gripping a device that measures body fat. To illustrate how this works, FIGURE 30.22 models an upper arm as part muscle and part fat, showing the resistivities of each. Nonconductive elements, such as skin and bone, have been ignored. This is obviously not a picture of the actual structure, but gathering all the fat tissue together and all the muscle tissue together is a model that predicts the arm's electrical character quite well.

A 0.87 mA current is recorded when a 0.60 V potential difference is applied across an upper arm having the dimensions shown in the figure. What are the percentages of muscle and fat in this person's upper arm?

FIGURE 30.22 A simple model for the resistance of an arm.

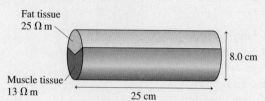

Fat tissue
25 Ω m

Muscle tissue
13 Ω m

8.0 cm

25 cm

MODEL Model the muscle and the fat as separate resistors connected to a 0.60 V battery. Assume the connecting wires to be ideal, with no "loss" of potential along the wires.

VISUALIZE FIGURE 30.23 shows the circuit, with the side-by-side muscle and fat resistors connected to the two terminals of the battery.

FIGURE 30.23 Circuit for passing current through the upper arm.

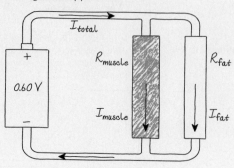

I_{total}

R_{muscle}

R_{fat}

0.60 V

I_{muscle}

I_{fat}

SOLVE The measured current of 0.87 mA is I_{total}, the current traveling from the battery to the arm and later back to the battery. This current splits at the junction between the two resistors. Kirchhoff's junction law, for the conservation of current, requires

$$I_{total} = I_{muscle} + I_{fat}$$

The current through each resistor can be found from Ohm's law: $I = \Delta V/R$. Each resistor has $\Delta V = 0.60$ V because each is connected to the battery terminals by lossless, ideal wires, but they have different resistances.

Let the fraction of muscle tissue be x; the fraction of fat is then $1 - x$. If the cross-section area of the upper arm is $A = \pi r^2$, then the muscle resistor has $A_{muscle} = xA$ while the fat resistor has $A_{fat} = (1 - x)A$. The resistances are related to the resistivities and the geometry by

$$R_{muscle} = \frac{\rho_{muscle}L}{A_{muscle}} = \frac{\rho_{muscle}L}{x\pi r^2}$$

$$R_{fat} = \frac{\rho_{fat}L}{A_{fat}} = \frac{\rho_{fat}L}{(1-x)\pi r^2}$$

The currents are thus

$$I_{muscle} = \frac{\Delta V}{R_{muscle}} = \frac{x\pi r^2 \Delta V}{\rho_{muscle}L} = 0.93x \text{ mA}$$

$$I_{fat} = \frac{\Delta V}{R_{fat}} = \frac{(1-x)\pi r^2 \Delta V}{\rho_{fat}L} = 0.48(1-x) \text{ mA}$$

The sum of these is the total current:

$$I_{total} = 0.87\text{mA} = 0.93x \text{ mA} + 0.48(1-x) \text{ mA}$$

$$= (0.48 + 0.45x) \text{ mA}$$

Solving, we find $x = 0.87$. This subject's upper arm is 87% muscle tissue, 13% fat tissue.

ASSESS The percentages seem reasonable for a healthy adult. A real measurement of body fat requires a more detailed model of the human body, because the current passes through both arms and across the chest, but the principles are the same.

SUMMARY

The goal of Chapter 30 has been to learn how and why charge moves through a conductor as what we call a current.

General Principles

Current is a nonequilibrium motion of charges sustained by an electric field. Nonuniform surface charge density creates an electric field in a wire. The electric field pushes the electron current i_e in a direction opposite to $\vec{E}$. The conventional current I is in the direction in which positive charge *seems* to move.

Conservation of Current

The current is the same at any two points in a wire.
At a junction,

$$\sum I_{in} = \sum I_{out}$$

This is **Kirchhoff's junction law.**

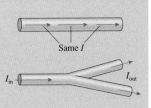

Electron current

i_e = rate of electron flow

$N_e = i_e \Delta t$

Conventional current

I = rate of charge flow = $e i_e$

$Q = I \Delta t$

Current density

$J = I/A$

Important Concepts

Sea of electrons

Conduction electrons move freely around the positive ions that form the atomic lattice.

Conduction

An electric field causes a slow drift at speed v_d to be superimposed on the rapid but random thermal motions of the electrons.

Collisions of electrons with the ions transfer energy to the atoms. This makes the wire warm and lightbulbs glow. More collisions mean a higher resistivity ρ and a lower conductivity σ.

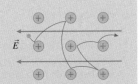

The drift speed is $v_d = \dfrac{e\tau}{m} E$, where τ is the mean time between collisions.

The electron current is related to the drift speed by

$$i_e = n_e A v_d$$

where n_e is the electron density.

An electric field E in a conductor causes a current density $J = n_e e v_d = \sigma E$, where the conductivity is

$$\sigma = \frac{n_e e^2 \tau}{m}$$

The resistivity is $\rho = 1/\sigma$.

Applications

Resistors

A potential difference ΔV_{wire} between the ends of a wire creates an electric field inside the wire:

$$E_{wire} = \frac{\Delta V_{wire}}{L}$$

The electric field causes a current in the direction of decreasing potential.

The size of the current is

$$I = \frac{\Delta V_{wire}}{R}$$

where $R = \dfrac{\rho L}{A}$ is the wire's **resistance.**
This is **Ohm's law.**

Terms and Notation

current, I	ampere, A	conductivity, σ	Ohm's law
drift speed, v_d	current density, J	resistivity, ρ	resistor
electron current, i_e	law of conservation of current	superconductivity	ideal wire
mean time between	junction	resistance, R	ideal insulator
collisions, τ	Kirchhoff's junction law	ohm, Ω	

CONCEPTUAL QUESTIONS

1. Suppose a time machine has just brought you forward from 1750 (post-Newton but pre-electricity) and you've been shown the lightbulb demonstration of FIGURE Q30.1. Do observations or *simple* measurements you might make—measurements that must make sense to you with your 1700s knowledge—prove that something is *flowing* through the wires? Or might you advance an alternative hypothesis for why the bulb is glowing? If your answer to the first question is yes, state what observations and/or measurements are relevant and the reasoning from which you can infer that something must be flowing. If not, can you offer an alternative hypothesis about why the bulb glows that could be tested?

FIGURE Q30.1

2. Consider a lightbulb circuit such as the one in FIGURE Q30.1.
   a. From the simple observations and measurements you can make on this circuit, can you distinguish a current composed of positive charge carriers from a current composed of negative charge carriers? If so, describe how you can tell which it is. If not, why not?
   b. One model of current is the motion of discrete charged particles. Another model is that current is the flow of a continuous charged fluid. Do simple observations and measurements on this circuit provide evidence in favor of either one of these models? If so, describe how.

3. The electron drift speed in a wire is exceedingly slow—typically only a fraction of a millimeter per second. Yet when you turn on a flashlight switch, the light comes on almost instantly. Resolve this apparent paradox.

4. Is FIGURE Q30.4 a possible surface charge distribution for a current-carrying wire? If so, in which direction is the current? If not, why not?

FIGURE Q30.4

5. What is the difference between current and current density?

6. All the wires in FIGURE Q30.6 are made of the same material and have the same diameter. Rank in order, from largest to smallest, the currents I_a to I_d. Explain.

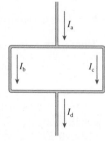

FIGURE Q30.6

7. Both batteries in FIGURE Q30.7 are identical and all lightbulbs are the same. Rank in order, from brightest to least bright, the brightness of bulbs a to c. Explain.

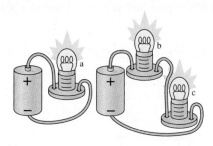

FIGURE Q30.7

8. Both batteries in FIGURE Q30.8 are identical and all lightbulbs are the same. Rank in order, from brightest to least bright, the brightness of bulbs a to c. Explain.

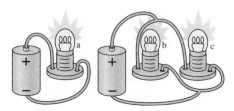

FIGURE Q30.8

9. The wire in FIGURE Q30.9 consists of two segments of different diameters but made from the same metal. The current in segment 1 is I_1.

FIGURE Q30.9

   a. Compare the currents in the two segments. That is, is I_2 greater than, less than, or equal to I_1? Explain.
   b. Compare the current densities J_1 and J_2 in the two segments.
   c. Compare the electric field strengths E_1 and E_2 in the two segments.
   d. Compare the drift speeds $(v_d)_1$ and $(v_d)_2$ in the two segments.

10. The current in a wire is doubled. What happens to (a) the current density, (b) the conduction-electron density, (c) the mean time between collisions, and (d) the electron drift speed? Are each of these doubled, halved, or unchanged? Explain.

11. The wires in FIGURE Q30.11 are all made of the same material. Rank in order, from largest to smallest, the resistances R_a to R_e of these wires. Explain.

FIGURE Q30.11

12. Which, if any, of these statements are true? (More than one may be true.) Explain.
    a. A battery supplies the energy to a circuit.
    b. A battery is a source of potential difference; the potential difference between the terminals of the battery is always the same.
    c. A battery is a source of current; the current leaving the battery is always the same.

EXERCISES AND PROBLEMS

Problems labeled integrate material from earlier chapters.

Exercises

Section 30.1 The Electron Current

1. || The electron drift speed in a 1.0-mm-diameter gold wire is 5.0×10^{-5} m/s. How long does it take 1 mole of electrons to flow through a cross section of the wire?
2. || 1.0×10^{20} electrons flow through a cross section of a 2.0-mm-diameter iron wire in 5.0 s. What is the electron drift speed?
3. || Electrons flow through a 1.6-mm-diameter aluminum wire at 2.0×10^{-4} m/s. How many electrons move through a cross section of the wire each day?
4. || 1.0×10^{16} electrons flow through a cross section of silver wire in 320 μs with a drift speed of 8.0×10^{-4} m/s. What is the diameter of the wire?

Section 30.2 Creating a Current

5. | The electron drift speed is 2.0×10^{-4} m/s in a metal with a mean time between collisions of 5.0×10^{-14} s. What is the electric field strength?
6. || a. How many conduction electrons are there in a 1.0-mm-diameter gold wire that is 10 cm long?
   b. How far must the sea of electrons in the wire move to deliver -32 nC of charge to an electrode?
7. || The mean time between collisions in iron is 4.2×10^{-15} s. What electron current is driven through a 1.8-mm-diameter iron wire by a 0.065 V/m electric field?
8. || A 2.0×10^{-3} V/m electric field creates a 3.5×10^{17} electrons/s current in a 1.0-mm-diameter aluminum wire. What are (a) the drift speed and (b) the mean time between collisions for electrons in this wire?

Section 30.3 Current and Current Density

9. | The wires leading to and from a 0.12-mm-diameter lightbulb filament are 1.5 mm in diameter. The wire to the filament carries a current with a current density of 4.5×10^5 A/m^2. What are (a) the current and (b) the current density in the filament?
10. || The current in a 100 watt lightbulb is 0.85 A. The filament inside the bulb is 0.25 mm in diameter.
    a. What is the current density in the filament?
    b. What is the electron current in the filament?
11. || In an integrated circuit, the current density in a 2.5-μm-thick $\times$ 75-μm-wide gold film is 7.5×10^5 A/m^2. How much charge flows through the film in 15 min?
12. | When a nerve cell fires, charge is transferred across the cell membrane to change the cell's potential from negative to positive. For a typical nerve cell, 9.0 pC of charge flows in a time of 0.50 ms. What is the average current through the cell membrane?
    BIO
13. | The current in an electric hair dryer is 10.0 A. How many electrons flow through the hair dryer in 5.0 min?
14. || 2.0×10^{13} electrons flow through a transistor in 1.0 ms. What is the current through the transistor?
15. | In an ionic solution, 5.0×10^{15} positive ions with charge $+2e$ pass to the right each second while 6.0×10^{15} negative ions with charge $-e$ pass to the left. What is the current in the solution?
16. | A hollow copper wire with an inner diameter of 1.0 mm and an outer diameter of 2.0 mm carries a current of 10 A. What is the current density in the wire?
17. || The current in a 2.0 mm $\times$ 2.0 mm square aluminum wire is 2.5 A. What are (a) the current density and (b) the electron drift speed?

Section 30.4 Conductivity and Resistivity

18. | What is the mean time between collisions for electrons in an aluminum wire and in an iron wire?
19. | The electric field in a 2.0 mm $\times$ 2.0 mm square aluminum wire is 0.012 V/m. What is the current in the wire?
20. | A 15-cm-long nichrome wire is connected across the terminals of a 1.5 V battery.
    a. What is the electric field inside the wire?
    b. What is the current density inside the wire?
    c. If the current in the wire is 2.0 A, what is the wire's diameter?
21. || A 3.0-mm-diameter wire carries a 12 A current when the electric field is 0.085 V/m. What is the wire's resistivity?
22. | A 0.0075 V/m electric field creates a 3.9 mA current in a 1.0-mm-diameter wire. What material is the wire made of?
23. || A 0.50-mm-diameter silver wire carries a 20 mA current. What are (a) the electric field and (b) the electron drift speed in the wire?
24. | The two segments of the wire in FIGURE EX30.24 have equal diameters but different conductivities σ_1 and σ_2. Current I passes through this wire. If the conductivities have the ratio $\sigma_2/\sigma_1 = 2$, what is the ratio E_2/E_1 of the electric field strengths in the two segments of the wire?

FIGURE EX30.24

25. | A metal cube 1.0 cm on each side is sandwiched between two electrodes. The electrodes create a 0.0050 V/m electric field in the metal. A current of 9.0 A passes through the cube, from the positive electrode to the negative electrode. Identify the metal.

Section 30.5 Resistance and Ohm's Law

26. | A 1.5 V battery provides 0.50 A of current.
    a. At what rate (C/s) is charge lifted by the charge escalator?
    b. How much work does the charge escalator do to lift 1.0 C of charge?
    c. What is the power output of the charge escalator?
27. || Wires 1 and 2 are made of the same metal. Wire 2 has twice the length and twice the diameter of wire 1. What are the ratios (a) ρ_2/ρ_1 of the resistivities and (b) R_2/R_1 of the resistances of the two wires?
28. | What is the resistance of
    a. A 2.0-m-long gold wire that is 0.20 mm in diameter?
    b. A 10-cm-long piece of carbon with a 1.0 mm $\times$ 1.0 mm square cross section?
29. || A 10-m-long wire with a diameter of 0.80 mm has a resistance of 1.1 Ω. Of what material is the wire made?

30. | The electric field inside a 30-cm-long copper wire is 5.0 mV/m. What is the potential difference between the ends of the wire?

31. | a. How long must a 0.60-mm-diameter aluminum wire be to have a 0.50 A current when connected to the terminals of a 1.5 V flashlight battery?
    b. What is the current if the wire is half this length?

32. ‖ The terminals of a 0.70 V watch battery are connected by a 100-m-long gold wire with a diameter of 0.10 mm. What is the current in the wire?

33. | BIO The femoral artery is the large artery that carries blood to the leg. What is the resistance of a 20-cm-long column of blood in a 1.0-cm-diameter femoral artery? The conductivity of blood is 0.63 $\Omega^{-1}\,\mathrm{m}^{-1}$.

34. ‖ Pencil "lead" is actually carbon. What is the current if a 9.0 V potential difference is applied between the ends of a 0.70-mm-diameter, 6.0-cm-long lead from a mechanical pencil?

35. ‖ The resistance of a very fine aluminum wire with a 10 μm × 10 μm square cross section is 1000 Ω. A 1000 Ω resistor is made by wrapping this wire in a spiral around a 3.0-mm-diameter glass core. How many turns of wire are needed?

36. | FIGURE EX30.36 is a current-versus-potential-difference graph for a material. What is the material's resistance?

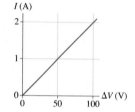

FIGURE EX30.36

37. ‖ A circuit calls for a 0.50-mm-diameter copper wire to be stretched between two points. You don't have any copper wire, but you do have aluminum wire in a wide variety of diameters. What diameter aluminum wire will provide the same resistance?

Problems

38. ‖ For what electric field strength would the current in a 2.0-mm-diameter nichrome wire be the same as the current in a 1.0-mm-diameter aluminum wire in which the electric field strength is 0.0080 V/m?

39. ‖ You've been asked to determine whether a new material your company has made is ohmic and, if so, to measure its electrical conductivity. Taking a 0.50 mm × 1.0 mm × 45 mm sample, you wire the ends of the long axis to a power supply and then measure the current for several different potential differences. Your data are as follows:

Voltage (V)	Current (A)
0.200	0.47
0.400	1.06
0.600	1.53
0.800	1.97

Use an appropriate graph of the data to determine whether the material is ohmic and, if so, its conductivity.

40. ‖ The electron beam inside a television picture tube is 0.40 mm in diameter and carries a current of 50 μA. This electron beam impinges on the inside of the picture tube screen.
    a. How many electrons strike the screen each second?
    b. What is the current density in the electron beam?

c. The electrons move with a velocity of 4.0×10^7 m/s. What electric field strength is needed to accelerate electrons from rest to this velocity in a distance of 5.0 mm?
d. Each electron transfers its kinetic energy to the picture tube screen upon impact. What is the *power* delivered to the screen by the electron beam?

41. ‖ FIGURE P30.41 shows a 4.0-cm-wide plastic film being wrapped onto a 2.0-cm-diameter roller that turns at 90 rpm. The plastic has a uniform surface charge density of -2.0 nC/cm^2.

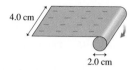

FIGURE P30.41

    a. What is the current of the moving film?
    b. How long does it take the roller to accumulate a charge of $-10\ \mu$C?

42. ‖ A sculptor has asked you to help electroplate gold onto a brass statue. You know that the charge carriers in the ionic solution are gold ions, and you've calculated that you must deposit 0.50 g of gold to reach the necessary thickness. How much current do you need, in mA, to plate the statue in 3.0 hours?

43. ‖ In a classic model of the hydrogen atom, the electron moves around the proton in a circular orbit of radius 0.053 nm.
    a. What is the electron's orbital frequency?
    b. What is the effective current of the electron?

44. | BIO The biochemistry that takes place inside cells depends on various elements, such as sodium, potassium, and calcium, that are dissolved in water as ions. These ions enter cells through narrow pores in the cell membrane known as *ion channels*. Each ion channel, which is formed from a specialized protein molecule, is selective for one type of ion. Measurements with microelectrodes have shown that a 0.30-nm-diameter potassium ion (K$^+$) channel carries a current of 1.8 pA.
    a. How many potassium ions pass through if the ion channel opens for 1.0 ms?
    b. What is the current density in the ion channel?

45. ‖ The starter motor of a car engine draws a current of 150 A from the battery. The copper wire to the motor is 5.0 mm in diameter and 1.2 m long. The starter motor runs for 0.80 s until the car engine starts.
    a. How much charge passes through the starter motor?
    b. How far does an electron travel along the wire while the starter motor is on?

46. | A car battery is rated at 90 A h, meaning that it can supply a 90 A current for 1 h before being completely discharged. If you leave your headlights on until the battery is completely dead, how much charge leaves the battery?

47. ‖ BIO Variations in the resistivity of blood can give valuable clues about changes in various properties of the blood. Suppose a medical device attaches two electrodes into a 1.5-mm-diameter vein at positions 5.0 cm apart. What is the blood resistivity if a 9.0 V potential difference causes a 230 μA current through the blood in the vein?

48. ‖ BIO The conducting path between the right hand and the left hand can be modeled as a 10-cm-diameter, 160-cm-long cylinder. The average resistivity of the interior of the human body is 5.0 Ω m. Dry skin has a much higher resistivity, but skin resistance can be made negligible by soaking the hands in salt water. If skin resistance is neglected, what potential difference between the hands is needed for a lethal shock of 100 mA across the chest? Your result shows that even small potential differences can produce dangerous currents when the skin is wet.

49. || You need to design a 1.0 A fuse that "blows" if the current exceeds 1.0 A. The fuse material in your stockroom melts at a current density of 500 A/cm². What diameter wire of this material will do the job?

50. || A hollow metal cylinder has inner radius a, outer radius b, length L, and conductivity σ. The current I is *radially* outward from the inner surface to the outer surface.
    a. Find an expression for the electric field strength inside the metal as a function of the radius r from the cylinder's axis.
    b. Evaluate the electric field strength at the inner and outer surfaces of an iron cylinder if $a = 1.0$ cm, $b = 2.5$ cm, $L = 10$ cm, and $I = 25$ A.

51. || A hollow metal sphere has inner radius a, outer radius b, and conductivity σ. The current I is *radially* outward from the inner surface to the outer surface.
    a. Find an expression for the electric field strength inside the metal as a function of the radius r from the center.
    b. Evaluate the electric field strength at the inner and outer surfaces of a copper sphere if $a = 1.0$ cm, $b = 2.5$ cm, and $I = 25$ A.

52. || The total amount of charge in coulombs that has entered a wire at time t is given by the expression $Q = 4t - t^2$, where t is in seconds and $t \geq 0$.
    a. Find an expression for the current in the wire at time t.
    b. Graph I versus t for the interval $0 \leq t \leq 4$ s.

53. || The total amount of charge that has entered a wire at time t is given by the expression $Q = (20\text{ C})(1 - e^{-t/(2.0\text{ s})})$, where t is in seconds and $t \geq 0$.
    a. Find an expression for the current in the wire at time t.
    b. What is the maximum value of the current?
    c. Graph I versus t for the interval $0 \leq t \leq 10$ s.

54. || The current in a wire at time t is given by the expression $I = (2.0\text{ A})e^{-t/(2.0\ \mu s)}$, where t is in microseconds and $t \geq 0$.
    a. Find an expression for the total amount of charge (in coulombs) that has entered the wire at time t. The initial conditions are $Q = 0$ C at $t = 0\ \mu s$.
    b. Graph Q versus t for the interval $0 \leq t \leq 10\ \mu s$.

55. || The two wires in FIGURE P30.55 are made of the same material. What are the current and the electron drift speed in the 2.0-mm-diameter segment of the wire?

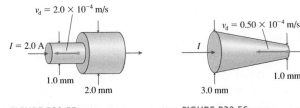

FIGURE P30.55   FIGURE P30.56

56. || What is the electron drift speed at the 3.0-mm-diameter end (the left end) of the wire in FIGURE P30.56?

57. | What diameter should the nichrome wire in FIGURE P30.57 be in order for the electric field strength to be the same in both wires?

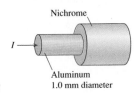

FIGURE P30.57

58. || An aluminum wire consists of the three segments shown in FIGURE P30.58. The current in the top segment is 10 A. For each of these three segments, find the
    a. Current I.
    b. Current density J.
    c. Electric field E.
    d. Drift velocity v_d.
    e. Electron current i.
    Place your results in a table for easy viewing.

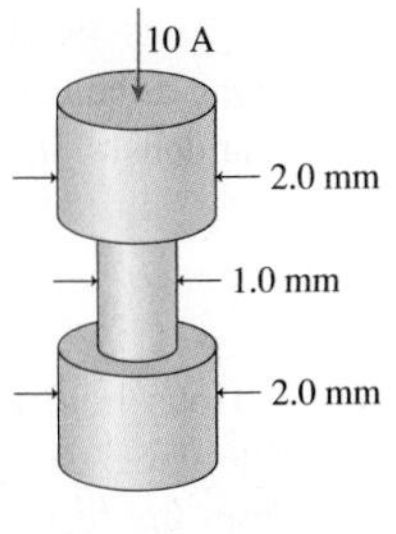

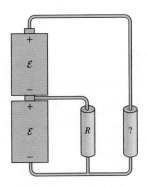

FIGURE P30.58

59. || What electric field strength is needed to create a 5.0 A current in a 2.0-mm-diameter iron wire?

60. || A 20-cm-long hollow nichrome tube of inner diameter 2.8 mm, outer diameter 3.0 mm is connected to a 3.0 V battery. What is the current in the tube?

61. || The batteries in FIGURE P30.61 are identical. Both resistors have equal currents. What is the resistance of the resistor on the right?

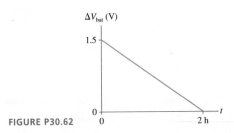

FIGURE P30.61

62. || A 1.5 V flashlight battery is connected to a wire with a resistance of 3.0 Ω. FIGURE P30.62 shows the battery's potential difference as a function of time. What is the total charge lifted by the charge escalator?

FIGURE P30.62

63. || Two 10-cm-diameter metal plates 1.0 cm apart are charged to ± 12.5 nC. They are suddenly connected together by a 0.224-mm-diameter copper wire stretched taut from the center of one plate to the center of the other.
    a. What is the maximum current in the wire?
    b. Does the current increase with time, decrease with time, or remain steady? Explain.
    c. What is the total amount of energy dissipated in the wire?

64. || A long, round wire has resistance R. What will the wire's resistance be if you stretch it to twice its initial length?

65. ‖ FIGURE P30.65 shows the potential along a tungsten wire. What is the current density in the wire?

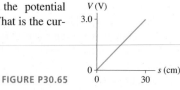

FIGURE P30.65

66. ‖ Household wiring often uses 2.0-mm-diameter copper wires. The wires can get rather long as they snake through the walls from the fuse box to the farthest corners of your house. What is the potential difference across a 20-m-long, 2.0-mm-diameter copper wire carrying an 8.0 A current?

67. ‖ You've decided to protect your house by placing a 5.0-m-tall iron lightning rod next to the house. The top is sharpened to a point and the bottom is in good contact with the ground. From your research, you've learned that lightning bolts can carry up to 50 kA of current and last up to 50 μs.
   a. How much charge is delivered by a lightning bolt with these parameters?
   b. You don't want the potential difference between the top and bottom of the lightning rod to exceed 100 V. What minimum diameter must the rod have?

Challenge Problems

68. The conductive tissues of the upper leg can be modeled as a
   BIO 40-cm-long, 12-cm-diameter cylinder of muscle and fat. The resistivities of muscle and fat are 13 Ω m and 25 Ω m, respectively. One person's upper leg is 82% muscle, 18% fat. What current is measured if a 1.5 V potential difference is applied between the person's hip and knee?

69. The current supplied by a battery slowly decreases as the battery runs down. Suppose that the current as a function of time is $I = (0.75 \text{ A})e^{-t/(6 \text{ h})}$. What is the total number of electrons transported from the positive electrode to the negative electrode by the charge escalator from the time the battery is first used until it is completely dead?

70. The electric field in a current-carrying wire can be modeled as the electric field at the midpoint between two charged rings. Model a 3.0-mm-diameter aluminum wire as two 3.0-mm-diameter rings 2.0 mm apart. What is the current in the wire after 20 electrons are transferred from one ring to the other?

71. A 5.0-mm-diameter proton beam carries a total current of 1.5 mA. The current density in the proton beam, which increases with distance from the center, is given by $J = J_{\text{edge}}(r/R)$, where R is the radius of the beam and J_{edge} is the current density at the edge.
   a. How many protons per second are delivered by this proton beam?
   b. Determine the value of J_{edge}.

72. A metal wire connecting the terminals of a battery with potential difference ΔV_{bat} gets warm as it draws a current I.
   a. What is ΔU, the change in potential energy of charge Q as it passes through the wire?
   b. Where does this energy go?
   c. Power is the *rate* of transfer of energy. Based on your answer to part a, find an expression for the power supplied by the battery to warm the wire.
   d. What power does a 1.5 V battery supply to a wire drawing a 1.2 A current?

73. FIGURE CP30.73 shows a wire that is made of two equal-diameter segments with conductivities σ_1 and σ_2. When current I passes through the wire, a thin layer of charge appears at the boundary between the segments.

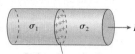

Surface charge density η

FIGURE CP30.73

   a. Find an expression for the surface charge density η on the boundary. Give your result in terms of I, σ_1, σ_2, and the wire's cross-section area A.
   b. A 1.0-mm-diameter wire made of copper and iron segments carries a 5.0 A current. How much charge accumulates at the boundary between the segments?

STOP TO THINK ANSWERS

Stop to Think 30.1: $i_c > i_b > i_a > i_d$. The electron current is proportional to $r^2 v_d$. Changing r by a factor of 2 has more influence than changing v_d by a factor of 2.

Stop to Think 30.2: The electrons don't have to move from the switch to the bulb, which could take hours. Because the wire between the switch and the bulb is already full of electrons, a flow of electrons from the switch into the wire immediately causes electrons to flow from the other end of the wire into the lightbulb.

Stop to Think 30.3: $E_d > E_b > E_e > E_a = E_c$. The electric field strength depends on the *difference* in the charge on the two wires. The electric fields of the rings in a and c are opposed to each other, so the net field is zero. The rings in d have the largest charge *difference*.

Stop to Think 30.4: 1 A into the junction. The total current entering the junction must equal the total current leaving the junction.

Stop to Think 30.5: $J_b > J_a = J_d > J_c$. The current density $J = I/\pi r^2$ is independent of the conductivity σ, so a and d are the same. Changing r by a factor of 2 has more influence than changing I by a factor of 2.

Stop to Think 30.6: $I_a = I_b = I_c = I_d$. Conservation of current requires $I_a = I_b$. The current in each wire is $I = \Delta V_{\text{wire}}/R$. All the wires have the same resistance because they are identical, and they all have the same potential difference because each is connected directly to the battery, which is a *source of potential.*

31 Fundamentals of Circuits

A microprocessor, the heart of a powerful computer, is a complex device. Even so, a microprocessor operates on the basis of just a few fundamental physical principles.

▶ **Looking Ahead** The goal of Chapter 31 is to understand the fundamental physical principles that govern electric circuits.

DC Circuits

Circuits—from a simple lightbulb to a supercomputer—are based on the controlled motion of charges. You will learn about the fundamental physical principles by which circuits operate.

This chapter will focus on **DC circuits,** meaning *direct current*, in which potentials and currents are steady. Chapter 35 will extend these ideas to AC circuits in which the potential difference oscillates sinusoidally.

◀ **Looking Back**
Section 29.2 Sources of potential

Analyzing Circuits

Circuits consist of many elements—batteries, resistors, capacitors, and more—connected together. Two basic tools will help you find the potential difference across and current through each element:

- Kirchhoff's junction law.
- Kirchhoff's loop law.

Also important will be Ohm's law, for resistors, and the properties of batteries and capacitors.

Energy and Power

Circuits do things by using energy. You'll learn to calculate *power*, the rate at which the battery supplies energy to a circuit and the rate at which a resistor dissipates it.

The power delivered by these photovoltaic cells is the product of their emf and the current they deliver. One solar panel provides about 200 W at midday on a sunny day.

Circuit Diagrams

You will learn how to use symbols of circuit elements to draw a **circuit diagram.** This is a logical picture of how the circuit elements are related rather than a literal picture of how they look.

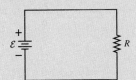

This is the circuit diagram of a simple circuit in which a resistor is connected to a battery.

Combining Resistors

Resistors often occur **in series** or **in parallel.**

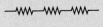 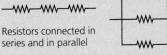

Resistors connected in series and in parallel

You'll learn that these combinations of resistors can be "simplified" by replacing them with one **equivalent resistor.**

◀ **Looking Back**
Sections 30.3–30.5 Current, resistance, and Ohm's law

RC Circuits

A capacitor is charged or discharged by current through a resistor. These important circuits are called *RC* **circuits.** Applications range from defibrillators to timing circuits.

You'll learn that the capacitor charge decays exponentially. The time to decay to e^{-1} of the initial value is called the time constant τ.

◀ **Looking Back**
Section 29.5 Capacitors

31.1 Circuit Elements and Diagrams

FIGURE 31.1 An electric circuit.

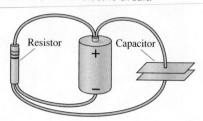

FIGURE 31.1 shows an electric circuit in which a resistor and a capacitor are connected by wires to a battery. To understand the functioning of this circuit, we do not need to know whether the wires are bent or straight, or whether the battery is to the right or to the left of the resistor. The literal picture of Figure 31.1 provides many irrelevant details. It is customary when describing or analyzing circuits to use a more abstract picture called a **circuit diagram**. A circuit diagram is a *logical* picture of what is connected to what.

A circuit diagram also replaces pictures of the circuit elements with symbols. FIGURE 31.2 shows the basic symbols that we will need. The longer line at one end of the battery symbol represents the positive terminal of the battery. Notice that a lightbulb, like a wire or a resistor, has two "ends," and current passes *through* the bulb. It is often useful to think of a lightbulb as a resistor that gives off light when a current is present. A lightbulb filament is not a perfectly ohmic material, but the resistance of a *glowing* lightbulb remains reasonably constant if you don't change ΔV by much.

FIGURE 31.2 A library of basic symbols used for electric circuit drawings.

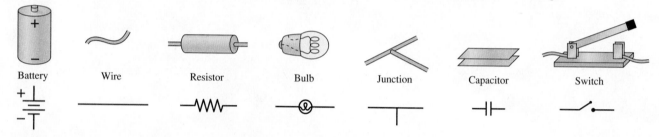

FIGURE 31.3 A circuit diagram for the circuit of Figure 31.1.

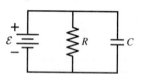

FIGURE 31.3 is a circuit diagram of the circuit shown in Figure 31.1. Notice how the circuit elements are labeled. The battery's emf $\mathcal{E}$ is shown beside the battery, and + and − symbols, even though somewhat redundant, are shown beside the terminals. We would use numerical values for $\mathcal{E}$, R, and C if we knew them. The wires, which in practice may bend and curve, are shown as straight-line connections between the circuit elements.

STOP TO THINK 31.1 Which of these diagrams represent the same circuit?

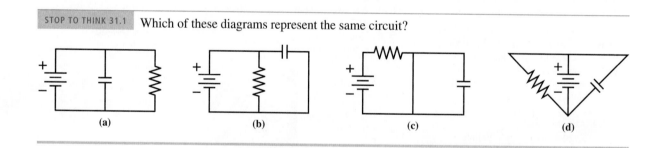

(a) (b) (c) (d)

31.2 Kirchhoff's Laws and the Basic Circuit

We are now ready to begin analyzing circuits. To analyze a circuit means to find:

1. The potential difference across each circuit component.
2. The current in each circuit component.

FIGURE 31.4 Kirchhoff's junction law.

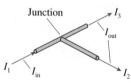

Junction law: $I_1 = I_2 + I_3$

Because charge and current are conserved, the total current into the junction of FIGURE 31.4 must equal the total current leaving the junction. That is,

$$\sum I_{\text{in}} = \sum I_{\text{out}} \qquad (31.1)$$

This statement, which you met in Chapter 30, is **Kirchhoff's junction law.**

An important property of the electric potential is that the sum of the potential differences around any loop or closed path is zero. This is a statement of energy conservation, because a charge that moves around a closed path and returns to its starting point has $\Delta U = 0$. We apply this idea to the circuit of FIGURE 31.5 by adding all of the potential differences *around* the loop formed by the circuit. Doing so gives

$$\Delta V_{\text{loop}} = \sum (\Delta V)_i = 0 \qquad (31.2)$$

where $(\Delta V)_i$ is the potential difference of the ith component in the loop. This statement, introduced in Chapter 29, is **Kirchhoff's loop law.**

Kirchhoff's loop law can be true only if at least one of the $(\Delta V)_i$ is negative. To apply the loop law, we need to explicitly identify which potential differences are positive and which are negative.

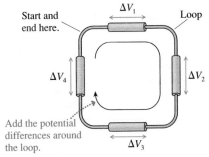

FIGURE 31.5 Kirchhoff's loop law.

Loop law: $\Delta V_1 + \Delta V_2 + \Delta V_3 + \Delta V_4 = 0$

TACTICS
BOX 31.1 Using Kirchhoff's loop law

❶ **Draw a circuit diagram.** Label all known and unknown quantities.

❷ **Assign a direction to the current.** Draw and label a current arrow I to show your choice.

- If you know the actual current direction, choose that direction.

- If you don't know the actual current direction, make an arbitrary choice. All that will happen if you choose wrong is that your value for I will end up negative.

❸ **"Travel" around the loop.** Start at any point in the circuit, then go all the way around the loop in the direction you assigned to the current in step 2. As you go through each circuit element, ΔV is interpreted to mean

$$\Delta V = V_{\text{downstream}} - V_{\text{upstream}}$$

- For an ideal battery in the negative-to-positive direction:

$$\Delta V_{\text{bat}} = +\mathcal{E}$$

- For an ideal battery in the positive-to-negative direction:

$$\Delta V_{\text{bat}} = -\mathcal{E}$$

- For a resistor: $\quad \Delta V_{\text{res}} = -\Delta V_R = -IR$

❹ **Apply the loop law:** $\sum (\Delta V)_i = 0$

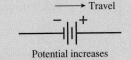

→ Travel

Potential increases

→ Travel

Potential decreases

→ I

Potential decreases

Exercises 4–7

NOTE ▶ Ohm's law gives us only the *magnitude* $\Delta V_R = IR$ of the potential difference across a resistor. Kirchhoff's law requires us to recognize that the electric potential inside a resistor *decreases* in the direction of the current. Thus $\Delta V_{\text{res}} = V_{\text{downstream}} - V_{\text{upstream}} = -\Delta V_R$. ◀

The Basic Circuit

The most basic electric circuit is a single resistor connected to the two terminals of a battery. FIGURE 31.6a on the next page shows a literal picture of the circuit elements and the connecting wires; FIGURE 31.6b is the circuit diagram. Notice that this is a **complete circuit,** forming a continuous path between the battery terminals.

FIGURE 31.6 The basic circuit of a resistor connected to a battery.

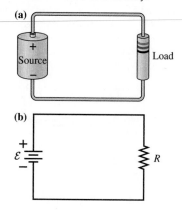

FIGURE 31.7 Analysis of the basic circuit using Kirchhoff's loop law.

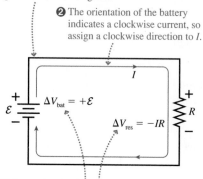

The resistor might be a known resistor, such as "a 10 Ω resistor," or it might be some other resistive device, such as a lightbulb. Regardless of what the resistor is, it is called the **load.** The battery is called the **source.**

FIGURE 31.7 shows the use of Kirchhoff's loop law to analyze this circuit. Two things are worth noting:

1. This circuit has no junctions, so the current I is the same in all four sides of the circuit. Kirchhoff's junction law is not needed.
2. We're assuming the ideal-wire model, in which there are *no* potential differences along the connecting wires.

Kirchhoff's loop law, with two circuit elements, is

$$\Delta V_{\text{loop}} = \sum (\Delta V)_i = \Delta V_{\text{bat}} + \Delta V_{\text{res}} = 0 \qquad (31.3)$$

Let's look at each of the two terms in Equation 31.3:

1. The potential *increases* as we travel through the battery on our clockwise journey around the loop. We enter the negative terminal and, farther downstream, exit the positive terminal after having gained potential $\mathcal{E}$. Thus

$$\Delta V_{\text{bat}} = +\mathcal{E}$$

2. The potential of a conductor *decreases* in the direction of the current, which we've indicated with the + and − signs in Figure 31.7. Thus

$$\Delta V_{\text{res}} = V_{\text{downstream}} - V_{\text{upstream}} = -IR$$

NOTE ▶ Determining which potential differences are positive and which are negative is perhaps *the* most important step in circuit analysis. ◀

With this information, the loop equation becomes

$$\mathcal{E} - IR = 0 \qquad (31.4)$$

We can solve the loop equation to find that the current in the circuit is

$$I = \frac{\mathcal{E}}{R} \qquad (31.5)$$

We can then use the current to find that the magnitude of the resistor's potential difference is

$$\Delta V_R = IR = \mathcal{E} \qquad (31.6)$$

This result should come as no surprise. The potential energy that the charges gain in the battery is subsequently lost as they "fall" through the resistor.

NOTE ▶ The current that the battery delivers depends jointly on the emf of the battery and the resistance of the load. ◀

EXAMPLE 31.1 **Two resistors and two batteries**

Analyze the circuit shown in FIGURE 31.8.

a. Find the current in and the potential difference across each resistor.
b. Draw a graph showing how the potential changes around the circuit, starting from $V = 0$ V at the negative terminal of the 6 V battery.

MODEL Assume ideal connecting wires and ideal batteries, for which $\Delta V_{\text{bat}} = \mathcal{E}$.

FIGURE 31.8 Circuit for Example 31.1.

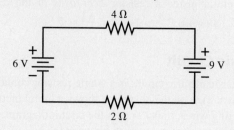

VISUALIZE In FIGURE 31.9, we've redrawn the circuit and defined $\mathcal{E}_1$, $\mathcal{E}_2$, R_1, and R_2. Because there are no junctions, the current is the same through *each* component in the circuit. With some thought, we might deduce whether the current is cw or ccw, but we do not need to know in advance of our analysis. We will choose a clockwise direction and solve for the value of I. If our solution is positive, then the current really is cw. If the solution should turn out to be negative, we will know that the current is ccw.

FIGURE 31.9 Analyzing the circuit.

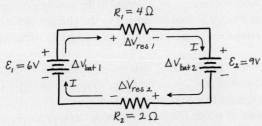

SOLVE a. How do we deal with *two* batteries? Can charge flow "backward" through a battery, from positive to negative? Consider the charge escalator analogy. Left to itself, a charge escalator lifts charge from lower to higher potential. But it *is* possible to run down an up escalator, as many of you have probably done. If two escalators are placed "head to head," whichever is stronger will, indeed, force the charge to run down the up escalator of the other battery. The current in a battery *can* be from positive to negative if driven in that direction by a larger emf from a second battery. Indeed, this is how rechargeable batteries are recharged.

Kirchhoff's loop law, going clockwise from the negative terminal of battery 1, is

$$\Delta V_{\text{closed loop}} = \sum (\Delta V)_i = \Delta V_{\text{bat 1}} + \Delta V_{\text{res 1}}$$

$$+ \Delta V_{\text{bat 2}} + \Delta V_{\text{res 2}} = 0$$

All the signs are $+$ because this is a formal statement of *adding* potential differences around the loop. Next we can evaluate each ΔV. As we go cw, the charges *gain* potential in battery 1 but *lose* potential in battery 2. Thus $\Delta V_{\text{bat 1}} = +\mathcal{E}_1$ and $\Delta V_{\text{bat 2}} = -\mathcal{E}_2$. There is a *loss* of potential in traveling through each resistor, because we're traversing them in the

direction we assigned to the current, so $\Delta V_{\text{res 1}} = -IR_1$ and $\Delta V_{\text{res 2}} = -IR_2$. Thus Kirchhoff's loop law becomes

$$\sum (\Delta V)_i = \mathcal{E}_1 - IR_1 - \mathcal{E}_2 - IR_2$$

$$= \mathcal{E}_1 - \mathcal{E}_2 - I(R_1 + R_2) = 0$$

We can solve this equation to find the current in the loop:

$$I = \frac{\mathcal{E}_1 - \mathcal{E}_2}{R_1 + R_2} = \frac{6 \text{ V} - 9 \text{ V}}{4 \text{ }\Omega + 2 \text{ }\Omega} = -0.50 \text{ A}$$

The value of I is negative; hence the actual current in this circuit is 0.50 A *counterclockwise*. You perhaps anticipated this from the orientation of the 9 V battery with its larger emf.

b. The potential difference across the 4 Ω resistor is

$$\Delta V_{\text{res 1}} = -IR_1 = -(-0.50 \text{ A})(4 \text{ }\Omega) = +2.0 \text{ V}$$

Because the current is actually ccw, the resistor's potential *increases* in the cw direction of our travel around the loop. Similarly, the potential difference across the 2 Ω resistor is $\Delta V_{\text{res 2}} = 1.0$ V. FIGURE 31.10 shows the potential experienced by charges flowing around the circuit. The distance s is measured from the 6 V battery's negative terminal, and we have chosen to let $V = 0$ V at that point. The potential ends at the value from which it started.

FIGURE 31.10 A graphical presentation of how the potential changes around the loop.

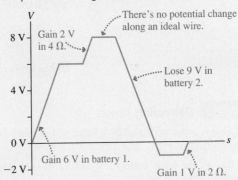

ASSESS Notice how the potential *drops* 9 V upon passing through battery 2 in the cw direction. It then gains 1 V upon passing through R_2 to end at the starting potential.

STOP TO THINK 31.2 What is ΔV across the unspecified circuit element? Does the potential increase or decrease when traveling through this element in the direction assigned to I?

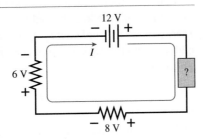

31.3 Energy and Power

FIGURE 31.11 Which lightbulb is brighter?

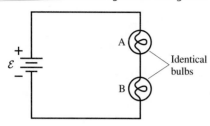

FIGURE 31.11 Which lightbulb is brighter?

The circuit of FIGURE 31.11 has two identical lightbulbs, A and B. Which is brighter? Or are they equally bright? Think about this before going on.

You might have been tempted to say that A is brighter. After all, the current gets to A first, so A might "use up" some of the current and leave less for B. But this would violate the laws of conservation of charge and conservation of current. There are no junctions between A and B, so the current through the two bulbs must be the same. Hence the bulbs are equally bright.

It's not current that the bulbs use up, it's *energy*. Because a battery supplies a potential difference, it also supplies energy to a circuit. The charge escalator is an energy-transfer process, transferring chemical energy E_{chem} stored in the battery to the potential energy U of the charges. That energy is then dissipated as the charges move through the wires and resistors, increasing their thermal energy until, in the case of the lightbulb filaments, they glow.

A charge gains potential energy $\Delta U = q \Delta V_{bat}$ as it moves up the charge escalator in the battery. For an ideal battery, with $\Delta V_{bat} = \mathcal{E}$, the battery supplies energy $\Delta U = q\mathcal{E}$ as it lifts charge q from the negative to the positive terminal.

It is useful to know the *rate* at which the battery supplies energy to the charges. Recall from Chapter 11 that the rate at which energy is transferred is *power*, measured in joules per second or *watts*. If energy $\Delta U = q\mathcal{E}$ is transferred to charge q, then the *rate* at which energy is transferred from the battery to the moving charges is

$$P_{bat} = \text{rate of energy transfer} = \frac{dU}{dt} = \frac{dq}{dt}\mathcal{E} \qquad (31.7)$$

But dq/dt, the rate at which charge moves through the battery, is the current I. Hence the power supplied by a battery, or the rate at which the battery (or any other source of emf) transfers energy to the charges passing through it, is

$$P_{bat} = I\mathcal{E} \qquad \text{(power delivered by an emf)} \qquad (31.8)$$

$I\mathcal{E}$ has units of J/s, or W.

EXAMPLE 31.2 | **Delivering power**

A 90 Ω load is connected to a 120 V battery. How much power is delivered by the battery?

SOLVE This is our basic battery-and-resistor circuit, which we analyzed earlier. In this case

$$I = \frac{\mathcal{E}}{R} = \frac{120 \text{ V}}{90 \text{ }\Omega} = 1.33 \text{ A}$$

Thus the power delivered by the battery is

$$P_{bat} = I\mathcal{E} = (1.33 \text{ A})(120 \text{ V}) = 160 \text{ W}$$

FIGURE 31.12 A current-carrying resistor dissipates power because the electric force does work on the charges.

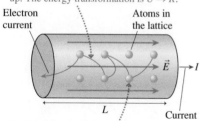

The electric field causes electrons to speed up. The energy transformation is $U \rightarrow K$.

Electron current

Atoms in the lattice

$\vec{E}$ → I

L

Current

Collisions transfer energy to the lattice. The energy transformation is $K \rightarrow E_{th}$.

P_{bat} is the energy transferred per second from the battery's store of chemicals to the moving charges that make up the current. But what happens to this energy? Where does it end up? FIGURE 31.12, a section of a current-carrying resistor, reminds you of our microscopic model of conduction. The electrons accelerate in the electric field, then collide with atoms in the lattice. The acceleration phase is a transformation of potential to kinetic energy. The collisions then transfer the electron's kinetic energy to the *thermal* energy of the lattice. The potential energy was acquired in the battery, from the conversion of chemical energy, so the entire energy-transfer process looks like

$$E_{chem} \rightarrow U \rightarrow K \rightarrow E_{th}$$

The net result is that **the battery's chemical energy is transferred to the thermal energy of the resistors,** raising their temperature.

Suppose the average distance between collisions is d. The electric force $\vec{F} = q\vec{E}$ exerted on charge q does work as it pushes the charge through distance d. The field is constant inside the resistor, so the work is simply

$$W = F\,\Delta s = qEd \qquad (31.9)$$

According to the work-kinetic energy theorem, this work increases the kinetic energy of charge q by $\Delta K = W = qEd$. This kinetic energy is transferred to the lattice when charge q collides with a lattice atom, causing the energy of the lattice to increase by

$$\Delta E_{\text{per collision}} = \Delta K = qEd$$

Collisions occur over and over as the charge makes its way through the resistor. After many such collisions, the total energy that charge q transfers while traveling distance L, the length of the resistor, is

$$\Delta E_{\text{th}} = qEL \qquad (31.10)$$

But EL is the potential difference ΔV_{R} between the two ends of the resistor. Thus *each* charge q, as it travels the length of the resistor, transfers energy to the atomic lattice in the amount

$$\Delta E_{\text{th}} = q\,\Delta V_{\text{R}} \qquad (31.11)$$

The *rate* at which energy is transferred from the current to the resistor is thus

$$P_{\text{R}} = \frac{dE_{\text{th}}}{dt} = \frac{dq}{dt}\Delta V_{\text{R}} = I\,\Delta V_{\text{R}} \qquad (31.12)$$

We say that this power—so many joules per second—is *dissipated* by the resistor as charge flows through it. The resistor, in turn, transfers this energy to the air and to the circuit board on which it is mounted, causing the circuit and all its surroundings to heat up.

From our analysis of the basic circuit, in which a single resistor is connected to a battery, we learned that $\Delta V_{\text{R}} = \mathcal{E}$. That is, the potential difference across the resistor is exactly the emf supplied by the battery. But then Equations 31.8 and 31.12, for P_{bat} and P_{R}, are numerically equal, and we find that

$$P_{\text{R}} = P_{\text{bat}} \qquad (31.13)$$

The answer to the question "What happens to the energy supplied by the battery?" is "The battery's chemical energy is transformed into the thermal energy of the resistor." The *rate* at which the battery supplies energy is exactly equal to the *rate* at which the resistor dissipates energy. This is, of course, exactly what we would have expected from energy conservation.

EXAMPLE 31.3   **The power of light**

How much current is "drawn" by a 100 W lightbulb connected to a 120 V outlet?

MODEL Most household appliances, such as a 100 W lightbulb or a 1500 W hair dryer, have a power rating. The rating does *not* mean that these appliances *always* dissipate that much power. These appliances are intended for use at a standard household voltage of 120 V, and their rating is the power they will dissipate *if* operated with a potential difference of 120 V. Their power consumption will differ from the rating if they are operated at any other potential difference.

SOLVE Because the lightbulb is operating as intended, it will dissipate 100 W of power. Thus

$$I = \frac{P_{\text{R}}}{\Delta V_{\text{R}}} = \frac{100\text{ W}}{120\text{ V}} = 0.833\text{ A}$$

ASSESS A current of 0.833 A in this lightbulb transfers 100 J/s to the thermal energy of the filament, which, in turn, dissipates 100 J/s as heat and light to its surroundings.

A resistor obeys Ohm's law, $\Delta V_R = IR$. (Remember that Ohm's law gives only the *magnitude* of ΔV_R.) This gives us two alternative ways of writing the power dissipated by a resistor. We can either substitute IR for ΔV_R or substitute $\Delta V_R/R$ for I. Thus

$$P_R = I\,\Delta V_R = I^2 R = \frac{(\Delta V_R)^2}{R} \qquad \text{(power dissipated by a resistor)} \qquad (31.14)$$

If the same current I passes through several resistors in series, then $P_R = I^2 R$ tells us that most of the power will be dissipated by the largest resistance. This is why a lightbulb filament glows but the connecting wires do not. Essentially *all* of the power supplied by the battery is dissipated by the high-resistance lightbulb filament and essentially no power is dissipated by the low-resistance wires. The filament gets very hot, but the wires do not.

EXAMPLE 31.4 | **The power of sound**

Most loudspeakers are designed to have a resistance of 8 Ω. If an 8 Ω loudspeaker is connected to a stereo amplifier with a rating of 100 W, what is the maximum possible current to the loudspeaker?

MODEL The rating of an amplifier is the *maximum* power it can deliver. Most of the time it delivers far less, but the maximum might be reached for brief, intense sounds like cymbal crashes.

SOLVE The loudspeaker is a resistive load. The maximum current to the loudspeaker occurs when the amplifier delivers maximum power $P_{max} = (I_{max})^2 R$. Thus

$$I_{max} = \sqrt{\frac{P_{max}}{R}} = \sqrt{\frac{100\text{ W}}{8\ \Omega}} = 3.5\text{ A}$$

Kilowatt Hours

The energy dissipated (i.e., transformed into thermal energy) by a resistor during time Δt is $E_{th} = P_R \Delta t$. The product of watts and seconds is joules, the SI unit of energy. However, your local electric company prefers to use a different unit, the *kilowatt hour,* to measure the energy you use each month.

A load that consumes P_R kW of electricity for Δt hours has used $P_R \Delta t$ **kilowatt hours** of energy, abbreviated kWh. For example, a 4000 W electric water heater uses 40 kWh of energy in 10 hours. A 1500 W hair dryer uses 0.25 kWh of energy in 10 minutes. Despite the rather unusual name, a kilowatt hour is a unit of energy. A homework problem will let you find the conversion factor from kilowatt hours to joules.

Your monthly electric bill specifies the number of kilowatt hours you used last month. This is the amount of energy that the electric company delivered to you, via an electric current, and that you transformed into light and thermal energy inside your home. The cost of electricity varies throughout the country, but the average cost of electricity in the United States is approximately 10¢ per kWh ($0.10/kWh). Thus it costs about $4.00 to run your water heater for 10 hours, about 2.5¢ to dry your hair.

The electric meter on the side of your house or apartment records the kilowatt hours of electric energy that you use.

STOP TO THINK 31.3 Rank in order, from largest to smallest, the powers P_a to P_d dissipated in resistors a to d.

(a) $+\ \Delta V\ -$ R
(b) $+\ 2\Delta V\ -$ R
(c) $+\ \Delta V\ -$ $2R$
(d) $+\ \Delta V\ -$ $\frac{1}{2}R$

31.4 Series Resistors

Consider the three lightbulbs in FIGURE 31.13. The batteries are identical and the bulbs are identical. You learned in the previous section that B and C are equally bright, because of conservation of current, but how does the brightness of B compare to that of A? Think about this before going on.

FIGURE 31.14a shows two resistors placed end to end between points a and b. Resistors that are aligned end to end, *with no junctions between them*, are called **series resistors** or, sometimes, resistors "in series." Because there are no junctions and because current is conserved, the current I must be the same through each of these resistors. That is, the current out of the last resistor in a series is equal to the current into the first resistor.

The potential differences across the two resistors are $\Delta V_1 = IR_1$ and $\Delta V_2 = IR_2$. The total potential difference ΔV_{ab} between points a and b is the sum of the individual potential differences:

$$\Delta V_{ab} = \Delta V_1 + \Delta V_2 = IR_1 + IR_2 = I(R_1 + R_2) \qquad (31.15)$$

FIGURE 31.14 Replacing two series resistors with an equivalent resistor.

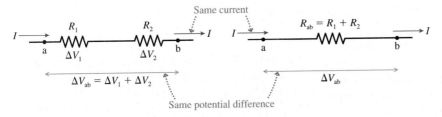

(a) Two resistors in series

(b) An equivalent resistor

Suppose, as in FIGURE 31.14b, we replaced the two resistors with a single resistor having current I and potential difference $\Delta V_{ab} = \Delta V_1 + \Delta V_2$. We can then use Ohm's law to find that the resistance R_{ab} between points a and b is

$$R_{ab} = \frac{\Delta V_{ab}}{I} = \frac{I(R_1 + R_2)}{I} = R_1 + R_2 \qquad (31.16)$$

Because the battery has to establish the same potential difference across the load and provide the same current in both cases, the two resistors R_1 and R_2 act exactly the same as a *single* resistor of value $R_1 + R_2$. We can say that the single resistor R_{ab} is *equivalent* to the two resistors in series.

There was nothing special about having only two resistors. If we have N resistors in series, their **equivalent resistance** is

$$R_{eq} = R_1 + R_2 + \cdots + R_N \qquad \text{(series resistors)} \qquad (31.17)$$

The current and the power output of the battery will be unchanged if the N series resistors are replaced by the single resistor R_{eq}. The key idea in this analysis is that **resistors in series all have the same current.**

NOTE ▶ Compare this idea to what you learned in Chapter 29 about capacitors in series. The end-to-end connections are the same, but the equivalent capacitance is *not* the sum of the individual capacitances. ◀

Now we can answer the lightbulb question posed at the beginning of this section. Suppose the resistance of each lightbulb is R. The battery drives current $I_A = \mathcal{E}/R$ through bulb A. Bulbs B and C are in series, with an equivalent resistance $R_{eq} = 2R$, but the battery has the same emf $\mathcal{E}$. Thus the current through bulbs B and C is $I_{B+C} = \mathcal{E}/R_{eq} = \mathcal{E}/2R = \frac{1}{2}I_A$. Bulb B has only half the current of bulb A, so B is dimmer.

Many people predict that A and B should be equally bright. It's the same battery, so shouldn't it provide the same current to both circuits? No! A battery is a source of emf, *not* a source of current. In other words, the battery's emf is the same no matter how the battery is used. When you buy a 1.5 V battery you're buying a device that provides a specified amount of potential difference, not a specified amount of current. The battery does provide the current to the circuit, but the *amount* of current depends on the resistance of the load. Your 1.5 V battery causes 1 A to pass through a 1.5 Ω load but only 0.1 A to pass through a 15 Ω load. As an analogy, think about a water

FIGURE 31.13 How does the brightness of bulb B compare to that of A?

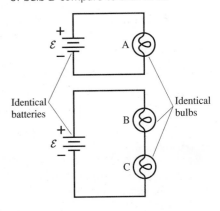

faucet. The pressure in the water main underneath the street is a fixed and unvarying quantity set by the water company, but the amount of water coming out of a faucet depends on how far you open it. A faucet opened slightly has a "high resistance," so only a little water flows. A wide-open faucet has a "low resistance," and the water flow is large.

In summary, **a battery provides a fixed and unvarying emf (potential difference). It does *not* provide a fixed and unvarying current. The amount of current depends jointly on the battery's emf *and* the resistance of the circuit attached to the battery.**

EXAMPLE 31.5 | **A series resistor circuit**

a. What is the current in the circuit of FIGURE 31.15a?
b. Draw a graph of potential versus position in the circuit, going cw from $V = 0$ V at the battery's negative terminal.

MODEL The three resistors are end to end, with no junctions between them, and thus are in series. Assume ideal connecting wires and an ideal battery.

SOLVE a. The battery "acts" the same—it provides the same current at the same potential difference—if we replace the three series resistors by their equivalent resistance

$$R_{eq} = 15\ \Omega + 4\ \Omega + 8\ \Omega = 27\ \Omega$$

This is shown as an equivalent circuit in FIGURE 31.15b. Now we have a circuit with a single battery and a single resistor, for which we know the current to be

$$I = \frac{\mathcal{E}}{R_{eq}} = \frac{9\ \text{V}}{27\ \Omega} = 0.333\ \text{A}$$

b. $I = 0.333$ A is the current in each of the three resistors in the original circuit. Thus the potential differences across the resistors are $\Delta V_{res\ 1} = -IR_1 = -5.0$ V, $\Delta V_{res\ 2} = -IR_2 = -1.3$ V, and $\Delta V_{res\ 3} = -IR_3 = -2.7$ V for the 15 Ω, the 4 Ω, and the 8 Ω resistors, respectively. FIGURE 31.15c shows that the potential increases by 9 V due to the battery's emf, then decreases by 9 V in three steps.

FIGURE 31.15 Analyzing a circuit with series resistors.

(a)
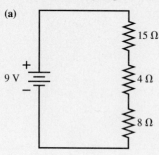

(b)
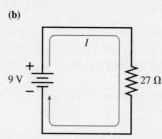

(c)
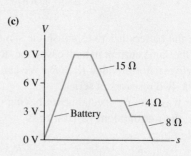

FIGURE 31.16 An ammeter measures the current in a circuit element.

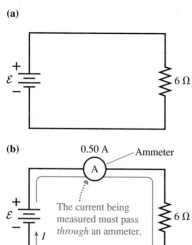

Ammeters

A device that measures the current in a circuit element is called an **ammeter**. Because charge flows *through* circuit elements, an ammeter must be placed *in series* with the circuit element whose current is to be measured.

FIGURE 31.16a shows a simple one-resistor circuit with an unknown emf $\mathcal{E}$. We can measure the current in the circuit by inserting the ammeter as shown in FIGURE 31.16b. Notice that we have to *break the connection* between the battery and the resistor in order to insert the ammeter. Now the current in the resistor has to first pass through the ammeter.

Because the ammeter is now in series with the resistor, the total resistance seen by the battery is $R_{eq} = 6\ \Omega + R_{ammeter}$. In order that the ammeter measure the current without changing the current, the ammeter's resistance must, in this case, be $\ll 6\ \Omega$. Indeed, an ideal ammeter has $R_{ammeter} = 0\ \Omega$ and thus has no effect on the current. Real ammeters come very close to this ideal.

The ammeter in Figure 31.16b reads 0.50 A, meaning that the current through the 6 Ω resistor is $I = 0.50$ A. Thus the resistor's potential difference is $\Delta V_R = IR = 3.0$ V. If the ammeter is ideal, with no resistance and thus no potential difference across it, then, from Kirchhoff's loop law, the battery's emf is $\mathcal{E} = \Delta V_R = 3.0$ V.

STOP TO THINK 31.4   What are the current and the potential at points a to e?

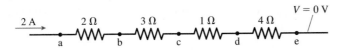

31.5 Real Batteries

Let's look at how real batteries differ from the ideal battery we have been assuming. Real batteries, like ideal batteries, separate charge and create a potential difference. However, real batteries also provide a slight resistance to the charges on the charge escalator. They have what is called an **internal resistance,** which is symbolized by r. FIGURE 31.17 shows both an ideal and a real battery.

From our vantage point outside a battery, we cannot see $\mathcal{E}$ and r separately. To the user, the battery provides a potential difference ΔV_{bat} called the **terminal voltage.** $\Delta V_{bat} = \mathcal{E}$ for an ideal battery, but the presence of the internal resistance affects ΔV_{bat}. Suppose the current in the battery is I. As charges travel from the negative to the positive terminal, they gain potential $\mathcal{E}$ but *lose* potential $\Delta V_{int} = -Ir$ due to the internal resistance. Thus the terminal voltage of the battery is

$$\Delta V_{bat} = \mathcal{E} - Ir \leq \mathcal{E} \tag{31.18}$$

Only when $I = 0$, meaning that the battery is not being used, is $\Delta V_{bat} = \mathcal{E}$.

FIGURE 31.18 shows a single resistor R connected to the terminals of a battery having emf $\mathcal{E}$ and internal resistance r. Resistances R and r are in series, so we can replace them, for the purpose of circuit analysis, with a single equivalent resistor $R_{eq} = R + r$. Hence the current in the circuit is

$$I = \frac{\mathcal{E}}{R_{eq}} = \frac{\mathcal{E}}{R + r} \tag{31.19}$$

If $r \ll R$, so that the internal resistance of the battery is negligible, then $I \approx \mathcal{E}/R$, exactly the result we found before. But the current decreases significantly as r increases.

FIGURE 31.17 An ideal battery and a real battery.

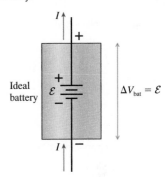

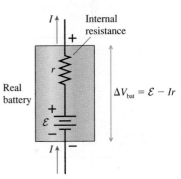

FIGURE 31.18 A single resistor connected to a real battery is in series with the battery's internal resistance, giving $R_{eq} = R + r$.

Although physically separated, the internal resistance r is electrically in series with R.

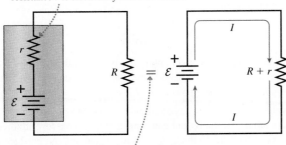

This means the two circuits are equivalent.

We can use Ohm's law to find that the potential difference across the load resistor R is

$$\Delta V_R = IR = \frac{R}{R + r}\mathcal{E} \tag{31.20}$$

Similarly, the potential difference across the terminals of the battery is

$$\Delta V_{bat} = \mathcal{E} - Ir = \mathcal{E} - \frac{r}{R + r}\mathcal{E} = \frac{R}{R + r}\mathcal{E} \tag{31.21}$$

The potential difference across the resistor is equal to the potential difference between the *terminals* of the battery, where the resistor is attached, *not* equal to the battery's emf. Notice that $\Delta V_{bat} = \mathcal{E}$ only if $r = 0$ (an ideal battery with no internal resistance).

EXAMPLE 31.6 Lighting up a flashlight

A 6 Ω flashlight bulb is powered by a 3 V battery with an internal resistance of 1 Ω. What are the power dissipation of the bulb and the terminal voltage of the battery?

MODEL Assume ideal connecting wires but not an ideal battery.

VISUALIZE The circuit diagram looks like Figure 31.18. R is the resistance of the bulb's filament.

SOLVE Equation 31.19 gives us the current:

$$I = \frac{\mathcal{E}}{R + r} = \frac{3\text{ V}}{6\text{ Ω} + 1\text{ Ω}} = 0.43\text{ A}$$

This is 15% less than the 0.5 A an ideal battery would supply. The potential difference across the resistor is $\Delta V_R = IR = 2.6$ V, thus the power dissipation is

$$P_R = I \Delta V_R = 1.1\text{ W}$$

The battery's terminal voltage is

$$\Delta V_{bat} = \frac{R}{R + r}\mathcal{E} = \frac{6\text{ Ω}}{6\text{ Ω} + 1\text{ Ω}} 3\text{ V} = 2.6\text{ V}$$

ASSESS 1 Ω is a typical internal resistance for a flashlight battery. The internal resistance causes the battery's terminal voltage to be 0.4 V less than its emf in this circuit.

A Short Circuit

FIGURE 31.19 The short-circuit current of a battery.

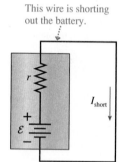

This wire is shorting out the battery.

In **FIGURE 31.19** we've replaced the resistor with an ideal wire having $R_{wire} = 0$ Ω. When a connection of very low or zero resistance is made between two points in a circuit that are normally separated by a higher resistance, we have what is called a **short circuit.** The wire in Figure 31.19 is *shorting out* the battery.

If the battery were ideal, shorting it with an ideal wire ($R = 0$ Ω) would cause the current to be $I = \mathcal{E}/0 = \infty$. The current, of course, cannot really become infinite. Instead, the battery's internal resistance r becomes the only resistance in the circuit. If we use $R = 0$ Ω in Equation 31.19, we find that the *short-circuit current* is

$$I_{short} = \frac{\mathcal{E}}{r} \tag{31.22}$$

A 3 V battery with 1 Ω internal resistance generates a short circuit current of 3 A. This is the *maximum possible current* that this battery can produce. Adding any external resistance R will decrease the current to a value less than 3 A.

EXAMPLE 31.7 A short-circuited battery

What is the short-circuit current of a 12 V car battery with an internal resistance of 0.020 Ω? What happens to the power supplied by the battery?

SOLVE The short-circuit current is

$$I_{short} = \frac{\mathcal{E}}{r} = \frac{12\text{ V}}{0.02\text{ Ω}} = 600\text{ A}$$

Power is generated by chemical reactions in the battery and dissipated by the load resistance. But with a short-circuited battery, the load resistance is *inside* the battery! The "shorted" battery has to dissipate power $P = I^2 r = 7200$ W *internally.*

ASSESS This value is realistic. Car batteries are designed to drive the starter motor, which has a very small resistance and can draw a current of a few hundred amps. That is why the battery cables are so thick. A shorted car battery can produce an *enormous* amount of current. The normal response of a shorted car battery is to explode; it simply cannot dissipate this much power. Shorting a flashlight battery can make it rather hot, but your life is not in danger. Although the voltage of a car battery is relatively small, a car battery can be dangerous and should be treated with great respect.

Most of the time a battery is used under conditions in which $r \ll R$ and the internal resistance is negligible. The ideal battery model is fully justified in that case. Thus we will assume that batteries are ideal *unless stated otherwise.* But keep in mind that batteries (and other sources of emf) do have an internal resistance, and this internal resistance limits the current of the battery.

31.6 Parallel Resistors

FIGURE 31.20 is another lightbulb puzzle. Initially the switch is open. The current is the same through bulbs A and B, because of conservation of current, and they are equally bright. Bulb C is not glowing. What happens to the brightness of A and B when the switch is closed? And how does the brightness of C then compare to that of A and B? Think about this before going on.

FIGURE 31.21a shows two resistors aligned side by side with their ends connected at c and d. Resistors connected *at both ends* are called **parallel resistors** or, sometimes, resistors "in parallel." The left ends of both resistors are at the same potential V_c. Likewise, the right ends are at the same potential V_d. Thus the potential *differences* ΔV_1 and ΔV_2 are the *same* and are simply ΔV_{cd}.

Kirchhoff's junction law applies at the junctions. The input current I splits into currents I_1 and I_2 at the left junction. On the right, the two currents are recombined into current I. According to the junction law,

$$I = I_1 + I_2 \tag{31.23}$$

We can apply Ohm's law to each resistor, along with $\Delta V_1 = \Delta V_2 = \Delta V_{cd}$, to find that the current is

$$I = \frac{\Delta V_1}{R_1} + \frac{\Delta V_2}{R_2} = \frac{\Delta V_{cd}}{R_1} + \frac{\Delta V_{cd}}{R_2} = \Delta V_{cd}\left(\frac{1}{R_1} + \frac{1}{R_2}\right) \tag{31.24}$$

Suppose, as in FIGURE 31.21b, we replaced the two resistors with a single resistor having current I and potential difference ΔV_{cd}. This resistor is equivalent to the original two because the battery has to establish the same potential difference and provide the same current in either case. A second application of Ohm's law shows that the resistance between points c and d is

$$R_{cd} = \frac{\Delta V_{cd}}{I} = \left(\frac{1}{R_1} + \frac{1}{R_2}\right)^{-1} \tag{31.25}$$

The single resistor R_{cd} draws the same current as resistors R_1 and R_2, so, as far as the battery is concerned, resistor R_{cd} is *equivalent* to the two resistors in parallel.

There is nothing special about having chosen two resistors to be in parallel. If we have N resistors in parallel, the *equivalent resistance* is

$$R_{eq} = \left(\frac{1}{R_1} + \frac{1}{R_2} + \cdots + \frac{1}{R_N}\right)^{-1} \quad \text{(parallel resistors)} \tag{31.26}$$

The behavior of the circuit will be unchanged if the N parallel resistors are replaced by the single resistor R_{eq}. The key idea of this analysis is that **resistors in parallel all have the same potential difference.**

NOTE ▶ Don't forget to take the inverse—the -1 exponent in Equation 31.26—after adding the inverses of all the resistances. ◀

FIGURE 31.20 What happens to the brightness of the bulbs when the switch is closed?

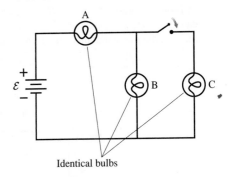

Identical bulbs

FIGURE 31.21 Replacing two parallel resistors with an equivalent resistor.

(a) Two resistors in parallel

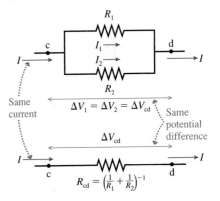

$R_{cd} = \left(\frac{1}{R_1} + \frac{1}{R_2}\right)^{-1}$

(b) An equivalent resistor

Two identical resistors*

In series	$R_{eq} = 2R$
In parallel	$R_{eq} = \dfrac{R}{2}$

*$R_1 = R_2 = R$

EXAMPLE 31.8 **A parallel resistor circuit**

The three resistors of FIGURE 31.22 are connected to a 9 V battery. Find the potential difference across and the current through each resistor.

FIGURE 31.22 Parallel resistor circuit of Example 31.8.

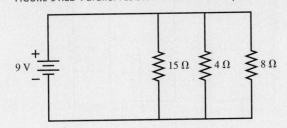

MODEL The resistors are in parallel. Assume an ideal battery and ideal connecting wires.

SOLVE The three parallel resistors can be replaced by a single equivalent resistor

$$R_{eq} = \left(\frac{1}{15\ \Omega} + \frac{1}{4\ \Omega} + \frac{1}{8\ \Omega}\right)^{-1} = (0.4417\ \Omega^{-1})^{-1} = 2.26\ \Omega$$

The equivalent circuit is shown in FIGURE 31.23a on the next page, from which we find the current to be

$$I = \frac{\mathcal{E}}{R_{eq}} = \frac{9\ \text{V}}{2.26\ \Omega} = 3.98\ \text{A}$$

Continued

The potential difference across R_{eq} is $\Delta V_{eq} = \mathcal{E} = 9.0$ V. Now we have to be careful. Current I divides at the junction into the smaller currents I_1, I_2, and I_3 shown in FIGURE 31.23b. However, the division is *not* into three equal currents. According to Ohm's law, resistor i has current $I_i = \Delta V_i/R_i$. Because the three resistors are each connected to the battery by ideal wires, as is the equivalent resistor, their potential differences are equal:

$$\Delta V_1 = \Delta V_2 = \Delta V_3 = \Delta V_{eq} = 9.0 \text{ V}$$

Thus the currents are

$$I_1 = \frac{9 \text{ V}}{15 \text{ }\Omega} = 0.60 \text{ A} \qquad I_2 = \frac{9 \text{ V}}{4 \text{ }\Omega} = 2.25 \text{ A}$$

$$I_3 = \frac{9 \text{ V}}{8 \text{ }\Omega} = 1.13 \text{ A}$$

ASSESS The *sum* of the three currents is 3.98 A, as required by Kirchhoff's junction law.

FIGURE 31.23 The parallel resistors can be replaced by a single equivalent resistor.

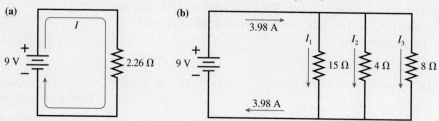

Summary of series and parallel resistors

	I	ΔV
Series	Same	Add
Parallel	Add	Same

The result of Example 31.8 seems surprising. The equivalent of a parallel combination of 15 Ω, 4 Ω, and 8 Ω was found to be 2.26 Ω. How can the equivalent of a group of resistors be *less* than any single resistance in the group? Shouldn't more resistors imply more resistance? The answer is yes for resistors in series but not for resistors in parallel. Even though a resistor is an obstacle to the flow of charge, parallel resistors provide more pathways for charge to get through. Consequently, the equivalent of several resistors in parallel is always *less* than any single resistor in the group.

Complex combinations of resistors can often be reduced to a single equivalent resistance through a step-by-step application of the series and parallel rules. The final example in this section illustrates this idea.

EXAMPLE 31.9 **A combination of resistors**

What is the equivalent resistance of the group of resistors shown in FIGURE 31.24?

FIGURE 31.24 A combination of resistors.

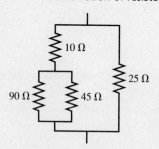

MODEL This circuit contains both series and parallel resistors.

SOLVE Reduction to a single equivalent resistance is best done in a series of steps, with the circuit being redrawn after each step. The procedure is shown in FIGURE 31.25. Note that the 10 Ω and 25 Ω

resistors are *not* in parallel. They are connected at their top ends but not at their bottom ends. Resistors must be connected to each other at *both* ends to be in parallel. Similarly, the 10 Ω and 45 Ω resistors are *not* in series because of the junction between them. If the original group of four resistors occurred within a larger circuit, they could be replaced with a single 15.4 Ω resistor without having any effect on the rest of the circuit.

FIGURE 31.25 The combination is reduced to a single equivalent resistor.

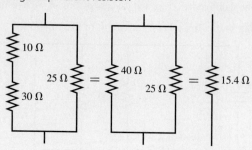

To return to the lightbulb question that began this section, FIGURE 31.26 has redrawn the circuit with each bulb shown as a resistance R. Initially, before the switch is closed, bulbs A and B are in series with equivalent resistance $2R$. The current from the battery is

$$I_{\text{before}} = \frac{\mathcal{E}}{2R} = \frac{1}{2}\frac{\mathcal{E}}{R}$$

This is the current in both bulbs.

Closing the switch places bulbs B and C in parallel. The equivalent resistance of two identical resistors in parallel is $R_{\text{eq}} = \frac{1}{2}R$. This equivalent resistance of B and C is in series with bulb A; hence the total resistance of the circuit is $\frac{3}{2}R$ and the current leaving the battery is

$$I_{\text{after}} = \frac{\mathcal{E}}{3R/2} = \frac{2}{3}\frac{\mathcal{E}}{R} > I_{\text{before}}$$

Closing the switch *decreases* the circuit resistance and thus *increases* the current leaving the battery.

All the charge flows through A, so A *increases* in brightness when the switch is closed. The current I_{after} then splits at the junction. Bulbs B and C have equal resistance, so the current splits equally. The current in B is $\frac{1}{3}(\mathcal{E}/R)$, which is *less* than I_{before}. Thus B *decreases* in brightness when the switch is closed. Bulb C has the same brightness as bulb B.

Voltmeters

A device that measures the potential difference across a circuit element is called a **voltmeter**. Because potential difference is measured *across* a circuit element, from one side to the other, a voltmeter is placed in *parallel* with the circuit element whose potential difference is to be measured.

FIGURE 31.27a shows a simple circuit in which a 17 Ω resistor is connected across a 9 V battery with an unknown internal resistance. By connecting a voltmeter across the resistor, as shown in FIGURE 31.27b, we can measure the potential difference across the resistor. Unlike an ammeter, using a voltmeter does *not* require us to break the connections.

Because the voltmeter is now in parallel with the resistor, the total resistance seen by the battery is $R_{\text{eq}} = (1/17\ \Omega + 1/R_{\text{voltmeter}})^{-1}$. In order that the voltmeter measure the voltage without changing the voltage, the voltmeter's resistance must, in this case, be $\gg 17\ \Omega$. Indeed, an *ideal voltmeter* has $R_{\text{voltmeter}} = \infty\ \Omega$, and thus has no effect on the voltage. Real voltmeters come very close to this ideal, and we will always assume them to be so.

The voltmeter in Figure 31.27b reads 8.5 V. This is less than $\mathcal{E}$ because of the battery's internal resistance. Equation 31.20 found an expression for the resistor's potential difference ΔV_R. That equation is easily solved for the internal resistance r:

$$r = \frac{\mathcal{E} - \Delta V_R}{\Delta V_R}R = \frac{0.5\text{ V}}{8.5\text{ V}}\,17\ \Omega = 1.0\ \Omega$$

Here a voltmeter reading was the one piece of experimental data we needed in order to determine the battery's internal resistance.

FIGURE 31.26 The lightbulbs of Figure 31.20 with the switch open and closed.

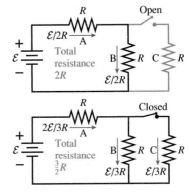

FIGURE 31.27 A voltmeter measures the potential difference across an element.

(a)

(b)

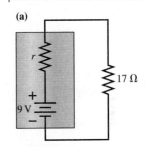

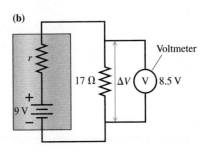

STOP TO THINK 31.5   Rank in order, from brightest to dimmest, the identical bulbs A to D.

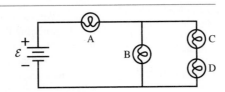

31.7 Resistor Circuits

We can use the information in this chapter to analyze a variety of more complex but more realistic circuits.

PROBLEM-SOLVING
STRATEGY 31.1 **Resistor circuits**

MODEL Assume that wires are ideal and, where appropriate, that batteries are ideal.

VISUALIZE Draw a circuit diagram. Label all known and unknown quantities.

SOLVE Base your mathematical analysis on Kirchhoff's laws and on the rules for series and parallel resistors.

- Step by step, reduce the circuit to the smallest possible number of equivalent resistors.
- Write Kirchhoff's loop law for each independent loop in the circuit.
- Determine the current through and the potential difference across the equivalent resistors.
- Rebuild the circuit, using the facts that the current is the same through all resistors in series and the potential difference is the same for all parallel resistors.

ASSESS Use two important checks as you rebuild the circuit.

- Verify that the sum of the potential differences across series resistors matches ΔV for the equivalent resistor.
- Verify that the sum of the currents through parallel resistors matches I for the equivalent resistor.

Exercise 26

EXAMPLE 31.10 **Analyzing a complex circuit**

Find the current through and the potential difference across each of the four resistors in the circuit shown in FIGURE 31.28.

FIGURE 31.28 A complex resistor circuit.

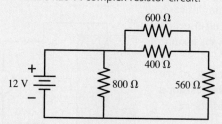

MODEL Assume an ideal battery, with no internal resistance, and ideal connecting wires.

VISUALIZE Figure 31.28 shows the circuit diagram. We'll keep redrawing the diagram as we analyze the circuit.

SOLVE First, we break the circuit down, step by step, into one with a single resistor. FIGURE 31.29a shows this done in three steps. The final battery-and-resistor circuit is our basic circuit, with current

$$I = \frac{\mathcal{E}}{R} = \frac{12 \text{ V}}{400 \text{ }\Omega} = 0.030 \text{ A} = 30 \text{ mA}$$

The potential difference across the 400 Ω resistor is $\Delta V_{400} = \Delta V_{bat} = \mathcal{E} = 12$ V.

Second, we rebuild the circuit, step by step, finding the currents and potential differences at each step. FIGURE 31.29b repeats the steps of Figure 31.29a exactly, but in reverse order. The 400 Ω resistor came from two 800 Ω resistors in parallel. Because $\Delta V_{400} = 12$ V, it must be true that each $\Delta V_{800} = 12$ V. The current through each 800 Ω is then $I = \Delta V/R = 15$ mA. The checkpoint is to note that 15 mA + 15 mA = 30 mA.

The right 800 Ω resistor was formed by 240 Ω and 560 Ω in series. Because $I_{800} = 15$ mA, it must be true that $I_{240} = I_{560} = 15$ mA. The potential difference across each is $\Delta V = IR$, so $\Delta V_{240} = 3.6$ V and $\Delta V_{560} = 8.4$ V. Here the checkpoint is to note that 3.6 V + 8.4 V = 12 V = ΔV_{800}, so the potential differences add as they should.

Finally, the 240 Ω resistor came from 600 Ω and 400 Ω in parallel, so they each have the same 3.6 V potential difference as their 240 Ω equivalent. The currents are $I_{600} = 6$ mA and $I_{400} = 9$ mA. Note that 6 mA + 9 mA = 15 mA, which is our third checkpoint. We now know all currents and potential differences.

ASSESS We *checked our work* at each step of the rebuilding process by verifying that currents summed properly at junctions and that potential differences summed properly along a series of resistances. This "check as you go" procedure is extremely important. It provides you, the problem solver, with a built-in error finder that will immediately inform you if a mistake has been made.

FIGURE 31.29 The step-by-step circuit analysis.

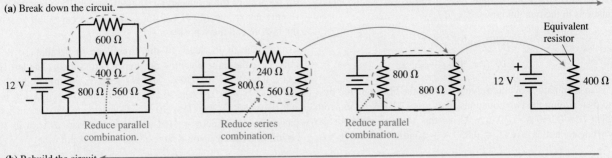

(a) Break down the circuit.

Equivalent resistor

Reduce parallel combination.   Reduce series combination.   Reduce parallel combination.

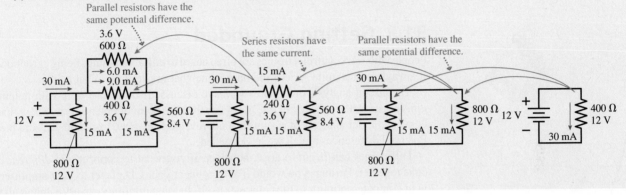

(b) Rebuild the circuit.

Parallel resistors have the same potential difference.

Series resistors have the same current.

Parallel resistors have the same potential difference.

EXAMPLE 31.11 **Analyzing a two-loop circuit**

Find the current through and the potential difference across the 100 Ω resistor in the circuit of **FIGURE 31.30**.

FIGURE 31.30 A two-loop circuit.

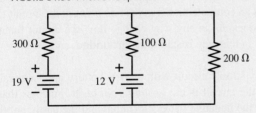

MODEL Assume ideal batteries and ideal connecting wires.

VISUALIZE Figure 31.30 shows the circuit diagram. None of the resistors are connected in series or in parallel, so this circuit cannot be reduced to a simpler circuit.

SOLVE Kirchhoff's loop law applies to *any* loop. To analyze a multiloop problem, we need to write a loop-law equation for each loop. **FIGURE 31.31** redraws the circuit and defines clockwise currents I_1 in the left loop and I_2 in the right loop. But what about the middle branch? Let's assign a downward current I_3 to the middle branch. If we apply Kirchhoff's junction law $\sum I_{\text{in}} = \sum I_{\text{out}}$ to the junction above the 100 Ω resistor, as shown in the blow-up of Figure 31.31, we see that $I_1 = I_2 + I_3$ and thus $I_3 = I_1 - I_2$. If I_3 ends up being a positive number, then the current in the middle branch really is downward. A negative I_3 will signify an upward current.

FIGURE 31.31 Applying Kirchhoff's laws.

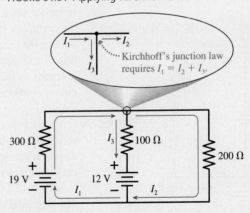

Kirchhoff's junction law requires $I_1 = I_2 + I_3$.

Kirchhoff's loop law for the left loop, going clockwise from the lower-left corner, is

$$\sum (\Delta V)_i = 19\text{ V} - (300\text{ }\Omega)I_1 - (100\text{ }\Omega)I_3 - 12\text{ V} = 0$$

We're traveling through the 100 Ω resistor in the direction of I_3, the "downhill" direction, so the potential decreases. The 12 V battery is traversed positive to negative, so there we have $\Delta V = -\mathcal{E} = -12$ V. For the right loop, we're going to travel "uphill" through the 100 Ω resistor, opposite to I_3, and gain potential. Thus the loop law for the right loop is

$$\sum (\Delta V)_i = 12\text{ V} + (100\text{ }\Omega)I_3 - (200\text{ }\Omega)I_2 = 0$$

Continued

If we substitute $I_3 = I_1 - I_2$ and then rearrange the terms, we find that the two independent loops have given us two simultaneous equations in the two unknowns I_1 and I_2:

$$400I_1 - 100I_2 = 7$$

$$-100I_1 + 300I_2 = 12$$

We can eliminate I_2 by multiplying through the first equation by 3 and then adding the two equations. This gives $1100I_1 = 33$, from which $I_1 = 0.030$ A $= 30$ mA. Using this value in either of the two loop equations gives $I_2 = 0.050$ A $= 50$ mA. Because $I_2 > I_1$,

the current through the 100 Ω resistor is $I_3 = I_1 - I_2 = -20$ mA, or, because of the minus sign, 20 mA upward. The potential difference across the 100 Ω resistor is $\Delta V_{100\,\Omega} = I_3 R = 2.0$ V, with the bottom end more positive.

ASSESS The three "legs" of the circuit are in parallel, so they must have the same potential difference across them. The left leg has $\Delta V = 19$ V $- (0.030$ A$)(300\ \Omega) = 10$ V, the middle leg has $\Delta V = 12$ V $- (0.020$ A$)(100\ \Omega) = 10$ V, and the right leg has $\Delta V = (0.050$ A$)(200\ \Omega) = 10$ V. Consistency checks such as these are very important. Had we made a numerical error in our circuit analysis, we would have caught it at this point.

31.8 Getting Grounded

People who work with electronics are often heard to talk about things being "grounded." It always sounds quite serious, perhaps somewhat mysterious. What is it?

The circuit analysis procedures we have discussed so far deal only with potential *differences*. Although we are free to choose the zero point of potential anywhere that is convenient, our analysis of circuits has not revealed any need to establish a zero point. Potential differences are all we have needed.

Difficulties can begin to arise, however, if you want to connect two *different* circuits together. Perhaps you would like to connect your CD player to your amplifier or your computer monitor to the computer itself. Incompatibilities can arise unless all the circuits to be connected have a *common* reference point for the potential.

You learned previously that the earth itself is a conductor. Suppose we have two circuits. If we connect *one* point of each circuit to the earth by an ideal wire, and we also agree to call the potential of the earth $V_{earth} = 0$ V, then both circuits have a common reference point. But notice something very important: *one* wire connects the circuit to the earth, but there is not a second wire returning to the circuit. That is, the wire connecting the circuit to the earth is not part of a complete circuit, so there is *no current* in this wire! Because the wire is an equipotential, it gives one point in the circuit the same potential as the earth, but it does *not* in any way change how the circuit functions. A circuit connected to the earth in this way is said to be **grounded**, and the wire is called the *ground wire*.

FIGURE 31.32a shows a fairly simple circuit with a 10 V battery and two resistors in series. The symbol beneath the circuit is the *ground symbol*. It indicates that a wire has been connected between the negative battery terminal and the earth, but the presence of the ground wire does not affect the circuit's behavior. The total resistance is $8\ \Omega + 12\ \Omega = 20\ \Omega$, so the current in the loop is $I = (10$ V$)/(20\ \Omega) = 0.50$ A. The potential differences across the two resistors are found, using Ohm's law, to be $\Delta V_8 = 4$ V and $\Delta V_{12} = 6$ V. These are the same values that we would find if the ground wire were *not* present. So what has grounding the circuit accomplished?

FIGURE 31.32b shows the actual potential at several points in the circuit. By definition, $V_{earth} = 0$ V. The negative battery terminal and the bottom of the 12 Ω resistor are connected by ideal wires to the earth, so *the* potential at these two points must also be zero. The positive terminal of the battery is 10 V more positive than the negative terminal, so $V_{neg} = 0$ V implies $V_{pos} = +10$ V. Similarly, the fact that the potential *decreases* by 6 V as charge flows through the 12 Ω resistor now implies that *the* potential at the junction of the resistors must be $+6$ V. The potential difference across the 8 Ω resistor is 4 V, so the top has to be at $+10$ V. This agrees with the potential at the positive battery terminal, as it must because these two points are connected by an ideal wire.

All that grounding the circuit does is allow us to have *specific values* for the potential at each point in the circuit. Now we can say "The voltage at the resistor junction is 6 V," whereas before all we could say was "There is a 6 V potential difference across the 12 Ω resistor."

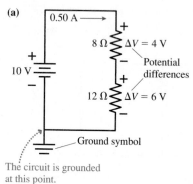

The circular prong of a three-prong plug is a connection to ground.

FIGURE 31.32 A circuit that is grounded at one point.

(a)

0.50 A →

8 Ω $\Delta V = 4$ V

+

10 V

−

Potential differences

12 Ω $\Delta V = 6$ V

Ground symbol

The circuit is grounded at this point.

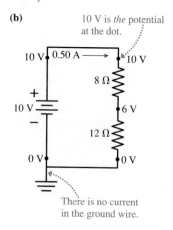

(b)

10 V is *the* potential at the dot.

10 V 0.50 A → 10 V

8 Ω

+

10 V   6 V

−

12 Ω

0 V   0 V

There is no current in the ground wire.

There is one important lesson from this: **Being grounded does not affect the circuit's behavior under normal conditions.** You cannot use "because it is grounded" to *explain* anything about a circuit's behavior.

We added "under normal conditions" because there is one exception. Most circuits are enclosed in a case of some sort that is held away from the circuit with insulators. Sometimes a circuit breaks or malfunctions in such a way that the case comes into electrical contact with the circuit. If the circuit uses high voltage, or even ordinary 120 V household voltage, anyone touching the case could be injured or killed by electrocution. To prevent this, many appliances or electrical instruments have the case itself grounded. Grounding ensures that the potential of the case will always remain at 0 V and be safe. If a malfunction occurs that connects the case to the circuit, a large current will pass through the ground wire to the earth and cause a fuse to blow. This is the *only* time the ground wire would ever have a current, and it is *not* a normal operation of the circuit.

EXAMPLE 31.12 **A grounded circuit**

Suppose the circuit of Figure 31.32 were grounded at the junction between the two resistors instead of at the bottom. Find the potential at each corner of the circuit.

VISUALIZE **FIGURE 31.33** shows the new circuit. (It is customary to draw the ground symbol so that its "point" is always down.)

FIGURE 31.33 Circuit of Figure 31.32 grounded at the point between the resistors.

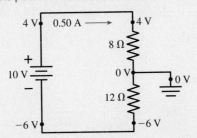

SOLVE Changing the ground point does not affect the circuit's behavior. The current is still 0.50 A, and the potential differences across the two resistors are still 4 V and 6 V. All that has happened is that we have moved the $V = 0$ V reference point. Because the earth has $V_{earth} = 0$ V, the junction itself now has a potential of 0 V. The potential decreases by 4 V as charge flows through the 8 Ω resistor. Because it *ends* at 0 V, the potential at the top of the 8 Ω resistor must be $+4$ V. Similarly, the potential decreases by 6 V through the 12 Ω resistor. Because it *starts* at 0 V, the bottom of the 12 Ω resistor must be at -6 V. The negative battery terminal is at the same potential as the bottom of the 12 Ω resistor, because they are connected by a wire, so $V_{neg} = -6$ V. Finally, the potential increases by 10 V as the charge flows through the battery, so $V_{pos} = +4$ V, in agreement, as it should be, with the potential at the top of the 8 Ω resistor.

ASSESS A negative voltage means only that the potential at that point is less than the potential at some other point that we chose to call $V = 0$ V. Only potential *differences* are physically meaningful, and only potential differences enter into Ohm's law: $I = \Delta V/R$. The potential difference across the 12 Ω resistor in this example is 6 V, decreasing from top to bottom, regardless of which point we choose to call $V = 0$ V.

31.9 *RC* Circuits

Thus far we've considered only circuits in which the current is steady and continuous. There are many circuits in which the time dependence of the current is a crucial feature. Charging and discharging a capacitor is an important example.

FIGURE 31.34a shows a charged capacitor, a switch, and a resistor. The capacitor has charge Q_0 and potential difference $\Delta V_0 = Q_0/C$. There is no current, so the potential difference across the resistor is zero. Then, at $t = 0$, the switch closes and the capacitor begins to discharge through the resistor. A circuit such as this, with resistors and capacitors, is called an **RC circuit.**

How long does the capacitor take to discharge? How does the current through the resistor vary as a function of time? To answer these questions, **FIGURE 31.34b** shows the circuit at some point in time after the switch was closed.

Kirchhoff's loop law is valid for any circuit, not just circuits with batteries. The loop law applied to the circuit of Figure 31.34b, going around the loop cw, is

$$\Delta V_{cap} + \Delta V_{res} = \frac{Q}{C} - IR = 0 \qquad (31.27)$$

Q and I in this equation are the *instantaneous* values of the capacitor charge and the resistor current.

FIGURE 31.34 An *RC* circuit.

(a) Before the switch closes

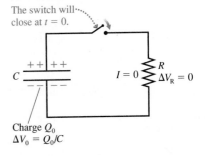

(b) After the switch closes

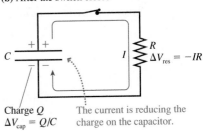

The rear flasher on a bike helmet flashes on and off. The timing is controlled by an *RC* circuit.

The current *I* is the rate at which charge flows through the resistor: $I = dq/dt$. But the charge flowing through the resistor is charge that was *removed* from the capacitor. That is, an infinitesimal charge dq flows through the resistor when the capacitor charge *decreases* by dQ. Thus $dq = -dQ$, and the resistor current is related to the instantaneous capacitor charge by

$$I = -\frac{dQ}{dt} \tag{31.28}$$

Now *I* is positive when *Q* is decreasing, as we would expect. The reasoning that has led to Equation 31.28 is rather subtle but very important. You'll see the same reasoning later in other contexts.

If we substitute Equation 31.28 into Equation 31.27 and then divide by *R*, the loop law for the *RC* circuit becomes

$$\frac{dQ}{dt} + \frac{Q}{RC} = 0 \tag{31.29}$$

Equation 31.29 is a first-order differential equation for the capacitor charge *Q*, but one that we can solve by direct integration. First, we rearrange Equation 31.29 to get all the charge terms on one side of the equation:

$$\frac{dQ}{Q} = -\frac{1}{RC} dt$$

The product *RC* is a constant for any particular circuit.

The capacitor charge was Q_0 at $t = 0$ when the switch was closed. We want to integrate from these starting conditions to charge *Q* at a later time *t*. That is,

$$\int_{Q_0}^{Q} \frac{dQ}{Q} = -\frac{1}{RC} \int_0^t dt \tag{31.30}$$

Both are well-known integrals, giving

$$\ln Q \Big|_{Q_0}^{Q} = \ln Q - \ln Q_0 = \ln\left(\frac{Q}{Q_0}\right) = -\frac{t}{RC}$$

We can solve for the capacitor charge *Q* by taking the exponential of both sides, then multiplying by Q_0. Doing so gives

$$Q = Q_0 e^{-t/RC} \tag{31.31}$$

Notice that $Q = Q_0$ at $t = 0$, as expected.

The argument of an exponential function must be dimensionless, so the quantity *RC* must have dimensions of time. It is useful to define the **time constant** τ of the *RC* circuit as

$$\tau = RC \tag{31.32}$$

We can then write Equation 31.31 as

$$Q = Q_0 e^{-t/\tau} \tag{31.33}$$

And because the capacitor voltage is directly proportional to the charge, it also decays exponentially as

$$\Delta V_C = \Delta V_0 e^{-t/\tau} \tag{31.34}$$

The meaning of Equation 31.33 is easier to understand if we portray it graphically. FIGURE 31.35a shows the capacitor charge as a function of time. The charge decays exponentially, starting from Q_0 at $t = 0$ and asymptotically approaching zero as $t \to \infty$. The time constant τ is the time at which the charge has decreased to e^{-1} (about 37%) of its initial value. At time $t = 2\tau$, the charge has decreased to e^{-2} (about 13%) of its initial value. A voltage graph would have the same shape.

NOTE ▶ The *shape* of the graph of *Q* is always the same, regardless of the specific value of the time constant τ. ◀

FIGURE 31.35 The decay curves of the capacitor charge and the resistor current.

(a)

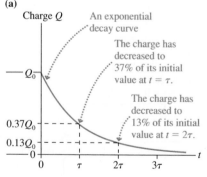

(b)

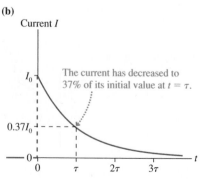

We find the resistor current by using Equation 31.28:

$$I = -\frac{dQ}{dt} = \frac{Q_0}{\tau}e^{-t/\tau} = \frac{Q_0}{RC}e^{-t/\tau} = \frac{\Delta V_0}{R}e^{-t/\tau} = I_0 e^{-t/\tau} \qquad (31.35)$$

where $I_0 = \Delta V_0/R$ is the initial current, immediately after the switch closes. FIGURE 31.35b is a graph of the resistor current versus t. You can see that the current undergoes the same exponential decay, with the same time constant, as the capacitor charge.

NOTE ▶ There's no specific time at which the capacitor has been discharged, because Q approaches zero asymptotically, but the charge and current have dropped to less than 1% of their initial values at $t = 5\tau$. Thus 5τ is a reasonable answer to the question "How long does it take to discharge a capacitor?" ◀

EXAMPLE 31.13 **Measuring capacitance**

To determine the capacitance of an unmarked capacitor, you set up the circuit shown in FIGURE 31.36. After holding the switch in position a for several seconds, you suddenly flip it—at a time you choose to call $t = 0$ s—to position b while monitoring the resistor voltage with a voltmeter. Your measurements are as follows:

Time (s)	Voltage (V)
0.0	9.0
2.0	5.4
4.0	2.7
6.0	1.6
8.0	1.0

FIGURE 31.36 An *RC* circuit for measuring capacitance.

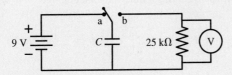

What is the capacitance? And what was the resistor current 5.0 s after the switch changed position?

MODEL The battery charges the capacitor to 9.0 V. Then, when the switch is flipped to position b, the capacitor discharges through the 25,000 Ω resistor with time constant $\tau = RC$.

SOLVE With the switch in position b, the resistor is in parallel with the capacitor and both have the same potential difference $\Delta V_R = \Delta V_C = Q/C$ at all times. The capacitor charge decays exponentially as

$$Q = Q_0 e^{-t/\tau}$$

Consequently, the resistor (and capacitor) voltage also decays exponentially:

$$\Delta V_R = \frac{Q_0}{C}e^{-t/\tau} = \Delta V_0 e^{-t/\tau}$$

where $\Delta V_0 = 9.0$ V is the potential difference at the instant the switch closes. To analyze exponential decays, we take the natural logarithm of both sides. This gives

$$\ln(\Delta V_R) = \ln(\Delta V_0) + \ln(e^{-t/\tau}) = \ln(\Delta V_0) - \frac{1}{\tau}t$$

This result tells us that a graph of $\ln(\Delta V_R)$ versus t—a *semi-log graph*—should be linear with y-intercept $\ln(\Delta V_0)$ and slope $-1/\tau$. If this turns out to be true, we can determine τ and hence C from an experimental measurement of the slope.

FIGURE 31.37 is a graph of $\ln(\Delta V_R)$ versus t. It is, indeed, linear with a negative slope. From the y-intercept of the best-fit line, we find $\Delta V_0 = e^{2.20} = 9.0$ V, as expected. This gives us confidence in our analysis. Using the slope, we find

$$\tau = -\frac{1}{\text{slope}} = -\frac{1}{-0.28 \text{ s}^{-1}} = 3.6 \text{ s}$$

With this, we can calculate

$$C = \frac{\tau}{R} = \frac{3.6 \text{ s}}{25,000 \text{ }\Omega} = 1.4 \times 10^{-4} \text{ F} = 140 \text{ }\mu\text{F}$$

The initial current is $I_0 = (9.0 \text{ V})/(25,000 \text{ }\Omega) = 360 \text{ }\mu\text{A}$. Current also decays exponentially with the same time constant, so the current after 5.0 s is

$$I = I_0 e^{-t/\tau} = (360 \text{ }\mu\text{A})e^{-(5.0 \text{ s})/(3.6 \text{ s})} = 90 \text{ }\mu\text{A}$$

FIGURE 31.37 A semi-log graph of the data.

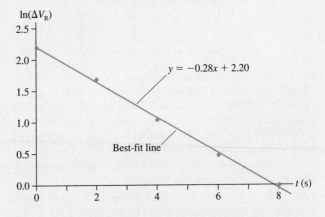

ASSESS The time constant of an exponential decay can be estimated as the time required to decay to one-third of the initial value. Looking at the data, we see that the voltage drops to one-third of the initial 9.0 V in just under 4 s. This is consistent with the more precise $\tau = 3.6$ s, so we have confidence in our results.

FIGURE 31.38 A circuit for charging a capacitor.

(a)

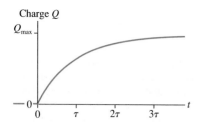

Switch closes at $t = 0$ s.

(b)

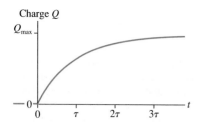

Charge Q

Charging a Capacitor

FIGURE 31.38a shows a circuit that charges a capacitor. After the switch is closed, the battery's charge escalator moves charge from the bottom electrode of the capacitor to the top electrode. The resistor, by limiting the current, slows the process but doesn't stop it. The capacitor charges until $\Delta V_C = \mathcal{E}$; then the charging current ceases. The full charge of the capacitor is $Q_{max} = C(\Delta V_C)_{max} = C\mathcal{E}$.

As a homework problem, you can show that the capacitor charge at time t is

$$Q = Q_{max}(1 - e^{-t/\tau}) \qquad (31.36)$$

where again $\tau = RC$. This "upside-down decay" to Q_{max} is shown graphically in **FIGURE 31.38b**. *RC* circuits that alternately charge and discharge a capacitor are at the heart of time-keeping circuits in computers and other digital electronics.

STOP TO THINK 31.6 The time constant for the discharge of this capacitor is

a. 5 s.
b. 4 s.
c. 2 s.
d. 1 s.
e. The capacitor doesn't discharge because the resistors cancel each other.

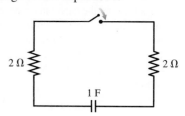

CHALLENGE EXAMPLE 31.14 | **Energy dissipated during a capacitor discharge**

The switch in **FIGURE 31.39** has been in position a for a long time. It is suddenly switched to position b for 1.0 s, then back to a. How much energy is dissipated by the 5500 Ω resistor?

FIGURE 31.39 Circuit of a switched capacitor.

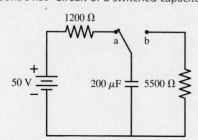

MODEL With the switch in position a, the capacitor charges through the 1200 Ω resistor with time constant $\tau_{charge} = (1200\ \Omega)(2.0 \times 10^{-4}\ \text{F}) = 0.24$ s. Because the switch has been in position a for a "long time," which we interpret as being much longer than 0.24 s, we will assume that the capacitor is fully charged to 50 V when the switch is changed to position b. The capacitor then discharges through the 5500 Ω resistor until the switch is returned to position a. Assume ideal wires.

SOLVE Let $t = 0$ s be the time when the switch is moved from a to b, initiating the discharge. The battery and 1200 Ω resistor are irrelevant during the discharge, so the circuit looks like that of Figure 31.34b. The time constant is $\tau = (5500\ \Omega)(2.0 \times 10^{-4}\ \text{F}) = 1.1$ s, so the capacitor voltage decreases from 50 V at $t = 0$ s to

$$\Delta V_C = (50\ \text{V})e^{-(1.0\ \text{s})/(1.1\ \text{s})} = 20\ \text{V}$$

at $t = 1.0$ s.

There are two ways to determine the energy dissipated in the resistor. We learned in Section 31.3 that a resistor dissipates energy at the rate $dE/dt = P_R = I^2 R$. The current decays exponentially as $I = I_0 \exp(-t/\tau)$, with $I_0 = \Delta V_0/R = 9.09$ mA. We can find the energy dissipated during a time T by integrating:

$$\Delta E = \int_0^T I^2 R\, dt = I_0^2 R \int_0^T e^{-2t/\tau}\, dt = -\frac{1}{2}\tau I_0^2 R e^{-2t/\tau}\Big|_0^T$$

$$= \frac{1}{2}\tau I_0^2 R \left(1 - e^{-2T/\tau}\right)$$

The 2 in the exponent appears because we squared the expression for *I*. Evaluating for $T = 1.0$ s, we find

$$\Delta E = \frac{1}{2}(1.1\ \text{s})(0.00909\ \text{A})^2 (5500\ \Omega)\left(1 - e^{-(2.0\ \text{s})/(1.1\ \text{s})}\right) = 0.21\ \text{J}$$

Alternatively, we can use the known capacitor voltages at $t = 0$ s and $t = 1.0$ s and $U_C = \frac{1}{2}C(\Delta V_C)^2$ to calculate the energy stored in the capacitor at these times:

$$U_C\,(t = 0.0\ \text{s}) = \frac{1}{2}(2.0 \times 10^{-4}\ \text{F})(50\ \text{V})^2 = 0.25\ \text{J}$$

$$U_C\,(t = 1.0\ \text{s}) = \frac{1}{2}(2.0 \times 10^{-4}\ \text{F})(20\ \text{V})^2 = 0.04\ \text{J}$$

The capacitor has lost $\Delta E = 0.21$ J of energy, and this energy was dissipated by the current through the resistor.

ASSESS Not every problem can be solved two ways, but doing so when it's possible gives us great confidence in our result.

SUMMARY

The goal of Chapter 31 has been to understand the fundamental physical principles that govern electric circuits.

General Strategy

MODEL Assume that wires and, where appropriate, batteries are ideal.

VISUALIZE Draw a circuit diagram. Label all known and unknown quantities.

SOLVE Base the solution on Kirchhoff's laws.

- Reduce the circuit to the smallest possible number of equivalent resistors.
- Write one loop equation for each independent loop.
- Find the current and the potential difference.
- Rebuild the circuit to find I and ΔV for each resistor.

ASSESS Verify that

- The sum of potential differences across series resistors matches ΔV for the equivalent resistor.
- The sum of the currents through parallel resistors matches I for the equivalent resistor.

Kirchhoff's loop law

For a closed loop:

- Assign a direction to the current I.
- $\sum (\Delta V)_i = 0$

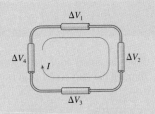

Kirchhoff's junction law

For a junction:

- $\sum I_{in} = \sum I_{out}$

Important Concepts

Ohm's Law

A potential difference ΔV between the ends of a conductor with resistance R creates a current

$$I = \frac{\Delta V}{R}$$

Signs of ΔV for Kirchhoff's loop law

$\Delta V_{bat} = +\mathcal{E}$ $\Delta V_{bat} = -\mathcal{E}$ $\Delta V_{res} = -IR$

The energy used by a circuit is supplied by the emf $\mathcal{E}$ of the battery through the energy transformations

$$E_{chem} \rightarrow U \rightarrow K \rightarrow E_{th}$$

The battery *supplies* energy at the rate

$$P_{bat} = I\mathcal{E}$$

The resistors *dissipate* energy at the rate

$$P_R = I\Delta V_R = I^2 R = \frac{(\Delta V_R)^2}{R}$$

Applications

Series resistors

$$R_{eq} = R_1 + R_2 + R_3 + \cdots$$

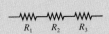

Parallel resistors

$$R_{eq} = \left(\frac{1}{R_1} + \frac{1}{R_2} + \frac{1}{R_3} + \cdots\right)^{-1}$$

RC circuits

The discharge of a capacitor through a resistor satisfies:

$$Q = Q_0 e^{-t/\tau}$$

$$I = -\frac{dQ}{dt} = \frac{Q_0}{\tau} e^{-t/\tau} = I_0 e^{-t/\tau}$$

where $\tau = RC$ is the **time constant.**

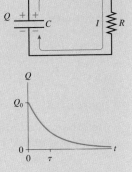

Terms and Notation

circuit diagram	source	internal resistance, r	grounded
Kirchhoff's junction law	kilowatt hour, kWh	terminal voltage, ΔV_{bat}	RC circuit
Kirchhoff's loop law	series resistors	short circuit	time constant, τ
complete circuit	equivalent resistance, R_{eq}	parallel resistors	
load	ammeter	voltmeter	

CONCEPTUAL QUESTIONS

1. Rank in order, from largest to smallest, the currents I_a to I_d through the four resistors in FIGURE Q31.1.

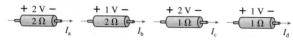

FIGURE Q31.1

2. The tip of a flashlight bulb is touching the top of the 3 V battery in FIGURE Q31.2. Does the bulb light? Why or why not?

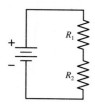

FIGURE Q31.2   FIGURE Q31.3

3. The wire is broken on the right side of the circuit in FIGURE Q31.3. What is the potential difference ΔV_{12} between points 1 and 2? Explain.

4. The circuit of FIGURE Q31.4 has two resistors, with $R_1 > R_2$. Which of the two resistors dissipates the larger amount of power? Explain.

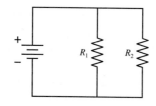

FIGURE Q31.4   FIGURE Q31.5

5. The circuit of FIGURE Q31.5 has two resistors, with $R_1 > R_2$. Which of the two resistors dissipates the larger amount of power? Explain.

6. Rank in order, from largest to smallest, the powers P_a to P_d dissipated by the four resistors in FIGURE Q31.6.

FIGURE Q31.6

7. Are the two resistors in FIGURE Q31.7 in series or in parallel? Explain.   FIGURE Q31.7

8. A battery with internal resistance r is connected to a load resistance R. If R is increased, does the terminal voltage of the battery increase, decrease, or stay the same? Explain.

9. Initially bulbs A and B in FIGURE Q31.9 are glowing. What happens to each bulb if the switch is closed? Does it get brighter, stay the same, get dimmer, or go out? Explain.

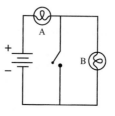

FIGURE Q31.9

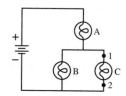

FIGURE Q31.10

10. Bulbs A, B, and C in FIGURE Q31.10 are identical, and all are glowing.
    a. Rank in order, from most to least, the brightnesses of the three bulbs. Explain.
    b. Suppose a wire is connected between points 1 and 2. What happens to each bulb? Does it get brighter, stay the same, get dimmer, or go out? Explain.

11. Bulbs A and B in FIGURE Q31.11 are identical, and both are glowing. Bulb B is removed from its socket. Does the potential difference ΔV_{12} between points 1 and 2 increase, stay the same, decrease, or become zero? Explain.

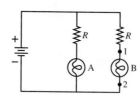

FIGURE Q31.11

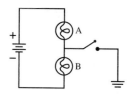

FIGURE Q31.12

12. Bulbs A and B in FIGURE Q31.12 are identical, and both are glowing. What happens to each bulb when the switch is closed? Does its brightness increase, stay the same, decrease, or go out? Explain.

13. FIGURE Q31.13 shows the voltage as a function of time of a capacitor as it is discharged (separately) through three different resistors. Rank in order, from largest to smallest, the values of the resistances R_1 to R_3.

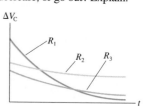

FIGURE Q31.13

EXERCISES AND PROBLEMS

Problems labeled ▓ integrate material from earlier chapters.

Exercises

Section 31.1 Circuit Elements and Diagrams

1. | Draw a circuit diagram for the circuit of FIGURE EX31.1.

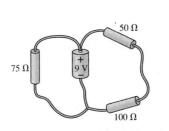

FIGURE EX31.1

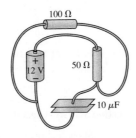

FIGURE EX31.2

2. | Draw a circuit diagram for the circuit of FIGURE EX31.2.

Section 31.2 Kirchhoff's Laws and the Basic Circuit

3. ‖ In FIGURE EX31.3, what is the current in the wire to the right of the junction? Does the charge in this wire flow to the right or to the left?

FIGURE EX31.3

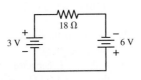

FIGURE EX31.4

4. | a. What are the magnitude and direction of the current in the 18 Ω resistor in FIGURE EX31.4?
   b. Draw a graph of the potential as a function of the distance traveled through the circuit, traveling cw from $V = 0$ V at the lower left corner.

5. | a. What are the magnitude and direction of the current in the 10 Ω resistor in FIGURE EX31.5?
   b. Draw a graph of the potential as a function of the distance traveled through the circuit, traveling cw from $V = 0$ V at the lower left corner.

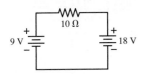

FIGURE EX31.5

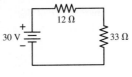

FIGURE EX31.6

6. | What is the potential difference across each resistor in FIGURE EX31.6?

Section 31.3 Energy and Power

7. | What is the resistance of a 1500 W (120 V) hair dryer? What is the current in the hair dryer when it is used?

8. | How much power is dissipated by each resistor in FIGURE EX31.8?

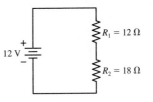

FIGURE EX31.8

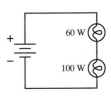

FIGURE EX31.9

9. ‖ A 60 W lightbulb and a 100 W lightbulb are placed one after the other in the circuit of FIGURE EX31.9. The battery's emf is large enough that both bulbs are glowing. Which is the true statement?
   A. The 60 W bulb is brighter.
   B. Both bulbs are equally bright.
   C. The 100 W bulb is brighter.
   D. There's not enough information to tell which bulb is brighter.

10. ‖ A standard 100 W (120 V) lightbulb contains a 7.0-cm-long tungsten filament. The high-temperature resistivity of tungsten is 9.0×10^{-7} Ω m. What is the diameter of the filament?

11. ‖ A typical American family uses 1000 kWh of electricity a month.
    a. What is the average current in the 120 V power line to the house?
    b. On average, what is the resistance of a household?

12. | A waterbed heater uses 450 W of power. It is on 25% of the time, off 75%. What is the annual cost of electricity at a billing rate of $0.12/kWh?

Section 31.4 Series Resistors

Section 31.5 Real Batteries

13. | Two of the three resistors in FIGURE EX31.13 are unknown but equal. The total resistance between points a and b is 200 Ω. What is the value of R?

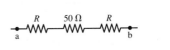

FIGURE EX31.13

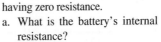

FIGURE EX31.14

14. | What is the value of resistor R in FIGURE EX31.14?

15. | The battery in FIGURE EX31.15 is short-circuited by an ideal ammeter having zero resistance.
    a. What is the battery's internal resistance?
    b. How much power is dissipated inside the battery?

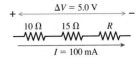

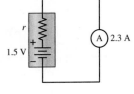

FIGURE EX31.15

16. ‖ The voltage across the terminals of a 9.0 V battery is 8.5 V when the battery is connected to a 20 Ω load. What is the battery's internal resistance?

17. ‖ Compared to an ideal battery, by what percentage does the battery's internal resistance reduce the potential difference across the 20 Ω resistor in FIGURE EX31.17?

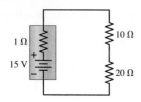

FIGURE EX31.17

Section 31.6 Parallel Resistors

18. ‖ A metal wire of resistance R is cut into two pieces of equal length. The two pieces are connected together side by side. What is the resistance of the two connected wires?
19. | Two of the three resistors in FIGURE EX31.19 are unknown but equal. The total resistance between points a and b is 75 Ω. What is the value of R?

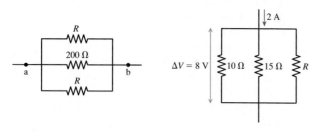

FIGURE EX31.19   FIGURE EX31.20

20. | What is the value of resistor R in FIGURE EX31.20?
21. | What is the equivalent resistance between points a and b in FIGURE EX31.21?

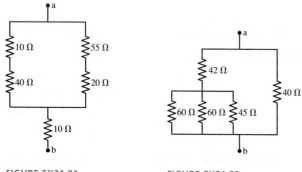

FIGURE EX31.21   FIGURE EX31.22

22. | What is the equivalent resistance between points a and b in FIGURE EX31.22?
23. | What is the equivalent resistance between points a and b in FIGURE EX31.23?

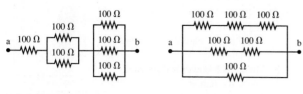

FIGURE EX31.23   FIGURE EX31.24

24. | What is the equivalent resistance between points a and b in FIGURE EX31.24?

Section 31.8 Getting Grounded

25. ‖ In FIGURE EX31.25, what is the value of the potential at points a and b?

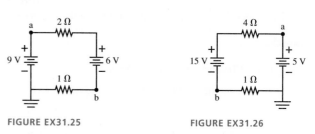

FIGURE EX31.25   FIGURE EX31.26

26. ‖‖ In FIGURE EX31.26, what is the value of the potential at points a and b?

Section 31.9 *RC* Circuits

27. | Show that the product RC has units of s.
28. | What is the time constant for the discharge of the capacitors in FIGURE EX31.28?

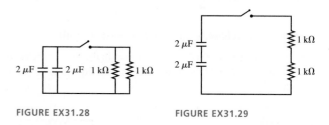

FIGURE EX31.28   FIGURE EX31.29

29. ‖ What is the time constant for the discharge of the capacitors in FIGURE EX31.29?
30. ‖ A 10 μF capacitor initially charged to 20 μC is discharged through a 1.0 kΩ resistor. How long does it take to reduce the capacitor's charge to 10 μC?
31. | The switch in FIGURE EX31.31 has been in position a for a long time. It is changed to position b at $t = 0$ s. What are the charge Q on the capacitor and the current I through the resistor (a) immediately after the switch is closed? (b) at $t = 50$ μs? (c) at $t = 200$ μs?

FIGURE EX31.31

32. ‖ What value resistor will discharge a 1.0 μF capacitor to 10% of its initial charge in 2.0 ms?
33. ‖ A capacitor is discharged through a 100 Ω resistor. The discharge current decreases to 25% of its initial value in 2.5 ms. What is the value of the capacitor?

Problems

34. ‖ The five identical bulbs in FIGURE P31.34 are all glowing. The battery is ideal. What is the order of brightness of the bulbs, from brightest to dimmest? Some may be equal.
A. P = S > Q = R = T
B. P = S = T > Q = R
C. P > S = T > Q = R
D. P > Q = R > S = T

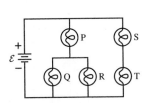

FIGURE P31.34

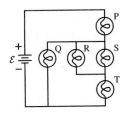

FIGURE P31.35

35. ‖ The five identical bulbs in FIGURE P31.35 are all glowing. The battery is ideal. What is the order of brightness of the bulbs, from brightest to dimmest? Some may be equal.
   A. $P = T > Q = R = S$
   B. $P > Q = R = S > T$
   C. $P = T > Q > R = S$
   D. $P > Q > T > R = S$

36. ‖‖ Two 75 W (120 V) lightbulbs are wired in series, then the combination is connected to a 120 V supply. How much power is dissipated by each bulb?

37. ‖‖ The corroded contacts in a lightbulb socket have 5.0 Ω resistance. How much actual power is dissipated by a 100 W (120 V) lightbulb screwed into this socket?

38. ‖ An electric eel develops a 450 V potential difference between
BIO   its head and tail. The eel can stun a fish or other prey by using this potential difference to drive a 0.80 A current pulse for 1.0 ms. What are (a) the energy delivered by this pulse and (b) the total charge that flows?

39. ‖ You have a 2.0 Ω resistor, a 3.0 Ω resistor, a 6.0 Ω resistor, and a 6.0 V battery. Draw a diagram of a circuit in which all three resistors are used and the battery delivers 9.0 W of power.

40. ‖ You have three 12 Ω resistors. Draw diagrams showing how you could arrange all three so that their equivalent resistance is (a) 4.0 Ω, (b) 8.0 Ω, (c) 18 Ω, and (d) 36 Ω.

41. ‖ What is the equivalent resistance between points a and b in FIGURE P31.41?

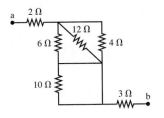

FIGURE P31.41

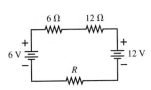

FIGURE P31.42

42. ‖ There is a current of 0.25 A in the circuit of FIGURE P31.42. What is the power dissipated by R?

43. ‖‖ A variable resistor R is connected across the terminals of a battery. FIGURE P31.43 shows the current in the circuit as R is varied. What are the emf and internal resistance of the battery?

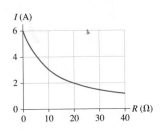

FIGURE P31.43

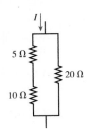

FIGURE P31.44

44. ‖ The 10 Ω resistor in FIGURE P31.44 is dissipating 40 W of power. How much power are the other two resistors dissipating?

45. ‖‖ What are the emf and internal resistance of the battery in FIGURE P31.45?

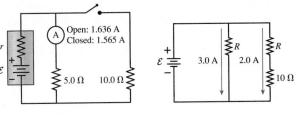

FIGURE P31.45   **FIGURE P31.46**

46. ‖ What is the emf of the battery in FIGURE P31.46?

47. ‖‖ A 2.5 V battery with 0.70 Ω internal resistance is connected in parallel with a 1.5 V battery having 0.30 Ω internal resistance. That is, their positive terminals are connected by a wire and their negative terminals are connected by a wire. What is the terminal voltage of the 2.5 V battery?

48. ‖‖ a. Load resistor R is attached to a battery of emf $\mathcal{E}$ and internal resistance r. For what value of the resistance R, in terms of $\mathcal{E}$ and r, will the power dissipated by the load resistor be a maximum?
   b. What is the maximum power that the load can dissipate if the battery has $\mathcal{E} = 9.0$ V and $r = 1.0$ Ω?
   c. *Why* should the power dissipated by the load have a maximum value? Explain.

Hint: What happens to the power dissipation when R is either very small or very large?

49. ‖ The ammeter in FIGURE P31.49 reads 3.0 A. Find I_1, I_2, and $\mathcal{E}$.

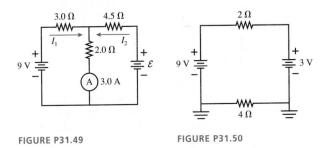

FIGURE P31.49   **FIGURE P31.50**

50. ‖‖ What is the current in the 2 Ω resistor in FIGURE P31.50?

51. ‖ It seems hard to justify spending $5 for a compact fluorescent lightbulb when an ordinary incandescent bulb costs 50¢. To see if this makes sense, compare a 60 W incandescent bulb lasting 1000 hours to a 15 W compact fluorescent bulb having a lifetime of 10,000 hours. Both bulbs produce the same amount of visible light and are interchangeable. If electricity costs $0.10/kWh, what is the total cost—purchase plus energy—to obtain 10,000 hours of light from each type of bulb? This is called the *life-cycle cost*.

52. ‖‖ A refrigerator has a 1000 W compressor, but the compressor runs only 20% of the time.
   a. If electricity costs $0.10/kWh, what is the monthly (30 day) cost of running the refrigerator?
   b. A more energy-efficient refrigerator with an 800 W compressor costs $100 more. If you buy the more expensive refrigerator, how many months will it take to recover your additional cost?

53. | For an ideal battery ($r = 0\ \Omega$), closing the switch in FIGURE P31.53 does not affect the brightness of bulb A. In practice, bulb A dims *just a little* when the switch closes. To see why, assume that the 1.50 V battery has an internal resistance $r = 0.50\ \Omega$ and that the resistance of a glowing bulb is $R = 6.00\ \Omega$.
   a. What is the current through bulb A when the switch is open?
   b. What is the current through bulb A after the switch has closed?
   c. By what percentage does the current through A change when the switch is closed?

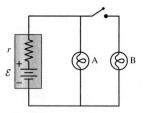

FIGURE P31.53

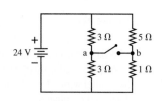

FIGURE P31.54

54. | What are the battery current I_{bat} and the potential difference ΔV_{ab} between points a and b when the switch in FIGURE P31.54 is (a) open and (b) closed?

55. ‖ The circuit in FIGURE P31.55 is called a *voltage divider*. What value of R will make $V_{out} = V_{in}/10$?

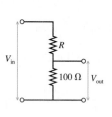

FIGURE P31.55

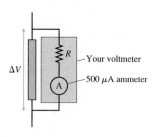

FIGURE P31.56

56. ‖ A circuit you're building needs a voltmeter that goes from 0 V to a full-scale reading of 5.0 V. Unfortunately, the only meter in the storeroom is an *ammeter* that goes from 0 μA to a full-scale reading of 500 μA. Fortunately, you've just finished a physics class, and you realize that you can convert this meter to a voltmeter by putting a resistor in series with it, as shown in FIGURE P31.56. You've measured that the resistance of the ammeter is 50.0 Ω, not the 0 Ω of an ideal ammeter. What value of R must you use so that the meter will go to full scale when the potential difference across the object being measured is 5.0 V?

57. ‖ A circuit you're building needs an ammeter that goes from 0 mA to a full-scale reading of 50 mA. Unfortunately, the only ammeter in the storeroom goes from 0 μA to a full-scale reading of only 500 μA. Fortunately, you've just finished a physics class, and you realize that you can make this ammeter work by putting a resistor in parallel with it, as shown in FIGURE P31.57. You've measured that the resistance of the ammeter is 50.0 Ω, not the 0 Ω of an ideal ammeter.
   a. What value of R must you use so that the meter will go to full scale when the current I is 50 mA?
   b. What is the effective resistance of your ammeter?

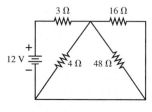

FIGURE P31.57

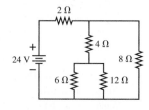

FIGURE P31.58

58. ‖ For the circuit shown in FIGURE P31.58, find the current through and the potential difference across each resistor. Place your results in a table for ease of reading.

59. ‖ For the circuit shown in FIGURE P31.59, find the current through and the potential difference across each resistor. Place your results in a table for ease of reading.

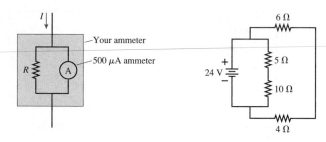

FIGURE P31.59   FIGURE P31.60

60. ‖ For the circuit shown in FIGURE P31.60, find the current through and the potential difference across each resistor. Place your results in a table for ease of reading.

61. ‖ For the circuit shown in FIGURE P31.61, find the current through and the potential difference across each resistor. Place your results in a table for ease of reading.

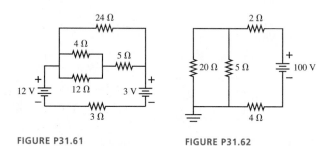

FIGURE P31.61   FIGURE P31.62

62. ‖ What is the current through the 20 Ω resistor in FIGURE P31.62?

63. ‖ What is the current through the 10 Ω resistor in FIGURE P31.63? Is the current from left to right or right to left?

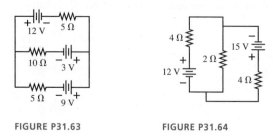

FIGURE P31.63   FIGURE P31.64

64. ‖ What power is dissipated by the 2 Ω resistor in FIGURE P31.64?

65. ‖ For what emf $\mathcal{E}$ does the 200 Ω resistor in FIGURE P31.65 dissipate no power? Should the emf be oriented with its positive terminal at the top or at the bottom?

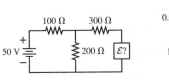

FIGURE P31.65

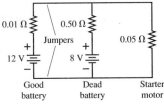

FIGURE P31.66

66. ‖ A 12 V car battery dies not so much because its voltage drops but because chemical reactions increase its internal resistance. A good battery connected with jumper cables can both start the engine and recharge the dead battery. Consider the automotive circuit of FIGURE P31.66.
   a. How much current could the good battery alone drive through the starter motor?
   b. How much current is the dead battery alone able to drive through the starter motor?
   c. With the jumper cables attached, how much current passes through the starter motor?
   d. With the jumper cables attached, how much current passes through the dead battery, and in which direction?

67. ‖ How much current flows through the bottom wire in FIGURE P31.67, and in which direction?

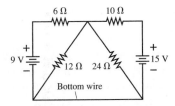

FIGURE P31.67

68. ‖ The capacitor in an RC circuit is discharged with a time constant of 10 ms. At what time after the discharge begins are (a) the charge on the capacitor reduced to half its initial value and (b) the energy stored in the capacitor reduced to half its initial value?

69. ‖ A circuit you're using discharges a 20 μF capacitor through an unknown resistor. After charging the capacitor, you close a switch at $t = 0$ s and then monitor the resistor current with an ammeter. Your data are as follows:

Time (s)	Current (μA)
0.5	890
1.0	640
1.5	440
2.0	270
2.5	200

Use an appropriate graph of the data to determine (a) the resistance and (b) the initial capacitor voltage.

70. ‖ A 150 μF defibrillator capacitor is charged to 1500 V. When
BIO fired through a patient's chest, it loses 95% of its charge in 40 ms. What is the resistance of the patient's chest?

71. ‖ A 50 μF capacitor that had been charged to 30 V is discharged through a resistor. FIGURE P31.71 shows the capacitor voltage as a function of time. What is the value of the resistance?

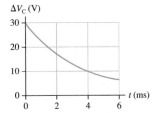

FIGURE P31.71

72. ‖ A 0.25 μF capacitor is charged to 50 V. It is then connected in series with a 25 Ω resistor and a 100 Ω resistor and allowed to discharge completely. How much energy is dissipated by the 25 Ω resistor?

73. ‖ The capacitor in FIGURE P31.73 begins to charge after the switch closes at $t = 0$ s.
   a. What is ΔV_C a very long time after the switch has closed?
   b. What is Q_{max} in terms of $\mathcal{E}$, R, and C?
   c. In this circuit, does $I = +dQ/dt$ or $-dQ/dt$? Explain.
   d. Find an expression for the current I at time t. Graph I from $t = 0$ to $t = 5\tau$.

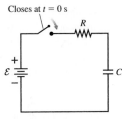

FIGURE P31.73

74. ‖ The capacitors in FIGURE P31.74 are charged and the switch closes at $t = 0$ s. At what time has the current in the 8 Ω resistor decayed to half the value it had immediately after the switch was closed?

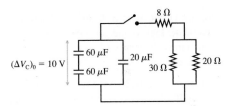

FIGURE P31.74

Challenge Problems

75. You've made the finals of the Science Olympics! As one of your tasks, you're given 1.0 g of aluminum and asked to make a wire, using all the aluminum, that will dissipate 7.5 W when connected to a 1.5 V battery. What length and diameter will you choose for your wire?

76. The switch in FIGURE CP31.76 has been closed for a very long time.
   a. What is the charge on the capacitor?
   b. The switch is opened at $t = 0$ s. At what time has the charge on the capacitor decreased to 10% of its initial value?

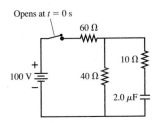

FIGURE CP31.76

77. A capacitor-charging circuit has a time constant of 40 ms. When the switch is closed, the initial current to the 50 μF capacitor is 65 mA. What is the capacitor's voltage after 20 ms? Assume the capacitor was completely uncharged when the switch closed.

78. The capacitor in Figure 31.38a begins to charge after the switch closes at $t = 0$ s. Analyze this circuit and show that $Q = Q_{max}(1 - e^{-t/\tau})$, where $Q_{max} = C\mathcal{E}$.

79. The switch in Figure 31.38a closes at $t = 0$ s and, after a very long time, the capacitor is fully charged. Find expressions for (a) the total energy supplied by the battery as the capacitor is being charged, (b) total energy dissipated by the resistor as the capacitor is being charged, and (c) the energy stored in the capacitor when it is fully charged. Your expressions will be in terms of $\mathcal{E}$, R, and C. (d) Do your results for parts a to c show that energy is conserved? Explain.

80. An *oscillator circuit* is important to many applications. A simple oscillator circuit can be built by adding a neon gas tube to an RC circuit, as shown in **FIGURE CP31.80**. Gas is normally a good insulator, and the resistance of the gas tube is essentially infinite when the light is off. This allows the capacitor to charge. When the capacitor voltage reaches a value V_{on}, the electric field inside the tube becomes strong enough to ionize the neon gas. Visually, the tube lights with an orange glow. Electrically, the ionization of the gas provides a very-low-resistance path through the tube. The capacitor very rapidly (we can think of it as instantaneously) discharges through the tube and the capacitor voltage drops. When the capacitor voltage has dropped to a value V_{off}, the electric field inside the tube becomes too weak to sustain the ionization and the neon light turns off. The capacitor then starts to charge again. The capacitor voltage oscillates between V_{off}, when it starts charging, and V_{on}, when the light comes on to discharge it.

a. Show that the oscillation period is

$$T = RC \ln\left(\frac{\mathcal{E} - V_{off}}{\mathcal{E} - V_{on}}\right)$$

b. A neon gas tube has $V_{on} = 80$ V and $V_{off} = 20$ V. What resistor value should you choose to go with a 10 μF capacitor and a 90 V battery to make a 10 Hz oscillator?

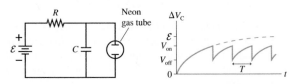

FIGURE CP31.80

81. A 2.0-m-long, 1.0-mm-diameter wire has a variable resistivity given by

$$\rho(x) = (2.5 \times 10^{-6})\left[1 + \left(\frac{x}{1.0 \text{ m}}\right)^2\right] \Omega \text{ m}$$

where x is measured from one end of the wire. What is the current if this wire is connected to the terminals of a 9.0 V battery?

STOP TO THINK ANSWERS

Stop to Think 31.1: a, b, and d. These three are the same circuit because the logic of the connections is the same. In c, the functioning of the circuit is changed by the extra wire connecting the two sides of the capacitor.

Stop to Think 31.2: ΔV increases by 2 V in the direction of I. Kirchhoff's loop law, starting on the left side of the battery, is then $+12$ V $+ 2$ V $- 8$ V $- 6$ V $= 0$ V.

Stop to Think 31.3: $P_b > P_d > P_a > P_c$. The power dissipated by a resistor is $P_R = (\Delta V_R)^2/R$. Increasing R decreases P_R; increasing ΔV_R increases P_R. But the potential has a larger effect because P_R depends on the square of ΔV_R.

Stop to Think 31.4: $I = 2$ A for all. $V_a = 20$ V, $V_b = 16$ V, $V_c = 10$ V, $V_d = 8$ V, $V_e = 0$ V. Current is conserved. The potential is 0 V on the right and increases by IR for each resistor going to the left.

Stop to Think 31.5: A > B > C = D. All the current from the battery goes through A, so it is brightest. The current divides at the junction, but not equally. Because B is in parallel with C + D but has half the resistance, twice as much current travels through B as through C + D. So B is dimmer than A but brighter than C and D. C and D are equal because of conservation of current.

Stop to Think 31.6: b. The two 2 Ω resistors are in series and equivalent to a 4 Ω resistor. Thus $\tau = RC = 4$ s.

32 The Magnetic Field

The aurora occurs when high-energy charged particles from the sun are steered into the upper atmosphere by the earth's magnetic field.

▶ **Looking Ahead** The goal of Chapter 32 is to learn how to calculate and use the magnetic field.

Magnetic Fields

Magnetism has been known since antiquity. Whereas electricity is understood in terms of electric charges, magnetism is based on **magnetic poles.** You'll learn how to use the **magnetic field,** with symbol $\vec{B}$, to work with the long-range interactions of magnetism.

Iron filings, like little compasses, show the shape of the magnetic field.

This bar magnet—a *dipole*, with a north and a south pole—is a permanent magnet.

A loop of current also creates a dipole magnetic field.

One of our key tasks will be to understand the connection between electromagnets and permanent magnets.

Compasses work because the earth is a large magnet. It is an electromagnet, with circulating currents in its molten iron core.

Magnetic Forces

Magnetic fields exert forces on *moving* charged particles. The force is perpendicular to the plane of $\vec{v}$ and $\vec{B}$.

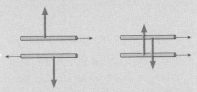

Currents are moving charged particles. You'll learn that currents create magnetic fields, and currents exert magnetic forces on each other. Opposite currents repel, parallel currents attract.

Magnetic Torque

Magnetic forces exert a *torque* on a current traveling around a closed loop.

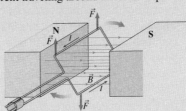

You'll learn that motors work because of magnetic torque.

◀ **Looking Back**
Sections 12.5 and 12.10 Torque and the vector cross product

Motion of Charges

The magnetic force causes charged particles to move in circular orbits in a magnetic field. This **cyclotron motion** has many important applications, from particle accelerators to the aurora.

Magnetism is three dimensional. You'll learn how to represent vectors perpendicular to a plane. Here the ×'s show a magnetic field into the page.

◀ **Looking Back**
Sections 8.2–8.3 Circular motion

Magnetic Materials

Iron and a few other materials exhibit pronounced magnetic properties, including the ability to form permanent magnets. You'll learn that **ferromagnetism** arises because electrons have an inherent magnetic moment called **electron spin.**

This hard disk is made of nickel, a magnetic material. It stores digital data—1's and 0's—in the alignment of microscopic **magnetic domains.**

32.1 Magnetism

We began our investigation of electricity in Chapter 25 by looking at the results of simple experiments with charged rods. We'll do the same with magnetism.

Discovering magnetism

Experiment 1

If a bar magnet is taped to a piece of cork and allowed to float in a dish of water, it always turns to align itself in an approximate north-south direction. The end of a magnet that points north is called the *north-seeking pole,* or simply the **north pole.** The other end is the **south pole.**

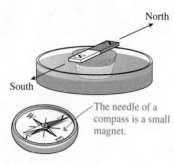

North

South

The needle of a compass is a small magnet.

Experiment 2

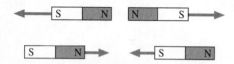

If the north pole of one magnet is brought near the north pole of another magnet, they repel each other. Two south poles also repel each other, but the north pole of one magnet exerts an attractive force on the south pole of another magnet.

Experiment 3

The north pole of a bar magnet attracts one end of a compass needle and repels the other. Apparently the compass needle itself is a little bar magnet with a north pole and a south pole.

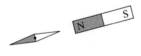

Experiment 4

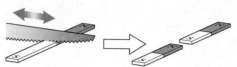

Cutting a bar magnet in half produces two weaker but still complete magnets, each with a north pole and a south pole. No matter how small the magnets are cut, even down to microscopic sizes, each piece remains a complete magnet with two poles.

Experiment 5

Magnets can pick up some objects, such as paper clips, but not all. If an object is attracted to one end of a magnet, it is also attracted to the other end. Most materials, including copper (a penny), aluminum, glass, and plastic, experience no force from a magnet.

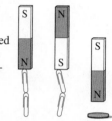

Experiment 6

A magnet does not affect an electroscope. A charged rod exerts a weak *attractive* force on *both* ends of a magnet. However, the force is the same as the force on a metal bar that isn't a magnet, so it is simply a polarization force like the ones we studied in Chapter 25. Other than polarization forces, charges have *no effects* on magnets.

No effect

What do these experiments tell us?

1. First, magnetism is not the same as electricity. **Magnetic poles and electric charges share some similar behavior, but they are not the same.**
2. Magnetism is a long-range force. Paper clips leap upward to a magnet.
3. Magnets have two poles, called north and south poles, and thus are **magnetic dipoles.** Two like poles exert repulsive forces on each other; two opposite poles attract. The behavior is *analogous* to electric charges, but, as noted, magnetic poles and electric charges are *not* the same. Unlike charges, isolated north or south poles do not exist.
4. The poles of a bar magnet can be identified by using it as a compass. The poles of other magnets, such as flat refrigerator magnets, can be identified by testing them against a bar magnet. A pole that attracts a known north pole and repels a known south pole must be a south magnetic pole.
5. Materials that are attracted to a magnet are called **magnetic materials.** The most common magnetic material is iron. Magnetic materials are attracted to *both* poles of a magnet. This attraction is analogous to how neutral objects are attracted to both positively and negatively charged rods by the polarization force. The difference is that *all* neutral objects are attracted to a charged rod whereas only a few materials are attracted to a magnet.

Our goal is to develop a theory of magnetism to explain these observations.

Compasses and Geomagnetism

The north pole of a compass needle is attracted toward the geographic north pole of the earth. Apparently the earth itself is a large magnet, as shown in **FIGURE 32.1**. The reasons for the earth's magnetism are complex, but geophysicists think that the earth's magnetic poles arise from currents in its molten iron core. Two interesting facts about the earth's magnetic field are (1) that the magnetic poles are offset slightly from the geographic poles of the earth's rotation axis, and (2) that the geographic north pole is actually a *south* magnetic pole! You should be able to use what you have learned thus far to convince yourself that this is the case.

FIGURE 32.1 The earth is a large magnet.

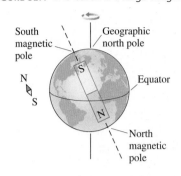

STOP TO THINK 32.1 Does the compass needle rotate clockwise (cw), counterclockwise (ccw), or not at all?

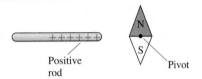

Positive rod

Pivot

32.2 The Discovery of the Magnetic Field

As electricity began to be seriously studied in the 18th century, some scientists speculated that there might be a connection between electricity and magnetism. Interestingly, the link between electricity and magnetism was discovered *in the midst of a classroom lecture demonstration* in 1819 by the Danish scientist Hans Christian Oersted. Oersted was using a battery—a fairly recent invention—to produce a large current in a wire. By chance, a compass was sitting next to the wire, and Oersted noticed that the current caused the compass needle to turn. In other words, the compass responded as if a magnet had been brought near.

Oersted had long been interested in a possible connection between electricity and magnetism, so the significance of this serendipitous observation was immediately apparent to him. Oersted's discovery that **magnetism is caused by an electric current** is illustrated in **FIGURE 32.2**. Part c of the figure demonstrates an important **right-hand rule** that relates the orientation of the compass needles to the direction of the current.

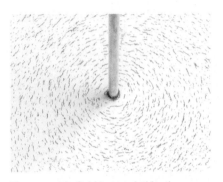

The magnetic field is revealed by the pattern of iron filings around a current-carrying wire.

FIGURE 32.2 Response of compass needles to a current in a straight wire.

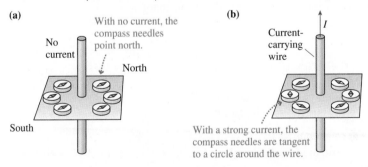

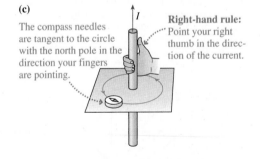

Magnetism is more demanding than electricity in requiring a three-dimensional perspective of the sort shown in Figure 32.2. But since two-dimensional figures are easier to draw, we will make as much use of them as we can. Consequently, we will often need to indicate field vectors or currents that are perpendicular to the page. **FIGURE 32.3** shows the notation we will use. **FIGURE 32.4** on the next page demonstrates this notation by showing the compasses around a current that is directed into the page. To use the right-hand rule, point your right thumb in the direction of the current (into the page). Your fingers will curl cw, and that is the direction in which the north poles of the compass needles point.

FIGURE 32.3 The notation for vectors and currents perpendicular to the page.

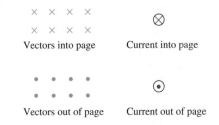

FIGURE 32.4 The orientation of the compasses around a current is given by the right-hand rule.

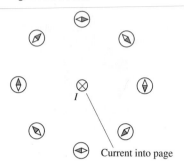

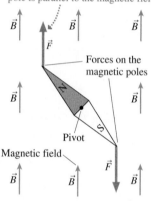

Current into page

FIGURE 32.5 The magnetic field exerts forces on the poles of a compass, causing the needle to align with the field.

The magnetic force on the north pole is parallel to the magnetic field.

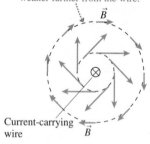

Forces on the magnetic poles

Pivot

Magnetic field

FIGURE 32.6 The magnetic field around a current-carrying wire.

(a) The magnetic field vectors are tangent to circles around the wire, pointing in the direction given by the right-hand rule. The field is weaker farther from the wire.

$\vec{B}$

Current-carrying wire $\vec{B}$

(b) Magnetic field lines are circles.

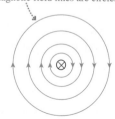

The Magnetic Field

We introduced the idea of a *field* as a way to understand the long-range electric force. Although this idea appeared rather far-fetched, it turned out to be very useful. We need a similar idea to understand the long-range force exerted by a current on a compass needle. Let us define the **magnetic field** $\vec{B}$ as having the following properties:

1. A magnetic field is created at *all* points in space surrounding a current-carrying wire.
2. The magnetic field at each point is a vector. It has both a magnitude, which we call the *magnetic field strength B*, and a direction.
3. The magnetic field exerts forces on magnetic poles. The force on a north pole is parallel to $\vec{B}$; the force on a south pole is opposite $\vec{B}$.

FIGURE 32.5 shows a compass needle in a magnetic field. The field vectors are shown at several points, but keep in mind that the field is present at *all* points in space. A magnetic force is exerted on each of the two poles of the compass, parallel to $\vec{B}$ for the north pole and opposite $\vec{B}$ for the south pole. This pair of opposite forces exerts a torque on the needle, rotating the needle until it is parallel to the magnetic field at that point.

Notice that the north pole of the compass needle, when it reaches the equilibrium position, is in the direction of the magnetic field. Thus a compass needle can be used as a probe of the magnetic field, just as a charge was a probe of the electric field. **Magnetic forces cause a compass needle to become aligned parallel to a magnetic field, with the north pole of the compass showing the direction of the magnetic field at that point.**

Look back at the compass alignments around the current-carrying wire in Figure 32.4. Because compass needles align with the magnetic field, the magnetic field at each point must be tangent to a circle around the wire. **FIGURE 32.6a** shows the magnetic field by drawing field vectors. Notice that the field is weaker (shorter vectors) at greater distances from the wire.

Another way to picture the field is with the use of **magnetic field lines**. These are imaginary lines drawn through a region of space so that

- A tangent to a field line is in the direction of the magnetic field, and
- The field lines are closer together where the magnetic field strength is larger.

FIGURE 32.6b shows the magnetic field lines around a current-carrying wire. Notice that magnetic field lines form loops, with no beginning or ending point. This is in contrast to electric field lines, which stop and start on charges.

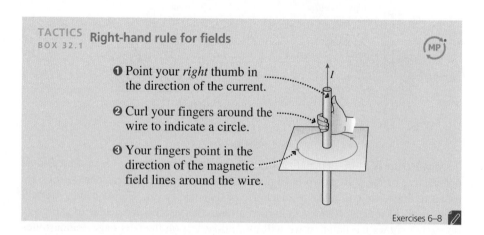

TACTICS BOX 32.1 **Right-hand rule for fields**

❶ Point your *right* thumb in the direction of the current.

❷ Curl your fingers around the wire to indicate a circle.

❸ Your fingers point in the direction of the magnetic field lines around the wire.

Exercises 6–8

NOTE ▶ The magnetic field of a current-carrying wire is very different from the electric field of a charged wire. The electric field of a charged wire points radially outward (positive wire) or inward (negative wire). ◀

Two Kinds of Magnetism?

You might be concerned that we have introduced two kinds of magnetism. We opened this chapter discussing permanent magnets and their forces. Then, without warning, we switched to the magnetic forces caused by a current. It is not at all obvious that these forces are the same kind of magnetism as that exhibited by stationary chunks of metal called "magnets." Perhaps there are two different types of magnetic forces, one having to do with currents and the other being responsible for permanent magnets. One of the major goals for our study of magnetism is to see that these two quite different ways of producing magnetic effects are really just two different aspects of a *single* magnetic force.

STOP TO THINK 32.2   The magnetic field at position P points

• P

a. Up.   b. Down.
c. Into the page.   d. Out of the page.

32.3 The Source of the Magnetic Field: Moving Charges

Figure 32.6 is a qualitative picture of the wire's magnetic field. Our first task is to turn that picture into a quantitative description. Because current in a wire generates a magnetic field, and a current is a collection of moving charges, our starting point is the idea that **moving charges are the source of the magnetic field.** FIGURE 32.7 shows a charged particle q moving with velocity $\vec{v}$. The magnetic field of this moving charge is found to be

$$\vec{B}_{\text{point charge}} = \left(\frac{\mu_0}{4\pi} \frac{qv\sin\theta}{r^2}, \text{ direction given by the right-hand rule} \right) \quad (32.1)$$

where r is the distance from the charge and θ is the angle between $\vec{v}$ and $\vec{r}$.

   Equation 32.1 is called the **Biot-Savart law** for a point charge (rhymes with *Leo* and *bazaar*), named for two French scientists whose investigations were motivated by Oersted's observations. It is analogous to Coulomb's law for the electric field of a point charge. Notice that the Biot-Savart law, like Coulomb's law, is an inverse-square law. However, the Biot-Savart law is somewhat more complex than Coulomb's law because the magnetic field depends on the angle θ between the charge's velocity and the line to the point where the field is evaluated.

   NOTE ▶ A moving charge has both a magnetic field *and* an electric field. What you know about electric fields has not changed. ◀

   The SI unit of magnetic field strength is the **tesla,** abbreviated as T. The tesla is defined as

$$1 \text{ tesla} = 1 \text{ T} \equiv 1 \text{ N/A m}$$

You will see later in the chapter that this definition is based on the magnetic force on a current-carrying wire. One tesla is quite a large field; most magnetic fields are a small fraction of a tesla. Table 32.1 lists a few magnetic field strengths.

   The constant μ_0 in Equation 32.1 is called the **permeability constant.** Its value is

$$\mu_0 = 4\pi \times 10^{-7} \text{ T m/A} = 1.257 \times 10^{-6} \text{ T m/A}$$

This constant plays a role in magnetism similar to that of the permittivity constant ϵ_0 in electricity.

FIGURE 32.7 The magnetic field of a moving point charge.

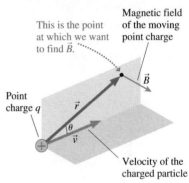

TABLE 32.1 Typical magnetic field strengths

Field source	Field strength (T)
Surface of the earth	5×10^{-5}
Refrigerator magnet	5×10^{-3}
Laboratory magnet	0.1 to 1
Superconducting magnet	10

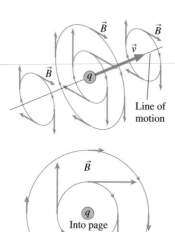

The right-hand rule for finding the direction of $\vec{B}$ is similar to the rule used for a current-carrying wire: Point your right thumb in the direction of $\vec{v}$. The magnetic field vector $\vec{B}$ is perpendicular to the plane of $\vec{r}$ and $\vec{v}$, pointing in the direction in which your fingers curl. In other words, the $\vec{B}$ vectors are tangent to circles drawn about the charge's line of motion. **FIGURE 32.8** shows a more complete view of the magnetic field of a moving positive charge. Notice that $\vec{B}$ is zero along the line of motion, where $\theta = 0°$ or $180°$, due to the $\sin\theta$ term in Equation 32.1.

NOTE ▶ The vector arrows in Figure 32.8 would have the same lengths but be reversed in direction for a negative charge. ◀

The requirement that a charge be moving to generate a magnetic field is explicit in Equation 32.1. If the speed v of the particle is zero, the magnetic field (but not the electric field!) is zero. This helps to emphasize a fundamental distinction between electric and magnetic fields: **All charges create electric fields, but only *moving* charges create magnetic fields.**

EXAMPLE 32.1 **The magnetic field of a proton**

A proton moves with velocity $\vec{v} = 1.0 \times 10^7\,\hat{\imath}$ m/s. As it passes the origin, what is the magnetic field at the (x, y, z) positions (1 mm, 0 mm, 0 mm), (0 mm, 1 mm, 0 mm), and (1 mm, 1 mm, 0 mm)?

MODEL The magnetic field is that of a moving charged particle.

VISUALIZE **FIGURE 32.9** shows the geometry. The first point is on the x-axis, directly in front of the proton, with $\theta_1 = 0°$. The second point is on the y-axis, with $\theta_2 = 90°$, and the third is in the xy-plane.

FIGURE 32.9 The magnetic field of Example 32.1.

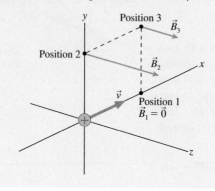

SOLVE Position 1, which is along the line of motion, has $\theta_1 = 0°$. Thus $\vec{B}_1 = \vec{0}$. Position 2 (at 0 mm, 1 mm, 0 mm) is at distance $r_2 = 1$ mm $= 0.001$ m. Equation 32.1, the Biot-Savart law, gives us the magnetic field strength at this point as

$$B = \frac{\mu_0}{4\pi}\frac{qv\sin\theta_2}{r_2^2}$$

$$= \frac{4\pi \times 10^{-7}\,\text{T m/A}\,(1.60 \times 10^{-19}\,\text{C})(1.0 \times 10^7\,\text{m/s})\sin 90°}{4\pi \qquad\qquad (0.0010\,\text{m})^2}$$

$$= 1.60 \times 10^{-13}\,\text{T}$$

According to the right-hand rule, the field points in the positive z-direction. Thus

$$\vec{B}_2 = 1.60 \times 10^{-13}\,\hat{k}\,\text{T}$$

where $\hat{k}$ is the unit vector in the positive z-direction. The field at position 3, at (1 mm, 1 mm, 0 mm), also points in the z-direction, but it is weaker than at position 2 both because r is larger *and* because θ is smaller. From geometry we know $r_3 = \sqrt{2}$ mm $= 0.00141$ m and $\theta_3 = 45°$. Another calculation using Equation 32.1 gives

$$\vec{B}_3 = 0.57 \times 10^{-13}\,\hat{k}\,\text{T}$$

ASSESS The magnetic field of a single moving charge is *very* small.

Superposition

The Biot-Savart law is the starting point for generating all magnetic fields, just as our earlier expression for the electric field of a point charge was the starting point for generating all electric fields. Magnetic fields, like electric fields, have been found experimentally to obey the principle of superposition. If there are n moving point charges, the net magnetic field is given by the vector sum

$$\vec{B}_{\text{total}} = \vec{B}_1 + \vec{B}_2 + \cdots + \vec{B}_n \qquad (32.2)$$

where each individual $\vec{B}$ is calculated with Equation 32.1. The principle of superposition will be the basis for calculating the magnetic fields of several important current distributions.

The Vector Cross Product

In Chapter 25, we found that the electric field of a point charge could be written concisely and accurately as

$$\vec{E} = \frac{1}{4\pi\epsilon_0} \frac{q}{r^2} \hat{r}$$

where $\hat{r}$ is a *unit vector* that points from the charge to the point at which we wish to calculate the field. Unit vector $\hat{r}$ expresses the idea "away from q."

The unit vector $\hat{r}$ also allows us to write the Biot-Savart law more concisely, but we'll need to use the form of vector multiplication called the *cross product*. To remind you, FIGURE 32.10 shows two vectors, $\vec{C}$ and $\vec{D}$, with angle α between them. The **cross product** of $\vec{C}$ and $\vec{D}$ is defined to be the vector

$$\vec{C} \times \vec{D} = (CD\sin\alpha, \text{direction given by the right-hand rule}) \quad (32.3)$$

The symbol $\times$ between the vectors is *required* to indicate a cross product.

NOTE ▶ The cross product of two vectors and the right-hand rule used to determine the direction of the cross product were introduced in Section 12.10 to describe torque and angular momentum. If you omitted that section, you will want to turn to it now to read about the cross product. A review is worthwhile even if you did learn about the cross product earlier. ◀

The Biot-Savart law, Equation 32.1, can be written in terms of the cross product as

$$\vec{B}_{\text{point charge}} = \frac{\mu_0}{4\pi} \frac{q\vec{v} \times \hat{r}}{r^2} \quad \text{(magnetic field of a point charge)} \quad (32.4)$$

where unit vector $\hat{r}$, shown in FIGURE 32.11, points from charge q to the point at which we want to evaluate the field. This expression for the magnetic field $\vec{B}$ has magnitude $(\mu_0/4\pi)qv\sin\theta/r^2$ (because the magnitude of $\hat{r}$ is 1) and points in the correct direction (given by the right-hand rule), so it agrees completely with Equation 32.1.

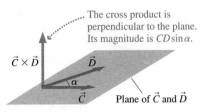

FIGURE 32.10 The cross product $\vec{C} \times \vec{D}$ is a vector perpendicular to the plane of vectors $\vec{C}$ and $\vec{D}$.

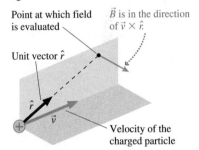

FIGURE 32.11 Unit vector $\hat{r}$ defines the direction from the moving charge to the point at which we want to evaluate the magnetic field.

EXAMPLE 32.2 **The magnetic field direction of a moving electron**

The electron in FIGURE 32.12 is moving to the right. What is the direction of the electron's magnetic field at the position indicated with a dot?

FIGURE 32.12
A moving electron.

VISUALIZE Because the charge is negative, the magnetic field points in the direction of $-(\vec{v} \times \hat{r})$, or opposite the direction of $\vec{v} \times \hat{r}$. Unit vector $\hat{r}$ points from the charge toward the dot. We can use the right-hand rule to find that $\vec{v} \times \hat{r}$ points *into* the page. Thus the electron's magnetic field at the dot points *out* of the page.

 STOP TO THINK 32.3 The positive charge is moving straight out of the page. What is the direction of the magnetic field at the position of the dot?

$\vec{v}$ out of page

a. Up   b. Down   c. Left   d. Right

32.4 The Magnetic Field of a Current

In practice we're more interested in the magnetic field of a current—a collection of moving charges—than in the very small magnetic fields of individual charges. The Biot-Savart law and the principle of superposition will be our primary tools for calculating magnetic fields. First, however, it will be useful to rewrite the Biot-Savart law in terms of current.

FIGURE 32.13 Relating the charge velocity $\vec{v}$ to the current I.

(a) Charge ΔQ in a small length Δs of a current-carrying wire

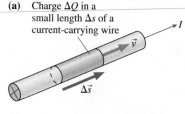

(b)

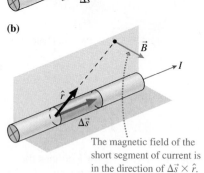

The magnetic field of the short segment of current is in the direction of $\Delta \vec{s} \times \hat{r}$.

FIGURE 32.13a shows a current-carrying wire. The wire as a whole is electrically neutral, but current I represents the motion of positive charge carriers through the wire. Suppose the small amount of moving charge ΔQ spans the small length Δs. The charge has velocity $\vec{v} = \Delta \vec{s}/\Delta t$, where the vector $\Delta \vec{s}$, which is parallel to $\vec{v}$, is the charge's displacement vector. If ΔQ is small enough to treat as a point charge, the magnetic field it creates at a point in space is proportional to $(\Delta Q)\vec{v}$. We can write $(\Delta Q)\vec{v}$ in terms of the wire's current I as

$$(\Delta Q)\vec{v} = \Delta Q \frac{\Delta \vec{s}}{\Delta t} = \frac{\Delta Q}{\Delta t} \Delta \vec{s} = I \Delta \vec{s} \tag{32.5}$$

where we used the definition of current, $I = \Delta Q/\Delta t$.

If we replace $q\vec{v}$ in the Biot-Savart law with $I \Delta \vec{s}$, we find that the magnetic field of a very short segment of wire carrying current I is

$$\vec{B}_{\text{current segment}} = \frac{\mu_0}{4\pi} \frac{I \Delta \vec{s} \times \hat{r}}{r^2} \tag{32.6}$$

(magnetic field of a very short segment of current)

Equation 32.6 is still the Biot-Savart law, only now written in terms of current rather than the motion of an individual charge. FIGURE 32.13b shows the direction of the current segment's magnetic field as determined by using the right-hand rule.

Equation 32.6 is the basis of a strategy for calculating the magnetic field of a current-carrying wire. You will recognize that it is the same basic strategy you learned for calculating the electric field of a continuous distribution of charge. The goal is to break a problem down into small steps that are individually manageable.

PROBLEM-SOLVING
STRATEGY 32.1 **The magnetic field of a current**

MODEL Model the wire as a simple shape, such as a straight line or a loop.

VISUALIZE For the pictorial representation:

❶ Draw a picture and establish a coordinate system.
❷ Identify the point P at which you want to calculate the magnetic field.
❸ Divide the current-carrying wire into small segments for which you *already know* how to determine $\vec{B}$. This is usually, though not always, a division into very short segments of length Δs.
❹ Draw the magnetic field vector for one or two segments. This will help you identify distances and angles that need to be calculated.
❺ Look for symmetries that simplify the field. You may conclude that some components of $\vec{B}$ are zero.

SOLVE The mathematical representation is $\vec{B}_{\text{net}} = \Sigma \vec{B}_i$.

- Use superposition to form an algebraic expression for *each* of the three components of $\vec{B}$ (unless you are sure one or more is zero) at point P.
- Let the (x, y, z)-coordinates of the point remain as variables.
- Express all angles and distances in terms of the coordinates.
- Let $\Delta s \rightarrow ds$ and the sum become an integral. Think carefully about the integration limits for this variable; they will depend on the boundaries of the wire and on the coordinate system you have chosen to use. Carry out the integration and simplify the results as much as possible.

ASSESS Check that your result is consistent with any limits for which you know what the field should be.

EXAMPLE 32.3 **The magnetic field of a long, straight wire**

A long, straight wire carries current I in the positive x-direction. Find the magnetic field at a point that is distance d from the wire.

MODEL Because the wire is "long," let's model it as being infinitely long.

VISUALIZE **FIGURE 32.14** illustrates the steps in the problem-solving strategy. We've chosen a coordinate system with point P on the y-axis. We've then divided the wire into small segments, labeled with index i, each containing a small amount ΔQ of *moving charge*. Unit vector $\hat{r}$ and angle θ_i are shown for segment i. You should use the right-hand rule to convince yourself that $\vec{B}_i$ points *out of the page*, in the positive z-direction. This is the direction no matter where segment i happens to be along the x-axis. Consequently, B_x (the component of $\vec{B}$ parallel to the wire) and B_y (the component of $\vec{B}$ straight away from the wire) are zero. The only component of $\vec{B}$ we need to evaluate is B_z, the component tangent to a circle around the wire.

FIGURE 32.14 Calculating the magnetic field of a long, straight wire carrying current I.

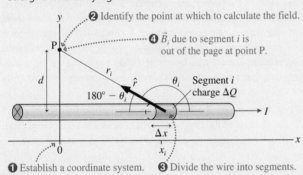

❷ Identify the point at which to calculate the field.

❹ $\vec{B}_i$ due to segment i is out of the page at point P.

Segment i charge ΔQ

❶ Establish a coordinate system.   ❸ Divide the wire into segments.

SOLVE We can use the Biot-Savart law to find the field $(B_i)_z$ of segment i. The cross product $\Delta \vec{s}_i \times \hat{r}$ has magnitude $(\Delta x)(1)\sin\theta_i$, hence

$$(B_i)_z = \frac{\mu_0}{4\pi} \frac{I\,\Delta x \sin\theta_i}{r_i^2} = \frac{\mu_0}{4\pi} \frac{I\sin\theta_i}{r_i^2}\Delta x = \frac{\mu_0}{4\pi} \frac{I\sin\theta_i}{x_i^2 + d^2}\Delta x$$

where we wrote the distance r_i in terms of x_i and d. We also need to express θ_i in terms of x_i and d. Because $\sin(180° - \theta) = \sin\theta$, this is

$$\sin\theta_i = \sin(180° - \theta_i) = \frac{d}{r_i} = \frac{d}{\sqrt{x_i^2 + d^2}}$$

With this expression for $\sin\theta_i$, the magnetic field of segment i is

$$(B_i)_z = \frac{\mu_0}{4\pi} \frac{Id}{(x_i^2 + d^2)^{3/2}}\Delta x$$

Now we're ready to sum the magnetic fields of all the segments. The superposition is a vector sum, but in this case only the z-components are nonzero. Summing all the $(B_i)_z$ gives

$$B_{\text{wire}} = \frac{\mu_0 Id}{4\pi} \sum_i \frac{\Delta x}{(x_i^2 + d^2)^{3/2}} \rightarrow \frac{\mu_0 Id}{4\pi} \int_{-\infty}^{\infty} \frac{dx}{(x^2 + d^2)^{3/2}}$$

Only at the very last step did we convert the sum to an integral. Then our model of the wire as being infinitely long sets the integration limits at $\pm\infty$. This is a standard integral that can be found in Appendix A or with integration software. Evaluation gives

$$B_{\text{wire}} = \frac{\mu_0 Id}{4\pi} \frac{x}{d^2(x^2 + d^2)^{1/2}} \Big|_{-\infty}^{\infty} = \frac{\mu_0}{2\pi} \frac{I}{d}$$

This is the magnitude of the field. The field direction is determined by using the right-hand rule. We can combine these two pieces of information to write

$$\vec{B}_{\text{wire}} = \left(\frac{\mu_0}{2\pi} \frac{I}{d}, \text{ tangent to a circle around the wire} \right)$$
$$\text{in the right-hand direction}$$

ASSESS **FIGURE 32.15** shows the magnetic field of a current-carrying wire. Compare this to Figure 32.2 and convince yourself that the direction shown agrees with the right-hand rule.

FIGURE 32.15 The magnetic field of a long, straight wire carrying current I.

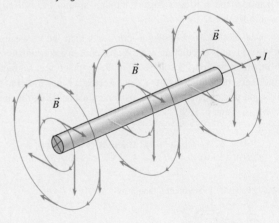

NOTE ▶ The difficulty magnetic field calculations present is not doing the integration itself, which is the last step, but setting up the calculation and knowing *what* to integrate. The purpose of the problem-solving strategy is to guide you through the process of setting up the integral. ◀

EXAMPLE 32.4 **The magnetic field strength near a heater wire**

A 1.0-m-long, 1.0-mm-diameter nichrome heater wire is connected to a 12 V battery. What is the magnetic field strength 1.0 cm away from the wire?

MODEL 1 cm is much less than the 1 m length of the wire, so model the wire as infinitely long.

SOLVE The current through the wire is $I = \Delta V_{\text{bat}}/R$, where the wire's resistance R is

$$R = \frac{\rho L}{A} = \frac{\rho L}{\pi r^2} = 1.91\ \Omega$$

Continued

The nichrome resistivity $\rho = 1.50 \times 10^{-6}\ \Omega\,\text{m}$ was taken from Table 30.2. Thus the current is $I = (12\ \text{V})/(1.91\ \Omega) = 6.28\ \text{A}$. The magnetic field strength at distance $d = 1.0\ \text{cm} = 0.010\ \text{m}$ from the wire is

$$B_{\text{wire}} = \frac{\mu_0\ I}{2\pi\ d} = (2.0 \times 10^{-7}\ \text{T m/A})\frac{6.28\ \text{A}}{0.010\ \text{m}}$$
$$= 1.3 \times 10^{-4}\ \text{T}$$

ASSESS The magnetic field of the wire is slightly more than twice the strength of the earth's magnetic field.

Motors, loudspeakers, metal detectors, and many other devices generate magnetic fields with *coils* of wire. The simplest coil is a single-turn circular loop of wire. A circular loop of wire with a circulating current is called a **current loop**.

EXAMPLE 32.5 The magnetic field of a current loop

FIGURE 32.16a shows a current loop, a circular loop of wire with radius R that carries current I. Find the magnetic field of the current loop at distance z on the axis of the loop.

FIGURE 32.16 A current loop.

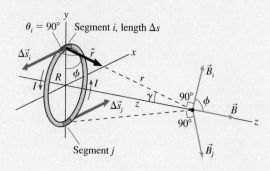

(a) A practical current loop   (b) An ideal current loop

MODEL Real coils need wires to bring the current in and out, but we'll model the coil as a current moving around the full circle shown in FIGURE 32.16b.

VISUALIZE FIGURE 32.17 shows a loop for which we've assumed that the current is circulating ccw. We've chosen a coordinate system in which the loop lies at $z = 0$ in the xy-plane. Let segment i be the segment at the top of the loop. Vector $\Delta\vec{s}_i$ is parallel to the x-axis and unit vector $\hat{r}$ is in the yz-plane, thus angle θ_i, the angle between $\Delta\vec{s}_i$ and $\hat{r}$, is $90°$.

FIGURE 32.17 Calculating the magnetic field of a current loop.

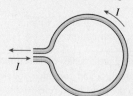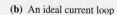

The direction of $\vec{B}_i$, the magnetic field due to the current in segment i, is given by the cross product $\Delta\vec{s}_i \times \hat{r}$. $\vec{B}_i$ must be perpendicular to $\Delta\vec{s}_i$ and perpendicular to $\hat{r}$. You should convince yourself that $\vec{B}_i$ in Figure 32.17 points in the correct direction. Notice that the y-component of $\vec{B}_i$ is canceled by the y-component of magnetic field $\vec{B}_j$ due to the current segment at the bottom of the loop, $180°$ away. In fact, *every* current segment on the loop can be paired with a segment $180°$ away, on the opposite side of the loop, such that the x- and y-components of $\vec{B}$ cancel and the components of $\vec{B}$ parallel to the z-axis add. In other words, the symmetry of the loop requires the on-axis magnetic field to point along the z-axis. Knowing that we need to sum only the z-components will simplify our calculation.

SOLVE We can use the Biot-Savart law to find the z-component $(B_i)_z = B_i \cos\phi$ of the magnetic field of segment i. The cross product $\Delta\vec{s}_i \times \hat{r}$ has magnitude $(\Delta s)(1)\sin 90° = \Delta s$, thus

$$(B_i)_z = \frac{\mu_0}{4\pi}\frac{I\,\Delta s}{r^2}\cos\phi = \frac{\mu_0 I \cos\phi}{4\pi(z^2 + R^2)}\Delta s$$

where we used $r = (z^2 + R^2)^{1/2}$. You can see, because $\phi + \gamma = 90°$, that angle ϕ is also the angle between $\hat{r}$ and the radius of the loop. Hence $\cos\phi = R/r$, and

$$(B_i)_z = \frac{\mu_0 IR}{4\pi(z^2 + R^2)^{3/2}}\Delta s$$

The final step is to sum the magnetic fields due to all the segments:

$$B_{\text{loop}} = \sum_i (B_i)_z = \frac{\mu_0 IR}{4\pi(z^2 + R^2)^{3/2}}\sum_i \Delta s$$

In this case, unlike the straight wire, none of the terms multiplying Δs depends on the position of segment i, so all these terms can be factored out of the summation. We're left with a summation that adds up the lengths of all the small segments. But this is just the total length of the wire, which is the circumference $2\pi R$. Thus the on-axis magnetic field of a current loop is

$$B_{\text{loop}} = \frac{\mu_0 IR}{4\pi(z^2 + R^2)^{3/2}}2\pi R = \frac{\mu_0}{2}\frac{IR^2}{(z^2 + R^2)^{3/2}}$$

In practice, current often passes through a *coil* consisting of N *turns* of wire. If the turns are all very close together, so that the magnetic field of each is essentially the same, then the magnetic field of a coil is N times the magnetic field of a current loop. The magnetic field at the center ($z = 0$) of an N-turn coil is

$$B_{\text{coil center}} = \frac{\mu_0}{2}\frac{NI}{R} \qquad (32.7)$$

EXAMPLE 32.6 **Matching the earth's magnetic field**

What current is needed in a 5-turn, 10-cm-diameter coil to cancel the earth's magnetic field at the center of the coil?

MODEL One way to create a zero-field region of space is to generate a magnetic field equal to the earth's field but pointing in the opposite direction. The vector sum of the two fields is zero.

VISUALIZE FIGURE 32.18 shows a five-turn coil of wire. The magnetic field is five times that of a single current loop.

SOLVE The earth's magnetic field, from Table 32.1, is 5×10^{-5} T. We can use Equation 32.7 to find that the current needed to generate a 5×10^{-5} T field is

$$I = \frac{2RB}{\mu_0 N} = \frac{2(0.050 \text{ m})(5.0 \times 10^{-5} \text{ T})}{5(4\pi \times 10^{-7} \text{ T m/A})} = 0.80 \text{ A}$$

FIGURE 32.18 A coil of wire.

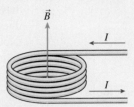

ASSESS A 0.80 A current is easily produced. Although there are better ways to cancel the earth's field than using a simple coil, this illustrates the idea.

32.5 Magnetic Dipoles

We were able to calculate the on-axis magnetic field of a current loop, but determining the field at off-axis points requires either numerical integrations or an experimental mapping of the field. FIGURE 32.19 shows the full magnetic field of a current loop. This is a field with *rotational symmetry,* so to picture the full three-dimensional field, imagine FIGURE 32.19a rotated about the axis of the loop. FIGURE 32.19b shows the magnetic field in the plane of the loop as seen from the right. There is a clear sense, seen in the photo of FIGURE 32.19c, that the magnetic field leaves the loop on one side, "flows" around the outside, then returns to the loop.

FIGURE 32.19 The magnetic field of a current loop.

(a) Cross section through the current loop   **(b)** The current loop seen from the right   **(c)** A photo of iron filings

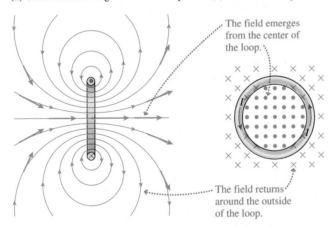
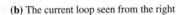
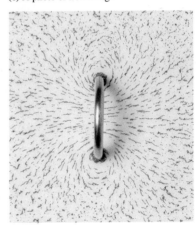

The field emerges from the center of the loop.

The field returns around the outside of the loop.

There are two versions of the right-hand rule that you can use to determine which way a loop's field points. Try these in Figure 32.19. Being able to quickly ascertain the field direction of a current loop is an important skill.

TACTICS BOX 32.2 **Finding the magnetic field direction of a current loop**

Use either of the following methods to find the magnetic field direction:

❶ Point your right thumb in the direction of the current at any point on the loop and let your fingers curl through the center of the loop. Your fingers are then pointing in the direction in which $\vec{B}$ leaves the loop.

❷ Curl the fingers of your right hand around the loop in the direction of the current. Your thumb is then pointing in the direction in which $\vec{B}$ leaves the loop.

Exercises 18–20

A Current Loop Is a Magnetic Dipole

A current loop has two distinct sides. Bar magnets and flat refrigerator magnets also have two distinct sides or ends, so you might wonder if current loops are related to these permanent magnets. Consider the following experiments with a current loop. Notice that we're using a simplified picture that shows the magnetic field only in the plane of the loop.

Investigating current loops

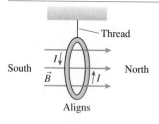

South   North

Aligns

A current loop hung by a thread aligns itself with the magnetic field pointing north.

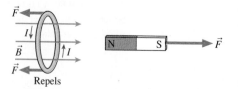

Repels

The north pole of a permanent magnet repels the side of a current loop from which the magnetic field is emerging.

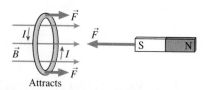

Attracts

The south pole of a permanent magnet attracts the side of a current loop from which the magnetic field is emerging.

These investigations show that **a current loop is a magnet,** just like a permanent magnet. A magnet created by a current in a coil of wire is called an **electromagnet.** An electromagnet picks up small pieces of iron, influences a compass needle, and acts in every way like a permanent magnet.

In fact, FIGURE 32.20 shows that a flat permanent magnet and a current loop generate the same magnetic field—the field of a magnetic dipole. For both, **you can identify the north pole as the face or end** *from which* **the magnetic field emerges.** The magnetic fields of both point *into* the south pole.

FIGURE 32.20 A current loop has magnetic poles and generates the same magnetic field as a flat permanent magnet.

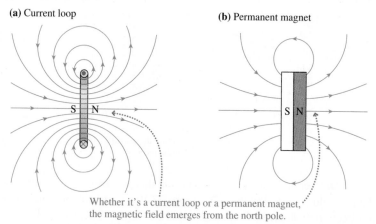

(a) Current loop   **(b)** Permanent magnet

Whether it's a current loop or a permanent magnet, the magnetic field emerges from the north pole.

One of the goals of this chapter is to show that magnetic forces exerted by currents and magnetic forces exerted by permanent magnets are just two different aspects of a single magnetism. We've now found a strong connection between permanent magnets and current loops, and this connection will turn out to be a big piece of the puzzle.

The Magnetic Dipole Moment

The expression for the electric field of an electric dipole was considerably simplified when we considered the field at distances significantly larger than the size of the charge separation s. The on-axis field of an electric dipole when $z \gg s$ is

$$\vec{E}_{\text{dipole}} = \frac{1}{4\pi\epsilon_0}\frac{2\vec{p}}{z^3}$$

where the electric dipole moment $\vec{p} = (qs$, from negative to positive charge).

The on-axis magnetic field of a current loop is

$$B_{\text{loop}} = \frac{\mu_0}{2}\frac{IR^2}{(z^2+R^2)^{3/2}}$$

If z is much larger than the diameter of the current loop, $z \gg R$, we can make the approximation $(z^2+R^2)^{3/2} \to z^3$. Then the loop's field is

$$B_{\text{loop}} \approx \frac{\mu_0}{2}\frac{IR^2}{z^3} = \frac{\mu_0}{4\pi}\frac{2(\pi R^2)I}{z^3} = \frac{\mu_0}{4\pi}\frac{2AI}{z^3} \qquad (32.8)$$

where $A = \pi R^2$ is the area of the loop.

A more advanced treatment of current loops shows that, if z is much larger than the size of the loop, Equation 32.8 is the on-axis magnetic field of a current loop of *any* shape, not just a circular loop. The shape of the loop affects the nearby field, but the distant field depends only on the current I and the area A enclosed within the loop. With this in mind, let's define the **magnetic dipole moment** $\vec{\mu}$ of a current loop enclosing area A to be

$$\vec{\mu} = (AI, \text{ from the south pole to the north pole})$$

The SI units of the magnetic dipole moment are $A\,m^2$.

NOTE ▶ Don't confuse the magnetic dipole moment $\vec{\mu}$ with the constant μ_0 in the Biot-Savart law. ◀

The magnetic dipole moment, like the electric dipole moment, is a vector. It has the same direction as the on-axis magnetic field. Thus the right-hand rule for determining the direction of $\vec{B}$ also shows the direction of $\vec{\mu}$. FIGURE 32.21 shows the magnetic dipole moment of a circular current loop.

Because the on-axis magnetic field of a current loop points in the same direction as $\vec{\mu}$, we can combine Equation 32.8 and the definition of $\vec{\mu}$ to write the on-axis field of a magnetic dipole as

$$\vec{B}_{\text{dipole}} = \frac{\mu_0}{4\pi}\frac{2\vec{\mu}}{z^3} \quad \text{(on the axis of a magnetic dipole)} \qquad (32.9)$$

If you compare $\vec{B}_{\text{dipole}}$ to $\vec{E}_{\text{dipole}}$, you can see that the magnetic field of a magnetic dipole has the same basic shape as the electric field of an electric dipole.

A permanent magnet also has a magnetic dipole moment and its on-axis magnetic field is given by Equation 32.9 when z is much larger than the size of the magnet. Equation 32.9 and laboratory measurements of the on-axis magnetic field can be used to determine a permanent magnet's dipole moment.

FIGURE 32.21 The magnetic dipole moment of a circular current loop.

The magnetic dipole moment is perpendicular to the loop, in the direction of the right-hand rule. The magnitude of $\vec{\mu}$ is AI.

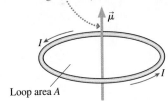

Loop area A

EXAMPLE 32.7 **Measuring current in a superconducting loop**

You'll learn in Chapter 33 that a current can be *induced* in a closed loop of wire. If the loop happens to be made of a superconducting material, with zero resistance, the induced current will—in principle—persist forever. The current cannot be measured with an ammeter because any real ammeter has resistance that will quickly stop the current. Instead, physicists measure the persistent current in a superconducting loop by measuring its magnetic field. In one experiment, the axial magnetic field of a 3.0-mm-diameter superconducting loop is measured at several distances from the center of the loop, yielding the following data:

Distance (cm)	Magnetic field (μT)
1.0	130
1.5	35
2.0	19
2.5	9
3.0	5

Continued

To measure such small magnetic fields requires careful shielding from the earth's magnetic field. Based on these data, what is the current in the loop?

MODEL The measurements are made far enough from the loop in comparison to its radius ($z \gg R$) that we can approximate the loop as a magnetic dipole rather than using the exact expression for the on-axis field of a current loop.

SOLVE The axial magnetic field strength of a dipole is

$$B = \frac{\mu_0}{4\pi} \frac{2\mu}{z^3} = \frac{\mu_0}{4\pi} \frac{2\pi R^2 I}{z^3} = \frac{\mu_0 R^2 I}{2} \frac{1}{z^3}$$

where we used $\mu = AI = \pi R^2 I$ for the magnetic dipole moment of a circular loop of radius R. If we graph B versus $1/z^3$ the result should be a straight line whose slope can be used to determine I.

The graph of **FIGURE 32.22** is a straight line passing through the origin, as expected. The best-fit line has slope $129\,\mu\text{T cm}^3$, where the rather unusual units are determined by rise over run. Converting to SI units, we find

$$\text{slope} = 129\,\mu\text{T cm}^3 \times \frac{10^{-6}\,\text{T}}{1\,\mu\text{T}} \times \left(\frac{1\,\text{m}}{100\,\text{cm}}\right)^3$$

$$= 1.29 \times 10^{-10}\,\text{T m}^3$$

With this, we can determine the current:

$$I = \frac{2}{\mu_0 R^2} \times \text{slope} = 91\,\text{A}$$

FIGURE 32.22 A graph of the data.

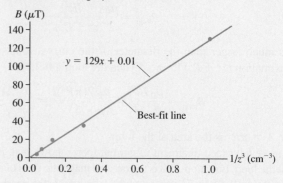

ASSESS This would be a very large current for ordinary wire. An important property of superconducting wires is their ability to carry current that would melt an ordinary wire.

STOP TO THINK 32.4 What is the current direction in this loop? And which side of the loop is the north pole?

a. Current cw; north pole on top
b. Current cw; north pole on bottom
c. Current ccw; north pole on top
d. Current ccw; north pole on bottom

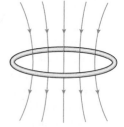

32.6 Ampère's Law and Solenoids

In principle, the Biot-Savart law can be used to calculate the magnetic field of any current distribution. In practice, the integrals are difficult to evaluate for anything other than very simple situations. We faced a similar situation for calculating electric fields, but we discovered an alternative method—Gauss's law—for calculating the electric field of charge distributions with a high degree of symmetry.

Likewise, there's an alternative method, called *Ampère's law,* for calculating the magnetic fields of current distributions with a high degree of symmetry. Ampère's law, like Gauss's law, doesn't work in all situations, but it is simple and elegant where it does. Whereas Gauss's law is written in terms of a surface integral, Ampère's law is based on the mathematical procedure called a *line integral.*

FIGURE 32.23 Integrating along a line from i to f.

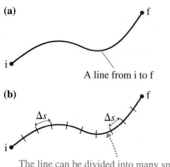

The line can be divided into many small segments. The sum of all the Δs's is the length l of the line.

Line Integrals

We've flirted with the idea of a line integral ever since introducing the concept of work in Chapter 11, but now we need to take a more serious look at what a line integral represents and how it is used. **FIGURE 32.23a** shows a curved line that goes from an initial point i to a final point f.

Suppose, as shown in FIGURE 32.23b, we divide the line into many small segments of length Δs. The first segment is Δs_1, the second is Δs_2, and so on. The sum of all the Δs's is the length l of the line between i and f. We can write this mathematically as

$$l = \sum_k \Delta s_k \rightarrow \int_i^f ds \qquad (32.10)$$

where, in the last step, we let $\Delta s \rightarrow ds$ and the sum become an integral.

This integral is called a **line integral.** All we've done is to subdivide a line into infinitely many infinitesimal pieces, then add them up. This is exactly what you do in calculus when you evaluate an integral such as $\int x \, dx$. In fact, an integration along the x-axis *is* a line integral, one that happens to be along a straight line. Figure 32.23 differs only in that the line is curved. The underlying idea in both cases is that an integral is just a fancy way of doing a sum.

The line integral of Equation 32.10 is not terribly exciting. FIGURE 32.24a makes things more interesting by allowing the line to pass through a magnetic field. FIGURE 32.24b again divides the line into small segments, but this time $\Delta \vec{s}_k$ is the displacement vector of segment k. The magnetic field at this point in space is $\vec{B}_k$.

Suppose we were to evaluate the dot product $\vec{B}_k \cdot \Delta \vec{s}_k$ at each segment, then add the values of $\vec{B}_k \cdot \Delta \vec{s}_k$ due to every segment. Doing so, and again letting the sum become an integral, we have

$$\sum_k \vec{B}_k \cdot \Delta \vec{s}_k \rightarrow \int_i^f \vec{B} \cdot d\vec{s} = \text{the line integral of } \vec{B} \text{ from i to f}$$

Once again, the integral is just a shorthand way to say: Divide the line into lots of little pieces, evaluate $\vec{B}_k \cdot \Delta \vec{s}_k$ for each piece, then add them up.

Although this process of evaluating the integral could be difficult, the only line integrals we'll need to deal with fall into two simple cases. If the magnetic field is *everywhere perpendicular* to the line, then $\vec{B} \cdot d\vec{s} = 0$ at every point along the line and the integral is zero. If the magnetic field is *everywhere tangent* to the line *and* has the same magnitude B at every point, then $\vec{B} \cdot d\vec{s} = B \, ds$ at every point and

$$\int_i^f \vec{B} \cdot d\vec{s} = \int_i^f B \, ds = B \int_i^f ds = Bl \qquad (32.11)$$

We used Equation 32.10 in the last step to integrate ds along the line.

Tactics Box 32.3 summarizes these two situations.

FIGURE 32.24 Integrating $\vec{B}$ along a line from i to f.

(a)

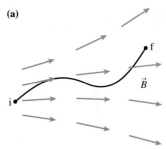

The line passes through a magnetic field.

(b)

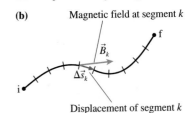

Magnetic field at segment k

Displacement of segment k

TACTICS
BOX 32.3 **Evaluating line integrals**   (MP)

❶ If $\vec{B}$ is everywhere perpendicular to a line, the line integral of $\vec{B}$ is

$$\int_i^f \vec{B} \cdot d\vec{s} = 0$$

❷ If $\vec{B}$ is everywhere tangent to a line of length l *and* has the same magnitude B at every point, then

$$\int_i^f \vec{B} \cdot d\vec{s} = Bl$$

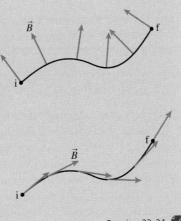

Exercises 23–24

Ampère's Law

FIGURE 32.25 Integrating the magnetic field around a wire.

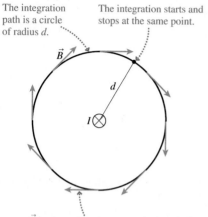

The integration path is a circle of radius d.

The integration starts and stops at the same point.

$\vec{B}$ is everywhere tangent to the integration path and has constant magnitude.

FIGURE 32.25 shows a wire carrying current I into the page and the magnetic field at distance d. The magnetic field of a current-carrying wire is everywhere tangent to a circle around the wire *and* has the same magnitude $\mu_0 I/2\pi d$ at all points on the circle. According to Tactics Box 32.3, these conditions allow us to easily evaluate the line integral of $\vec{B}$ along a circular path around the wire. Suppose we were to integrate the magnetic field *all the way around* the circle. That is, the initial point i of the integration path and the final point f will be the same point. This would be a line integral around a *closed curve,* which is denoted

$$\oint \vec{B} \cdot d\vec{s}$$

The little circle on the integral sign indicates that the integration is performed around a closed curve. The notation has changed, but the meaning has not.

Because $\vec{B}$ is tangent to the circle *and* of constant magnitude at every point on the circle, we can use Option 2 from Tactics Box 32.3 to write

$$\oint \vec{B} \cdot d\vec{s} = Bl = B(2\pi d) \tag{32.12}$$

where, in this case, the path length l is the circumference $2\pi d$ of the circle. The magnetic field strength of a current-carrying wire is $B = \mu_0 I/2\pi d$, thus

$$\oint \vec{B} \cdot d\vec{s} = \mu_0 I \tag{32.13}$$

The interesting result is that the line integral of $\vec{B}$ around the current-carrying wire is independent of the radius of the circle. Any circle, from one touching the wire to one far away, would give the same result. The integral depends only on the amount of current passing *through* the circle that we integrated around.

This is reminiscent of Gauss's law. In our investigation of Gauss's law, we started with the observation that the electric flux Φ_e through a sphere surrounding a point charge depends only on the amount of charge inside, not on the radius of the sphere. After examining several cases, we concluded that the shape of the surface wasn't relevant. The electric flux through *any* closed surface enclosing total charge Q_{in} turned out to be $\Phi_e = Q_{in}/\epsilon_0$.

Although we'll skip the details, the same type of reasoning that we used to prove Gauss's law shows that the result of Equation 32.13

- Is independent of the shape of the curve around the current.
- Is independent of where the current passes through the curve.
- Depends only on the total amount of current through the area enclosed by the integration path.

FIGURE 32.26 Using Ampère's law.

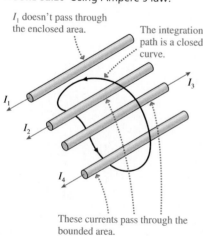

I_1 doesn't pass through the enclosed area.

The integration path is a closed curve.

These currents pass through the bounded area.

Thus whenever total current $I_{through}$ passes through an area bounded by a *closed curve,* the line integral of the magnetic field around the curve is

$$\oint \vec{B} \cdot d\vec{s} = \mu_0 I_{through} \tag{32.14}$$

This result for the magnetic field is known as **Ampère's law.**

To make practical use of Ampère's law, we need to determine which currents are positive and which are negative. The right-hand rule is once again the proper tool. If you curl your right fingers around the closed path in the direction in which you are going to integrate, then any current passing though the bounded area in the direction of your thumb is a positive current. Any current in the opposite direction is a negative current. In FIGURE 32.26, for example, currents I_2 and I_4 are positive, I_3 is negative. Thus $I_{through} = I_2 - I_3 + I_4$.

NOTE ▶ The integration path of Ampère's law is a mathematical curve through space. It does not have to match a physical surface or boundary, although it could if we want it to. ◀

In one sense, Ampère's law doesn't tell us anything new. After all, we derived Ampère's law from the Biot-Savart law for the magnetic field of a current. But in another sense, Ampère's law is more important than the Biot-Savart law because it states a very general property about magnetic fields. We will use Ampère's law to find the magnetic fields of some important current distributions that have a high degree of symmetry.

EXAMPLE 32.8 **The magnetic field inside a current-carrying wire**

A wire of radius R carries current I. Find the magnetic field inside the wire at distance $r < R$ from the axis.

MODEL Assume the current density is uniform over the cross section of the wire.

VISUALIZE **FIGURE 32.27** shows a cross section through the wire. The wire has cylindrical symmetry, with all the charges moving parallel to the wire, so the magnetic field *must* be tangent to circles that are concentric with the wire. We don't know how the strength of the magnetic field depends on the distance from the center—that's what we're going to find—but the symmetry of the situation dictates the *shape* of the magnetic field.

FIGURE 32.27 Using Ampère's law inside a current-carrying wire.

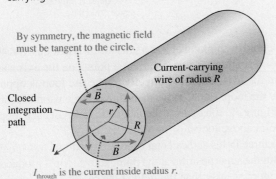

By symmetry, the magnetic field must be tangent to the circle.

Current-carrying wire of radius R

Closed integration path

$\vec{B}$

r

R

I

$\vec{B}$

I_{through} is the current inside radius r.

SOLVE To find the field strength at radius r, we draw a circle of radius r. The amount of current passing through this circle is

$$I_{\text{through}} = JA_{\text{circle}} = \pi r^2 J$$

where J is the current density. Our assumption of a uniform current density allows us to use the full current I passing through a wire of radius R to find that

$$J = \frac{I}{A} = \frac{I}{\pi R^2}$$

Thus the current through the circle of radius r is

$$I_{\text{through}} = \frac{r^2}{R^2}I$$

Let's integrate $\vec{B}$ around the circumference of this circle. According to Ampère's law,

$$\oint \vec{B} \cdot d\vec{s} = \mu_0 I_{\text{through}} = \frac{\mu_0 r^2}{R^2}I$$

We know from the symmetry of the wire that $\vec{B}$ is everywhere tangent to the circle *and* has the same magnitude at all points on the circle. Consequently, the line integral of $\vec{B}$ around the circle can be evaluated using Option 2 of Tactics Box 32.3:

$$\oint \vec{B} \cdot d\vec{s} = Bl = 2\pi r B$$

where $l = 2\pi r$ is the path length. If we substitute this expression into Ampère's law, we find that

$$2\pi r B = \frac{\mu_0 r^2}{R^2}I$$

Solving for B, we find that the magnetic field strength at radius r *inside* a current-carrying wire is

$$B = \frac{\mu_0 I}{2\pi R^2}r$$

ASSESS The magnetic field strength increases linearly with distance from the center of the wire until, at the surface of the wire, $B = \mu_0 I/2\pi R$ matches our earlier solution for the magnetic field outside a current-carrying wire. This agreement at $r = R$ gives us confidence in our result. The magnetic field strength both inside and outside the wire is shown graphically in **FIGURE 32.28**.

FIGURE 32.28 Graphical representation of the magnetic field of a current-carrying wire.

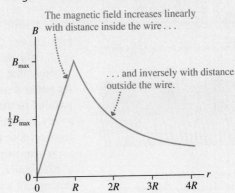

The magnetic field increases linearly with distance inside the wire . . .

. . . and inversely with distance outside the wire.

B

B_{max}

$\frac{1}{2}B_{\text{max}}$

0

0 R $2R$ $3R$ $4R$ r

The Magnetic Field of a Solenoid

FIGURE 32.29 A solenoid.

In our study of electricity, we made extensive use of the idea of a uniform electric field: a field that is the same at every point in space. We found that two closely spaced, parallel charged plates generate a uniform electric field between them, and this was one reason we focused so much attention on the parallel-plate capacitor.

Similarly, there are many applications of magnetism for which we would like to generate a **uniform magnetic field,** a field having the same magnitude and the same direction at every point within some region of space. None of the sources we have looked at thus far produces a uniform magnetic field.

In practice, a uniform magnetic field is generated with a **solenoid.** A solenoid, shown in FIGURE 32.29, is a helical coil of wire with the same current I passing through each loop in the coil. Solenoids may have hundreds or thousands of coils, often called *turns,* sometimes wrapped in several layers.

FIGURE 32.30 Using superposition to find the magnetic field of a stack of current loops.

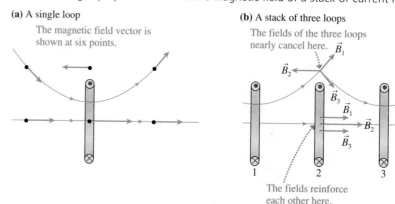

(a) A single loop

The magnetic field vector is shown at six points.

(b) A stack of three loops

The fields of the three loops nearly cancel here. $\vec{B}_1$

$\vec{B}_2$

$\vec{B}_3$

$\vec{B}_1$

$\vec{B}_2$

$\vec{B}_3$

1   2   3

The fields reinforce each other here.

We can understand a solenoid by thinking of it as a stack of current loops. FIGURE 32.30a shows the magnetic field of a single current loop at three points on the axis and three points equally distant from the axis. The field directly above the loop is opposite in direction to the field inside the loop. FIGURE 32.30b then shows three parallel loops. We can use information from Figure 32.30b to draw the magnetic fields of each loop at the center of loop 2 and at a point above loop 2.

The superposition of the three fields at the center of loop 2 produces a *stronger* field than that of loop 2 alone. But the superposition at the point above loop 2 produces a net magnetic field that is very much weaker than the field at the center of the loop. We've used only three current loops to illustrate the idea, but these tendencies are reinforced by including more loops. With many current loops along the same axis, **the field in the center is strong and roughly parallel to the axis, whereas the field outside the loops is very close to zero.**

FIGURE 32.31a is a photo of the magnetic field of a short solenoid. You can see that the magnetic field inside the coils is nearly uniform (i.e., the field lines are nearly parallel) and the field outside is much weaker. Our goal of producing a uniform magnetic field can be achieved by increasing the number of coils until we have an *ideal solenoid* that is infinitely long and in which the coils are as close together as possible. As FIGURE 32.31b shows, **the magnetic field inside an ideal solenoid is uniform and parallel to the axis; the magnetic field outside is zero.** No real solenoid is ideal, but a very uniform magnetic field can be produced near the center of a tightly wound solenoid whose length is much larger than its diameter.

We can use Ampère's law to calculate the field of an ideal solenoid. FIGURE 32.32 shows a cross section through an infinitely long solenoid. The integration path that we'll use is a rectangle of width l, enclosing N turns of the solenoid coil. Because this is a mathematical curve, not a physical boundary, there's no difficulty with letting it protrude through the

FIGURE 32.31 The magnetic field of a solenoid.

(a) A short solenoid

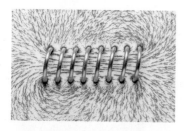

(b)

$\vec{B} = \vec{0}$

$\vec{B}$

$\vec{B} = \vec{0}$

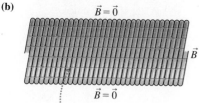

The magnetic field is uniform inside this section of an ideal, infinitely long solenoid. The magnetic field outside the solenoid is zero.

wall of the solenoid wherever we wish. The solenoid's magnetic field direction, given by the right-hand rule, is left to right, so we'll integrate around this path in the ccw direction.

Each of the N wires enclosed by the integration path carries current I, so the total current passing through the rectangle is $I_{\text{through}} = NI$. Ampère's law is thus

$$\oint \vec{B} \cdot d\vec{s} = \mu_0 I_{\text{through}} = \mu_0 NI \tag{32.15}$$

The line integral around this path is the sum of the line integrals along each side. Along the bottom, where $\vec{B}$ is parallel to $d\vec{s}$ and of constant value B, the integral is simply Bl. The integral along the top is zero because the magnetic field outside an ideal solenoid is zero.

The left and right sides sample the magnetic field both inside and outside the solenoid. The magnetic field outside is zero, and the interior magnetic field is everywhere *perpendicular* to the line of integration. Consequently, as we recognized in Option 1 of Tactics Box 32.3, the line integral is zero.

Only the integral along the bottom path is nonzero, leading to

$$\oint \vec{B} \cdot d\vec{s} = Bl = \mu_0 NI$$

Thus the strength of the uniform magnetic field inside a solenoid is

$$B_{\text{solenoid}} = \frac{\mu_0 NI}{l} = \mu_0 nI \tag{32.16}$$

where $n = N/l$ is the number of turns per unit length. Measurements that need a uniform magnetic field are often conducted inside a solenoid, which can be built quite large.

FIGURE 32.32 A closed path inside and outside an ideal solenoid.

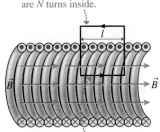

EXAMPLE 32.9 Generating an MRI magnetic field

A 1.0-m-long MRI solenoid generates a 1.2 T magnetic field. To produce such a large field, the solenoid is wrapped with superconducting wire that can carry a 100 A current. How many turns of wire does the solenoid need?

MODEL Assume that the solenoid is ideal.

SOLVE Generating a magnetic field with a solenoid is a trade-off between current and turns of wire. A larger current requires fewer turns, but the resistance of ordinary wires causes them to overheat if the current is too large. For a superconducting wire that can

carry 100 A with no resistance, we can use Equation 32.16 to find the required number of turns:

$$N = \frac{lB}{\mu_0 I} = \frac{(1.0\text{ m})(1.2\text{ T})}{(4\pi \times 10^{-7}\text{ T m/A})(100\text{ A})} = 9500\text{ turns}$$

ASSESS The solenoid coil requires a large number of turns, but that's not surprising for generating a very strong field. If the wires are 1 mm in diameter, there would be 10 layers with approximately 1000 turns per layer.

The magnetic field of a finite-length solenoid is approximately uniform *inside* the solenoid and weak, but not zero, outside. As FIGURE 32.33 shows, the magnetic field outside the solenoid looks like that of a bar magnet. Thus a solenoid is an electromagnet, and you can use the right-hand rule to identify the north-pole end. A solenoid with many turns and a large current can be a very powerful magnet.

FIGURE 32.33 The magnetic fields of a finite-length solenoid and of a bar magnet.

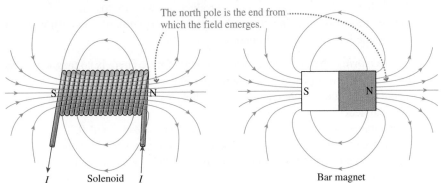

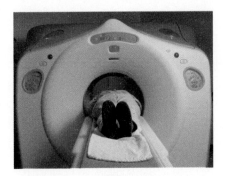

This patient is undergoing magnetic resonance imaging (MRI). The large cylinder surrounding the patient contains a solenoid that is wound with superconducting wire to generate a strong uniform magnetic field.

32.7 The Magnetic Force on a Moving Charge

FIGURE 32.34 Ampère's experiment to observe the forces between parallel current-carrying wires.

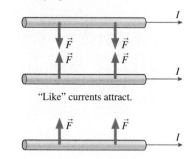

"Like" currents attract.

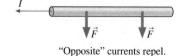

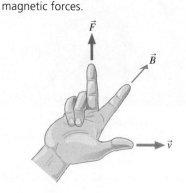

"Opposite" currents repel.

It's time to switch our attention from how magnetic fields are generated to how magnetic fields exert forces and torques. Oersted discovered that a current passing through a wire causes a magnetic torque to be exerted on a nearby compass needle. Upon hearing of Oersted's discovery, André-Marie Ampère, for whom the SI unit of current is named, reasoned that the current was acting like a magnet and, if this were true, that two current-carrying wires should exert magnetic forces on each other.

To find out, Ampère set up two parallel wires that could carry large currents either in the same direction or in opposite (or "antiparallel") directions. **FIGURE 32.34** shows the outcome of his experiment. Notice that, for currents, "likes" attract and "opposites" repel. This is the opposite of what would have happened had the wires been charged and thus exerting electric forces on each other. Ampère's experiment showed that **a magnetic field exerts a force on a current.**

Magnetic Force

Because a current consists of moving charges, Ampère's experiment implied that a magnetic field exerts a force on a *moving* charge. This is true, although the exact form of the force law was not discovered until later in the 19th century. The magnetic force turns out to depend not only on the charge and the charge's velocity, but also on how the velocity vector is oriented relative to the magnetic field. **FIGURE 32.35** shows the outcome of three experiments to observe the magnetic force.

FIGURE 32.35 The relationship among $\vec{v}$, $\vec{B}$, and $\vec{F}$.

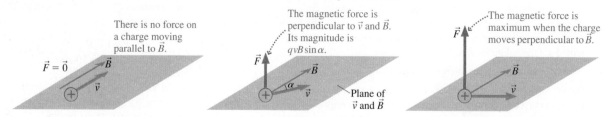

If you compare the experiment in the middle of Figure 32.35 to Figure 32.10, you'll see that the relationship among $\vec{v}$, $\vec{B}$, and $\vec{F}$ is exactly the same as the geometric relationship among $\vec{C}$, $\vec{D}$, and $\vec{C} \times \vec{D}$. The magnetic force on a charge q as it moves through a magnetic field $\vec{B}$ with velocity $\vec{v}$ can be written

$$\vec{F}_{\text{on } q} = q\vec{v} \times \vec{B} = (qvB \sin \alpha, \text{ direction of right-hand rule}) \quad (32.17)$$

FIGURE 32.36 The right-hand rule for magnetic forces.

where α is the angle between $\vec{v}$ and $\vec{B}$.

The right-hand rule is that of the cross product, shown in **FIGURE 32.36**. Notice that **the magnetic force on a moving charged particle is perpendicular to both $\vec{v}$ and $\vec{B}$.**

The magnetic force has several important properties:

1. Only a *moving* charge experiences a magnetic force. There is no magnetic force on a charge at rest ($\vec{v} = \vec{0}$) in a magnetic field.
2. There is no force on a charge moving parallel ($\alpha = 0°$) or antiparallel ($\alpha = 180°$) to a magnetic field.
3. When there is a force, the force is perpendicular to *both* $\vec{v}$ and $\vec{B}$.
4. The force on a negative charge is in the direction *opposite* to $\vec{v} \times \vec{B}$.
5. For a charge moving perpendicular to $\vec{B}$ ($\alpha = 90°$), the magnitude of the magnetic force is $F = |q|vB$.

FIGURE 32.37 shows the relationship among $\vec{v}$, $\vec{B}$, and $\vec{F}$ for four moving charges. (The *source* of the magnetic field isn't shown, only the field itself.) You can see the inherent three-dimensionality of magnetism, with the force perpendicular to both $\vec{v}$ and $\vec{B}$. The magnetic force is very different from the electric force, which is parallel to the electric field.

FIGURE 32.37 Magnetic forces on moving charges.

EXAMPLE 32.10 | The magnetic force on an electron

A long wire carries a 10 A current from left to right. An electron 1.0 cm above the wire is traveling to the right at a speed of 1.0×10^7 m/s. What are the magnitude and the direction of the magnetic force on the electron?

MODEL The magnetic field is that of a long, straight wire.

VISUALIZE FIGURE 32.38 shows the current and an electron moving to the right. The right-hand rule tells us that the wire's magnetic

FIGURE 32.38 An electron moving parallel to a current-carrying wire.

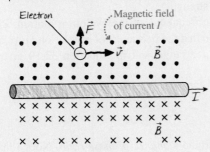

field above the wire is out of the page, so the electron is moving perpendicular to the field.

SOLVE The electron charge is negative, thus the direction of the force is opposite the direction of $\vec{v} \times \vec{B}$. The right-hand rule shows that $\vec{v} \times \vec{B}$ points down, toward the wire, so $\vec{F}$ points up, away from the wire. The magnitude of the force is $|q|vB = evB$. The field is that of a long, straight wire:

$$B = \frac{\mu_0 I}{2\pi d} = 2.0 \times 10^{-4}\,\text{T}$$

Thus the magnitude of the force on the electron is

$$F = evB = (1.60 \times 10^{-19}\,\text{C})(1.0 \times 10^7\,\text{m/s})(2.0 \times 10^{-4}\,\text{T})$$
$$= 3.2 \times 10^{-16}\,\text{N}$$

The force on the electron is $\vec{F} = (3.2 \times 10^{-16}\,\text{N, up})$.

ASSESS This force will cause the electron to curve away from the wire.

We can draw an interesting and important conclusion at this point. You have seen that the magnetic field is *created by* moving charges. Now you also see that magnetic forces are *exerted on* moving charges. Thus it appears that **magnetism is an interaction between moving charges.** Any two charges, whether moving or stationary, interact with each other through the electric field. In addition, two *moving* charges interact with each other through the magnetic field.

Cyclotron Motion

Many important applications of magnetism involve the motion of charged particles in a magnetic field. Older television picture tubes use magnetic fields to steer electrons through a vacuum from the electron gun to the screen. Microwave generators, which are used in applications ranging from ovens to radar, use a device called a *magnetron* in which electrons oscillate rapidly in a magnetic field.

You've just seen that there is no force on a charge that has velocity $\vec{v}$ parallel or antiparallel to a magnetic field. Consequently, **a magnetic field has no effect on a charge moving parallel or antiparallel to the field.** To understand the motion of charged particles in magnetic fields, we need to consider only motion *perpendicular* to the field.

FIGURE 32.39 shows a positive charge q moving with a velocity $\vec{v}$ in a plane that is perpendicular to a *uniform* magnetic field $\vec{B}$. According to the right-hand rule, the

FIGURE 32.39 Cyclotron motion of a charged particle moving in a uniform magnetic field.

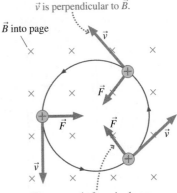

The magnetic force is always perpendicular to $\vec{v}$, causing the particle to move in a circle.

magnetic force on this particle is *perpendicular* to the velocity $\vec{v}$. A force that is always perpendicular to $\vec{v}$ changes the *direction* of motion, by deflecting the particle sideways, but it cannot change the particle's speed. Thus **a particle moving perpendicular to a uniform magnetic field undergoes uniform circular motion at constant speed.** This motion is called the **cyclotron motion** of a charged particle in a magnetic field.

NOTE ▶ A negative charge will orbit in the opposite direction from that shown in Figure 32.39 for a positive charge. ◀

You've seen many analogies to cyclotron motion earlier in this text. For a mass moving in a circle at the end of a string, the tension force is always perpendicular to $\vec{v}$. For a satellite moving in a circular orbit, the gravitational force is always perpendicular to $\vec{v}$. Now, for a charged particle moving in a magnetic field, it is the magnetic force of strength $F = qvB$ that points toward the center of the circle and causes the particle to have a centripetal acceleration.

Newton's second law for circular motion, which you learned in Chapter 8, is

$$F = qvB = ma_r = \frac{mv^2}{r} \qquad (32.18)$$

Thus the radius of the cyclotron orbit is

$$r_{cyc} = \frac{mv}{qB} \qquad (32.19)$$

The inverse dependence on B indicates that the size of the orbit can be decreased by increasing the magnetic field strength.

We can also determine the frequency of the cyclotron motion. Recall from your earlier study of circular motion that the frequency of revolution f is related to the speed and radius by $f = v/2\pi r$. A rearrangement of Equation 32.19 gives the **cyclotron frequency:**

$$f_{cyc} = \frac{qB}{2\pi m} \qquad (32.20)$$

where the ratio q/m is the particle's *charge-to-mass ratio.* Notice that the cyclotron frequency depends on the charge-to-mass ratio and the magnetic field strength but *not* on the charge's velocity.

Electrons undergoing circular motion in a magnetic field. You can see the electrons' path because they collide with a low-density gas that then emits light.

EXAMPLE 32.11 **The radius of cyclotron motion**

In FIGURE 32.40, an electron is accelerated from rest through a potential difference of 500 V, then injected into a uniform magnetic field. Once in the magnetic field, it completes half a revolution in 2.0 ns. What is the radius of its orbit?

FIGURE 32.40 An electron is accelerated, then injected into a magnetic field.

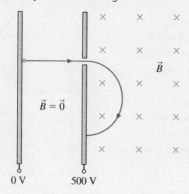

MODEL Energy is conserved as the electron is accelerated by the potential difference. The electron then undergoes cyclotron motion in the magnetic field, although it completes only half a revolution before hitting the back of the acceleration electrode.

SOLVE The electron accelerates from rest ($v_i = 0$ m/s) at $V_i = 0$ V to speed v_f at $V_f = 500$ V. We can use conservation of energy $K_f + qV_f = K_i + qV_i$ to find the speed v_f with which it enters the magnetic field:

$$\frac{1}{2}mv_f^2 + (-e)V_f = 0 + 0$$

$$v_f = \sqrt{\frac{2eV_f}{m}} = \sqrt{\frac{2(1.60 \times 10^{-19}\,\text{C})(500\,\text{V})}{9.11 \times 10^{-31}\,\text{kg}}}$$

$$= 1.33 \times 10^7\,\text{m/s}$$

The cyclotron radius in the magnetic field is $r_{cyc} = mv/eB$, but we first need to determine the field strength. Were it not for the electrode, the electron would undergo circular motion with period $T = 4.0$ ns. Hence the cyclotron frequency is $f = 1/T = 2.5 \times 10^8$ Hz. We can use the cyclotron frequency to determine that the magnetic field strength is

$$B = \frac{2\pi m f_{cyc}}{e} = \frac{2\pi(9.11 \times 10^{-31} \text{ kg})(2.50 \times 10^8 \text{ Hz})}{1.60 \times 10^{-19} \text{ C}}$$

$$= 8.94 \times 10^{-3} \text{ T}$$

Thus the radius of the electron's orbit is

$$r_{cyc} = \frac{mv}{qB} = 8.5 \times 10^{-3} \text{ m} = 8.5 \text{ mm}$$

FIGURE 32.41a shows a more general situation in which the charged particle's velocity $\vec{v}$ is neither parallel nor perpendicular to $\vec{B}$. The component of $\vec{v}$ parallel to $\vec{B}$ is not affected by the field, so the charged particle spirals around the magnetic field lines in a helical trajectory. The radius of the helix is determined by $\vec{v}_\perp$, the component of $\vec{v}$ perpendicular to $\vec{B}$.

FIGURE 32.41 In general, charged particles spiral along helical trajectories around the magnetic field lines. This motion is responsible for the earth's aurora.

(a) Charged particles spiral around the magnetic field lines.

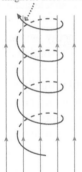

(b) The earth's magnetic field leads particles into the atmosphere near the poles, causing the aurora.

(c) The aurora

The motion of charged particles in a magnetic field is responsible for the earth's aurora. High-energy particles and radiation streaming out from the sun, called the *solar wind,* create ions and electrons as they strike molecules high in the atmosphere. Some of these charged particles become trapped in the earth's magnetic field, creating what is known as the *Van Allen radiation belt.*

As **FIGURE 32.41b** shows, the electrons spiral along the magnetic field lines until the field leads them into the atmosphere. The shape of the earth's magnetic field is such that most electrons enter the atmosphere near the magnetic poles. There they collide with oxygen and nitrogen atoms, exciting the atoms and causing them to emit auroral light seen in **FIGURE 32.41c**.

STOP TO THINK 32.5   An electron moves perpendicular to a magnetic field. What is the direction of $\vec{B}$?

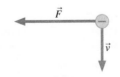

a. Left
d. Right
b. Up
e. Down
c. Into the page
f. Out of the page

The Cyclotron

Physicists studying the structure of the atomic nucleus and of elementary particles usually use a device called a *particle accelerator.* The first practical particle accelerator,

FIGURE 32.42 A cyclotron.

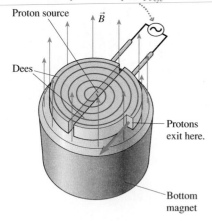

FIGURE 32.42 A cyclotron.

The potential ΔV oscillates at the cyclotron frequency f_{cyc}.

Proton source

$\vec{B}$

Dees

Protons exit here.

Bottom magnet

invented in the 1930s, was the **cyclotron.** Cyclotrons remain important for many applications of nuclear physics, such as the creation of radioisotopes for medicine.

A cyclotron, shown in FIGURE 32.42, consists of an evacuated chamber within a large, uniform magnetic field. Inside the chamber are two hollow conductors shaped like the letter D and hence called "dees." The dees are made of copper, which doesn't affect the magnetic field; are open along the straight sides; and are separated by a small gap. A charged particle, typically a proton, is injected into the magnetic field from a source near the center of the cyclotron, and it begins to move in and out of the dees in a circular cyclotron orbit.

The cyclotron operates by taking advantage of the fact that the cyclotron frequency f_{cyc} of a charged particle is independent of the particle's speed. An *oscillating* potential difference ΔV is connected across the dees and adjusted until its frequency is exactly the cyclotron frequency. There is almost no electric field inside the dees (you learned in Chapter 27 that the electric field inside a hollow conductor is zero), but a strong electric field points from the positive to the negative dee in the gap between them.

Suppose the proton emerges into the gap from the positive dee. The electric field in the gap *accelerates* the proton across the gap into the negative dee, and it gains kinetic energy $e\Delta V$. A half cycle later, when it next emerges into the gap, the potential of the dees (whose potential difference is oscillating at f_{cyc}) will have changed sign. The proton will *again* be emerging from the positive dee and will *again* accelerate across the gap and gain kinetic energy $e\Delta V$.

This pattern will continue orbit after orbit. The proton's kinetic energy increases by $2e\Delta V$ every orbit, so after N orbits its kinetic energy is $K = 2Ne\Delta V$ (assuming that its initial kinetic energy was near zero). The radius of its orbit increases as it speeds up; hence the proton follows the *spiral* path shown in Figure 32.42 until it finally reaches the outer edge of the dee. It is then directed out of the cyclotron and aimed at a target. Although ΔV is modest, usually a few hundred volts, the fact that the proton can undergo many thousands of orbits before reaching the outer edge allows it to acquire a very large kinetic energy.

The Hall Effect

A charged particle moving through a vacuum is deflected sideways, perpendicular to $\vec{v}$, by a magnetic field. In 1879, a graduate student named Edwin Hall showed that the same is true for the charges moving through a conductor as part of a current. This phenomenon—now called the **Hall effect**—is used to gain information about the charge carriers in a conductor. It is also the basis of a widely used technique for measuring magnetic field strengths.

FIGURE 32.43a shows a magnetic field perpendicular to a flat, current-carrying conductor. You learned in Chapter 30 that the charge carriers move through a conductor at the drift speed v_d. Their motion is perpendicular to $\vec{B}$, so each charge carrier experiences a magnetic force $F_B = ev_dB$ perpendicular to both $\vec{B}$ and the current I. However, for the first time we have a situation in which it *does* matter whether the charge carriers are positive or negative.

FIGURE 32.43 The charge carriers in a current are deflected to one surface of a conductor, creating the Hall voltage ΔV_H.

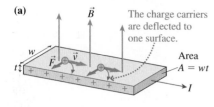

(a) $\vec{B}$

The charge carriers are deflected to one surface.

w

t

$\vec{F}$ $\vec{v}$

Area $A = wt$

I

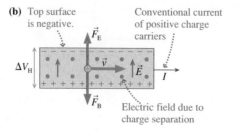

(b) Top surface is negative.

Conventional current of positive charge carriers

$\vec{F}_E$

ΔV_H

$\vec{v}$

$\vec{E}$

I

$\vec{F}_B$

Electric field due to charge separation

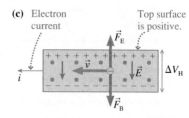

(c) Electron current

Top surface is positive.

$\vec{F}_E$

i

$\vec{v}$

$\vec{E}$

ΔV_H

$\vec{F}_B$

FIGURE 32.43b, with the field out of the page, shows that positive charge carriers moving in the direction of I are pushed toward the bottom surface of the conductor. This creates an excess positive charge on the bottom surface and leaves an excess negative charge on the top. FIGURE 32.43c, where the electrons in an electron current i move opposite the direction of I, shows that electrons would be pushed toward the bottom surface. (Be sure to use the right-hand rule and the sign of the electron charge to confirm the deflections shown in these figures.) Thus the sign of the excess charge on the bottom surface is the same as the sign of the charge carriers. Experimentally, the bottom surface is negative when the conductor is a metal, and this is one more piece of evidence that the charge carriers in metals are electrons.

Electrons are deflected toward the bottom surface once the current starts flowing, but the process can't continue indefinitely. As excess charge accumulates on the top and bottom surfaces, it acts like the charge on the plates of a capacitor, creating a potential difference ΔV between the two surfaces and an electric field $E = \Delta V/w$ *inside* the conductor of width w. This electric field increases until the upward electric force $\vec{F}_E$ on the charge carriers exactly balances the downward magnetic force $\vec{F}_B$. Once the forces are balanced, a steady state is reached in which the charge carriers move in the direction of the current and no additional charge is deflected to the surface.

The steady-state condition, in which $F_B = F_E$, is

$$F_B = ev_d B = F_E = eE = e\frac{\Delta V}{w} \tag{32.21}$$

Thus the steady-state potential difference between the two surfaces of the conductor, which is called the **Hall voltage** ΔV_H, is

$$\Delta V_H = wv_d B \tag{32.22}$$

You learned in Chapter 30 that the drift speed is related to the current density J by $J = nev_d$, where n is the charge-carrier density (charge carriers per m^3). Thus

$$v_d = \frac{J}{ne} = \frac{I/A}{ne} = \frac{I}{wtne} \tag{32.23}$$

where $A = wt$ is the cross-section area of the conductor. If we use this expression for v_d in Equation 32.22, we find that the Hall voltage is

$$\Delta V_H = \frac{IB}{tne} \tag{32.24}$$

The Hall voltage is very small for metals in laboratory-sized magnetic fields, typically in the microvolt range. Even so, measurements of the Hall voltage in a known magnetic field are used to determine the charge-carrier density n. Interestingly, the Hall voltage is larger for *poor* conductors that have smaller charge-carrier densities. A laboratory probe for measuring magnetic field strengths, called a *Hall probe*, measures ΔV_H for a poor conductor whose charge-carrier density is known. The magnetic field is then determined from Equation 32.24.

EXAMPLE 32.12 | **Measuring the magnetic field**

A Hall probe consists of a strip of the metal bismuth that is 0.15 mm thick and 5.0 mm wide. Bismuth is a poor conductor with charge-carrier density 1.35×10^{25} m^{-3}. The Hall voltage on the probe is 2.5 mV when the current through it is 1.5 A. What is the strength of the magnetic field, and what is the electric field strength inside the bismuth?

VISUALIZE The bismuth strip looks like Figure 32.43a. The thickness is $t = 1.5 \times 10^{-4}$ m and the width is $w = 5.0 \times 10^{-3}$ m.

SOLVE Equation 32.24 gives the Hall voltage. We can rearrange the equation to find that the magnetic field is

$$B = \frac{tne}{I}\Delta V_H$$

$$= \frac{(1.5 \times 10^{-4}\ \text{m})(1.35 \times 10^{25}\ \text{m}^{-3})(1.60 \times 10^{-19}\ \text{C})}{1.5\ \text{A}}0.0025\ \text{V}$$

$$= 0.54\ \text{T}$$

The electric field created inside the bismuth by the excess charge on the surface is

$$E = \frac{\Delta V_H}{w} = \frac{0.0025\ \text{V}}{5.0 \times 10^{-3}\ \text{m}} = 0.50\ \text{V/m}$$

ASSESS 0.54 T is a fairly typical strength for a laboratory magnet.

32.8 Magnetic Forces on Current-Carrying Wires

FIGURE 32.44 Magnetic force on a current-carrying wire.

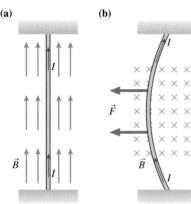

(a)   (b)

There's no force on a current parallel to a magnetic field.   A current perpendicular to the field experiences a force in the direction of the right-hand rule.

Ampère's observation of magnetic forces between current-carrying wires motivated us to look at the magnetic forces on moving charges. We're now ready to apply that knowledge to Ampère's experiment. As a first step, let us find the force exerted by a uniform magnetic field on a long, straight wire carrying current I through the field. As FIGURE 32.44a shows, there's *no force* on a current-carrying wire *parallel* to a magnetic field. This shouldn't be surprising; it follows from the fact that there is no force on a charged particle moving parallel to $\vec{B}$.

FIGURE 32.44b shows a wire *perpendicular* to the magnetic field. By the right-hand rule, each charge in the current has a force of magnitude qvB directed to the left. Consequently, the entire length of wire within the magnetic field experiences a force to the left, perpendicular to both the current direction and the field direction.

To find the magnitude of the force, we must relate the current I in the wire to the charge q moving through the wire. FIGURE 32.45 shows a segment of wire of length l carrying current I. The current I, by definition, is the amount of moving charge q in this segment of wire divided by the time Δt it takes the charge to flow through the segment: $I = q/\Delta t$. The time required is $\Delta t = l/v$, giving

$$q = I\Delta t = I\frac{l}{v}$$

FIGURE 32.45 Two ways to think of a current.

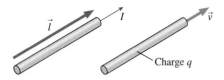

A current consists of charge carriers q moving with velocity $\vec{v}$.

Thus $Il = qv$. If we define vector $\vec{l}$ to have magnitude l and point in the direction of $\vec{v}$, the direction of current, then $I\vec{l} = q\vec{v}$. Substituting this for $q\vec{v}$ in the force equation $\vec{F} = q\vec{v} \times \vec{B}$, we find that the magnetic force on a current-carrying wire is

$$\vec{F}_{\text{wire}} = I\vec{l} \times \vec{B} = (IlB \sin\alpha, \text{ direction of right-hand rule}) \quad (32.25)$$

where α is the angle between $\vec{l}$ (the direction of the current) and $\vec{B}$. As an aside, you can see from Equation 32.25 that the magnetic field B must have units of N/A m. This is why we defined 1 T = 1 N/A m in Section 32.3.

NOTE ▶ The familiar right-hand rule applies to a current-carrying wire. Point your right thumb in the direction of the current (parallel to $\vec{l}$) and your index finger in the direction of $\vec{B}$. Your middle finger is then pointing in the direction of the force $\vec{F}$ on the wire. ◀

EXAMPLE 32.13 **Magnetic levitation**

The 0.10 T uniform magnetic field of FIGURE 32.46 is horizontal, parallel to the floor. A straight segment of 1.0-mm-diameter copper wire, also parallel to the floor, is perpendicular to the magnetic field. What current through the wire, and in which direction, will allow the wire to "float" in the magnetic field?

FIGURE 32.46 Magnetic levitation.

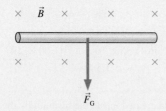

MODEL The wire will float in the magnetic field if the magnetic force on the wire points upward and has magnitude mg, allowing it to balance the downward gravitational force.

SOLVE We can use the right-hand rule to determine which current direction experiences an upward force. With $\vec{B}$ pointing away from us, the direction of the current needs to be from left to right. The forces will balance when

$$F = IlB = mg = \rho(\pi r^2 l)g$$

where $\rho = 8920$ kg/m³ is the density of copper. The length of the wire cancels, leading to

$$I = \frac{\rho\pi r^2 g}{B} = \frac{(8920 \text{ kg/m}^3)\pi(0.00050 \text{ m})^2 (9.80 \text{ m/s}^2)}{0.10 \text{ T}}$$
$$= 0.69 \text{ A}$$

A 0.69 A current from left to right will levitate the wire in the magnetic field.

ASSESS A 0.69 A current is quite reasonable, but this idea is useful only if we can get the current into and out of this segment of wire. In practice, we could do so with wires that come in from below the page. These input and output wires would be parallel to $\vec{B}$ and not experience a magnetic force. Although this example is very simple, it is the basis for applications such as magnetic levitation trains.

Force Between Two Parallel Wires

Now consider Ampère's experimental arrangement of two parallel wires of length l, distance d apart. FIGURE 32.47a shows the currents I_1 and I_2 in the same direction; FIGURE 32.47b shows the currents in opposite directions. We will assume that the wires are sufficiently long to allow us to use the earlier result for the magnetic field of a long, straight wire: $B = \mu_0 I/2\pi d$.

FIGURE 32.47 Magnetic forces between parallel current-carrying wires.

(a) Currents in same direction

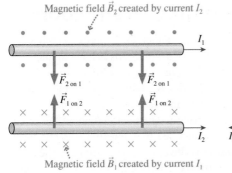

(b) Currents in opposite directions

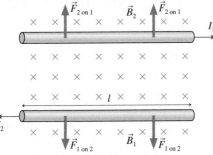

As Figure 32.47a shows, the current I_2 in the lower wire creates a magnetic field $\vec{B}_2$ at the position of the upper wire. $\vec{B}_2$ points out of the page, perpendicular to current I_1. **It is field $\vec{B}_2$, due to the lower wire, that exerts a magnetic force on the upper wire.** Using the right-hand rule, you can see that the force on the upper wire is downward, thus attracting it toward the lower wire. The field of the lower current is not a uniform field, but it is the *same* at all points along the upper wire because the two wires are parallel. Consequently, we can use the field of a long, straight wire to determine the magnetic force exerted by the lower wire on the upper wire when they are separated by distance d:

$$F_{\text{parallel wires}} = I_1 l B_2 = I_1 l \frac{\mu_0 I_2}{2\pi d} = \frac{\mu_0 l I_1 I_2}{2\pi d} \qquad (32.26)$$

As an exercise, you should convince yourself that the current in the upper wire exerts an upward-directed magnetic force on the lower wire with exactly the same magnitude. You should also convince yourself, using the right-hand rule, that the forces are repulsive and tend to push the wires apart if the two currents are in opposite directions.

Thus two parallel wires exert equal but opposite forces on each other, as required by Newton's third law. **Parallel wires carrying currents in the same direction attract each other; parallel wires carrying currents in opposite directions repel each other.**

EXAMPLE 32.14 | **A current balance**

Two stiff, 50-cm-long, parallel wires are connected at the ends by metal springs. Each spring has an unstretched length of 5.0 cm and a spring constant of 0.025 N/m. The wires push each other apart when a current travels around the loop. How much current is required to stretch the springs to lengths of 6.0 cm?

MODEL Two parallel wires carrying currents in opposite directions exert repulsive magnetic forces on each other.

VISUALIZE FIGURE 32.48 shows the "circuit." The springs are conductors, allowing a current to travel around the loop. In equilibrium, the repulsive magnetic forces between the wires are balanced by the restoring forces $F_{\text{sp}} = k\Delta y$ of the springs.

FIGURE 32.48 The current-carrying wires of Example 32.14.

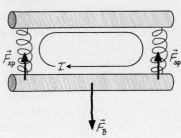

Continued

SOLVE Figure 32.48 shows the forces on the lower wire. The net force is zero, hence the magnetic force is $F_B = 2F_{sp}$. The force between the wires is given by Equation 32.26 with $I_1 = I_2 = I$:

$$F_B = \frac{\mu_0 l I^2}{2\pi d} = 2F_{sp} = 2k\Delta y$$

where k is the spring constant and $\Delta y = 1.0$ cm is the amount by which each spring stretches. Solving for the current, we find

$$I = \sqrt{\frac{4\pi k d \Delta y}{\mu_0 l}} = 17 \text{ A}$$

ASSESS Devices in which a magnetic force balances a mechanical force are called *current balances*. They can be used to make very accurate current measurements.

32.9 Forces and Torques on Current Loops

You have seen that a current loop is a magnetic dipole, much like a permanent magnet. We will now look at some important features of how current loops behave in magnetic fields. This discussion will be largely qualitative, but it will highlight some of the important properties of magnets and magnetic fields. We will use these ideas in the next section to make the connection between electromagnets and permanent magnets.

FIGURE 32.49a shows two current loops. Using what we just learned about the forces between parallel and antiparallel currents, you can see that **parallel current loops exert attractive magnetic forces on each other if the currents circulate in the same direction; they repel each other when the currents circulate in opposite directions.**

FIGURE 32.49 Two alternative but equivalent ways to view magnetic forces.

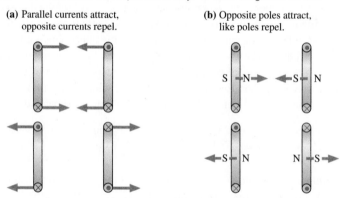

(a) Parallel currents attract, opposite currents repel.

(b) Opposite poles attract, like poles repel.

We can think of these forces in terms of magnetic poles. Recall that the north pole of a current loop is the side from which the magnetic field emerges, which you can determine with the right-hand rule. FIGURE 32.49b shows the north and south magnetic poles of the current loops. When the currents circulate in the same direction, a north and a south pole face each other and exert attractive forces on each other. When the currents circulate in opposite directions, the two like poles repel each other.

Here, at last, we have a real connection to the behavior of magnets that opened our discussion of magnetism—namely, that like poles repel and opposite poles attract. Now we have an *explanation* for this behavior, at least for electromagnets. **Magnetic poles attract or repel because the moving charges in one current exert attractive or repulsive magnetic forces on the moving charges in the other current.** Our tour through interacting moving charges is finally starting to show some practical results!

Now let's consider what happens to a current loop in a magnetic field. FIGURE 32.50 shows a square current loop in a uniform magnetic field along the z-axis. As we've learned, the field exerts magnetic forces on the currents in each of the four sides of the loop. Their directions are given by the right-hand rule. Forces $\vec{F}_{front}$ and $\vec{F}_{back}$ are opposite to each other and cancel. Forces $\vec{F}_{top}$ and $\vec{F}_{bottom}$ also add to give no net force, but because $\vec{F}_{top}$ and $\vec{F}_{bottom}$ don't act along the same line they will *rotate* the loop by exerting a torque on it.

Recall that torque is the magnitude of the force F multiplied by the moment arm d, the distance between the pivot point and the line of action. Both forces have the same moment arm $d = \frac{1}{2}l\sin\theta$, hence the torque on the loop—a torque exerted by the magnetic field—is

$$\tau = 2Fd = 2(IlB)(\tfrac{1}{2}l\sin\theta) = (Il^2)B\sin\theta = \mu B\sin\theta \qquad (32.27)$$

where $\mu = Il^2 = IA$ is the loop's magnetic dipole moment.

Although we derived Equation 32.27 for a square loop, the result is valid for a current loop of any shape. Notice that Equation 32.27 looks like another example of a cross product. We earlier defined the magnetic dipole moment vector $\vec{\mu}$ to be a vector perpendicular to the current loop in a direction given by the right-hand rule. Figure 32.50 shows that θ is the angle between $\vec{B}$ and $\vec{\mu}$, hence the torque on a magnetic dipole is

$$\vec{\tau} = \vec{\mu} \times \vec{B} \qquad (32.28)$$

The torque is zero when the magnetic dipole moment $\vec{\mu}$ is aligned parallel or antiparallel to the magnetic field, and is maximum when $\vec{\mu}$ is perpendicular to the field. It is this magnetic torque that causes a compass needle—a magnetic moment—to rotate until it is aligned with the magnetic field.

An Electric Motor

The torque on a current loop in a magnetic field is the basis for how an electric motor works. As FIGURE 32.51 shows, the *armature* of a motor is a coil of wire wound on an axle. When a current passes through the coil, the magnetic field exerts a torque on the armature and causes it to rotate. If the current were steady, the armature would oscillate back and forth around the equilibrium position until (assuming there's some friction or damping) it stopped with the plane of the coil perpendicular to the field. To keep the motor turning, a device called a *commutator* reverses the current direction in the coils every 180°. (Notice that the commutator is split, so the positive terminal of the battery sends current into whichever wire touches the right half of the commutator.) The current reversal prevents the armature from ever reaching an equilibrium position, so the magnetic torque keeps the motor spinning as long as there is a current.

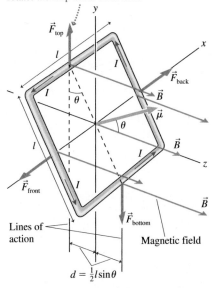

FIGURE 32.50 A uniform magnetic field exerts a torque on a current loop.

$\vec{F}_{top}$ and $\vec{F}_{bottom}$ exert a torque that rotates the loop about the x-axis.

FIGURE 32.51 A simple electric motor.

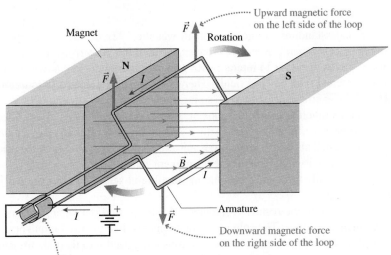

The commutator reverses the current in the loop every half cycle so that the force is always upward on the left side of the loop.

STOP TO THINK 32.6   What is the current direction in the loop?

a. Out of the page at the top of the loop, into the page at the bottom
b. Out of the page at the bottom of the loop, into the page at the top

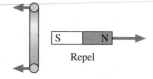

Repel

32.10 Magnetic Properties of Matter

Our theory has focused mostly on the magnetic properties of currents, yet our everyday experience is mostly with permanent magnets. We have seen that current loops and solenoids have magnetic poles and exhibit behaviors like those of permanent magnets, but we still lack a specific connection between electromagnets and permanent magnets. The goal of this section is to complete our understanding by developing an atomic-level view of the magnetic properties of matter.

Atomic Magnets

FIGURE 32.52 A classical orbiting electron is a tiny magnetic dipole.

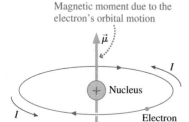

Magnetic moment due to the electron's orbital motion

$\vec{\mu}$

Nucleus

I

I

Electron

A plausible explanation for the magnetic properties of materials is the orbital motion of the atomic electrons. **FIGURE 32.52** shows a simple, classical model of an atom in which a negative electron orbits a positive nucleus. In this picture of the atom, the electron's motion is that of a current loop! It is a microscopic current loop, to be sure, but a current loop nonetheless. Consequently, an orbiting electron acts as a tiny magnetic dipole, with a north pole and a south pole. You can think of the magnetic dipole as an atomic-size magnet.

However, the atoms of most elements contain many electrons. Unlike the solar system, where all of the planets orbit in the same direction, electron orbits are arranged to oppose each other: one electron moves counterclockwise for every electron that moves clockwise. Thus the magnetic moments of individual orbits tend to cancel each other and the *net* magnetic moment is either zero or very small.

The cancellation continues as the atoms are joined into molecules and the molecules into solids. When all is said and done, the net magnetic moment of any bulk matter due to the orbiting electrons is so small as to be negligible. There are various subtle magnetic effects that can be observed under laboratory conditions, but orbiting electrons cannot explain the very strong magnetic effects of a piece of iron.

The Electron Spin

FIGURE 32.53 Magnetic moment of the electron.

The arrow represents the inherent magnetic moment of the electron.

The key to understanding atomic magnetism was the 1922 discovery that electrons have an *inherent magnetic moment*. Perhaps this shouldn't be surprising. An electron has a *mass,* which allows it to interact with gravitational fields, and a *charge,* which allows it to interact with electric fields. There's no reason an electron shouldn't also interact with magnetic fields, and to do so it comes with a magnetic moment.

An electron's inherent magnetic moment, shown in **FIGURE 32.53**, is often called the electron *spin* because, in a classical picture, a spinning ball of charge would have a magnetic moment. This classical picture is not a realistic portrayal of how the electron really behaves, but its inherent magnetic moment makes it seem *as if* the electron were spinning. While it may not be spinning in a literal sense, an electron really is a microscopic magnet.

We must appeal to the results of quantum physics to find out what happens in an atom with many electrons. The spin magnetic moments, like the orbital magnetic moments, tend to oppose each other as the electrons are placed into their shells, causing the net magnetic moment of a *filled* shell to be zero. However, atoms containing an odd number of electrons must have at least one valence electron with an unpaired spin. These atoms have net magnetic moment due to the electron's spin.

But atoms with magnetic moments don't necessarily form a solid with magnetic properties. For most elements, the magnetic moments of the atoms are randomly arranged when the atoms join together to form a solid. As FIGURE 32.54 shows, this random arrangement produces a solid whose net magnetic moment is very close to zero. This agrees with our common experience that most materials are not magnetic.

Ferromagnetism

It happens that in iron, and a few other substances, the spins interact with each other in such a way that atomic magnetic moments tend to all line up in the *same* direction, as shown in FIGURE 32.55. Materials that behave in this fashion are called **ferromagnetic,** with the prefix *ferro* meaning "iron-like."

In ferromagnetic materials, the individual magnetic moments add together to create a *macroscopic* magnetic dipole. The material has a north and a south magnetic pole, generates a magnetic field, and aligns parallel to an external magnetic field. In other words, it is a magnet!

FIGURE 32.54 The random magnetic moments of the atoms in a typical solid.

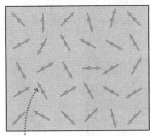

The atomic magnetic moments due to unpaired spins point in random directions. The sample has no net magnetic moment.

FIGURE 32.55 The aligned atomic magnetic moments in a ferromagnetic material.

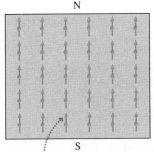

The atomic magnetic moments are aligned. The sample has north and south magnetic poles.

FIGURE 32.56 Magnetic domains in a ferromagnetic material. The net magnetic dipole is nearly zero.

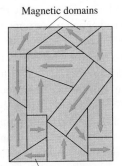

Magnetic domains

Magnetic moment of the domain

Although iron is a magnetic material, a typical piece of iron is not a strong permanent magnet. You need not worry that a steel nail, which is mostly iron and is easily lifted with a magnet, will leap from your hands and pin itself against the hammer because of its own magnetism. It turns out, as shown in FIGURE 32.56, that a piece of iron is divided into small regions, typically less than $100\ \mu$m in size, called **magnetic domains.** The magnetic moments of all the iron atoms within each domain are perfectly aligned, so each individual domain, like Figure 32.55, is a strong magnet.

However, the various magnetic domains that form a larger solid, such as you might hold in your hand, are randomly arranged. Their magnetic dipoles largely cancel, much like the cancellation that occurs on the atomic scale for nonferromagnetic substances, so the solid as a whole has only a small magnetic moment. That is why the nail is not a strong permanent magnet.

Induced Magnetic Dipoles

If a ferromagnetic substance is subjected to an *external* magnetic field, the external field exerts a torque on the magnetic dipole of each domain. The torque causes many of the domains to rotate and become aligned with the external field, just as a compass needle aligns with a magnetic field, although internal forces between the domains generally prevent the alignment from being perfect. In addition, atomic-level forces between the spins can cause the *domain boundaries* to move. Domains that are aligned along the external field become larger at the expense of domains that are opposed to the field. These changes in the size and orientation of the domains cause the material to develop a *net magnetic dipole* that is aligned with the external field. This magnetic dipole has been *induced* by the external field, so it is called an **induced magnetic dipole.**

NOTE ▶ The induced magnetic dipole is analogous to the polarization forces and induced electric dipoles that you studied in Chapter 26. ◀

FIGURE 32.57 The magnetic field of the solenoid creates an induced magnetic dipole in the iron.

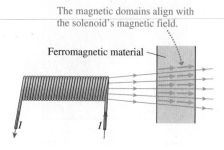

The magnetic domains align with the solenoid's magnetic field.

Ferromagnetic material

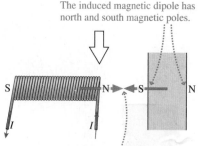

The induced magnetic dipole has north and south magnetic poles.

The attractive force between the opposite poles pulls the ferromagnetic material toward the solenoid.

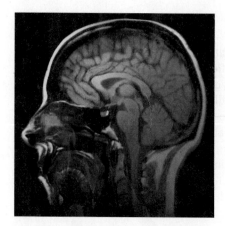

Magnetic resonance imaging, or MRI, uses the magnetic properties of atoms as a noninvasive probe of the human body.

FIGURE 32.57 shows a ferromagnetic material near the end of a solenoid. The magnetic moments of the domains align with the solenoid's field, creating an induced magnetic dipole whose south pole faces the solenoid's north pole. Consequently, the magnetic force between the poles pulls the ferromagnetic object to the electromagnet.

The fact that a magnet attracts and picks up ferromagnetic objects was one of the basic observations about magnetism with which we started the chapter. Now we have an *explanation* of how it works, based on three ideas:

1. Electrons are microscopic magnets due to their spin.
2. A ferromagnetic material in which the spins are aligned is organized into magnetic domains.
3. The individual domains align with an external magnetic field to produce an induced magnetic dipole moment for the entire object.

The object's magnetic dipole may not return to zero when the external field is removed because some domains remain "frozen" in the alignment they had in the external field. Thus a ferromagnetic object that has been in an external field may be left with a net magnetic dipole moment after the field is removed. In other words, the object has become a **permanent magnet.** A permanent magnet is simply a ferromagnetic material in which a majority of the magnetic domains are aligned with each other to produce a net magnetic dipole moment.

Whether or not a ferromagnetic material can be made into a permanent magnet depends on the internal crystalline structure of the material. *Steel* is an alloy of iron with other elements. An alloy of mostly iron with the right percentages of chromium and nickel produces *stainless steel,* which has virtually no magnetic properties at all because its particular crystalline structure is not conducive to the formation of domains. A very different steel alloy called Alnico V is made with 51% iron, 24% cobalt, 14% nickel, 8% aluminum, and 3% copper. It has extremely prominent magnetic properties and is used to make high-quality permanent magnets. You can see from the complex formula that developing good magnetic materials requires a lot of engineering skill as well as a lot of patience!

So we've come full circle. One of our initial observations about magnetism was that a permanent magnet can exert forces on some materials but not others. The *theory* of magnetism that we then proceeded to develop was about the interactions between moving charges. What moving charges had to do with permanent magnets was not obvious. But finally, by considering magnetic effects at the atomic level, we found that properties of permanent magnets and magnetic materials can be traced to the interactions of vast numbers of electron spins.

STOP TO THINK 32.7   Which magnet or magnets induced this magnetic dipole?

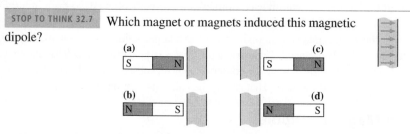

CHALLENGE EXAMPLE 32.15 **Designing a loudspeaker**

A loudspeaker consists of a paper cone wrapped at the bottom with several turns of fine wire. As FIGURE 32.58 shows, this coil sits in a narrow gap between the poles of a circular magnet. To produce sound, the amplifier drives a current through the coil. The magnetic field then exerts a force on this current, pushing the cone and thus pushing the air to create a sound wave. An ideal speaker would experience only forces from the magnetic field, thus responding only to the current from the amplifier. Real speakers are balanced so as to come close to this ideal unless driven very hard.

FIGURE 32.58 The coil and magnet of a loudspeaker.

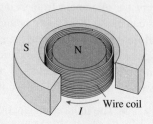

Consider a 5.5 g loudspeaker cone with a 5.0-cm-diameter, 20-turn coil having a resistance of 8.0 Ω. There is a 0.18 T field in the gap between the poles. These values are typical of the loudspeakers found in car stereo systems. What is the oscillation amplitude of this speaker if driven by a 100 Hz oscillatory voltage from the amplifier with a peak value of 12 V?

MODEL Model the loudspeaker as ideal, responding only to magnetic forces. These forces cause the cone to accelerate. We'll use kinematics to relate the acceleration to the displacement.

VISUALIZE **FIGURE 32.59** shows the coil in the gap between the magnet poles. Magnetic fields go from north to south poles, so the field is radially outward. Consequently, the field at all points is perpendicular to the circular current. According to the right-hand rule, the magnetic force on the current is into or out of the page, depending on whether the current is counterclockwise or clockwise, respectively.

FIGURE 32.59 The magnetic field in the gap, from north to south, is perpendicular to the current.

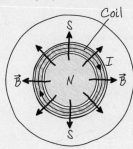

SOLVE We can write the output voltage of the amplifier as $\Delta V = V_0 \cos \omega t$, where $V_0 = 12$ V is the peak voltage and $\omega = 2\pi f = 628$ rad/s is the angular frequency at 100 Hz. The voltage drives current

$$I = \frac{\Delta V}{R} = \frac{V_0 \cos \omega t}{R}$$

through the coil, where R is the coil's resistance. This causes the oscillating in-and-out force that drives the speaker cone back and forth. Even though the coil isn't a straight wire, the fact that the magnetic field is everywhere perpendicular to the current means that we can calculate the magnetic force as $F = IlB$ where l is the total length of the wire in the coil. The circumference of the coil is $\pi(0.050$ m$) = 0.157$ m so 20 turns gives $l = 3.1$ m. The cone responds to the force by accelerating with $a = F/m$. Combining these pieces, we find the cone's acceleration is

$$a = \frac{IlB}{m} = \frac{V_0 lB \cos \omega t}{mR} = a_{max} \cos \omega t$$

It is straightforward to evaluate $a_{max} = 152$ m/s^2.

From kinematics, $a = dv/dt$ and $v = dx/dt$. We need to integrate twice to find the displacement. First,

$$v = \int a \, dt = a_{max} \int \cos \omega t \, dt = \frac{a_{max}}{\omega} \sin \omega t$$

The integration constant is zero because we know, from simple harmonic motion, that the average velocity is zero. Integrating again, we get

$$x = \int v \, dt = \frac{a_{max}}{\omega} \int \sin \omega t \, dt = -\frac{a_{max}}{\omega^2} \cos \omega t$$

where the integration constant is again zero if we assume the oscillation takes place around the origin. The minus sign tells us that the displacement and acceleration are out of phase. The *amplitude* of the oscillation, which we seek, is

$$A = \frac{a_{max}}{\omega^2} = \frac{152 \text{ m/s}^2}{(628 \text{ rad/s})^2} = 3.8 \times 10^{-4} \text{ m} = 0.38 \text{ mm}$$

ASSESS If you've ever placed your hand on a loudspeaker cone, you know that you can feel a slight vibration. An amplitude of 0.38 mm is consistent with this observation. The fact that the amplitude increases with the inverse square of the frequency explains why you can sometimes *see* the cone vibrating with an amplitude of several millimeters for low-frequency bass notes.

SUMMARY

The goal of Chapter 32 has been to learn how to calculate and use the magnetic field.

General Principles

At its most fundamental level, magnetism is an interaction between moving charges. The magnetic field of one moving charge exerts a force on another moving charge.

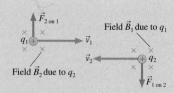

Magnetic Fields

The Biot-Savart law

- A **point charge**, $\vec{B} = \dfrac{\mu_0}{4\pi}\dfrac{q\vec{v}\times\hat{r}}{r^2}$

- A **short current element**, $\vec{B} = \dfrac{\mu_0}{4\pi}\dfrac{I\Delta\vec{s}\times\hat{r}}{r^2}$

To find the magnetic field of a current:

- Divide the wire into many short segments.

- Find the field of each segment Δs.

- Find $\vec{B}$ by summing the fields of all Δs, usually as an integral.

An alternative method for fields with a high degree of symmetry is Ampère's law:

$$\oint \vec{B}\cdot d\vec{s} = \mu_0 I_{\text{through}}$$

where I_{through} is the current through the area bounded by the integration path.

Magnetic Forces

The magnetic force on a moving charge is

$$\vec{F} = q\vec{v}\times\vec{B}$$

The force is perpendicular to $\vec{v}$ and $\vec{B}$.

The magnetic force on a current-carrying wire is

$$\vec{F} = I\vec{l}\times\vec{B}$$

$\vec{F} = \vec{0}$ for a charge or current moving parallel to $\vec{B}$.

The magnetic torque on a magnetic dipole is

$$\vec{\tau} = \vec{\mu}\times\vec{B}$$

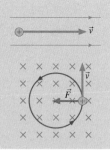

Applications

Wire

$B = \dfrac{\mu_0}{2\pi}\dfrac{I}{d}$

Loop

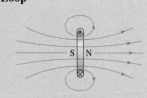

Solenoid

$B = \dfrac{\mu_0 NI}{l}$

Flat magnet

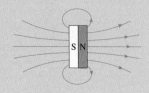

Right-hand rule

Point your right thumb in the direction of I. Your fingers curl in the direction of $\vec{B}$. For a dipole, $\vec{B}$ emerges from the side that is the north pole.

Charged-particle motion

No force if $\vec{v}$ is parallel to $\vec{B}$

Circular motion at the cyclotron frequency $f_{\text{cyc}} = qB/2\pi m$ if $\vec{v}$ is perpendicular to $\vec{B}$

Parallel wires and current loops

Parallel currents attract.
Opposite currents repel.

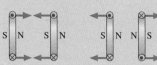

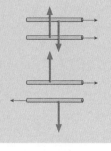

Terms and Notation

north pole	Biot-Savart law	line integral	Hall effect
south pole	tesla, T	Ampère's law	Hall voltage, ΔV_{H}
magnetic dipole	permeability constant, μ_0	uniform magnetic field	ferromagnetic
magnetic material	cross product	solenoid	magnetic domain
right-hand rule	current loop	cyclotron motion	induced magnetic dipole
magnetic field, $\vec{B}$	electromagnet	cyclotron frequency, f_{cyc}	permanent magnet
magnetic field lines	magnetic dipole moment, $\vec{\mu}$	cyclotron	

CONCEPTUAL QUESTIONS

1. The lightweight glass sphere in FIGURE Q32.1 hangs by a thread. The north pole of a bar magnet is brought near the sphere.
   a. Suppose the sphere is electrically neutral. Is it attracted to, repelled by, or not affected by the magnet? Explain.
   b. Answer the same question if the sphere is positively charged.

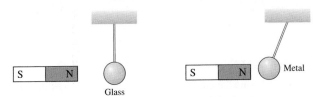

FIGURE Q32.1　　　　FIGURE Q32.2

2. The metal sphere in FIGURE Q32.2 hangs by a thread. When the north pole of a magnet is brought near, the sphere is strongly attracted to the magnet. Then the magnet is reversed and its south pole is brought near the sphere. How does the sphere respond? Explain.

3. You have two electrically neutral metal cylinders that exert strong attractive forces on each other. You have no other metal objects. Can you determine if *both* of the cylinders are magnets, or if one is a magnet and the other is just a piece of iron? If so, how? If not, why not?

4. What is the current direction in the wire of FIGURE Q32.4? Explain.

FIGURE Q32.4　　　　　　FIGURE Q32.5

5. What is the current direction in the wire of FIGURE Q32.5? Explain.

6. What is the *initial* direction of deflection for the charged particles entering the magnetic fields shown in FIGURE Q32.6?

FIGURE Q32.6

7. What is the *initial* direction of deflection for the charged particles entering the magnetic fields shown in FIGURE Q32.7?

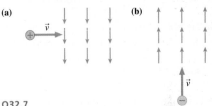

FIGURE Q32.7

8. Determine the magnetic field direction that causes the charged particles shown in FIGURE Q32.8 to experience the indicated magnetic force.

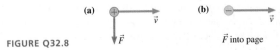

FIGURE Q32.8

9. Determine the magnetic field direction that causes the charged particles shown in FIGURE Q32.9 to experience the indicated magnetic force.

FIGURE Q32.9　　$\vec{F}$ out of page　　$\vec{v}$ into page

10. You have a horizontal cathode-ray tube (CRT) for which the controls have been adjusted such that the electron beam *should* make a single spot of light exactly in the center of the screen. You observe, however, that the spot is deflected to the right. It is possible that the CRT is broken. But as a clever scientist, you realize that your laboratory might be in either an electric or a magnetic field. Assuming that you do not have a compass, any magnets, or any charged rods, how can you use the CRT itself to determine whether the CRT is broken, is in an electric field, or is in a magnetic field? You cannot remove the CRT from the room.

11. The south pole of a bar magnet is brought toward the current loop of FIGURE Q32.11. Does the bar magnet attract, repel, or have no effect on the loop? Explain.

FIGURE Q32.11

12. Give a step-by-step explanation, using both words and pictures, of how a permanent magnet can pick up a piece of nonmagnetized iron.

EXERCISES AND PROBLEMS

Problems labeled   integrate material from earlier chapters.

Exercises

Section 32.3 The Source of the Magnetic Field: Moving Charges

1. | Points 1 and 2 in FIGURE EX32.1 are the same distance from the wires as the point where $B = 2.0$ mT. What are the strength and direction of $\vec{B}$ at points 1 and 2?

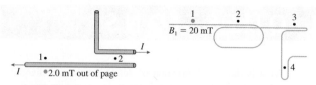

FIGURE EX32.1   FIGURE EX32.2

2. | What is the magnetic field strength at points 2 to 4 in FIGURE EX32.2? Assume that the wires overlap closely and that points 1 to 4 are equally distant from the wires.

3. ‖ A proton moves along the x-axis with $v_x = 1.0 \times 10^7$ m/s. As it passes the origin, what are the strength and direction of the magnetic field at the (x, y, z) positions (a) (1 cm, 0 cm, 0 cm), (b) (0 cm, 1 cm, 0 cm), and (c) (0 cm, −2 cm, 0 cm)?

4. ‖ An electron moves along the z-axis with $v_z = 2.0 \times 10^7$ m/s. As it passes the origin, what are the strength and direction of the magnetic field at the (x, y, z) positions (a) (1 cm, 0 cm, 0 cm), (b) (0 cm, 0 cm, 1 cm), and (c) (0 cm, 1 cm, 1 cm)?

5. ‖ What is the magnetic field at the position of the dot in FIGURE EX32.5? Give your answer as a vector.

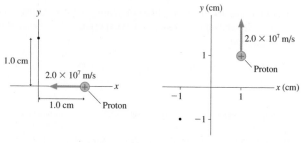

FIGURE EX32.5   FIGURE EX32.6

6. ‖ What is the magnetic field at the position of the dot in FIGURE EX32.6? Give your answer as a vector.

7. ‖ A proton is passing the origin. The magnetic field at the (x, y, z) position (1 mm, 0 mm, 0 mm) is $1.0 \times 10^{-13} \hat{j}$ T. The field at (0 mm, 1 mm, 0 mm) is $-1.0 \times 10^{-13} \hat{i}$ T. What are the speed and direction of the proton?

Section 32.4 The Magnetic Field of a Current

8. | What currents are needed to generate the magnetic field strengths of Table 32.1 at a point 1.0 cm from a long, straight wire?

9. | At what distances from a very thin, straight wire carrying a 10 A current would the magnetic field strengths of Table 32.1 be generated?

10. ‖ The element niobium, which is a metal, is a superconductor (i.e., no electrical resistance) at temperatures below 9 K. However, the superconductivity is destroyed if the magnetic field at the surface of the metal reaches or exceeds 0.10 T. What is the maximum current in a straight, 3.0-mm-diameter superconducting niobium wire?

11. ‖ The magnetic field at the center of a 1.0-cm-diameter loop is 2.5 mT.
    a. What is the current in the loop?
    b. A long straight wire carries the same current you found in part a. At what distance from the wire is the magnetic field 2.5 mT?

12. | A wire carries current I into the junction shown in FIGURE EX32.12. What is the magnetic field at the dot?

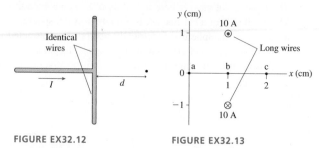

FIGURE EX32.12   FIGURE EX32.13

13. | What are the magnetic fields at points a to c in FIGURE EX32.13? Give your answers as vectors.

14. ‖ What are the magnetic field strength and direction at points a to c in FIGURE EX32.14?

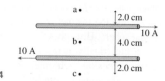

FIGURE EX32.14

Section 32.5 Magnetic Dipoles

15. | The on-axis magnetic field strength 10 cm from a small bar magnet is 5.0 μT.
    a. What is the bar magnet's magnetic dipole moment?
    b. What is the on-axis field strength 15 cm from the magnet?

16. ‖ A 100 A current circulates around a 2.0-mm-diameter superconducting ring.
    a. What is the ring's magnetic dipole moment?
    b. What is the on-axis magnetic field strength 5.0 cm from the ring?

17. ‖ A small, square loop carries a 25 A current. The on-axis magnetic field strength 50 cm from the loop is 7.5 nT. What is the edge length of the square?

18. ‖ The earth's magnetic dipole moment is 8.0×10^{22} A m².
    a. What is the magnetic field strength on the surface of the earth at the earth's north magnetic pole? How does this compare to the value in Table 32.1? You can assume that the current loop is deep inside the earth.
    b. Astronauts discover an earth-size planet without a magnetic field. To create a magnetic field with the same strength as earth's, they propose running a current through a wire around the equator. What size current would be needed?

Section 32.6 Ampère's Law and Solenoids

19. | What is the line integral of $\vec{B}$ between points i and f in FIGURE EX32.19?

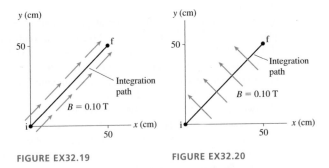

FIGURE EX32.19   FIGURE EX32.20

20. | What is the line integral of $\vec{B}$ between points i and f in FIGURE EX32.20?

21. ‖ The value of the line integral of $\vec{B}$ around the closed path in FIGURE EX32.21 is 3.77×10^{-6} T m. What is I_3?

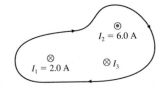

$I_2 = 6.0$ A

$I_1 = 2.0$ A $\otimes I_3$

FIGURE EX32.21

22. ‖ The value of the line integral of $\vec{B}$ around the closed path in FIGURE EX32.22 is 1.38×10^{-5} T m. What are the direction (in or out of the page) and magnitude of I_3?

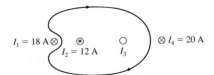

$I_1 = 18$ A $\otimes$ $\odot$ $\bigcirc$ $\otimes I_4 = 20$ A
              $I_2 = 12$ A I_3

FIGURE EX32.22

23. ‖ What is the line integral of $\vec{B}$ between points i and f in FIGURE EX32.23?

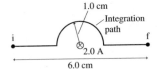

1.0 cm
Integration path
i   f
2.0 A
6.0 cm

FIGURE EX32.23

24. ‖ Magnetic resonance imaging needs a magnetic field strength
BIO of 1.5 T. The solenoid is 1.8 m long and 75 cm in diameter. It is tightly wound with a single layer of 2.0-mm-diameter superconducting wire. What size current is needed?

25. ‖ A 2.0-cm-diameter, 15-cm-long solenoid is tightly wound with one layer of wire. A 2.5 A current through the wire generates a 3.0 mT magnetic field inside the solenoid. What is the diameter of the wire, in mm?

Section 32.7 The Magnetic Force on a Moving Charge

26. | A proton moves in the magnetic field $\vec{B} = 0.50\,\hat{\imath}$ T with a speed of 1.0×10^7 m/s in the directions shown in FIGURE EX32.26. For each, what is magnetic force $\vec{F}$ on the proton? Give your answers in component form.

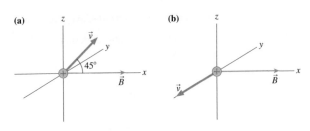

FIGURE EX32.26

27. ‖ An electron moves in the magnetic field $\vec{B} = 0.50\,\hat{\imath}$ T with a speed of 1.0×10^7 m/s in the directions shown in FIGURE EX32.27. For each, what is magnetic force $\vec{F}$ on the electron? Give your answers in component form.

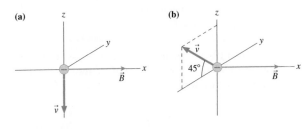

FIGURE EX32.27

28. ‖ To five significant figures, what are the cyclotron frequencies in a 3.0000 T magnetic field of the ions (a) O_2^+, (b) N_2^+, and (c) CO^+? The atomic masses are shown in the table; the mass of the missing electron is less than 0.001 u and is not relevant at this level of precision. Although N_2^+ and CO^+ both have a nominal molecular mass of 28, they are easily distinguished by virtue of their slightly different cyclotron frequencies. Use the following constants: $1\ \text{u} = 1.6605 \times 10^{-27}$ kg, $e = 1.6022 \times 10^{-19}$ C.

Atomic masses

^{12}C	12.000
^{14}N	14.003
^{16}O	15.995

29. | Radio astronomers detect electromagnetic radiation at 45 MHz from an interstellar gas cloud. They suspect this radiation is emitted by electrons spiraling in a magnetic field. What is the magnetic field strength inside the gas cloud?

30. | For your senior project, you would like to build a cyclotron that will accelerate protons to 10% of the speed of light. The largest vacuum chamber you can find is 50 cm in diameter. What magnetic field strength will you need?

31. | The Hall voltage across a conductor in a 55 mT magnetic field is 1.9 μV. When used with the same current in a different magnetic field, the voltage across the conductor is 2.8 μV. What is the strength of the second field?

32. | Test instruments to measure magnetic field strengths are often based on the Hall effect. In one instrument, the "probe" is a 1.0-mm-thick, 6.0-mm-wide semiconductor with a charge-carrier density of 2.1×10^{21} m^{-3}, much less than the charge-carrier density in a conductor. Passing a 60 mA current through the probe generates a Hall voltage of 120 mV. What is the magnetic field strength?

Section 32.8 Magnetic Forces on Current-Carrying Wires

33. | What magnetic field strength and direction will levitate the 2.0 g wire in FIGURE EX32.33?

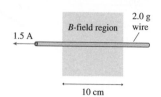

FIGURE EX32.33

34. | The right edge of the circuit in FIGURE EX32.34 extends into a 50 mT uniform magnetic field. What are the magnitude and direction of the net force on the circuit?

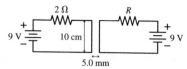

FIGURE EX32.34

35. || The two 10-cm-long parallel wires in FIGURE EX32.35 are separated by 5.0 mm. For what value of the resistor R will the force between the two wires be 5.4×10^{-5} N?

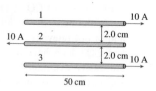

FIGURE EX32.35

36. | What is the net force (magnitude and direction) on each wire in FIGURE EX32.36?

FIGURE EX32.36

Section 32.9 Forces and Torques on Current Loops

37. || A square current loop 5.0 cm on each side carries a 500 mA current. The loop is in a 1.2 T uniform magnetic field. The axis of the loop, perpendicular to the plane of the loop, is 30° away from the field direction. What is the magnitude of the torque on the current loop?

38. | A small bar magnet experiences a 0.020 N m torque when the axis of the magnet is at 45° to a 0.10 T magnetic field. What is the magnitude of its magnetic dipole moment?

39. || a. What is the magnitude of the torque on the current loop in FIGURE EX32.39?
    b. What is the loop's equilibrium orientation?

FIGURE EX32.39

Problems

40. | Although the evidence is weak, there has been concern in recent years over possible health effects from the magnetic fields generated by electric transmission lines. A typical high-voltage transmission line is 20 m above the ground and carries a 200 A current at a potential of 110 kV.
    a. What is the magnetic field strength on the ground directly under such a transmission line?
    b. What percentage is this of the earth's magnetic field of 50 μT?

41. | A biophysics experiment uses a very sensitive magnetic field probe to determine the current associated with a nerve impulse traveling along an axon. If the peak field strength 1.0 mm from an axon is 8.0 pT, what is the peak current carried by the axon?

42. || A long wire carrying a 5.0 A current perpendicular to the xy-plane intersects the x-axis at $x = -2.0$ cm. A second, parallel wire carrying a 3.0 A current intersects the x-axis at $x = +2.0$ cm. At what point or points on the x-axis is the magnetic field zero if (a) the two currents are in the same direction and (b) the two currents are in opposite directions?

43. || The two insulated wires in FIGURE P32.43 cross at a 30° angle but do not make electrical contact. Each wire carries a 5.0 A current. Points 1 and 2 are each 4.0 cm from the intersection and equally distant from both wires. What are the magnitude and direction of the magnetic fields at points 1 and 2?

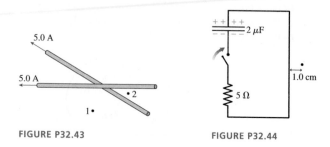

FIGURE P32.43   FIGURE P32.44

44. || The capacitor in FIGURE P32.44 is charged to 50 V. The switch closes at $t = 0$ s. Draw a graph showing the magnetic field strength as a function of time at the position of the dot. On your graph indicate the maximum field strength, and provide an appropriate numerical scale on the horizontal axis.

45. || At what distance on the axis of a current loop is the magnetic field half the strength of the field at the center of the loop? Give your answer as a multiple of R.

46. || Find an expression for the magnetic field strength at the center (point P) of the circular arc in FIGURE P32.46.

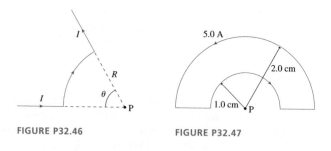

FIGURE P32.46   FIGURE P32.47

47. || What are the strength and direction of the magnetic field at point P in FIGURE P32.47?

48. || What are the strength and direction of the magnetic field at the center of the loop in FIGURE P32.48?

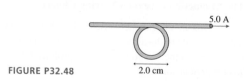

FIGURE P32.48   2.0 cm

49. ‖ Your employer asks you to build a 20-cm-long solenoid with an interior field of 5.0 mT. The specifications call for a single layer of wire, wound with the coils as close together as possible. You have two spools of wire available. Wire with a #18 gauge has a diameter of 1.02 mm and has a maximum current rating of 6 A. Wire with a #26 gauge is 0.41 mm in diameter and can carry up to 1 A. Which wire should you use, and what current will you need?

50. ‖ The magnetic field strength at the north pole of a 2.0-cm-diameter, 8-cm-long Alnico magnet is 0.10 T. To produce the same field with a solenoid of the same size, carrying a current of 2.0 A, how many turns of wire would you need? Does this seem feasible?

51. ‖ The earth's magnetic field, with a magnetic dipole moment of 8.0×10^{22} A m^2, is generated by currents within the molten iron of the earth's outer core. Suppose we model the core current as a 3000-km-diameter current loop made from a 1000-km-diameter "wire." The loop diameter is measured from the centers of this very fat wire.
   a. What is the current in the current loop?
   b. What is the current density J in the current loop?
   c. To decide whether this is a large or a small current density, compare it to the current density of a 1.0 A current in a 1.0-mm-diameter wire.

52. ‖ BIO Weak magnetic fields can be measured at the surface of the brain. Although the currents causing these fields are quite complicated, we can estimate their size by modeling them as a current loop around the equator of a 16-cm-diameter (the width of a typical head) sphere. What current is needed to produce a 3.0 pT field—the strength measured for one subject—at the pole of this sphere?

53. ‖ BIO The heart produces a weak magnetic field that can be used to diagnose certain heart problems. It is a dipole field produced by a current loop in the outer layers of the heart.
   a. It is estimated that the field at the center of the heart is 90 pT. What current must circulate around an 8.0-cm-diameter loop, about the size of a human heart, to produce this field?
   b. What is the magnitude of the heart's magnetic dipole moment?

54. ‖ Two identical coils are parallel to each other on the same axis. They are separated by a distance equal to their radius. They each have N turns and carry equal currents I in the same direction.
   a. Find an expression for the magnetic field strength at the midpoint between the loops.
   b. Calculate the field strength if the loops are 10 cm in diameter, have 10 turns, and carry a 1.0 A current.

55. ‖ Use the Biot-Savart law to find the magnetic field strength at the center of the semicircle in FIGURE P32.55.

FIGURE P32.55

56. ‖ The *toroid* of FIGURE P32.56 is a coil of wire wrapped around a doughnut-shaped ring (a *torus*) made of nonconducting material. Toroidal magnetic fields are used to confine fusion plasmas.
   a. From symmetry, what must be the *shape* of the magnetic field in this toroid? Explain.
   b. Consider a toroid with N closely spaced turns carrying current I. Use Ampère's law to find an expression for the magnetic field strength at a point inside the torus at distance r from the axis.
   c. Is a toroidal magnetic field a uniform field? Explain.

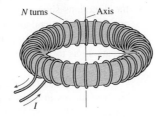

FIGURE P32.56

57. ‖ A long, hollow wire has inner radius R_1 and outer radius R_2. The wire carries current I uniformly distributed across the area of the wire. Use Ampère's law to find an expression for the magnetic field strength in the three regions $0 < r < R_1$, $R_1 < r < R_2$, and $R_2 < r$.

58. ‖ An electron orbits in a 5.0 mT field with angular momentum 8.0×10^{-26} kg m^2/s. What is the diameter of the orbit?

59. ‖ A proton moving in a uniform magnetic field with $\vec{v}_1 = 1.00 \times 10^6 \hat{\imath}$ m/s experiences force $\vec{F}_1 = 1.20 \times 10^{-16} \hat{k}$ N. A second proton with $\vec{v}_2 = 2.00 \times 10^6 \hat{\jmath}$ m/s experiences $\vec{F}_2 = -4.16 \times 10^{-16} \hat{k}$ N in the same field. What is $\vec{B}$? Give your answer as a magnitude and an angle measured ccw from the $+x$-axis.

60. ‖ An electron travels with speed 1.0×10^7 m/s between the two parallel charged plates shown in FIGURE P32.60. The plates are separated by 1.0 cm and are charged by a 200 V battery. What magnetic field strength and direction will allow the electron to pass between the plates without being deflected?

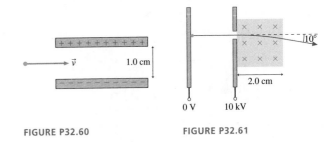

FIGURE P32.60   **FIGURE P32.61**

61. ‖ An electron in a cathode-ray tube is accelerated through a potential difference of 10 kV, then passes through the 2.0-cm-wide region of uniform magnetic field in FIGURE P32.61. What field strength will deflect the electron by 10°?

62. ‖ The microwaves in a microwave oven are produced in a special tube called a *magnetron*. The electrons orbit the magnetic field at 2.4 GHz, and as they do so they emit 2.4 GHz electromagnetic waves.
   a. What is the magnetic field strength?
   b. If the maximum diameter of the electron orbit before the electron hits the wall of the tube is 2.5 cm, what is the maximum electron kinetic energy?

63. ‖ An antiproton (same properties as a proton except that $q = -e$) is moving in the combined electric and magnetic fields of FIGURE P32.63. What are the magnitude and direction of the antiproton's acceleration at this instant?

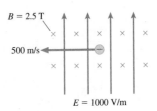

FIGURE P32.63

64. ‖ a. A 65-cm-diameter cyclotron uses a 500 V oscillating potential difference between the dees. What is the maximum kinetic energy of a proton if the magnetic field strength is 0.75 T?
   b. How many revolutions does the proton make before leaving the cyclotron?

65. ||| FIGURE P32.65 shows a *mass spectrometer,* an analytical instrument used to identify the various molecules in a sample by measuring their charge-to-mass ratio q/m. The sample is ionized, the positive ions are accelerated (starting from rest) through a potential difference ΔV, and they then enter a region of uniform magnetic field. The field bends the ions into circular trajectories, but after just half a circle they either strike the wall or pass through a small opening to a detector. As the accelerating voltage is slowly increased, different ions reach the detector and are measured. Consider a mass spectrometer with a 200.00 mT magnetic field and an 8.0000 cm spacing between the entrance and exit holes. To five significant figures, what accelerating potential differences ΔV are required to detect the ions (a) O_2^+, (b) N_2^+, and (c) CO^+? See Exercise 28 for atomic masses; the mass of the missing electron is less than 0.001 u and is not relevant at this level of precision. Although N_2^+ and CO^+ both have a nominal molecular mass of 28, they are easily distinguished by virtue of their slightly different accelerating voltages. Use the following constants: $1\text{ u} = 1.6605 \times 10^{-27}$ kg, $e = 1.6022 \times 10^{-19}$ C.

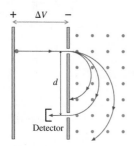

FIGURE P32.65

66. || A Hall-effect probe to measure magnetic field strengths needs to be calibrated in a known magnetic field. Although it is not easy to do, magnetic fields can be precisely measured by measuring the cyclotron frequency of protons. A testing laboratory adjusts a magnetic field until the proton's cyclotron frequency is 10.0 MHz. At this field strength, the Hall voltage on the probe is 0.543 mV when the current through the probe is 0.150 mA. Later, when an unknown magnetic field is measured, the Hall voltage at the same current is 1.735 mV. What is the strength of this magnetic field?

67. || The 10-turn loop of wire shown in FIGURE P32.67 lies in a horizontal plane, parallel to a uniform horizontal magnetic field, and carries a 2.0 A current. The loop is free to rotate about a nonmagnetic axle through the center. A 50 g mass hangs from one edge of the loop. What magnetic field strength will prevent the loop from rotating about the axle?

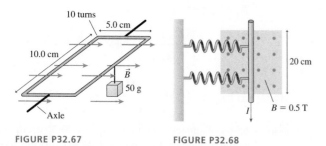

FIGURE P32.67   FIGURE P32.68

68. || The two springs in FIGURE P32.68 each have a spring constant of 10 N/m. They are compressed by 1.0 cm when a current passes through the wire. How big is the current?

69. || Magnetic fields are sometimes measured by balancing magnetic forces against known mechanical forces. Your task is to measure the strength of a horizontal magnetic field using a 12-cm-long rigid metal rod that hangs from two nonmagnetic springs, one at each end, with spring constants 1.3 N/m. You first position the rod to be level and perpendicular to the field, whose direction you determined with a compass. You then connect the ends of the rod to wires that run parallel to the field and thus experience no forces. Finally, you measure the downward deflection of the rod, stretching the springs, as you pass current through it. Your data are as follows:

Current (A)	Deflection (mm)
1.0	4
2.0	9
3.0	12
4.0	15
5.0	21

Use an appropriate graph of the data to determine the magnetic field strength.

70. || A conducting bar of length l and mass m rests at the left end of the two frictionless rails of length d in FIGURE P32.70. A uniform magnetic field of strength B points upward.
   a. In which direction, into or out of the page, will a current through the conducting bar cause the bar to experience a force to the right?
   b. Find an expression for the bar's speed as it leaves the rails at the right end.

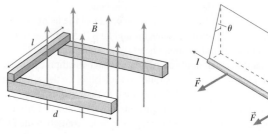

FIGURE P32.70   FIGURE P32.71

71. || a. In FIGURE P32.71, a long, straight, current-carrying wire of linear mass density μ is suspended by threads. A magnetic field perpendicular to the wire exerts a horizontal force that deflects the wire to an equilibrium angle θ. Find an expression for the strength and direction of the magnetic field $\vec{B}$.
   b. What $\vec{B}$ deflects a 55 g/m wire to a 12° angle when the current is 10 A?

72. || FIGURE P32.72 is a cross section through three long wires with linear mass density 50 g/m. They each carry equal currents in the directions shown. The lower two wires are 4.0 cm apart and are attached to a table. What current I will allow the upper wire to "float" so as to form an equilateral triangle with the lower wires?

FIGURE P32.72

73. || In the semiclassical Bohr model of the hydrogen atom, the electron moves in a circular orbit of radius 5.3×10^{-11} m with speed 2.2×10^6 m/s. According to this model, what is the magnetic field at the center of a hydrogen atom?
   Hint: Determine the *average* current of the orbiting electron.

74. ‖ A wire along the x-axis carries current I in the negative x-direction through the magnetic field

$$\vec{B} = \begin{cases} B_0 \dfrac{x}{l} \hat{k} & 0 \le x \le l \\ 0 & \text{elsewhere} \end{cases}$$

a. Draw a graph of B versus x over the interval $-\frac{3}{2}l < x < \frac{3}{2}l$.
b. Find an expression for the net force $\vec{F}_{\text{net}}$ on the wire.
c. Find an expression for the net torque on the wire about the point $x = 0$.

75. ‖ A *nonuniform* magnetic field exerts a net force on a current loop of radius R. FIGURE P32.75 shows a magnetic field that is diverging from the end of a bar magnet. The magnetic field at the position of the current loop makes an angle θ with respect to the vertical.

a. Find an expression for the net magnetic force on the current.
b. Calculate the force if $R = 2.0$ cm, $I = 0.50$ A, $B = 200$ mT, and $\theta = 20°$.

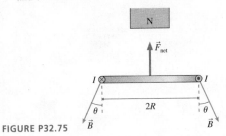

FIGURE P32.75

Challenge Problems

76. You have a 1.0-m-long copper wire. You want to make an N-turn current loop that generates a 1.0 mT magnetic field at the center when the current is 1.0 A. You must use the entire wire. What will be the diameter of your coil?

77. a. Derive an expression for the magnetic field strength at distance d from the center of a straight wire of finite length l that carries current I.
b. Determine the field strength at the center of a current-carrying *square* loop having sides of length $2R$.
c. Compare your answer to part b to the field at the center of a *circular* loop of diameter $2R$. Do so by computing the ratio $B_{\text{square}}/B_{\text{circle}}$.

78. A flat, circular disk of radius R is uniformly charged with total charge Q. The disk spins at angular velocity ω about an axis through its center. What is the magnetic field strength at the center of the disk?

79. A long, straight conducting wire of radius R has a nonuniform current density $J = J_0 r/R$, where J_0 is a constant. The wire carries total current I.
a. Find an expression for J_0 in terms of I and R.
b. Find an expression for the magnetic field strength inside the wire at radius r.
c. At the boundary, $r = R$, does your solution match the known field outside a long, straight current-carrying wire?

80. The coaxial cable shown in FIGURE CP32.80 consists of a solid inner conductor of radius R_1 surrounded by a hollow, very thin outer conductor of radius R_2. The two carry equal currents I, but in *opposite* directions. The current density is uniformly distributed over each conductor.

a. Find expressions for three magnetic fields: within the inner conductor, in the space between the conductors, and outside the outer conductor.
b. Draw a graph of B versus r from $r = 0$ to $r = 2R_2$ if $R_1 = \frac{1}{3}R_2$.

FIGURE CP32.80

81. An infinitely wide flat sheet of charge flows out of the page in FIGURE CP32.81. The current per unit width along the sheet (amps per meter) is given by the linear current density J_s.
a. What is the *shape* of the magnetic field? To answer this question, you may find it helpful to approximate the current sheet as many parallel, closely spaced current-carrying wires. Give your answer as a picture showing magnetic field vectors.
b. Find the magnetic field strength at distance d above or below the current sheet.

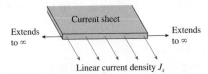

FIGURE CP32.81

82. The uniform 30 mT magnetic field in FIGURE CP32.82 points in the positive z-direction. An electron enters the region of magnetic field with a speed of 5.0×10^6 m/s and at an angle of $30°$ above the xy-plane. Find the radius r and the pitch p of the electron's spiral trajectory.

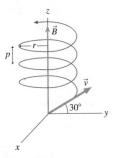

FIGURE CP32.82

STOP TO THINK ANSWERS

Stop to Think 32.1: Not at all. The charge exerts weak, attractive polarization forces on both ends of the compass needle, but in this configuration the forces will balance and have no net effect.

Stop to Think 32.2: d. Point your right thumb in the direction of the current and curl your fingers around the wire.

Stop to Think 32.3: b. Point your right thumb out of the page, in the direction of $\vec{v}$. Your fingers are pointing down as they curl around the left side.

Stop to Think 32.4: b. The right-hand rule gives a downward $\vec{B}$ for a clockwise current. The north pole is on the side from which the field emerges.

Stop to Think 32.5: c. For a field pointing into the page, $\vec{v} \times \vec{B}$ is to the right. But the electron is negative, so the force is in the direction of $-(\vec{v} \times \vec{B})$.

Stop to Think 32.6: b. Repulsion indicates that the south pole of the loop is on the right, facing the bar magnet; the north pole is on the left. Then the right-hand rule gives the current direction.

Stop to Think 32.7: a or c. Any magnetic field to the right, whether leaving a north pole or entering a south pole, will align the magnetic domains as shown.

33 Electromagnetic Induction

Electromagnetic induction is the physics that underlies many modern technologies, from the generation of electricity to data storage.

▶ **Looking Ahead** The goal of Chapter 33 is to understand and apply electromagnetic induction.

Magnetic Flux

A key idea will be the amount of magnetic field passing through a loop. This is called the **magnetic flux.**

You'll find that the magnetic flux depends on the strength of the magnetic field, the area of the loop, and the angle between them.

◀ **Looking Back**
Sections 27.2–27.3 Electric flux

Connecting *E* and *B*

We previously found that a current generates a magnetic field. In fact, the connection between electric and magnetic fields is much more profound.

You'll learn that pushing a magnet into a coil of wire, or pulling it out, causes an **induced current** in the wire. The process is called **electromagnetic induction.**

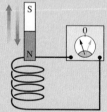

◀ **Looking Back**
Chapter 32 Magnetic fields and forces

Applications

Electromagnetic induction has many applications. You'll learn about using **inductors**—coils of wire that store magnetic energy—in circuits.

You'll also learn how a generator, such as this **generator** turned by windmill blades, transforms mechanical energy into electric energy.

Lenz's Law

Lenz's law says that a current is induced in a closed loop *if and only if the magnetic flux through the loop is changing.* Simply having a magnetic flux doesn't do anything; the flux has to *change.*

You'll learn how to use Lenz's law to determine the direction of an induced current.

◀ **Looking Back**
Section 29.2 Sources of potential

Faraday's Law

You'll learn to use **Faraday's law,** the most important law connecting electric and magnetic fields.

The magnetic flux through this loop is increasing as the loop moves into the field. Faraday's law allows us to compute the induced emf and the induced current. The current direction is given by Lenz's law.

Induced Fields

At its most fundamental level, Faraday's law tells us that a changing magnetic field creates an **induced electric field.** It is the induced electric field that then creates the induced current in a conducting loop.

An increasing magnetic field (blue) creates an electric field (red) that circulates in closed loops. You'll learn to calculate the strength of the induced field.

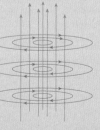

33.1 Induced Currents

Oersted's 1820 discovery that a current creates a magnetic field generated enormous excitement. One question scientists hoped to answer was whether the converse of Oersted's discovery was true: that is, can a magnet be used to create a current?

The breakthrough came in 1831 when the American science teacher Joseph Henry and the English scientist Michael Faraday each discovered the process we now call *electromagnetic induction*. Faraday—whom you met in Chapter 25 as the inventor of the concept of a *field*—was the first to publish his findings, so today we study Faraday's law rather than Henry's law.

Faraday's 1831 discovery, like Oersted's, was a happy combination of an unplanned event and a mind that was ready to recognize its significance. Faraday was experimenting with two coils of wire wrapped around an iron ring, as shown in FIGURE 33.1. He had hoped that the magnetic field generated in the coil on the left would induce a magnetic field in the iron, and that the magnetic field in the iron might then somehow create a current in the circuit on the right.

Like all his previous attempts, this technique failed to generate a current. But Faraday happened to notice that the needle of the current meter jumped ever so slightly at the instant he closed the switch in the circuit on the left. After the switch was closed, the needle immediately returned to zero. The needle again jumped when he later opened the switch, but this time in the opposite direction. Faraday recognized that the motion of the needle indicated a current in the circuit on the right, but a momentary current only during the brief interval when the current on the left was starting or stopping.

Faraday's observations, coupled with his mental picture of field lines, led him to suggest that a current is generated only if the magnetic field through the coil is *changing*. This explains why all the previous attempts to generate a current with static magnetic fields had been unsuccessful. Faraday set out to test this hypothesis.

FIGURE 33.1 Faraday's discovery.

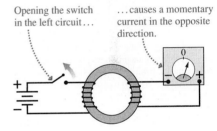

Closing the switch in the left circuit . . . . . . causes a momentary current in the right circuit.

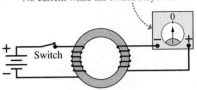

No current while the switch stays closed

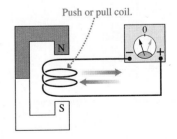

Opening the switch in the left circuit . . . . . . causes a momentary current in the opposite direction.

Faraday investigates electromagnetic induction

Faraday placed one coil directly above the other, without the iron ring. There was no current in the lower circuit while the switch was in the closed position, but a momentary current appeared whenever the switch was opened or closed.

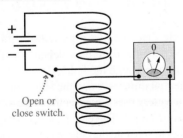

Opening or closing the switch creates a momentary current.

He pushed a bar magnet into a coil of wire. This action caused a momentary deflection of the current-meter needle, although *holding* the magnet inside the coil had no effect. A quick withdrawal of the magnet deflected the needle in the other direction.

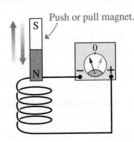

Push or pull magnet.

Pushing the magnet into the coil or pulling it out creates a momentary current.

Must the magnet move? Faraday created a momentary current by rapidly pulling a coil of wire out of a magnetic field. Pushing the coil *into* the magnet caused the needle to deflect in the opposite direction.

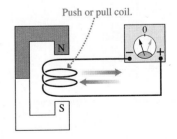

Push or pull coil.

Pushing the coil into the magnet or pulling it out creates a momentary current.

Faraday found that there is a current in a coil of wire if and only if the magnetic field passing through the coil is *changing*. This is an informal statement of what we'll soon call *Faraday's law*. The current in a circuit due to a changing magnetic field is called an **induced current**. An induced current is not caused by a battery; it is a completely new way to generate a current.

33.2 Motional emf

We'll start our investigation of electromagnetic induction by looking at situations in which the magnetic field is fixed while the circuit moves or changes. Consider a conductor of length l that moves with velocity $\vec{v}$ through a perpendicular uniform magnetic field $\vec{B}$, as shown in FIGURE 33.2. The charge carriers inside the wire—assumed to be positive—also move with velocity $\vec{v}$, so they each experience a magnetic force $\vec{F}_B = q\vec{v} \times \vec{B}$ of strength $F_B = qvB$. This force causes the charge carriers to move, separating the positive and negative charges. The separated charges then create an electric field inside the conductor.

FIGURE 33.2 The magnetic force on the charge carriers in a moving conductor creates an electric field inside the conductor.

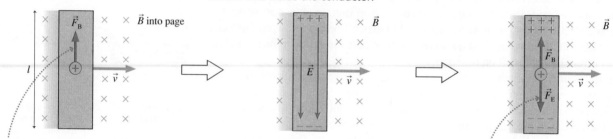

Charge carriers in the wire experience an upward force of magnitude $F_B = qvB$. Being free to move, positive charges flow upward (or, if you prefer, negative charges downward).

The charge separation creates an electric field in the conductor. $\vec{E}$ increases as more charge flows.

The charge flow continues until the downward electric force $\vec{F}_E$ is large enough to balance the upward magnetic force $\vec{F}_B$. Then the net force on a charge is zero and the current ceases.

The charge carriers continue to separate until the electric force $F_E = qE$ exactly balances the magnetic force $F_B = qvB$, creating an equilibrium situation. This balance happens when the electric field strength is

$$E = vB \tag{33.1}$$

In other words, **the magnetic force on the charge carriers in a moving conductor creates an electric field $E = vB$ inside the conductor.**

The electric field, in turn, creates an electric potential difference between the two ends of the moving conductor. FIGURE 33.3a defines a coordinate system in which $\vec{E} = -vB\hat{j}$. Using the connection between the electric field and the electric potential,

$$\Delta V = V_{top} - V_{bottom} = -\int_0^l E_y\,dy = -\int_0^l (-vB)\,dy = vlB \tag{33.2}$$

Thus **the motion of the wire through a magnetic field *induces* a potential difference vlB between the ends of the conductor.** The potential difference depends on the strength of the magnetic field and on the wire's speed through the field.

There's an important analogy between this potential difference and the potential difference of a battery. FIGURE 33.3b reminds you that a battery uses a nonelectric force—the charge escalator—to separate positive and negative charges. The emf $\mathcal{E}$ of the battery was defined as the work performed per charge (W/q) to separate the charges. An isolated battery, with no current, has a potential difference $\Delta V_{bat} = \mathcal{E}$. We could refer to a battery, where the charges are separated by chemical reactions, as a source of *chemical emf.*

The moving conductor develops a potential difference because of the work done by magnetic forces to separate the charges. You can think of the moving conductor as a "battery" that stays charged only as long as it keeps moving but "runs down" if it stops. The emf of the conductor is due to its motion, rather than to chemical reactions inside, so we can define the **motional emf** of a conductor moving with velocity $\vec{v}$ perpendicular to a magnetic field $\vec{B}$ to be

$$\mathcal{E} = vlB \tag{33.3}$$

FIGURE 33.3 Generating an emf.

(a) Magnetic forces separate the charges and cause a potential difference between the ends. This is a motional emf.

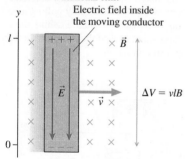

(b) Chemical reactions separate the charges and cause a potential difference between the ends. This is a chemical emf.

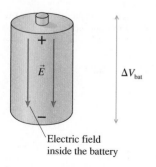

Electric field inside the battery

STOP TO THINK 33.1 A square conductor moves through a uniform magnetic field. Which of the figures shows the correct charge distribution on the conductor?

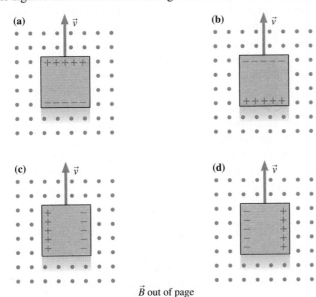

$\vec{B}$ out of page

EXAMPLE 33.1 **Measuring the earth's magnetic field**

It is known that the earth's magnetic field over northern Canada points straight down. The crew of a Boeing 747 aircraft flying at 260 m/s over northern Canada finds a 0.95 V potential difference between the wing tips. The wing span of a Boeing 747 is 65 m. What is the magnetic field strength there?

MODEL The wing is a conductor moving through a magnetic field, so there is a motional emf.

SOLVE The magnetic field is perpendicular to the velocity, so we can use Equation 33.3 to find

$$B = \frac{\mathcal{E}}{vL} = \frac{0.95 \text{ V}}{(260 \text{ m/s})(65 \text{ m})} = 5.6 \times 10^{-5} \text{ T}$$

ASSESS Chapter 32 noted that the earth's magnetic field is roughly 5×10^{-5} T. The field is somewhat stronger than this near the magnetic poles, somewhat weaker near the equator.

EXAMPLE 33.2 **Potential difference along a rotating bar**

A metal bar of length l rotates with angular velocity ω about a pivot at one end of the bar. A uniform magnetic field $\vec{B}$ is perpendicular to the plane of rotation. What is the potential difference between the ends of the bar?

VISUALIZE **FIGURE 33.4** is a pictorial representation of the bar. The magnetic forces on the charge carriers will cause the outer end to be positive with respect to the pivot.

FIGURE 33.4 Pictorial representation of a metal bar rotating in a magnetic field.

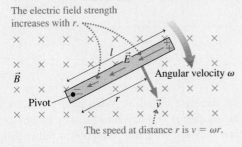

The electric field strength increases with r.

$\vec{B}$

Pivot

Angular velocity ω

The speed at distance r is $v = \omega r$.

SOLVE Even though the bar is rotating, rather than moving in a straight line, the velocity of each charge carrier is perpendicular to $\vec{B}$. Consequently, the electric field created inside the bar is exactly that given in Equation 33.1, $E = vB$. But v, the speed of the charge carrier, now depends on its distance from the pivot. Recall that in rotational motion the tangential speed at radius r from the center of rotation is $v = \omega r$. Thus the electric field at distance r from the pivot is $E = \omega r B$. The electric field increases in strength as you move outward along the bar.

The electric field $\vec{E}$ points toward the pivot, so its radial component is $E_r = -\omega r B$. If we integrate outward from the center, the potential difference between the ends of the bar is

$$\Delta V = V_{\text{tip}} - V_{\text{pivot}} = -\int_0^l E_r \, dr$$

$$= -\int_0^l (-\omega r B) \, dr = \omega B \int_0^l r \, dr = \frac{1}{2} \omega l^2 B$$

ASSESS $\frac{1}{2}\omega l$ is the speed at the midpoint of the bar. Thus ΔV is $v_{\text{mid}} l B$, which seems reasonable.

Induced Current in a Circuit

The moving conductor of Figure 33.2 had an emf, but it couldn't sustain a current because the charges had nowhere to go. It's like a battery that is disconnected from a circuit. We can change this by including the moving conductor in a circuit.

FIGURE 33.5 shows a conducting wire sliding with speed v along a U-shaped conducting rail. We'll assume that the rail is attached to a table and cannot move. The wire and the rail together form a closed conducting loop—a circuit.

Suppose a magnetic field $\vec{B}$ is perpendicular to the plane of the circuit. Charges in the moving wire will be pushed to the ends of the wire by the magnetic force, just as they were in Figure 33.2, but now the charges can continue to flow around the circuit. That is, the moving wire acts like a battery in a circuit.

The current in the circuit is an *induced current.* In this example, the induced current is counterclockwise (ccw). If the total resistance of the circuit is R, the induced current is given by Ohm's law as

$$I = \frac{\mathcal{E}}{R} = \frac{vlB}{R} \qquad (33.4)$$

In this situation, the induced current is due to magnetic forces on moving charges.

We've assumed that the wire is moving along the rail at constant speed. It turns out that we must apply a continuous pulling force $\vec{F}_{pull}$ to make this happen. FIGURE 33.6 shows why. The moving wire, which now carries induced current I, is in a magnetic field. You learned in Chapter 32 that a magnetic field exerts a force on a current-carrying wire. According to the right-hand rule, the magnetic force $\vec{F}_{mag}$ on the moving wire points to the left. This "magnetic drag" will cause the wire to slow down and stop *unless* we exert an equal but opposite pulling force $\vec{F}_{pull}$ to keep the wire moving.

The magnitude of the magnetic force on a current-carrying wire was found in Chapter 32 to be $F_{mag} = IlB$. Using that result, along with Equation 33.4 for the induced current, we find that the force required to pull the wire with a constant speed v is

$$F_{pull} = F_{mag} = IlB = \left(\frac{vlB}{R}\right)lB = \frac{vl^2B^2}{R} \qquad (33.5)$$

FIGURE 33.5 A current is induced in the circuit as the wire moves through a magnetic field.

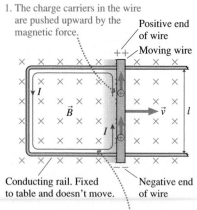

1. The charge carriers in the wire are pushed upward by the magnetic force.

Positive end of wire

Moving wire

Conducting rail. Fixed to table and doesn't move.

Negative end of wire

2. The charge carriers flow around the conducting loop as an induced current.

FIGURE 33.6 A pulling force is needed to move the wire to the right.

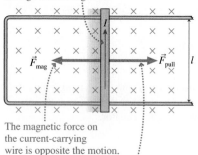

The induced current flows through the moving wire.

The magnetic force on the current-carrying wire is opposite the motion.

A pulling force to the right must balance the magnetic force to keep the wire moving at constant speed. This force does work on the wire.

STOP TO THINK 33.2 Is there an induced current in this circuit? If so, what is its direction?

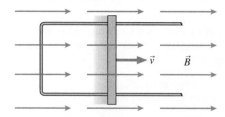

Energy Considerations

The environment must do work on the wire to pull it. What happens to the energy transferred to the wire by this work? Is energy conserved as the wire moves along the rail? It will be easier to answer this question if we think about power rather than work. Power is the *rate* at which work is done on the wire. You learned in Chapter 11 that the power exerted by a force pushing or pulling an object with velocity v is $P = Fv$. The power provided to the circuit by pulling on the wire is

$$P_{\text{input}} = F_{\text{pull}}v = \frac{v^2l^2B^2}{R} \qquad (33.6)$$

This is the rate at which energy is added to the circuit by the pulling force.

But the circuit also dissipates energy by transforming electric energy into the thermal energy of the wires and components, heating them up. The power dissipated by current I as it passes through resistance R is $P = I^2R$. Equation 33.4 for the induced current I gives us the power dissipated by the circuit of Figure 33.5:

$$P_{\text{dissipated}} = I^2R = \frac{v^2l^2B^2}{R} \qquad (33.7)$$

You can see that Equations 33.6 and 33.7 are identical. **The rate at which work is done on the circuit exactly balances the rate at which energy is dissipated.** Thus *energy is conserved.*

If you have to *pull* on the wire to get it to move to the right, you might think that it would spring back to the left on its own. FIGURE 33.7 shows the same circuit with the wire moving to the left. In this case, you must *push* the wire to the left to keep it moving. The magnetic force is always opposite to the wire's direction of motion.

In both Figure 33.6, where the wire is pulled, and Figure 33.7, where it is pushed, a mechanical force is used to create a current. In other words, we have a conversion of *mechanical* energy to *electric* energy. A device that converts mechanical energy to electric energy is called a **generator.** The slide-wire circuits of Figures 33.6 and 33.7 are simple examples of a generator. We will look at more practical examples of generators later in the chapter.

We can summarize our analysis as follows:

1. Pulling or pushing the wire through the magnetic field at speed v creates a motional emf $\mathcal{E}$ in the wire and induces a current $I = \mathcal{E}/R$ in the circuit.
2. To keep the wire moving at constant speed, a pulling or pushing force must balance the magnetic force on the wire. This force does work on the circuit.
3. The work done by the pulling or pushing force exactly balances the energy dissipated by the current as it passes through the resistance of the circuit.

FIGURE 33.7 A pushing force is needed to move the wire to the left.

1. The magnetic force on the charge carriers is down, so the induced current flows clockwise.

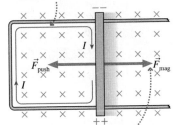

2. The magnetic force on the current-carrying wire is to the right.

EXAMPLE 33.3   **Lighting a bulb**

FIGURE 33.8 shows a circuit consisting of a flashlight bulb, rated 3.0 V/1.5 W, and ideal wires with no resistance. The right wire of the circuit, which is 10 cm long, is pulled at constant speed v through a perpendicular magnetic field of strength 0.10 T.

a. What speed must the wire have to light the bulb to full brightness?

b. What force is needed to keep the wire moving?

FIGURE 33.8 Circuit of Example 33.3.

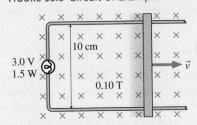

MODEL Treat the moving wire as a source of motional emf.

VISUALIZE The magnetic force on the charge carriers, $\vec{F}_B = q\vec{v} \times \vec{B}$, causes a counterclockwise (ccw) induced current.

SOLVE a. The bulb's rating of 3.0 V/1.5 W means that at full brightness it will dissipate 1.5 W at a potential difference of 3.0 V. Because the power is related to the voltage and current by $P = I\Delta V$, the current causing full brightness is

$$I = \frac{P}{\Delta V} = \frac{1.5 \text{ W}}{3.0 \text{ V}} = 0.50 \text{ A}$$

The bulb's resistance—the total resistance of the circuit—is

$$R = \frac{\Delta V}{I} = \frac{3.0 \text{ V}}{0.50 \text{ A}} = 6.0 \text{ } \Omega$$

Equation 33.4 gives the speed needed to induce this current:

$$v = \frac{IR}{lB} = \frac{(0.50 \text{ A})(6.0 \text{ } \Omega)}{(0.10 \text{ m})(0.10 \text{ T})} = 300 \text{ m/s}$$

You can confirm from Equation 33.6 that the input power at this speed is 1.5 W.

Continued

b. From Equation 33.5, the pulling force must be

$$F_{\text{pull}} = \frac{vl^2B^2}{R} = 5.0 \times 10^{-3} \text{ N}$$

You can also obtain this result from $F_{\text{pull}} = P/v$.

ASSESS Example 33.1 showed that high speeds are needed to produce significant potential difference. Thus 300 m/s is not surprising. The pulling force is not very large, but even a small force can deliver large amounts of power $P = Fv$ when v is large.

FIGURE 33.9 Eddy currents.

(a) Eddy currents are induced when a metal sheet is pulled through a magnetic field.

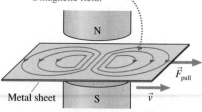

(b) The magnetic force on the eddy currents is opposite in direction to $\vec{v}$.

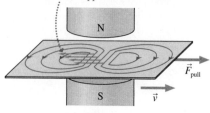

FIGURE 33.10 Magnetic braking system.

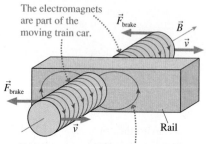

Eddy currents are induced in the rail. Magnetic forces between the eddy currents and the electromagnets slow the train.

Eddy Currents

These ideas have interesting implications. Consider pulling a *sheet* of metal through a magnetic field, as shown in FIGURE 33.9a. The metal, we will assume, is not a magnetic material, so it experiences no magnetic force if it is at rest. The charge carriers in the metal experience a magnetic force as the sheet is dragged between the pole tips of the magnet. A current is induced, just as in the loop of wire, but here the currents do not have wires to define their path. As a consequence, two "whirlpools" of current begin to circulate in the metal. These spread-out current whirlpools in a solid metal are called **eddy currents.**

FIGURE 33.9b shows the magnetic force on the eddy current as it passes between the pole tips. This force is to the left, acting as a retarding force. Thus **an external force is required to pull a metal through a magnetic field.** If the pulling force ceases, the retarding magnetic force quickly causes the metal to decelerate until it stops. Similarly, a force is required to push a sheet of metal *into* a magnetic field.

Eddy currents are often undesirable. The power dissipation of eddy currents can cause unwanted heating, and the magnetic forces on eddy currents mean that extra energy must be expended to move metals in magnetic fields. But eddy currents also have important useful applications. A good example is magnetic braking.

The moving train car has an electromagnet that straddles the rail, as shown in FIGURE 33.10. During normal travel, there is no current through the electromagnet and no magnetic field. To stop the car, a current is switched into the electromagnet. The current creates a strong magnetic field that passes *through* the rail, and the motion of the rail relative to the magnet induces eddy currents in the rail. The magnetic force between the electromagnet and the eddy currents acts as a braking force on the magnet and, thus, on the car. Magnetic braking systems are very efficient, and they have the added advantage that they heat the rail rather than the brakes.

STOP TO THINK 33.3 A square loop of copper wire is pulled through a region of magnetic field. Rank in order, from strongest to weakest, the pulling forces $\vec{F}_a$, $\vec{F}_b$, $\vec{F}_c$, and $\vec{F}_d$ that must be applied to keep the loop moving at constant speed.

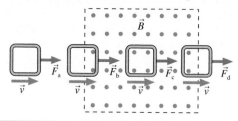

33.3 Magnetic Flux

Faraday found that a current is induced when the amount of magnetic field passing through a coil or a loop of wire changes. And that's exactly what happens as the slide wire moves down the rail in Figure 33.5! As the circuit expands, more magnetic field passes through. It's time to define more clearly what we mean by "the amount of field passing through a loop."

Imagine holding a rectangular loop of wire in front of a fan, as shown in FIGURE 33.11. The amount of air that flows through the loop depends on the effective area of the loop as seen along the direction of flow. You can see from the figure that the effective area (i.e., as seen facing the fan) is

$$A_{eff} = ab\cos\theta = A\cos\theta \qquad (33.8)$$

where $A = ab$ is the area of the loop and θ is the tilt angle of the loop. A loop perpendicular to the flow, with $\theta = 0°$, has $A_{eff} = A$, the full area of the loop. No air at all flows through the loop if it is tilted 90°, and you can see that $A_{eff} = 0$ in this case.

FIGURE 33.11 The amount of air flowing through a loop depends on the effective area of the loop.

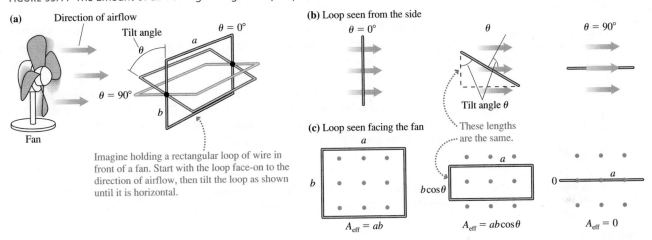

We can apply this idea to a magnetic field passing through a loop. FIGURE 33.12 shows a loop of area $A = ab$ in a uniform magnetic field. Think of the field vectors, seen here from behind, as if they were arrows shot into the page. The density of arrows (arrows per m²) is proportional to the strength B of the magnetic field; a stronger field would be represented by arrows packed closer together. The number of arrows passing through a loop of wire depends on two factors:

1. The density of arrows, which is proportional to B, and
2. The effective area $A_{eff} = A\cos\theta$ of the loop.

FIGURE 33.12 Magnetic field through a loop that is tilted at various angles.

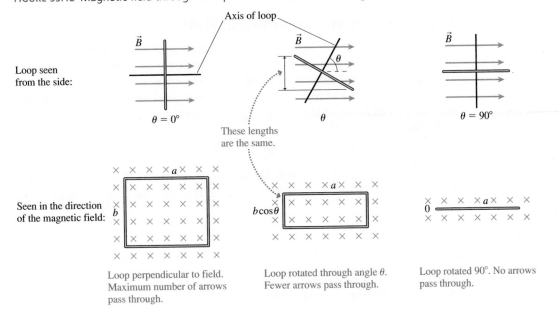

The angle θ is the angle between the magnetic field and the axis of the loop. The maximum number of arrows passes through the loop when it is perpendicular to the magnetic field ($\theta = 0°$). No arrows pass through the loop if it is tilted 90°.

With this in mind, let's define the **magnetic flux Φ_m** as

$$\Phi_m = A_{eff}B = AB\cos\theta \qquad (33.9)$$

The magnetic flux measures the amount of magnetic field passing through a loop of area A if the loop is tilted at angle θ from the field. The SI unit of magnetic flux is the **weber.** From Equation 33.9 you can see that

$$1\text{ weber} = 1\text{ Wb} = 1\text{ T m}^2$$

Equation 33.9 is reminiscent of the vector dot product: $\vec{A} \cdot \vec{B} = AB\cos\theta$. With that in mind, let's define an **area vector** $\vec{A}$ to be a vector *perpendicular* to the loop, with magnitude equal to the area A of the loop. Vector $\vec{A}$ has units of m^2. FIGURE 33.13a shows the area vector $\vec{A}$ for a circular loop of area A.

FIGURE 33.13b shows a magnetic field passing through a loop. The angle between vectors $\vec{A}$ and $\vec{B}$ is the same angle used in Equations 33.8 and 33.9 to define the effective area and the magnetic flux. So Equation 33.9 really is a dot product, and we can define the magnetic flux more concisely as

$$\Phi_m = \vec{A} \cdot \vec{B} \qquad (33.10)$$

Writing the flux as a dot product helps make clear how angle θ is defined: θ is the angle between the magnetic field and the axis of the loop.

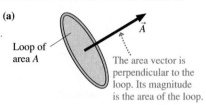

FIGURE 33.13 Magnetic flux can be defined in terms of an area vector $\vec{A}$.

(a)

Loop of area A

The area vector is perpendicular to the loop. Its magnitude is the area of the loop.

(b)

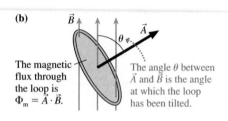

The magnetic flux through the loop is $\Phi_m = \vec{A} \cdot \vec{B}$.

The angle θ between $\vec{A}$ and $\vec{B}$ is the angle at which the loop has been tilted.

EXAMPLE 33.4 **A circular loop in a magnetic field**

FIGURE 33.14 is an edge view of a 10-cm-diameter circular loop in a uniform 0.050 T magnetic field. What is the magnetic flux through the loop?

SOLVE Angle θ is the angle between the loop's area vector $\vec{A}$, which is perpendicular to the plane of the loop, and the magnetic field $\vec{B}$. In this case, $\theta = 60°$, not the 30° angle shown in the figure. Vector $\vec{A}$ has magnitude $A = \pi r^2 = 7.85 \times 10^{-3}$ m^2. Thus the magnetic flux is

$$\Phi_m = \vec{A} \cdot \vec{B} = AB\cos\theta = 2.0 \times 10^{-4}\text{ Wb}$$

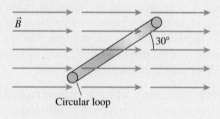

FIGURE 33.14 A circular loop in a magnetic field.

Circular loop

Magnetic Flux in a Nonuniform Field

Equation 33.10 for the magnetic flux assumes that the field is uniform over the area of the loop. We can calculate the flux in a nonuniform field, one where the field strength changes from one edge of the loop to the other, but we'll need to use calculus.

FIGURE 33.15 shows a loop in a nonuniform magnetic field. Imagine dividing the loop into many small pieces of area dA. The infinitesimal flux $d\Phi_m$ through one such area, where the magnetic field is $\vec{B}$, is

$$d\Phi_m = \vec{B} \cdot d\vec{A} \qquad (33.11)$$

The total magnetic flux through the loop is the sum of the fluxes through each of the small areas. We find that sum by integrating. Thus the total magnetic flux through the loop is

$$\Phi_m = \int_{\text{area of loop}} \vec{B} \cdot d\vec{A} \qquad (33.12)$$

Equation 33.12 is a more general definition of magnetic flux. It may look rather formidable, so we'll illustrate its use with an example.

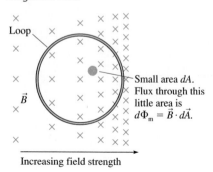

FIGURE 33.15 A loop in a nonuniform magnetic field.

Loop

Small area dA. Flux through this little area is $d\Phi_m = \vec{B} \cdot d\vec{A}$.

$\vec{B}$

Increasing field strength

EXAMPLE 33.5 **Magnetic flux from the current in a long straight wire**

The 1.0 cm × 4.0 cm rectangular loop of FIGURE 33.16 is 1.0 cm away from a long straight wire. The wire carries a current of 1.0 A. What is the magnetic flux through the loop?

FIGURE 33.16 A loop next to a current carrying wire.

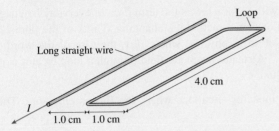

MODEL We'll treat the wire as if it were infinitely long. The magnetic field strength of a wire decreases with distance from the wire, so the field is *not* uniform over the area of the loop.

FIGURE 33.17 Calculating the magnetic flux through the loop.

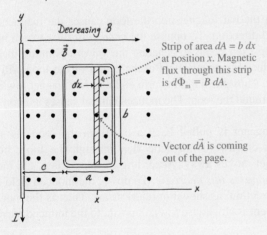

VISUALIZE Using the right-hand rule, we see that the field, as it circles the wire, is perpendicular to the plane of the loop. FIGURE 33.17 redraws the loop with the field coming out of the page and establishes a coordinate system.

SOLVE Let the loop have dimensions a and b, as shown, with the near edge at distance c from the wire. The magnetic field varies with distance x from the wire, but the field is constant along a line parallel to the wire. This suggests dividing the loop into many narrow rectangular strips of length b and width dx, each forming a small area $dA = b\,dx$. The magnetic field has the same strength at all points within this small area. One such strip is shown in the figure at position x.

The area vector $d\vec{A}$ is perpendicular to the strip (coming out of the page), which makes it parallel to $\vec{B}$ ($\theta = 0°$). Thus the infinitesimal flux through this little area is

$$d\Phi_m = \vec{B} \cdot d\vec{A} = B\,dA = B(b\,dx) = \frac{\mu_0 I b}{2\pi x}dx$$

where, from Chapter 32, we've used $B = \mu_0 I/2\pi x$ as the magnetic field at distance x from a long straight wire. Integrating "over the area of the loop" means to integrate from the near edge of the loop at $x = c$ to the far edge at $x = c + a$. Thus

$$\Phi_m = \frac{\mu_0 I b}{2\pi}\int_c^{c+a}\frac{dx}{x} = \frac{\mu_0 I b}{2\pi}\ln x \Big|_c^{c+a} = \frac{\mu_0 I b}{2\pi}\ln\left(\frac{c+a}{c}\right)$$

Evaluating for $a = c = 0.010$ m, $b = 0.040$ m, and $I = 1.0$ A gives

$$\Phi_m = 5.5 \times 10^{-9} \text{ Wb}$$

ASSESS The flux measures how much of the wire's magnetic field passes through the loop, but we had to integrate, rather than simply using Equation 33.10, because the field is stronger at the near edge of the loop than at the far edge.

33.4 Lenz's Law

We started out by looking at a situation in which a moving wire caused a loop to expand in a magnetic field. This is one way to change the magnetic flux through the loop. But Faraday found that a current can be induced by any change in the magnetic flux, no matter how it's accomplished.

For example, a momentary current is induced in the loop of FIGURE 33.18 as the bar magnet is pushed toward the loop, increasing the flux through the loop. Pulling the magnet back out of the loop causes the current meter to deflect in the opposite direction. The conducting wires aren't moving, so this is not a motional emf. Nonetheless, the induced current is very real.

The German physicist Heinrich Lenz began to study electromagnetic induction after learning of Faraday's discovery. Three years later, in 1834, Lenz announced a rule for determining the direction of the induced current. We now call his rule **Lenz's law,** and it can be stated as follows:

FIGURE 33.18 Pushing a bar magnet toward the loop induces a current.

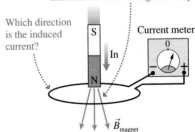

> **Lenz's law** There is an induced current in a closed, conducting loop if and only if the magnetic flux through the loop is changing. The direction of the induced current is such that the induced magnetic field opposes the *change* in the flux.

Lenz's law is rather subtle, and it takes some practice to see how to apply it.

NOTE ▶ One difficulty with Lenz's law is the term *flux*. In everyday language, the word *flux* already implies that something is changing. Think of the phrase, "The situation is in flux." Not so in physics, where *flux*, the root of the word *flow*, means "passes through." A steady magnetic field through a loop creates a steady, *un*changing magnetic flux. ◀

Lenz's law tells us to look for situations where the flux is *changing*. This can happen in three ways.

1. The magnetic field through the loop changes (increases or decreases),
2. The loop changes in area or angle, or
3. The loop moves into or out of a magnetic field.

Lenz's law depends on the idea that an induced current generates its own magnetic field $\vec{B}_{induced}$. This is the *induced magnetic field* of Lenz's law. You learned in Chapter 32 how to use the right-hand rule to determine the direction of this induced magnetic field.

In Figure 33.18, pushing the bar magnet into the loop causes the magnetic flux to *increase* in the downward direction. To oppose the *change* in flux, which is what Lenz's law requires, the loop itself needs to generate the *upward*-pointing magnetic field of FIGURE 33.19. The induced magnetic field at the center of the loop will point upward if the current is ccw. Thus pushing the north end of a bar magnet toward the loop induces a ccw current around the loop. The induced current ceases as soon as the magnet stops moving.

Now suppose the bar magnet is pulled back away from the loop, as shown in FIGURE 33.20a. There is a downward magnetic flux through the loop, but the flux *decreases* as the magnet moves away. According to Lenz's law, the induced magnetic field of the loop *opposes this decrease*. To do so, the induced field needs to point in the *downward* direction, as shown in FIGURE 33.20b. Thus as the magnet is withdrawn, the induced current is clockwise (cw), opposite to the induced current of Figure 33.19.

FIGURE 33.19 The induced current is ccw.

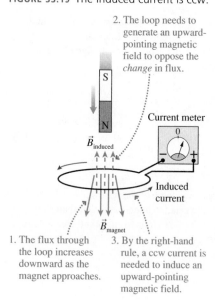

2. The loop needs to generate an upward-pointing magnetic field to oppose the *change* in flux.

1. The flux through the loop increases downward as the magnet approaches.

3. By the right-hand rule, a ccw current is needed to induce an upward-pointing magnetic field.

FIGURE 33.20 Pulling the magnet away induces a cw current.

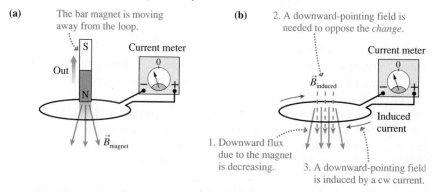

(a) The bar magnet is moving away from the loop.

(b) 2. A downward-pointing field is needed to oppose the *change*.

1. Downward flux due to the magnet is decreasing.

3. A downward-pointing field is induced by a cw current.

NOTE ▶ Notice that the magnetic field of the bar magnet is pointing downward in both Figures 33.19 and 33.20. It is not the *flux* due to the magnet that the induced current opposes, but the *change* in the flux. This is a subtle but critical distinction.

If the induced current opposed the flux itself, the current in both Figures 33.19 and 33.20 would be ccw to generate an upward magnetic field. But that's not what happens. When the field of the magnet points down and is increasing, the induced current opposes the increase by generating an upward field. When the field of the magnet points down but is decreasing, the induced current opposes the decrease by generating a downward field. ◄

FIGURE 33.21 shows six basic situations. The magnetic field can point either up or down through the loop. For each, the flux can either increase, hold steady, or decrease in strength. These observations form the basis for a set of rules about using Lenz's law.

FIGURE 33.21 The induced current for six different situations.

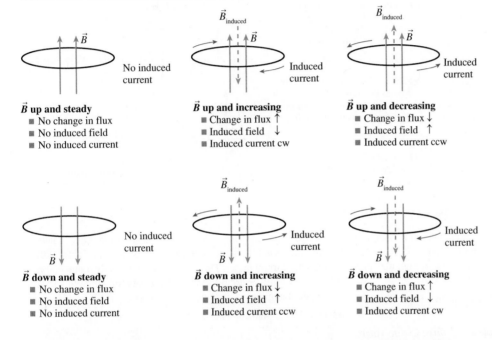

$\vec{B}$ **up and steady**
- No change in flux
- No induced field
- No induced current

$\vec{B}$ **up and increasing**
- Change in flux ↑
- Induced field ↓
- Induced current cw

$\vec{B}$ **up and decreasing**
- Change in flux ↓
- Induced field ↑
- Induced current ccw

$\vec{B}$ **down and steady**
- No change in flux
- No induced field
- No induced current

$\vec{B}$ **down and increasing**
- Change in flux ↓
- Induced field ↑
- Induced current ccw

$\vec{B}$ **down and decreasing**
- Change in flux ↑
- Induced field ↓
- Induced current cw

TACTICS
BOX 33.1   Using Lenz's law   (MP)

❶ **Determine the direction of the applied magnetic field.** The field must pass through the loop.

❷ **Determine how the flux is changing.** Is it increasing, decreasing, or staying the same?

❸ **Determine the direction of an induced magnetic field that will oppose the *change* in the flux.**

- Increasing flux: the induced magnetic field points opposite the applied magnetic field.

- Decreasing flux: the induced magnetic field points in the same direction as the applied magnetic field.

- Steady flux: there is no induced magnetic field.

❹ **Determine the direction of the induced current.** Use the right-hand rule to determine the current direction in the loop that generates the induced magnetic field you found in step 3.

Exercises 10–14

Let's look at some examples.

EXAMPLE 33.6 | **Lenz's law 1**

FIGURE 33.22 shows two loops, one above the other. The upper loop has a battery and a switch that has been closed for a long time. How does the lower loop respond when the switch is opened in the upper loop?

MODEL We'll use the right-hand rule to find the magnetic fields of current loops.

SOLVE FIGURE 33.23 shows the four steps of using Lenz's law. Opening the switch induces a ccw current in the lower loop. This is a momentary current, lasting only until the magnetic field of the upper loop drops to zero.

ASSESS The conclusion is consistent with Figure 33.21.

FIGURE 33.22 The two loops of Example 33.6.

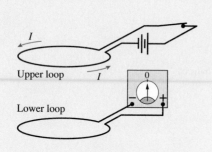

Upper loop

Lower loop

FIGURE 33.23 Applying Lenz's law.

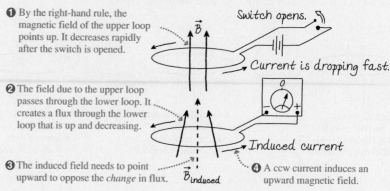

❶ By the right-hand rule, the magnetic field of the upper loop points up. It decreases rapidly after the switch is opened.

❷ The field due to the upper loop passes through the lower loop. It creates a flux through the lower loop that is up and decreasing.

❸ The induced field needs to point upward to oppose the *change* in flux.

❹ A ccw current induces an upward magnetic field.

EXAMPLE 33.7 | **Lenz's law 2**

FIGURE 33.24 shows two coils wrapped side by side on a cylinder. When the switch for coil 1 is closed, does the induced current in coil 2 pass from right to left or from left to right through the current meter?

MODEL We'll use the right-hand rule to find the magnetic field of a coil.

VISUALIZE It is very important to look at the *direction* in which a coil is wound around the cylinder. Notice that the two coils in Figure 33.24 are wound in opposite directions.

SOLVE FIGURE 33.25 shows the four steps of using Lenz's law. Closing the switch induces a current that passes from right to left through the current meter. The induced current is only momentary. It lasts only until the field from coil 1 reaches full strength and is no longer changing.

ASSESS The conclusion is consistent with Figure 33.21.

FIGURE 33.24 The two solenoids of Example 33.7.

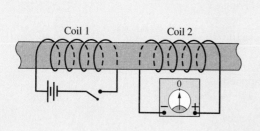

Coil 1   Coil 2

FIGURE 33.25 Applying Lenz's law.

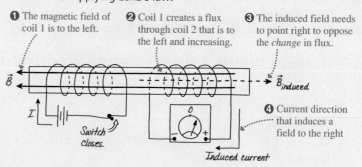

❶ The magnetic field of coil 1 is to the left.

❷ Coil 1 creates a flux through coil 2 that is to the left and increasing.

❸ The induced field needs to point right to oppose the *change* in flux.

❹ Current direction that induces a field to the right

STOP TO THINK 33.4 A current-carrying wire is pulled away from a conducting loop in the direction shown. As the wire is moving, is there a cw current around the loop, a ccw current, or no current?

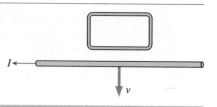

33.5 Faraday's Law

Charges don't start moving spontaneously. A current requires an emf to provide the energy. We started our analysis of induced currents with circuits in which a *motional emf* can be understood in terms of magnetic forces on moving charges. But we've also seen that a current can be induced by changing the magnetic field through a stationary circuit, a circuit in which there is no motion. There *must* be an emf in this circuit, even though the mechanism for this emf is not yet clear.

The emf associated with a changing magnetic flux, regardless of what causes the change, is called an **induced emf** $\mathcal{E}$. Then, if there is a complete circuit having resistance R, a current

$$I_{\text{induced}} = \frac{\mathcal{E}}{R} \qquad (33.13)$$

is established in the wire as a *consequence* of the induced emf. The direction of the current is given by Lenz's law. The last piece of information we need is the size of the induced emf $\mathcal{E}$.

The research of Faraday and others eventually led to the discovery of the basic law of electromagnetic induction, which we now call **Faraday's law.** It states:

> **Faraday's law** An emf $\mathcal{E}$ is induced around a closed loop if the magnetic flux through the loop changes. The magnitude of the emf is
>
> $$\mathcal{E} = \left| \frac{d\Phi_{\text{m}}}{dt} \right| \qquad (33.14)$$
>
> and the direction of the emf is such as to drive an induced current in the direction given by Lenz's law.

In other words, the induced emf is the *rate of change* of the magnetic flux through the loop.

As a corollary to Faraday's law, an N-turn coil of wire in a changing magnetic field acts like N batteries in series. The induced emf of each of the coils adds, so the induced emf of the entire coil is

$$\mathcal{E}_{\text{coil}} = N \left| \frac{d\Phi_{\text{per coil}}}{dt} \right| \qquad \text{(Faraday's law for an } N\text{-turn coil)} \qquad (33.15)$$

As a first example of using Faraday's law, return to the situation of Figure 33.5, where a wire moves through a magnetic field by sliding on a U-shaped conducting rail. FIGURE 33.26 shows the circuit again. The magnetic field $\vec{B}$ is perpendicular to the plane of the conducting loop, so $\theta = 0°$ and the magnetic flux is $\Phi = AB$, where A is the area of the loop. If the slide wire is distance x from the end, the area is $A = xl$ and the flux at that instant of time is

$$\Phi_{\text{m}} = AB = xlB \qquad (33.16)$$

The flux through the loop increases as the wire moves. According to Faraday's law, the induced emf is

$$\mathcal{E} = \left| \frac{d\Phi_{\text{m}}}{dt} \right| = \frac{d}{dt}(xlB) = \frac{dx}{dt}lB = vlB \qquad (33.17)$$

where the wire's velocity is $v = dx/dt$. We can now use Equation 33.13 to find that the induced current is

$$I = \frac{\mathcal{E}}{R} = \frac{vlB}{R} \qquad (33.18)$$

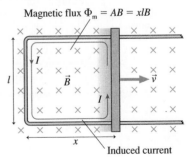

FIGURE 33.26 The magnetic flux through the loop increases as the slide wire moves.

Magnetic flux $\Phi_{\text{m}} = AB = xlB$

The flux is increasing into the loop, so the induced magnetic field opposes this increase by pointing out of the loop. This requires a ccw induced current in the loop. Faraday's law leads us to the conclusion that the loop will have a ccw induced current $I = vlB/R$. This is exactly the conclusion we reached in Section 33.2, where we analyzed the situation from the perspective of magnetic forces on moving charge carriers. Faraday's law confirms what we already knew but, at least in this case, doesn't seem to offer anything new.

Using Faraday's Law

Most electromagnetic induction problems can be solved with a four-step strategy.

> **PROBLEM-SOLVING STRATEGY 33.1** **Electromagnetic induction**
>
> **MODEL** Make simplifying assumptions about wires and magnetic fields.
>
> **VISUALIZE** Draw a picture or a circuit diagram. Use Lenz's law to determine the direction of the induced current.
>
> **SOLVE** The mathematical representation is based on Faraday's law
>
> $$\mathcal{E} = \left| \frac{d\Phi_m}{dt} \right|$$
>
> For an N-turn coil, multiply by N. The size of the induced current is $I = \mathcal{E}/R$.
>
> **ASSESS** Check that your result has the correct units, is reasonable, and answers the question.
>
> Exercise 18

EXAMPLE 33.8 **Electromagnetic induction in a solenoid**

A 2.0-cm-diameter loop of wire with a resistance of 0.010 Ω is placed in the center of the solenoid seen in FIGURE 33.27a. The solenoid is 4.0 cm in diameter, 20 cm long, and wrapped with 1000 turns of wire. FIGURE 33.27b shows the current through the

FIGURE 33.27 A loop inside a solenoid.

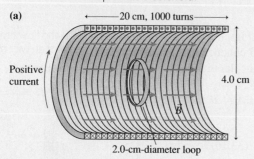

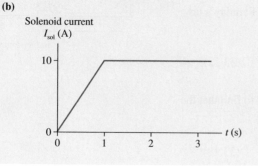

solenoid as a function of time as the solenoid is "powered up." A positive current is defined to be cw when seen from the left. Find the current in the loop as a function of time and show the result as a graph.

MODEL The solenoid's length is much greater than its diameter, so the field near the center should be nearly uniform.

VISUALIZE The magnetic field of the solenoid creates a magnetic flux through the loop of wire. The solenoid current is always positive, meaning that it is cw as seen from the left. Consequently, from the right-hand rule, the magnetic field inside the solenoid always points to the right. During the first second, while the solenoid current is increasing, the flux through the loop is to the right and increasing. To oppose the *change* in the flux, the loop's induced magnetic field must point to the left. Thus, again using the right-hand rule, the induced current must flow ccw as seen from the left. This is a *negative* current. There's no *change* in the flux for $t > 1$ s, so the induced current is zero.

SOLVE Now we're ready to use Faraday's law to find the magnitude of the current. Because the field is uniform inside the solenoid and perpendicular to the loop ($\theta = 0°$), the flux is $\Phi_m = AB$, where $A = \pi r^2 = 3.14 \times 10^{-4}$ m² is the area of the loop (*not* the area of the solenoid). The field of a long solenoid of length l was found in Chapter 32 to be

$$B = \frac{\mu_0 N I_{sol}}{l}$$

The flux when the solenoid current is I_{sol} is thus

$$\Phi_m = \frac{\mu_0 A N I_{sol}}{l}$$

The changing flux creates an induced emf $\mathcal{E}$ that is given by Faraday's law:

$$\mathcal{E} = \left|\frac{d\Phi_m}{dt}\right| = \frac{\mu_0 A N}{l}\left|\frac{dI_{sol}}{dt}\right| = 2.0 \times 10^{-6}\left|\frac{dI_{sol}}{dt}\right|$$

From the slope of the graph, we find

$$\left|\frac{dI_{sol}}{dt}\right| = \begin{cases} 10 \text{ A/s} & 0.0 \text{ s} < t < 1.0 \text{ s} \\ 0 & 1.0 \text{ s} < t < 3.0 \text{ s} \end{cases}$$

Thus the induced emf is

$$\mathcal{E} = \begin{cases} 2.0 \times 10^{-5} \text{ V} & 0.0 \text{ s} < t < 1.0 \text{ s} \\ 0 \text{ V} & 1.0 \text{ s} < t < 3.0 \text{ s} \end{cases}$$

Finally, the current induced in the loop is

$$I_{loop} = \frac{\mathcal{E}}{R} = \begin{cases} -2.0 \text{ mA} & 0.0 \text{ s} < t < 1.0 \text{ s} \\ 0 \text{ mA} & 1.0 \text{ s} < t < 3.0 \text{ s} \end{cases}$$

where the negative sign comes from Lenz's law. This result is shown in FIGURE 33.28.

FIGURE 33.28 The induced current in the loop.

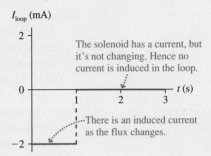

EXAMPLE 33.9 **Current induced by an MRI machine**

The body is a conductor, so rapid magnetic field changes in an MRI machine can induce currents in the body. To estimate the size of these currents, and any biological hazard they might impose, consider the "loop" of muscle tissue shown in FIGURE 33.29. This might be muscle circling the bone of your arm or thigh. Although muscle is not a great conductor—its resistivity is 1.5 Ω m—we can consider it to be a conducting loop with a rather high resistance. Suppose the magnetic field along the axis of the loop drops from 1.6 T to 0 T in 0.30 s, which is about the largest possible rate of change for an MRI solenoid. What current will be induced?

FIGURE 33.29 Edge view of a loop of muscle tissue in a magnetic field.

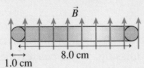

MODEL Model the muscle as a current loop. Assume that B decreases linearly with time.

SOLVE The magnetic field is parallel to the axis of the loop, with $\theta = 0°$, so the magnetic flux through the loop is $\Phi_m = AB = \pi r^2 B$. The flux changes with time because B changes. According to Faraday's law, the magnitude of the induced emf is

$$\mathcal{E} = \left|\frac{d\Phi_m}{dt}\right| = \pi r^2\left|\frac{dB}{dt}\right|$$

The rate at which the magnetic field changes is

$$\frac{dB}{dt} = \frac{\Delta B}{\Delta t} = \frac{-1.60 \text{ T}}{0.30 \text{ s}} = -5.3 \text{ T/s}$$

dB/dt is negative because the field is decreasing, but all we need for Faraday's law is the absolute value. Thus

$$\mathcal{E} = \pi r^2\left|\frac{dB}{dt}\right| = \pi(0.040 \text{ m})^2(5.3 \text{ T/s}) = 0.027 \text{ V}$$

To find the current, we need to know the resistance of the loop. Recall, from Chapter 30, that a conductor with resistivity ρ, length L, and cross-section area A has resistance $R = \rho L/A$. The length is the circumference of the loop, $L = 0.25$ m, and we can use the 1.0 cm diameter of the "wire" to find $A = 7.9 \times 10^{-5}$ m². With these values, we can compute $R = 4700$ Ω. As a result, the induced current is

$$I = \frac{\mathcal{E}}{R} = \frac{0.027 \text{ V}}{4700 \text{ }\Omega} = 5.7 \times 10^{-6} \text{ A} = 5.7 \text{ }\mu\text{A}$$

ASSESS This is a very small current. Power—the rate of energy dissipation in the muscle—is

$$P = I^2 R = (5.7 \times 10^{-6} \text{ A})^2(4700 \text{ }\Omega) = 1.5 \times 10^{-7} \text{ W}$$

The current is far too small to notice, and the tiny energy dissipation will certainly not heat the tissue.

What Does Faraday's Law Tell Us?

The induced current in the slide-wire circuit of Figure 33.26 can be understood as a motional emf due to magnetic forces on moving charges. We had not anticipated this kind of current in Chapter 32, but it takes no new laws of physics to understand it. The induced currents in Examples 33.8 and 33.9 are different. We cannot explain these induced currents on the basis of previous laws or principles. This is new physics.

Faraday recognized that all induced currents are associated with a changing magnetic flux. There are two fundamentally different ways to change the magnetic flux through a conducting loop:

1. The loop can expand, contract, or rotate, creating a motional emf.
2. The magnetic field can change.

We can see both of these if we write Faraday's law as

$$\mathcal{E} = \left| \frac{d\Phi_m}{dt} \right| = \left| \vec{B} \cdot \frac{d\vec{A}}{dt} + \vec{A} \cdot \frac{d\vec{B}}{dt} \right| \qquad (33.19)$$

The first term on the right side represents a motional emf. The magnetic flux changes because the loop itself is changing. This term includes not only situations like the slide-wire circuit, where the area A changes, but also loops that rotate in a magnetic field. The physical area of a rotating loop does not change, but the area *vector* $\vec{A}$ does. The loop's motion causes magnetic forces on the charge carriers in the loop.

The second term on the right side is the new physics in Faraday's law. It says that an emf can also be created simply by changing a magnetic field, even if nothing is moving. This was the case in Examples 33.8 and 33.9. Faraday's law tells us that the induced emf is simply the rate of change of the magnetic flux through the loop, *regardless* of what causes the flux to change.

STOP TO THINK 33.5   A conducting loop is halfway into a magnetic field. Suppose the magnetic field begins to increase rapidly in strength. What happens to the loop?

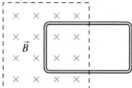

a. The loop is pushed upward, toward the top of the page.
b. The loop is pushed downward, toward the bottom of the page.
c. The loop is pulled to the left, into the magnetic field.
d. The loop is pushed to the right, out of the magnetic field.
e. The tension in the wires increases but the loop does not move.

FIGURE 33.30 An induced electric field creates a current in the loop.

(a)

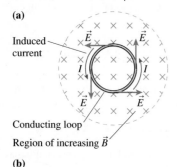

Induced current

Conducting loop

Region of increasing $\vec{B}$

(b)

Induced electric field $\vec{E}$

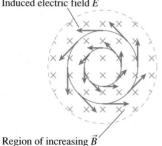

Region of increasing $\vec{B}$

33.6 Induced Fields

Faraday's law is a tool for calculating the strength of an induced current, but one important piece of the puzzle is still missing. What *causes* the current? That is, what *force* pushes the charges around the loop against the resistive forces of the metal? The agents that exert forces on charges are electric fields and magnetic fields. Magnetic forces are responsible for motional emfs, but magnetic forces cannot explain the current induced in a *stationary* loop by a changing magnetic field.

FIGURE 33.30a shows a conducting loop in an increasing magnetic field. According to Lenz's law, there is an induced current in the ccw direction. Something has to act on the charge carriers to make them move, so we infer that there must be an *electric* field tangent to the loop at all points. This electric field is *caused* by the changing magnetic field and is called an **induced electric field.** The induced electric field is the mechanism that creates a current inside a stationary loop when there's a changing magnetic field.

The conducting loop isn't necessary. The space in which the magnetic field is changing is filled with the pinwheel pattern of induced electric fields shown in FIGURE 33.30b. Charges will move if a conducting path is present, but the induced electric field is there as a direct consequence of the changing magnetic field.

But this is a rather peculiar electric field. All the electric fields we have examined until now have been created by charges. Electric field vectors pointed away from positive charges and toward negative charges. An electric field created by charges is called a **Coulomb electric field.** The induced electric field of Figure 33.30b is caused not by charges but by a changing magnetic field. It is called a **non-Coulomb electric field.**

So it appears that there are two different ways to create an electric field:

1. A Coulomb electric field is created by positive and negative charges.
2. A non-Coulomb electric field is created by a changing magnetic field.

Both exert a force $\vec{F} = q\vec{E}$ on a charge, and both create a current in a conductor. However, the origins of the fields are very different. FIGURE 33.31 is a quick summary of the two ways to create an electric field.

We first introduced the idea of a field as a way of thinking about how two charges exert long-range forces on each other through the emptiness of space. The field may have seemed like a useful pictorial representation of charge interactions, but we had little evidence that fields are *real,* that they actually exist. Now we do. The electric field has shown up in a completely different context, independent of charges, as the explanation of the very real existence of induced currents.

The electric field is not just a pictorial representation; it is real.

Calculating the Induced Field

The induced electric field is peculiar in another way: It is nonconservative. Recall that a force is conservative if it does no net work on a particle moving around a closed path. "Uphills" are balanced by "downhills." We can associate a potential energy with a conservative force, hence we have gravitational potential energy for the conservative gravitational force and electric potential energy for the conservative electric force of charges (a Coulomb electric field).

But a charge moving around a closed path in the induced electric field of Figure 33.30 is always being pushed *in the same direction* by the electric force $\vec{F} = q\vec{E}$. There's never any negative work to balance the positive work, so the net work done in going around a closed path is not zero. Because it's nonconservative, we cannot associate an electric potential with an induced electric field. Only the Coulomb field of charges has an electric potential.

However, we can associate the induced field with the emf of Faraday's law. The emf was defined as the work required per unit charge to separate the charge. That is,

$$\mathcal{E} = \frac{W}{q} \tag{33.20}$$

In batteries, a familiar source of emf, this work is done by chemical forces. But the emf that appears in Faraday's law arises when work is done by the force of an induced electric field.

If a charge q moves through a small displacement $d\vec{s}$, the small amount of work done by the electric field is $dW = \vec{F} \cdot d\vec{s} = q\vec{E} \cdot d\vec{s}$. The emf of Faraday's law is an emf around a *closed curve* through which the magnetic flux Φ_m is changing. The work done by the induced electric field as charge q moves around a closed curve is

$$W_{\text{closed curve}} = q \oint \vec{E} \cdot d\vec{s} \tag{33.21}$$

where the integration symbol with the circle is the same as the one we used in Ampère's law to indicate an integral around a closed curve. If we use this work in Equation 33.20, we find that the emf around a closed loop is

$$\mathcal{E} = \frac{W_{\text{closed curve}}}{q} = \oint \vec{E} \cdot d\vec{s} \tag{33.22}$$

FIGURE 33.31 Two ways to create an electric field.

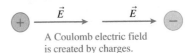

A Coulomb electric field is created by charges.

A non-Coulomb electric field is created by a changing magnetic field.

If we restrict ourselves to situations such as Figure 33.30 where the loop is perpendicular to the magnetic field and only the field is changing, we can write Faraday's law as $\mathcal{E} = |d\Phi_m/dt| = A|dB/dt|$. Consequently

$$\oint \vec{E} \cdot d\vec{s} = A\left|\frac{dB}{dt}\right| \tag{33.23}$$

Equation 33.23 is an alternative statement of Faraday's law that relates the induced electric field to the changing magnetic field.

The solenoid in FIGURE 33.32a provides a good example of the connection between $\vec{E}$ and $\vec{B}$. If there were a conducting loop inside the solenoid, we could use Lenz's law to determine that the direction of the induced current would be clockwise. But Faraday's law, in the form of Equation 33.23, tells us that **an induced electric field is present whether there's a conducting loop or not.** The electric field is induced simply due to the fact that $\vec{B}$ is changing.

FIGURE 33.32 The induced electric field circulates around the changing magnetic field inside a solenoid.

(a) The current through the solenoid is increasing.

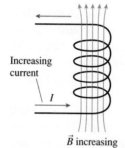

(b) The induced electric field circulates around the magnetic field lines.

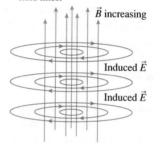

(c) Top view into the solenoid. $\vec{B}$ is coming out of the page.

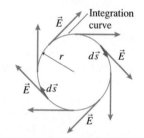

The shape and direction of the induced electric field have to be such that it *could* drive a current around a conducting loop, if one were present, and it has to be consistent with the cylindrical symmetry of the solenoid. The only possible choice, shown in FIGURE 33.32b, is an electric field that circulates clockwise around the magnetic field lines.

NOTE ▶ Circular electric field lines violate the Chapter 26 rule that electric field lines have to start and stop on charges. However, that rule applied only to Coulomb fields created by source charges. An induced electric field is a non-Coulomb field created not by source charges but by a changing magnetic field. Without source charges, induced electric field lines *must* form closed loops. ◀

To use Faraday's law, choose a *clockwise* circle of radius r as the closed curve for evaluating the integral. FIGURE 33.32c shows that the electric field vectors are everywhere tangent to the curve, so the line integral of $\vec{E}$ is

$$\oint \vec{E} \cdot d\vec{s} = El = 2\pi r E \tag{33.24}$$

where $l = 2\pi r$ is the length of the closed curve. This is exactly like the integrals we did for Ampère's law in Chapter 32.

If we stay inside the solenoid ($r < R$), the flux passes through area $A = \pi r^2$ and Equation 33.24 becomes

$$\oint \vec{E} \cdot d\vec{s} = 2\pi r E = A\left|\frac{dB}{dt}\right| = \pi r^2 \left|\frac{dB}{dt}\right| \tag{33.25}$$

Thus the strength of the induced electric field inside the solenoid is

$$E_{\text{inside}} = \frac{r}{2}\left|\frac{dB}{dt}\right| \qquad (33.26)$$

This result shows very directly that the induced electric field is created by a *changing* magnetic field. A constant $\vec{B}$, with $dB/dt = 0$, would give $E = 0$.

EXAMPLE 33.10 **An induced electric field**

A 4.0-cm-diameter solenoid is wound with 2000 turns per meter. The current through the solenoid oscillates at 60 Hz with an amplitude of 2.0 A. What is the maximum strength of the induced electric field inside the solenoid?

MODEL Assume that the magnetic field inside the solenoid is uniform.

VISUALIZE The electric field lines are concentric circles around the magnetic field lines, as was shown in Figure 33.32b. They reverse direction twice every period as the current oscillates.

SOLVE You learned in Chapter 32 that the magnetic field strength inside a solenoid with n turns per meter is $B = \mu_0 nI$. In this case, the current through the solenoid is $I = I_0 \sin \omega t$, where $I_0 = 2.0$ A is the peak current and $\omega = 2\pi(60 \text{ Hz}) = 377$ rad/s. Thus the induced electric field strength at radius r is

$$E = \frac{r}{2}\left|\frac{dB}{dt}\right| = \frac{r}{2}\frac{d}{dt}(\mu_0 nI_0 \sin \omega t) = \frac{1}{2}\mu_0 nr\omega I_0 \cos \omega t$$

The field strength is maximum at maximum radius ($r = R$) *and* at the instant when $\cos \omega t = 1$. That is,

$$E_{\max} = \frac{1}{2}\mu_0 nR\omega I_0 = 0.019 \text{ V/m}$$

ASSESS This field strength, although not large, is similar to the field strength that the emf of a battery creates in a wire. Hence this induced electric field can drive a substantial induced current through a conducting loop *if* a loop is present. But the induced electric field exists inside the solenoid whether or not there is a conducting loop.

Occasionally it is useful to have a version of Faraday's law without the absolute value signs. The essence of Lenz's law is that the emf $\mathcal{E}$ opposes the *change* in Φ_m. Mathematically, this means that $\mathcal{E}$ must be opposite in sign to dB/dt. Consequently, we can write Faraday's law as

$$\mathcal{E} = \oint \vec{E} \cdot d\vec{s} = -\frac{d\Phi_m}{dt} \qquad (33.27)$$

For practical applications, it's always easier to calculate just the magnitude of the emf with Faraday's law and to use Lenz's law to find the direction of the emf or the induced current. However, the mathematically rigorous version of Faraday's law in Equation 33.27 will prove to be useful when we combine it with other equations, in Chapter 34, to predict the existence of electromagnetic waves.

Maxwell's Theory of Electromagnetic Waves

In 1855, less than two years after receiving his undergraduate degree, the Scottish physicist James Clerk Maxwell presented a paper titled "On Faraday's Lines of Force." In this paper, he began to sketch out how Faraday's pictorial ideas about fields could be given a rigorous mathematical basis. Maxwell was troubled by a certain lack of symmetry. Faraday had found that a changing magnetic field creates an induced electric field, a non-Coulomb electric field not tied to charges. But what, Maxwell began to wonder, about a changing *electric* field?

To complete the symmetry, Maxwell proposed that a changing electric field creates an **induced magnetic field,** a new kind of magnetic field not tied to the existence of currents. **FIGURE 33.33** shows a region of space where the *electric* field is increasing. This region of space, according to Maxwell, is filled with a pinwheel pattern of induced magnetic fields. The induced magnetic field looks like the induced electric field, with $\vec{E}$ and $\vec{B}$ interchanged, except that—for technical reasons explored in the

FIGURE 33.33 Maxwell hypothesized the existence of induced magnetic fields.

A changing magnetic field creates an induced electric field.

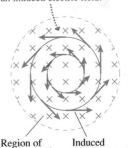

Region of increasing $\vec{B}$ Induced electric field $\vec{E}$

A changing electric field creates an induced magnetic field.

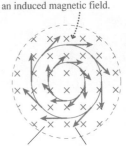

Region of increasing $\vec{E}$ Induced magnetic field $\vec{B}$

next chapter—the induced $\vec{B}$ points the opposite way from the induced $\vec{E}$. Although there was no experimental evidence that induced magnetic fields existed, Maxwell went ahead and included them in his electromagnetic field theory. This was an inspired hunch, soon to be vindicated.

Maxwell soon realized that it might be possible to establish self-sustaining electric and magnetic fields that would be entirely independent of any charges or currents. That is, a changing electric field $\vec{E}$ creates a magnetic field $\vec{B}$, which then changes in just the right way to recreate the electric field, which then changes in just the right way to again recreate the magnetic field, and so on. The fields are continually recreated through electromagnetic induction without any reliance on charges or currents.

Maxwell was able to predict that electric and magnetic fields would be able to sustain themselves, free from charges and currents, if they took the form of an **electromagnetic wave.** The wave would have to have a very specific geometry, shown in FIGURE 33.34, in which $\vec{E}$ and $\vec{B}$ are perpendicular to each other as well as perpendicular to the direction of travel. That is, an electromagnetic wave would be a *transverse* wave.

Furthermore, Maxwell's theory predicted that the wave would travel with speed

$$v_{\text{em wave}} = \frac{1}{\sqrt{\epsilon_0 \mu_0}}$$

where ϵ_0 is the permittivity constant from Coulomb's law and μ_0 is the permeability constant from the law of Biot and Savart. Maxwell computed that an electromagnetic wave, if it existed, would travel with speed $v_{\text{em wave}} = 3.00 \times 10^8$ m/s.

We don't know Maxwell's immediate reaction, but it must have been both shock and excitement. His predicted speed for electromagnetic waves, a prediction that came directly from his theory, was none other than the speed of light! This agreement could be just a coincidence, but Maxwell didn't think so. Making a bold leap of imagination, Maxwell concluded that **light is an electromagnetic wave.**

It took 25 more years for Maxwell's predictions to be tested. In 1886, the German physicist Heinrich Hertz discovered how to generate and transmit radio waves. Two years later, in 1888, he was able to show that radio waves travel at the speed of light. Maxwell, unfortunately, did not live to see his triumph. He had died in 1879, at the age of 48.

Chapter 34 will develop some of the mathematical details of Maxwell's theory and show how the ideas contained in Faraday's law lead to electromagnetic waves.

FIGURE 33.34 A self-sustaining electromagnetic wave.

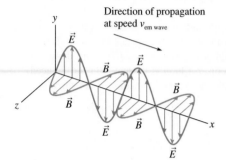

A generator inside a hydroelectric dam uses electromagnetic induction to convert the mechanical energy of a spinning turbine into electric energy.

33.7 Induced Currents: Three Applications

There are many applications of Faraday's law and induced currents in modern technology. In this section we will look at three: generators, transformers, and metal detectors.

Generators

A generator is a device that transforms mechanical energy into electric energy. FIGURE 33.35 shows a generator in which a coil of wire, perhaps spun by a windmill, rotates in a magnetic field. Both the field and the area of the loop are constant, but the magnetic flux through the loop changes continuously as the loop rotates. The induced current is removed from the rotating loop by *brushes* that press up against rotating *slip rings.*

The flux through the coil is

$$\Phi_{\text{m}} = \vec{A} \cdot \vec{B} = AB\cos\theta = AB\cos\omega t \qquad (33.28)$$

where ω is the angular frequency ($\omega = 2\pi f$) with which the coil rotates. The induced emf is given by Faraday's law,

$$\mathcal{E}_{coil} = -N\frac{d\Phi_m}{dt} = -ABN\frac{d}{dt}(\cos\omega t) = \omega ABN \sin\omega t \qquad (33.29)$$

where N is the number of turns on the coil. Here it's best to use the signed version of Faraday's law to see how $\mathcal{E}_{coil}$ alternates between positive and negative.

Because the emf alternates in sign, the current through resistor R alternates back and forth in direction. Hence the generator of Figure 33.35 is an alternating-current generator, producing what we call an *AC voltage.*

FIGURE 33.35 An alternating-current generator.

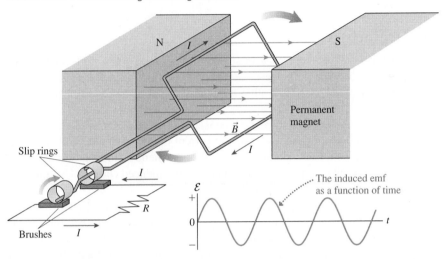

EXAMPLE 33.11 An AC generator

A coil with area 2.0 m² rotates in a 0.010 T magnetic field at a frequency of 60 Hz. How many turns are needed to generate a peak voltage of 160 V?

SOLVE The coil's maximum voltage is found from Equation 33.29:

$$\mathcal{E}_{max} = \omega ABN = 2\pi f ABN$$

The number of turns needed to generate $\mathcal{E}_{max} = 160$ V is

$$N = \frac{\mathcal{E}_{max}}{2\pi fAB} = \frac{160\text{ V}}{2\pi(60\text{ Hz})(2.0\text{ m}^2)(0.010\text{ T})} = 21\text{ turns}$$

ASSESS A 0.010 T field is modest, so you can see that generating large voltages is not difficult with large (2 m²) coils. Commercial generators use water flowing through a dam, rotating windmill blades, or turbines spun by expanding steam to rotate the generator coils. Work is required to rotate the coil, just as work was required to pull the slide wire in Section 33.2, because the magnetic field exerts retarding forces on the currents in the coil. Thus a generator is a device that turns motion (mechanical energy) into a current (electric energy). A generator is the opposite of a motor, which turns a current into motion.

Transformers

FIGURE 33.36 shows two coils wrapped on an iron core. The left coil is called the **primary coil.** It has N_1 turns and is driven by an oscillating voltage $V_1\cos\omega t$. The magnetic field of the primary follows the iron core and passes through the right coil, which has N_2 turns and is called the **secondary coil.** The alternating current through the primary coil causes an oscillating magnetic flux through the secondary coil and, hence, an induced emf. The induced emf of the secondary coil is delivered to the load as the oscillating voltage $V_2\cos\omega t$.

The changing magnetic field inside the iron core is inversely proportional to the number of turns on the primary coil: $B \propto 1/N_1$. (This relation is a consequence of the coil's inductance, an idea discussed in the next section.) According to Faraday's law, the emf induced in the secondary coil is directly proportional to its number of turns:

FIGURE 33.36 A transformer.

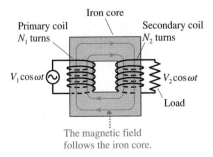

Transformers are essential for transporting electric energy from the power plant to cities and homes.

FIGURE 33.37 A metal detector.

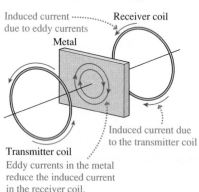

Induced current due to eddy currents
Receiver coil
Metal
Induced current due to the transmitter coil
Transmitter coil
Eddy currents in the metal reduce the induced current in the receiver coil.

$\mathcal{E}_{\text{sec}} \propto N_2$. Combining these two proportionalities, the secondary voltage of an ideal transformer is related to the primary voltage by

$$V_2 = \frac{N_2}{N_1} V_1 \qquad (33.30)$$

Depending on the ratio N_2/N_1, the voltage V_2 across the load can be *transformed* to a higher or a lower voltage than V_1. Consequently, this device is called a **transformer.** Transformers are widely used in the commercial generation and transmission of electricity. A *step-up transformer,* with $N_2 \gg N_1$, boosts the voltage of a generator up to several hundred thousand volts. Delivering power with smaller currents at higher voltages reduces losses due to the resistance of the wires. High-voltage transmission lines carry electric power to urban areas, where *step-down transformers* ($N_2 \ll N_1$) lower the voltage to 120 V.

Metal Detectors

Metal detectors, such as those used in airports for security, seem fairly mysterious. How can they detect the presence of *any* metal—not just magnetic materials such as iron—but not detect plastic or other materials? Metal detectors work because of induced currents.

A metal detector, shown in FIGURE 33.37, consists of two coils: a *transmitter coil* and a *receiver coil.* A high-frequency alternating current in the transmitter coil generates an alternating magnetic field along the axis. This magnetic field creates a changing flux through the receiver coil and causes an alternating induced current. The transmitter and receiver are similar to a transformer.

Suppose a piece of metal is placed between the transmitter and the receiver. The alternating magnetic field through the metal induces eddy currents in a plane parallel to the transmitter and receiver coils. The receiver coil then responds to the *superposition* of the transmitter's magnetic field and the magnetic field of the eddy currents. Because the eddy currents attempt to prevent the flux from changing, in accordance with Lenz's law, the net field at the receiver *decreases* when a piece of metal is inserted between the coils. Electronic circuits detect the current decrease in the receiver coil and set off an alarm. Eddy currents can't flow in an insulator, so this device detects only metals.

33.8 Inductors

Capacitors are useful circuit elements because they store potential energy U_C in the electric field. Similarly, a coil of wire can be a useful circuit element because it stores energy in the magnetic field. Using as an analogy the definition of capacitance as the charge-to-voltage ratio, $C = Q/\Delta V$, let's define the **inductance** L of a coil as its flux-to-current ratio:

$$L = \frac{\Phi_m}{I} \qquad (33.31)$$

Strictly speaking, this is called *self-inductance* because the flux we're considering is the magnetic flux the solenoid creates in itself when there is a current.

The SI unit of inductance is the **henry,** named in honor of Joseph Henry, defined as

$$1 \text{ henry} = 1 \text{ H} \equiv 1 \text{ Wb/A} = 1 \text{ T m}^2/\text{A}$$

Practical inductances are typically millihenries (mH) or microhenries (μH).

A coil of wire used in a circuit for the purpose of providing inductance is called an **inductor.** An *ideal inductor* is one for which the wire forming the coil has no electric resistance. The circuit symbol for an inductor is —⁀⁀⁀⁀—.

It's not hard to find the inductance of a solenoid. In Chapter 32 we found that the magnetic field inside an ideal solenoid having N turns and length l is

$$B = \frac{\mu_0 N I}{l}$$

The magnetic flux through one turn of the coil is $\Phi_{\text{per turn}} = AB$, where A is the cross-section area of the solenoid. The total magnetic flux through all N turns is

$$\Phi_{\text{m}} = N\Phi_{\text{per turn}} = \frac{\mu_0 N^2 A}{l} I \qquad (33.32)$$

Thus the inductance of the solenoid, using the definition of Equation 33.31, is

$$L_{\text{solenoid}} = \frac{\Phi_{\text{m}}}{I} = \frac{\mu_0 N^2 A}{l} \qquad (33.33)$$

The inductance of a solenoid depends only on its geometry, not at all on the current. You may recall that the capacitance of two parallel plates depends only on their geometry, not at all on their potential difference.

EXAMPLE 33.12 **The length of an inductor**

An inductor is made by tightly wrapping 0.30-mm-diameter wire around a 4.0-mm-diameter cylinder. What length cylinder has an inductance of 10 μH?

SOLVE The cross-section area of the solenoid is $A = \pi r^2$. If the wire diameter is d, the number of turns of wire on a cylinder of length l is $N = l/d$. Thus the inductance is

$$L = \frac{\mu_0 N^2 A}{l} = \frac{\mu_0 (l/d)^2 \pi r^2}{l} = \frac{\mu_0 \pi r^2 l}{d^2}$$

The length needed to give inductance $L = 1.0 \times 10^{-5}$ H is

$$l = \frac{d^2 L}{\mu_0 \pi r^2} = \frac{(0.00030 \text{ m})^2 (1.0 \times 10^{-5} \text{ H})}{(4\pi \times 10^{-7} \text{ T m/A})\pi (0.0020 \text{ m})^2}$$

$$= 0.057 \text{ m} = 5.7 \text{ cm}$$

The Potential Difference Across an Inductor

An inductor is not very interesting when the current through it is steady. If the inductor is ideal, with $R = 0\ \Omega$, the potential difference due to a steady current is zero. Inductors become important circuit elements when currents are changing. FIGURE 33.38a shows a steady current into the left side of an inductor. The solenoid's magnetic field passes through the coils of the solenoid, establishing a flux.

In FIGURE 33.38b, the current into the solenoid is increasing. This creates an increasing flux to the left. According to Lenz's law, an induced current in the coils will oppose this increase by creating an induced magnetic field pointing to the right. This requires the induced current to be *opposite* the current into the solenoid. This induced current will carry positive charge carriers to the left until a potential difference is established across the solenoid.

You saw a similar situation in Section 33.2. The induced current in a conductor moving through a magnetic field carried positive charge carriers to the top of the wire and established a potential difference across the conductor. The induced current in the moving wire was due to magnetic forces on the moving charges. Now, in Figure 33.38b, the induced current is due to the non-Coulomb electric field induced by the changing magnetic field. Nonetheless, the outcome is the same: a potential difference across the conductor.

We can use Faraday's law to find the potential difference. The emf induced in a coil is

$$\mathcal{E}_{\text{coil}} = N\left|\frac{d\Phi_{\text{per turn}}}{dt}\right| = \left|\frac{d\Phi_{\text{m}}}{dt}\right| \qquad (33.34)$$

FIGURE 33.38 Increasing the current through an inductor.

(a)

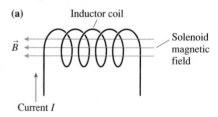

Inductor coil

Solenoid magnetic field

Current I

(b) The induced current is opposite the solenoid current.

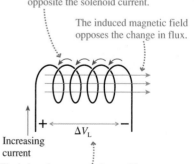

The induced magnetic field opposes the change in flux.

Increasing current

ΔV_{L}

The induced current carries positive charge carriers to the left and establishes a potential difference across the inductor.

FIGURE 33.39 Decreasing the current through an inductor.

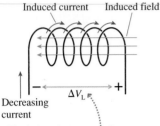

Decreasing current

ΔV_L

The induced current carries positive charge carriers to the right. The potential difference is opposite that of Figure 33.38b.

where $\Phi_m = N\Phi_{\text{per turn}}$ is the total flux through all the coils. The inductance was defined such that $\Phi_m = LI$, so Equation 33.34 becomes

$$\mathcal{E}_{\text{coil}} = L\left|\frac{dI}{dt}\right| \tag{33.35}$$

The induced emf is directly proportional to the *rate of change* of current through the coil. We'll consider the appropriate sign in a moment, but Equation 33.35 gives us the size of the potential difference that is developed across a coil as the current through the coil changes. Note that $\mathcal{E}_{\text{coil}} = 0$ for a steady, unchanging current.

FIGURE 33.39 shows the same inductor, but now the current (still *in* to the left side) is decreasing. To oppose the decrease in flux, the induced current is in the *same* direction as the input current. The induced current carries charge to the right and establishes a potential difference opposite that in Figure 33.38b.

NOTE ▶ Notice that the induced current does not oppose the current through the inductor, which is from left to right in both Figures 33.38 and 33.39. Instead, in accordance with Lenz's law, the induced current opposes the *change* in the current in the solenoid. The practical result is that it is hard to change the current through an inductor. Any effort to increase or decrease the current is met with opposition in the form of an opposing induced current. You can think of the current in an inductor as having inertia, trying to continue what it was doing without change. ◀

Before we can use inductors in a circuit we need to establish a rule about signs that is consistent with our earlier circuit analysis. FIGURE 33.40 first shows current I passing through a resistor. You learned in Chapter 31 that the potential difference across a resistor is $\Delta V_{\text{res}} = -\Delta V_R = -IR$, where the minus sign indicates that the potential *decreases* in the direction of the current.

We'll use the same convention for an inductor. The potential difference across an inductor, *measured along the direction of the current,* is

FIGURE 33.40 The potential difference across a resistor and an inductor.

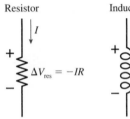

Resistor

Inductor

$\Delta V_{\text{res}} = -IR$

$\Delta V_L = -L\dfrac{dI}{dt}$

The potential always decreases.

The potential decreases if the current is increasing.

The potential increases if the current is decreasing.

$$\Delta V_L = -L\frac{dI}{dt} \tag{33.36}$$

If the current is increasing ($dI/dt > 0$), the input side of the inductor is more positive than the output side and the potential decreases in the direction of the current ($\Delta V_L < 0$). This was the situation in Figure 33.38b. If the current is decreasing ($dI/dt < 0$), the input side is more negative and the potential increases in the direction of the current ($\Delta V_L > 0$). This was the situation in Figure 33.39.

The potential difference across an inductor can be very large if the current changes very abruptly (large dI/dt). FIGURE 33.41 shows an inductor connected across a battery. There is a large current through the inductor, limited only by the internal resistance of the battery. Suppose the switch is suddenly opened. A very large induced voltage is created across the inductor as the current rapidly drops to zero. This potential difference (plus ΔV_{bat}) appears across the gap of the switch as it is opened. A large potential difference across a small gap often creates a spark.

FIGURE 33.41 Creating sparks.

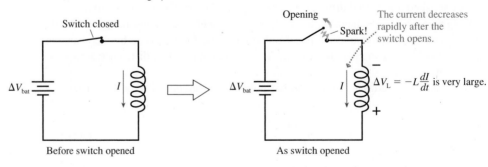

Switch closed

ΔV_{bat}

I

Before switch opened

Opening

Spark!

The current decreases rapidly after the switch opens.

ΔV_{bat}

I

$\Delta V_L = -L\dfrac{dI}{dt}$ is very large.

As switch opened

Indeed, this is exactly how the spark plugs in your car work. The car's generator sends a current through the *coil,* which is a big inductor. When a switch is suddenly opened, breaking the current, the induced voltage, typically a few thousand volts, appears across the terminals of the spark plug, creating the spark that ignites the gasoline. Older cars use a *distributor* to open and close an actual switch; more recent cars have *electronic ignition* in which the mechanical switch has been replaced by a transistor.

EXAMPLE 33.13 **Large voltage across an inductor**

A 1.0 A current passes through a 10 mH inductor coil. What potential difference is induced across the coil if the current drops to zero in 5.0 μs?

MODEL Assume this is an ideal inductor, with $R = 0\ \Omega$, and that the current decrease is linear with time.

SOLVE The rate of current decrease is

$$\frac{dI}{dt} \approx \frac{\Delta I}{\Delta t} = \frac{-1.0\ \text{A}}{5.0 \times 10^{-6}\ \text{s}} = -2.0 \times 10^5\ \text{A/s}$$

The induced voltage is

$$\Delta V_{\text{L}} = -L\frac{dI}{dt} \approx -(0.010\ \text{H})(-2.0 \times 10^5\ \text{A/s}) = 2000\ \text{V}$$

ASSESS Inductors may be physically small, but they can pack a punch if you try to change the current through them too quickly.

STOP TO THINK 33.6   The potential at a is higher than the potential at b. Which of the following statements about the inductor current I could be true?

a. I is from a to b and steady.
b. I is from a to b and increasing.
c. I is from a to b and decreasing.
d. I is from b to a and steady.
e. I is from b to a and increasing.
f. I is from b to a and decreasing.

$V_{\text{a}} > V_{\text{b}}$

a ⎯⎯•⎯⎯⎯⎯⎯⎯⎯•⎯⎯ b

Energy in Inductors and Magnetic Fields

Recall that electric power is $P_{\text{elec}} = I\Delta V$. As current passes through an inductor, for which $\Delta V_{\text{L}} = -L(dI/dt)$, the electric power is

$$P_{\text{elec}} = I\Delta V_{\text{L}} = -LI\frac{dI}{dt} \qquad (33.37)$$

P_{elec} is negative because the current is *losing* electric energy. That energy is being transferred to the inductor, which is *storing* energy U_{L} at the rate

$$\frac{dU_{\text{L}}}{dt} = +LI\frac{dI}{dt} \qquad (33.38)$$

where we've noted that power is the rate of change of energy.

We can find the total energy stored in an inductor by integrating Equation 33.38 from $I = 0$, where $U_{\text{L}} = 0$, to a final current I. Doing so gives

$$U_{\text{L}} = L\int_0^I I\,dI = \frac{1}{2}LI^2 \qquad (33.39)$$

The potential energy stored in an inductor depends on the square of the current through it. Notice the analogy with the energy $U_{\text{C}} = \frac{1}{2}C(\Delta V)^2$ stored in a capacitor.

In working with circuits we say that the energy is "stored in the inductor." Strictly speaking, the energy is stored in the inductor's magnetic field, analogous to how a capacitor stores energy in the electric field. We can use the inductance of a solenoid, Equation 33.33, to relate the inductor's energy to the magnetic field strength:

$$U_L = \frac{1}{2}LI^2 = \frac{\mu_0 N^2 A}{2l}I^2 = \frac{1}{2\mu_0}Al\left(\frac{\mu_0 NI}{l}\right)^2 \tag{33.40}$$

We made the last rearrangement in Equation 33.40 because $\mu_0 NI/l$ is the magnetic field inside the solenoid. Thus

$$U_L = \frac{1}{2\mu_0}AlB^2 \tag{33.41}$$

But Al is the volume inside the solenoid. Dividing by Al, the magnetic field *energy density* inside the solenoid (energy per m³) is

$$u_B = \frac{1}{2\mu_0}B^2 \tag{33.42}$$

We've derived this expression for energy density based on the properties of a solenoid, but it turns out to be the correct expression for the energy density anywhere there's a magnetic field. Compare this to the energy density of an electric field $u_E = \frac{1}{2}\epsilon_0 E^2$ that we found in Chapter 29.

Energy in electric and magnetic fields

Electric fields	Magnetic fields
A capacitor stores energy	An inductor stores energy
$U_C = \frac{1}{2}C(\Delta V)^2$	$U_L = \frac{1}{2}LI^2$
Energy density in the field is	Energy density in the field is
$u_E = \frac{\epsilon_0}{2}E^2$	$u_B = \frac{1}{2\mu_0}B^2$

EXAMPLE 33.14 **Energy stored in an inductor**

The 10 μH inductor of Example 33.12 was 5.7 cm long and 4.0 mm in diameter. Suppose it carries a 100 mA current. What are the energy stored in the inductor, the magnetic energy density, and the magnetic field strength?

SOLVE The stored energy is

$$U_L = \frac{1}{2}LI^2 = \frac{1}{2}(1.0 \times 10^{-5}\,\text{H})(0.10\,\text{A})^2 = 5.0 \times 10^{-8}\,\text{J}$$

The solenoid volume is $(\pi r^2)l = 7.16 \times 10^{-7}\,\text{m}^3$. Using this gives the energy density of the magnetic field:

$$u_B = \frac{5.0 \times 10^{-8}\,\text{J}}{7.16 \times 10^{-7}\,\text{m}^3} = 0.070\,\text{J/m}^3$$

From Equation 33.42, the magnetic field with this energy density is

$$B = \sqrt{2\mu_0 u_B} = 4.2 \times 10^{-4}\,\text{T}$$

33.9 *LC* Circuits

Telecommunication—radios, televisions, cell phones—is based on electromagnetic signals that *oscillate* at a well-defined frequency. These oscillations are generated and detected by a simple circuit consisting of an inductor and a capacitor in parallel. This is called an **LC circuit**. In this section we will learn why an *LC* circuit oscillates and determine the oscillation frequency.

FIGURE 33.42 shows a capacitor with initial charge Q_0, an inductor, and a switch. The switch has been open for a long time, so there is no current in the circuit. Then, at $t = 0$, the switch is closed. How does the circuit respond? Let's think it through qualitatively before getting into the mathematics.

As FIGURE 33.43 shows, the inductor provides a conducting path for discharging the capacitor. However, the discharge current has to pass through the inductor, and, as we've seen, an inductor resists changes in current. Consequently, the current doesn't stop when the capacitor charge reaches zero.

A block attached to a stretched spring is a useful mechanical analogy. Closing the switch to discharge the capacitor is like releasing the block. The block doesn't stop when it reaches the origin; its momentum keeps it going until the spring is fully compressed. Likewise, the current continues until it has recharged the capacitor with the opposite polarization. This process repeats over and over, charging the capacitor first one way, then the other. That is, the charge and current *oscillate*.

FIGURE 33.42 An *LC* circuit.

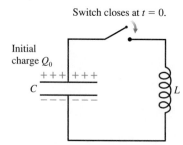

Switch closes at $t = 0$.

Initial charge Q_0

FIGURE 33.43 The capacitor charge oscillates much like a block attached to a spring.

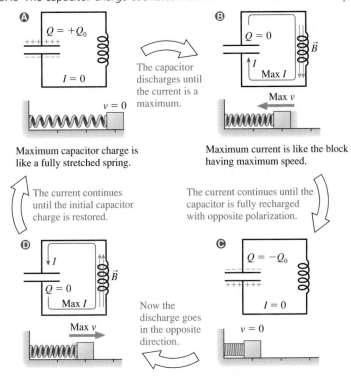

The capacitor discharges until the current is a maximum.

Maximum capacitor charge is like a fully stretched spring.

Maximum current is like the block having maximum speed.

The current continues until the initial capacitor charge is restored.

The current continues until the capacitor is fully recharged with opposite polarization.

Now the discharge goes in the opposite direction.

·The goal of our circuit analysis will be to find expressions showing how the capacitor charge Q and the inductor current I change with time. As always, our starting point for circuit analysis is Kirchhoff's voltage law, which says that all the potential differences around a closed loop must sum to zero. Choosing a cw direction for I, Kirchhoff's law is

$$\Delta V_C + \Delta V_L = 0 \qquad (33.43)$$

The potential difference across a capacitor is $\Delta V_C = Q/C$, and we found the potential difference across an inductor in Equation 33.36. Using these, Kirchhoff's law becomes

$$\frac{Q}{C} - L\frac{dI}{dt} = 0 \qquad (33.44)$$

Equation 33.44 has two unknowns, Q and I. We can eliminate one of the unknowns by finding another relation between Q and I. Current is the rate at which charge moves, $I = dq/dt$, but the charge flowing through the inductor is charge that was *removed* from the capacitor. That is, an infinitesimal charge dq flows through the inductor when the capacitor charge changes by $dQ = -dq$. Thus the current through the inductor is related to the charge on the capacitor by

$$I = -\frac{dQ}{dt} \qquad (33.45)$$

Now I is positive when Q is decreasing, as we would expect. This is a subtle but important step in the reasoning.

Equations 33.44 and 33.45 are two equations in two unknowns. To solve them, we'll first take the time derivative of Equation 33.45:

$$\frac{dI}{dt} = \frac{d}{dt}\left(-\frac{dQ}{dt}\right) = -\frac{d^2Q}{dt^2} \qquad (33.46)$$

We can substitute this result into Equation 33.44:

$$\frac{Q}{C} + L\frac{d^2Q}{dt^2} = 0 \qquad (33.47)$$

Now we have an equation for the capacitor charge Q.

A cell phone is actually a very sophisticated two-way radio that communicates with the nearest base station via high-frequency radio waves—roughly 1000 MHz. As in any radio or communications device, the transmission frequency is established by the oscillating current in an *LC* circuit.

Equation 33.47 is a second-order differential equation for Q. Fortunately, it is an equation we've seen before and already know how to solve. To see this, we rewrite Equation 33.47 as

$$\frac{d^2Q}{dt^2} = -\frac{1}{LC}Q \qquad (33.48)$$

Recall, from Chapter 14, that the equation of motion for an undamped mass on a spring is

$$\frac{d^2x}{dt^2} = -\frac{k}{m}x \qquad (33.49)$$

Equation 33.48 is *exactly the same equation,* with x replaced by Q and k/m replaced by $1/LC$. This should be no surprise because we've already seen that a mass on a spring is a mechanical analog of the LC circuit.

We know the solution to Equation 33.49. It is simple harmonic motion $x(t) = x_0 \cos \omega t$ with angular frequency $\omega = \sqrt{k/m}$. Thus the solution to Equation 33.48 must be

$$Q(t) = Q_0 \cos \omega t \qquad (33.50)$$

where Q_0 is the initial charge, at $t = 0$, and the angular frequency is

$$\omega = \sqrt{\frac{1}{LC}} \qquad (33.51)$$

The charge on the upper plate of the capacitor oscillates back and forth between $+Q_0$ and $-Q_0$ (the opposite polarization) with period $T = 2\pi/\omega$.

As the capacitor charge oscillates, so does the current through the inductor. Using Equation 33.45 gives the current through the inductor:

$$I = -\frac{dQ}{dt} = \omega Q_0 \sin \omega t = I_{max} \sin \omega t \qquad (33.52)$$

where $I_{max} = \omega Q_0$ is the maximum current.

An LC circuit is an *electric oscillator,* oscillating at frequency $f = \omega/2\pi$. FIGURE 33.44 shows graphs of the capacitor charge Q and the inductor current I as functions of time. Notice that Q and I are 90° out of phase. The current is zero when the capacitor is fully charged, as expected, and the charge is zero when the current is maximum.

FIGURE 33.44 The oscillations of an LC circuit.

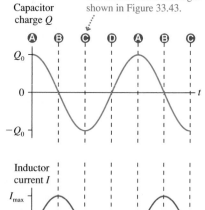

The letters match the stages shown in Figure 33.43.

EXAMPLE 33.15 | **An AM radio oscillator**

You have a 1.0 mH inductor. What capacitor should you choose to make an oscillator with a frequency of 920 kHz? (This frequency is near the center of the AM radio band.)

SOLVE The angular frequency is $\omega = 2\pi f = 5.78 \times 10^6$ rad/s. Using Equation 33.51 for ω gives the required capacitor:

$$C = \frac{1}{\omega^2 L} = \frac{1}{(5.78 \times 10^6 \text{ rad/s})^2 (0.0010 \text{ H})}$$

$$= 3.0 \times 10^{-11} \text{ F} = 30 \text{ pF}$$

An LC circuit, like a mass on a spring, wants to respond only at its natural oscillation frequency $\omega = 1/\sqrt{LC}$. In Chapter 14 we defined a strong response at the natural frequency as a *resonance,* and resonance is the basis for all telecommunications. The input circuit in radios, televisions, and cell phones is an LC circuit driven by the signal picked up by the antenna. This signal is the superposition of hundreds of sinusoidal waves at different frequencies, one from each transmitter in the area, but the circuit responds only to the *one* signal that matches the circuit's natural frequency. That particular signal generates a large-amplitude current that can be further amplified and decoded to become the output that you hear.

33.10 *LR* Circuits

A circuit consisting of an inductor, a resistor, and (perhaps) a battery is called an **LR circuit**. FIGURE 33.45a is an example of an *LR* circuit. We'll assume that the switch has been in position a for such a long time that the current is steady and unchanging. There's no potential difference across the inductor, because $dI/dt = 0$, so it simply acts like a piece of wire. The current flowing around the circuit is determined entirely by the battery and the resistor: $I_0 = \Delta V_{bat}/R$.

What happens if, at $t = 0$, the switch is suddenly moved to position b? With the battery no longer in the circuit, you might expect the current to stop immediately. But the inductor won't let that happen. The current will continue for some period of time as the inductor's magnetic field drops to zero. In essence, the energy stored in the inductor allows it to act like a battery for a short period of time. Our goal is to determine how the current decays after the switch is moved.

NOTE ▶ It's important not to open switches in inductor circuits because they'll spark, as Figure 33.41 showed. The unusual switch in Figure 33.45 is designed to make the new contact just before breaking the old one. ◀

FIGURE 33.45b shows the circuit after the switch is changed. Our starting point, once again, is Kirchhoff's voltage law. The potential differences around a closed loop must sum to zero. For this circuit, Kirchhoff's law is

$$\Delta V_{res} + \Delta V_L = 0 \tag{33.53}$$

The potential differences in the direction of the current are $\Delta V_{res} = -IR$ for the resistor and $\Delta V_L = -L(dI/dt)$ for the inductor. Substituting these into Equation 33.53 gives

$$-RI - L\frac{dI}{dt} = 0 \tag{33.54}$$

We're going to need to integrate to find the current I as a function of time. Before doing so, we rearrange Equation 33.54 to get all the current terms on one side of the equation and all the time terms on the other:

$$\frac{dI}{I} = -\frac{R}{L}dt = -\frac{dt}{(L/R)} \tag{33.55}$$

We know that the current at $t = 0$, when the switch was moved, was I_0. We want to integrate from these starting conditions to current I at the unspecified time t. That is,

$$\int_{I_0}^{I}\frac{dI}{I} = -\frac{1}{(L/R)}\int_0^t dt \tag{33.56}$$

Both are common integrals, giving

$$\ln I\Big|_{I_0}^{I} = \ln I - \ln I_0 = \ln\left(\frac{I}{I_0}\right) = -\frac{t}{(L/R)} \tag{33.57}$$

We can solve for the current I by taking the exponential of both sides, then multiplying by I_0. Doing so gives I, the current as a function of time:

$$I = I_0 e^{-t/(L/R)} \tag{33.58}$$

Notice that $I = I_0$ at $t = 0$, as expected.

The argument of the exponential function must be dimensionless, so L/R must have dimensions of time. If we define the **time constant** τ of the *LR* circuit to be

$$\tau = \frac{L}{R} \tag{33.59}$$

FIGURE 33.45 An *LR* circuit.

(a)

The switch has been in this position for a long time. At $t = 0$ it is moved to position b.

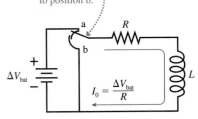

$$I_0 = \frac{\Delta V_{bat}}{R}$$

(b)

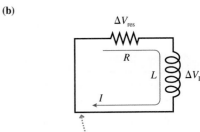

This is the circuit with the switch in position b. The inductor prevents the current from stopping instantly.

FIGURE 33.46 The current decay in an *LR* circuit.

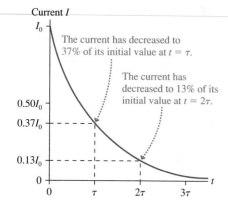

Current *I*

The current has decreased to 37% of its initial value at $t = \tau$.

The current has decreased to 13% of its initial value at $t = 2\tau$.

then we can write Equation 33.58 as

$$I = I_0 e^{-t/\tau} \qquad (33.60)$$

The time constant is the time at which the current has decreased to e^{-1} (about 37%) of its initial value. We can see this by computing the current at the time $t = \tau$:

$$I(\text{at } t = \tau) = I_0 e^{-\tau/\tau} = e^{-1} I_0 = 0.37 I_0 \qquad (33.61)$$

Thus the time constant for an *LR* circuit functions in exactly the same way as the time constant for the *RC* circuit we analyzed in Chapter 31. At time $t = 2\tau$, the current has decreased to $e^{-2} I_0$, or about 13% of its initial value.

The current is graphed in FIGURE 33.46. You can see that the current decays exponentially. The *shape* of the graph is always the same, regardless of the specific value of the time constant τ.

EXAMPLE 33.16 | **Exponential decay in an *LR* circuit**

The switch in FIGURE 33.47 has been in position a for a long time. It is changed to position b at $t = 0$ s.

a. What is the current in the circuit at $t = 5.0$ μs?
b. At what time has the current decayed to 1% of its initial value?

FIGURE 33.47 The *LR* circuit of Example 33.16.

The switch moves from a to b at $t = 0$.

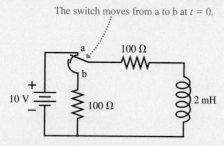

MODEL This is an *LR* circuit. We'll assume ideal wires and an ideal inductor.

VISUALIZE The two resistors will be in series after the switch is thrown.

SOLVE Before the switch is thrown, while $\Delta V_L = 0$, the current is $I_0 = (10 \text{ V})/(100 \ \Omega) = 0.10 \text{ A} = 100 \text{ mA}$. This will be the initial current after the switch is thrown because the current through an

inductor can't change instantaneously. The circuit resistance after the switch is thrown is $R = 200 \ \Omega$, so the time constant is

$$\tau = \frac{L}{R} = \frac{2.0 \times 10^{-3} \text{ H}}{200 \ \Omega} = 1.0 \times 10^{-5} \text{ s} = 10 \ \mu\text{s}$$

a. The current at $t = 5.0$ μs is

$$I = I_0 e^{-t/\tau} = (100 \text{ mA}) e^{-(5.0 \mu s)/(10 \mu s)} = 61 \text{ mA}$$

b. To find the time at which a particular current is reached we need to go back to Equation 33.57 and solve for t:

$$t = -\frac{L}{R} \ln\left(\frac{I}{I_0}\right) = -\tau \ln\left(\frac{I}{I_0}\right)$$

The time at which the current has decayed to 1 mA (1% of I_0) is

$$t = -(10 \ \mu\text{s}) \ln\left(\frac{1 \text{ mA}}{100 \text{ mA}}\right) = 46 \ \mu\text{s}$$

ASSESS For all practical purposes, the current has decayed away in ≈ 50 μs. The inductance in this circuit is not large, so a short decay time is not surprising.

STOP TO THINK 33.7 Rank in order, from largest to smallest, the time constants τ_a, τ_b, and τ_c of these three circuits.

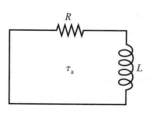

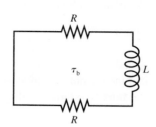

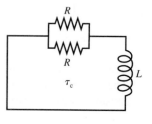

CHALLENGE EXAMPLE 33.17 | **Induction heating**

Induction heating uses induced currents to heat metal objects to high temperatures for applications such as surface hardening, brazing, or even melting. To illustrate the idea, consider a copper wire formed into a 4.0 cm × 4.0 cm square loop and placed in a magnetic field—perpendicular to the plane of the loop—that oscillates with 0.010 T amplitude at a frequency of 1000 Hz. What is the wire's initial temperature rise, in °C/min?

MODEL The changing magnetic flux through the loop will induce a current that, because of the wire's resistance, will heat the wire. Eventually, when the wire gets hot, heat loss through radiation and/or convection will limit the temperature rise, but initially we can consider the temperature change due only to the heating by the current. Assume that the wire's diameter is much less than the 4.0 cm width of the loop.

VISUALIZE **FIGURE 33.48** shows the copper loop in the magnetic field. The wire's cross-section area A is unknown, but our assumption of a thin wire means that the loop has a well-defined area L^2. Values of copper's resistivity, density, and specific heat were taken from tables inside the back cover of the book. We've used subscripts to distinguish between mass density ρ_{mass} and resistivity ρ_{elec}, a potentially confusing duplication of symbols.

FIGURE 33.48 A copper wire being heated by induction.

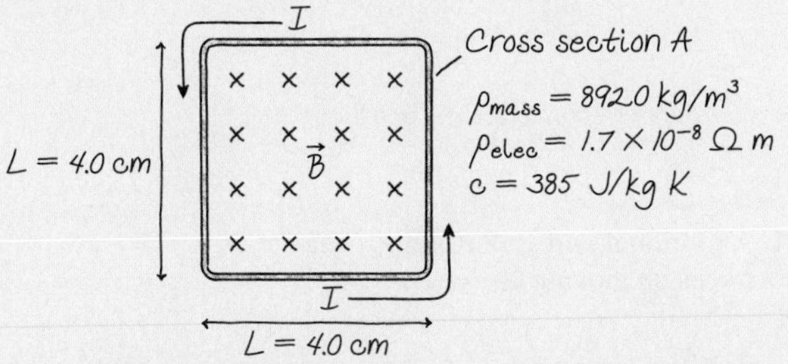

SOLVE Power dissipation by a current, $P = I^2R$, heats the wire. As long as heat losses are negligible, we can use the heating rate and the wire's specific heat c to calculate the rate of temperature change. Our first task is to find the induced current. According to Faraday's law,

$$I = \frac{\mathcal{E}}{R} = -\frac{1}{R}\frac{d\Phi_m}{dt} = -\frac{L^2}{R}\frac{dB}{dt}$$

where R is the loop's resistance and $\Phi_m = L^2B$ is the magnetic flux through a loop of area L^2. The oscillating magnetic field can be written $B = B_0 \cos \omega t$, with $B_0 = 0.010$ T and $\omega = 2\pi \times 1000$ Hz = 6280 rad/s. Thus

$$\frac{dB}{dt} = -\omega B_0 \sin \omega t$$

from which we find that the induced current oscillates as

$$I = \frac{\omega B_0 L^2}{R} \sin \omega t$$

As the current oscillates, the power dissipation in the wire is

$$P = I^2R = \frac{\omega^2 B_0^2 L^4}{R} \sin^2 \omega t$$

The power dissipation also oscillates, but very rapidly in comparison to a temperature rise that we expect to occur over seconds or minutes. Consequently, we are justified in replacing the oscillating P with its *average* value P_{avg}. Recall that the time average of the function $\sin^2 \omega t$ is $\frac{1}{2}$, a result that can be proven by integration or justified by noticing that a graph of $\sin^2 \omega t$ oscillates symmetrically between 0 and 1. Thus the average power dissipation in the wire is

$$P_{avg} = \frac{\omega^2 B_0^2 L^4}{2R}$$

Recall that power is the *rate* of energy transfer. In this case, the power dissipated in the wire is the wire's heating rate: $dQ/dt = P_{avg}$, where here Q is heat, not charge. Using $Q = mc\Delta T$, from thermodynamics, we can write

$$\frac{dQ}{dt} = mc\frac{dT}{dt} = P_{avg} = \frac{\omega^2 B_0^2 L^4}{2R}$$

To complete the calculation, we need the mass and resistance of the wire. The wire's total length is $4L$, and its cross-section area is A. Thus

$$m = \rho_{mass}V = 4\rho_{mass}LA$$

$$R = \frac{\rho_{elec}(4L)}{A} = \frac{4\rho_{elec}L}{A}$$

Substituting these into the heating equation, we have

$$4\rho_{mass}LAc\frac{dT}{dt} = \frac{\omega^2 B_0^2 L^3 A}{8\rho_{elec}}$$

Interestingly, the wire's cross-section area cancels. The wire's temperature initially increases at the rate

$$\frac{dT}{dt} = \frac{\omega^2 B_0^2 L^2}{32\rho_{elec}\rho_{mass}c}$$

All the terms on the right-hand side are known. Evaluating, we find

$$\frac{dT}{dt} = 3.3 \text{ K/s} = 200°\text{C/min}$$

ASSESS This is a rapid but realistic temperature rise for a small object, although the rate of increase will slow as the object begins losing heat to the environment through radiation and/or convection. Induction heating can increase an object's temperature by several hundred degrees in a few minutes.

SUMMARY

The goal of Chapter 33 has been to understand and apply electromagnetic induction.

General Principles

Faraday's Law

MODEL Make simplifying assumptions.

VISUALIZE Use Lenz's law to determine the direction of the **induced current.**

SOLVE The **induced emf** is

$$\mathcal{E} = \left| \frac{d\Phi_m}{dt} \right|$$

Multiply by N for an N-turn coil.
The size of the induced current is $I = \mathcal{E}/R$.

ASSESS Is the result reasonable?

Lenz's Law

There is an induced current in a closed conducting loop if and only if the magnetic flux through the loop is changing. The direction of the induced current is such that the induced magnetic field opposes the *change* in the flux.

Magnetic flux

Magnetic flux measures the amount of magnetic field passing through a surface.

$$\Phi_m = \vec{A} \cdot \vec{B} = AB\cos\theta$$

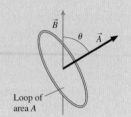

Loop of area A

Important Concepts

Three ways to change the flux

1. A loop moves into or out of a magnetic field.

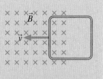

2. The loop changes area or rotates.

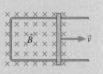

3. The magnetic field through the loop increases or decreases.

Two ways to create an induced current

1. A **motional emf** is due to magnetic forces on moving charge carriers.

2. An induced electric field is due to a changing magnetic field.

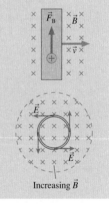

Increasing $\vec{B}$

Applications

Inductors

Solenoid inductance $L_{\text{solenoid}} = \dfrac{\mu_0 N^2 A}{l}$

Potential difference $\Delta V_L = -L\dfrac{dI}{dt}$

Energy stored $U_L = \frac{1}{2}LI^2$

Magnetic energy density $u_B = \dfrac{1}{2\mu_0}B^2$

LC circuit

Oscillates at $\omega = \sqrt{\dfrac{1}{LC}}$

LR circuit

Exponential change with $\tau = \dfrac{L}{R}$

Terms and Notation

electromagnetic induction	area vector, $\vec{A}$	induced magnetic field	inductor
induced current	Lenz's law	electromagnetic wave	LC circuit
motional emf	induced emf, $\mathcal{E}$	primary coil	LR circuit
generator	Faraday's law	secondary coil	time constant, τ
eddy current	induced electric field	transformer	
magnetic flux, Φ_m	Coulomb electric field	inductance, L	
weber, Wb	non-Coulomb electric field	henry, H	

CONCEPTUAL QUESTIONS

1. What is the direction of the induced current in FIGURE Q33.1?

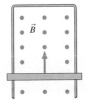

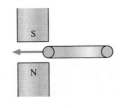

FIGURE Q33.1   FIGURE Q33.2

2. You want to insert a loop of copper wire between the two permanent magnets in FIGURE Q33.2. Is there an attractive magnetic force that tends to *pull* the loop in, like a magnet pulls on a paper clip? Or do you need to *push* the loop in against a repulsive force? Explain.

3. A vertical, rectangular loop of copper wire is half in and half out of the horizontal magnetic field in FIGURE Q33.3. (The field is zero beneath the dashed line.) The loop is released and starts to fall. Is there a net magnetic force on the loop? If so, in which direction? Explain.

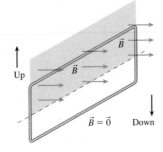

FIGURE Q33.3

4. Does the loop of wire in FIGURE Q33.4 have a clockwise current, a counterclockwise current, or no current under the following circumstances? Explain.
   a. The magnetic field points out of the page and is increasing.
   b. The magnetic field points out of the page and is constant.
   c. The magnetic field points out of the page and is decreasing.

FIGURE Q33.4   FIGURE Q33.5

5. The two loops of wire in FIGURE Q33.5 are stacked one above the other. Does the upper loop have a clockwise current, a counterclockwise current, or no current at the following times? Explain.
   a. Before the switch is closed.
   b. Immediately after the switch is closed.
   c. Long after the switch is closed.
   d. Immediately after the switch is reopened.

6. FIGURE Q33.6 shows a bar magnet being pushed toward a conducting loop from below, along the axis of the loop.
   a. What is the current direction in the loop? Explain.
   b. Is there a magnetic force on the loop? If so, in which direction? Explain.
   Hint: A current loop is a magnetic dipole.
   c. Is there a force on the magnet? If so, in which direction?

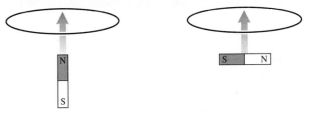

FIGURE Q33.6   FIGURE Q33.7

7. A bar magnet is pushed toward a loop of wire as shown in FIGURE Q33.7. Is there a current in the loop? If so, in which direction? If not, why not?

8. FIGURE Q33.8 shows a bar magnet, a coil of wire, and a current meter. Is the current through the meter right to left, left to right, or zero for the following circumstances? Explain.
   a. The magnet is inserted into the coil.
   b. The magnet is held at rest inside the coil.
   c. The magnet is withdrawn from the left side of the coil.

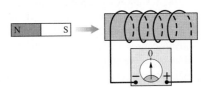

FIGURE Q33.8

9. Is the magnetic field strength in FIGURE Q33.9 increasing, decreasing, or steady? Explain.

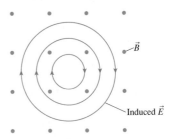

FIGURE Q33.9

10. An inductor with a 2.0 A current stores energy. At what current will the stored energy be twice as large?

11. a. Can you tell which of the inductors in FIGURE Q33.11 has the larger current through it? If so, which one? Explain.
    b. Can you tell through which inductor the current is changing more rapidly? If so, which one? Explain.
    c. If the current enters the inductor from the bottom, can you tell if the current is increasing, decreasing, or staying the same? If so, which? Explain.

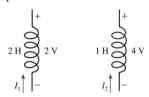

FIGURE Q33.11

12. An *LC* circuit oscillates at a frequency of 2000 Hz. What will the frequency be if the inductance is quadrupled?

13. Rank in order, from largest to smallest, the three time constants τ_a to τ_c for the three circuits in FIGURE Q33.13. Explain.

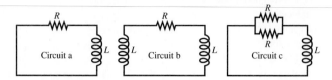

FIGURE Q33.13

14. For the circuit of FIGURE Q33.14:
    a. What is the battery current immediately after the switch closes? Explain.
    b. What is the battery current after the switch has been closed a long time? Explain.

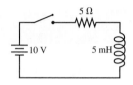

FIGURE Q33.14

EXERCISES AND PROBLEMS

Problems labled [ ] integrate material from earlier chapters.

Exercises

Section 33.2 Motional emf

1. | The earth's magnetic field strength is 5.0×10^{-5} T. How fast would you have to drive your car to create a 1.0 V motional emf along your 1.0-m-long radio antenna? Assume that the motion of the antenna is perpendicular to $\vec{B}$.

2. | A potential difference of 0.050 V is developed across the 10-cm-long wire of FIGURE EX33.2 as it moves through a magnetic field perpendicular to the page. What are the strength and direction (in or out) of the magnetic field?

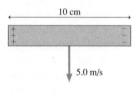

FIGURE EX33.2

3. ‖ A 10-cm-long wire is pulled along a U-shaped conducting rail in a perpendicular magnetic field. The total resistance of the wire and rail is 0.20 Ω. Pulling the wire at a steady speed of 4.0 m/s causes 4.0 W of power to be dissipated in the circuit.
   a. How big is the pulling force?
   b. What is the strength of the magnetic field?

Section 33.3 Magnetic Flux

4. | What is the magnetic flux through the loop shown in FIGURE EX33.4?

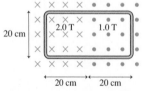

FIGURE EX33.4

5. ‖ FIGURE EX33.5 shows a 2.0-cm-diameter solenoid passing through the center of a 6.0-cm-diameter loop. The magnetic field inside the solenoid is 0.20 T. What is the magnetic flux through the loop when it is perpendicular to the solenoid and when it is tilted at a 60° angle?

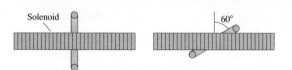

FIGURE EX33.5

6. ‖ What is the magnetic flux through the loop shown in FIGURE EX33.6?

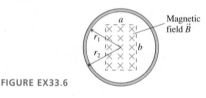

FIGURE EX33.6

Section 33.4 Lenz's Law

7. | There is a cw induced current in the conducting loop shown in FIGURE EX33.7. Is the magnetic field inside the loop increasing in strength, decreasing in strength, or steady?

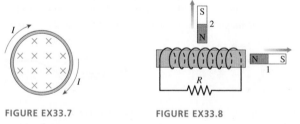

FIGURE EX33.7   FIGURE EX33.8

8. | A solenoid is wound as shown in FIGURE EX33.8.
   a. Is there an induced current as magnet 1 is moved away from the solenoid? If so, what is the current direction through resistor *R*?
   b. Is there an induced current as magnet 2 is moved away from the solenoid? If so, what is the current direction through resistor *R*?

9. ‖ The current in the solenoid of FIGURE EX33.9 is increasing. The solenoid is surrounded by a conducting loop. Is there a current in the loop? If so, is the loop current cw or ccw?

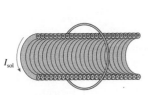

FIGURE EX33.9

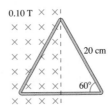

FIGURE EX33.10

10. | The metal equilateral triangle in FIGURE EX33.10, 20 cm on each side, is halfway into a 0.10 T magnetic field.
   a. What is the magnetic flux through the triangle?
   b. If the magnetic field strength decreases, what is the direction of the induced current in the triangle?

Section 33.5 Faraday's Law

11. | FIGURE EX33.11 shows a 10-cm-diameter loop in three different magnetic fields. The loop's resistance is 0.20 Ω. For each, what are the size and direction of the induced current?

(a) B increasing at 0.50 T/s

(b) B decreasing at 0.50 T/s

(c) B decreasing at 0.50 T/s

FIGURE EX33.11

12. | The loop in FIGURE EX33.12 is being pushed into the 0.20 T magnetic field at 50 m/s. The resistance of the loop is 0.10 Ω. What are the direction and the magnitude of the current in the loop?

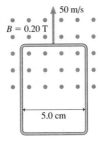

FIGURE EX33.12

13. ‖ A 1000-turn coil of wire 1.0 cm in diameter is in a magnetic field that increases from 0.10 T to 0.30 T in 10 ms. The axis of the coil is parallel to the field. What is the emf of the coil?

14. | The resistance of the loop in FIGURE EX33.14 is 0.20 Ω. Is the magnetic field strength increasing or decreasing? At what rate (T/s)?

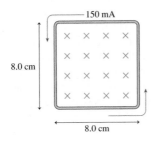

FIGURE EX33.14

Section 33.6 Induced Fields

15. ‖ FIGURE EX33.15 shows the current as a function of time through a 20-cm-long, 4.0-cm-diameter solenoid with 400 turns. Draw a graph of the induced electric field strength as a function of time at a point 1.0 cm from the axis of the solenoid.

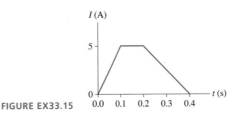

FIGURE EX33.15

16. ‖ The magnetic field inside a 5.0-cm-diameter solenoid is 2.0 T and decreasing at 4.0 T/s. What is the electric field strength inside the solenoid at a point (a) on the axis and (b) 2.0 cm from the axis?

17. ‖ The magnetic field in FIGURE EX33.17 is decreasing at the rate 0.10 T/s. What is the acceleration (magnitude and direction) of a proton initially at rest at points a to d?

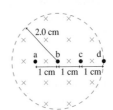

FIGURE EX33.17

Section 33.8 Inductors

18. | What is the potential difference across a 10 mH inductor if the current through the inductor drops from 150 mA to 50 mA in 10 μs? What is the direction of this potential difference? That is, does the potential increase or decrease along the direction of the current?

19. | The maximum allowable potential difference across a 200 mH inductor is 400 V. You need to raise the current through the inductor from 1.0 A to 3.0 A. What is the minimum time you should allow for changing the current?

20. | A 100 mH inductor whose windings have a resistance of 4.0 Ω is connected across a 12 V battery having an internal resistance of 2.0 Ω. How much energy is stored in the inductor?

21. ‖ How much energy is stored in a 3.0-cm-diameter, 12-cm-long solenoid that has 200 turns of wire and carries a current of 0.80 A?

Section 33.9 *LC* Circuits

22. ‖ An FM radio station broadcasts at a frequency of 100 MHz. What inductance should be paired with a 10 pF capacitor to build a receiver circuit for this station?

23. | A 2.0 mH inductor is connected in parallel with a variable capacitor. The capacitor can be varied from 100 pF to 200 pF. What is the range of oscillation frequencies for this circuit?

24. | An MRI machine needs to detect signals that oscillate at
BIO very high frequencies. It does so with an *LC* circuit containing a 15 mH coil. To what value should the capacitance be set to detect a 450 MHz signal?

Section 33.10 *LR* Circuits

25. ‖ What value of resistor R gives the circuit in FIGURE EX33.25 a time constant of 25 μs?

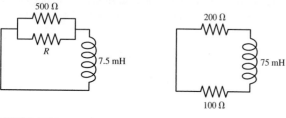

FIGURE EX33.25   FIGURE EX33.26

26. ‖ At $t = 0$ s, the current in the circuit in FIGURE EX33.26 is I_0. At what time is the current $\frac{1}{2}I_0$?

Problems

27. ‖ FIGURE P33.27 shows a 10 cm × 10 cm square bent at a 90° angle. A uniform 0.050 T magnetic field points downward at a 45° angle. What is the magnetic flux through the loop?

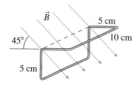

FIGURE P33.27

28. ‖ A 5.0-cm-diameter coil has 20 turns and a resistance of 0.50 Ω. A magnetic field perpendicular to the coil is $B = 0.020t + 0.010t^2$, where B is in tesla and t is in seconds.
    a. Find an expression for the induced current $I(t)$ as a function of time.
    b. Evaluate I at $t = 5$ s and $t = 10$ s.

29. ‖ A 20 cm × 20 cm square loop has a resistance of 0.10 Ω. A magnetic field perpendicular to the loop is $B = 4t - 2t^2$, where B is in tesla and t is in seconds. What is the current in the loop at $t = 0.0$ s, $t = 1.0$ s, and $t = 2.0$ s?

30. ‖ A 100-turn, 2.0-cm-diameter coil is at rest in a horizontal plane. A uniform magnetic field 60° away from vertical increases from 0.50 T to 1.50 T in 0.60 s. What is the induced emf in the coil?

31. ‖ A 100-turn, 8.0-cm-diameter coil is made of 0.50-mm-diameter copper wire. A magnetic field is parallel to the axis of the coil. At what rate must B increase to induce a 2.0 A current in the coil?

32. ‖ A circular loop made from a flexible, conducting wire is shrinking. Its radius as a function of time is $r = r_0e^{-\beta t}$. The loop is perpendicular to a steady, uniform magnetic field B. Find an expression for the induced emf in the loop at time t.

33. ‖ A 10 cm × 10 cm square loop lies in the xy-plane. The magnetic field in this region of space is $B = (0.30t\,\hat{i} + 0.50t^2\,\hat{k})$ T, where t is in s. What is the emf induced in the loop at (a) $t = 0.5$ s and (b) $t = 1.0$ s?

34. ‖ A 20 cm × 20 cm square loop of wire lies in the xy-plane with its bottom edge on the x-axis. The resistance of the loop is 0.50 Ω. A magnetic field parallel to the z-axis is given by $B = 0.80y^2t$, where B is in tesla, y in meters, and t in seconds. What is the size of the induced current in the loop at $t = 0.50$ s?

35. ‖‖ A 2.0 cm × 2.0 cm square loop of wire with resistance 0.010 Ω has one edge parallel to a long straight wire. The near edge of the loop is 1.0 cm from the wire. The current in the wire is increasing at the rate of 100 A/s. What is the current in the loop?

36. ‖‖ The rectangular loop in FIGURE P33.36 has 0.020 Ω resistance. What is the induced current in the loop at this instant?

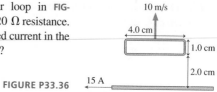

FIGURE P33.36

37. ‖ FIGURE P33.37 shows a 4.0-cm-diameter loop with resistance 0.10 Ω around a 2.0-cm-diameter solenoid. The solenoid is 10 cm long, has 100 turns, and carries the current shown in the graph. A positive current is cw when seen from the left. Find the current in the loop at (a) $t = 0.5$ s, (b) $t = 1.5$ s, and (c) $t = 2.5$ s.

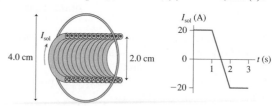

FIGURE P33.37

38. ‖‖ FIGURE P33.38 shows a 1.0-cm-diameter loop with $R = 0.50\ \Omega$ inside a 2.0-cm-diameter solenoid. The solenoid is 8.0 cm long, has 120 turns, and carries the current shown in the graph. A positive current is cw when seen from the left. Determine the current in the loop at $t = 0.010$ s.

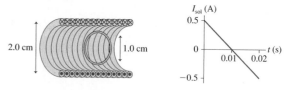

FIGURE P33.38

39. ‖ FIGURE P33.39 shows two 20-turn coils tightly wrapped on the same 2.0-cm-diameter cylinder with 1.0-mm-diameter wire. The current through coil 1 is shown in the graph. Determine the current in coil 2 at (a) $t = 0.05$ s and (b) $t = 0.25$ s. A positive current is into the page at the top of a loop. Assume that the magnetic field of coil 1 passes entirely through coil 2.

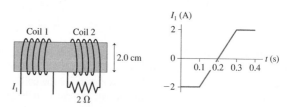

FIGURE P33.39

40. ‖‖ A 50-turn, 4.0-cm-diameter coil with $R = 0.50\ \Omega$ surrounds a 2.0-cm-diameter solenoid. The solenoid is 20 cm long and has 200 turns. The 60 Hz current through the solenoid is $I_{sol} = (0.50\ \text{A})\sin(2\pi ft)$. Find an expression for I_{coil}, the induced current in the coil as a function of time.

41. ‖ A loop antenna, such as is used on older televisions to pick up UHF broadcasts, is 25 cm in diameter. The plane of the loop is perpendicular to the oscillating magnetic field of a 150 MHz electromagnetic wave. The magnetic field through the loop is $B = (20\ \text{nT})\sin\omega t$.

a. What is the maximum emf induced in the antenna?

b. What is the maximum emf if the loop is turned 90° to be perpendicular to the oscillating electric field?

42. ‖ A 40-turn, 4.0-cm-diameter coil with $R = 0.40\ \Omega$ surrounds a 3.0-cm-diameter solenoid. The solenoid is 20 cm long and has 200 turns. The 60 Hz current through the solenoid is $I = I_0 \sin(2\pi ft)$. What is I_0 if the maximum induced current in the coil is 0.20 A?

43. ‖ Electricity is distributed from electrical substations to neighborhoods at 15,000 V. This is a 60 Hz oscillating (AC) voltage. Neighborhood transformers, seen on utility poles, step this voltage down to the 120 V that is delivered to your house.

a. How many turns does the primary coil on the transformer have if the secondary coil has 100 turns?

b. No energy is lost in an ideal transformer, so the output power P_{out} from the secondary coil equals the input power P_{in} to the primary coil. Suppose a neighborhood transformer delivers 250 A at 120 V. What is the current in the 15,000 V line from the substation?

44. ‖‖ A small, 2.0-mm-diameter circular loop with $R = 0.020\ \Omega$ is at the center of a large 100-mm-diameter circular loop. Both loops lie in the same plane. The current in the outer loop changes from $+1.0$ A to -1.0 A in 0.10 s. What is the induced current in the inner loop?

45. ‖ The square loop shown in FIGURE P33.45 moves into a 0.80 T magnetic field at a constant speed of 10 m/s. The loop has a resistance of $0.10\ \Omega$, and it enters the field at $t = 0$ s.

a. Find the induced current in the loop as a function of time. Give your answer as a graph of I versus t from $t = 0$ s to $t = 0.020$ s.

b. What is the maximum current? What is the position of the loop when the current is maximum?

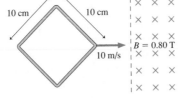

FIGURE P33.45

46. ‖ The L-shaped conductor in FIGURE P33.46 moves at 10 m/s across a stationary L-shaped conductor in a 0.10 T magnetic field. The two vertices overlap, so that the enclosed area is zero, at $t = 0$ s. The conductor has a resistance of 0.010 ohms *per meter*.

a. What is the direction of the induced current?

b. Find expressions for the induced emf and the induced current as functions of time.

c. Evaluate $\mathcal{E}$ and I at $t = 0.10$ s.

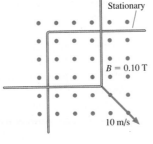

FIGURE P33.46

47. ‖ A 4.0-cm-long slide wire moves outward with a speed of 100 m/s in a 1.0 T magnetic field. (See Figure 33.26.) At the instant the circuit forms a 4.0 cm × 4.0 cm square, with $R = 0.010\ \Omega$ on each side, what are

a. The induced emf?

b. The induced current?

c. The potential difference between the two ends of the moving wire?

48. ‖ A 20-cm-long, zero-resistance slide wire moves outward, on zero-resistance rails, at a steady speed of 10 m/s in a 0.10 T magnetic field. (See Figure 33.26.) On the opposite side, a $1.0\ \Omega$ carbon resistor completes the circuit by connecting the two rails. The mass of the resistor is 50 mg.

a. What is the induced current in the circuit?

b. How much force is needed to pull the wire at this speed?

c. If the wire is pulled for 10 s, what is the temperature increase of the carbon? The specific heat of carbon is 710 J/kg K.

49. ‖ Your camping buddy has an idea for a light to go inside your tent. He happens to have a powerful (and heavy!) horseshoe magnet that he bought at a surplus store. This magnet creates a 0.20 T field between two pole tips 10 cm apart. His idea is to build the hand-cranked generator shown in FIGURE P33.49. He thinks you can make enough current to fully light a $1.0\ \Omega$ lightbulb rated at 4.0 W. That's not super bright, but it should be plenty of light for routine activities in the tent.

a. Find an expression for the induced current as a function of time if you turn the crank at frequency f. Assume that the semicircle is at its highest point at $t = 0$ s.

b. With what frequency will you have to turn the crank for the maximum current to fully light the bulb? Is this feasible?

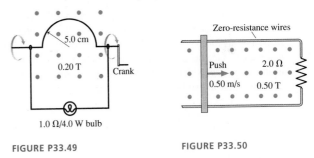

FIGURE P33.49   **FIGURE P33.50**

50. ‖ The 10-cm-wide, zero-resistance slide wire shown in FIGURE P33.50 is pushed toward the $2.0\ \Omega$ resistor at a steady speed of 0.50 m/s. The magnetic field strength is 0.50 T.

a. How big is the pushing force?

b. How much power does the pushing force supply to the wire?

c. What are the direction and magnitude of the induced current?

d. How much power is dissipated in the resistor?

51. ‖ One way to determine a magnetic field strength is to measure the emf induced in a rotating coil. To calibrate a large magnet in your laboratory, you attach a 2.0-cm-diameter, 100-turn coil to the end of a motor-driven shaft, place the coil between the pole tips of the magnet, and rotate it at different frequencies. The emf oscillates, so you use a voltmeter that measures its amplitude. The table shows your data:

Frequency (Hz)	Voltage (mV)
10	380
15	610
20	780
25	1020
30	1160

Use an appropriate graph of the data to determine the magnetic field strength.

52. ‖ You've decided to make the magnetic projectile launcher shown in FIGURE P33.52 for your science project. An aluminum

bar of length *l* slides along metal rails through a magnetic field *B*. The switch closes at *t* = 0 s, while the bar is at rest, and a battery of emf $\mathcal{E}_{bat}$ starts a current flowing around the loop. The battery has internal resistance *r*. The resistance of the rails and the bar are effectively zero.

a. Show that the bar reaches a terminal speed v_{term}, and find an expression for v_{term}.

b. Evaluate v_{term} for $\mathcal{E}_{bat}$ = 1.0 V, *r* = 0.10 Ω, *l* = 6.0 cm, and *B* = 0.50 T.

FIGURE P33.52

53. ||| A slide wire of length *l*, mass *m*, and resistance *R* slides down a U-shaped metal track that is tilted upward at angle θ. The track has zero resistance and no friction. A vertical magnetic field *B* fills the loop formed by the track and the slide wire.

a. Find an expression for the induced current *I* when the slide wire moves at speed *v*.

b. Show that the slide wire reaches a terminal speed v_{term}, and find an expression for v_{term}.

54. || FIGURE P33.54 shows a U-shaped conducting rail that is oriented vertically in a horizontal magnetic field. The rail has no electric resistance and does not move. A slide wire with mass *m* and resistance *R* can slide up and down without friction while maintaining electrical contact with the rail. The slide wire is released from rest.

a. Show that the slide wire reaches a terminal speed v_{term}, and find an expression for v_{term}.

b. Determine the value of v_{term} if *l* = 20 cm, *m* = 10 g, *R* = 0.10 Ω, and *B* = 0.50 T.

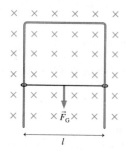

FIGURE P33.54

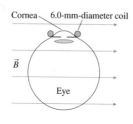

FIGURE P33.55

55. || Experiments to study vision often need to track the movements
BIO of a subject's eye. One way of doing so is to have the subject sit in a magnetic field while wearing special contact lenses with a coil of very fine wire circling the edge. A current is induced in the coil each time the subject rotates his eye. Consider the experiment of FIGURE P33.55 in which a 20-turn, 6.0-mm-diameter coil of wire circles the subject's cornea while a 1.0 T magnetic field is directed as shown. The subject begins by looking straight ahead. What emf is induced in the coil if the subject shifts his gaze by 5° in 0.20 s?

56. || A 10-turn coil of wire having a diameter of 1.0 cm and a resistance of 0.20 Ω is in a 1.0 mT magnetic field, with the coil oriented for maximum flux. The coil is connected to an uncharged 1.0 μF capacitor rather than to a current meter. The coil is quickly pulled out of the magnetic field. Afterward, what is the voltage across the capacitor?

Hint: Use *I* = *dq*/*dt* to relate the *net* change of flux to the amount of charge that flows to the capacitor.

57. ||| The magnetic field at one place on the earth's surface is 55 μT in strength and tilted 60° down from horizontal. A 200-turn coil having a diameter of 4.0 cm and a resistance of 2.0 Ω is connected to a 1.0 μF capacitor rather than to a current meter. The coil is held in a horizontal plane and the capacitor is discharged. Then the coil is quickly rotated 180° so that the side that had been facing up is now facing down. Afterward, what is the voltage across the capacitor? See the Hint in Problem 56.

58. || The magnetic field inside a 4.0-cm-diameter superconducting solenoid varies sinusoidally between 8.0 T and 12.0 T at a frequency of 10 Hz.

a. What is the maximum electric field strength at a point 1.5 cm from the solenoid axis?

b. What is the value of *B* at the instant *E* reaches its maximum value?

59. || Equation 33.26 is an expression for the induced electric field inside a solenoid (*r* < *R*). Find an expression for the induced electric field outside a solenoid (*r* > *R*) in which the magnetic field is changing at the rate *dB*/*dt*.

60. || A solenoid inductor has an emf of 0.20 V when the current through it changes at the rate 10.0 A/s. A steady current of 0.10 A produces a flux of 5.0 μWb per turn. How many turns does the inductor have?

61. || a. What is the magnetic energy density at the center of a 4.0-cm-diameter loop carrying a current of 1.0 A?

b. What current in a straight wire gives the magnetic energy density you found in part a at a point 2.0 cm from the wire?

62. | MRI (magnetic resonance imaging) is a medical technique
BIO that produces detailed "pictures" of the interior of the body. The patient is placed into a solenoid that is 40 cm in diameter and 1.0 m long. A 100 A current creates a 5.0 T magnetic field inside the solenoid. To carry such a large current, the solenoid wires are cooled with liquid helium until they become supercon- ducting (no electric resistance).

a. How much magnetic energy is stored in the solenoid? Assume that the magnetic field is uniform within the solenoid and quickly drops to zero outside the solenoid.

b. How many turns of wire does the solenoid have?

63. | One possible concern with MRI (see Problem 62) is turning
BIO the magnetic field on or off too quickly. Bodily fluids are con- ductors, and a changing magnetic field could cause electric cur- rents to flow through the patient. Suppose a typical patient has a maximum cross-section area of 0.060 m². What is the smallest time interval in which a 5.0 T magnetic field can be turned on or off if the induced emf around the patient's body must be kept to less than 0.10 V?

64. || FIGURE P33.64 shows the current through a 10 mH inductor. Draw a graph showing the potential difference ΔV_L across the inductor for these 6 ms.

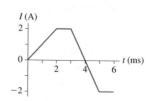

FIGURE P33.64

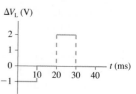

FIGURE P33.65

65. || FIGURE P33.65 shows the potential difference across a 50 mH inductor. The current through the inductor at *t* = 0 s is 0.20 A.

Draw a graph showing the current through the inductor from $t = 0$ s to $t = 40$ ms.

66. ‖ The current through inductance L is given by $I = I_0 \sin \omega t$.
    a. Find an expression for the potential difference ΔV_L across the inductor.
    b. The maximum voltage across the inductor is 0.20 V when $L = 50 \ \mu\text{H}$ and $f = 500$ kHz. What is I_0?

67. ‖ The current through inductance L is given by $I = I_0 e^{-t/\tau}$.
    a. Find an expression for the potential difference ΔV_L across the inductor.
    b. Evaluate ΔV_L at $t = 0$, 1, 2, and 3 ms if $L = 20$ mH, $I_0 = 50$ mA, and $\tau = 1.0$ ms.

68. ‖ An LC circuit is built with a 20 mH inductor and an 8.0 pF capacitor. The capacitor voltage has its maximum value of 25 V at $t = 0$ s.
    a. How long is it until the capacitor is first fully discharged?
    b. What is the inductor current at that time?

69. ‖ An LC circuit has a 10 mH inductor. The current has its maximum value of 0.60 A at $t = 0$ s. A short time later the capacitor reaches its maximum potential difference of 60 V. What is the value of the capacitance?

70. ‖ An electric oscillator is made with a 0.10 μF capacitor and a 1.0 mH inductor. The capacitor is initially charged to 5.0 V. What is the maximum current through the inductor as the circuit oscillates?

71. ‖‖ In recent years it has been possible to buy a 1.0 F capacitor. This is an enormously large amount of capacitance. Suppose you want to build a 1.0 Hz oscillator with a 1.0 F capacitor. You have a spool of 0.25-mm-diameter wire and a 4.0-cm-diameter plastic cylinder. How long must your inductor be if you wrap it with 2 layers of closely spaced turns?

72. ‖ For your final exam in electronics, you're asked to build an LC circuit that oscillates at 10 kHz. In addition, the maximum current must be 0.10 A and the maximum energy stored in the capacitor must be 1.0×10^{-5} J. What values of inductance and capacitance must you use?

73. ‖ The switch in FIGURE P33.73 has been in position 1 for a long time. It is changed to position 2 at $t = 0$ s.
    a. What is the maximum current through the inductor?
    b. What is the first time at which the current is maximum?

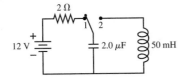

FIGURE P33.73

74. ‖ The 300 μF capacitor in FIGURE P33.74 is initially charged to 100 V, the 1200 μF capacitor is uncharged, and the switches are both open.
    a. What is the maximum voltage to which you can charge the 1200 μF capacitor by the proper closing and opening of the two switches?
    b. How would you do it? Describe the sequence in which you would close and open switches and the times at which you would do so. The first switch is closed at $t = 0$ s.

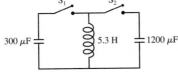

FIGURE P33.74

75. ‖ The switch in FIGURE P33.75 has been open for a long time. It is closed at $t = 0$ s.
    a. What is the current through the battery immediately after the switch is closed?
    b. What is the current through the battery after the switch has been closed a long time?

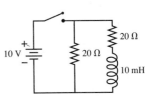

FIGURE P33.75

76. ‖ The switch in FIGURE P33.76 has been open for a long time. It is closed at $t = 0$ s. What is the current through the 20 Ω resistor
    a. immediately after the switch is closed?
    b. after the switch has been closed a long time?
    c. immediately after the switch is reopened?

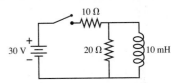

FIGURE P33.76

77. ‖ The switch in FIGURE P33.77 has been open for a long time. It is closed at $t = 0$ s.
    a. After the switch has been closed for a long time, what is the current in the circuit? Call this current I_0.
    b. Find an expression for the current I as a function of time. Write your expression in terms of I_0, R, and L.
    c. Sketch a current-versus-time graph from $t = 0$ s until the current is no longer changing.

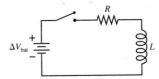

FIGURE P33.77

78. ‖ To determine the inductance of an unmarked inductor, you set up the circuit shown in FIGURE P33.78. After moving the switch from a to b at $t = 0$ s, you monitor the resistor voltage with an oscilloscope. Your data are as follows:

Time (μs)	Voltage (V)
0	9.0
10	6.7
20	4.6
30	3.2
40	2.5

Use an appropriate graph of the data to determine the inductance.

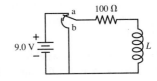

FIGURE P33.78

Challenge Problems

79. The metal wire in FIGURE CP33.79 moves with speed v parallel to a straight wire that is carrying current I. The distance between the two wires is d. Find an expression for the potential difference between the two ends of the moving wire.

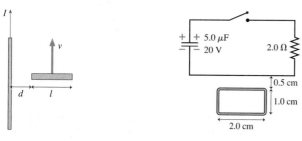

FIGURE CP33.79

80. A rectangular metal loop with 0.050 Ω resistance is placed next to one wire of the RC circuit shown in FIGURE CP33.80. The capacitor is charged to 20 V with the polarity shown, then the switch is closed at $t = 0$ s.
    a. What is the direction of current in the loop for $t > 0$ s?
    b. What is the current in the loop at $t = 5.0 \, \mu s$? Assume that only the circuit wire next to the loop is close enough to produce a significant magnetic field.

81. A closed, square loop is formed with 40 cm of wire having $R = 0.10 \, \Omega$, as shown in FIGURE CP33.81. A 0.50 T magnetic field is perpendicular to the loop. At $t = 0$ s, two diagonally opposite corners of the loop begin to move apart at 0.293 m/s.
    a. How long does it take the loop to collapse to a straight line?
    b. Find an expression for the induced current I as a function of time while the loop is collapsing. Assume that the sides remain straight lines during the collapse.
    c. Evaluate I at four or five times during the collapse, then draw a graph of I versus t.

FIGURE CP33.81

82. Let's look at the details of eddy-current braking. A square loop, length l on each side, is shot with velocity v_0 into a uniform magnetic field B. The field is perpendicular to the plane of the loop. The loop has mass m and resistance R, and it enters the field at $t = 0$ s. Assume that the loop is moving to the right along the x-axis and that the field begins at $x = 0$ m.
    a. Find an expression for the loop's velocity as a function of time as it enters the magnetic field. You can ignore gravity, and you can assume that the back edge of the loop has not entered the field.
    b. Calculate and draw a graph of v over the interval $0 \, \text{s} \leq t \leq 0.04$ s for the case that $v_0 = 10$ m/s, $l = 10$ cm, $m = 1.0$ g, $R = 0.0010 \, \Omega$, and $B = 0.10$ T. The back edge of the loop does not reach the field during this time interval.

83. An 8.0 cm × 8.0 cm square loop is halfway into a magnetic field perpendicular to the plane of the loop. The loop's mass is 10 g and its resistance is 0.010 Ω. A switch is closed at $t = 0$ s, causing the magnetic field to increase from 0 to 1.0 T in 0.010 s.
    a. What is the induced current in the square loop?
    b. With what speed is the loop "kicked" away from the magnetic field?
    Hint: What is the impulse on the loop?

84. A 2.0-cm-diameter solenoid is wrapped with 1000 turns per meter. 0.50 cm from the axis, the strength of an induced electric field is 5.0×10^{-4} V/m. What is the rate dI/dt with which the current through the solenoid is changing?

85. High-frequency signals are often transmitted along a *coaxial cable*, such as the one shown in FIGURE CP33.85. For example, the cable TV hookup coming into your home is a coaxial cable. The signal is carried on a wire of radius r_1 while the outer conductor of radius r_2 is grounded. A soft, flexible insulating material fills the space between them, and an insulating plastic coating goes around the outside.
    a. Find an expression for the inductance per meter of a coaxial cable. To do so, consider the flux through a rectangle of length l that spans the gap between the inner and outer conductors.
    b. Evaluate the inductance per meter of a cable having $r_1 = 0.50$ mm and $r_2 = 3.0$ mm.

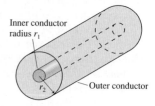

FIGURE CP33.85

FIGURE CP33.80

STOP TO THINK ANSWERS

Stop to Think 33.1: d. According to the right-hand rule, the magnetic force on a positive charge carrier is to the right.

Stop to Think 33.2: No. The charge carriers in the wire move parallel to $\vec{B}$. There's no magnetic force on a charge moving parallel to a magnetic field.

Stop to Think 33.3: $F_b = F_d > F_a = F_c$. $\vec{F}_a$ is zero because there's no field. $\vec{F}_c$ is also zero because there's no current around the loop. The charge carriers in both the right and left edges are pushed to the bottom of the loop, creating a motional emf but no current. The currents at b and d are in opposite directions, but the forces on the segments in the field are both to the left and of equal magnitude.

Stop to Think 33.4: Clockwise. The wire's magnetic field as it passes through the loop is into the page. The flux through the loop decreases

into the page as the wire moves away. To oppose this decrease, the induced magnetic field needs to point into the page.

Stop to Think 33.5: d. The flux is increasing into the loop. To oppose this increase, the induced magnetic field needs to point out of the page. This requires a ccw induced current. Using the right-hand rule, the magnetic force on the current in the left edge of the loop is to the right, away from the field. The magnetic forces on the top and bottom segments of the loop are in opposite directions and cancel each other.

Stop to Think 33.6: b or f. The potential decreases in the direction of increasing current and increases in the direction of decreasing current.

Stop to Think 33.7: $\tau_c > \tau_a > \tau_b$. $\tau = L/R$, so smaller total resistance gives a larger time constant. The parallel resistors have total resistance $R/2$. The series resistors have total resistance $2R$.

34 Electromagnetic Fields and Waves

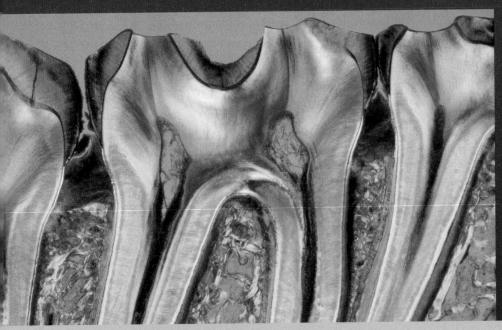

A thin section of molar teeth seen in polarized light. The rainbow of colors arises because different biological materials have different effects on the light's polarization.

▶ **Looking Ahead** The goal of Chapter 34 is to study the properties of electromagnetic fields and waves.

Maxwell's Theory of Electromagnetism

All of electricity and magnetism is based on four equations for the fields, called **Maxwell's equations,** and one equation that tells us how charges respond to fields.

Gauss's law: Charged particles create electric fields.

Faraday's law: Electric fields can also be created by changing magnetic fields.

Gauss's law for magnetism: There are no isolated magnetic poles.

Ampère-Maxwell law: Magnetic fields can be created by currents or by changing electric fields.

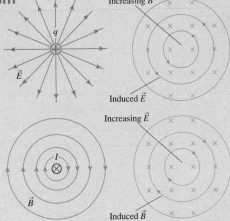

Electric and magnetic fields exert forces on charged particles. When combined, the net force on charge q is called the **Lorentz force law:**

$$\vec{F} = q(\vec{E} + \vec{v} \times \vec{B})$$

◀ **Looking Back**
Section 27.4 Gauss's law
Section 32.6 Ampère's law
Section 33.5 Faraday's law

Electromagnetic Waves

You'll learn that Maxwell's equations predict the existence of self-sustaining **electromagnetic waves** that travel through space without the presence of charges or currents.

- $\vec{E}$ and $\vec{B}$ are perpendicular to each other and to the direction of travel.
- In vacuum $v_{em} = 1/\sqrt{\epsilon_0 \mu_0} = c$, the speed of light.

Electromagnetic waves are often **polarized,** meaning that the electric field always oscillates in the same plane. You'll learn to calculate the intensity of light transmitted through a polarizing filter.

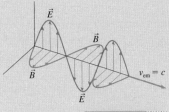

No light is transmitted through *crossed polarizers* whose axes are perpendicular to each other.

Field Transformations

Electric and magnetic fields turn out not to be separate, independent entities. Whether the field at a point is electric or magnetic depends on your motion relative to the charges and currents.

You'll learn how to transform the fields measured in reference frame A to a second frame B moving relative to A.

What are the fields of this charge?

◀ **Looking Back**
Section 4.4 Relative motion

34.1 *E* or *B*? It Depends on Your Perspective

FIGURE 34.1 Brittney carries a charge past Alec.

(a)

Charge *q* moves with velocity $\vec{v}$ relative to Alec.

(b)

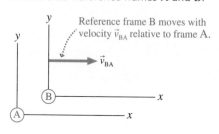

Charge *q* moves through a magnetic field established by Alec.

It seems clear, after the last nine chapters, that charges create electric fields and that moving charges, or currents, create magnetic fields. But consider FIGURE 34.1a, where Brittney, carrying charge *q*, runs past Alec with velocity $\vec{v}$. Alec sees a moving charge, and he knows that this charge creates a magnetic field. But from Brittney's perspective, the charge is at rest. Stationary charges don't create magnetic fields, so Brittney claims that the magnetic field is zero. Is there, or is there not, a magnetic field?

Or what about the situation in FIGURE 34.1b? Now Brittney is carrying the charge through a magnetic field that Alec has created. Alec sees a charge moving in a magnetic field, so he knows there's a force $\vec{F} = q\vec{v} \times \vec{B}$ on the charge. But for Brittney the charge is still at rest. Stationary charges don't experience magnetic forces, so Brittney claims that $\vec{F} = \vec{0}$.

Now, we may be a bit uncertain about magnetic fields, but surely there can be no disagreement over forces. After all, forces cause observable and measurable effects, so Alec and Brittney should be able to agree on whether or not the charge experiences a force. Further, if Brittney runs with constant velocity, then both Alec and Brittney are in *inertial reference frames*. You learned in Chapter 4 that these are the reference frames in which Newton's laws are valid, so we can't say that there's anything abnormal or unusual about Alec's and Brittney's observations. We have a paradox.

This paradox has arisen because magnetic fields and forces depend on velocity, but we haven't looked at the issue of velocity *with respect to what* or velocity *as measured by whom*. The resolution of this paradox will lead us to the conclusion that $\vec{E}$ and $\vec{B}$ are not, as we've been assuming, separate and independent entities. They are closely intertwined.

Reference Frames

FIGURE 34.2 Reference frames A and B.

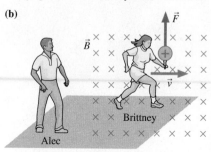

We introduced reference frames and relative motion in Chapter 4. To remind you, FIGURE 34.2 shows two reference frames labeled A and B. You can think of these as the reference frames in which Alec and Brittney, respectively, are at rest. Frame B moves with velocity $\vec{v}_{BA}$ with respect to frame A. That is, an observer (Alec) at rest in A sees the origin of B (Brittney) go past with velocity $\vec{v}_{BA}$. Of course, Brittney would say that Alec has velocity $\vec{v}_{AB} = -\vec{v}_{BA}$ relative to her reference frame. There's no implication that either frame is "at rest." All we know is that the two reference frames move relative to each other. We will stipulate that both reference frames are inertial reference frames, so $\vec{v}_{BA}$ is constant.

FIGURE 34.3 shows a charged particle C. Experimenters in frame A measure the motion of the particle and find that its velocity *relative to frame A* is $\vec{v}_{CA}$. At the same instant, experimenters in B find that the particle's velocity *relative to frame B* is $\vec{v}_{CB}$. In Chapter 4, we found that $\vec{v}_{CA}$ and $\vec{v}_{CB}$ are related by

$$\vec{v}_{CA} = \vec{v}_{CB} + \vec{v}_{BA} \tag{34.1}$$

FIGURE 34.3 The particle's velocity is measured in both frame A and frame B.

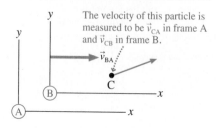

Equation 34.1, the *Galilean transformation of velocity*, tells us that the velocity of the particle relative to reference frame A is its velocity relative to frame B plus (vector addition!) the velocity of frame B relative to frame A.

Suppose the charged particle in Figure 34.3 is accelerating, as it would if acted on by a net force. How does its acceleration $\vec{a}_{CA}$, as measured by experimenters in frame A, compare to the acceleration $\vec{a}_{CB}$ measured in frame B? We can answer this question by taking the time derivative of Equation 34.1:

$$\frac{d\vec{v}_{CA}}{dt} = \frac{d\vec{v}_{CB}}{dt} + \frac{d\vec{v}_{BA}}{dt}$$

The derivatives of $\vec{v}_{CA}$ and $\vec{v}_{CB}$ are the particle's accelerations $\vec{a}_{CA}$ and $\vec{a}_{CB}$ in frames A and B, respectively. But $\vec{v}_{BA}$ is a *constant* velocity, so $d\vec{v}_{BA}/dt = \vec{0}$. Thus the Galilean transformation of acceleration is simply

$$\vec{a}_{CA} = \vec{a}_{CB} \qquad (34.2)$$

Brittney and Alec may measure different positions and velocities for a particle, but they *agree* on its acceleration. And if they agree on its acceleration, they must, by using Newton's second law, agree on the force acting on the particle. That is, **experimenters in all inertial reference frames agree about the force acting on a particle.** This conclusion is the key to understanding how different experimenters see electric and magnetic fields.

The Transformation of Electric and Magnetic Fields

Imagine that Alec has measured the electric field $\vec{E}_A$ and the magnetic field $\vec{B}_A$ in reference frame A. Our investigations thus far give us no reason to think that Brittney's measurements of the fields will differ from Alec's. After all, it seems like the fields are just "there," waiting to be measured. Thus our expectation is that Brittney, in frame B, will measure $\vec{E}_B = \vec{E}_A$ and $\vec{B}_B = \vec{B}_A$.

To find out if this is true, Alec establishes a region of space with a uniform magnetic field $\vec{B}_A$ but no electric field ($\vec{E}_A = \vec{0}$). Then, as shown in FIGURE 34.4, he shoots a positive charge q through the magnetic field. At an instant when q is moving horizontally with velocity $\vec{v}_{CA}$, Alec observes that the particle experiences force $\vec{F}_A = q(\vec{E}_A + \vec{v}_{CA} \times \vec{B}_A) = q\vec{v}_{CA} \times \vec{B}_A$. The direction of the force, given by the right-hand rule, is straight up.

Suppose that Brittney, in frame B, runs alongside the charge with the same velocity: $\vec{v}_{BA} = \vec{v}_{CA}$. To her, in frame B, the charge is at rest. Nonetheless, because both experimenters must agree about forces, Brittney *must* observe the same upward force on the charge that Alec observed. But there is *no* magnetic force on a stationary charge, so how can this be?

Because Brittney sees a stationary charge being acted on by an upward force, her only possible conclusion is that there is an upward-pointing *electric field*. After all, the electric field was initially defined in terms of the force experienced by a stationary charge. If the electric field in frame B is $\vec{E}_B$, then the force on the charge is $\vec{F}_B = q\vec{E}_B$. But we know that $\vec{F}_B = \vec{F}_A$, and Alec has already measured $\vec{F}_A = q\vec{v}_{CA} \times \vec{B}_A = q\vec{v}_{BA} \times \vec{B}_A$. Thus we're led to the conclusion that

$$\vec{E}_B = \vec{v}_{BA} \times \vec{B}_A \qquad (34.3)$$

As Brittney runs past Alec, she finds that at least part of Alec's magnetic field has become an electric field! **Whether a field is seen as "electric" or "magnetic" depends on the motion of the reference frame relative to the sources of the field.**

FIGURE 34.5 shows the situation from Brittney's perspective. There is a force on charge q, the same force that Alec measured in Figure 34.4, but Brittney attributes this force to an electric field rather than a magnetic field. (Brittney needs a moving charge to measure magnetic forces, so we'll need a different experiment to see whether or not there's a magnetic field in frame B.)

More generally, suppose that an experimenter in reference frame A creates both an electric field $\vec{E}_A$, and a magnetic field $\vec{B}_A$. A charge moving in A with velocity $\vec{v}_{CA}$ experiences the force $\vec{F}_A = q(\vec{E}_A + \vec{v}_{CA} \times \vec{B}_A)$ shown in FIGURE 34.6a on the next page. The charge is at rest in a reference frame B that moves with velocity $\vec{v}_{BA} = \vec{v}_{CA}$ so the force in B can be due only to an electric field: $\vec{F}_B = q\vec{E}_B$. Equating the forces, because experimenters in all inertial reference frames agree about forces, we find that

$$\vec{E}_B = \vec{E}_A + \vec{v}_{BA} \times \vec{B}_A \qquad (34.4)$$

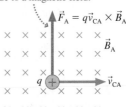

FIGURE 34.4 A charged particle moves through a magnetic field in reference frame A and experiences a magnetic force.

In A, the force on q is due to a magnetic field.

$\vec{F}_A = q\vec{v}_{CA} \times \vec{B}_A$

The situation in frame A

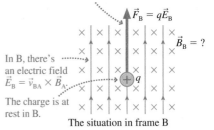

FIGURE 34.5 In frame B, the charge experiences an electric force.

In B, the force on q is due to an electric field.

$\vec{F}_B = q\vec{E}_B$

$\vec{B}_B = ?$

In B, there's an electric field $\vec{E}_B = \vec{v}_{BA} \times \vec{B}_A$.

The charge is at rest in B.

The situation in frame B

Equation 34.4 transforms the electric and magnetic fields measured in reference frame A into the electric field measured in a frame B that moves relative to A with velocity $\vec{v}_{BA}$. FIGURE 34.6b shows the outcome. Although we used a charge as a probe to find Equation 34.4, the equation is strictly about fields in different reference frames; it makes no mention of charges.

FIGURE 34.6 A charge in reference frame A experiences electric and magnetic forces. The charge experiences the same force in frame B, but it is due only to an electric field.

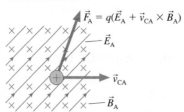

(a) The electric and magnetic fields in frame A

$$\vec{F}_A = q(\vec{E}_A + \vec{v}_{CA} \times \vec{B}_A)$$

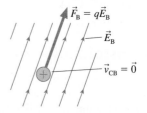

(b) The electric field in frame B, where the charged particle is at rest

$$\vec{F}_B = q\vec{E}_B$$

EXAMPLE 34.1   **Transforming the electric field**

A laboratory experimenter has created the parallel electric and magnetic fields $\vec{E} = 10{,}000\,\hat{\imath}$ V/m and $\vec{B} = 0.10\,\hat{\imath}$ T. A proton is shot into these fields with velocity $\vec{v} = 1.0 \times 10^5\,\hat{\jmath}$ m/s. What is the electric field in the proton's reference frame?

MODEL Let the laboratory be reference frame A and a frame moving with the proton be reference frame B. The relative velocity is $\vec{v}_{BA} = 1.0 \times 10^5\,\hat{\jmath}$ m/s.

VISUALIZE FIGURE 34.7 shows the geometry. The laboratory fields, now labeled A, are parallel to the x-axis while $\vec{v}_{BA}$ is in the y-direction. Thus $\vec{v}_{BA} \times \vec{B}_A$ points in the negative z-direction.

SOLVE $\vec{v}_{BA}$ and $\vec{B}_A$ are perpendicular, so the magnitude of $\vec{v}_{BA} \times \vec{B}_A$ is $(1.0 \times 10^5$ m/s$)(0.10$ T$)(\sin 90°) = 10{,}000$ V/m. Thus the electric field in frame B, the proton's frame, is

$$\vec{E}_B = \vec{E}_A + \vec{v}_{BA} \times \vec{B}_A = (10{,}000\,\hat{\imath} - 10{,}000\,\hat{k})\text{ V/m}$$

$$= (14{,}000 \text{ V/m}, 45° \text{ below the } x\text{-axis})$$

FIGURE 34.7 Finding electric field $\vec{E}_B$.

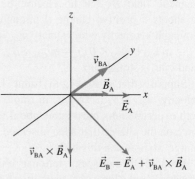

ASSESS The force on the proton is the same in both reference frames. But in the proton's reference frame that force is due entirely to an electric field tilted 45° below the x-axis.

To find a transformation equation for the magnetic field, FIGURE 34.8a shows charge q at rest in reference frame A. Alec measures the fields of a stationary point charge, which we know are

$$\vec{E}_A = \frac{1}{4\pi\epsilon_0}\frac{q}{r^2}\hat{r} \qquad \vec{B}_A = \vec{0}$$

What are the fields at this point in space as measured by Brittney in frame B? We can use Equation 34.4 to find $\vec{E}_B$. Because $\vec{B}_A = \vec{0}$, the electric field in frame B is

$$\vec{E}_B = \vec{E}_A = \frac{1}{4\pi\epsilon_0}\frac{q}{r^2}\hat{r} \qquad (34.5)$$

In other words, Coulomb's law is still valid in a frame in which the point charge is moving.

But Brittney also measures a magnetic field $\vec{B}_B$, because, as seen in FIGURE 34.8b, charge q is moving in reference frame B. The magnetic field of a moving point charge is given by the Biot-Savart law:

$$\vec{B}_B = \frac{\mu_0}{4\pi}\frac{q}{r^2}\vec{v}_{CB} \times \hat{r} = -\frac{\mu_0}{4\pi}\frac{q}{r^2}\vec{v}_{BA} \times \hat{r} \qquad (34.6)$$

where we used the fact that the charge's velocity in frame B is $\vec{v}_{CB} = -\vec{v}_{BA}$.

It will be useful to rewrite Equation 34.6 as

$$\vec{B}_B = -\frac{\mu_0}{4\pi}\frac{q}{r^2}\vec{v}_{BA}\times\hat{r} = -\epsilon_0\mu_0\vec{v}_{BA}\times\left(\frac{1}{4\pi\epsilon_0}\frac{q}{r^2}\hat{r}\right)$$

The expression in parentheses is simply $\vec{E}_A$, the electric field in frame A, so we have

$$\vec{B}_B = -\epsilon_0\mu_0\vec{v}_{BA}\times\vec{E}_A \qquad (34.7)$$

Equation 34.7 expresses the remarkable idea that **the Biot-Savart law for the magnetic field of a moving point charge is nothing other than the Coulomb electric field of a stationary point charge transformed into a moving reference frame.**

We will assert without proof that if the experimenters in frame A create a magnetic field $\vec{B}_A$ in addition to the electric field $\vec{E}_A$, then the magnetic field $\vec{B}_B$ measured in frame B is

$$\vec{B}_B = \vec{B}_A - \epsilon_0\mu_0\vec{v}_{BA}\times\vec{E}_A \qquad (34.8)$$

This is a general transformation matching Equation 34.4 for the electric field $\vec{E}_B$.

Notice something interesting. The constant μ_0 has units of T m/A; those of ϵ_0 are C²/N m². By definition, 1 T = 1 N/A m and 1 A = 1 C/s. Consequently, the units of $\epsilon_0\mu_0$ turn out to be s²/m². In other words, the quantity $1/\sqrt{\epsilon_0\mu_0}$, with units of m/s, is a speed. But what speed? The constants are well known from measurements of static electric and magnetic fields, so we can compute

$$\frac{1}{\sqrt{\epsilon_0\mu_0}} = \frac{1}{\sqrt{(8.85\times10^{-12}\,\text{C}^2/\text{N m}^2)(1.26\times10^{-6}\,\text{T m/A})}} = 3.00\times10^8\,\text{m/s}$$

Of all the possible values you might get from evaluating $1/\sqrt{\epsilon_0\mu_0}$, what are the chances it equals c, the speed of light? It is not a random coincidence. In Section 34.5 we'll show that electric and magnetic fields can exist as a *traveling wave*, and that the wave speed is predicted by the theory to be none other than

$$v_{em} = c = \frac{1}{\sqrt{\epsilon_0\mu_0}} \qquad (34.9)$$

For now, we'll go ahead and write $\epsilon_0\mu_0 = 1/c^2$. With this, our **Galilean field transformation equations** are

$$\vec{E}_B = \vec{E}_A + \vec{v}_{BA}\times\vec{B}_A$$
$$\vec{B}_B = \vec{B}_A - \frac{1}{c^2}\vec{v}_{BA}\times\vec{E}_A \qquad (34.10)$$

where $\vec{v}_{BA}$ is the velocity of reference frame B relative to frame A and where, to reiterate, the fields are measured *at the same point in space* by experimenters *at rest* in each reference frame.

NOTE ▶ We'll see shortly that these equations are valid only if $v_{BA} \ll c$. ◀

We can no longer believe that electric and magnetic fields have a separate, independent existence. Changing from one reference frame to another mixes and rearranges the fields. Different experimenters watching an event will agree on the outcome, such as the deflection of a charged particle, but they will ascribe it to different combinations of fields. Our conclusion is that **there is a single electromagnetic field that presents different faces, in terms of $\vec{E}$ and $\vec{B}$, to different viewers.**

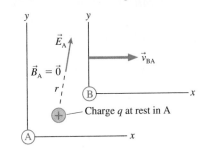

FIGURE 34.8 A charge at rest in frame A is moving in frame B and creates a magnetic field $\vec{B}_B$.

(a) In frame A, the static charge creates an electric field but no magnetic field.

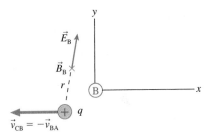

(b) In frame B, the moving charge creates both an electric and a magnetic field.

Two views of a magnet

The 1.0 T field of a large laboratory magnet points straight up. A rocket flies past the laboratory, parallel to the ground, at 1000 m/s. What are the fields between the magnet's pole tips as measured—very quickly!—by scientists on the rocket?

MODEL Let the laboratory be reference frame A and a frame moving with the rocket be reference frame B.

VISUALIZE FIGURE 34.9 shows the magnet and establishes the coordinate systems. The relative velocity is $\vec{v}_{BA} = 1000\,\hat{\imath}$ m/s.

FIGURE 34.9 The rocket and the magnet.

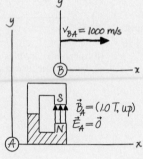

SOLVE The fields in the laboratory reference frame are $\vec{E}_A = \vec{0}$ and $\vec{B}_A = 1.0\,\hat{\jmath}$ T. Transforming the fields to the rocket's reference frame gives first, for the electric field,

$$\vec{E}_B = \vec{E}_A + \vec{v}_{BA} \times \vec{B}_A = \vec{v}_{BA} \times \vec{B}_A$$

From the right-hand rule, $\vec{v}_{BA} \times \vec{B}_A$ is out of the page, in the z-direction. $\vec{v}_{BA}$ and $\vec{B}_A$ are perpendicular, so

$$\vec{E}_B = v_{BA}B_A\hat{k} = 1000\,\hat{k}\ \text{V/m}$$

Similarly, for the magnetic field,

$$\vec{B}_B = \vec{B}_A - \frac{1}{c^2}\vec{v}_{BA} \times \vec{E}_A = \vec{B}_A = 1.0\,\hat{\jmath}\ \text{T}$$

Thus the rocket scientists measure

$$\vec{E}_B = 1000\,\hat{k}\ \text{V/m} \quad \text{and} \quad \vec{B}_B = 1.0\,\hat{\jmath}\ \text{T}$$

Almost Relativity

FIGURE 34.10 Two charges moving parallel to each other.

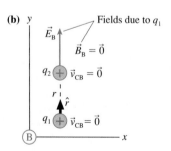

(a) Fields seen in frame A

(b) Fields seen in frame B

FIGURE 34.10a shows two positive charges moving side by side through frame A with velocity $\vec{v}_{CA}$. Charge q_1 creates an electric field and a magnetic field at the position of charge q_2. These are

$$\vec{E}_A = \frac{1}{4\pi\epsilon_0}\frac{q_1}{r^2}\hat{\jmath} \quad \text{and} \quad \vec{B}_A = \frac{\mu_0}{4\pi}\frac{q_1 v_{CA}}{r^2}\hat{k}$$

where r is the distance between the charges, and we've used $\hat{r} = \hat{\jmath}$ and $\vec{v} \times \hat{r} = v\hat{k}$.

How are the fields seen in frame B, which moves with $\vec{v}_{BA} = \vec{v}_{CA}$ and in which the charges are at rest? From the field transformation equations,

$$\vec{B}_B = \vec{B}_A - \frac{1}{c^2}\vec{v}_{BA} \times \vec{E}_A = \frac{\mu_0}{4\pi}\frac{q_1 v_{CA}}{r^2}\hat{k} - \frac{1}{c^2}\left(v_{CA}\hat{\imath} \times \frac{1}{4\pi\epsilon_0}\frac{q_1}{r^2}\hat{\jmath}\right)$$

$$= \frac{\mu_0}{4\pi}\frac{q_1 v_{CA}}{r^2}\left(1 - \frac{1}{\epsilon_0\mu_0 c^2}\right)\hat{k} \tag{34.11}$$

where we used $\hat{\imath} \times \hat{\jmath} = \hat{k}$. But $\epsilon_0\mu_0 = 1/c^2$, so the term in parentheses is zero and thus $\vec{B}_B = \vec{0}$. This result was expected because q_1 is at rest in frame B and shouldn't create a magnetic field.

The transformation of the electric field is similar:

$$\vec{E}_B = \vec{E}_A + \vec{v}_{BA} \times \vec{B}_A = \frac{1}{4\pi\epsilon_0}\frac{q_1}{r^2}\hat{\jmath} + v_{BA}\hat{\imath} \times \frac{\mu_0}{4\pi}\frac{q_1 v_{CA}}{r^2}\hat{k}$$

$$= \frac{1}{4\pi\epsilon_0}\frac{q_1}{r^2}(1 - \epsilon_0\mu_0 v_{BA}^2)\hat{\jmath} = \frac{1}{4\pi\epsilon_0}\frac{q_1}{r^2}\left(1 - \frac{v_{BA}^2}{c^2}\right)\hat{\jmath} \tag{34.12}$$

where we used $\hat{\imath} \times \hat{k} = -\hat{\jmath}$, $\vec{v}_{CA} = \vec{v}_{BA}$, and $\epsilon_0\mu_0 = 1/c^2$. FIGURE 34.10b shows the charges and fields in frame B.

But now we have a problem. In frame B where the two charges are at rest and separated by distance r, the electric field due to charge q_1 should be simply

$$\vec{E}_B = \frac{1}{4\pi\epsilon_0}\frac{q_1}{r^2}\hat{j}$$

The field transformation equations have given a "wrong" result for the electric field $\vec{E}_B$.

It turns out that the field transformations of Equations 34.10, which are based on Galilean relativity, aren't quite right. We would need Einstein's relativity—a topic that we'll take up in Chapter 36—to give the correct transformations. However, the Galilean field transformations in Equations 34.10 are equivalent to the relativistically correct transformations when $v \ll c$, in which case $v^2/c^2 \ll 1$. You can see that the two expressions for $\vec{E}_B$ do, in fact, agree if v_{BA}^2/c^2 can be neglected.

Thus our use of the field transformation equations has an additional rule: Set v^2/c^2 to zero. This is an acceptable rule for speeds $v < 10^7$ m/s. Even with this limitation, our investigation has provided us with a deeper understanding of electric and magnetic fields.

STOP TO THINK 34.1   The first diagram shows electric and magnetic fields in reference frame A. Which diagram shows the fields in frame B?

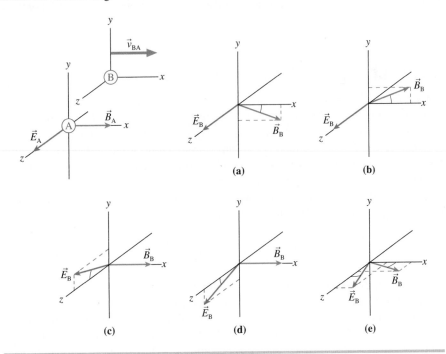

Faraday's Law Revisited

The transformation of electric and magnetic fields gives us new insight into Faraday's law. FIGURE 34.11a on the next page shows a reference frame A in which a conducting loop is moving with velocity $\vec{v}$ into a magnetic field. You learned in Chapter 33 that the magnetic field exerts a magnetic force $\vec{F}_B = q\vec{v} \times \vec{B} = (qvB, \text{upward})$ on the charges in the leading edge of the wire, creating an emf $\mathcal{E} = vLB$ and an induced current in the loop. We called this a *motional emf*.

How do things appear to an experimenter who is in frame B that moves with the loop at velocity $\vec{v}_{BA} = \vec{v}$ and for whom the loop is at rest? We have learned the important lesson that experimenters in different inertial reference frames agree about the outcome of any experiment; hence an experimenter in frame B agrees that there is an

FIGURE 34.11 A motional emf as seen in two different reference frames.

(a) Laboratory frame A

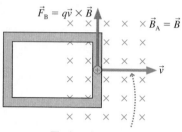

The loop is moving to the right.

(b) Loop frame B   The induced electric field points up.

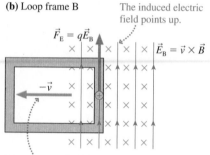

The magnetic field is moving to the left.

induced current in the loop. But the charges are at rest in frame B so there cannot be any magnetic force on them. How is the emf established in frame B?

We can use the field transformations to determine that the fields in frame B are

$$\vec{E}_B = \vec{E}_A + \vec{v} \times \vec{B}_A = \vec{v} \times \vec{B}$$
$$\vec{B}_B = \vec{B}_A - \frac{1}{c^2}\vec{v} \times \vec{E}_A = \vec{B} \tag{34.13}$$

where we used the fact that $\vec{E}_A = \vec{0}$ in frame A.

An experimenter in the loop's frame sees not only a magnetic field but also the electric field $\vec{E}_B$ shown in **FIGURE 34.11b**. The magnetic field exerts no force on the charges, because they're at rest in this frame, but the electric field does. The force on charge q is $\vec{F}_E = q\vec{E}_B = q\vec{v} \times \vec{B} = (qvB,$ upward). This is the same force as was measured in the laboratory frame, so it will cause the same emf and the same current. The outcome is identical, as we knew it had to be, but the experimenter in B attributes the emf to an electric field whereas the experimenter in A attributes it to a magnetic field.

Field $\vec{E}_B$ is, in fact, the *induced electric field* of Faraday's law. Faraday's law, fundamentally, is a statement that **a changing magnetic field creates an electric field.** But only in frame B, the frame of the loop, is the magnetic field changing. Thus the induced electric field is seen in the loop's frame but not in the laboratory frame.

34.2 The Field Laws Thus Far

Let's remind ourselves where we are in terms of discovering laws about the electromagnetic field. Gauss's law, which you studied in Chapter 27, states a very general property of the electric field. It says that charges create electric fields in such a way that the electric flux of the field is the same through *any* closed surface surrounding the charges. **FIGURE 34.12** illustrates this idea by showing the field lines passing through a Gaussian surface enclosing a charge.

The mathematical statement of Gauss's law for the electric field says that for any *closed* surface enclosing total charge Q_{in}, the net electric flux through the surface is

$$(\Phi_e)_{\text{closed surface}} = \oint \vec{E} \cdot d\vec{A} = \frac{Q_{in}}{\epsilon_0} \tag{34.14}$$

The circle on the integral sign indicates that the integration is over a closed surface. Gauss's law is the first of what will turn out to be four *field equations*.

There's an analogous equation for magnetic fields, an equation we implied in Chapter 32—where we noted that isolated north or south poles do not exist—but didn't explicitly write it down. **FIGURE 34.13** shows a Gaussian surface around a magnetic dipole. Magnetic field lines form continuous curves, without starting or stopping, so every field line leaving the surface at some point must reenter it at another. Consequently, the net magnetic flux over a *closed* surface is zero.

We've shown only one surface and one magnetic field, but this conclusion turns out to be a general property of magnetic fields. Because every north pole is accompanied by a south pole, we can't enclose a "net pole" within a surface. Thus Gauss's law for magnetic fields is

$$(\Phi_m)_{\text{closed surface}} = \oint \vec{B} \cdot d\vec{A} = 0 \tag{34.15}$$

FIGURE 34.12 A Gaussian surface enclosing a charge.

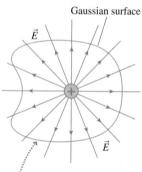

There is a net electric flux through this surface that encloses a charge.

FIGURE 34.13 There is no net flux through a Gaussian surface around a magnetic dipole.

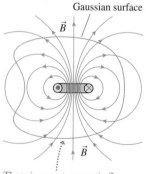

There is no net magnetic flux through this closed surface.

Equation 34.14 is the mathematical statement that Coulomb electric field lines start and stop on charges. Equation 34.15 is the mathematical statement that magnetic field lines form closed loops; they don't start or stop (i.e., there are no isolated magnetic poles).

These two versions of Gauss's law are important statements about what types of fields can and cannot exist. They will become two of Maxwell's equations.

The third field law we've established is Faraday's law:

$$\mathcal{E} = \oint \vec{E} \cdot d\vec{s} = -\frac{d\Phi_m}{dt} \tag{34.16}$$

where the line integral of $\vec{E}$ is around the closed curve that bounds the surface through which the magnetic flux Φ_m is calculated. Equation 34.16 is the mathematical statement that an electric field (and thus an emf $\mathcal{E}$) can also be created by a changing magnetic field. The correct use of Faraday's law requires a convention for determining when fluxes are positive and negative. The sign convention will be given in the next section, where we discuss the fourth and last field equation—an analogous equation for magnetic fields.

34.3 The Displacement Current

We introduced Ampère's law in Chapter 32 as an alternative to the Biot-Savart law for calculating the magnetic field of a current. Whenever total current $I_{through}$ passes through an area bounded by a closed curve, the line integral of the magnetic field around the curve is

$$\oint \vec{B} \cdot d\vec{s} = \mu_0 I_{through} \tag{34.17}$$

FIGURE 34.14 illustrates the geometry of Ampère's law. The sign of each current can be determined by using Tactics Box 34.1. In this case, $I_{through} = I_1 - I_2$.

FIGURE 34.14 Ampère's law relates the line integral of $\vec{B}$ around curve C to the current passing through surface S.

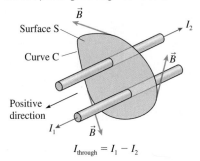

$$I_{through} = I_1 - I_2$$

TACTICS BOX 34.1 **Determining the signs of flux and current** (MP)

❶ For a surface S bounded by a closed curve C, choose either the clockwise (cw) or counterclockwise (ccw) direction around C.
❷ Curl the fingers of your *right* hand around the curve in the chosen direction, with your thumb perpendicular to the surface. Your thumb defines the positive direction.

■ A flux Φ through the surface is positive if the field is in the same direction as your thumb, negative if the field is in the opposite direction.

■ A current through the surface in the direction of your thumb is positive, in the direction opposite your thumb is negative.

Exercises 4–6

Ampère's law is the formal statement that **currents create magnetic fields.** Although Ampère's law can be used to calculate magnetic fields in situations with a high degree of symmetry, it is more important as a statement about what types of magnetic field can and cannot exist.

Something Is Missing

Nothing restricts the bounded surface of Ampère's law to being flat. It's not hard to see that any current passing through surface S_1 in FIGURE 34.15 on the next page must also pass through the curved surface S_2. To interpret Ampère's law properly, we have to say that the current $I_{through}$ is the net current passing through *any* surface S that is bounded by curve C.

FIGURE 34.15 The *net* current passing through the flat surface S₁ also passes through the curved surface S₂.

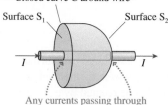
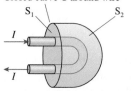

Any currents passing through S₁ must also pass through S₂.

Even in this case, the *net* current through S₁, namely zero, matches the net current through S₂.

FIGURE 34.16 There is no current through surface S₂ as the capacitor charges, but there is a changing electric flux.

(a) Cross section through a closed curve C around the wire

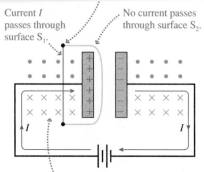

Current *I* passes through surface S₁.

No current passes through surface S₂.

This is the magnetic field of the current *I* that is charging the capacitor.

(b)

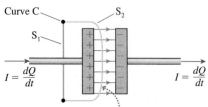

Curve C   S₂   S₁ $I = \dfrac{dQ}{dt}$ $I = \dfrac{dQ}{dt}$

The electric flux Φ_e through surface S₂ increases as the capacitor charges.

But this leads to an interesting puzzle. FIGURE 34.16a shows a capacitor being charged. Current *I*, from the left, brings positive charge to the left capacitor plate. The same current carries charges away from the right capacitor plate, leaving the right plate negatively charged. This is a perfectly ordinary current in a conducting wire, and you can use the right-hand rule to verify that its magnetic field is as shown.

Curve C is a closed curve encircling the wire on the left. The current passes through surface S₁, a flat surface across C, and we could use Ampère's law to find that the magnetic field is that of a straight wire. But what happens if we try to use surface S₂ to determine $I_{through}$? Ampère's law says that we can consider *any* surface bounded by curve C, and surface S₂ certainly qualifies. But *no* current passes through S₂. Charges are brought to the left plate of the capacitor and charges are removed from the right plate, but *no* charge moves across the gap between the plates. Surface S₁ has $I_{through} = I$, but surface S₂ has $I_{through} = 0$. Another dilemma!

It would appear that Ampère's law is either wrong or incomplete. Maxwell was the first to recognize the seriousness of this problem. He noted that there may be no current passing through S₂, but, as FIGURE 34.16b shows, there is an electric flux Φ_e through S₂ due to the electric field inside the capacitor. Furthermore, this flux is *changing* with time as the capacitor charges and the electric field strength grows. Faraday had discovered the significance of a changing magnetic flux, but no one had considered a changing electric flux.

The current *I* passes through S₁, so Ampère's law applied to S₁ gives

$$\oint \vec{B} \cdot d\vec{s} = \mu_0 I_{through} = \mu_0 I$$

We believe this result because it gives the correct magnetic field for a current-carrying wire. Now the line integral depends only on the magnetic field at points on curve C, so its value won't change if we choose a different surface S to evaluate the current. The problem is with the right side of Ampère's law, which would incorrectly give zero if applied to surface S₂. We need to modify the right side of Ampère's law to recognize that an electric flux rather than a current passes through S₂.

The electric flux between two capacitor plates of surface area A is

$$\Phi_e = EA$$

The capacitor's electric field is $E = Q/\epsilon_0 A$; hence the flux is actually independent of the plate size:

$$\Phi_e = \frac{Q}{\epsilon_0 A} A = \frac{Q}{\epsilon_0} \qquad (34.18)$$

The *rate* at which the electric flux is changing is

$$\frac{d\Phi_e}{dt} = \frac{1}{\epsilon_0} \frac{dQ}{dt} = \frac{I}{\epsilon_0} \qquad (34.19)$$

where we used $I = dQ/dt$. The flux is changing with time at a rate directly proportional to the charging current *I*.

Equation 34.19 suggests that the quantity $\epsilon_0(d\Phi_e/dt)$ is in some sense "equivalent" to current I. Maxwell called the quantity

$$I_{\text{disp}} = \epsilon_0 \frac{d\Phi_e}{dt} \tag{34.20}$$

the **displacement current.** He had started with a fluid-like model of electric and magnetic fields, and the displacement current was analogous to the displacement of a fluid. The fluid model has since been abandoned, but the name lives on despite the fact that nothing is actually being displaced.

Maxwell hypothesized that the displacement current was the "missing" piece of Ampère's law, so he modified Ampère's law to read

$$\oint \vec{B} \cdot d\vec{s} = \mu_0(I_{\text{through}} + I_{\text{disp}}) = \mu_0\left(I_{\text{through}} + \epsilon_0 \frac{d\Phi_e}{dt}\right) \tag{34.21}$$

Equation 34.21 is now known as the Ampère-Maxwell law. When applied to Figure 34.16b, the Ampère-Maxwell law gives

$$S_1: \quad \oint \vec{B} \cdot d\vec{s} = \mu_0\left(I_{\text{through}} + \epsilon_0 \frac{d\Phi_e}{dt}\right) = \mu_0(I + 0) = \mu_0 I$$

$$S_2: \quad \oint \vec{B} \cdot d\vec{s} = \mu_0\left(I_{\text{through}} + \epsilon_0 \frac{d\Phi_e}{dt}\right) = \mu_0(0 + I) = \mu_0 I$$

where, for surface S_2, we used Equation 34.19 for $d\Phi_e/dt$. Surfaces S_1 and S_2 now both give the same result for the line integral of $\vec{B} \cdot d\vec{s}$ around the closed curve C.

NOTE ▶ The displacement current I_{disp} between the capacitor plates is numerically equal to the current I in the wires to and from the capacitor, so in some sense it allows "current" to be conserved all the way through the capacitor. Nonetheless, the displacement current is *not* a flow of charge. The displacement current is equivalent to a real current in that it creates the same magnetic field, but it does so with a changing electric flux rather than a flow of charge. ◀

The Induced Magnetic Field

Ordinary Coulomb electric fields are created by charges, but a second way to create an electric field is by having a changing magnetic field. That's Faraday's law. Ordinary magnetic fields are created by currents, but now we see that a second way to create a magnetic field is by having a changing electric field. Just as the electric field created by a changing $\vec{B}$ is called an induced electric field, the magnetic field created by a changing $\vec{E}$ is called an *induced magnetic field*.

FIGURE 34.17 shows the close analogy between induced electric fields, governed by Faraday's law, and induced magnetic fields, governed by the second term in the Ampère-Maxwell law. An increasing solenoid current causes an increasing magnetic field. The changing magnetic field, in turn, induces a circular electric field. The negative sign in Faraday's law dictates that the induced electric field direction is ccw when seen looking along the magnetic field direction.

An increasing capacitor charge causes an increasing electric field. The changing electric field, in turn, induces a circular magnetic field. But the sign of the Ampère-Maxwell law is positive, the opposite of the sign of Faraday's law, so the induced magnetic field direction is cw when you're looking along the electric field direction.

FIGURE 34.17 The close analogy between an induced electric field and an induced magnetic field.

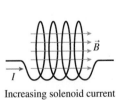

Faraday's law describes an induced electric field.

The Ampère-Maxwell law describes an induced magnetic field.

EXAMPLE 34.3 **The fields inside a charging capacitor**

A 2.0-cm-diameter parallel-plate capacitor with a 1.0 mm spacing is being charged at the rate 0.50 C/s. What is the magnetic field strength inside the capacitor at a point 0.50 cm from the axis?

MODEL The electric field inside a parallel-plate capacitor is uniform. As the capacitor is charged, the changing electric field induces a magnetic field.

VISUALIZE FIGURE 34.18 shows the fields. The induced magnetic field lines are circles concentric with the capacitor.

FIGURE 34.18 The magnetic field strength is found by integrating around a closed curve of radius r.

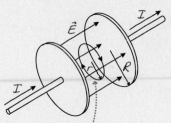

The magnetic field line is a circle concentric with the capacitor. The electric flux through this circle is $\pi r^2 E$.

SOLVE The electric field of a parallel-plate capacitor is $E = Q/\epsilon_0 A = Q/\epsilon_0 \pi R^2$. The electric flux through the circle of radius r (not the full flux of the capacitor) is

$$\Phi_e = \pi r^2 E = \pi r^2 \frac{Q}{\epsilon_0 \pi R^2} = \frac{r^2}{R^2} \frac{Q}{\epsilon_0}$$

Thus the Ampère-Maxwell law is

$$\oint \vec{B} \cdot d\vec{s} = \epsilon_0 \mu_0 \frac{d\Phi_e}{dt} = \epsilon_0 \mu_0 \frac{d}{dt}\left(\frac{r^2}{R^2} \frac{Q}{\epsilon_0} \right) = \mu_0 \frac{r^2}{R^2} \frac{dQ}{dt}$$

The magnetic field is everywhere tangent to the circle of radius r, so the integral of $\vec{B} \cdot d\vec{s}$ around the circle is simply $BL = 2\pi rB$. With this value for the line integral, the Ampère-Maxwell law becomes

$$2\pi rB = \mu_0 \frac{r^2}{R^2} \frac{dQ}{dt}$$

and thus

$$B = \frac{\mu_0}{2\pi} \frac{r}{R^2} \frac{dQ}{dt} = (2.0 \times 10^{-7}\,\text{T m/A}) \frac{0.0050\,\text{m}}{(0.010\,\text{m})^2}(0.50\,\text{C/s})$$

$$= 5.0 \times 10^{-6}\,\text{T}$$

If a changing magnetic field can induce an electric field and a changing electric field can induce a magnetic field, what happens when both fields change simultaneously? That is the question that Maxwell was finally able to answer after he modified Ampère's law to include the displacement current, and it is the subject to which we turn next.

STOP TO THINK 34.2 The electric field in four identical capacitors is shown as a function of time. Rank in order, from largest to smallest, the magnetic field strength at the outer edge of the capacitor at time T.

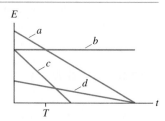

34.4 Maxwell's Equations

James Clerk Maxwell was a young, mathematically brilliant Scottish physicist. In 1855, barely 24 years old, he presented a paper to the Cambridge Philosophical Society entitled "On Faraday's Lines of Force." It had been 30 years and more since the major discoveries of Oersted, Ampère, Faraday, and others, but electromagnetism remained a loose collection of facts and "rules of thumb" without a consistent theory to link these ideas together.

Maxwell's goal was to synthesize this body of knowledge and to form a *theory* of electromagnetic fields. The critical step along the way was his recognition of the need to include a displacement-current term in Ampère's law.

Maxwell's theory of electromagnetism is embodied in four equations that we today call **Maxwell's equations.** These are

$$\oint \vec{E} \cdot d\vec{A} = \frac{Q_{in}}{\epsilon_0} \qquad\qquad \text{Gauss's law}$$

$$\oint \vec{B} \cdot d\vec{A} = 0 \qquad\qquad \text{Gauss's law for magnetism}$$

$$\oint \vec{E} \cdot d\vec{s} = -\frac{d\Phi_m}{dt} \qquad\qquad \text{Faraday's law}$$

$$\oint \vec{B} \cdot d\vec{s} = \mu_0 I_{through} + \epsilon_0 \mu_0 \frac{d\Phi_e}{dt} \qquad \text{Ampère-Maxwell law}$$

Maxwell's claim is that these four equations are a *complete* description of electric and magnetic fields. They tell us how fields are created by charges and currents, and also how fields can be induced by the changing of other fields. We need one more equation for completeness, an equation that tells us how matter responds to electromagnetic fields. The general force equation

$$\vec{F} = q(\vec{E} + \vec{v} \times \vec{B}) \qquad \text{(Lorentz force law)}$$

is known as the *Lorentz force law.* **Maxwell's equations for the fields, together with the Lorentz force law to tell us how matter responds to the fields, form the complete theory of electromagnetism.**

Maxwell's equations bring us to the pinnacle of classical physics. When combined with Newton's three laws of motion, his law of gravity, and the first and second laws of thermodynamics, we have all of classical physics—a total of just 11 equations.

While some physicists might quibble over whether all 11 are truly fundamental, the important point is not the exact number but how few equations we need to describe the overwhelming majority of our experience of the physical world. It seems as if we could have written them all on page 1 of this book and been finished, but it doesn't work that way. Each of these equations is the synthesis of a tremendous number of physical phenomena and conceptual developments. To know physics isn't just to know the equations, but to know what the equations *mean* and how they're used. That's why it's taken us so many chapters and so much effort to get to this point. Each equation is a shorthand way to summarize a book's worth of information!

Let's summarize the physical meaning of the five electromagnetic equations:

Classical physics

Newton's first law
Newton's second law
Newton's third law
Newton's law of gravity
Gauss's law
Gauss's law for magnetism
Faraday's law
Ampère-Maxwell law
Lorentz force law
First law of thermodynamics
Second law of thermodynamics

- **Gauss's law:** Charged particles create an electric field.
- **Faraday's law:** An electric field can also be created by a changing magnetic field.
- **Gauss's law for magnetism:** There are no isolated magnetic poles.
- **Ampère-Maxwell law, first half:** Currents create a magnetic field.
- **Ampère-Maxwell law, second half:** A magnetic field can also be created by a changing electric field.
- **Lorentz force law, first half:** An electric force is exerted on a charged particle in an electric field.
- **Lorentz force law, second half:** A magnetic force is exerted on a charge moving in a magnetic field.

These are the *fundamental ideas* of electromagnetism. Other important ideas, such as Ohm's law, Kirchhoff's laws, and Lenz's law, despite their practical importance, are not fundamental ideas. They can be derived from Maxwell's equations, sometimes with the addition of empirically based concepts such as resistance.

It's true that Maxwell's equations are mathematically more complex than Newton's laws and that their solution, for many problems of practical interest, requires advanced mathematics. Fortunately, we have the mathematical tools to get just far enough into Maxwell's equations to discover their most startling and revolutionary implication—the prediction of electromagnetic waves.

34.5 Electromagnetic Waves

It had been known since the early 19th century, from experiments on interference and diffraction, that light is a wave. We studied the wave properties of light in Part V, but at that time we were not able to determine just what is "waving."

Faraday speculated that light was somehow connected with electricity and magnetism, but Maxwell, using his equations of the electromagnetic field, was the first to understand that light is an oscillation of the electromagnetic field. Maxwell was able to predict that

- Electromagnetic waves can exist at any frequency, not just at the frequencies of visible light. This prediction was the harbinger of radio waves.
- All electromagnetic waves travel in a vacuum with the same speed, a speed that we now call the *speed of light.*

A general wave equation can be derived from Maxwell's equations, but the necessary mathematical techniques are beyond the level of this textbook. We'll adopt a simpler approach in which we *assume* an electromagnetic wave of a certain form and then show that it's consistent with Maxwell's equations. After all, the wave can't exist *unless* it's consistent with Maxwell's equations.

To begin, we're going to assume that electric and magnetic fields can exist independently of charges and currents in a *source-free* region of space. This is a very important assumption because it makes the statement that **fields are real entities.** They're not just cute pictures that tell us about charges and currents, but real things that can exist all by themselves. Our assertion is that the fields can exist in a self-sustaining mode in which a changing magnetic field creates an electric field (Faraday's law) that in turn changes in just the right way to re-create the original magnetic field (the Ampère-Maxwell law).

The source-free Maxwell's equations, with no charges or currents, are

$$\oint \vec{E} \cdot d\vec{A} = 0 \qquad \oint \vec{E} \cdot d\vec{s} = -\frac{d\Phi_m}{dt}$$
$$\oint \vec{B} \cdot d\vec{A} = 0 \qquad \oint \vec{B} \cdot d\vec{s} = \epsilon_0\mu_0\frac{d\Phi_e}{dt}$$
(34.22)

Any electromagnetic wave traveling in empty space must be consistent with these equations.

Let's postulate that an electromagnetic plane wave traveling with speed v_{em} has the characteristics shown in FIGURE 34.19. It's a useful picture, and one that you'll see in any textbook, but a picture that can be very misleading if you don't think about it carefully. $\vec{E}$ and $\vec{B}$ are *not* spatial vectors. That is, they don't stretch spatially in the y- or z-direction for a certain distance. Instead, these vectors are showing the values of the electric and magnetic fields along a single line, the x-axis. An $\vec{E}$ vector pointing in the y-direction says that *at this position* on the x-axis, where the vector's tail is, the electric field points in the y-direction and has a certain strength. Nothing is "reaching" to a point in space above the x-axis. In fact, this picture contains no information about the fields anywhere other than right on the x-axis.

However, we are assuming that this is a *plane wave,* which, you'll recall from Chapter 20, is a wave for which the fields are the same *everywhere* in any yz-plane, perpendicular to the x-axis. FIGURE 34.20a shows a small section of the xy-plane where, at this instant of time,

Large radar installations like this one are used to track rockets and missiles.

FIGURE 34.19 A sinusoidal electromagnetic wave.

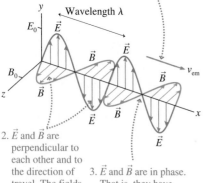

1. A sinusoidal wave with frequency f and wavelength λ travels with wave speed v_{em}.

2. $\vec{E}$ and $\vec{B}$ are perpendicular to each other and to the direction of travel. The fields have amplitudes E_0 and B_0.

3. $\vec{E}$ and $\vec{B}$ are in phase. That is, they have matching crests, troughs, and zeros.

$\vec{E}$ is pointing up and $\vec{B}$ is pointing toward you. The field strengths vary with x, the direction of travel, but not with y. As the wave moves forward, the fields that are now in the x_1-plane will soon arrive in the x_2-plane, and those now in the x_2-plane will move to x_3.

FIGURE 34.20b shows a section of the yz-plane that slices the x-axis at x_2. These fields are moving out of the page, coming toward you. The fields are the same everywhere in this plane, which is what we mean by a plane wave. If you watched a movie of the event, you would see the $\vec{E}$ and $\vec{B}$ fields at each point in this plane *oscillating* in time, but always synchronized with all the other points in the plane.

Gauss's Laws

Now that we understand the shape of the electromagnetic field, we can check its consistency with Maxwell's equations. This field is a sinusoidal wave, so the components of the fields are

$$
\begin{aligned}
E_x = 0 \quad & E_y = E_0 \sin\left(2\pi(x/\lambda - ft)\right) \quad E_z = 0 \\
B_x = 0 \quad & B_y = 0 \qquad\qquad\qquad\qquad B_z = B_0 \sin\left(2\pi(x/\lambda - ft)\right)
\end{aligned}
\tag{34.23}
$$

where E_0 and B_0 are the amplitudes of the oscillating electric and magnetic fields.

FIGURE 34.21 shows an imaginary box—a Gaussian surface—centered on the x-axis. Both electric and magnetic field vectors exist at each point in space, but the figure shows them separately for clarity. $\vec{E}$ oscillates along the y-axis, so all electric field lines enter and leave the box through the top and bottom surfaces; no electric field lines pass through the sides of the box.

FIGURE 34.21 A closed surface can be used to check Gauss's law for the electric and magnetic fields.

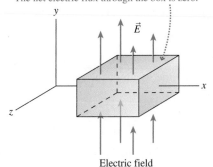

The net electric flux through the box is zero.

Electric field

The net magnetic flux through the box is zero.

Magnetic field

Because this is a plane wave, the magnitude of each electric field vector entering the bottom of the box is exactly matched by the electric field vector leaving the top. The electric flux through the top of the box is equal in magnitude but opposite in sign to the flux through the bottom, and the flux through the sides is zero. Thus the *net* electric flux is $\Phi_e = 0$. There is no charge inside the box because there are no sources in this region of space, so we also have $Q_{in} = 0$. Hence the electric field of a plane wave is consistent with the first of the source-free Maxwell's equations, Gauss's law.

The exact same argument applies to the magnetic field. The net magnetic flux is $\Phi_m = 0$; thus the magnetic field is consistent with the second of Maxwell's equations.

Faraday's Law

Faraday's law is concerned with the changing magnetic flux through a closed curve. We'll apply Faraday's law to a narrow rectangle in the xy-plane, shown in FIGURE 34.22, with height h and width Δx. We'll assume Δx to be so small that $\vec{B}$ is essentially constant over the width of the rectangle.

FIGURE 34.20 Interpreting the electromagnetic wave of Figure 34.19.

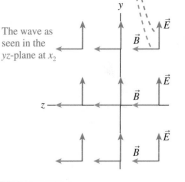

(a) Wave traveling to the right

The wave as seen in the xy-plane

$\vec{B}$ out of page

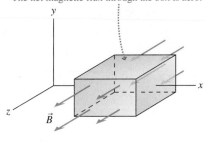

(b) Wave coming toward you

The wave as seen in the yz-plane at x_2

FIGURE 34.22 Faraday's law can be applied to a narrow rectangle in the xy-plane.

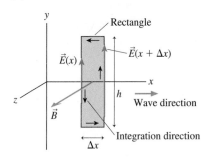

Rectangle

$\vec{E}(x)$

$\vec{E}(x + \Delta x)$

h

Wave direction

$\vec{B}$

Integration direction

Δx

The magnetic field $\vec{B}$ points in the z-direction, perpendicular to the rectangle. The magnetic flux through the rectangle is $\Phi_m = B_z A_{\text{rectangle}} = B_z h \Delta x$, hence the flux *changes* at the rate

$$\frac{d\Phi_m}{dt} = \frac{d}{dt}(B_z h \Delta x) = \frac{\partial B_z}{\partial t} h \Delta x \tag{34.24}$$

The ordinary derivative dB_z/dt, which is the full rate of change of B from all possible causes, becomes a partial derivative $\partial B_z/\partial t$ in this situation because the change in magnetic flux is due entirely to the change of B with time and not at all to the spatial variation of B.

According to our sign convention, we have to go around the rectangle in a ccw direction to make the flux positive. Thus we must also use a ccw direction to evaluate the line integral

$$\oint \vec{E} \cdot d\vec{s} = \int_{\text{right}} \vec{E} \cdot d\vec{s} + \int_{\text{top}} \vec{E} \cdot d\vec{s} + \int_{\text{left}} \vec{E} \cdot d\vec{s} + \int_{\text{bottom}} \vec{E} \cdot d\vec{s} \tag{34.25}$$

The electric field $\vec{E}$ points in the y-direction, hence $\vec{E} \cdot d\vec{s} = 0$ at all points on the top and bottom edges, and these two integrals are zero.

Along the left edge of the loop, at position x, $\vec{E}$ has the same value at every point. Figure 34.22 shows that the direction of $\vec{E}$ is *opposite* to $d\vec{s}$, thus $\vec{E} \cdot d\vec{s} = -E_y(x)\,ds$. On the right edge of the loop, at position $x + \Delta x$, $\vec{E}$ is *parallel* to $d\vec{s}$ and $\vec{E} \cdot d\vec{s} = E_y(x + \Delta x)\,ds$. Thus the line integral of $\vec{E} \cdot d\vec{s}$ around the rectangle is

$$\oint \vec{E} \cdot d\vec{s} = -E_y(x)h + E_y(x + \Delta x)h = [E_y(x + \Delta x) - E_y(x)]h \tag{34.26}$$

NOTE ▶ $E_y(x)$ indicates that E_y is a function of the position x. It is *not* E_y multiplied by x. ◀

You learned in calculus that the derivative of the function $f(x)$ is

$$\frac{df}{dx} = \lim_{\Delta x \to 0}\left[\frac{f(x + \Delta x) - f(x)}{\Delta x}\right]$$

We've assumed that Δx is very small. If we now let the width of the rectangle go to zero, $\Delta x \to 0$, Equation 34.26 becomes

$$\oint \vec{E} \cdot d\vec{s} = \frac{\partial E_y}{\partial x} h \Delta x \tag{34.27}$$

We've used a partial derivative because E_y is a function of both position x and time t.

Now, using Equations 34.24 and 34.27, we can write Faraday's law as

$$\oint \vec{E} \cdot d\vec{s} = \frac{\partial E_y}{\partial x} h \Delta x = -\frac{d\Phi_m}{dt} = -\frac{\partial B_z}{\partial t} h \Delta x$$

The area $h\Delta x$ of the rectangle cancels, and we're left with

$$\frac{\partial E_y}{\partial x} = -\frac{\partial B_z}{\partial t} \tag{34.28}$$

Equation 34.28, which compares the rate at which E_y varies with position to the rate at which B_z varies with time, is a *required condition* that an electromagnetic wave must satisfy to be consistent with Maxwell's equations. We can use Equations 34.23 for E_y and B_z to evaluate the partial derivatives:

$$\frac{\partial E_y}{\partial x} = \frac{2\pi E_0}{\lambda} \cos\left(2\pi(x/\lambda - ft)\right)$$

$$\frac{\partial B_z}{\partial t} = -2\pi f B_0 \cos\left(2\pi(x/\lambda - ft)\right)$$

Thus the required condition of Equation 34.28 is

$$\frac{\partial E_y}{\partial x} = \frac{2\pi E_0}{\lambda}\cos\left(2\pi(x/\lambda - ft)\right) = -\frac{\partial B_z}{\partial t} = 2\pi fB_0\cos\left(2\pi(x/\lambda - ft)\right)$$

Canceling the many common factors, and multiplying by λ, we're left with

$$E_0 = (\lambda f)B_0 = v_{em}B_0 \qquad (34.29)$$

where we used the fact that $\lambda f = v$ for any sinusoidal wave.

Equation 34.29, which came from applying Faraday's law, tells us that the field amplitudes E_0 and B_0 of an electromagnetic wave are not arbitrary. **Once the amplitude B_0 of the magnetic field wave is specified, the electric field amplitude E_0 must be $E_0 = v_{em}B_0$.** Otherwise the fields won't satisfy Maxwell's equations.

The Ampère-Maxwell Law

We have one equation to go, but this one will now be easier. The Ampère-Maxwell law is concerned with the changing electric flux through a closed curve. FIGURE 34.23 shows a very narrow rectangle of width Δx and length l in the xz-plane. The electric field is perpendicular to this rectangle; hence the electric flux through it is $\Phi_e = E_y A_{rectangle} = E_y l\Delta x$. This flux is changing at the rate

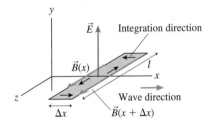

FIGURE 34.23 The Ampère-Maxwell law can be applied to a narrow rectangle in the xz-plane.

$$\frac{d\Phi_e}{dt} = \frac{d}{dt}(E_y l\Delta x) = \frac{\partial E_y}{\partial t}l\Delta x \qquad (34.30)$$

The line integral of $\vec{B}\cdot d\vec{s}$ around this closed rectangle is calculated just like the line integral of $\vec{E}\cdot d\vec{s}$ in Figure 34.22. $\vec{B}$ is perpendicular to $d\vec{s}$ on the narrow ends, so $\vec{B}\cdot d\vec{s} = 0$. The field at *all* points on the left edge, at position x, is $\vec{B}(x)$, and this field is parallel to $d\vec{s}$ to make $\vec{B}\cdot d\vec{s} = B_z(x)\,ds$. Similarly, $\vec{B}\cdot d\vec{s} = -B_z(x + \Delta x)\,ds$ at all points on the right edge, where $\vec{B}$ is opposite to $d\vec{s}$.

Thus, if we let $\Delta x \to 0$,

$$\oint \vec{B}\cdot d\vec{s} = B_z(x)l - B_z(x + \Delta x)l = -[B_z(x + \Delta x) - B_z(x)]l$$
$$= -\frac{\partial B_z}{\partial x}l\Delta x \qquad (34.31)$$

Equations 34.30 and 34.31 can now be used in the Ampère-Maxwell law:

$$\oint \vec{B}\cdot d\vec{s} = -\frac{\partial B_z}{\partial x}l\Delta x = \epsilon_0\mu_0\frac{d\Phi_e}{dt} = \epsilon_0\mu_0\frac{\partial E_y}{\partial t}l\Delta x$$

The area of the rectangle cancels, and we're left with

$$\frac{\partial B_z}{\partial x} = -\epsilon_0\mu_0\frac{\partial E_y}{\partial t} \qquad (34.32)$$

Equation 34.32 is a second required condition that the fields must satisfy. If we again evaluate the partial derivatives, using Equations 34.23 for E_y and B_z, we find

$$\frac{\partial E_y}{\partial t} = -2\pi fE_0\cos\left(2\pi(x/\lambda - ft)\right)$$

$$\frac{\partial B_z}{\partial x} = \frac{2\pi B_0}{\lambda}\cos\left(2\pi(x/\lambda - ft)\right)$$

With these, Equation 34.32 becomes

$$\frac{\partial B_z}{\partial x} = \frac{2\pi B_0}{\lambda}\cos\left(2\pi(x/\lambda - ft)\right) = -\epsilon_0\mu_0\frac{\partial E_y}{\partial t} = 2\pi\epsilon_0\mu_0 fE_0\cos\left(2\pi(x/\lambda - ft)\right)$$

A final round of cancellations and another use of $\lambda f = v_{em}$ leave us with

$$E_0 = \frac{B_0}{\epsilon_0 \mu_0 \lambda f} = \frac{B_0}{\epsilon_0 \mu_0 v_{em}} \tag{34.33}$$

The last of Maxwell's equations gives us another constraint between E_0 and B_0.

The Speed of Light

But how can Equation 34.29, which required $E_0 = v_{em}B_0$, and Equation 34.33 both be true at the same time? The one and only way is if

$$\frac{1}{\epsilon_0 \mu_0 v_{em}} = v_{em}$$

from which we find

$$v_{em} = \frac{1}{\sqrt{\epsilon_0 \mu_0}} = 3.00 \times 10^8 \text{ m/s} = c \tag{34.34}$$

This is a remarkable conclusion. The constants ϵ_0 and μ_0 are from electrostatics and magnetostatics, where they determine the size of $\vec{E}$ and $\vec{B}$ due to point charges. Coulomb's law and the Biot-Savart law, where ϵ_0 and μ_0 first appeared, have nothing to do with waves. Yet Maxwell's theory of electromagnetism ends up predicting that electric and magnetic fields can form a self-sustaining electromagnetic wave *if* that wave travels at the specific speed $v_{em} = 1/\sqrt{\epsilon_0 \mu_0}$. No other speed will satisfy Maxwell's equations.

We've made no assumption about the frequency of the wave, so apparently all electromagnetic waves, regardless of their frequency, travel (in vacuum) at the same speed $v_{em} = 1/\sqrt{\epsilon_0 \mu_0}$. We call this speed c, the "speed of light," but it applies equally well from low-frequency radio waves to ultrahigh-frequency x rays.

STOP TO THINK 34.3 An electromagnetic wave is propagating in the positive x-direction. At this instant of time, what is the direction of $\vec{E}$ at the center of the rectangle?

a. In the positive x-direction b. In the negative x-direction
c. In the positive y-direction d. In the negative y-direction
e. In the positive z-direction f. In the negative z-direction

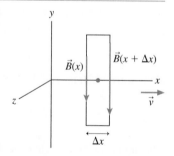

34.6 Properties of Electromagnetic Waves

We've demonstrated that one very specific sinusoidal wave is consistent with Maxwell's equations. It's possible to show that *any* electromagnetic wave, whether it's sinusoidal or not, must satisfy four basic conditions:

1. The fields $\vec{E}$ and $\vec{B}$ are perpendicular to the direction of propagation $\vec{v}_{em}$. Thus an electromagnetic wave is a transverse wave.
2. $\vec{E}$ and $\vec{B}$ are perpendicular to each other in a manner such that $\vec{E} \times \vec{B}$ is in the direction of $\vec{v}_{em}$.
3. The wave travels in vacuum at speed $v_{em} = 1/\sqrt{\epsilon_0 \mu_0} = c$.
4. $E = cB$ at any point on the wave.

In this section, we'll look at some other properties of electromagnetic waves.

Energy and Intensity

Waves transfer energy. Ocean waves erode beaches, sound waves set your eardrums vibrating, and light from the sun warms the earth. The energy flow of an electromagnetic wave is described by the **Poynting vector** $\vec{S}$, defined as

$$\vec{S} \equiv \frac{1}{\mu_0}\vec{E} \times \vec{B} \qquad (34.35)$$

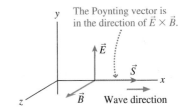

FIGURE 34.24 The Poynting vector.

The Poynting vector, shown in FIGURE 34.24, has two important properties:

1. The Poynting vector points in the direction in which an electromagnetic wave is traveling. You can see this by looking back at Figure 34.19.
2. It is straightforward to show that the units of S are W/m², or power (joules per second) per unit area. Thus the magnitude S of the Poynting vector measures the rate of energy transfer per unit area of the wave.

Because $\vec{E}$ and $\vec{B}$ of an electromagnetic wave are perpendicular to each other, and $E = cB$, the magnitude of the Poynting vector is

$$S = \frac{EB}{\mu_0} = \frac{E^2}{c\mu_0} = c\epsilon_0 E^2$$

The Poynting vector is a function of time, oscillating from zero to $S_{max} = E_0^2/c\mu_0$ and back to zero twice during each period of the wave's oscillation. That is, the energy flow in an electromagnetic wave is not smooth. It "pulses" as the electric and magnetic fields oscillate in intensity. We're unaware of this pulsing because the electromagnetic waves that we can sense—light waves—have such high frequencies.

Of more interest is the *average* energy transfer, averaged over one cycle of oscillation, which is the wave's **intensity** I. In our earlier study of waves, we defined the intensity of a wave to be $I = P/A$, where P is the power (energy transferred per second) of a wave that impinges on area A. Because $E = E_0 \sin\left(2\pi(x/\lambda - ft)\right)$, and the average over one period of $\sin^2\left(2\pi(x/\lambda - ft)\right)$ is $\frac{1}{2}$, the intensity of an electromagnetic wave is

$$I = \frac{P}{A} = S_{avg} = \frac{1}{2c\mu_0}E_0^2 = \frac{c\epsilon_0}{2}E_0^2 \qquad (34.36)$$

Equation 34.36 relates the intensity of an electromagnetic wave, a quantity that is easily measured, to the amplitude of the wave's electric field.

The intensity of a plane wave, with constant electric field amplitude E_0, would not change with distance. But a plane wave is an idealization; there are no true plane waves in nature. You learned in Chapter 20 that, to conserve energy, the intensity of a wave far from its source decreases with the inverse square of the distance. If a source with power P_{source} emits electromagnetic waves *uniformly* in all directions, the electromagnetic wave intensity at distance r from the source is

$$I = \frac{P_{source}}{4\pi r^2} \qquad (34.37)$$

Equation 34.37 simply expresses the recognition that the energy of the wave is spread over a sphere of surface area $4\pi r^2$.

EXAMPLE 34.4 **Fields of a cell phone**

A digital cell phone broadcasts a 0.60 W signal at a frequency of 1.9 GHz. What are the amplitudes of the electric and magnetic fields at a distance of 10 cm, about the distance to the center of the user's brain?

MODEL Treat the cell phone as a point source of electromagnetic waves.

Continued

SOLVE The intensity of a 0.60 W point source at a distance of 10 cm is

$$I = \frac{P_{source}}{4\pi r^2} = \frac{0.60 \text{ W}}{4\pi(0.10 \text{ m})^2} = 4.78 \text{ W/m}^2$$

We can find the electric field amplitude from the intensity:

$$E_0 = \sqrt{\frac{2I}{c\epsilon_0}} = \sqrt{\frac{2(4.78 \text{ W/m}^2)}{(3.00 \times 10^8 \text{ m/s})(8.85 \times 10^{-12} \text{ C}^2/\text{N m}^2)}}$$

$$= 60 \text{ V/m}$$

The amplitudes of the electric and magnetic fields are related by the speed of light. This allows us to compute

$$B_0 = \frac{E_0}{c} = 2.0 \times 10^{-7} \text{ T}$$

ASSESS The electric field amplitude is modest; the magnetic field amplitude is very small. This implies that the interaction of electromagnetic waves with matter is mostly due to the electric field.

STOP TO THINK 34.4   An electromagnetic wave is traveling in the positive y-direction. The electric field at one instant of time is shown at one position. The magnetic field at this position points

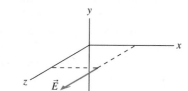

a. In the positive x-direction.   b. In the negative x-direction.
c. In the positive y-direction.   d. In the negative y-direction.
e. Toward the origin.   f. Away from the origin.

Radiation Pressure

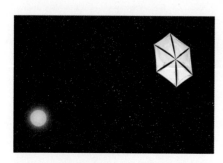

Artist's conception of a future spacecraft powered by radiation pressure from the sun.

Electromagnetic waves transfer not only energy but also momentum. An object gains momentum when it absorbs electromagnetic waves, much as a ball at rest gains momentum when struck by a ball in motion.

Suppose we shine a beam of light on an object that completely absorbs the light energy. If the object absorbs energy during a time interval Δt, its momentum changes by

$$\Delta p = \frac{\text{energy absorbed}}{c}$$

This is a consequence of Maxwell's theory, which we'll state without proof.

The momentum change implies that the light is exerting a force on the object. Newton's second law, in terms of momentum, is $F = \Delta p/\Delta t$. The radiation force due to the beam of light is

$$F = \frac{\Delta p}{\Delta t} = \frac{(\text{energy absorbed})/\Delta t}{c} = \frac{P}{c}$$

where P is the power (joules per second) of the light.

It's more interesting to consider the force exerted on an object per unit area, which is called the **radiation pressure** p_{rad}. The radiation pressure on an object that absorbs all the light is

$$p_{rad} = \frac{F}{A} = \frac{P/A}{c} = \frac{I}{c} \tag{34.38}$$

where I is the intensity of the light wave. The subscript on p_{rad} is important in this context to distinguish the radiation pressure from the momentum p.

EXAMPLE 34.5 **Solar sailing**

A low-cost way of sending spacecraft to other planets would be to use the radiation pressure on a solar sail. The intensity of the sun's electromagnetic radiation at distances near the earth's orbit is about 1300 W/m². What size sail would be needed to accelerate a 10,000 kg spacecraft toward Mars at 0.010 m/s²?

MODEL Assume that the solar sail is perfectly absorbing.

SOLVE The force that will create a 0.010 m/s² acceleration is $F = ma = 100$ N. We can use Equation 34.38 to find the sail

area that, by absorbing light, will receive a 100 N force from the sun:

$$A = \frac{cF}{I} = \frac{(3.00 \times 10^8 \text{ m/s})(100 \text{ N})}{1300 \text{ W/m}^2} = 2.3 \times 10^7 \text{ m}^2$$

ASSESS If the sail is a square, it would need to be 4.8 km × 4.8 km, or roughly 3 mi × 3 mi. This is large, but not entirely out of the question with thin films that can be unrolled in space. But how will the crew return from Mars?

Antennas

We've seen that an electromagnetic wave is self-sustaining, independent of charges or currents. However, charges and currents are needed at the *source* of an electromagnetic wave. We'll take a brief look at how an electromagnetic wave is generated by an antenna.

FIGURE 34.25 is the electric field of an electric dipole. If the dipole is vertical, the electric field $\vec{E}$ at points along a horizontal line is also vertical. Reversing the dipole, by switching the charges, reverses $\vec{E}$. If the charges were to oscillate back and forth, switching position at frequency f, then $\vec{E}$ would oscillate in a vertical plane. The changing $\vec{E}$ would then create an induced magnetic field $\vec{B}$, which could then create an $\vec{E}$, which could then create a $\vec{B}, \ldots$, and an electromagnetic wave at frequency f would radiate out into space.

FIGURE 34.25 An electric dipole creates an electric field that reverses direction if the dipole charges are switched.

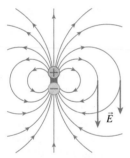
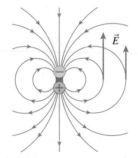

Positive charge on top   Negative charge on top

This is exactly what an **antenna** does. FIGURE 34.26 shows two metal wires attached to the terminals of an oscillating voltage source. The figure shows an instant when the top wire is negative and the bottom is positive, but these will reverse in half a cycle. The wire is basically an oscillating dipole, and it creates an oscillating electric field. The oscillating $\vec{E}$ induces an oscillating $\vec{B}$, and they take off as an electromagnetic wave at speed $v_{em} = c$. The wave does need oscillating charges as a *wave source*, but once created it is self-sustaining and independent of the source. The antenna might be destroyed, but the wave could travel billions of light years across the universe, bearing the legacy of James Clerk Maxwell.

FIGURE 34.26 An antenna generates a self-sustaining electromagnetic wave.

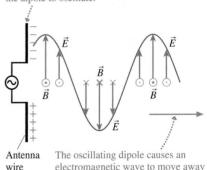

An oscillating voltage causes the dipole to oscillate.

Antenna wire — The oscillating dipole causes an electromagnetic wave to move away from the antenna at speed $v_{em} = c$.

STOP TO THINK 34.5 The amplitude of the oscillating electric field at your cell phone is 4.0 μV/m when you are 10 km east of the broadcast antenna. What is the electric field amplitude when you are 20 km east of the antenna?

a. 1.0 μV/m
b. 2.0 μV/m
c. 4.0 μV/m
d. There's not enough information to tell.

34.7 Polarization

The plane of the electric field vector $\vec{E}$ and the Poynting vector $\vec{S}$ (the direction of propagation) is called the **plane of polarization** of an electromagnetic wave. Figure 34.27 shows two electromagnetic waves moving along the x-axis. The electric field in FIGURE 34.27a oscillates vertically, so we would say that this wave is *vertically polarized*. Similarly the wave in FIGURE 34.27b is *horizontally polarized*. Other polarizations are possible, such as a wave polarized 30° away from horizontal.

FIGURE 34.27 The plane of polarization is the plane in which the electric field vector oscillates.

(a) Vertical polarization

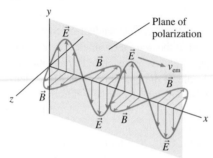

(b) Horizontal polarization

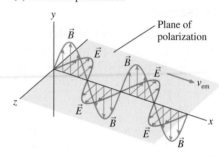

NOTE ▶ This use of the term "polarization" is completely independent of the idea of *charge polarization* that you learned about in Chapter 25. ◀

Some wave sources, such as lasers and radio antennas, emit *polarized* electromagnetic waves with a well-defined plane of polarization. By contrast, most natural sources of electromagnetic radiation are unpolarized, emitting waves whose electric fields oscillate randomly with all possible orientations.

A few natural sources are *partially polarized,* meaning that one direction of polarization is more prominent than others. The light of the sky at right angles to the sun is partially polarized because of how the sun's light scatters from air molecules to create skylight. Bees and other insects make use of this partial polarization to navigate. Light reflected from a flat, horizontal surface, such as a road or the surface of a lake, has a predominantly horizontal polarization. This is the rationale for using polarizing sunglasses.

The most common way of artificially generating polarized visible light is to send unpolarized light through a *polarizing filter.* The first widely used polarizing filter was invented by Edwin Land in 1928, while he was still an undergraduate student. He developed an improved version, called Polaroid, in 1938. Polaroid, as shown in FIGURE 34.28, is a plastic sheet containing very long organic molecules known as polymers. The sheets are formed in such a way that the polymers are all aligned to form a grid, rather like the metal bars in a barbecue grill. The sheet is then chemically treated to make the polymer molecules somewhat conducting.

As a light wave travels through Polaroid, the component of the electric field oscillating parallel to the polymer grid drives the conduction electrons up and down the molecules. The electrons absorb energy from the light wave, so the parallel component of $\vec{E}$ is absorbed in the filter. But the conduction electrons can't oscillate perpendicular to the molecules, so the component of $\vec{E}$ perpendicular to the polymer grid passes through without absorption. Thus the light wave emerging from a polarizing filter is polarized perpendicular to the polymer grid.

FIGURE 34.28 A polarizing filter.

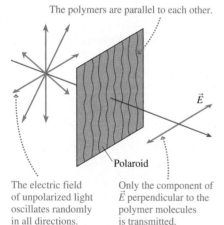

The polymers are parallel to each other.

Polaroid

The electric field of unpolarized light oscillates randomly in all directions.

Only the component of $\vec{E}$ perpendicular to the polymer molecules is transmitted.

Malus's Law

Suppose a *polarized* light wave of intensity I_0 approaches a polarizing filter. What is the intensity of the light that passes through the filter? FIGURE 34.29 shows that an oscillating electric field can be decomposed into components parallel and perpendicular to the polarizer's axis (i.e., the polarization direction transmitted by the polarizer). If we call the polarizer axis the y-axis, then the incident electric field is

$$\vec{E}_{\text{incident}} = E_\perp \hat{\imath} + E_\parallel \hat{\jmath} = E_0 \sin\theta\, \hat{\imath} + E_0 \cos\theta\, \hat{\jmath} \qquad (34.39)$$

where θ is the angle between the incident plane of polarization and the polarizer axis.

If the polarizer is ideal, meaning that light polarized parallel to the axis is 100% transmitted and light perpendicular to the axis is 100% blocked, then the electric field of the light transmitted by the filter is

$$\vec{E}_{\text{transmitted}} = E_\parallel \hat{\jmath} = E_0 \cos\theta\, \hat{\jmath} \qquad (34.40)$$

Because the intensity depends on the square of the electric field amplitude, you can see that the transmitted intensity is related to the incident intensity by

$$I_{\text{transmitted}} = I_0 \cos^2\theta \quad \text{(incident light polarized)} \qquad (34.41)$$

This result, which was discovered experimentally in 1809, is called **Malus's law.**

FIGURE 34.30a shows that Malus's law can be demonstrated with two polarizing filters. The first, called the *polarizer,* is used to produce polarized light of intensity I_0. The second, called the *analyzer,* is rotated by angle θ relative to the polarizer. As the photographs of **FIGURE 34.30b** show, the transmission of the analyzer is (ideally) 100% when $\theta = 0°$ and steadily decreases to zero when $\theta = 90°$. Two polarizing filters with perpendicular axes, called *crossed polarizers,* block all the light.

FIGURE 34.29 An incident electric field can be decomposed into components parallel and perpendicular to a polarizer's axis.

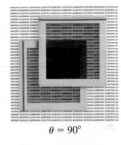

The incident light is polarized at angle θ with respect to the polarizer's axis. Only the component of $\vec{E}$ in the direction of the axis is transmitted.

FIGURE 34.30 The intensity of the transmitted light depends on the angle between the polarizing filters.

(a)
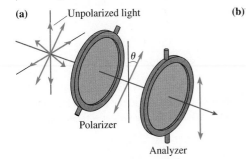
Unpolarized light / Polarizer / Analyzer

(b)
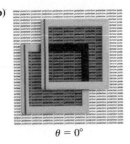 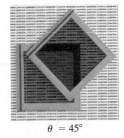 
$\theta = 0°$ $\theta = 45°$ $\theta = 90°$

Suppose the light incident on a polarizing filter is *unpolarized,* as is the light incident from the left on the polarizer in Figure 34.30a. The electric field of unpolarized light varies randomly through all possible values of θ. Because the *average* value of $\cos^2\theta$ is $\frac{1}{2}$, the intensity transmitted by a polarizing filter is

$$I_{\text{transmitted}} = \frac{1}{2}I_0 \quad \text{(incident light unpolarized)} \qquad (34.42)$$

In other words, a polarizing filter passes 50% of unpolarized light and blocks 50%.

In polarizing sunglasses, the polymer grid is aligned horizontally (when the glasses are in the normal orientation) so that the glasses transmit vertically polarized light. Most natural light is unpolarized, so the glasses reduce the light intensity by 50%. But *glare*—the reflection of the sun and the skylight from roads and other horizontal surfaces—has a strong horizontal polarization. This light is almost completely blocked by the Polaroid, so the sunglasses "cut glare" without affecting the main scene you wish to see.

You can test whether your sunglasses are polarized by holding them in front of you and rotating them as you look at the glare reflecting from a horizontal surface. Polarizing sunglasses substantially reduce the glare when the glasses are "normal" but not when the glasses are 90° from normal. (You can also test them against a pair of sunglasses known to be polarizing by seeing if all light is blocked when the lenses of the two pairs are crossed.)

The vertical polarizer blocks the horizontally polarized glare from the surface of the water.

STOP TO THINK 34.6   Unpolarized light of equal intensity is incident on four pairs of polarizing filters. Rank in order, from largest to smallest, the intensities I_a to I_d transmitted through the second polarizer of each pair.

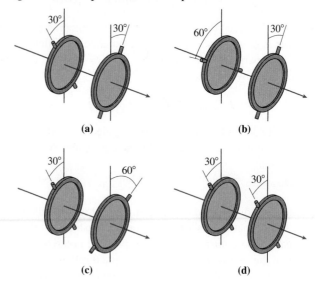

(a)   (b)

(c)   (d)

CHALLENGE EXAMPLE 34.6   **Light propulsion**

Future space rockets might propel themselves by firing laser beams, rather than exhaust gases, out the back. The acceleration would be small, but it could continue for months or years in the vacuum of space. Consider a 1200 kg unmanned space probe powered by a 15 MW laser. After one year, how far will it have traveled and how fast will it be going?

MODEL Assume the laser efficiency is so high that it can be powered for a year with a negligible mass of fuel.

SOLVE Light waves transfer not only energy but also momentum, which is how they exert a radiation-pressure force. We found that the radiation force of a light beam of power P is

$$F = \frac{P}{c}$$

From Newton's third law, the emitted light waves must exert an equal-but-opposite reaction force on the source of the light. In this case, the emitted light exerts a force of this magnitude on the space probe to which the laser is attached. This reaction force causes the probe to accelerate at

$$a = \frac{F}{m} = \frac{P}{mc} = \frac{15 \times 10^6 \text{ W}}{(1200 \text{ kg})(3.0 \times 10^8 \text{ m/s})}$$
$$= 4.2 \times 10^{-5} \text{ m/s}^2$$

As expected, the acceleration is extremely small. But one year is a large amount of time: $\Delta t = 3.15 \times 10^7$ s. After one year of acceleration,

$$v = a\Delta t = 1300 \text{ m/s}$$

$$d = \tfrac{1}{2}a(\Delta t)^2 = 2.1 \times 10^{10} \text{ m}$$

The space probe will have traveled 2.1×10^{10} m and will be going 1300 m/s.

ASSESS Even after a year, the speed is not exceptionally fast—only about 2900 mph. But the probe will have traveled a substantial distance, about 25% of the distance to Mars.

SUMMARY

The goal of Chapter 34 has been to study the properties of electromagnetic fields and waves.

General Principles

Maxwell's Equations

These equations govern electromagnetic fields:

$$\oint \vec{E} \cdot d\vec{A} = \frac{Q_{in}}{\epsilon_0}$$ Gauss's law

$$\oint \vec{B} \cdot d\vec{A} = 0$$ Gauss's law for magnetism

$$\oint \vec{E} \cdot d\vec{s} = -\frac{d\Phi_m}{dt}$$ Faraday's law

$$\oint \vec{B} \cdot d\vec{s} = \mu_0 I_{through} + \epsilon_0 \mu_0 \frac{d\Phi_e}{dt}$$ Ampère-Maxwell law

Maxwell's equations tell us that:

An electric field can be created by

- Charged particles
- A changing magnetic field

A magnetic field can be created by

- A current
- A changing electric field

Lorentz Force

This force law governs the interaction of charged particles with electromagnetic fields:

$$\vec{F} = q(\vec{E} + \vec{v} \times \vec{B})$$

- An electric field exerts a force on any charged particle.
- A magnetic field exerts a force on a moving charged particle.

Field Transformations

Fields measured in reference frame A to be $\vec{E}_A$ and $\vec{B}_A$ are found in frame B to be

$$\vec{E}_B = \vec{E}_A + \vec{v}_{BA} \times \vec{B}_A$$

$$\vec{B}_B = \vec{B}_A - \frac{1}{c^2}\vec{v}_{BA} \times \vec{E}_A$$

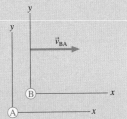

Important Concepts

Induced fields

An induced electric field is created by a changing magnetic field.

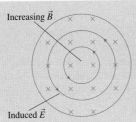

An induced magnetic field is created by a changing electric field.

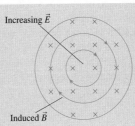

These fields can exist independently of charges and currents.

An electromagnetic wave is a self-sustaining electromagnetic field.

- An em wave is a transverse wave with $\vec{E}$, $\vec{B}$, and $\vec{v}_{em}$ mutually perpendicular.
- An em wave propagates with speed $v_{em} = c = 1/\sqrt{\epsilon_0\mu_0}$.
- The electric and magnetic field strengths are related by $E = cB$.
- The **Poynting vector** $\vec{S} = (\vec{E} \times \vec{B})/\mu_0$ is the energy transfer in the direction of travel.
- The wave **intensity** is $I = P/A = (1/2c\mu_0)E_0^2 = (c\epsilon_0/2)E_0^2$.

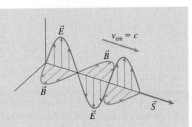

Applications

Polarization

The electric field and the Poynting vector define the **plane of polarization.** The intensity of polarized light transmitted through a polarizing filter is given by Malus's law:

$$I = I_0 \cos^2\theta$$

where θ is the angle between the electric field and the polarizer axis.

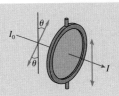

Terms and Notation

Galilean field transformation equations	Poynting vector, $\vec{S}$	antenna
displacement current	intensity, I	plane of polarization
Maxwell's equations	radiation pressure, p_{rad}	Malus's law

CONCEPTUAL QUESTIONS

1. Andre is flying his spaceship to the left through the laboratory magnetic field of FIGURE Q34.1.
   a. Does Andre see a magnetic field? If so, in which direction does it point?
   b. Does Andre see an electric field? If so, in which direction does it point?

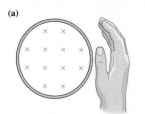

FIGURE Q34.1

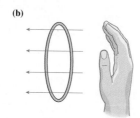

FIGURE Q34.2

2. Sharon drives her rocket through the magnetic field of FIGURE Q34.2 traveling to the right at a speed of 1000 m/s as measured by Bill. As she passes Bill, she shoots a positive charge backward at a speed of 1000 m/s relative to her.
   a. According to Sharon, what kind of force or forces act on the charge? In which directions? Explain.
   b. According to Bill, what kind of force or forces act on the charge? In which directions? Explain.

3. If you curl the fingers of your right hand as shown, are the electric fluxes in FIGURE Q34.3 positive or negative?

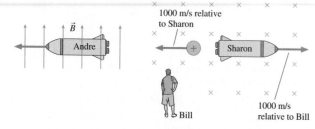

FIGURE Q34.3

4. What is the current through surface S in FIGURE Q34.4 if you curl your right fingers in the direction of the arrow?

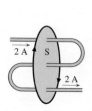

FIGURE Q34.4

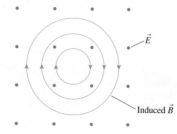

FIGURE Q34.5

5. Is the electric field strength in FIGURE Q34.5 increasing, decreasing, or not changing? Explain.

6. Do the situations in FIGURE Q34.6 represent possible electromagnetic waves? If not, why not?

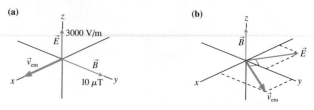

FIGURE Q34.6

7. In what directions are the electromagnetic waves traveling in FIGURE Q34.7?

FIGURE Q34.7

8. The intensity of an electromagnetic wave is 10 W/m^2. What will the intensity be if:
   a. The amplitude of the electric field is doubled?
   b. The amplitude of the magnetic field is doubled?
   c. The amplitudes of both the electric and the magnetic fields are doubled?
   d. The frequency is doubled?

9. Older televisions used a *loop antenna* like the one in FIGURE Q34.9. How does this antenna work?

FIGURE Q34.9

10. A vertically polarized electromagnetic wave passes through the five polarizers in FIGURE Q34.10. Rank in order, from largest to smallest, the transmitted intensities I_a to I_e.

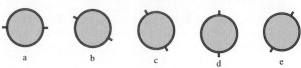

FIGURE Q34.10

EXERCISES AND PROBLEMS

Problems labeled [   ] integrate material from earlier chapters.

Exercises

Section 34.1 *E* or *B*? It Depends on Your Perspective

1. | A rocket cruises past a laboratory at 1.00×10^6 m/s in the positive *x*-direction just as a proton is launched with velocity (in the laboratory frame) $\vec{v} = (1.41 \times 10^6 \hat{\imath} + 1.41 \times 10^6 \hat{\jmath})$ m/s. What are the proton's speed and its angle from the *y*-axis in (a) the laboratory frame and (b) the rocket frame?

2. | **FIGURE EX34.2** shows the electric and magnetic field in frame A. A rocket in frame B travels parallel to one of the axes of the A coordinate system. Along which axis must the rocket travel, and in which direction, in order for the rocket scientists to measure (a) $B_B > B_A$, (b) $B_B = B_A$, and (c) $B_B < B_A$?

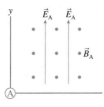

FIGURE EX34.2

3. ‖ Scientists in the laboratory create a uniform electric field $\vec{E} = 1.0 \times 10^6 \hat{k}$ V/m in a region of space where $\vec{B} = \vec{0}$. What are the fields in the reference frame of a rocket traveling in the positive *x*-direction at 1.0×10^6 m/s?

4. | Laboratory scientists have created the electric and magnetic fields shown in **FIGURE EX34.4**. These fields are also seen by scientists that zoom past in a rocket traveling in the *x*-direction at 1.0×10^6 m/s. According to the rocket scientists, what angle does the electric field make with the axis of the rocket?

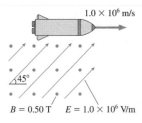

FIGURE EX34.4   **FIGURE EX34.5**

5. | A rocket zooms past the earth at $v = 2.0 \times 10^6$ m/s. Scientists on the rocket have created the electric and magnetic fields shown in **FIGURE EX34.5**. What are the fields measured by an earthbound scientist?

Section 34.2 The Field Laws Thus Far

Section 34.3 The Displacement Current

6. ‖ The magnetic field is uniform over each face of the box shown in **FIGURE EX34.6**. What are the magnetic field strength and direction on the front surface?

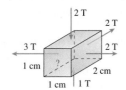

FIGURE EX34.6

7. | Show that the quantity $\epsilon_0 (d\Phi_e/dt)$ has units of current.

8. ‖ Show that the displacement current inside a parallel-plate capacitor can be written $C(dV_C/dt)$.

9. | What capacitance, in μF, has its potential difference increasing at 1.0×10^6 V/s when the displacement current in the capacitor is 1.0 A?

10. ‖ A 10-cm-diameter parallel-plate capacitor has a 1.0 mm spacing. The electric field between the plates is increasing at the rate 1.0×10^6 V/m s. What is the magnetic field strength (a) on the axis, (b) 3.0 cm from the axis, and (c) 7.0 cm from the axis?

11. ‖ A 5.0-cm-diameter parallel-plate capacitor has a 0.50 mm gap. What is the displacement current in the capacitor if the potential difference across the capacitor is increasing at 500,000 V/s?

Section 34.5 Electromagnetic Waves

12. | What is the electric field amplitude of an electromagnetic wave whose magnetic field amplitude is 2.0 mT?

13. | What is the magnetic field amplitude of an electromagnetic wave whose electric field amplitude is 10 V/m?

14. | The magnetic field of an electromagnetic wave in a vacuum is $B_z = (3.00 \, \mu\text{T}) \sin((1.00 \times 10^7)x - \omega t)$, where *x* is in m and *t* is in s. What are the wave's (a) wavelength, (b) frequency, and (c) electric field amplitude?

15. ‖ The electric field of an electromagnetic wave in a vacuum is $E_y = (20.0 \, \text{V/m}) \cos((6.28 \times 10^8)x - \omega t)$, where *x* is in m and *t* is in s. What are the wave's (a) wavelength, (b) frequency, and (c) magnetic field amplitude?

Section 34.6 Properties of Electromagnetic Waves

16. | A radio wave is traveling in the negative *y*-direction. What is the direction of $\vec{E}$ at a point where $\vec{B}$ is in the positive *x*-direction?

17. | a. What is the magnetic field amplitude of an electromagnetic wave whose electric field amplitude is 100 V/m?
    b. What is the intensity of the wave?

18. | A radio receiver can detect signals with electric field amplitudes as small as 300 μV/m. What is the intensity of the smallest detectable signal?

19. ‖ A helium-neon laser emits a 1.0-mm-diameter laser beam with a power of 1.0 mW. What are the amplitudes of the electric and magnetic fields of the light wave?

20. ‖ A 200 MW laser pulse is focused with a lens to a diameter of 2.0 μm.
    a. What is the laser beam's electric field amplitude at the focal point?
    b. What is the ratio of the laser beam's electric field to the electric field that keeps the electron bound to the proton of a hydrogen atom? The radius of the electron orbit is 0.053 nm.

21. ‖ A radio antenna broadcasts a 1.0 MHz radio wave with 25 kW of power. Assume that the radiation is emitted uniformly in all directions.
    a. What is the wave's intensity 30 km from the antenna?
    b. What is the electric field amplitude at this distance?

22. ‖ At what distance from a 10 W point source of electromagnetic waves is the magnetic field amplitude 1.0 μT?

23. | A 1000 W carbon-dioxide laser emits light with a wavelength of 10 μm into a 3.0-mm-diameter laser beam. What force does the laser beam exert on a completely absorbing target?

Section 34.7 Polarization

24. | FIGURE EX34.24 shows a vertically polarized radio wave of frequency 1.0×10^6 Hz traveling into the page. The maximum electric field strength is 1000 V/m. What are
    a. The maximum magnetic field strength?
    b. The magnetic field strength and direction at a point where $\vec{E} = $ (500 V/m, down)?

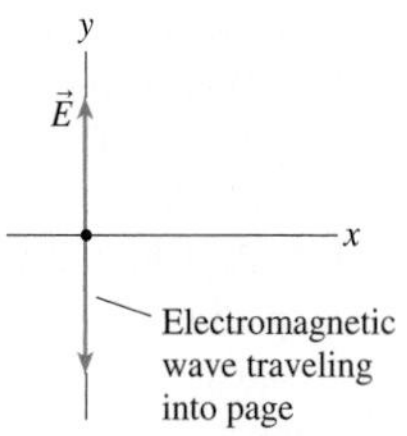

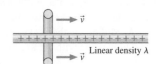

FIGURE EX34.24

Electromagnetic wave traveling into page

25. ‖ Only 25% of the intensity of a polarized light wave passes through a polarizing filter. What is the angle between the electric field and the axis of the filter?

26. ‖ A 200 mW vertically polarized laser beam passes through a polarizing filter whose axis is 35° from horizontal. What is the power of the laser beam as it emerges from the filter?

27. ‖ Unpolarized light with intensity 350 W/m² passes first through a polarizing filter with its axis vertical, then through a second polarizing filter. It emerges from the second filter with intensity 131 W/m². What is the angle from vertical of the axis of the second polarizing filter?

Problems

28. ‖ What is the force (magnitude and direction) on the proton in FIGURE P34.28? Give the direction as an angle cw or ccw from vertical.

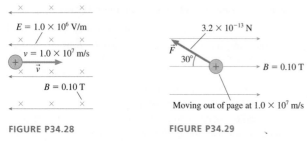

FIGURE P34.28

FIGURE P34.29

29. ‖ What are the electric field strength and direction at the position of the proton in FIGURE P34.29?

30. | What electric field strength and direction will allow the electron in FIGURE P34.30 to pass through this region of space without being deflected?

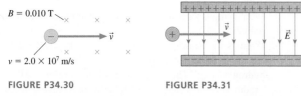

FIGURE P34.30

FIGURE P34.31

31. | A proton is fired with a speed of 1.0×10^6 m/s through the parallel-plate capacitor shown in FIGURE P34.31. The capacitor's electric field is $\vec{E} = (1.0 \times 10^5$ V/m, down).
    a. What magnetic field $\vec{B}$, both strength and direction, must be applied to allow the proton to pass through the capacitor with no change in speed or direction?

    b. Find the electric and magnetic fields in the proton's reference frame.
    c. How does an experimenter in the proton's frame explain that the proton experiences no force as the charged plates fly by?

32. ‖ An electron travels with $\vec{v} = 5.0 \times 10^6 \hat{\imath}$ m/s through a point in space where $\vec{E} = (2.0 \times 10^5 \hat{\imath} - 2.0 \times 10^5 \hat{\jmath})$ V/m and $\vec{B} = -0.10 \hat{k}$ T. What is the force on the electron?

33. ‖ A very long, 1.0-mm-diameter wire carries a 2.5 A current from left to right. Thin plastic insulation on the wire is positively charged with linear charge density 2.5 nC/cm. A mosquito 1.0 cm from the center of the wire would like to move in such a way as to experience an electric field but no magnetic field. How fast and which direction should she fly?

34. ‖ In FIGURE P34.34, a circular loop of radius r travels with speed v along a charged wire having linear charge density λ. The wire is at rest in the laboratory frame, and it passes through the center of the loop.
    a. What are $\vec{E}$ and $\vec{B}$ at a point on the loop as measured by a scientist in the laboratory? Include both strength and direction.
    b. What are the fields $\vec{E}$ and $\vec{B}$ at a point on the loop as measured by a scientist in the frame of the loop?
    c. Show that an experimenter in the loop's frame sees a current $I = \lambda v$ passing through the center of the loop.
    d. What electric and magnetic fields would an experimenter in the loop's frame calculate at distance r from the current of part c?
    e. Show that your fields of parts b and d are the same.

FIGURE P34.34

Linear density λ

35. ‖ The magnetic field inside a 4.0-cm-diameter superconducting solenoid varies sinusoidally between 8.0 T and 12.0 T at a frequency of 10 Hz.
    a. What is the maximum electric field strength at a point 1.5 cm from the solenoid axis?
    b. What is the value of B at the instant E reaches its maximum value?

36. ‖ A simple series circuit consists of a 150 Ω resistor, a 25 V battery, a switch, and a 2.5 pF parallel-plate capacitor (initially uncharged) with plates 5.0 mm apart. The switch is closed at $t = 0$ s.
    a. After the switch is closed, find the maximum electric flux and the maximum displacement current through the capacitor.
    b. Find the electric flux and the displacement current at $t = 0.50$ ns.

37. ‖ A wire with conductivity σ carries current I. The current is increasing at the rate dI/dt.
    a. Show that there is a displacement current in the wire equal to $(\epsilon_0/\sigma)(dI/dt)$.
    b. Evaluate the displacement current for a copper wire in which the current is increasing at 1.0×10^6 A/s.

38. ‖ A 10 A current is charging a 1.0-cm-diameter parallel-plate capacitor.
    a. What is the magnetic field strength at a point 2.0 mm radially from the center of the wire leading to the capacitor?
    b. What is the magnetic field strength at a point 2.0 mm radially from the center of the capacitor?

39. ‖ FIGURE P34.39 shows the voltage across a $0.10\ \mu F$ capacitor. Draw a graph showing the displacement current through the capacitor as a function of time.

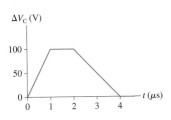

FIGURE P34.39

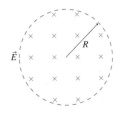

FIGURE P34.40

40. ‖ FIGURE P34.40 shows the electric field inside a cylinder of radius $R = 3.0$ mm. The field strength is increasing with time as $E = 1.0 \times 10^8 t^2$ V/m, where t is in s. The electric field outside the cylinder is always zero, and the field inside the cylinder was zero for $t < 0$.
   a. Find an expression for the electric flux Φ_e through the entire cylinder as a function of time.
   b. Draw a picture showing the magnetic field lines inside and outside the cylinder. Be sure to include arrowheads showing the field's direction.
   c. Find an expression for the magnetic field strength as a function of time at a distance $r < R$ from the center. Evaluate the magnetic field strength at $r = 2.0$ mm, $t = 2.0$ s.
   d. Find an expression for the magnetic field strength as a function of time at a distance $r > R$ from the center. Evaluate the magnetic field strength at $r = 4.0$ mm, $t = 2.0$ s.

41. ‖ A $1.0\ \mu F$ capacitor is discharged, starting at $t = 0$ s. The displacement current through the plates is $I_{disp} = (10\ A)\exp(-t/2.0\ \mu s)$. What was the capacitor's initial voltage $(\Delta V_C)_0$?

42. ‖ At one instant, the electric and magnetic fields at one point of an electromagnetic wave are $\vec{E} = (200\,\hat{i} + 300\,\hat{j} - 50\,\hat{k})$ V/m and $\vec{B} = B_0(7.3\,\hat{i} - 7.3\,\hat{j} + a\,\hat{k})\ \mu T$.
   a. What are the values of a and B_0?
   b. What is the Poynting vector at this time and position?

43. ‖ a. Show that u_E and u_B, the energy densities of the electric and magnetic fields, are equal to each other in an electromagnetic wave. In other words, show that the wave's energy is divided equally between the electric field and the magnetic field.
   b. What is the total energy density in an electromagnetic wave of intensity 1000 W/m²?

44. ‖ Assume that a 7.0-cm-diameter, 100 W lightbulb radiates all its energy as a single wavelength of visible light. Estimate the electric and magnetic field strengths at the surface of the bulb.

45. │ The intensity of sunlight reaching the earth is 1360 W/m².
   a. What is the power output of the sun?
   b. What is the intensity of sunlight on Mars?

46. ‖ A cube of water 10 cm on a side is placed in a microwave beam having $E_0 = 11$ kV/m. The microwaves illuminate one face of the cube, and the water absorbs 80% of the incident energy. How long will it take to raise the water temperature by 50°C? Assume that the water has no heat loss during this time.

47. ‖ A laser beam passes through a converging lens with a focal length f. At what distance past the lens has the laser beam's (a) intensity and (b) electric field strength increased by a factor of 4?

48. │ When the Voyager 2 spacecraft passed Neptune in 1989, it was 4.5×10^9 km from the earth. Its radio transmitter, with which it sent back data and images, broadcast with a mere 21 W of power. Assuming that the transmitter broadcast equally in all directions,
   a. What signal intensity was received on the earth?
   b. What electric field amplitude was detected?
   The received signal was somewhat stronger than your result because the spacecraft used a directional antenna, but not by much.

49. ‖ In reading the instruction manual that came with your garage-door opener, you see that the transmitter unit in your car produces a 250 mW signal and that the receiver unit is supposed to respond to a radio wave of the correct frequency if the electric field amplitude exceeds 0.10 V/m. You wonder if this is really true. To find out, you put fresh batteries in the transmitter and start walking away from your garage while opening and closing the door. Your garage door finally fails to respond when you're 42 m away. Are the manufacturer's claims true?

50. ‖ The maximum electric field strength in air is 3.0 MV/m. Stronger electric fields ionize the air and create a spark. What is the maximum power that can be delivered by a 1.0-cm-diameter laser beam propagating through air?

51. ‖ A LASIK vision-correction system uses a laser that emits
   BIO 10-ns-long pulses of light, each with 2.5 mJ of energy. The laser beam is focused to a 0.85-mm-diameter circle on the cornea. What is the electric field amplitude of the light wave at the cornea?

52. ‖ The intensity of sunlight reaching the earth is 1360 W/m². Assuming all the sunlight is absorbed, what is the radiation-pressure force on the earth? Give your answer (a) in newtons and (b) as a fraction of the sun's gravitational force on the earth.

53. ‖ For radio and microwaves, the depth of penetration into the
   BIO human body is proportional to $\lambda^{1/2}$. If 27 MHz radio waves penetrate to a depth of 14 cm, how far do 2.4 GHz microwaves penetrate?

54. ‖ A laser beam shines straight up onto a flat, black foil of mass m.
   a. Find an expression for the laser power P needed to levitate the foil.
   b. Evaluate P for a foil with a mass of 25 μg.

55. │ For a science project, you would like to horizontally suspend an 8.5 by 11 inch sheet of black paper in a vertical beam of light whose dimensions exactly match the paper. If the mass of the sheet is 1.0 g, what light intensity will you need?

56. ‖ You've recently read about a chemical laser that generates a 20-cm-diameter, 25 MW laser beam. One day, after physics class, you start to wonder if you could use the radiation pressure from this laser beam to launch small payloads into orbit. To see if this might be feasible, you do a quick calculation of the acceleration of a 20-cm-diameter, 100 kg, perfectly absorbing block. What speed would such a block have if pushed *horizontally* 100 m along a frictionless track by such a laser?

57. ‖ An 80 kg astronaut has gone outside his space capsule to do some repair work. Unfortunately, he forgot to lock his safety tether in place, and he has drifted 5.0 m away from the capsule. Fortunately, he has a 1000 W portable laser with fresh batteries that will operate it for 1.0 h. His only chance is to accelerate himself toward the space capsule by firing the laser in the opposite direction. He has a 10-h supply of oxygen. How long will it take him to reach safety?

58. ‖ Unpolarized light of intensity I_0 is incident on three polarizing filters. The axis of the first is vertical, that of the second is 45° from vertical, and that of the third is horizontal. What light intensity emerges from the third filter?

Challenge Problems

59. An electron travels with $\vec{v} = 5.0 \times 10^6 \hat{\imath}$ m/s through a point in space where $\vec{B} = 0.10 \hat{\jmath}$ T. The force on the electron at this point is $\vec{F} = (9.6 \times 10^{-14} \hat{\imath} - 9.6 \times 10^{-14} \hat{k})$ N. What is the electric field?

60. A 4.0-cm-diameter parallel-plate capacitor with a 1.0 mm spacing is charged to 1000 V. A switch closes at $t = 0$ s, and the capacitor is discharged through a wire with 0.20 Ω resistance.
    a. Find an expression for the magnetic field strength inside the capacitor at $r = 1.0$ cm as a function of time.
    b. Draw a graph of B versus t.

61. The radar system at an airport broadcasts 11 GHz microwaves with 150 kW of power. An approaching airplane with a 31 m² cross section is 30 km away. Assume that the radar broadcasts uniformly in all directions and that the airplane scatters microwaves uniformly in all directions. What is the electric field strength of the microwave signal received back at the airport 200 μs later?

62. Large quantities of dust should have been left behind after the creation of the solar system. Larger dust particles, comparable in size to soot and sand grains, are common. They create shooting stars when they collide with the earth's atmosphere. But very small dust particles are conspicuously absent. Astronomers believe that the very small dust particles have been blown out of the solar system by the sun. By comparing the forces on dust particles, determine the diameter of the smallest dust particles that can remain in the solar system over long periods of time. Assume that the dust particles are spherical, black, and have a density of 2000 kg/m³. The sun emits electromagnetic radiation with power 3.9×10^{26} W.

63. Consider current I passing through a resistor of radius r, length L, and resistance R.
    a. Determine the electric and magnetic fields at the surface of the resistor. Assume that the electric field is uniform throughout, including at the surface.
    b. Determine the strength and direction of the Poynting vector at the surface of the resistor.
    c. Show that the flux of the Poynting vector (i.e., the integral of $\vec{S} \cdot d\vec{A}$) over the surface of the resistor is I^2R. Then give an interpretation of this result.

64. Unpolarized light of intensity I_0 is incident on a stack of 7 polarizing filters, each with its axis rotated 15° cw with respect to the previous filter. What light intensity emerges from the last filter?

STOP TO THINK ANSWERS

Stop to Think 34.1: b. $\vec{v}_{AB}$ is parallel to $\vec{B}_A$ hence $\vec{v}_{AB} \times \vec{B}_A$ is zero. Thus $\vec{E}_B = \vec{E}_A$ and points in the positive z-direction. $\vec{v}_{AB} \times \vec{E}_A$ points down, in the negative y-direction, so $-\vec{v}_{AB} \times \vec{E}_A/c^2$ points in the positive y-direction and causes $\vec{B}_B$ to be angled upward.

Stop to Think 34.2: $B_c > B_a > B_d > B_b$. The induced magnetic field strength depends on the *rate* dE/dt at which the electric field is changing. Steeper slopes on the graph correspond to larger magnetic fields.

Stop to Think 34.3: e. $\vec{E}$ is perpendicular to $\vec{B}$ and to $\vec{v}$, so it can only be along the z-axis. According to the Ampère-Maxwell law, $d\Phi_e/dt$ has the same sign as the line integral of $\vec{B} \cdot d\vec{s}$ around the closed curve. The integral is positive for a cw integration. Thus, from the right-hand rule, $\vec{E}$ is either into the page (negative z-direction) and increasing, or out of the page (positive z-direction) and decreasing. We can see from the figure that B is decreasing in strength as the wave moves from left

to right, so E must also be decreasing. Thus $\vec{E}$ points along the positive z-axis.

Stop to Think 34.4: a. The Poynting vector $\vec{S} = (\vec{E} \times \vec{B})/\mu_0$ points in the direction of travel, which is the positive y-direction. $\vec{B}$ must point in the positive x-direction in order for $\vec{E} \times \vec{B}$ to point upward.

Stop to Think 34.5: b. The intensity along a line from the antenna decreases inversely with the square of the distance, so the intensity at 20 km is $\frac{1}{4}$ that at 10 km. But the intensity depends on the square of the electric field amplitude, or, conversely, E_0 is proportional to $I^{1/2}$. Thus E_0 at 20 km is $\frac{1}{2}$ that at 10 km.

Stop to Think 34.6: $I_d > I_a > I_b = I_c$. The intensity depends on $\cos^2\theta$, where θ is the angle *between* the axes of the two filters. The filters in d have $\theta = 0°$. The two filters in both b and c are crossed ($\theta = 90°$) and transmit no light at all.

35 AC Circuits

Transmission lines carry alternating current at voltages as high as 500,000 V.

▶ **Looking Ahead** The goal of Chapter 35 is to understand and apply basic techniques of AC circuit analysis.

AC Electricity

The wires that transport electricity across the country—the *grid*—use alternating current, called **AC.**

Transformers allow an oscillating voltage to be "stepped up" to a higher voltage so that power can be delivered using lower currents that don't overheat the wires. Smaller transformers bring the voltage down to 120 V.

Capacitors and Inductors

You'll learn that capacitors and inductors are much more useful in AC circuits than they were in DC circuits.

The peak current and peak voltage of a capacitor or an inductor are related by a resistance-like quantity called **reactance,** also measured in ohms. An inductor's reactance increases with frequency; that of a capacitor decreases.

◀ **Looking Back**
Section 29.5 Capacitors
Section 33.8 Inductors

RLC Circuits

A circuit that is especially important in communication electronics is the series ***RLC* circuit,** consisting of a resistor, capacitor, and inductor.

You'll learn that an *RLC* circuit exhibits *resonance*, allowing it to be tuned to a specific frequency.

Phasors

Voltages and currents oscillate, so the mathematics of AC circuits is similar to that of simple harmonic motion.

You'll learn a new way to represent oscillating quantities with rotating vectors called **phasors.** The instantaneous value of a phasor is its horizontal projection.

◀ **Looking Back**
Chapter 14 Simple harmonic motion and resonance

Filter Circuits

Simple circuits consisting of resistors and capacitors can act as *filters.*

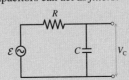

You'll see how this circuit transmits low frequencies to the output—the capacitor voltage— but blocks high frequencies. It is called a **low-pass filter.**

◀ **Looking Back**
Chapter 31 Circuit analysis

Phase and Power

The emf and the current of an AC circuit oscillate with the same frequency but usually not in phase with each other. You'll find that the *phase difference* limits an emf's ability to deliver power because the current and voltage aren't pushing and pulling together.

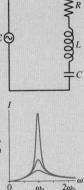

The power delivered to, say, a motor is reduced by a quantity called the **power factor.**

FIGURE 35.1 An oscillating emf can be represented as a graph or as a phasor diagram.

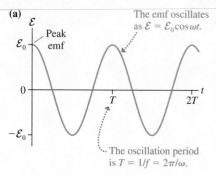

(a)

(b)

35.1 AC Sources and Phasors

One of the examples of Faraday's law cited in Chapter 33 was an electric generator. A turbine, which might be powered by expanding steam or falling water, causes a coil of wire to rotate in a magnetic field. As the coil spins, the emf and the induced current oscillate sinusoidally. The emf is alternately positive and negative, causing the charges to flow in one direction and then, a half cycle later, in the other. The oscillation frequency of the *grid* in North and South America is $f = 60$ Hz, whereas most of the rest of the world uses a 50 Hz oscillation.

The generator's peak emf—the peak voltage—is a fixed, unvarying quantity, so it might seem logical to call a generator an *alternating voltage source.* Nonetheless, circuits powered by a sinusoidal emf are called **AC circuits,** where AC stands for *alternating current.* By contrast, the steady-current circuits you studied in Chapter 31 are called **DC circuits,** for *direct current.*

AC circuits are not limited to the use of 50 Hz or 60 Hz power-line voltages. Audio, radio, television, and telecommunication equipment all make extensive use of AC circuits, with frequencies ranging from approximately 10^2 Hz in audio circuits to approximately 10^9 Hz in cell phones. These devices use *electrical oscillators* rather than generators to produce a sinusoidal emf, but the basic principles of circuit analysis are the same.

You can think of an AC generator or oscillator as a battery whose output voltage undergoes sinusoidal oscillations. The instantaneous emf of an AC generator or oscillator, shown graphically in FIGURE 35.1a, can be written

$$\mathcal{E} = \mathcal{E}_0 \cos \omega t \qquad (35.1)$$

where $\mathcal{E}_0$ is the peak or maximum emf and $\omega = 2\pi f$ is the angular frequency in radians per second. Recall that the units of emf are volts. As you can imagine, the mathematics of AC circuit analysis are going to be very similar to the mathematics of simple harmonic motion.

An alternative way to represent the emf and other oscillatory quantities is with the *phasor diagram* of FIGURE 35.1b. A **phasor** is a vector that rotates *counterclockwise* (ccw) around the origin at angular frequency ω. The length or magnitude of the phasor is the maximum value of the quantity. For example, the length of an emf phasor is $\mathcal{E}_0$. The angle ωt is the *phase angle,* an idea you learned about in Chapter 14, where we made a connection between circular motion and simple harmonic motion.

The quantity's instantaneous value, the value you would measure at time t, is the projection of the phasor onto the horizontal axis. This is also analogous to the connection between circular motion and simple harmonic motion. FIGURE 35.2 helps you visualize the phasor rotation by showing how the phasor corresponds to the more familiar graph at several specific points in the cycle.

FIGURE 35.2 The correspondence between a phasor and points on a graph.

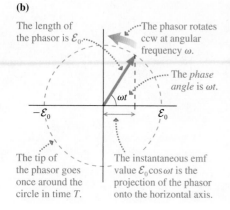

STOP TO THINK 35.1 The magnitude of the instantaneous value of the emf represented by this phasor is

a. Increasing.
b. Decreasing.
c. Constant.
d. It's not possible to tell without knowing t.

Resistor Circuits

In Chapter 31 you learned to analyze a circuit in terms of the current I, voltage V, and potential difference ΔV. Now, because the current and voltage are oscillating, we will use lowercase i to represent the *instantaneous* current through a circuit element and v for the circuit element's *instantaneous* voltage.

FIGURE 35.3 shows the instantaneous current i_R through a resistor R. The potential difference across the resistor, which we call the *resistor voltage* v_R, is given by Ohm's law:

$$v_R = i_R R \qquad (35.2)$$

FIGURE 35.4 shows a resistor R connected across an AC emf $\mathcal{E}$. Notice that the circuit symbol for an AC generator is ——. We can analyze this circuit in exactly the same way we analyzed a DC resistor circuit. Kirchhoff's loop law says that the sum of all the potential differences around a closed path is zero:

$$\sum \Delta V = \Delta V_{\text{source}} + \Delta V_{\text{res}} = \mathcal{E} - v_R = 0 \qquad (35.3)$$

The minus sign appears, just as it did in the equation for a DC circuit, because the potential *decreases* when we travel through a resistor in the direction of the current. We find from the loop law that $v_R = \mathcal{E} = \mathcal{E}_0 \cos \omega t$. This isn't surprising because the resistor is connected directly across the terminals of the emf.

The resistor voltage in an AC circuit can be written

$$v_R = V_R \cos \omega t \qquad (35.4)$$

where V_R is the peak or maximum voltage. You can see that $V_R = \mathcal{E}_0$ in the single-resistor circuit of Figure 35.4. Thus the current through the resistor is

$$i_R = \frac{v_R}{R} = \frac{V_R \cos \omega t}{R} = I_R \cos \omega t \qquad (35.5)$$

where $I_R = V_R/R$ is the peak current.

NOTE ▶ Ohm's law applies to both the instantaneous *and* peak currents and voltages. ◀

The resistor's instantaneous current and voltage are in phase, both oscillating as $\cos \omega t$. FIGURE 35.5 shows the voltage and the current simultaneously on a graph and as a phasor diagram. The fact that the current phasor is shorter than the voltage phasor has no significance. Current and voltage are measured in different units, so you can't compare the length of one to the length of the other. Showing the two different quantities on a single graph—a tactic that can be misleading if you're not careful—illustrates that they oscillate in phase and that their phasors rotate together at the same angle and frequency.

FIGURE 35.3 Instantaneous current i_R through a resistor.

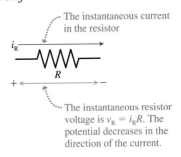

The instantaneous current in the resistor

The instantaneous resistor voltage is $v_R = i_R R$. The potential decreases in the direction of the current.

FIGURE 35.4 An AC resistor circuit.

This is the current direction when $\mathcal{E} > 0$. A half cycle later it will be in the opposite direction.

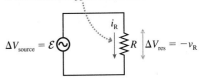

$\Delta V_{\text{source}} = \mathcal{E}$ $\Delta V_{\text{res}} = -v_R$

FIGURE 35.5 Graph and phasor diagram of the resistor current and voltage.

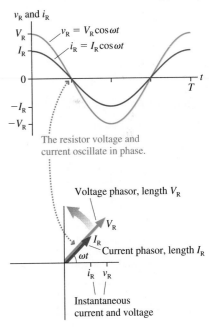

v_R and i_R

$v_R = V_R \cos \omega t$

$i_R = I_R \cos \omega t$

The resistor voltage and current oscillate in phase.

Voltage phasor, length V_R

V_R

I_R Current phasor, length I_R

ωt

i_R v_R

Instantaneous current and voltage

EXAMPLE 35.1 | **Finding resistor voltages**

In the circuit of FIGURE 35.6, what are (a) the peak voltage across each resistor and (b) the instantaneous resistor voltages at $t = 20$ ms?

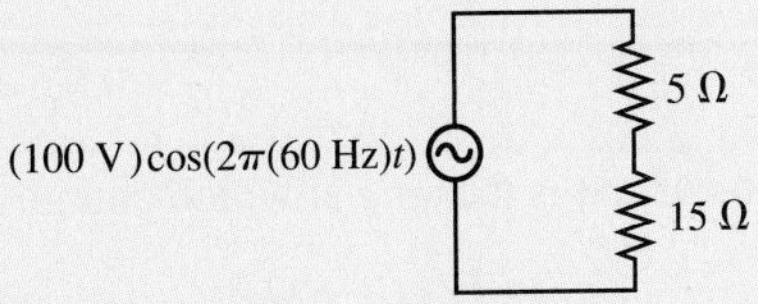

VISUALIZE Figure 35.6 shows the circuit diagram. The two resistors are in series.

SOLVE a. The equivalent resistance of the two series resistors is $R_{\text{eq}} = 5\ \Omega + 15\ \Omega = 20\ \Omega$. The instantaneous current through the equivalent resistance is

$$i_R = \frac{v_R}{R_{\text{eq}}} = \frac{\mathcal{E}_0 \cos \omega t}{R_{\text{eq}}} = \frac{(100\ \text{V}) \cos(2\pi(60\ \text{Hz})t)}{20\ \Omega}$$

$$= (5.0\ \text{A}) \cos(2\pi(60\ \text{Hz})t)$$

FIGURE 35.6 An AC resistor circuit.

$(100\ \text{V})\cos(2\pi(60\ \text{Hz})t)$ $5\ \Omega$ $15\ \Omega$

Continued

The peak current is $I_R = 5.0$ A, and this is also the peak current through the two resistors that form the 20 Ω equivalent resistance. Hence the peak voltage across each resistor is

$$V_R = I_R R = \begin{cases} 25 \text{ V} & 5 \text{ Ω resistor} \\ 75 \text{ V} & 15 \text{ Ω resistor} \end{cases}$$

b. The instantaneous current at $t = 0.020$ s is

$$i_R = (5.0 \text{ A}) \cos\left(2\pi(60 \text{ Hz})(0.020 \text{ s})\right) = 1.55 \text{ A}$$

The resistor voltages at this time are

$$v_R = i_R R = \begin{cases} 7.7 \text{ V} & 5 \text{ Ω resistor} \\ 23.2 \text{ V} & 15 \text{ Ω resistor} \end{cases}$$

ASSESS The sum of the instantaneous voltages, 30.9 V, is what you would find by calculating $\mathcal{E}$ at $t = 20$ ms. This self-consistency gives us confidence in the answer.

STOP TO THINK 35.2 The resistor whose voltage and current phasors are shown here has resistance R

a. > 1 Ω
b. < 1 Ω
c. It's not possible to tell.

35.2 Capacitor Circuits

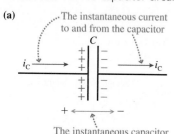

FIGURE 35.7 An AC capacitor circuit.

(a)

The instantaneous current to and from the capacitor

The instantaneous capacitor voltage is $v_C = q/C$. The potential decreases from + to −.

(b)

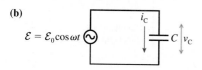

FIGURE 35.7a shows current i_C charging a capacitor with capacitance C. The instantaneous capacitor voltage is $v_C = q/C$, where $\pm q$ is the charge on the two capacitor plates at this instant. It is useful to compare Figure 35.7a to Figure 35.3 for a resistor.

FIGURE 35.7b, where capacitance C is connected across an AC source of emf $\mathcal{E}$, is the most basic capacitor circuit. The capacitor is in parallel with the source, so the capacitor voltage equals the emf: $v_C = \mathcal{E} = \mathcal{E}_0 \cos \omega t$. It will be useful to write

$$v_C = V_C \cos \omega t \tag{35.6}$$

where V_C is the peak or maximum voltage across the capacitor. You can see that $V_C = \mathcal{E}_0$ in this single-capacitor circuit.

To find the current to and from the capacitor, we first write the charge

$$q = Cv_C = CV_C \cos \omega t \tag{35.7}$$

The current is the *rate* at which charge flows through the wires, $i_C = dq/dt$, thus

$$i_C = \frac{dq}{dt} = \frac{d}{dt}(CV_C \cos \omega t) = -\omega CV_C \sin \omega t \tag{35.8}$$

We can most easily see the relationship between the capacitor voltage and current if we use the trigonometric identity $-\sin(x) = \cos(x + \pi/2)$ to write

$$i_C = \omega CV_C \cos\left(\omega t + \frac{\pi}{2}\right) \tag{35.9}$$

In contrast to a resistor, a capacitor's current and voltage are *not* in phase. In FIGURE 35.8a, a graph of the instantaneous voltage v_C and current i_C, you can see that the current peaks one-quarter of a period *before* the voltage peaks. The phase angle of the current phasor on the phasor diagram of FIGURE 35.8b is $\pi/2$ rad—a quarter of a circle—larger than the phase angle of the voltage phasor.

We can summarize this finding:

The AC current of a capacitor *leads* the capacitor voltage by $\pi/2$ rad, or 90°.

The current reaches its peak value I_C at the instant the capacitor is fully discharged and $v_C = 0$. The current is zero at the instant the capacitor is fully charged.

A simple harmonic oscillator provides a mechanical analogy of the 90° phase difference between current and voltage. You learned in Chapter 14 that the position and velocity of a simple harmonic oscillator are

$$x = A \cos \omega t$$

$$v = \frac{dx}{dt} = -\omega A \sin \omega t = -v_{max} \sin \omega t = v_{max} \cos\left(\omega t + \frac{\pi}{2}\right)$$

You can see that the velocity of an oscillator leads the position by 90° in the same way that the capacitor current leads the voltage.

Capacitive Reactance

We can use Equation 35.9 to see that the peak current to and from a capacitor is $I_C = \omega C V_C$. This relationship between the peak voltage and peak current looks much like Ohm's law for a resistor if we define the **capacitive reactance** X_C to be

$$X_C \equiv \frac{1}{\omega C} \qquad (35.10)$$

With this definition,

$$I_C = \frac{V_C}{X_C} \quad \text{or} \quad V_C = I_C X_C \qquad (35.11)$$

The units of reactance, like those of resistance, are ohms.

NOTE ▶ Reactance relates the *peak* voltage V_C and current I_C. But reactance differs from resistance in that it does *not* relate the instantaneous capacitor voltage and current because they are out of phase. That is, $v_C \neq i_C X_C$. ◀

A resistor's resistance R is independent of the emf frequency. In contrast, as FIGURE 35.9 shows, a capacitor's reactance X_C depends inversely on the frequency. The reactance becomes very large at low frequencies (i.e., the capacitor is a large impediment to current). This makes sense because $\omega = 0$ would be a nonoscillating DC circuit, and we know that a steady DC current cannot pass through a capacitor. The reactance decreases as the frequency increases until, at very high frequencies, $X_C \approx 0$ and the capacitor begins to act like an ideal wire. This result has important consequences for how capacitors are used in many circuits.

FIGURE 35.8 Graph and phasor diagrams of the capacitor current and voltage.

(a) i_C peaks $\frac{1}{4}T$ before v_C peaks. We say that the current *leads* the voltage by 90°.

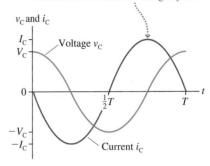

(b)
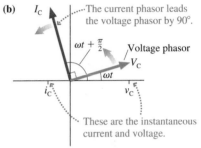

FIGURE 35.9 The capacitive reactance as a function of frequency.

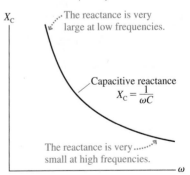

EXAMPLE 35.2 | **Capacitive reactance**

What is the capacitive reactance of a 0.10 μF capacitor at a 100 Hz audio frequency and at a 100 MHz FM-radio frequency?

SOLVE At 100 Hz,

$$X_C(\text{at 100 Hz}) = \frac{1}{\omega C} = \frac{1}{2\pi(100 \text{ Hz})(1.0 \times 10^{-7} \text{ F})} = 16,000 \text{ }\Omega$$

Increasing the frequency by a factor of 10^6 decreases X_C by a factor of 10^6, giving

$$X_C(\text{at 100 MHz}) = 0.016 \text{ }\Omega$$

ASSESS A capacitor with a substantial reactance at audio frequencies has virtually no reactance at FM-radio frequencies.

EXAMPLE 35.3 **Capacitor current**

A 10 μF capacitor is connected to a 1000 Hz oscillator with a peak emf of 5.0 V. What is the peak current to the capacitor?

VISUALIZE Figure 35.7b showed the circuit diagram. It is a simple one-capacitor circuit.

SOLVE The capacitive reactance at $\omega = 2\pi f = 6280$ rad/s is

$$X_C = \frac{1}{\omega C} = \frac{1}{(6280 \text{ rad/s})(10 \times 10^{-6} \text{ F})} = 16 \text{ }\Omega$$

The peak voltage across the capacitor is $V_C = \mathcal{E}_0 = 5.0$ V; hence the peak current is

$$I_C = \frac{V_C}{X_C} = \frac{5.0 \text{ V}}{16 \text{ }\Omega} = 0.31 \text{ A}$$

ASSESS Using reactance is just like using Ohm's law, but don't forget it applies to only the *peak* current and voltage, not the instantaneous values.

STOP TO THINK 35.3 What is the capacitive reactance of "no capacitor," just a continuous wire?

a. 0   b. ∞ c. Undefined

35.3 *RC* Filter Circuits

FIGURE 35.10 An *RC* circuit driven by an AC source.

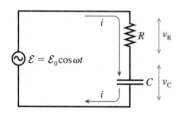

You learned in Chapter 31 that a resistance R causes a capacitor to be charged or discharged with time constant $\tau = RC$. We called this an RC circuit. Now that we've looked at resistors and capacitors individually, let's explore what happens if an RC circuit is driven continuously by an alternating current source.

FIGURE 35.10 shows a circuit in which a resistor R and capacitor C are in series with an emf $\mathcal{E}$ oscillating at angular frequency ω. Before launching into a formal analysis, let's try to understand qualitatively how this circuit will respond as the frequency is varied. If the frequency is very low, the capacitive reactance will be very large, and thus the peak current I_C will be very small. The peak current through the resistor is the same as the peak current to and from the capacitor (conservation of current requires $I_R = I_C$); hence we expect the resistor's peak voltage $V_R = I_R R$ to be very small at very low frequencies.

On the other hand, suppose the frequency is very high. Then the capacitive reactance approaches zero and the peak current, determined by the resistance alone, will be $I_R = \mathcal{E}_0/R$. The resistor's peak voltage $V_R = IR$ will approach the peak source voltage $\mathcal{E}_0$ at very high frequencies.

This reasoning leads us to expect that V_R will *increase* steadily from 0 to $\mathcal{E}_0$ as ω is increased from 0 to very high frequencies. Kirchhoff's loop law has to be obeyed, so the capacitor voltage V_C will *decrease* from $\mathcal{E}_0$ to 0 during the same change of frequency. A quantitative analysis will show us how this behavior can be used as a *filter*.

The goal of a quantitative analysis is to determine the peak current I and the two peak voltages V_R and V_C as functions of the emf amplitude $\mathcal{E}_0$ and frequency ω. Our analytic procedure is based on the fact that the instantaneous current i is the same for two circuit elements in series.

Using phasors to analyze an *RC* circuit

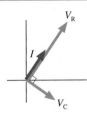

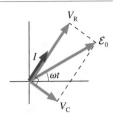

Begin by drawing a current phasor of length *I*. This is the starting point because the series circuit elements have the same current *i*. The angle at which the phasor is drawn is not relevant.

The current and voltage of a resistor are in phase, so draw a resistor voltage phasor of length V_R parallel to the current phasor *I*. The capacitor current leads the capacitor voltage by 90°, so draw a capacitor voltage phasor of length V_C that is 90° behind [i.e., clockwise (cw) from] the current phasor.

The series resistor and capacitor are in parallel with the emf, so their *instantaneous* voltages satisfy $v_R + v_C = \mathcal{E}$. This is a *vector* addition of phasors, so draw the emf phasor as the vector sum of the two voltage phasors. The emf is $\mathcal{E} = \mathcal{E}_0 \cos \omega t$, hence the emf phasor is at angle ωt.

The length of the emf phasor, $\mathcal{E}_0$, is the hypotenuse of a right triangle formed by the resistor and capacitor phasors. Thus $\mathcal{E}_0^2 = V_R^2 + V_C^2$.

The relationship $\mathcal{E}_0^2 = V_R^2 + V_C^2$ is based on the peak values, not the instantaneous values, because the peak values are the lengths of the sides of the right triangle. The peak voltages are related to the peak current *I* via $V_R = IR$ and $V_C = IX_C$, thus

$$\mathcal{E}_0^2 = V_R^2 + V_C^2 = (IR)^2 + (IX_C)^2 = (R^2 + X_C^2)I^2$$
$$= (R^2 + 1/\omega^2 C^2)I^2 \tag{35.12}$$

Consequently, the peak current in the *RC* circuit is

$$I = \frac{\mathcal{E}_0}{\sqrt{R^2 + X_C^2}} = \frac{\mathcal{E}_0}{\sqrt{R^2 + 1/\omega^2 C^2}} \tag{35.13}$$

Knowing *I* gives us the two peak voltages:

$$V_R = IR = \frac{\mathcal{E}_0 R}{\sqrt{R^2 + X_C^2}} = \frac{\mathcal{E}_0 R}{\sqrt{R^2 + 1/\omega^2 C^2}}$$

$$V_C = IX_C = \frac{\mathcal{E}_0 X_C}{\sqrt{R^2 + X_C^2}} = \frac{\mathcal{E}_0/\omega C}{\sqrt{R^2 + 1/\omega^2 C^2}} \tag{35.14}$$

Frequency Dependence

Our goal was to see how the peak current and voltages vary as functions of the frequency ω. Equations 35.13 and 35.14 are rather complex and best interpreted by looking at graphs. FIGURE 35.11 is a graph of V_R and V_C versus ω.

You can see that our qualitative predictions have been borne out. That is, V_R increases from 0 to $\mathcal{E}_0$ as ω is increased, while V_C decreases from $\mathcal{E}_0$ to 0. The explanation for this behavior is that the capacitive reactance X_C decreases as ω increases. For low frequencies, where $X_C \gg R$, the circuit is primarily capacitive. For high frequencies, where $X_C \ll R$, the circuit is primarily resistive.

The frequency at which $V_R = V_C$ is called the **crossover frequency** ω_c. The *crossover* frequency is easily found by setting the two expressions in Equations 35.14 equal to each other. The denominators are the same and cancel, as does $\mathcal{E}_0$, leading to

$$\omega_c = \frac{1}{RC} \tag{35.15}$$

In practice, $f_c = \omega_c/2\pi$ is also called the crossover frequency.

FIGURE 35.11 Graph of the resistor and capacitor peak voltages as functions of the emf angular frequency ω.

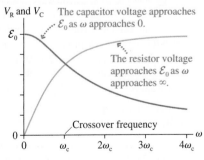

We'll leave it as a homework problem to show that $V_R = V_C = \mathcal{E}_0/\sqrt{2}$ when $\omega = \omega_c$. This may seem surprising. After all, shouldn't V_R and V_C add up to $\mathcal{E}_0$?

No! V_R and V_C are the *peak values* of oscillating voltages, not the instantaneous values. The instantaneous values do, indeed, satisfy $v_R + v_C = \mathcal{E}$ at all instants of time. But the resistor and capacitor voltages are out of phase with each other, as the phasor diagram shows, so the two circuit elements don't reach their peak values at the same time. The peak values are related by $\mathcal{E}_0^2 = V_R^2 + V_C^2$, and you can see that $V_R = V_C = \mathcal{E}_0/\sqrt{2}$ satisfies this equation.

NOTE ▶ It's very important in AC circuit analysis to make a clear distinction between instantaneous values and peak values of voltages and currents. Relationships that are true for one set of values may not be true for the other. ◀

Filters

FIGURE 35.12 Low-pass and high-pass filter circuits.

(a) Low-pass filter

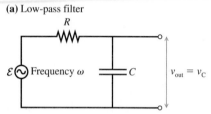

Transmits frequencies $\omega < \omega_c$ and blocks frequencies $\omega > \omega_c$

(b) High-pass filter

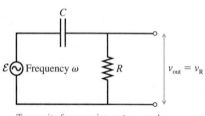

Transmits frequencies $\omega > \omega_c$ and blocks frequencies $\omega < \omega_c$

FIGURE 35.12a is the circuit we've just analyzed; the only difference is that the capacitor voltage v_C is now identified as the *output voltage* v_{out}. This is a voltage you might measure or, perhaps, send to an amplifier for use elsewhere in an electronic instrument. You can see from the capacitor voltage graph in Figure 35.11 that the peak output voltage is $V_{out} \approx \mathcal{E}_0$ if $\omega \ll \omega_c$, but $V_{out} \approx 0$ if $\omega \gg \omega_c$. In other words,

- If the frequency of an input signal is well below the crossover frequency, the input signal is transmitted with little loss to the output.
- If the frequency of an input signal is well above the crossover frequency, the input signal is strongly attenuated and the output is very nearly zero.

This circuit is called a **low-pass filter.**

The circuit of FIGURE 35.12b, which instead uses the resistor voltage v_R for the output v_{out}, is a **high-pass filter.** The output is $V_{out} \approx 0$ if $\omega \ll \omega_c$, but $V_{out} \approx \mathcal{E}_0$ if $\omega \gg \omega_c$. That is, an input signal whose frequency is well above the crossover frequency is transmitted without loss to the output.

Filter circuits are widely used in electronics. For example, a high-pass filter designed to have $f_c = 100$ Hz would pass the audio frequencies associated with speech ($f > 200$ Hz) while blocking 60 Hz "noise" that can be picked up from power lines. Similarly, the high-frequency hiss from old vinyl records can be attenuated with a low-pass filter, allowing the lower-frequency audio signal to pass.

A simple RC filter suffers from the fact that the crossover region where $V_R \approx V_C$ is fairly broad. More sophisticated filters have a sharper transition from off ($V_{out} \approx 0$) to on ($V_{out} \approx \mathcal{E}_0$), but they're based on the same principles as the RC filter analyzed here.

EXAMPLE 35.4 | **Designing a filter**

For a science project, you've built a radio to listen to AM radio broadcasts at frequencies near 1 MHz. The basic circuit is an antenna, which produces a very small oscillating voltage when it absorbs the energy of an electromagnetic wave, and an amplifier. Unfortunately, your neighbor's short-wave radio broadcast at 10 MHz interferes with your reception. Having just finished physics, you decide to solve this problem by placing a filter between the antenna and the amplifier. You happen to have a 500 pF capacitor. What frequency should you select as the filter's crossover frequency? What value of resistance will you need to build this filter?

MODEL You need a low-pass filter to block signals at 10 MHz while passing the lower-frequency AM signal at 1 MHz.

VISUALIZE The circuit will look like the low-pass filter in Figure 35.12a. The oscillating voltage generated by the antenna will be the emf, and v_{out} will be sent to the amplifier.

SOLVE You might think that a crossover frequency near 5 MHz, about halfway between 1 MHz and 10 MHz, would work best. But 5 MHz is a factor of 5 higher than 1 MHz while only a factor of 2 less than 10 MHz. A crossover frequency the same factor above 1 MHz as it is below 10 MHz will give the best results. In practice, choosing $f_c = 3$ MHz would be sufficient. You can then use Equation 35.15 to select the proper resistor value:

$$R = \frac{1}{\omega_c C} = \frac{1}{2\pi(3 \times 10^6 \text{ Hz})(500 \times 10^{-12} \text{ F})}$$

$$= 106 \ \Omega \approx 100 \ \Omega$$

ASSESS Rounding to 100 Ω is appropriate because the crossover frequency was determined to only one significant figure. Such "sloppy design" is adequate when the two frequencies you need to distinguish are well separated.

STOP TO THINK 35.4   Rank in order, from largest to smallest, the crossover frequencies $(\omega_c)_a$ to $(\omega_c)_d$ of these four circuits.

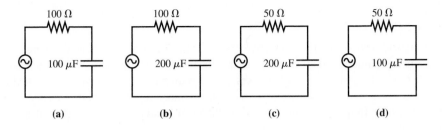

(a)   (b)   (c)   (d)

35.4 Inductor Circuits

FIGURE 35.13a shows the instantaneous current i_L through an inductor. If the current is changing, the instantaneous inductor voltage is

$$v_L = L\frac{di_L}{dt} \tag{35.16}$$

You learned in Chapter 33 that the potential decreases in the direction of the current if the current is increasing ($di_L/dt > 0$) and increases if the current is decreasing ($di_L/dt < 0$).

FIGURE 35.13b is the simplest inductor circuit. The inductor L is connected across the AC source, so the inductor voltage equals the emf: $v_L = \mathcal{E} = \mathcal{E}_0 \cos \omega t$. We can write

$$v_L = V_L \cos \omega t \tag{35.17}$$

where V_L is the peak or maximum voltage across the inductor. You can see that $V_L = \mathcal{E}_0$ in this single-inductor circuit.

We can find the inductor current i_L by integrating Equation 35.17. First, we use Equation 35.17 to write Equation 35.16 as

$$di_L = \frac{v_L}{L}dt = \frac{V_L}{L}\cos \omega t\, dt \tag{35.18}$$

Integrating gives

$$i_L = \frac{V_L}{L}\int \cos \omega t\, dt = \frac{V_L}{\omega L}\sin \omega t = \frac{V_L}{\omega L}\cos\left(\omega t - \frac{\pi}{2}\right)$$

$$= I_L \cos\left(\omega t - \frac{\pi}{2}\right) \tag{35.19}$$

where $I_L = V_L/\omega L$ is the peak or maximum inductor current.

NOTE ▶ Mathematically, Equation 35.19 could have an integration constant i_0. An integration constant would represent a constant DC current through the inductor, but there is no DC source of potential in an AC circuit. Hence, on physical grounds, we set $i_0 = 0$ for an AC circuit. ◀

We define the **inductive reactance,** analogous to the capacitive reactance, to be

$$X_L \equiv \omega L \tag{35.20}$$

FIGURE 35.13 Using an inductor in an AC circuit.

(a) The instantaneous current through the inductor

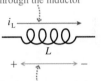

The instantaneous inductor voltage is $v_L = L(di_L/dt)$.

(b)

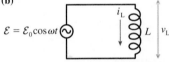

Then the peak current $I_L = V_L/\omega L$ and the peak voltage are related by

$$I_L = \frac{V_L}{X_L} \quad \text{or} \quad V_L = I_L X_L \qquad (35.21)$$

FIGURE 35.14 The inductive reactance as a function of frequency.

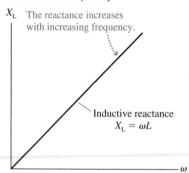

X_L The reactance increases with increasing frequency.

Inductive reactance $X_L = \omega L$

ω

FIGURE 35.14 shows that the inductive reactance increases as the frequency increases. This makes sense. Faraday's law tells us that the induced voltage across a coil increases as the time rate of change of $\vec{B}$ increases, and $\vec{B}$ is directly proportional to the inductor current. For a given peak current I_L, $\vec{B}$ changes more rapidly at higher frequencies than at lower frequencies, and thus V_L is larger at higher frequencies than at lower frequencies.

FIGURE 35.15a is a graph of the inductor voltage and current. You can see that the current peaks one-quarter of a period *after* the voltage peaks. The angle of the current phasor on the phasor diagram of **FIGURE 35.15b** is $\pi/2$ rad less than the angle of the voltage phasor. We can summarize this finding:

The AC current through an inductor *lags* the inductor voltage by $\pi/2$ rad, or 90°.

FIGURE 35.15 Graphs and phasor diagrams of the inductor current and voltage.

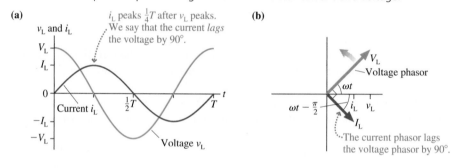

(a) i_L peaks $\frac{1}{4}T$ after v_L peaks. We say that the current *lags* the voltage by 90°.

v_L and i_L

V_L
I_L
0
$-I_L$
$-V_L$

Current i_L
$\frac{1}{2}T$
T
t

Voltage v_L

(b)

V_L
Voltage phasor
ωt
$\omega t - \frac{\pi}{2}$
i_L v_L
I_L

The current phasor lags the voltage phasor by 90°.

EXAMPLE 35.5 | **Current and voltage of an inductor**

A 25 μH inductor is used in a circuit that oscillates at 100 kHz. The current through the inductor reaches a peak value of 20 mA at $t = 5.0\ \mu$s. What is the peak inductor voltage, and when, closest to $t = 5.0\ \mu$s, does it occur?

MODEL The inductor current lags the voltage by 90°, or, equivalently, the voltage reaches its peak value one-quarter period *before* the current.

VISUALIZE The circuit looks like Figure 35.13b.

SOLVE The inductive reactance at $f = 100$ kHz is

$$X_L = \omega L = 2\pi(1.0 \times 10^5\ \text{Hz})(25 \times 10^{-6}\ \text{H}) = 16\ \Omega$$

Thus the peak voltage is $V_L = I_L X_L = (20\ \text{mA})(16\ \Omega) = 320\ \text{mV}$. The voltage peak occurs one-quarter period before the current peaks, and we know that the current peaks at $t = 5.0\ \mu$s. The period of a 100 kHz oscillation is 10.0 μs, so the voltage peaks at

$$t = 5.0\ \mu\text{s} - \frac{10.0\ \mu\text{s}}{4} = 2.5\ \mu\text{s}$$

35.5 The Series *RLC* Circuit

FIGURE 35.16 A series *RLC* circuit.

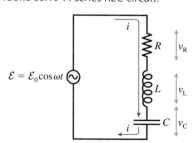

i
R v_R
$\mathcal{E} = \mathcal{E}_0 \cos \omega t$
L v_L
i
C v_C

The circuit of **FIGURE 35.16**, where a resistor, inductor, and capacitor are in series, is called a **series *RLC* circuit**. The series *RLC* circuit has many important applications because, as you will see, it exhibits resonance behavior.

The analysis, which is very similar to our analysis of the *RC* circuit in Section 35.3, will be based on a phasor diagram. Notice that the three circuit elements are in series with each other and, together, are in parallel with the emf. We can draw two conclusions that form the basis of our analysis:

1. The instantaneous current of all three elements is the same: $i = i_R = i_L = i_C$.
2. The sum of the instantaneous voltages matches the emf: $\mathcal{E} = v_R + v_L + v_C$.

Using phasors to analyze an *RLC* circuit

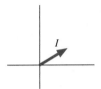

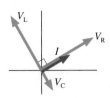

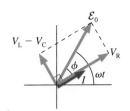

Begin by drawing a current phasor of length I. This is the starting point because the series circuit elements have the same current i.	The current and voltage of a resistor are in phase, so draw a resistor voltage phasor parallel to the current phasor I. The capacitor current leads the capacitor voltage by 90°, so draw a capacitor voltage phasor that is 90° behind the current phasor. The inductor current *lags* the voltage by 90°, so draw an inductor voltage phasor 90° ahead of the current phasor.	The instantaneous voltages satisfy $\mathcal{E} = v_R + v_L + v_C$. In terms of phasors, this is a *vector* addition. We can do the addition in two steps. Because the capacitor and inductor phasors are in opposite directions, their vector sum has length $V_L - V_C$. Adding the resistor phasor, at right angles, then gives the emf phasor $\mathcal{E}$ at angle ωt.	The length $\mathcal{E}_0$ of the emf phasor is the hypotenuse of a right triangle. Thus $$\mathcal{E}_0^2 = V_R^2 + (V_L - V_C)^2$$

If $V_L > V_C$, which we've assumed, then the instantaneous current i lags the emf by a phase angle ϕ. We can write the current, in terms of ϕ, as

$$i = I\cos(\omega t - \phi) \qquad (35.22)$$

Of course, there's no guarantee that V_L will be larger than V_C. If the opposite is true, $V_L < V_C$, the emf phasor is on the other side of the current phasor. Our analysis is still valid if we consider ϕ to be negative when i is ccw from $\mathcal{E}$. Thus ϕ can be anywhere between $-90°$ and $+90°$.

Now we can continue much as we did with the *RC* circuit. Based on the right triangle, $\mathcal{E}_0^2$ is

$$\mathcal{E}_0^2 = V_R^2 + (V_L - V_C)^2 = [R^2 + (X_L - X_C)^2]I^2 \qquad (35.23)$$

where we wrote each of the peak voltages in terms of the peak current I and a resistance or a reactance. Consequently, the peak current in the *RLC* circuit is

$$I = \frac{\mathcal{E}_0}{\sqrt{R^2 + (X_L - X_C)^2}} = \frac{\mathcal{E}_0}{\sqrt{R^2 + (\omega L - 1/\omega C)^2}} \qquad (35.24)$$

The three peak voltages, if you need them, are then found from $V_R = IR$, $V_L = IX_L$, and $V_C = IX_C$.

Impedance

The denominator of Equation 35.24 is called the **impedance** Z of the circuit:

$$Z = \sqrt{R^2 + (X_L - X_C)^2} \qquad (35.25)$$

Impedance, like resistance and reactance, is measured in ohms. The circuit's peak current can be written in terms of the source emf and the circuit impedance as

$$I = \frac{\mathcal{E}_0}{Z} \qquad (35.26)$$

Equation 35.26 is a compact way to write I, but it doesn't add anything new to Equation 35.24.

Phase Angle

It is often useful to know the phase angle ϕ between the emf and the current. You can see from FIGURE 35.17 that

$$\tan \phi = \frac{V_L - V_C}{V_R} = \frac{(X_L - X_C)I}{RI}$$

The current I cancels, and we're left with

$$\phi = \tan^{-1}\left(\frac{X_L - X_C}{R}\right) \qquad (35.27)$$

We can check that Equation 35.27 agrees with our analyses of single-element circuits. A resistor-only circuit has $X_L = X_C = 0$ and thus $\phi = \tan^{-1}(0) = 0$ rad. In other words, as we discovered previously, the emf and current are in phase. An AC inductor circuit has $R = X_C = 0$ and thus $\phi = \tan^{-1}(\infty) = \pi/2$ rad, agreeing with our earlier finding that the inductor current lags the voltage by 90°.

Other relationships can be found from the phasor diagram and written in terms of the phase angle. For example, it is useful to write the peak resistor voltage as

$$V_R = \mathcal{E}_0 \cos \phi \qquad (35.28)$$

Notice that the resistor voltage oscillates in phase with the emf only if $\phi = 0$ rad.

Resonance

Suppose we vary the emf frequency ω while keeping everything else constant. There is very little current at very low frequencies because the capacitive reactance $X_C = 1/\omega C$ is very large. Similarly, there is very little current at very high frequencies because the inductive reactance $X_L = \omega L$ becomes very large.

If I approaches zero at very low and very high frequencies, there should be some intermediate frequency where I is a maximum. Indeed, you can see from Equation 35.24 that the denominator will be a minimum, making I a maximum, when $X_L = X_C$, or

$$\omega L = \frac{1}{\omega C} \qquad (35.29)$$

The frequency ω_0 that satisfies Equation 35.29 is called the **resonance frequency**:

$$\omega_0 = \frac{1}{\sqrt{LC}} \qquad (35.30)$$

This is the frequency for *maximum current* in the series *RLC* circuit. The maximum current

$$I_{max} = \frac{\mathcal{E}_0}{R} \qquad (35.31)$$

is that of a purely resistive circuit because the impedance is $Z = R$ at resonance.

You'll recognize ω_0 as the oscillation frequency of the *LC* circuit we analyzed in Chapter 33. The current in an ideal *LC* circuit oscillates forever as energy is transferred back and forth between the capacitor and the inductor. This is analogous to an ideal, frictionless simple harmonic oscillator in which the energy is transformed back and forth between kinetic and potential.

Adding a resistor to the circuit is like adding damping to a mechanical oscillator. The emf is then a sinusoidal driving force, and the series *RLC* circuit is directly analogous to the driven, damped oscillator that you studied in Chapter 14. A mechanical oscillator exhibits *resonance* by having a large-amplitude response when the driving frequency matches the system's natural frequency. Equation 35.30 is the natural frequency of the series *RLC*

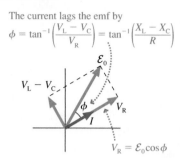

FIGURE 35.17 The current is not in phase with the emf.

The current lags the emf by
$$\phi = \tan^{-1}\left(\frac{V_L - V_C}{V_R}\right) = \tan^{-1}\left(\frac{X_L - X_C}{R}\right)$$

$V_R = \mathcal{E}_0 \cos \phi$

circuit, the frequency at which the current would like to oscillate. Consequently, the circuit has a large current response when the oscillating emf matches this frequency.

FIGURE 35.18 shows the peak current I of a series *RLC* circuit as the emf frequency ω is varied. Notice how the current increases until reaching a maximum at frequency ω_0, then decreases. This is the hallmark of a resonance.

As R decreases, causing the damping to decrease, the maximum current becomes larger and the curve in Figure 35.18 becomes narrower. You saw exactly the same behavior for a driven mechanical oscillator. The emf frequency must be very close to ω_0 in order for a lightly damped system to respond, but the response at resonance is very large.

For a different perspective, **FIGURE 35.19** graphs the instantaneous emf $\mathcal{E} = \mathcal{E}_0 \cos \omega t$ and current $i = I \cos(\omega t - \phi)$ for frequencies below, at, and above ω_0. The current and the emf are in phase at resonance ($\phi = 0$ rad) because the capacitor and inductor essentially cancel each other to give a purely resistive circuit. Away from resonance, the current decreases *and* begins to get out of phase with the emf. You can see, from Equation 35.27, that the phase angle ϕ is negative when $X_L < X_C$ (i.e., the frequency is below resonance) and positive when $X_L > X_C$ (the frequency is above resonance).

Resonance circuits are widely used in radio, television, and communication equipment because of their ability to respond to one particular frequency (or very narrow range of frequencies) while suppressing others. The selectivity of a resonance circuit improves as the resistance decreases, but the inherent resistance of the wires and the inductor coil keeps R from being 0 Ω.

FIGURE 35.18 A graph of the current I versus emf frequency for a series *RLC* circuit.

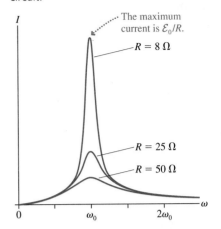

FIGURE 35.19 Graphs of the emf $\mathcal{E}$ and the current i at frequencies below, at, and above the resonance frequency ω_0.

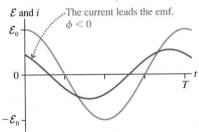

Below resonance: $\omega < \omega_0$

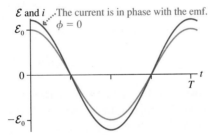

Resonance: $\omega = \omega_0$
Maximum current

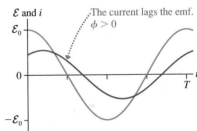

Above resonance: $\omega > \omega_0$

EXAMPLE 35.6 Designing a radio receiver

An AM radio antenna picks up a 1000 kHz signal with a peak voltage of 5.0 mV. The tuning circuit consists of a 60 μH inductor in series with a variable capacitor. The inductor coil has a resistance of 0.25 Ω, and the resistance of the rest of the circuit is negligible.

a. To what value should the capacitor be tuned to listen to this radio station?
b. What is the peak current through the circuit at resonance?
c. A stronger station at 1050 kHz produces a 10 mV antenna signal. What is the current at this frequency when the radio is tuned to 1000 kHz?

MODEL The inductor's 0.25 Ω resistance can be modeled as a resistance in series with the inductance, hence we have a series *RLC* circuit. The antenna signal at $\omega = 2\pi \times 1000$ kHz is the emf.

VISUALIZE The circuit looks like Figure 35.16.

SOLVE a. The capacitor needs to be tuned to where it and the inductor are resonant at $\omega_0 = 2\pi \times 1000$ kHz. The appropriate value is

$$C = \frac{1}{L\omega_0^2} = \frac{1}{(60 \times 10^{-6} \text{ H})(6.28 \times 10^6 \text{ rad/s})^2}$$
$$= 4.2 \times 10^{-10} \text{ F} = 420 \text{ pF}$$

b. $X_L = X_C$ at resonance, so the peak current is

$$I = \frac{\mathcal{E}_0}{R} = \frac{5.0 \times 10^{-3} \text{ V}}{0.25 \ \Omega} = 0.020 \text{ A} = 20 \text{ mA}$$

c. The 1050 kHz signal is "off resonance," so we need to compute $X_L = \omega L = 396 \ \Omega$ and $X_C = 1/\omega C = 361 \ \Omega$ at $\omega = 2\pi \times 1050$ kHz. The peak voltage of this signal is $\mathcal{E}_0 = 10$ mV. With these values, Equation 35.24 for the peak current is

$$I = \frac{\mathcal{E}_0}{\sqrt{R^2 + (X_L - X_C)^2}} = 0.28 \text{ mA}$$

ASSESS These are realistic values for the input stage of an AM radio. You can see that the signal from the 1050 kHz station is strongly suppressed when the radio is tuned to 1000 kHz.

A series *RLC* circuit has $V_C = 5.0$ V, $V_R = 7.0$ V, and $V_L = 9.0$ V. Is the frequency above, below, or equal to the resonance frequency?

35.6 Power in AC Circuits

A primary role of the emf is to supply energy. Some circuit devices, such as motors and lightbulbs, use the energy to perform useful tasks. Other circuit devices dissipate the energy as an increased thermal energy in the components and the surrounding air. Chapter 31 examined the topic of power in DC circuits. Now we can perform a similar analysis for AC circuits.

The emf supplies energy to a circuit at the rate

$$p_{\text{source}} = i\mathcal{E} \tag{35.32}$$

where i and $\mathcal{E}$ are the instantaneous current from and potential difference across the emf. We've used a lowercase p to indicate that this is the instantaneous power. We need to look at the power losses in individual circuit elements.

Resistors

A resistor dissipates energy at the rate

$$p_R = i_R v_R = i_R^2 R \tag{35.33}$$

We can use $i_R = I_R \cos \omega t$ to write the resistor's instantaneous power loss as

$$p_R = i_R^2 R = I_R^2 R \cos^2 \omega t \tag{35.34}$$

FIGURE 35.20 shows the instantaneous power graphically. You can see that, because the cosine is squared, the power oscillates twice during every cycle of the emf. The energy dissipation peaks both when $i_R = I_R$ and when $i_R = -I_R$.

In practice, we're more interested in the *average power* than in the instantaneous power. The **average power** P is the total energy dissipated per second. We can find P_R for a resistor by using the identity $\cos^2(x) = \frac{1}{2}(1 + \cos 2x)$ to write

$$P_R = I_R^2 R \cos^2 \omega t = I_R^2 R \left[\frac{1}{2}(1 + \cos 2\omega t) \right] = \frac{1}{2} I_R^2 R + \frac{1}{2} I_R^2 R \cos 2\omega t$$

The $\cos 2\omega t$ term oscillates positive and negative twice during each cycle of the emf. Its average, over one cycle, is zero. Thus the average power loss in a resistor is

$$P_R = \frac{1}{2} I_R^2 R \qquad \text{(average power loss in a resistor)} \tag{35.35}$$

It is useful to write Equation 35.35 as

$$P_R = \left(\frac{I_R}{\sqrt{2}} \right)^2 R = (I_{\text{rms}})^2 R \tag{35.36}$$

where the quantity

$$I_{\text{rms}} = \frac{I_R}{\sqrt{2}} \tag{35.37}$$

is called the **root-mean-square current,** or rms current, I_{rms}. Technically, an rms quantity is the square root of the average, or mean, of the quantity squared. For a sinusoidal oscillation, the rms value turns out to be the peak value divided by $\sqrt{2}$.

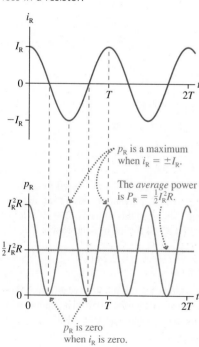

FIGURE 35.20 The instantaneous power loss in a resistor.

p_R is a maximum when $i_R = \pm I_R$.

The *average* power is $P_R = \frac{1}{2} I_R^2 R$.

p_R is zero when i_R is zero.

The rms current allows us to compare Equation 35.36 directly to the energy dissipated by a resistor in a DC circuit: $P = I^2 R$. You can see that the average power loss of a resistor in an AC circuit with $I_{rms} = 1$ A is the same as in a DC circuit with $I = 1$ A. **As far as power is concerned, an rms current is equivalent to an equal DC current.**

Similarly, we can define the root-mean-square voltage and emf:

$$V_{rms} = \frac{V_R}{\sqrt{2}} \qquad \mathcal{E}_{rms} = \frac{\mathcal{E}_0}{\sqrt{2}} \qquad (35.38)$$

The resistor's average power loss in terms of the rms quantities is

$$P_R = (I_{rms})^2 R = \frac{(V_{rms})^2}{R} = I_{rms} V_{rms} \qquad (35.39)$$

and the average power supplied by the emf is

$$P_{source} = I_{rms} \mathcal{E}_{rms} \qquad (35.40)$$

The single-resistor circuit that we analyzed in Section 35.1 had $V_R = \mathcal{E}$ or, equivalently, $V_{rms} = \mathcal{E}_{rms}$. You can see from Equations 35.39 and 35.40 that the power loss in the resistor exactly matches the power supplied by the emf. This must be the case in order to conserve energy.

NOTE ▶ Voltmeters, ammeters, and other AC measuring instruments are calibrated to give the rms value. An AC voltmeter would show that the "line voltage" of an electrical outlet in the United States is 120 V. This is $\mathcal{E}_{rms}$. The peak voltage $\mathcal{E}_0$ is larger by a factor of $\sqrt{2}$, or $\mathcal{E}_0 = 170$ V. The power-line voltage is sometimes specified as "120 V/60 Hz," showing the rms voltage and the frequency. ◀

The power rating on a lightbulb is its average power at $V_{rms} = 120$ V.

EXAMPLE 35.7 **Lighting a bulb**

A 100 W incandescent lightbulb is plugged into a 120 V/60 Hz outlet. What is the resistance of the bulb's filament? What is the peak current through the bulb?

MODEL The filament in a lightbulb acts as a resistor.

VISUALIZE FIGURE 35.21 is a simple one-resistor circuit.

FIGURE 35.21 An AC circuit with a lightbulb as a resistor.

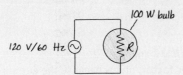

SOLVE A bulb labeled 100 W is designed to dissipate an average 100 W at $V_{rms} = 120$ V. We can use Equation 35.39 to find

$$R = \frac{(V_{rms})^2}{P_R} = \frac{(120 \text{ V})^2}{100 \text{ W}} = 144 \ \Omega$$

The rms current is then found from

$$I_{rms} = \frac{P_R}{V_{rms}} = \frac{100 \text{ W}}{120 \text{ V}} = 0.833 \text{ A}$$

The peak current is $I_R = \sqrt{2} I_{rms} = 1.18$ A.

ASSESS Calculations with rms values are just like the calculations for DC circuits.

Capacitors and Inductors

In Section 35.2, we found that the instantaneous current to a capacitor is $i_C = -\omega C V_C \sin \omega t$. Thus the instantaneous energy dissipation in a capacitor is

$$p_C = v_C i_C = (V_C \cos \omega t)(-\omega C V_C \sin \omega t) = -\frac{1}{2} \omega C V_C^2 \sin 2\omega t \quad (35.41)$$

where we used $\sin(2x) = 2\sin(x)\cos(x)$.

FIGURE 35.22 on the next page shows Equation 35.41 graphically. Energy is transferred into the capacitor (positive power) as it is charged, but, instead of being dissipated, as it would be by a resistor, the energy is stored as potential energy in the capacitor's electric field. Then, as the capacitor discharges, this energy is given back to the circuit. Power is the rate at which energy is *removed* from the circuit, hence p is negative as the capacitor transfers energy back into the circuit.

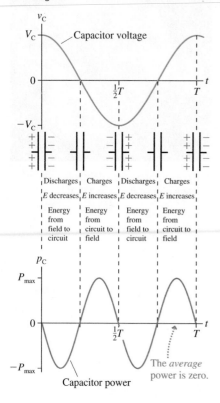

FIGURE 35.22 Energy flows into and out of a capacitor as it is charged and discharged.

Using a mechanical analogy, a capacitor is like an ideal, frictionless simple harmonic oscillator. Kinetic and potential energy are constantly being exchanged, but there is no dissipation because none of the energy is transformed into thermal energy. The important conclusion is that **a capacitor's average power loss is zero:** $P_C = 0$.

The same is true of an inductor. An inductor alternately stores energy in the magnetic field, as the current is increasing, then transfers energy back to the circuit as the current decreases. The instantaneous power oscillates between positive and negative, but **an inductor's average power loss is zero:** $P_L = 0$.

NOTE ▶ We're assuming ideal capacitors and inductors. Real capacitors and inductors inevitably have a small amount of resistance and dissipate a small amount of energy. However, their energy dissipation is negligible compared to that of the resistors in most practical circuits. ◀

The Power Factor

In an *RLC* circuit, energy is supplied by the emf and dissipated by the resistor. But an *RLC* circuit is unlike a purely resistive circuit in that the current is not in phase with the potential difference of the emf.

We found in Equation 35.22 that the instantaneous current in an *RLC* circuit is $i = I\cos(\omega t - \phi)$, where ϕ is the angle by which the current lags the emf. Thus the instantaneous power supplied by the emf is

$$p_{\text{source}} = i\mathcal{E} = (I\cos(\omega t - \phi))(\mathcal{E}_0 \cos\omega t) = I\mathcal{E}_0 \cos\omega t \cos(\omega t - \phi) \quad (35.42)$$

We can use the expression $\cos(x - y) = \cos(x)\cos(y) + \sin(x)\sin(y)$ to write the power as

$$p_{\text{source}} = (I\mathcal{E}_0 \cos\phi)\cos^2\omega t + (I\mathcal{E}_0 \sin\phi)\sin\omega t \cos\omega t \quad (35.43)$$

In our analysis of the power loss in a resistor and a capacitor, we found that the average of $\cos^2\omega t$ is $\frac{1}{2}$ and the average of $\sin\omega t \cos\omega t$ is zero. Thus we can immediately write that the *average* power supplied by the emf is

$$P_{\text{source}} = \frac{1}{2}I\mathcal{E}_0 \cos\phi = I_{\text{rms}}\mathcal{E}_{\text{rms}} \cos\phi \quad (35.44)$$

The rms values, you will recall, are $I/\sqrt{2}$ and $\mathcal{E}_0/\sqrt{2}$.

The term $\cos\phi$, called the **power factor,** arises because the current and the emf in a series *RLC* circuit are not in phase. Because the current and the emf aren't pushing and pulling together, the source delivers less energy to the circuit.

We'll leave it as a homework problem for you to show that the peak current in an *RLC* circuit can be written $I = I_{\text{max}}\cos\phi$, where $I_{\text{max}} = \mathcal{E}_0/R$ was given in Equation 35.31. In other words, the current term in Equation 35.44 is a function of the power factor. Consequently, the average power is

$$P_{\text{source}} = P_{\text{max}}\cos^2\phi \quad (35.45)$$

where $P_{\text{max}} = \frac{1}{2}I_{\text{max}}\mathcal{E}_0$ is the *maximum* power the source can deliver to the circuit.

The source delivers maximum power only when $\cos\phi = 1$. This is the case when $X_L - X_C = 0$, requiring either a purely resistive circuit or an *RLC* circuit operating at the resonance frequency ω_0. The average power loss is zero for a purely capacitive or purely inductive load with, respectively, $\phi = -90°$ or $\phi = +90°$, as found above.

Motors of various types, especially large industrial motors, use a significant fraction of the electric energy generated in industrialized nations. Motors operate most efficiently, doing the maximum work per second, when the power factor is as close to

Industrial motors use a significant fraction of the electric energy generated in the United States.

1 as possible. But motors are inductive devices, due to their electromagnet coils, and if too many motors are attached to the electric grid, the power factor is pulled away from 1. To compensate, the electric company places large capacitors throughout the transmission system. The capacitors dissipate no energy, but they allow the electric system to deliver energy more efficiently by keeping the power factor close to 1.

Finally, we found in Equation 35.28 that the resistor's peak voltage in an *RLC* circuit is related to the emf peak voltage by $V_R = \mathcal{E}_0 \cos \phi$ or, dividing both sides by $\sqrt{2}$, $V_{rms} = \mathcal{E}_{rms} \cos \phi$. We can use this result to write the energy loss in the resistor as

$$P_R = I_{rms} V_{rms} = I_{rms} \mathcal{E}_{rms} \cos \phi \qquad (35.46)$$

But this expression is P_{source}, as we found in Equation 35.44. Thus we see that the energy supplied to an *RLC* circuit by the emf is ultimately dissipated by the resistor.

STOP TO THINK 35.6 The emf and the current in a series *RLC* circuit oscillate as shown. Which of the following (perhaps more than one) would increase the rate at which energy is supplied to the circuit?

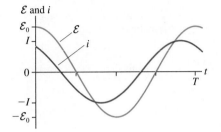

a. Increase $\mathcal{E}_0$
b. Increase L
c. Increase C
d. Decrease $\mathcal{E}_0$
e. Decrease L
f. Decrease C

CHALLENGE EXAMPLE 35.8 **Power in an *RLC* circuit**

An audio amplifier drives a series *RLC* circuit consisting of an 8.0 Ω loudspeaker, a 160 μF capacitor, and a 1.5 mH inductor. The amplifier output is 15.0 V rms at 500 Hz.

a. What power is delivered to the speaker?
b. What maximum power could the amplifier deliver, and how would the capacitor have to be changed for this to happen?

MODEL The emf and voltage of an *RLC* circuit are not in phase, and that affects the power delivered to the circuit. All the power is dissipated by the circuit's resistance, which in this case is the loudspeaker.

VISUALIZE The circuit looks like Figure 35.16.

SOLVE a. The power delivered by the emf is $P_{source} = I_{rms}\mathcal{E}_{rms} \cos \phi$, where ϕ is the phase angle between the emf and the current. In an AC circuit, the current is $I = \mathcal{E}/Z$, where Z is the impedance. To calculate Z, we need the reactances of the capacitor and inductor, and these, in turn, depend on the frequency. At 500 Hz, the angular frequency is $\omega = 2\pi(500 \text{ Hz}) = 3140 \text{ rad/s}$. With this, we can find

$$X_C = \frac{1}{\omega C} = \frac{1}{(3140 \text{ rad/s})(160 \times 10^{-6} \text{ F})} = 1.99 \ \Omega$$

$$X_L = \omega L = (3140 \text{ rad/s})(0.0015 \text{ H}) = 4.71 \ \Omega$$

Now we can calculate the impedance:

$$Z = \sqrt{R^2 + (X_L - X_C)^2} = 8.45 \ \Omega$$

and thus

$$I_{rms} = \frac{\mathcal{E}_{rms}}{Z} = \frac{15.0 \text{ V}}{8.45 \ \Omega} = 1.78 \text{ A}$$

Lastly, we need the phase angle between the emf and the current:

$$\phi = \tan^{-1}\left(\frac{X_L - X_C}{R}\right) = 18.8°$$

The power factor is $\cos(18.8°) = 0.947$, and thus the power delivered by the emf is

$$P_{source} = I_{rms}\mathcal{E}_{rms} \cos \phi = (1.78 \text{ A})(15.0 \text{ V})(0.947) = 25 \text{ W}$$

b. Maximum power is delivered when the current is in phase with the emf, making the power factor 1.00. This occurs when $X_C = X_L$, making the impedance $Z = R = 8.0 \ \Omega$ and the current $I_{rms} = \mathcal{E}_{rms}/R = 1.88 \text{ A}$. Then

$$P_{source} = I_{rms}\mathcal{E}_{rms} \cos \phi = (1.88 \text{ A})(15.0 \text{ V})(1.00) = 28 \text{ W}$$

To deliver maximum power, we need to change the capacitance to make $X_C = X_L = 4.71 \ \Omega$. The required capacitance is

$$C = \frac{1}{(3140 \text{ rad/s})(4.71 \ \Omega)} = 68 \ \mu\text{F}$$

So delivering maximum power requires lowering the capacitance from 160 μF to 68 μF.

ASSESS Changing the capacitor not only increases the power factor, it also increases the current. Both contribute to the higher power.

SUMMARY

The goal of Chapter 35 has been to understand and apply basic techniques of AC circuit analysis.

Important Concepts

AC circuits are driven by an emf

$$\mathcal{E} = \mathcal{E}_0 \cos \omega t$$

that oscillates with angular frequency $\omega = 2\pi f$.

Phasors can be used to represent the oscillating emf, current, and voltage.

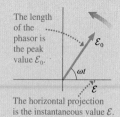

The length of the phasor is the peak value $\mathcal{E}_0$.

The horizontal projection is the instantaneous value $\mathcal{E}$.

Basic circuit elements

Element	i and v	Resistance/ reactance	I and V	Power
Resistor	In phase	R is fixed	$V = IR$	$I_{rms}V_{rms}$
Capacitor	i leads v by 90°	$X_C = 1/\omega C$	$V = IX_C$	0
Inductor	i lags v by 90°	$X_L = \omega L$	$V = IX_L$	0

For many purposes, especially calculating power, the **root-mean-square** (rms) quantities

$$V_{rms} = V/\sqrt{2} \qquad I_{rms} = I/\sqrt{2} \qquad \mathcal{E}_{rms} = \mathcal{E}_0/\sqrt{2}$$

are equivalent to the corresponding DC quantities.

Key Skills

Using phasor diagrams

- Start with a phasor (v or i) common to two or more circuit elements.

- The sum of instantaneous quantities is vector addition.

- Use the Pythagorean theorem to relate peak quantities.

For an RC circuit, shown here,

$$v_R + v_C = \mathcal{E}$$
$$V_R^2 + V_C^2 = \mathcal{E}_0^2$$

Kirchhoff's laws

Loop law The sum of the potential differences around a loop is zero.

Junction law The sum of currents entering a junction equals the sum leaving the junction.

Instantaneous and peak quantities

Instantaneous quantities v and i generally obey different relationships than peak quantities V and I.

Applications

RC filter circuits

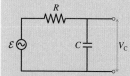

$$V_C = \frac{\mathcal{E}_0 X_C}{\sqrt{R^2 + X_C^2}}$$

$$V_C \rightarrow \mathcal{E}_0 \text{ as } \omega \rightarrow 0$$

A **low-pass filter** transmits low frequencies and blocks high frequencies.

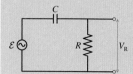

$$V_R = \frac{\mathcal{E}_0 R}{\sqrt{R^2 + X_C^2}}$$

$$V_R \rightarrow \mathcal{E}_0 \text{ as } \omega \rightarrow \infty$$

A **high-pass filter** transmits high frequencies and blocks low frequencies.

Series RLC circuits

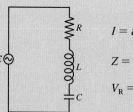

$I = \mathcal{E}_0/Z$ where Z is the **impedance**

$$Z = \sqrt{R^2 + (X_L - X_C)^2}$$

$$V_R = IR \qquad V_L = IX_L \qquad V_C = IX_C$$

When $\omega = \omega_0 = 1/\sqrt{LC}$ (the **resonance frequency**), the current in the circuit is a maximum $I_{max} = \mathcal{E}_0/R$.

In general, the current i lags behind $\mathcal{E}$ by the **phase angle** $\phi = \tan^{-1}\big((X_L - X_C)/R\big)$.

The power supplied by the emf is $P_{source} = I_{rms}\mathcal{E}_{rms}\cos\phi$, where $\cos\phi$ is called the **power factor.**

The power lost in the resistor is $P_R = I_{rms}V_{rms} = (I_{rms})^2 R$.

Terms and Notation

AC circuit	crossover frequency, ω_c	series *RLC* circuit	root-mean-square current, I_{rms}
DC circuit	low-pass filter	impedance, Z	power factor, $\cos\phi$
phasor	high-pass filter	resonance frequency, ω_0	
capacitive reactance, X_C	inductive reactance, X_L	average power, P	

CONCEPTUAL QUESTIONS

1. **FIGURE Q35.1** shows emf phasors a, b, and c.
   a. For each, what is the instantaneous value of the emf?
   b. At this instant, is the magnitude of each emf increasing, decreasing, or holding constant?

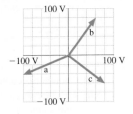

FIGURE Q35.1

2. A resistor is connected across an oscillating emf. The peak current through the resistor is 2.0 A. What is the peak current if:
   a. The resistance R is doubled?
   b. The peak emf $\mathcal{E}_0$ is doubled?
   c. The frequency ω is doubled?

3. A capacitor is connected across an oscillating emf. The peak current through the capacitor is 2.0 A. What is the peak current if:
   a. The capacitance C is doubled?
   b. The peak emf $\mathcal{E}_0$ is doubled?
   c. The frequency ω is doubled?

4. A low-pass *RC* filter has a crossover frequency $f_c = 200$ Hz. What is f_c if:
   a. The resistance R is doubled?
   b. The capacitance C is doubled?
   c. The peak emf $\mathcal{E}_0$ is doubled?

5. An inductor is connected across an oscillating emf. The peak current through the inductor is 2.0 A. What is the peak current if:
   a. The inductance L is doubled?
   b. The peak emf $\mathcal{E}_0$ is doubled?
   c. The frequency ω is doubled?

6. The resonance frequency of a series *RLC* circuit is 1000 Hz. What is the resonance frequency if:
   a. The resistance R is doubled?
   b. The inductance L is doubled?
   c. The capacitance C is doubled?
   d. The peak emf $\mathcal{E}_0$ is doubled?

7. In the series *RLC* circuit represented by the phasors of **FIGURE Q35.7**, is the emf frequency less than, equal to, or greater than the resonance frequency ω_0? Explain.

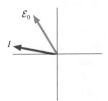

FIGURE Q35.7

8. The resonance frequency of a series *RLC* circuit is less than the emf frequency. Does the current lead or lag the emf? Explain.

9. The current in a series *RLC* circuit lags the emf by 20°. You cannot change the emf. What two different things could you do to the circuit that would increase the power delivered to the circuit by the emf?

10. The average power dissipated by a resistor is 4.0 W. What is P_R if:
    a. The resistance R is doubled while $\mathcal{E}_0$ is held fixed?
    b. The peak emf $\mathcal{E}_0$ is doubled while R is held fixed?
    c. Both are doubled simultaneously?

EXERCISES AND PROBLEMS

Problems labeled [ ] integrate material from earlier chapters.

Exercises

Section 35.1 AC Sources and Phasors

1. | The emf phasor in **FIGURE EX35.1** is shown at $t = 15$ ms.
   a. What is the angular frequency ω? Assume this is the first rotation.
   b. What is the instantaneous value of the emf?

FIGURE EX35.1

2. ‖ The emf phasor in **FIGURE EX35.2** is shown at $t = 2.0$ ms.
   a. What is the angular frequency ω? Assume this is the first rotation.
   b. What is the peak value of the emf?

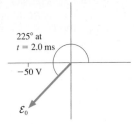

FIGURE EX35.2

3. | A 110 Hz source of emf has a peak voltage of 50 V. Draw the emf phasor at $t = 3.0$ ms.

4. ‖ Draw the phasor for the emf $\mathcal{E} = (170\ \text{V})\cos\big((2\pi \times 60\ \text{Hz})t\big)$ at $t = 60$ ms.

5. | A 200 Ω resistor is connected to an AC source with $\mathcal{E}_0 = 10$ V. What is the peak current through the resistor if the emf frequency is (a) 100 Hz? (b) 100 kHz?

6. | FIGURE EX35.6 shows voltage and current graphs for a resistor.
   a. What is the emf frequency f?
   b. What is the value of the resistance R?
   c. Draw the resistor's voltage and current phasors at $t = 15$ ms.

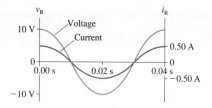

FIGURE EX35.6

Section 35.2 Capacitor Circuits

7. | A 0.30 μF capacitor is connected across an AC generator that produces a peak voltage of 10 V. What is the peak current to and from the capacitor if the emf frequency is (a) 100 Hz? (b) 100 kHz?

8. | The peak current to and from a capacitor is 10 mA. What is the peak current if
   a. The emf frequency is doubled?
   b. The emf peak voltage is doubled (at the original frequency)?

9. | A 20 nF capacitor is connected across an AC generator that produces a peak voltage of 5.0 V.
   a. At what frequency f is the peak current 50 mA?
   b. What is the instantaneous value of the emf at the instant when $i_C = I_C$?

10. ‖ A capacitor is connected to a 15 kHz oscillator. The peak current is 65 mA when the rms voltage is 6.0 V. What is the value of the capacitance C?

11. | A capacitor has a peak current of 330 μA when the peak voltage at 250 kHz is 2.2 V.
    a. What is the capacitance?
    b. If the peak voltage is held constant, what is the peak current at 500 kHz?

Section 35.3 RC Filter Circuits

12. | A high-pass RC filter is connected to an AC source with a peak voltage of 10.0 V. The peak capacitor voltage is 6.0 V. What is the resistor voltage?

13. | A high-pass RC filter with a crossover frequency of 1000 Hz uses a 100 Ω resistor. What is the value of the capacitor?

14. | A low-pass RC filter with a crossover frequency of 1000 Hz uses a 100 Ω resistor. What is the value of the capacitor?

15. ‖ What are V_R and V_C if the emf frequency in FIGURE EX35.15 is 10 kHz?

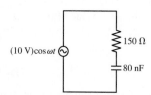

(10 V)cos ωt

150 Ω

80 nF

FIGURE EX35.15

16. | A low-pass filter consists of a 100 μF capacitor in series with a 159 Ω resistor. The circuit is driven by an AC source with a peak voltage of 5.00 V.

a. What is the crossover frequency f_c?
b. What is V_C when $f = \frac{1}{2}f_c$, f_c, and $2f_c$?

17. | A high-pass filter consists of a 1.59 μF capacitor in series with a 100 Ω resistor. The circuit is driven by an AC source with a peak voltage of 5.00 V.
   a. What is the crossover frequency f_c?
   b. What is V_R when $f = \frac{1}{2}f_c$, f_c, and $2f_c$?

Section 35.4 Inductor Circuits

18. | The peak current through an inductor is 10 mA. What is the peak current if
   a. The emf frequency is doubled?
   b. The emf peak voltage is doubled (at the original frequency)?

19. | A 20 mH inductor is connected across an AC generator that produces a peak voltage of 10 V. What is the peak current through the inductor if the emf frequency is (a) 100 Hz? (b) 100 kHz?

20. | An inductor is connected to a 15 kHz oscillator. The peak current is 65 mA when the rms voltage is 6.0 V. What is the value of the inductance L?

21. | A 500 μH inductor is connected across an AC generator that produces a peak voltage of 5.0 V.
   a. At what frequency f is the peak current 50 mA?
   b. What is the instantaneous value of the emf at the instant when $i_L = I_L$?

22. ‖ An inductor has a peak current of 330 μA when the peak voltage at 45 MHz is 2.2 V.
   a. What is the inductance?
   b. If the peak voltage is held constant, what is the peak current at 90 MHz?

Section 35.5 The Series RLC Circuit

23. | A series RLC circuit has a 200 kHz resonance frequency. What is the resonance frequency if
   a. The resistor value is doubled?
   b. The capacitor value is doubled?

24. | A series RLC circuit has a 200 kHz resonance frequency. What is the resonance frequency if the capacitor value is doubled and, at the same time, the inductor value is halved?

25. ‖ What capacitor in series with a 100 Ω resistor and a 20 mH inductor will give a resonance frequency of 1000 Hz?

26. | What inductor in series with a 100 Ω resistor and a 2.5 μF capacitor will give a resonance frequency of 1000 Hz?

27. | A series RLC circuit consists of a 50 Ω resistor, a 3.3 mH inductor, and a 480 nF capacitor. It is connected to an oscillator with a peak voltage of 5.0 V. Determine the impedance, the peak current, and the phase angle at frequencies (a) 3000 Hz, (b) 4000 Hz, and (c) 5000 Hz.

28. ‖ At what frequency f do a 1.0 μF capacitor and a 1.0 μH inductor have the same reactance? What is the value of the reactance at this frequency?

Section 35.6 Power in AC Circuits

29. | The heating element of a hair drier dissipates 1500 W when connected to a 120 V/60 Hz power line. What is its resistance?

30. ‖ A resistor dissipates 2.0 W when the rms voltage of the emf is 10.0 V. At what rms voltage will the resistor dissipate 10.0 W?

31. ‖ For what absolute value of the phase angle does a source deliver 75% of the maximum possible power to an RLC circuit?

32. ‖ The motor of an electric drill draws a 3.5 A rms current at the power-line voltage of 120 V rms. What is the motor's power if the current lags the voltage by 20°?

33. ‖ A series *RLC* circuit attached to a 120 V/60 Hz power line draws a 2.4 A rms current with a power factor of 0.87. What is the value of the resistor?

34. ‖ A series *RLC* circuit with a 100 Ω resistor dissipates 80 W when attached to a 120 V/60 Hz power line. What is the power factor?

Problems

35. ‖ a. For an *RC* circuit, find an expression for the angular frequency at which $V_R = \frac{1}{2}\mathcal{E}_0$.
   b. What is V_C at this frequency?

36. ‖ a. For an *RC* circuit, find an expression for the angular frequency at which $V_C = \frac{1}{2}\mathcal{E}_0$.
   b. What is V_R at this frequency?

37. ‖ a. Evaluate V_C in FIGURE P35.37 at emf frequencies 1, 3, 10, 30, and 100 kHz.
   b. Graph V_C versus frequency. Draw a smooth curve through your five points.

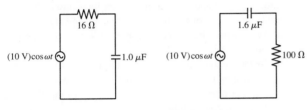

FIGURE P35.37   **FIGURE P35.38**

38. ‖ a. Evaluate V_R in FIGURE P35.38 at emf frequencies 100, 300, 1000, 3000, and 10,000 Hz.
   b. Graph V_R versus frequency. Draw a smooth curve through your five points.

39. ‖ For an *RC* filter circuit, show that $V_R = V_C = \mathcal{E}_0/\sqrt{2}$ at $\omega = \omega_c$.

40. ‖ When two capacitors are connected in parallel across a 10.0 V rms, 1.00 kHz oscillator, the oscillator supplies a total rms current of 545 mA. When the same two capacitors are connected to the oscillator in series, the oscillator supplies an rms current of 126 mA. What are the values of the two capacitors?

41. ‖ Show that Equation 35.27 for the phase angle ϕ of a series *RLC* circuit gives the correct result for a capacitor-only circuit.

42. ‖ a. What is the peak current supplied by the emf in FIGURE P35.42?
   b. What is the peak voltage across the 3.0 μF capacitor?

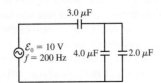

FIGURE P35.42

43. ‖ You have a resistor and a capacitor of unknown values. First, you charge the capacitor and discharge it through the resistor. By monitoring the capacitor voltage on an oscilloscope, you see that the voltage decays to half its initial value in 2.5 ms. You then use the resistor and capacitor to make a low-pass filter. What is the crossover frequency f_c?

44. ‖ FIGURE P35.44 shows a parallel *RC* circuit.
   a. Use a phasor-diagram analysis to find expressions for the peak currents I_R and I_C.
   Hint: What do the resistor and capacitor have in common? Use that as the initial phasor.
   b. Complete the phasor analysis by finding an expression for the peak emf current I.

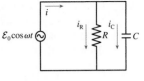

FIGURE P35.44

45. ‖ FIGURE P35.45 shows voltage and current graphs for a capacitor.
   a. What is the emf frequency f?
   b. What is the value of the capacitance C?

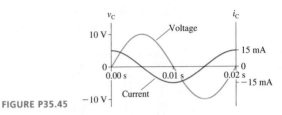

FIGURE P35.45

46. ‖ FIGURE P35.46 shows voltage and current graphs for an inductor.
   a. What is the emf frequency f?
   b. What is the value of the inductance L?

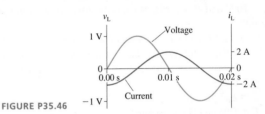

FIGURE P35.46

47. ‖ Use a phasor diagram to analyze the *RL* circuit of FIGURE P35.47. In particular,
   a. Find expressions for I, V_R, and V_L.
   b. What is V_R in the limits $\omega \to 0$ and $\omega \to \infty$?
   c. If the output is taken from the resistor, is this a low-pass or a high-pass filter? Explain.
   d. Find an expression for the crossover frequency ω_c.

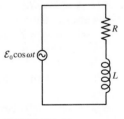

FIGURE P35.47

48. ‖ A series *RLC* circuit consists of a 100 Ω resistor, a 0.15 H inductor, and a 30 μF capacitor. It is attached to a 120 V/60 Hz power line. What are (a) the emf $\mathcal{E}_{rms}$, (b) the phase angle ϕ, and (c) the average power loss?

49. ‖ A series *RLC* circuit consists of a 25 Ω resistor, a 0.10 H inductor, and a 100 μF capacitor. It draws a 2.5 A rms current when attached to a 60 Hz source. What are (a) the emf $\mathcal{E}_{rms}$, (b) the phase angle ϕ, and (c) the average power loss?

50. ‖ For the circuit of FIGURE P35.50,
   a. What is the resonance frequency, in both rad/s and Hz?
   b. Find V_R and V_L at resonance.
   c. How can V_L be larger than $\mathcal{E}_0$? Explain.

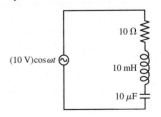

FIGURE P35.50

51. | For the circuit of FIGURE P35.51,
    a. What is the resonance frequency, in both rad/s and Hz?
    b. Find V_R and V_C at resonance.
    c. How can V_C be larger than $\mathcal{E}_0$? Explain.

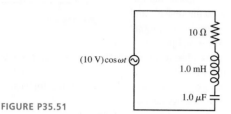

FIGURE P35.51

52. ‖ In FIGURE P35.52, what is the current supplied by the emf when (a) the frequency is very small and (b) the frequency is very large?

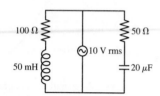

FIGURE P35.52

53. ‖ The current lags the emf by 30° in a series *RLC* circuit with $\mathcal{E}_0 = 10$ V and $R = 50$ Ω. What is the peak current through the circuit?

54. ‖ A series *RLC* circuit consists of a 50 Ω resistor, a 3.3 mH inductor, and a 480 nF capacitor. It is connected to a 5.0 kHz oscillator with a peak voltage of 5.0 V. What is the instantaneous current i when
    a. $\mathcal{E} = \mathcal{E}_0$?
    b. $\mathcal{E} = 0$ V and is decreasing?

55. ‖ A series *RLC* circuit consists of a 50 Ω resistor, a 3.3 mH inductor, and a 480 nF capacitor. It is connected to a 3.0 kHz oscillator with a peak voltage of 5.0 V. What is the instantaneous emf $\mathcal{E}$ when
    a. $i = I$?
    b. $i = 0$ A and is decreasing?
    c. $i = -I$?

56. ‖ A series *RLC* circuit consists of a 100 Ω resistor, a 10 mH inductor, and a 1.0 nF capacitor. It is connected to an oscillator with an rms voltage of 10 V. What is the power supplied to the circuit if (a) $\omega = \frac{1}{2}\omega_0$? (b) $\omega = \omega_0$? (c) $\omega = 2\omega_0$?

57. ‖ Show that the power factor of a series *RLC* circuit is $\cos\phi = R/Z$.

58. ‖ For a series *RLC* circuit, show that
    a. The peak current can be written $I = I_{max}\cos\phi$.
    b. The average power can be written $P_{source} = P_{max}\cos^2\phi$.

59. ‖ The tuning circuit in an FM radio receiver is a series *RLC* circuit with a 0.200 μH inductor.
    a. The receiver is tuned to a station at 104.3 MHz. What is the value of the capacitor in the tuning circuit?
    b. FM radio stations are assigned frequencies every 0.2 MHz, but two nearby stations cannot use adjacent frequencies. What is the maximum resistance the tuning circuit can have if the peak current at a frequency of 103.9 MHz, the closest frequency that can be used by a nearby station, is to be no more than 0.10% of the peak current at 104.3 MHz? The radio is still tuned to 104.3 MHz, and you can assume the two stations have equal strength.

60. ‖ A television channel is assigned the frequency range from 54 MHz to 60 MHz. A series *RLC* tuning circuit in a TV receiver resonates in the middle of this frequency range. The circuit uses a 16 pF capacitor.
    a. What is the value of the inductor?
    b. In order to function properly, the current throughout the frequency range must be at least 50% of the current at the resonance frequency. What is the minimum possible value of the circuit's resistance?

61. ‖ Lightbulbs labeled 40 W, 60 W, and 100 W are connected to a 120 V/60 Hz power line as shown in FIGURE P35.61. What is the rate at which energy is dissipated in each bulb?

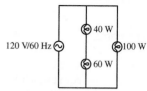

FIGURE P35.61

62. ‖ Commercial electricity is generated and transmitted as *three-phase electricity*. Instead of a single emf, three separate wires carry currents for the emfs $\mathcal{E}_1 = \mathcal{E}_0\cos\omega t$, $\mathcal{E}_2 = \mathcal{E}_0\cos(\omega t + 120°)$, and $\mathcal{E}_3 = \mathcal{E}_0\cos(\omega t - 120°)$ over three parallel wires, each of which supplies one-third of the power. This is why the long-distance transmission lines you see in the countryside have three wires. Suppose the transmission lines into a city supply a total of 450 MW of electric power, a realistic value.
    a. What would be the current in each wire if the transmission voltage were $\mathcal{E}_0 = 120$ V rms?
    b. In fact, transformers are used to step the transmission-line voltage up to 500 kV rms. What is the current in each wire?
    c. Big transformers are expensive. Why does the electric company use step-up transformers?

63. ‖ Commercial electricity is generated and transmitted as *three-phase electricity*. Instead of a single emf $\mathcal{E} = \mathcal{E}_0\cos\omega t$, three separate wires carry currents for the emfs $\mathcal{E}_1 = \mathcal{E}_0\cos\omega t$, $\mathcal{E}_2 = \mathcal{E}_0\cos(\omega t + 120°)$, and $\mathcal{E}_3 = \mathcal{E}_0\cos(\omega t - 120°)$. This is why the long-distance transmission lines you see in the countryside have three parallel wires, as do many distribution lines within a city.
    a. Draw a phasor diagram showing phasors for all three phases of a three-phase emf.
    b. Show that the sum of the three phases is zero, producing what is referred to as *neutral*. In *single-phase* electricity, provided by the familiar 120 V/60 Hz electric outlets in your home, one side of the outlet is neutral, as established at a nearby electrical substation. The other, called the *hot side*, is one of the three phases. (The round opening is connected to ground.)
    c. Show that the potential difference between any two of the phases has the rms value $\sqrt{3}\,\mathcal{E}_{rms}$, where $\mathcal{E}_{rms}$ is the familiar single-phase rms voltage. Evaluate this potential difference for $\mathcal{E}_{rms} = 120$ V. Some high-power home appliances, especially electric clothes dryers and hot-water heaters, are designed to operate between two of the phases rather than between one phase and neutral. Heavy-duty industrial motors are designed to operate from all three phases, but full three-phase power is rare in residential or office use.

64. ‖ A motor attached to a 120 V/60 Hz power line draws an 8.0 A current. Its average energy dissipation is 800 W.
    a. What is the power factor?
    b. What is the rms resistor voltage?
    c. What is the motor's resistance?
    d. How much series capacitance needs to be added to increase the power factor to 1.0?

Challenge Problems

65. The small transformers that power many consumer products produce a 12.0 V rms, 60 Hz emf. Design a circuit using resistors and capacitors that uses the transformer voltage as an input and produces a 6.0 V rms output that leads the input voltage by 45°.

66. FIGURE CP35.66 shows voltage and current graphs for a series *RLC* circuit.
    a. What is the resistance *R*?
    b. If $L = 200\ \mu H$, what is the resonance frequency?

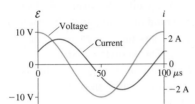

FIGURE CP35.66

67. You're the operator of a 15,000 V rms, 60 Hz electrical substation. When you get to work one day, you see that the station is delivering 6.0 MW of power with a power factor of 0.90.
    a. What is the rms current leaving the station?
    b. How much series capacitance should you add to bring the power factor up to 1.0?
    c. How much power will the station then be delivering?

68. a. Show that the average power loss in a series *RLC* circuit is

$$P_{avg} = \frac{\omega^2 \mathcal{E}_{rms}^2 R}{\omega^2 R^2 + L^2(\omega^2 - \omega_0^2)^2}$$

    b. Prove that the energy dissipation is a maximum at $\omega = \omega_0$.

69. a. Show that the peak inductor voltage in a series *RLC* circuit is maximum at frequency

$$\omega_L = \left(\frac{1}{\omega_0^2} - \frac{1}{2}R^2 C^2\right)^{-1/2}$$

    b. A series *RLC* circuit with $\mathcal{E}_0 = 10.0$ V consists of a 1.0 Ω resistor, a 1.0 μH inductor, and a 1.0 μF capacitor. What is V_L at $\omega = \omega_0$ and at $\omega = \omega_L$?

70. The telecommunication circuit shown in FIGURE CP35.70 has a parallel inductor and capacitor in series with a resistor.
    a. Use a phasor diagram to show that the peak current through the resistor is

$$I = \frac{\mathcal{E}_0}{\sqrt{R^2 + \left(\dfrac{1}{X_L} - \dfrac{1}{X_C}\right)^{-2}}}$$

    Hint: Start with the inductor phasor v_L.
    b. What is *I* in the limits $\omega \to 0$ and $\omega \to \infty$?
    c. What is the resonance frequency ω_0? What is *I* at this frequency?

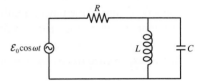

FIGURE CP35.70

71. Consider the parallel *RLC* circuit shown in FIGURE CP35.71.
    a. Show that the current drawn from the emf is

$$I = \mathcal{E}_0 \sqrt{\frac{1}{R^2} + \left(\frac{1}{\omega L} - \omega C\right)^2}$$

    Hint: Start with a phasor that is common to all three circuit elements.
    b. What is *I* in the limits $\omega \to 0$ and $\omega \to \infty$?
    c. Find the frequency for which *I* is a minimum.
    d. Sketch a graph of *I* versus ω.

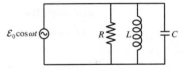

FIGURE CP35.71

STOP TO THINK ANSWERS

Stop to Think 35.1: a. The instantaneous emf value is the projection down onto the horizontal axis. The emf is negative but increasing in magnitude as the phasor, which rotates ccw, approaches the horizontal axis.

Stop to Think 35.2: c. Voltage and current are measured using different scales and units. You can't compare the length of a voltage phasor to the length of a current phasor.

Stop to Think 35.3: a. There is "no capacitor" when the separation between the two capacitor plates becomes zero and the plates touch. Capacitance *C* is inversely proportional to the plate spacing *d*, hence $C \to \infty$ as $d \to 0$. The capacitive reactance is inversely proportional to *C*, so $X_C \to 0$ as $C \to \infty$.

Stop to Think 35.4: $(\omega_c)_d > (\omega_c)_c = (\omega_c)_a > (\omega_c)_b$. The crossover frequency is $1/RC$.

Stop to Think 35.5: Above. $V_L > V_C$ tells us that $X_L > X_C$. This is the condition above resonance, where X_L is increasing with ω while X_C is decreasing.

Stop to Think 35.6: a, b, and **f.** You can always increase power by turning up the voltage. The current leads the emf, telling us that the circuit is primarily capacitive. The current can be brought into phase with the emf, thus maximizing the power, by decreasing *C* or increasing *L*.

VI Electricity and Magnetism

Mass and charge are the two most fundamental properties of matter. The first five parts of this text were investigations of the properties and interactions of masses. Part VI has been a study of the physics of charge—what charge is and how charges interact.

Electric and magnetic fields were introduced to enable us to understand the long-range forces of electricity and magnetism. The field concept is subtle, but it is an essential part of our modern understanding of the physical universe. One charge—the source charge—alters the space around it by creating an electric field and, if the charge is moving, a magnetic field. Other charges experience forces exerted *by the fields*. Thus the

electric and magnetic fields are the agents by which charges interact.

Faraday's discovery of electromagnetic induction led scientists to recognize that the fields are *real* and can exist independently of charges. The most vivid confirmation of this reality was Maxwell's discovery of electromagnetic waves—the quintessential electromagnetic phenomenon.

Part VI has introduced many new phenomena, concepts, and laws. The knowledge structure table draws together the major ideas about charges and fields, and it briefly summarizes some of the most important applications of electricity and magnetism.

KNOWLEDGE STRUCTURE VI **Electricity and Magnetism**

ESSENTIAL CONCEPTS Charge, dipole, field, potential, emf
BASIC GOALS How do charged particles interact?
What are the properties and characteristics of electromagnetic fields?

GENERAL PRINCIPLES

Coulomb's law
$$\vec{E}_{\text{point charge}} = \frac{1}{4\pi\epsilon_0}\frac{q}{r^2}\hat{r} = \left(\frac{1}{4\pi\epsilon_0}\frac{q}{r^2}, \text{away from } q\right)$$

Biot-Savart law
$$\vec{B}_{\text{point charge}} = \frac{\mu_0}{4\pi}\frac{q\vec{v}\times\hat{r}}{r^2} = \left(\frac{\mu_0}{4\pi}\frac{qv\sin\theta}{r^2}, \text{direction of right-hand rule}\right)$$

Faraday's law $\mathcal{E} = |d\Phi_m/dt|$ $I_{\text{induced}} = \mathcal{E}/R$ in the direction of Lenz's law

Lenz's law An induced current flows around a conducting loop in the direction such that the induced magnetic field opposes the *change* in the magnetic flux.

Lorentz force law $\vec{F}_{\text{on } q} = q(\vec{E} + \vec{v}\times\vec{B})$

Superposition The electric or magnetic field due to multiple charges is the vector sum of the field of each charge. This principle was used to derive the fields of many special charge distributions, such as wires, planes, and loops.

FIELD AND POTENTIAL The electric field of charges can also be described in terms of an electric potential *V*:

$$V_{\text{point charge}} = \frac{q}{4\pi\epsilon_0 r}$$

- The electric field is perpendicular to equipotential surfaces and in the direction of decreasing potential.
- The potential energy of charge *q* is $U = qV$. The total energy $K + qV$ of a group of charges is conserved.

ELECTROMAGNETIC WAVES All the properties of electromagnetic fields are summarized mathematically in four equations called *Maxwell's equations*. From Maxwell's equations we learn that electromagnetic fields can exist independently of charges as an *electromagnetic wave*.

- An em wave travels at speed $c = 1/\sqrt{\epsilon_0\mu_0}$.
- $\vec{E}$ and $\vec{B}$ are perpendicular to each other and to the direction of travel, with $E = cB$.

Electric and magnetic properties of materials

- Charges move through conductors but not through insulators.
- Conductors and insulators are *polarized* in an electric field.
- A magnetic moment in a magnetic field experiences a torque.

Applications to circuits

- Circuits obey Kirchhoff's loop law (conservation of energy) and junction law (conservation of current).
- Resistors control the current: $I = \Delta V/R$ (Ohm's law).
- Capacitors store charge $Q = C\Delta V$ and energy $V_C = \frac{1}{2}C(\Delta V_C)^2$.

Model of current and conductivity

- The charge carriers in metals are electrons.
- emf $\rightarrow$ electric field $\rightarrow$ current density $J = \sigma E \rightarrow I = JA$

The Telecommunications Revolution

In 1800, the year that Alessandro Volta invented the battery and Thomas Jefferson was elected president, the fastest a message could travel was the speed of a man or woman on horseback. News took three days to travel from New York to Boston, and well over a month to reach the frontier outpost of Cincinnati.

But Hans Oersted's 1820 discovery that a current creates a magnetic field introduced revolutionary changes to communications. The American scientist Joseph Henry, who shares with Faraday credit for the discovery of electromagnetic induction, saw a simple electromagnet in 1825. Inspired, he set about improving the device. By 1830, Henry was able to send current through more than a mile of wire to activate an electromagnet and strike a bell.

In 1835, Henry met an entrepreneur interested in the commercial development of electric technology—Samuel F. B. Morse. Morse was one of the most prominent American artists of the early 19th century, but he also had an abiding interest in technology. In the 1830s, he invented the famous code that bears his name—Morse code—and began to experiment with electromagnets.

With advice and encouragement from Henry, Morse developed the first practical telegraph. The first telegraph line, between Washington, D.C., and Baltimore, began operating in 1844; the first message sent was "What hath God wrought?" For the first time, long-distance communications could take place essentially instantaneously.

Telegraph communication advanced as quickly as wire could be strung, and a worldwide network had been established by 1875. But the telegraph didn't hold its monopoly for long, as other inventors began to think about using electromagnetic devices to transmit speech. The first to succeed was Alexander Graham Bell, who invented the telephone in 1876.

The telegraph and telephone provided electromagnetic communication over wires, but the discovery of electromagnetic waves opened up another possibility—wireless communication at the speed of light. Radio technology developed rapidly in the late 19th century, and in 1901 the Italian inventor Guglielmo Marconi sent and received the first transatlantic radio message. World War I prompted further development of radio, because of the need to communicate with military units as they moved about, and by 1925 more than 1000 radio stations were operating in the United States.

Radio and, later, television spanned the globe by 1960, but radio stations reached a few hundred miles at best, and television transmission was limited to each city. National broadcasts within the United States required the signal to be transmitted via microwave relays to local stations for rebroadcast. Network television shows were possible, but not live-from-the-scene broadcasts. Journalists had to film events, then return the film to the studio for broadcast. Television images from overseas could only be seen the next day, after film was flown back to the United States.

The first communications satellite was launched by NASA in 1960, followed two years later by a more practical satellite, Telstar, that used solar power to amplify signals received from earth and beam them back down. The first live transatlantic television transmission was made on July 11, 1962, and was broadcast throughout the United States.

Plans were made for a system of roughly 100 satellites, so that one would always be overhead, but another idea soon proved more practical. In 1945, 12 years before space flight began, the science-fiction writer Arthur C. Clarke proposed placing satellites in orbits 22,300 miles above the earth. A satellite at this altitude orbits with a 24-hour period, so from the ground it appears to hang stationary in space. We now call this a *geosynchronous orbit*. One such satellite would allow microwave communication between two points one-third of a world apart, so just three geosynchronous satellites would span the entire earth.

Much more energy is required to reach geosynchronous orbit than to reach low-earth orbit, but rocket technology was advancing faster than NASA could build Telstar satellites. The first commercial communications satellite was placed in geosynchronous orbit in 1965, and, for the first time, television images could be broadcast live to anywhere in the world. Today all of the world's intercontinental television and much of the intercontinental telephone traffic travel via microwaves to and from a cluster of these artificial stars floating high above the earth.

Today, in the 21st century, information and images span the world as quickly as or more quickly than they once moved through a small village. You can talk to friends or relatives anywhere around the globe, and each day's news brings live images from remote places. Telecommunication unites our world, and the technologies of telecommunications are direct descendants of Coulomb, Ampère, Oersted, Henry, and—most of all—Michael Faraday.

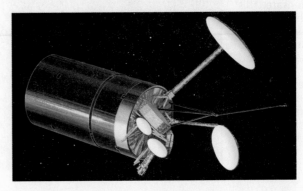

This INTELSAT telecommunications satellite is 12 m (40 ft) long.

ONE STEP BEYOND

The Telecommunications Revolution

PART VII
Relativity and Quantum Physics

This three-frame sequence shows a gas of a few thousand rubidium atoms condensing into a single quantum state known as a Bose-Einstein condensate. This phenomenon was predicted by Einstein in 1925 but not observed until 1995, when physicists learned how to use lasers to cool the atoms to temperatures below 200 nanokelvin.

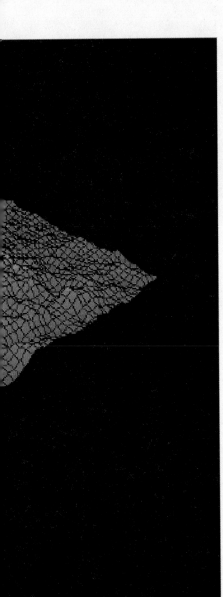

OVERVIEW

Contemporary Physics

Our journey into physics is nearing its end. We began roughly 350 years ago with Newton's discovery of the laws of motion. Part VI brought us to the end of the 19th century, just over 100 years ago. Along the way you've learned about the motion of particles, the conservation of energy, the physics of waves, and the electromagnetic interactions that hold atoms together and generate light waves. We begin the last phase of our journey with confidence.

Newton's mechanics and Maxwell's electromagnetism were the twin pillars of science at the end of the 19th century and the basis for much of engineering and applied science in the 20th century. Despite the successes of these theories, a series of discoveries starting around 1900 and continuing into the first few decades of the 20th century profoundly altered our understanding of the universe at the most fundamental level.

- Einstein's theory of relativity forced scientists to completely revise their concepts of space and time. Our exploration of these fascinating ideas will end with perhaps the most famous equation in physics: Einstein's $E = mc^2$.
- Experimenters found that the classical distinction between *particles* and *waves* breaks down at the atomic level. Light sometimes acts like a particle, while electrons and even entire atoms sometimes act like waves. We will need a new theory of light and matter—quantum physics—to explain these phenomena.

These two theories form the basis for physics as it is practiced today, and they are now having a significant impact on 21st-century engineering.

The complete theory of quantum physics, as it was developed in the 1920s, describes atomic particles in terms of an entirely new concept called a *wave function*. One of our most important tasks in Part VII will be to learn what a wave function is, what laws govern its behavior, and how to relate wave functions to experimental measurements. We will concentrate on one-dimensional models that, while not perfect, will be adequate for understanding the essential features of scanning tunneling microscopes, various semiconductor devices, radioactive decay, and other applications.

We'll complete our study of quantum physics with an introduction to atomic and nuclear physics. You will learn where the electron-shell model of chemistry comes from, how atoms emit and absorb light, what's inside the nucleus, and why some nuclei undergo radioactive decay.

The quantum world with its wave functions and probabilities can seem strange and mysterious, yet quantum physics gives the most definitive and accurate predictions of any physical theory ever devised. The contemporary perspective of quantum physics will be a fitting end to our journey.

NOTE ▶ This edition of *Physics for Scientists and Engineers* contains only Chapter 36, Relativity. The complete Part VII may be found either in the hardbound *Physics for Scientists and Engineers with Modern Physics* or in the softbound *Volume 5: Relativity and Quantum Physics.* ◀

36 Relativity

The Large Hadron Collider, the world's highest-energy particle accelerator, is built in a 27-km-circumference tunnel near Geneva, Switzerland. It accelerates protons to 99.999999% of the speed of light.

▶ **Looking Ahead** The goal of Chapter 36 is to understand how Einstein's theory of relativity changes our concepts of space and time.

Principle of Relativity

Einstein's theory of relativity is based on a simple-sounding principle: The laws of physics are the same in every inertial reference frame. This seemingly innocuous statement will force us to completely rethink our ideas of space and time.

The most well-known consequence of this principle is that light travels at the same speed c in all inertial reference frames.

You'll learn why it is that no object or information can travel faster than the speed of light.

Space

The physical length of an object is *less* when the object is moving in a reference frame than when it is at rest in that reference frame. This is **length contraction.**

To us, the Fermilab Accelerator is 3.9 miles in circumference. To protons in the accelerator, moving at 0.999999c, the circumference is only 30 feet.

Mass and Energy

You'll learn the significance of relativity's famous equation, $E = mc^2$. Mass can be transformed into energy, and energy into mass, as long as the total energy is conserved.

The sun is powered by the conversion of 4 billion kilograms of matter into energy every second. Even so, the sun will continue to shine for billions of years.

Reference Frames

You'll learn to work with **events** whose position in space and time of occurrence are measured by experimenters in different **inertial reference frames.**

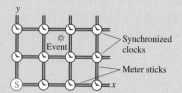
Synchronized clocks
Meter sticks
Event

◀ **Looking Back**
Section 4.4 Reference frames and relative velocity

Time

The time interval between ticks of a clock is *longer* when the clock is moving in a reference frame than when it is at rest in that reference frame. This is **time dilation.**

You'll learn about the **twin paradox.** If an astronaut travels to a distance star and back at a speed close to that of light, she'll be younger than her identical twin when she returns.

Applications of Relativity

Abstract though it may seem, relativity is important for modern technologies such as PET scans (positron-electron tomography) in medicine and nuclear energy. Relativity also underlies our understanding of the physics of stars and galaxies.

Your GPS device receives signals from precision clocks in orbiting satellites. The clocks must be corrected for relativistic effects in order for the GPS system to work.

36.1 Relativity: What's It All About?

What do you think of when you hear the phrase "theory of relativity"? A white-haired Einstein? $E = mc^2$? Black holes? Time travel? Perhaps you've heard that the theory of relativity is so complicated and abstract that only a handful of people in the whole world really understand it.

There is, without doubt, a certain mystique associated with relativity, an aura of the strange and exotic. The good news is that understanding the ideas of relativity is well within your grasp. Einstein's *special theory of relativity,* the portion of relativity we'll study, is not mathematically difficult at all. The challenge is conceptual because relativity questions deeply held assumptions about the nature of space and time. In fact, that's what relativity is all about—space and time.

What's Special About Special Relativity?

Einstein's first paper on relativity, in 1905, dealt exclusively with inertial reference frames, reference frames that move relative to each other with constant velocity. Ten years later, Einstein published a more encompassing theory of relativity that considered accelerated motion and its connection to gravity. The second theory, because it's more general in scope, is called *general relativity.* General relativity is the theory that describes black holes, curved spacetime, and the evolution of the universe. It is a fascinating theory but, alas, very mathematical and outside the scope of this textbook.

Motion at constant velocity is a "special case" of motion—namely, motion for which the acceleration is zero. Hence Einstein's first theory of relativity has come to be known as **special relativity.** It is special in the sense of being a restricted, special case of his more general theory, not special in the everyday sense meaning distinctive or exceptional. Special relativity, with its conclusions about time dilation and length contraction, is what we will study.

Albert Einstein (1879–1955) was one of the most influential thinkers in history.

36.2 Galilean Relativity

Relativity is the process of relating measurements in one reference frame to those in a different reference frame moving *relative to* the first. To appreciate and understand what is new in Einstein's theory, we need a firm grasp of the ideas of relativity that are embodied in Newtonian mechanics. Thus we begin with *Galilean relativity*.

Reference Frames

Suppose you're passing me as we both drive in the same direction along a freeway. My car's speedometer reads 55 mph while your speedometer shows 60 mph. Is 60 mph your "true" speed? That is certainly your speed relative to someone standing beside the road, but your speed relative to me is only 5 mph. Your speed is 120 mph relative to a driver approaching from the other direction at 60 mph.

An object does not have a "true" speed or velocity. The very definition of velocity, $v = \Delta x/\Delta t$, assumes the existence of a coordinate system in which, during some time interval Δt, the displacement Δx is measured. The best we can manage is to specify an object's velocity relative to, or with respect to, the coordinate system in which it is measured.

Let's define a **reference frame** to be a coordinate system in which experimenters equipped with meter sticks, stopwatches, and any other needed equipment make position and time measurements on moving objects. Three ideas are implicit in our definition of a reference frame:

- A reference frame extends infinitely far in all directions.
- The experimenters are at rest in the reference frame.
- The number of experimenters and the quality of their equipment are sufficient to measure positions and velocities to any level of accuracy needed.

The first two ideas are especially important. It is often convenient to say "the laboratory reference frame" or "the reference frame of the rocket." These are shorthand expressions for "a reference frame, infinite in all directions, in which the laboratory (or the rocket) and a set of experimenters happen to be at rest."

NOTE ▶ A reference frame is not the same thing as a "point of view." That is, each person or each experimenter does not have his or her own private reference frame. **All experimenters at rest relative to each other share the same reference frame.** ◀

FIGURE 36.1 shows two reference frames called S and S′. The coordinate axes in S are x, y, z and those in S′ are x′, y′, z′. Reference frame S′ moves with velocity v relative to S or, equivalently, S moves with velocity −v relative to S′. There's no implication that either reference frame is "at rest." Notice that the zero of time, when experimenters start their stopwatches, is the instant that the origins of S and S′ coincide.

We will restrict our attention to *inertial reference frames,* implying that the relative velocity v is constant. You should recall from Chapter 5 that an **inertial reference frame** is a reference frame in which Newton's first law, the law of inertia, is valid. In particular, an inertial reference frame is one in which an isolated particle, one on which there are no forces, either remains at rest or moves in a straight line at constant speed.

Any reference frame moving at constant velocity with respect to an inertial reference frame is itself an inertial reference frame. Conversely, a reference frame accelerating with respect to an inertial reference frame is *not* an inertial reference frame. Our restriction to reference frames moving with respect to each other at constant velocity—with no acceleration—is the "special" part of special relativity.

NOTE ▶ An inertial reference frame is an idealization. A true inertial reference frame would need to be floating in deep space, far from any gravitational influence. In practice, an earthbound laboratory is a good approximation of an inertial reference frame because the accelerations associated with the earth's rotation and motion around the sun are too small to influence most experiments. ◀

STOP TO THINK 36.1 Which of these is an inertial reference frame (or a very good approximation)?

a. Your bedroom
b. A car rolling down a steep hill
c. A train coasting along a level track
d. A rocket being launched
e. A roller coaster going over the top of a hill
f. A sky diver falling at terminal speed

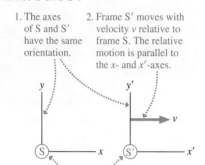

FIGURE 36.1 The standard reference frames S and S′.

1. The axes of S and S′ have the same orientation.
2. Frame S′ moves with velocity v relative to frame S. The relative motion is parallel to the x- and x′-axes.
3. The origins of S and S′ coincide at t = 0. This is our definition of t = 0.

4. Alternatively, frame S moves with velocity −v relative to frame S′.

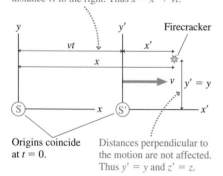

FIGURE 36.2 The position of an exploding firecracker is measured in reference frames S and S′.

At time t, the origin of S′ has moved distance vt to the right. Thus x = x′ + vt.

Origins coincide at t = 0.

Distances perpendicular to the motion are not affected. Thus y′ = y and z′ = z.

The Galilean Transformations

Suppose a firecracker explodes at time t. The experimenters in reference frame S determine that the explosion happened at position x. Similarly, the experimenters in S′ find that the firecracker exploded at x′ in their reference frame. What is the relationship between x and x′?

FIGURE 36.2 shows the explosion and the two reference frames. You can see from the figure that x = x′ + vt, thus

$$x = x' + vt \qquad x' = x - vt$$
$$y = y' \qquad \text{or} \qquad y' = y \qquad (36.1)$$
$$z = z' \qquad z' = z$$

These are the *Galilean transformations of position.* If you know a position measured by the experimenters in one inertial reference frame, you can calculate the position that would be measured by experimenters in any other inertial reference frame.

Suppose the experimenters in both reference frames now track the motion of the object in FIGURE 36.3 by measuring its position at many instants of time. The experimenters in S find that the object's velocity is $\vec{u}$. During the *same time interval* Δt, the experimenters in S′ measure the velocity to be $\vec{u}\,'$.

NOTE ▶ In this chapter, we will use v to represent the velocity of one reference frame relative to another. We will use $\vec{u}$ and $\vec{u}\,'$ to represent the velocities of objects with respect to reference frames S and S′. ◀

We can find the relationship between $\vec{u}$ and $\vec{u}\,'$ by taking the time derivatives of Equations 36.1 and using the definition $u_x = dx/dt$:

$$u_x = \frac{dx}{dt} = \frac{dx'}{dt} + v = u_x' + v$$

$$u_y = \frac{dy}{dt} = \frac{dy'}{dt} = u_y'$$

The equation for u_z is similar. The net result is

$$
\begin{array}{lcl}
u_x = u_x' + v & & u_x' = u_x - v \\
u_y = u_y' & \text{or} & u_y' = u_y \\
u_z = u_z' & & u_z' = u_z
\end{array}
\qquad (36.2)
$$

Equations 36.2 are the *Galilean transformations of velocity.* If you know the velocity of a particle in one inertial reference frame, you can find the velocity that would be measured by experimenters in any other inertial reference frame.

NOTE ▶ In Section 4.4 you learned the Galilean transformation of velocity as $\vec{v}_{CB} = \vec{v}_{CA} + \vec{v}_{AB}$, where $\vec{v}_{AB}$ means "the velocity of A relative to B." Equations 36.2 are equivalent for relative motion parallel to the x-axis but are written in a more formal notation that will be useful for relativity. ◀

FIGURE 36.3 The velocity of a moving object is measured in reference frames S and S′.

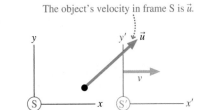

The object's velocity in frame S is $\vec{u}$.

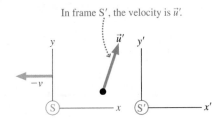

In frame S′, the velocity is $\vec{u}\,'$.

EXAMPLE 36.1 **The speed of sound**

An airplane is flying at speed 200 m/s with respect to the ground. Sound wave 1 is approaching the plane from the front, sound wave 2 is catching up from behind. Both waves travel at 340 m/s relative to the ground. What is the speed of each wave relative to the plane?

MODEL Assume that the earth (frame S) and the airplane (frame S′) are inertial reference frames. Frame S′, in which the airplane is at rest, moves at $v = 200$ m/s relative to frame S.

VISUALIZE FIGURE 36.4 shows the airplane and the sound waves.

FIGURE 36.4 Experimenters in the plane measure different speeds for the waves than do experimenters on the ground.

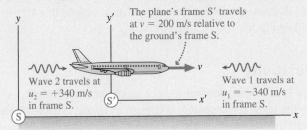

The plane's frame S′ travels at $v = 200$ m/s relative to the ground's frame S.

Wave 2 travels at $u_2 = +340$ m/s in frame S.

Wave 1 travels at $u_1 = -340$ m/s in frame S.

SOLVE The speed of a mechanical wave, such as a sound wave or a wave on a string, is its speed *relative to its medium.* Thus the *speed of sound* is the speed of a sound wave through a reference frame in which the air is at rest. This is reference frame S, where wave 1 travels with velocity $u_1 = -340$ m/s and wave 2 travels with velocity $u_2 = +340$ m/s. Notice that the Galilean transformations use *velocities,* with appropriate signs, not just speeds.

The airplane travels to the right with reference frame S′ at velocity v. We can use the Galilean transformations of velocity to find the velocities of the two sound waves in frame S′:

$$u_1' = u_1 - v = -340 \text{ m/s} - 200 \text{ m/s} = -540 \text{ m/s}$$

$$u_2' = u_2 - v = 340 \text{ m/s} - 200 \text{ m/s} = 140 \text{ m/s}$$

ASSESS This isn't surprising. If you're driving at 50 mph, a car coming the other way at 55 mph is approaching you at 105 mph. A car coming up behind you at 55 mph is gaining on you at the rate of only 5 mph. Wave speeds behave the same. Notice that a mechanical wave appears to be stationary to a person moving at the wave speed. To a surfer, the crest of the ocean wave remains at rest under his or her feet.

Ocean waves are approaching the beach at 10 m/s. A boat heading out to sea travels at 6 m/s. How fast are the waves moving in the boat's reference frame?

a. 16 m/s   b. 10 m/s   c. 6 m/s   d. 4 m/s

The Galilean Principle of Relativity

FIGURE 36.5 Experimenters in both reference frames test Newton's second law by measuring the force on a particle and its acceleration.

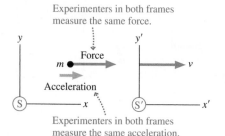

Experimenters in both frames measure the same force.

Experimenters in both frames measure the same acceleration.

Experimenters in reference frames S and S' measure different values for position and velocity. What about the force on and the acceleration of the particle in FIGURE 36.5? The strength of a force can be measured with a spring scale. The experimenters in reference frames S and S' both see the *same reading* on the scale (assume the scale has a bright digital display easily seen by all experimenters), so both conclude that the force is the same. That is, $F' = F$.

We can compare the accelerations measured in the two reference frames by taking the time derivative of the velocity transformation equation $u' = u - v$. (We'll assume, for simplicity, that the velocities and accelerations are all in the x-direction.) The relative velocity v between the two reference frames is *constant,* with $dv/dt = 0$, thus

$$a' = \frac{du'}{dt} = \frac{du}{dt} = a \qquad (36.3)$$

Experimenters in reference frames S and S' measure different values for an object's position and velocity, but they *agree* on its acceleration.

If $F = ma$ in reference frame S, then $F' = ma'$ in reference frame S'. Stated another way, if Newton's second law is valid in one inertial reference frame, then it is valid in all inertial reference frames. Because other laws of mechanics, such as the conservation laws, follow from Newton's laws of motion, we can state this conclusion as the *Galilean principle of relativity:*

> **Galilean principle of relativity** The laws of mechanics are the same in all inertial reference frames.

The Galilean principle of relativity is easy to state, but to understand it we must understand what is and is not "the same." To take a specific example, consider the law of conservation of momentum. FIGURE 36.6a shows two particles about to collide. Their total momentum in frame S, where particle 2 is at rest, is $P_i = 9.0$ kg m/s. This is an isolated system, hence the law of conservation of momentum tells us that the momentum after the collision will be $P_f = 9.0$ kg m/s.

FIGURE 36.6 Total momentum measured in two reference frames.

(a) Collision seen in frame S

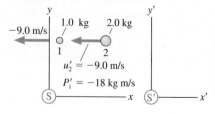

(b) Collision seen in frame S'

FIGURE 36.6b has used the velocity transformation to look at the same particles in frame S' in which particle 1 is at rest. The initial momentum in S' is $P_i' = -18$ kg m/s. Thus it is not the *value* of the momentum that is the same in all inertial reference frames. Instead, the Galilean principle of relativity tells us that the *law* of momentum conservation is the same in all inertial reference frames. If $P_f = P_i$ in frame S, then it must be true that $P_f' = P_i'$ in frame S'. Consequently, we can conclude that P_f' will be -18 kg m/s after the collision in S'.

Using Galilean Relativity

The principle of relativity is concerned with the laws of mechanics, not with the values that are needed to satisfy the laws. If momentum is conserved in one inertial reference frame, it is conserved in all inertial reference frames. Even so, a problem may be easier to solve in one reference frame than in others.

Elastic collisions provide a good example of using reference frames. You learned in Chapter 10 how to calculate the outcome of a perfectly elastic collision between two particles in the reference frame in which particle 2 is initially at rest. We can use that information together with the Galilean transformations to solve elastic-collision problems in any inertial reference frame.

TACTICS
BOX 36.1   **Analyzing elastic collisions**

❶ Transform the initial velocities of particles 1 and 2 from frame S to reference frame S′ in which particle 2 is at rest.

❷ The outcome of the collision in S′ is given by

$$u'_{1f} = \frac{m_1 - m_2}{m_1 + m_2} u'_{1i}$$

$$u'_{2f} = \frac{2m_1}{m_1 + m_2} u'_{1i}$$

❸ Transform the two final velocities from frame S′ back to frame S.

Exercises 4–5

EXAMPLE 36.2   **An elastic collision**

A 300 g ball moving to the right at 2.0 m/s has a perfectly elastic collision with a 100 g ball moving to the left at 4.0 m/s. What are the direction and speed of each ball after the collision?

MODEL The velocities are measured in the laboratory frame, which we call frame S.

VISUALIZE **FIGURE 36.7a** shows both the balls and a reference frame S′ in which ball 2 is at rest.

SOLVE The three steps of Tactics Box 36.1 are illustrated in **FIGURE 36.7b**. We're given u_{1i} and u_{2i}. The Galilean transformations of these velocities to frame S′, using $v = -4.0$ m/s, are

$$u'_{1i} = u_{1i} - v = (2.0 \text{ m/s}) - (-4.0 \text{ m/s}) = 6.0 \text{ m/s}$$

$$u'_{2i} = u_{2i} - v = (-4.0 \text{ m/s}) - (-4.0 \text{ m/s}) = 0 \text{ m/s}$$

The 100 g ball is at rest in frame S′, which is what we wanted. The velocities after the collision are

$$u'_{1f} = \frac{m_1 - m_2}{m_1 + m_2} u'_{1i} = 3.0 \text{ m/s}$$

$$u'_{2f} = \frac{2m_1}{m_1 + m_2} u'_{1i} = 9.0 \text{ m/s}$$

We've finished the collision analysis, but we're not done because these are the post-collision velocities in frame S′. Another application of the Galilean transformations tells us that the post-collision velocities in frame S are

$$u_{1f} = u'_{1f} + v = (3.0 \text{ m/s}) + (-4.0 \text{ m/s}) = -1.0 \text{ m/s}$$

$$u_{2f} = u'_{2f} + v = (9.0 \text{ m/s}) + (-4.0 \text{ m/s}) = 5.0 \text{ m/s}$$

Thus the 300 g ball rebounds to the left at a speed of 1.0 m/s and the 100 g ball is knocked to the right at a speed of 5.0 m/s.

FIGURE 36.7 Using reference frames to solve an elastic-collision problem.

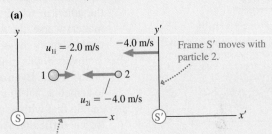

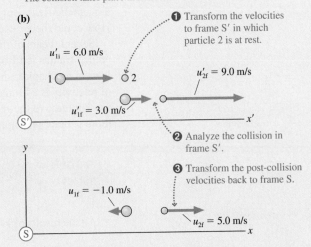

ASSESS You can easily verify that momentum is conserved: $P_f = P_i = 0.20$ kg m/s. The calculations in this example were easy. The important point of this example, and one worth careful thought, is the *logic* of what we did and why we did it.

36.3 Einstein's Principle of Relativity

The 19th century was an era of optics and electromagnetism. Thomas Young demonstrated in 1801 that light is a wave, and by midcentury scientists had devised techniques for measuring the speed of light. Faraday discovered electromagnetic induction in 1831, setting in motion a series of events leading to Maxwell's conclusion, in 1864, that light is an electromagnetic wave.

If light is a wave, what is the medium in which it travels? This was perhaps *the* most important scientific question of the second half of the 19th century. The medium in which light waves were assumed to travel was called the **ether.** Experiments to measure the speed of light were assumed to be measuring its speed through the ether. But just what *is* the ether? What are its properties? Can we collect a jar full of ether to study? Despite the significance of these questions, efforts to detect the ether or measure its properties kept coming up empty handed.

Maxwell's theory of electromagnetism didn't help the situation. The crowning success of Maxwell's theory was his prediction that light waves travel with speed

$$c = \frac{1}{\sqrt{\epsilon_0 \mu_0}} = 3.00 \times 10^8 \text{ m/s}$$

This is a very specific prediction with no wiggle room. The difficulty with such a specific prediction was the implication that Maxwell's laws of electromagnetism are valid *only* in the reference frame of the ether. After all, as **FIGURE 36.8** shows, the light speed should certainly be larger or smaller than c in a reference frame moving through the ether, just as the sound speed is different to someone moving through the air.

As the 19th century closed, it appeared that Maxwell's theory did not obey the classical principle of relativity. There was just one reference frame, the reference frame of the ether, in which the laws of electromagnetism seemed to be true. And to make matters worse, the fact that no one had been able to detect the ether meant that no one could identify the one reference frame in which Maxwell's equations "worked."

It was in this muddled state of affairs that a young Albert Einstein made his mark on the world. Even as a teenager, Einstein had wondered how a light wave would look to someone "surfing" the wave, traveling alongside the wave at the wave speed. You can do that with a water wave or a sound wave, but light waves seemed to present a logical difficulty. An electromagnetic wave sustains itself by virtue of the fact that a changing magnetic field induces an electric field and a changing electric field induces a magnetic field. But to someone moving with the wave, *the fields would not change.* How could there be an electromagnetic wave under these circumstances?

Several years of thinking about the connection between electromagnetism and reference frames led Einstein to the conclusion that *all* the laws of physics, not just the laws of mechanics, should obey the principle of relativity. In other words, the principle of relativity is a fundamental statement about the nature of the physical universe. Thus we can remove the restriction in the Galilean principle of relativity and state a much more general principle:

> **Principle of relativity** All the laws of physics are the same in all inertial reference frames.

All the results of Einstein's theory of relativity flow from this one simple statement.

The Constancy of the Speed of Light

If Maxwell's equations of electromagnetism are laws of physics, and there's every reason to think they are, then, according to the principle of relativity, Maxwell's equations must be true in *every* inertial reference frame. On the surface this seems to be an

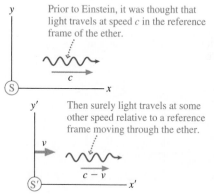

FIGURE 36.8 It seems as if the speed of light should differ from c in a reference frame moving through the ether.

Prior to Einstein, it was thought that light travels at speed c in the reference frame of the ether.

Then surely light travels at some other speed relative to a reference frame moving through the ether.

innocuous statement, equivalent to saying that the law of conservation of momentum is true in every inertial reference frame. But follow the logic:

1. Maxwell's equations are true in all inertial reference frames.
2. Maxwell's equations predict that electromagnetic waves, including light, travel at speed $c = 3.00 \times 10^8$ m/s.
3. Therefore, **light travels at speed c in all inertial reference frames.**

FIGURE 36.9 shows the implications of this conclusion. *All* experimenters, regardless of how they move with respect to each other, find that *all* light waves, regardless of the source, travel in their reference frame with the *same* speed c. If Cathy's velocity toward Bill and away from Amy is $v = 0.9c$, Cathy finds, by making measurements in her reference frame, that the light from Bill approaches her at speed c, not at $c + v = 1.9c$. And the light from Amy, which left Amy at speed c, catches up from behind at speed c *relative to Cathy,* not the $c - v = 0.1c$ you would have expected.

Although this prediction goes against all shreds of common sense, the experimental evidence for it is strong. Laboratory experiments are difficult because even the highest laboratory speed is insignificant in comparison to c. In the 1930s, however, physicists R. J. Kennedy and E. M. Thorndike realized that they could use the earth itself as a laboratory. The earth's speed as it circles the sun is about 30,000 m/s. The *relative* velocity of the earth in January differs by 60,000 m/s from its velocity in July, when the earth is moving in the opposite direction. Kennedy and Thorndike were able to use a very sensitive and stable interferometer to show that the numerical values of the speed of light in January and July differ by less than 2 m/s.

More recent experiments have used unstable elementary particles, called π mesons, that decay into high-energy photons of light. The π mesons, created in a particle accelerator, move through the laboratory at 99.975% the speed of light, or $v = 0.99975c$, as they emit photons at speed c in the π meson's reference frame. As FIGURE 36.10 shows, you would expect the photons to travel through the laboratory with speed $c + v = 1.99975c$. Instead, the measured speed of the photons in the laboratory was, within experimental error, 3.00×10^8 m/s.

In summary, *every* experiment designed to compare the speed of light in different reference frames has found that light travels at 3.00×10^8 m/s in every inertial reference frame, regardless of how the reference frames are moving with respect to each other.

How Can This Be?

You're in good company if you find this impossible to believe. Suppose I shot a ball forward at 50 m/s while driving past you at 30 m/s. You would certainly see the ball traveling at 80 m/s relative to you and the ground. What we're saying with regard to light is equivalent to saying that the ball travels at 50 m/s relative to my car and *at the same time* travels at 50 m/s relative to the ground, even though the car is moving across the ground at 30 m/s. It seems logically impossible.

You might think that this is merely a matter of semantics. If we can just get our definitions and use of words straight, then the mystery and confusion will disappear. Or perhaps the difficulty is a confusion between what we "see" versus what "really happens." In other words, a better analysis, one that focuses on what really happens, would find that light "really" travels at different speeds in different reference frames.

Alas, what "really happens" is that light travels at 3.00×10^8 m/s in every inertial reference frame, regardless of how the reference frames are moving with respect to each other. It's not a trick. There remains only one way to escape the logical contradictions.

The definition of velocity is $u = \Delta x/\Delta t$, the ratio of a distance traveled to the time interval in which the travel occurs. Suppose you and I both make measurements on an object as it moves, but you happen to be moving relative to me. Perhaps I'm standing on the corner, you're driving past in your car, and we're both trying to measure the velocity of a bicycle. Further, suppose we have agreed in advance to measure the position of the bicycle

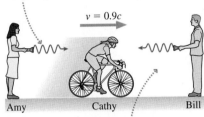

FIGURE 36.9 Light travels at speed c in all inertial reference frames, regardless of how the reference frames are moving with respect to the light source.

This light wave leaves Amy at speed c relative to Amy. It approaches Cathy at speed c relative to Cathy.

This light wave leaves Bill at speed c relative to Bill. It approaches Cathy at speed c relative to Cathy.

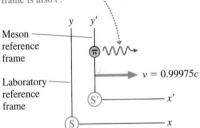

FIGURE 36.10 Experiments find that the photons travel through the laboratory with speed c, not the speed $1.99975c$ that you might expect.

A photon is emitted at speed c relative to the π meson. Measurements find that the photon's speed in the laboratory reference frame is also c.

first as it passes the tree in FIGURE 36.11, then later as it passes the lamppost. Your $\Delta x'$, the bicycle's displacement, differs from my Δx because of your motion relative to me, causing you to calculate a bicycle velocity u' in your reference frame that differs from its velocity u in my reference frame. This is just the Galilean transformations showing up again.

FIGURE 36.11 Measuring the velocity of an object by appealing to the basic definition $u = \Delta x/\Delta t$.

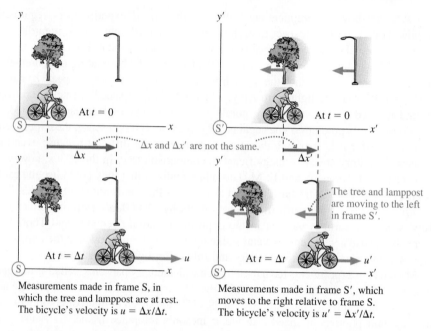

Measurements made in frame S, in which the tree and lamppost are at rest. The bicycle's velocity is $u = \Delta x/\Delta t$.

Measurements made in frame S', which moves to the right relative to frame S. The bicycle's velocity is $u' = \Delta x'/\Delta t$.

Now let's repeat the measurements, but this time let's measure the velocity of a light wave as it travels from the tree to the lamppost. Once again, your $\Delta x'$ differs from my Δx, and the obvious conclusion is that your light speed u' differs from my light speed u. The difference will be very small if you're driving past in your car, very large if you're flying past in a rocket traveling at nearly the speed of light. Although this conclusion seems obvious, it is wrong. Experiments show that, for a light wave, we'll get the *same* values: $u' = u$.

The only way this can be true is if your Δt is not the same as my Δt. If the time it takes the light to move from the tree to the lamppost in your reference frame, a time we'll now call $\Delta t'$, differs from the time Δt it takes the light to move from the tree to the lamppost in my reference frame, then we might find that $\Delta x'/\Delta t' = \Delta x/\Delta t$. That is, $u' = u$ even though you are moving with respect to me.

We've assumed, since the beginning of this textbook, that time is simply time. It flows along like a river, and all experimenters in all reference frames simply use it. For example, suppose the tree and the lamppost both have big clocks that we both can see. Shouldn't we be able to agree on the time interval Δt the light needs to move from the tree to the lamppost?

Perhaps not. It's demonstrably true that $\Delta x' \neq \Delta x$. It's experimentally verified that $u' = u$ for light waves. Something must be wrong with *assumptions* that we've made about the nature of time. The principle of relativity has painted us into a corner, and our only way out is to reexamine our understanding of time.

36.4 Events and Measurements

To question some of our most basic assumptions about space and time requires extreme care. We need to be certain that no assumptions slip into our analysis unnoticed. Our goal is to describe the motion of a particle in a clear and precise way, making the barest minimum of assumptions.

Events

The fundamental element of relativity is called an **event.** An event is a physical activity that takes place at a definite point in space and at a definite instant of time. An exploding firecracker is an event. A collision between two particles is an event. A light wave hitting a detector is an event.

Events can be observed and measured by experimenters in different reference frames. An exploding firecracker is as clear to you as you drive by in your car as it is to me standing on the street corner. We can quantify where and when an event occurs with four numbers: the coordinates (x, y, z) and the instant of time t. These four numbers, illustrated in **FIGURE 36.12**, are called the **spacetime coordinates** of the event.

The spatial coordinates of an event measured in reference frames S and S′ may differ. It now appears that the instant of time recorded in S and S′ may also differ. Thus the spacetime coordinates of an event measured by experimenters in frame S are (x, y, z, t) and the spacetime coordinates of the *same event* measured by experimenters in frame S′ are (x', y', z', t').

The motion of a particle can be described as a sequence of two or more events. We introduced this idea in the preceding section when we agreed to measure the velocity of a bicycle and then of a light wave by making measurements when the object passed the tree (first event) and when the object passed the lamppost (second event).

Measurements

Events are what "really happen," but how do we learn about an event? That is, how do the experimenters in a reference frame determine the spacetime coordinates of an event? This is a problem of *measurement.*

We defined a reference frame to be a coordinate system in which experimenters can make position and time measurements. That's a good start, but now we need to be more precise as to *how* the measurements are made. Imagine that a reference frame is filled with a cubic lattice of meter sticks, as shown in **FIGURE 36.13**. At every intersection is a clock, and all the clocks in a reference frame are *synchronized.* We'll return in a moment to consider how to synchronize the clocks, but assume for the moment it can be done.

Now, with our meter sticks and clocks in place, we can use a two-part measurement scheme:

■ The (x, y, z) coordinates of an event are determined by the intersection of the meter sticks closest to the event.
■ The event's time t is the time displayed on the clock nearest the event.

You can imagine, if you wish, that each event is accompanied by a flash of light to illuminate the face of the nearest clock and make its reading known.

Several important issues need to be noted:

1. The clocks and meter sticks in each reference frame are imaginary, so they have no difficulty passing through each other.
2. Measurements of position and time made in one reference frame must use only the clocks and meter sticks in that reference frame.
3. There's nothing special about the sticks being 1 m long and the clocks 1 m apart. The lattice spacing can be altered to achieve whatever level of measurement accuracy is desired.
4. We'll assume that the experimenters in each reference frame have assistants sitting beside every clock to record the position and time of nearby events.
5. Perhaps most important, t is the time at which the event *actually happens,* not the time at which an experimenter sees the event or at which information about the event reaches an experimenter.
6. All experimenters in one reference frame agree on the spacetime coordinates of an event. In other words, **an event has a unique set of spacetime coordinates in each reference frame.**

FIGURE 36.12 The location and time of an event are described by its spacetime coordinates.

An event has spacetime coordinates (x, y, z, t) in frame S and different spacetime coordinates (x', y', z', t') in frame S′.

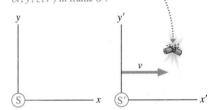

FIGURE 36.13 The spacetime coordinates of an event are measured by a lattice of meter sticks and clocks.

The spacetime coordinates of this event are measured by the nearest meter stick intersection and the nearest clock.

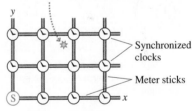

Reference frame S

Reference frame S′ has its own meter sticks and its own clocks.

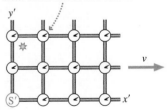

Reference frame S′

STOP TO THINK 36.3   A carpenter is working on a house two blocks away. You notice a slight delay between seeing the carpenter's hammer hit the nail and hearing the blow. At what time does the event "hammer hits nail" occur?

    a. At the instant you hear the blow
    b. At the instant you see the hammer hit
    c. Very slightly before you see the hammer hit
    d. Very slightly after you see the hammer hit

Clock Synchronization

It's important that all the clocks in a reference frame be **synchronized,** meaning that all clocks in the reference frame have the same reading at any one instant of time. Thus we need a method of synchronization. One idea that comes to mind is to designate the clock at the origin as the *master clock*. We could then carry this clock around to every clock in the lattice, adjust that clock to match the master clock, and finally return the master clock to the origin.

This would be a perfectly good method of clock synchronization in Newtonian mechanics, where time flows along smoothly, the same for everyone. But we've been driven to reexamine the nature of time by the possibility that time is different in reference frames moving relative to each other. Because the master clock would *move,* we cannot assume that the moving master clock would keep time in the same way as the stationary clocks.

We need a synchronization method that does not require moving the clocks. Fortunately, such a method is easy to devise. Each clock is resting at the intersection of meter sticks, so by looking at the meter sticks, the assistant knows, or can calculate, exactly how far each clock is from the origin. Once the distance is known, the assistant can calculate exactly how long a light wave will take to travel from the origin to each clock. For example, light will take 1.00 μs to travel to a clock 300 m from the origin.

NOTE ▶ It's handy for many relativity problems to know that the speed of light is $c = 300$ m/μs. ◀

To synchronize the clocks, the assistants begin by setting each clock to display the light travel time from the origin, but they don't start the clocks. Next, as FIGURE 36.14 shows, a light flashes at the origin and, simultaneously, the clock at the origin starts running from $t = 0$ s. The light wave spreads out in all directions at speed c. A photodetector on each clock recognizes the arrival of the light wave and, without delay, starts the clock. The clock had been preset with the light travel time, so each clock as it starts reads exactly the same as the clock at the origin. Thus all the clocks will be synchronized after the light wave has passed by.

FIGURE 36.14 Synchronizing clocks.

1. This clock is preset to 1.00 μs, the time it takes light to travel 300 m.

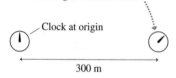

Clock at origin

300 m

2. At $t = 0$ s, a light flashes at the origin and the origin clock starts running. A very short time later, seen here, a light wave has begun to move outward.

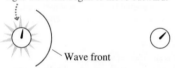

Wave front

3. The clock starts when the light wave reaches it. It is now synchronized with the origin clock.

Events and Observations

We noted above that t is the time the event *actually happens.* This is an important point, one that bears further discussion. Light waves take time to travel. Messages, whether they're transmitted by light pulses, telephone, or courier on horseback, take time to be delivered. An experimenter *observes* an event, such as an exploding firecracker, only *at a later time* when light waves reach his or her eyes. But our interest is in the event itself, not the experimenter's observation of the event. The time at which the experimenter sees the event or receives information about the event is not when the event actually occurred.

Suppose at $t = 0$ s a firecracker explodes at $x = 300$ m. The flash of light from the firecracker will reach an experimenter at the origin at $t_1 = 1.0$ μs. The sound of the explosion will reach a sightless experimenter at the origin at $t_2 = 0.88$ s. Neither of these is the time t_{event} of the explosion, although the experimenter can work backward from these times, using known wave speeds, to determine t_{event}. In this example, the spacetime coordinates of the event—the explosion—are (300 m, 0 m, 0 m, 0 s).

EXAMPLE 36.3 **Finding the time of an event**

Experimenter A in reference frame S stands at the origin looking in the positive *x*-direction. Experimenter B stands at *x* = 900 m looking in the negative *x*-direction. A firecracker explodes somewhere between them. Experimenter B sees the light flash at *t* = 3.0 μs. Experimenter A sees the light flash at *t* = 4.0 μs. What are the spacetime coordinates of the explosion?

MODEL Experimenters A and B are in the same reference frame and have synchronized clocks.

VISUALIZE FIGURE 36.15 shows the two experimenters and the explosion at unknown position *x*.

SOLVE The two experimenters observe light flashes at two different instants, but there's only one event. Light travels at 300 m/μs, so the additional 1.0 μs needed for the light to reach experimenter A implies that distance (*x* − 0 m) is 300 m longer than distance (900 m − *x*). That is,

$$(x - 0\text{ m}) = (900\text{ m} - x) + 300\text{ m}$$

This is easily solved to give *x* = 600 m as the position coordinate of the explosion. The light takes 1.0 μs to travel 300 m to experimenter B, 2.0 μs to travel 600 m to experimenter A. The light is

FIGURE 36.15 The light wave reaches the experimenters at different times. Neither of these is the time at which the event actually happened.

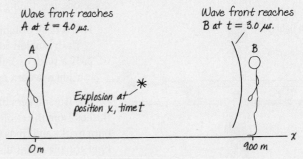

received at 3.0 μs and 4.0 μs, respectively; hence it was emitted by the explosion at *t* = 2.0 μs. The spacetime coordinates of the explosion are (600 m, 0 m, 0 m, 2.0 μs).

ASSESS Although the experimenters *see* the explosion at different times, they agree that the explosion actually *happened* at *t* = 2.0 μs.

Simultaneity

Two events 1 and 2 that take place at different positions x_1 and x_2 but at the *same time* $t_1 = t_2$, as measured in some reference frame, are said to be **simultaneous** in that reference frame. Simultaneity is determined by when the events actually happen, not when they are seen or observed. In general, simultaneous events are *not* seen at the same time because of the difference in light travel times from the events to an experimenter.

EXAMPLE 36.4 **Are the explosions simultaneous?**

An experimenter in reference frame S stands at the origin looking in the positive *x*-direction. At *t* = 3.0 μs she sees firecracker 1 explode at *x* = 600 m. A short time later, at *t* = 5.0 μs, she sees firecracker 2 explode at *x* = 1200 m. Are the two explosions simultaneous? If not, which firecracker exploded first?

MODEL Light from both explosions travels toward the experimenter at 300 m/μs.

SOLVE The experimenter *sees* two different explosions, but perceptions of the events are not the events themselves. When did the explosions *actually* occur? Using the fact that light travels at 300 m/μs, we can see that firecracker 1 exploded at $t_1 = 1.0$ μs and firecracker 2 also exploded at $t_2 = 1.0$ μs. The events *are* simultaneous.

STOP TO THINK 36.4   A tree and a pole are 3000 m apart. Each is suddenly hit by a bolt of lightning. Mark, who is standing at rest midway between the two, sees the two lightning bolts at the same instant of time. Nancy is at rest under the tree. Define event 1 to be "lightning strikes tree" and event 2 to be "lightning strikes pole." For Nancy, does event 1 occur before, after, or at the same time as event 2?

36.5 The Relativity of Simultaneity

We've now established a means for measuring the time of an event in a reference frame, so let's begin to investigate the nature of time. The following "thought experiment" is very similar to one suggested by Einstein.

FIGURE 36.16 on the next page shows a long railroad car traveling to the right with a velocity *v* that may be an appreciable fraction of the speed of light. A firecracker is tied

FIGURE 36.16 A railroad car traveling to the right with velocity *v*.

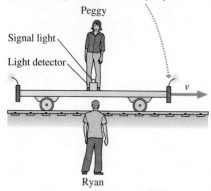

The firecrackers will make burn marks on the ground at the positions where they explode.

to each end of the car, just above the ground. Each firecracker is powerful enough so that, when it explodes, it will make a burn mark on the ground at the position of the explosion.

Ryan is standing on the ground, watching the railroad car go by. Peggy is standing in the exact center of the car with a special box at her feet. This box has two light detectors, one facing each way, and a signal light on top. The box works as follows:

1. If a flash of light is received at the detector facing right, as seen by Ryan, before a flash is received at the left detector, then the light on top of the box will turn green.
2. If a flash of light is received at the left detector before a flash is received at the right detector, or if two flashes arrive simultaneously, the light on top will turn red.

The firecrackers explode as the railroad car passes Ryan, and he sees the two light flashes from the explosions simultaneously. He then measures the distances to the two burn marks and finds that he was standing exactly halfway between the marks. Because light travels equal distances in equal times, Ryan concludes that the two explosions were simultaneous in his reference frame, the reference frame of the ground. Further, because he was midway between the two ends of the car, he was directly opposite Peggy when the explosions occurred.

FIGURE 36.17a shows the sequence of events in Ryan's reference frame. Light travels at speed *c* in all inertial reference frames, so, although the firecrackers were moving, the light waves are spheres centered on the burn marks. Ryan determines that the light wave coming from the right reaches Peggy and the box before the light wave coming from the left. Thus, according to Ryan, the signal light on top of the box turns green.

FIGURE 36.17 Exploding firecrackers seen in two different reference frames.

(a) The events in Ryan's frame

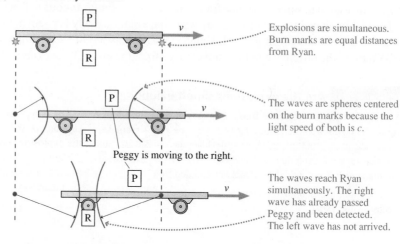

(b) The events in Peggy's frame

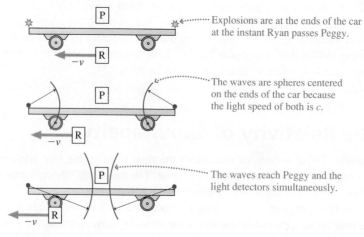

How do things look in Peggy's reference frame, a reference frame moving to the right at velocity v relative to the ground? As FIGURE 36.17b shows, Peggy sees Ryan moving to the left with speed v. Light travels at speed c in all inertial reference frames, so the light waves are spheres centered on the ends of the car. If the explosions are simultaneous, as Ryan has determined, the two light waves reach her and the box simultaneously. Thus, according to Peggy, the signal light on top of the box turns red!

Now the light on top must be either green or red. *It can't be both!* Later, after the railroad car has stopped, Ryan and Peggy can place the box in front of them. Either it has a red light or a green light. Ryan can't see one color while Peggy sees the other. Hence we have a paradox. It's impossible for Peggy and Ryan both to be right. But who is wrong, and why?

What do we know with absolute certainty?

1. Ryan detected the flashes simultaneously.
2. Ryan was halfway between the firecrackers when they exploded.
3. The light from the two explosions traveled toward Ryan at equal speeds.

The conclusion that the explosions were simultaneous in Ryan's reference frame is unassailable. The light is green.

Peggy, however, made an assumption. It's a perfectly ordinary assumption, one that seems sufficiently obvious that you probably didn't notice, but an assumption nonetheless. Peggy assumed that the explosions were simultaneous.

Didn't Ryan find them to be simultaneous? Indeed, he did. Suppose we call Ryan's reference frame S, the explosion on the right event R, and the explosion on the left event L. Ryan found that $t_R = t_L$. But Peggy has to use a different set of clocks, the clocks in her reference frame S$'$, to measure the times t_R' and t_L' at which the explosions occurred. The fact that $t_R = t_L$ in frame S does *not* allow us to conclude that $t_R' = t_L'$ in frame S$'$.

In fact, in frame S$'$ the right firecracker must explode *before* the left firecracker. Figure 36.17b, with its assumption about simultaneity, was incorrect. FIGURE 36.18 shows the situation in Peggy's reference frame, with the right firecracker exploding first. Now the wave from the right reaches Peggy and the box first, as Ryan had concluded, and the light on top turns green.

One of the most disconcerting conclusions of relativity is that **two events occurring simultaneously in reference frame S are *not* simultaneous in any reference frame S$'$ moving relative to S.** This is called the **relativity of simultaneity**.

The two firecrackers *really* explode at the same instant of time in Ryan's reference frame. And the right firecracker *really* explodes first in Peggy's reference frame. It's not a matter of when they see the flashes. Our conclusion refers to the times at which the explosions actually occur.

The paradox of Peggy and Ryan contains the essence of relativity, and it's worth careful thought. First, review the logic until you're certain that there *is* a paradox, a logical impossibility. Then convince yourself that the only way to resolve the paradox is to abandon the assumption that the explosions are simultaneous in Peggy's reference frame. If you understand the paradox and its resolution, you've made a big step toward understanding what relativity is all about.

FIGURE 36.18 The real sequence of events in Peggy's reference frame.

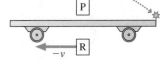

STOP TO THINK 36.5   A tree and a pole are 3000 m apart. Each is hit by a bolt of lightning. Mark, who is standing at rest midway between the two, sees the two lightning bolts at the same instant of time. Nancy is flying her rocket at $v = 0.5c$ in the direction from the tree toward the pole. The lightning hits the tree just as she passes by it. Define event 1 to be "lightning strikes tree" and event 2 to be "lightning strikes pole." For Nancy, does event 1 occur before, after, or at the same time as event 2?

36.6 Time Dilation

The principle of relativity has driven us to the logical conclusion that time is not the same for two reference frames moving relative to each other. Our analysis thus far has been mostly qualitative. It's time to start developing some quantitative tools that will allow us to compare measurements in one reference frame to measurements in another reference frame.

FIGURE 36.19a shows a special clock called a *light clock*. The light clock is a box with a light source at the bottom and a mirror at the top, separated by distance *h*. The light source emits a very short pulse of light that travels to the mirror and reflects back to a light detector beside the source. The clock advances one "tick" each time the detector receives a light pulse, and it immediately, with no delay, causes the light source to emit the next light pulse.

Our goal is to compare two measurements of the interval between two ticks of the clock: one taken by an experimenter standing next to the clock and the other by an experimenter moving with respect to the clock. To be specific, FIGURE 36.19b shows the clock at rest in reference frame S'. We call this the **rest frame** of the clock. Reference frame S' moves to the right with velocity *v* relative to reference frame S.

Relativity requires us to measure *events,* so let's define event 1 to be the emission of a light pulse and event 2 to be the detection of that light pulse. Experimenters in both reference frames are able to measure where and when these events occur *in their frame*. In frame S, the time interval $\Delta t = t_2 - t_1$ is one tick of the clock. Similarly, one tick in frame S' is $\Delta t' = t'_2 - t'_1$.

To be sure we have a clear understanding of the relativity result, let's first do a classical analysis. In frame S', the clock's rest frame, the light travels straight up and down, a total distance 2*h*, at speed *c*. The time interval is $\Delta t' = 2h/c$.

FIGURE 36.20a shows the operation of the light clock as seen in frame S. The clock is moving to the right at speed *v* in S, thus the mirror moves distance $\frac{1}{2}v(\Delta t)$ during the time $\frac{1}{2}(\Delta t)$ in which the light pulse moves from the source to the mirror. The distance traveled by the light during this interval is $\frac{1}{2}u_{\text{light}}(\Delta t)$, where u_{light} is the speed of light in frame S. You can see from the vector addition in FIGURE 36.20b that the speed of light in frame S is $u_{\text{light}} = (c^2 + v^2)^{1/2}$. (Remember, this is a classical analysis in which the speed of light *does* depend on the motion of the reference frame relative to the light source.)

The Pythagorean theorem applied to the right triangle in Figure 36.20a is

$$h^2 + \left(\frac{1}{2}v\,\Delta t\right)^2 = \left(\frac{1}{2}u_{\text{light}}\Delta t\right)^2 = \left(\frac{1}{2}\sqrt{c^2 + v^2}\,\Delta t\right)^2$$

$$= \left(\frac{1}{2}c\,\Delta t\right)^2 + \left(\frac{1}{2}v\,\Delta t\right)^2 \tag{36.4}$$

The term $\left(\frac{1}{2}v\,\Delta t\right)^2$ is common to both sides and cancels. Solving for Δt gives $\Delta t = 2h/c$, identical to $\Delta t'$. In other words, a classical analysis finds that the clock ticks at exactly the same rate in both frame S and frame S'. This shouldn't be surprising. There's only one kind of time in classical physics, measured the same by all experimenters independent of their motion.

The principle of relativity changes only one thing, but that change has profound consequences. According to the principle of relativity, light travels at the same speed in *all* inertial reference frames. In frame S', the rest frame of the clock, the light simply goes straight up and back. The time of one tick,

$$\Delta t' = \frac{2h}{c} \tag{36.5}$$

is unchanged from the classical analysis.

FIGURE 36.19 The ticking of a light clock can be measured by experimenters in two different reference frames.

(a) A light clock

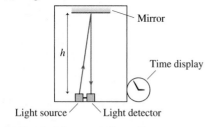

(b) The clock is at rest in frame S'.

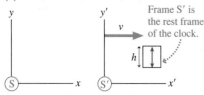

FIGURE 36.20 A classical analysis of the light clock.

(a)

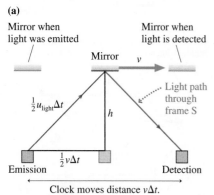

(b)

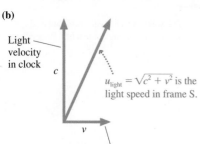

FIGURE 36.21 shows the light clock as seen in frame S. The difference from Figure 36.20a is that the light now travels along the hypotenuse at speed c. We can again use the Pythagorean theorem to write

$$h^2 + \left(\frac{1}{2}v\,\Delta t\right)^2 = \left(\frac{1}{2}c\,\Delta t\right)^2 \qquad (36.6)$$

Solving for Δt gives

$$\Delta t = \frac{2h/c}{\sqrt{1 - v^2/c^2}} = \frac{\Delta t'}{\sqrt{1 - v^2/c^2}} \qquad (36.7)$$

The time interval between two ticks in frame S is *not* the same as in frame S'.

It's useful to define $\beta = v/c$, the velocity as a fraction of the speed of light. For example, a reference frame moving with $v = 2.4 \times 10^8$ m/s has $\beta = 0.80$. In terms of β, Equation 36.7 is

$$\Delta t = \frac{\Delta t'}{\sqrt{1 - \beta^2}} \qquad (36.8)$$

NOTE ▶ The expression $(1 - v^2/c^2)^{1/2} = (1 - \beta^2)^{1/2}$ occurs frequently in relativity. The value of the expression is 1 when $v = 0$, and it steadily decreases to 0 as $v \rightarrow c$ (or $\beta \rightarrow 1$). The square root is an imaginary number if $v > c$, which would make Δt imaginary in Equation 36.8. Time intervals certainly have to be real numbers, suggesting that $v > c$ is not physically possible. One of the predictions of the theory of relativity, as you've undoubtedly heard, is that nothing can travel faster than the speed of light. Now you can begin to see why. We'll examine this topic more closely in Section 36.9. In the meantime, we'll require v to be less than c. ◀

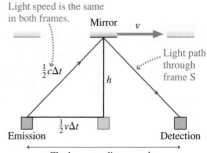

FIGURE 36.21 A light clock analysis in which the speed of light is the same in all reference frames.

Proper Time

Frame S' has one important distinction. It is the *one and only* inertial reference frame in which the light clock is at rest. Consequently, it is the one and only inertial reference frame in which the times of both events—the emission of the light and the detection of the light—are measured by the *same* reference-frame clock. You can see that the light pulse in Figure 36.19a starts and ends at the same position. In Figure 36.21, the emission and detection take place at different positions in frame S and must be measured by different reference-frame clocks, one at each position.

The time interval between two events that occur at the *same position* is called the **proper time** $\Delta\tau$. Only one inertial reference frame measures the proper time, and it does so with a single clock that is present at both events. An inertial reference frame moving with velocity $v = \beta c$ relative to the proper-time frame must use two clocks to measure the time interval: one at the position of the first event, the other at the position of the second event. The time interval between the two events in this frame is

$$\Delta t = \frac{\Delta\tau}{\sqrt{1 - \beta^2}} \geq \Delta\tau \qquad \text{(time dilation)} \qquad (36.9)$$

The "stretching out" of the time interval implied by Equation 36.9 is called **time dilation.** Time dilation is sometimes described by saying that "moving clocks run slow." This is not an accurate statement because it implies that some reference frames are "really" moving while others are "really" at rest. The whole point of relativity is that all inertial reference frames are equally valid, that all we know about reference frames is how they move relative to each other. A better description of time dilation is the statement that **the time interval between two ticks is the shortest in the reference frame in which the clock is at rest.** The time interval between two ticks is longer (i.e., the clock "runs slower") when it is measured in any reference frame in which the clock is moving.

NOTE ▶ Equation 36.9 was derived using a light clock because the operation of a light clock is clear and easy to analyze. But the conclusion is really about time itself. *Any* clock, regardless of how it operates, behaves the same. ◀

EXAMPLE 36.5 | **From the sun to Saturn**

Saturn is 1.43×10^{12} m from the sun. A rocket travels along a line from the sun to Saturn at a constant speed of $0.9c$ relative to the solar system. How long does the journey take as measured by an experimenter on earth? As measured by an astronaut on the rocket?

MODEL Let the solar system be in reference frame S and the rocket be in reference frame S′ that travels with velocity $v = 0.9c$ relative to S. Relativity problems must be stated in terms of *events*. Let event 1 be "the rocket and the sun coincide" (the experimenter on earth says that the rocket passes the sun; the astronaut on the rocket says that the sun passes the rocket) and event 2 be "the rocket and Saturn coincide."

FIGURE 36.22 Pictorial representation of the trip as seen in frames S and S′.

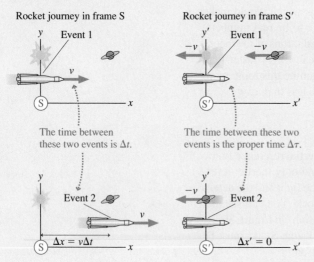

Rocket journey in frame S

Event 1

The time between these two events is Δt.

Event 2

$\Delta x = v\Delta t$

Rocket journey in frame S′

Event 1

The time between these two events is the proper time $\Delta \tau$.

Event 2

$\Delta x' = 0$

VISUALIZE **FIGURE 36.22** shows the two events as seen from the two reference frames. Notice that the two events occur at the *same position* in S′, the position of the rocket, and consequently can be measured by *one* clock.

SOLVE The time interval measured in the solar system reference frame, which includes the earth, is simply

$$\Delta t = \frac{\Delta x}{v} = \frac{1.43 \times 10^{12} \text{ m}}{0.9 \times (3.00 \times 10^8 \text{ m/s})} = 5300 \text{ s}$$

Relativity hasn't abandoned the basic definition $v = \Delta x / \Delta t$, although we do have to be sure that Δx and Δt are measured in just one reference frame and refer to the same two events.

How are things in the rocket's reference frame? The two events occur at the *same position* in S′ and can be measured by *one* clock, the clock at the origin. Thus the time measured by the astronauts is the *proper time* $\Delta \tau$ between the two events. We can use Equation 36.9 with $\beta = 0.9$ to find

$$\Delta \tau = \sqrt{1 - \beta^2} \, \Delta t = \sqrt{1 - 0.9^2} \, (5300 \text{ s}) = 2310 \text{ s}$$

ASSESS The time interval measured between these two events by the astronauts is less than half the time interval measured by experimenters on earth. The difference has nothing to do with when earthbound astronomers *see* the rocket pass the sun and Saturn. Δt is the time interval from when the rocket actually passes the sun, as measured by a clock at the sun, until it actually passes Saturn, as measured by a synchronized clock at Saturn. The interval between *seeing* the events from earth, which would have to allow for light travel times, would be something other than 5300 s. Δt and $\Delta \tau$ are different because *time is different* in two reference frames moving relative to each other.

STOP TO THINK 36.6 Molly flies her rocket past Nick at constant velocity v. Molly and Nick both measure the time it takes the rocket, from nose to tail, to pass Nick. Which of the following is true?

a. Both Molly and Nick measure the same amount of time.
b. Molly measures a shorter time interval than Nick.
c. Nick measures a shorter time interval than Molly.

Experimental Evidence

Is there any evidence for the crazy idea that clocks moving relative to each other tell time differently? Indeed, there's plenty. An experiment in 1971 sent an atomic clock around the world on a jet plane while an identical clock remained in the laboratory. This was a difficult experiment because the traveling clock's speed was so small compared to c, but measuring the small differences between the time intervals was just barely within the capabilities of atomic clocks. It was also a more complex experiment than we've analyzed because the clock accelerated as it moved around a circle. The scientists found that, upon its return, the eastbound clock, traveling faster than the laboratory on a rotating earth, was 60 ns behind the stay-at-home clock, which was exactly as predicted by relativity.

Very detailed studies have been done on unstable particles called *muons* that are created at the top of the atmosphere, at a height of about 60 km, when high-energy cosmic rays collide with air molecules. It is well known, from laboratory studies, that stationary muons decay with a *half-life* of 1.5 μs. That is, half the muons decay within 1.5 μs, half of those remaining decay in the next 1.5 μs, and so on. The decays can be used as a clock.

The muons travel down through the atmosphere at very nearly the speed of light. The time needed to reach the ground, assuming $v \approx c$, is $\Delta t \approx (60,000 \text{ m})/(3 \times 10^8 \text{ m/s}) = 200 \ \mu$s. This is 133 half-lives, so the fraction of muons reaching the ground should be $\approx \left(\frac{1}{2}\right)^{133} = 10^{-40}$. That is, only 1 out of every 10^{40} muons should reach the ground. In fact, experiments find that about 1 in 10 muons reach the ground, an experimental result that differs by a factor of 10^{39} from our prediction!

The discrepancy is due to time dilation. In FIGURE 36.23, the two events "muon is created" and "muon hits ground" take place at two different places in the earth's reference frame. However, these two events occur at the *same position* in the muon's reference frame. (The muon is like the rocket in Example 36.5.) Thus the muon's internal clock measures the proper time. The time-dilated interval $\Delta t = 200 \ \mu$s in the earth's reference frame corresponds to a proper time $\Delta \tau \approx 5 \ \mu$s in the muon's reference frame. That is, in the muon's reference frame it takes only 5 μs from creation at the top of the atmosphere until the ground runs into it. This is 3.3 half-lives, so the fraction of muons reaching the ground is $\left(\frac{1}{2}\right)^{3.3} = 0.1$, or 1 out of 10. We wouldn't detect muons at the ground at all if not for time dilation.

The details are beyond the scope of this textbook, but dozens of high-energy particle accelerators around the world that study quarks and other elementary particles have been designed and built on the basis of Einstein's theory of relativity. The fact that they work exactly as planned is strong testimony to the reality of time dilation.

FIGURE 36.23 We wouldn't detect muons at the ground if not for time dilation.

A muon travels ≈450 m in 1.5 μs. We would not detect muons at ground level if the half-life of a moving muon were 1.5 μs.

Muon is created.

Because of time dilation, the half-life of a high-speed muon is long enough in the earth's reference frame for about 1 in 10 muons to reach the ground.

Muon hits ground.

The Twin Paradox

The most well-known relativity paradox is the twin paradox. George and Helen are twins. On their 25th birthday, Helen departs on a starship voyage to a distant star. Let's imagine, to be specific, that her starship accelerates almost instantly to a speed of 0.95c and that she travels to a star that is 9.5 light years (9.5 ly) from earth. Upon arriving, she discovers that the planets circling the star are inhabited by fierce aliens, so she immediately turns around and heads home at 0.95c.

A **light year,** abbreviated ly, is the distance that light travels in one year. A light year is vastly larger than the diameter of the solar system. The distance between two neighboring stars is typically a few light years. For our purpose, we can write the speed of light as $c = 1$ ly/year. That is, light travels 1 light year per year.

This value for c allows us to determine how long, according to George and his fellow earthlings, it takes Helen to travel out and back. Her total distance is 19 ly and, due to her rapid acceleration and rapid turn-around, she travels essentially the entire distance at speed $v = 0.95c = 0.95$ ly/year. Thus the time she's away, as measured by George, is

$$\Delta t_G = \frac{19 \text{ ly}}{0.95 \text{ ly/year}} = 20 \text{ years} \tag{36.10}$$

George will be 45 years old when his sister Helen returns with tales of adventure.

While she's away, George takes a physics class and studies Einstein's theory of relativity. He realizes that time dilation will make Helen's clocks run more slowly than his clocks, which are at rest relative to him. Her heart—a clock—will beat fewer

The global positioning system (GPS), which allows you to pinpoint your location anywhere in the world to within a few meters, uses a set of orbiting satellites. Because of their motion, the atomic clocks on these satellites keep time differently from clocks on the ground. To determine an accurate position, the software in your GPS receiver must carefully correct for time-dilation effects.

times and the minute hand on her watch will go around fewer times. In other words, she's aging more slowly than he is. Although she is his twin, she will be younger than he is when she returns.

Calculating Helen's age is not hard. We simply have to identify Helen's clock, because it's always with Helen as she travels, as the clock that measures proper time $\Delta\tau$. From Equation 36.9,

$$\Delta t_{\mathrm{H}} = \Delta\tau = \sqrt{1 - \beta^2}\, \Delta t_{\mathrm{G}} = \sqrt{1 - 0.95^2}\ (20\ \text{years}) = 6.25\ \text{years} \quad (36.11)$$

George will have just celebrated his 45th birthday as he welcomes home his 31-year-and-3-month-old twin sister.

This may be unsettling because it violates our commonsense notion of time, but it's not a paradox. There's no logical inconsistency in this outcome. So why is it called "the twin paradox"?

Helen, knowing that she had quite of bit of time to kill on her journey, brought along several physics books to read. As she learns about relativity, she begins to think about George and her friends back on earth. Relative to her, they are all moving away at $0.95c$. Later they'll come rushing toward her at $0.95c$. Time dilation will cause their clocks to run more slowly than her clocks, which are at rest relative to her. In other words, as FIGURE 36.24 shows, Helen concludes that people on earth are aging more slowly than she is. Alas, she will be much older than they when she returns.

Finally, the big day arrives. Helen lands back on earth and steps out of the starship. George is expecting Helen to be younger than he is. Helen is expecting George to be younger than she is.

Here's the paradox! It's logically impossible for each to be younger than the other at the time they are reunited. Where, then, is the flaw in our reasoning? It seems to be a symmetrical situation—Helen moves relative to George and George moves relative to Helen—but symmetrical reasoning has led to a conundrum.

But are the situations really symmetrical? George goes about his business day after day without noticing anything unusual. Helen, on the other hand, experiences three distinct periods during which the starship engines fire, she's crushed into her seat, and free dust particles that had been floating inside the starship are no longer, in the starship's reference frame, at rest or traveling in a straight line at constant speed. In other words, George spends the entire time in an inertial reference frame, *but Helen does not*. The situation is *not* symmetrical.

The principle of relativity applies *only* to inertial reference frames. Our discussion of time dilation was for inertial reference frames. Thus George's analysis and calculations are correct. Helen's analysis and calculations are *not* correct because she was trying to apply an inertial reference frame result while traveling in a noninertial reference frame.

Helen is younger than George when she returns. This is strange, but not a paradox. It is a consequence of the fact that time flows differently in two reference frames moving relative to each other.

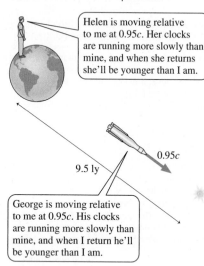

FIGURE 36.24 The twin paradox.

Helen is moving relative to me at $0.95c$. Her clocks are running more slowly than mine, and when she returns she'll be younger than I am.

9.5 ly

$0.95c$

George is moving relative to me at $0.95c$. His clocks are running more slowly than mine, and when I return he'll be younger than I am.

36.7 Length Contraction

We've seen that relativity requires us to rethink our idea of time. Now let's turn our attention to the concepts of space and distance. Consider the rocket that traveled from the sun to Saturn in Example 36.5. FIGURE 36.25a shows the rocket moving with velocity v through the solar system reference frame S. We define $L = \Delta x = x_{\mathrm{Saturn}} - x_{\mathrm{sun}}$ as the distance between the sun and Saturn in frame S or, more generally, the *length* of the spatial interval between two points. The rocket's speed is $v = L/\Delta t$, where Δt is the time measured in frame S for the journey from the sun to Saturn.

FIGURE 36.25 L and L' are the distances between the sun and Saturn in frames S and S'.

(a) Reference frame S: The solar system is stationary.

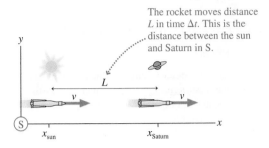

The rocket moves distance L in time Δt. This is the distance between the sun and Saturn in S.

(b) Reference frame S': The rocket is stationary.

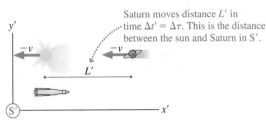

Saturn moves distance L' in time $\Delta t' = \Delta \tau$. This is the distance between the sun and Saturn in S'.

FIGURE 36.25b shows the situation in reference frame S', where the rocket is at rest. The sun and Saturn move to the left at speed $v = L'/\Delta t'$, where $\Delta t'$ is the time measured in frame S' for Saturn to travel distance L'.

Speed v is the relative speed between S and S' and is the same for experimenters in both reference frames. That is,

$$v = \frac{L}{\Delta t} = \frac{L'}{\Delta t'} \tag{36.12}$$

The time interval $\Delta t'$ measured in frame S' is the proper time $\Delta \tau$ because both events occur at the same position in frame S' and can be measured by one clock. We can use the time-dilation result, Equation 36.9, to relate $\Delta \tau$ measured by the astronauts to Δt measured by the earthbound scientists. Then Equation 36.12 becomes

$$\frac{L}{\Delta t} = \frac{L'}{\Delta \tau} = \frac{L'}{\sqrt{1 - \beta^2}\,\Delta t} \tag{36.13}$$

The Δt cancels, and the distance L' in frame S' is

$$L' = \sqrt{1 - \beta^2}\, L \tag{36.14}$$

Surprisingly, we find that **the distance between two objects in reference frame S' is *not the same* as the distance between the same two objects in reference frame S.**

Frame S, in which the distance is L, has one important distinction. It is the *one and only* inertial reference frame in which the objects are at rest. Experimenters in frame S can take all the time they need to measure L because the two objects aren't going anywhere. The distance L between two objects, or two points on one object, measured in the reference frame in which the objects are at rest is called the **proper length** ℓ. Only one inertial reference frame can measure the proper length.

We can use the proper length ℓ to write Equation 36.14 as

$$L' = \sqrt{1 - \beta^2}\,\ell \leq \ell \tag{36.15}$$

This "shrinking" of the distance between two objects, as measured by an experiment moving with respect to the objects, is called **length contraction.** Although we derived length contraction for the distance between two distinct objects, it applies equally well to the length of any physical object that stretches between two points along the x- and x'-axes. The length of an object is greatest in the reference frame in which the object is at rest. The object's length is less (i.e., the length is contracted) when it is measured in any reference frame in which the object is moving.

The Stanford Linear Accelerator (SLAC) is a 2-mi-long electron accelerator. The accelerator's length is less than 1 m in the reference frame of the electrons.

EXAMPLE 36.6 **The distance from the sun to Saturn**

In Example 36.5 a rocket traveled along a line from the sun to Saturn at a constant speed of 0.9c relative to the solar system. The Saturn-to-sun distance was given as 1.43×10^{12} m. What is the distance between the sun and Saturn in the rocket's reference frame?

MODEL Saturn and the sun are, at least approximately, at rest in the solar system reference frame S. Thus the given distance is the proper length ℓ.

SOLVE We can use Equation 36.15 to find the distance in the rocket's frame S':

$$L' = \sqrt{1 - \beta^2}\,\ell = \sqrt{1 - 0.9^2}\,(1.43 \times 10^{12}\text{ m})$$
$$= 0.62 \times 10^{12}\text{ m}$$

ASSESS The sun-to-Saturn distance measured by the astronauts is less than half the distance measured by experimenters on earth. L' and ℓ are different because *space is different* in two reference frames moving relative to each other.

FIGURE 36.26 Carmen and Dan each measure the length of the other's meter stick as they move relative to each other.

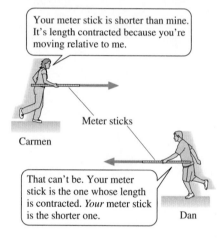

Your meter stick is shorter than mine. It's length contracted because you're moving relative to me.

Meter sticks

Carmen

That can't be. Your meter stick is the one whose length is contracted. *Your* meter stick is the shorter one.

Dan

FIGURE 36.27 Distance d is the same in both coordinate systems.

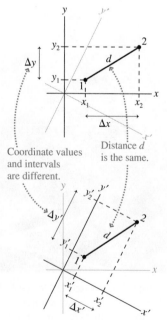

Measurements in the xy-system

Coordinate values and intervals are different.

Distance d is the same.

Measurements in the $x'y'$-system

The conclusion that space is different in reference frames moving relative to each other is a direct consequence of the fact that time is different. Experimenters in both reference frames agree on the relative velocity v, leading to Equation 36.12: $v = L/\Delta t = L'/\Delta t'$. We had already learned that $\Delta t' < \Delta t$ because of time dilation. Thus L' *has* to be less than L. That is the only way experimenters in the two reference frames can reconcile their measurements.

To be specific, the earthly experimenters in Examples 36.5 and 36.6 find that the rocket takes 5300 s to travel the 1.43×10^{12} m between the sun and Saturn. The rocket's speed is $v = L/\Delta t = 2.7 \times 10^8$ m/s $= 0.9c$. The astronauts in the rocket find that it takes only 2310 s for Saturn to reach them after the sun has passed by. But there's no conflict, because they also find that the distance is only 0.62×10^{12} m. Thus Saturn's speed toward them is $v = L'/\Delta t' = (0.62 \times 10^{12}\text{ m})/(2310\text{ s}) = 2.7 \times 10^8$ m/s $= 0.9c$.

Another Paradox?

Carmen and Dan are in their physics lab room. They each select a meter stick, lay the two side by side, and agree that the meter sticks are exactly the same length. Then, for an extra-credit project, they go outside and run past each other, in opposite directions, at a relative speed $v = 0.9c$. FIGURE 36.26 shows their experiment and a portion of their conversation.

Now, Dan's meter stick can't be both longer and shorter than Carmen's meter stick. Is this another paradox? No! Relativity allows us to compare the *same* events as they're measured in two different reference frames. This did lead to a real paradox when Peggy rolled past Ryan on the train. There the signal light on the box turns green (a single event) or it doesn't, and Peggy and Ryan have to agree about it. But the events by which Dan measures the length (in Dan's frame) of Carmen's meter stick are *not the same events* as those by which Carmen measures the length (in Carmen's frame) of Dan's meter stick.

There's no conflict between their measurements. In Dan's reference frame, Carmen's meter stick has been length contracted and is less than 1 m in length. In Carmen's reference frame, Dan's meter stick has been length contracted and is less than 1m in length. If this weren't the case, if both agreed that one of the meter sticks was shorter than the other, then we could tell which reference frame was "really" moving and which was "really" at rest. But the principle of relativity doesn't allow us to make that distinction. Each is moving relative to the other, so each should make the same measurement for the length of the other's meter stick.

The Spacetime Interval

Forget relativity for a minute and think about ordinary geometry. FIGURE 36.27 shows two ordinary coordinate systems. They are identical except for the fact that one has been rotated relative to the other. A student using the xy-system would measure coordinates (x_1, y_1) for point 1 and (x_2, y_2) for point 2. A second student, using the $x'y'$-system, would measure (x'_1, y'_1) and (x'_2, y'_2).

The students soon find that none of their measurements agree. That is, $x_1 \neq x_1'$ and so on. Even the intervals are different: $\Delta x \neq \Delta x'$ and $\Delta y \neq \Delta y'$. Each is a perfectly valid coordinate system, giving no reason to prefer one over the other, but each yields different measurements.

Is there *anything* on which the two students can agree? Yes, there is. The distance d between points 1 and 2 is independent of the coordinates. We can state this mathematically as

$$d^2 = (\Delta x)^2 + (\Delta y)^2 = (\Delta x')^2 + (\Delta y')^2 \qquad (36.16)$$

The quantity $(\Delta x)^2 + (\Delta y)^2$ is called an **invariant** in geometry because it has the same value in any Cartesian coordinate system.

Returning to relativity, is there an invariant in the spacetime coordinates, some quantity that has the *same value* in all inertial reference frames? There is, and to find it let's return to the light clock of Figure 36.21. FIGURE 36.28 shows the light clock as seen in reference frames S' and S''. The speed of light is the same in both frames, even though both are moving with respect to each other and with respect to the clock.

Notice that the clock's height h is common to both reference frames. Thus

$$h^2 = \left(\frac{1}{2}c\,\Delta t'\right)^2 - \left(\frac{1}{2}\Delta x'\right)^2 = \left(\frac{1}{2}c\,\Delta t''\right)^2 - \left(\frac{1}{2}\Delta x''\right)^2 \qquad (36.17)$$

The factor $\frac{1}{2}$ cancels, allowing us to write

$$c^2(\Delta t')^2 - (\Delta x')^2 = c^2(\Delta t'')^2 - (\Delta x'')^2 \qquad (36.18)$$

Let us define the **spacetime interval** s between two events to be

$$s^2 = c^2(\Delta t)^2 - (\Delta x)^2 \qquad (36.19)$$

What we've shown in Equation 36.18 is that **the spacetime interval s has the same value in all inertial reference frames.** That is, the spacetime interval between two events is an invariant. It is a value that all experimenters, in all reference frames, can agree upon.

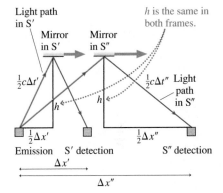

FIGURE 36.28 The light clock seen by experimenters in reference frames S' and S''.

EXAMPLE 36.7 **Using the spacetime interval**

A firecracker explodes at the origin of an inertial reference frame. Then, 2.0 μs later, a second firecracker explodes 300 m away. Astronauts in a passing rocket measure the distance between the explosions to be 200 m. According to the astronauts, how much time elapses between the two explosions?

MODEL The spacetime coordinates of two events are measured in two different inertial reference frames. Call the reference frame of the ground S and the reference frame of the rocket S'. The spacetime interval between these two events is the same in both reference frames.

SOLVE The spacetime interval (or, rather, its square) in frame S is

$$s^2 = c^2(\Delta t)^2 - (\Delta x)^2 = (600\ \text{m})^2 - (300\ \text{m})^2 = 270{,}000\ \text{m}^2$$

where we used $c = 300\ \text{m}/\mu\text{s}$ to determine that $c\,\Delta t = 600$ m. The spacetime interval has the same value in frame S'. Thus

$$s^2 = 270{,}000\ \text{m}^2 = c^2(\Delta t')^2 - (\Delta x')^2$$
$$= c^2(\Delta t')^2 - (200\ \text{m})^2$$

This is easily solved to give $\Delta t' = 1.85$ μs.

ASSESS The two events are closer together in both space and time in the rocket's reference frame than in the reference frame of the ground.

Einstein's legacy, according to popular culture, was the discovery that "everything is relative." But it's not so. Time intervals and space intervals may be relative, as were the intervals Δx and Δy in the purely geometric analogy with which we opened this section, but some things are *not* relative. In particular, the spacetime interval s between two events is not relative. It is a well-defined number, agreed on by experimenters in each and every inertial reference frame.

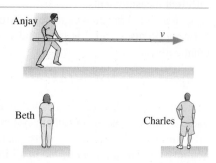

STOP TO THINK 36.7   Beth and Charles are at rest relative to each other. Anjay runs past at velocity v while holding a long pole parallel to his motion. Anjay, Beth, and Charles each measure the length of the pole at the instant Anjay passes Beth. Rank in order, from largest to smallest, the three lengths L_A, L_B, and L_C.

36.8 The Lorentz Transformations

The Galilean transformation $x' = x - vt$ of classical relativity lets us calculate the position x' of an event in frame S' if we know its position x in frame S. Classical relativity, of course, assumes that $t' = t$. Is there a similar transformation in relativity that would allow us to calculate an event's spacetime coordinates (x', t') in frame S' if we know their values (x, t) in frame S? Such a transformation would need to satisfy three conditions:

1. Agree with the Galilean transformations in the low-speed limit $v \ll c$.
2. Transform not only spatial coordinates but also time coordinates.
3. Ensure that the speed of light is the same in all reference frames.

FIGURE 36.29 The spacetime coordinates of an event are measured in inertial reference frames S and S'.

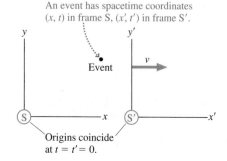

An event has spacetime coordinates (x, t) in frame S, (x', t') in frame S'.

Event

Origins coincide at $t = t' = 0$.

We'll continue to use reference frames in the standard orientation of **FIGURE 36.29**. The motion is parallel to the x- and x'-axes, and we *define* $t = 0$ and $t' = 0$ as the instant when the origins of S and S' coincide.

The requirement that a new transformation agree with the Galilean transformation when $v \ll c$ suggests that we look for a transformation of the form

$$x' = \gamma(x - vt) \quad \text{and} \quad x = \gamma(x' + vt') \tag{36.20}$$

where γ is a dimensionless function of velocity that satisfies $\gamma \to 1$ as $v \to 0$.

To determine γ, we consider the following two events:

Event 1: A flash of light is emitted from the origin of both reference frames $(x = x' = 0)$ at the instant they coincide $(t = t' = 0)$.

Event 2: The light strikes a light detector. The spacetime coordinates of this event are (x, t) in frame S and (x', t') in frame S'.

Light travels at speed c in both reference frames, so the positions of event 2 are $x = ct$ in S and $x' = ct'$ in S'. Substituting these expressions for x and x' into Equation 36.20 gives

$$ct' = \gamma(ct - vt) = \gamma(c - v)t$$
$$ct = \gamma(ct' + vt') = \gamma(c + v)t' \tag{36.21}$$

We solve the first equation for t', by dividing by c, then substitute this result for t' into the second:

$$ct = \gamma(c + v)\frac{\gamma(c - v)t}{c} = \gamma^2(c^2 - v^2)\frac{t}{c}$$

The t cancels, leading to

$$\gamma^2 = \frac{c^2}{c^2 - v^2} = \frac{1}{1 - v^2/c^2}$$

Thus the γ that "works" in the proposed transformation of Equation 36.20 is

$$\gamma = \frac{1}{\sqrt{1 - v^2/c^2}} = \frac{1}{\sqrt{1 - \beta^2}} \qquad (36.22)$$

You can see that $\gamma \rightarrow 1$ as $v \rightarrow 0$, as expected.

The transformation between t and t' is found by requiring that $x = x$ if you use Equation 36.20 to transform a position from S to S$'$ and then back to S. The details will be left for a homework problem. Another homework problem will let you demonstrate that the y and z measurements made perpendicular to the relative motion are not affected by the motion. We tacitly assumed this condition in our analysis of the light clock.

The full set of equations are called the **Lorentz transformations.** They are

$$
\begin{array}{ll}
x' = \gamma(x - vt) & x = \gamma(x' + vt') \\
y' = y & y = y' \\
z' = z & z = z' \\
t' = \gamma(t - vx/c^2) & t = \gamma(t' + vx'/c^2)
\end{array}
\qquad (36.23)
$$

The Lorentz transformations transform the spacetime coordinates of *one* event. Compare these to the Galilean transformation equations in Equations 36.1.

NOTE ▶ These transformations are named after the Dutch physicist H. A. Lorentz, who derived them prior to Einstein. Lorentz was close to discovering special relativity, but he didn't recognize that our concepts of space and time have to be changed before these equations can be properly interpreted. ◀

Using Relativity

Relativity is phrased in terms of *events;* hence relativity problems are solved by interpreting the problem statement in terms of specific events.

PROBLEM-SOLVING
STRATEGY 36.1 **Relativity**

MODEL Frame the problem in terms of events, things that happen at a specific place and time.

VISUALIZE A pictorial representation defines the reference frames.

- Sketch the reference frames, showing their motion relative to each other.
- Show events. Identify objects that are moving with respect to the reference frames.
- Identify any proper time intervals and proper lengths. These are measured in an object's rest frame.

SOLVE The mathematical representation is based on the Lorentz transformations, but not every problem requires the full transformation equations.

- Problems about time intervals can often be solved using time dilation: $\Delta t = \gamma \Delta \tau$.
- Problems about distances can often be solved using length contraction: $L = \ell/\gamma$.

ASSESS Are the results consistent with Galilean relativity when $v \ll c$?

Ryan and Peggy revisited

Peggy is standing in the center of a long, flat railroad car that has firecrackers tied to both ends. The car moves past Ryan, who is standing on the ground, with velocity $v = 0.8c$. Flashes from the exploding firecrackers reach him simultaneously 1.0 μs after the instant that Peggy passes him, and he later finds burn marks on the track 300 m to either side of where he had been standing.

a. According to Ryan, what is the distance between the two explosions, and at what times do the explosions occur relative to the time that Peggy passes him?

b. According to Peggy, what is the distance between the two explosions, and at what times do the explosions occur relative to the time that Ryan passes her?

MODEL Let the explosion on Ryan's right, the direction in which Peggy is moving, be event R. The explosion on his left is event L.

VISUALIZE Peggy and Ryan are in inertial reference frames. As FIGURE 36.30 shows, Peggy's frame S′ is moving with $v = 0.8c$ relative to Ryan's frame S. We've defined the reference frames such that Peggy and Ryan are at the origins. The instant they pass, by definition, is $t = t' = 0$ s. The two events are shown in Ryan's reference frame.

FIGURE 36.30 A pictorial representation of the reference frames and events.

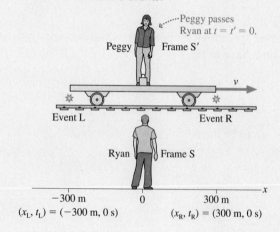

Peggy passes Ryan at $t = t' = 0$.

Peggy   Frame S′

v

Event L   Event R

Ryan   Frame S

−300 m   0   300 m x
$(x_L, t_L) = (-300 \text{ m}, 0 \text{ s})$ $(x_R, t_R) = (300 \text{ m}, 0 \text{ s})$

SOLVE a. The two burn marks tell Ryan that the distance between the explosions was $L = 600$ m. Light travels at $c = 300$ m/μs, and the burn marks are 300 m on either side of him, so Ryan can determine that each explosion took place 1.0 μs before he saw the flash. But this was the instant of time that Peggy passed him, so Ryan concludes that the explosions were simultaneous with each other and with Peggy's passing him. The spacetime coordinates of the two events in frame S are $(x_R, t_R) = (300 \text{ m}, 0 \text{ μs})$ and $(x_L, t_L) = (-300 \text{ m}, 0 \text{ μs})$.

b. We already know, from our qualitative analysis in Section 36.5, that the explosions are *not* simultaneous in Peggy's reference frame. Event R happens before event L in S′, but we don't know how they compare to the time at which Ryan passes Peggy. We can now use the Lorentz transformations to relate the spacetime coordinates of these events as measured by Ryan to the spacetime coordinates as measured by Peggy. Using $v = 0.8c$, we find that γ is

$$\gamma = \frac{1}{\sqrt{1 - v^2/c^2}} = \frac{1}{\sqrt{1 - 0.8^2}} = 1.667$$

For event L, the Lorentz transformations are

$$x'_L = 1.667((-300 \text{ m}) - (0.8c)(0 \text{ μs})) = -500 \text{ m}$$

$$t'_L = 1.667((0 \text{ μs}) - (0.8c)(-300 \text{ m})/c^2) = 1.33 \text{ μs}$$

And for event R,

$$x'_R = 1.667((300 \text{ m}) - (0.8c)(0 \text{ μs})) = 500 \text{ m}$$

$$t'_R = 1.667((0 \text{ μs}) - (0.8c)(300 \text{ m})/c^2) = -1.33 \text{ μs}$$

According to Peggy, the two explosions occur 1000 m apart. Furthermore, the first explosion, on the right, occurs 1.33 μs before Ryan passes her at $t' = 0$ s. The second, on the left, occurs 1.33 μs after Ryan goes by.

ASSESS Events that are simultaneous in frame S are *not* simultaneous in frame S′. The results of the Lorentz transformations agree with our earlier qualitative analysis.

A follow-up discussion of Example 36.8 is worthwhile. Because Ryan moves at speed $v = 0.8c = 240$ m/μs relative to Peggy, he moves 320 m during the 1.33 μs between the first explosion and the instant he passes Peggy, then another 320 m before the second explosion. Gathering this information together, FIGURE 36.31 shows the sequence of events in Peggy's reference frame.

The firecrackers define the ends of the railroad car, so the 1000 m distance between the explosions in Peggy's frame is the car's length L' in frame S′. The car is at rest in frame S′, hence length L' is the proper length: $\ell = 1000$ m. Ryan is measuring the length of a moving object, so he should see the car length contracted to

$$L = \sqrt{1 - \beta^2}\,\ell = \frac{\ell}{\gamma} = \frac{1000 \text{ m}}{1.667} = 600 \text{ m}$$

And, indeed, that is exactly the distance Ryan measured between the burn marks.

Finally, we can calculate the spacetime interval s between the two events. According to Ryan,

$$s^2 = c^2(\Delta t^2) - (\Delta x)^2 = c^2(0 \ \mu s)^2 - (600 \ \text{m})^2 = -(600 \ \text{m})^2$$

Peggy computes the spacetime interval to be

$$s^2 = c^2(\Delta t')^2 - (\Delta x')^2 = c^2(2.67 \ \mu s)^2 - (1000 \ \text{m})^2 = -(600 \ \text{m})^2$$

Their calculations of the spacetime interval agree, showing that s really is an invariant, but notice that s itself is an imaginary number.

Length

We've already introduced the idea of length contraction, but we didn't precisely define just what we mean by the *length* of a moving object. The length of an object at rest is clear because we can take all the time we need to measure it with meter sticks, surveying tools, or whatever we need. But how can we give clear meaning to the length of a moving object?

A reasonable definition of an object's length is the distance $L = \Delta x = x_R - x_L$ between the right and left ends when the positions x_R and x_L are measured *at the same time t*. In other words, length is the distance spanned by the object at *one instant* of time. Measuring an object's length requires *simultaneous* measurements of two positions (i.e., two events are required); hence the result won't be known until the information from two spatially separated measurements can be brought together.

FIGURE 36.32 shows an object traveling through reference frame S with velocity v. The object is at rest in reference frame S′ that travels with the object at velocity v; hence the length in frame S′ is the proper length ℓ. That is, $\Delta x' = x'_R - x'_L = \ell$ in frame S′.

At time t, an experimenter (and his or her assistants) in frame S makes simultaneous measurements of the positions x_R and x_L of the ends of the object. The difference $\Delta x = x_R - x_L = L$ is the length in frame S. The Lorentz transformations of x_R and x_L are

$$\begin{aligned} x'_R &= \gamma(x_R - vt) \\ x'_L &= \gamma(x_L - vt) \end{aligned} \qquad (36.24)$$

where, it is important to note, t is the *same* for both because the measurements are simultaneous.

Subtracting the second equation from the first, we find

$$x'_R - x'_L = \ell = \gamma(x_R - x_L) = \gamma L = \frac{L}{\sqrt{1 - \beta^2}}$$

Solving for L, we find, in agreement with Equation 36.15, that

$$L = \sqrt{1 - \beta^2}\, \ell \qquad (36.25)$$

This analysis has accomplished two things. First, by giving a precise definition of length, we've put our length-contraction result on a firmer footing. Second, we've had good practice at relativistic reasoning using the Lorentz transformation.

NOTE ▶ Length contraction does not tell us how an object would *look*. The visual appearance of an object is determined by light waves that arrive simultaneously at the eye. These waves left points on the object at different times (i.e., *not* simultaneously) because they had to travel different distances to the eye. The analysis needed to determine an object's visual appearance is considerably more complex. Length and length contraction are concerned only with the *actual* length of the object at one instant of time. ◀

FIGURE 36.31 The sequence of events as seen in Peggy's reference frame.

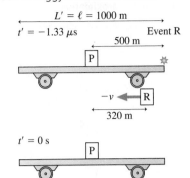

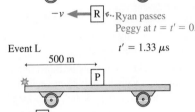

FIGURE 36.32 The length of an object is the distance between *simultaneous* measurements of the positions of the end points.

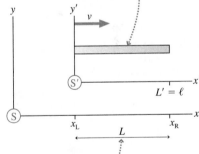

Because the object is moving in frame S, simultaneous measurements of its ends must be made to find its length L in frame S.

The Binomial Approximation

You've met the binomial approximation earlier in this text and in your calculus class. The binomial approximation is useful when we need to calculate a relativistic expression for a nonrelativistic velocity $v \ll c$. Because $v^2/c^2 \ll 1$ in these cases, we can write

$$\text{If } v \ll c: \begin{cases} \sqrt{1 - \beta^2} = (1 - v^2/c^2)^{1/2} \approx 1 - \dfrac{1}{2}\dfrac{v^2}{c^2} \\ \gamma = \dfrac{1}{\sqrt{1 - \beta^2}} = (1 - v^2/c^2)^{-1/2} \approx 1 + \dfrac{1}{2}\dfrac{v^2}{c^2} \end{cases} \quad (36.26)$$

The following example illustrates the use of the binomial approximation.

EXAMPLE 36.9 **The shrinking school bus**

An 8.0-m-long school bus drives past at 30 m/s. By how much is its length contracted?

MODEL The school bus is at rest in an inertial reference frame S′ moving at velocity $v = 30$ m/s relative to the ground frame S. The given length, 8.0 m, is the proper length ℓ in frame S′.

SOLVE In frame S, the school bus is length contracted to

$$L = \sqrt{1 - \beta^2}\,\ell$$

The bus's velocity v is much less than c, so we can use the binomial approximation to write

$$L \approx \left(1 - \frac{1}{2}\frac{v^2}{c^2}\right)\ell = \ell - \frac{1}{2}\frac{v^2}{c^2}\ell$$

The *amount* of the length contraction is

$$\ell - L = \frac{1}{2}\frac{v^2}{c^2}\ell = \frac{1}{2}\left(\frac{30 \text{ m/s}}{3.0 \times 10^8 \text{ m/s}}\right)^2 (8.0 \text{ m})$$

$$= 4.0 \times 10^{-14} \text{ m} = 40 \text{ fm}$$

where 1 fm = 1 femtometer = 10^{-15} m.

ASSESS The bus "shrinks" by only slightly more than the diameter of the nucleus of an atom. It's no wonder that we're not aware of length contraction in our everyday lives. If you had tried to calculate this number exactly, your calculator would have shown $\ell - L = 0$ because the difference between ℓ and L shows up only in the 14th decimal place. A scientific calculator determines numbers to 10 or 12 decimal places, but that isn't sufficient to show the difference. The binomial approximation provides an invaluable tool for finding the very tiny difference between two numbers that are nearly identical.

The Lorentz Velocity Transformations

FIGURE 36.33 The velocity of a moving object is measured to be u in frame S and u' in frame S′.

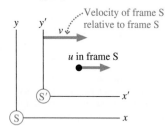

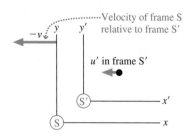

FIGURE 36.33 shows an object that is moving in both reference frame S and reference frame S′. Experimenters in frame S determine that the object's velocity is u, while experimenters in frame S′ find it to be u'. For simplicity, we'll assume that the object moves parallel to the x- and x′-axes.

The Galilean velocity transformation $u' = u - v$ was found by taking the time derivative of the position transformation. We can do the same with the Lorentz transformation if we take the derivative with respect to the time in each frame. Velocity u' in frame S′ is

$$u' = \frac{dx'}{dt'} = \frac{d(\gamma(x - vt))}{d(\gamma(t - vx/c^2))} \quad (36.27)$$

where we've used the Lorentz transformations for position x' and time t'.

Carrying out the differentiation gives

$$u' = \frac{\gamma(dx - v\,dt)}{\gamma(dt - v\,dx/c^2)} = \frac{dx/dt - v}{1 - v(dx/dt)/c^2} \quad (36.28)$$

But dx/dt is u, the object's velocity in frame S, leading to

$$u' = \frac{u - v}{1 - uv/c^2} \quad (36.29)$$

You can see that Equation 36.29 reduces to the Galilean transformation $u' = u - v$ when $v \ll c$, as expected.

The transformation from S′ to S is found by reversing the sign of v. Altogether,

$$u' = \frac{u - v}{1 - uv/c^2} \quad \text{and} \quad u = \frac{u' + v}{1 + u'v/c^2} \qquad (36.30)$$

Equations 36.30 are the Lorentz velocity transformation equations.

NOTE ▶ It is important to distinguish carefully between v, which is the relative velocity between two reference frames, and u and u', which are the velocities of an *object* as measured in the two different reference frames. ◀

EXAMPLE 36.10 **A really fast bullet**

A rocket flies past the earth at $0.90c$. As it goes by, the rocket fires a bullet in the forward direction at $0.95c$ with respect to the rocket. What is the bullet's speed with respect to the earth?

MODEL The rocket and the earth are inertial reference frames. Let the earth be frame S and the rocket be frame S′. The velocity of frame S′ relative to frame S is $v = 0.90c$. The bullet's velocity in frame S′ is $u' = 0.95c$.

SOLVE We can use the Lorentz velocity transformation to find

$$u = \frac{u' + v}{1 + u'v/c^2} = \frac{0.95c + 0.90c}{1 + (0.95c)(0.90c)/c^2} = 0.997c$$

The bullet's speed with respect to the earth is 99.7% of the speed of light.

NOTE ▶ Many relativistic calculations are much easier when velocities are specified as a fraction of c. ◀

ASSESS In Newtonian mechanics, the Galilean transformation of velocity would give $u = 1.85c$. Now, despite the very high speed of the rocket and of the bullet with respect to the rocket, the bullet's speed with respect to the earth remains less than c. This is yet more evidence that objects cannot travel faster than the speed of light.

Suppose the rocket in Example 36.10 fired a laser beam in the forward direction as it traveled past the earth at velocity v. The laser beam would travel away from the rocket at speed $u' = c$ in the rocket's reference frame S′. What is the laser beam's speed in the earth's frame S? According to the Lorentz velocity transformation, it must be

$$u = \frac{u' + v}{1 + u'v/c^2} = \frac{c + v}{1 + cv/c^2} = \frac{c + v}{1 + v/c} = \frac{c + v}{(c + v)/c} = c \qquad (36.31)$$

Light travels at speed c in both frame S and frame S′. This important consequence of the principle of relativity is "built into" the Lorentz transformations.

36.9 Relativistic Momentum

In Newtonian mechanics, the total momentum of a system is a conserved quantity. Further, as we've seen, the law of conservation of momentum, $P_f = P_i$, is true in all inertial reference frames *if* the particle velocities in different reference frames are related by the Galilean velocity transformations.

The difficulty, of course, is that the Galilean transformations are not consistent with the principle of relativity. It is a reasonable approximation when all velocities are very much less than c, but the Galilean transformations fail dramatically as velocities approach c. It's not hard to show that $P'_f \neq P'_i$ if the particle velocities in frame S′ are related to the particle velocities in frame S by the Lorentz transformations.

There are two possibilities:

1. The so-called law of conservation of momentum is not really a law of physics. It is approximately true at low velocities but fails as velocities approach the speed of light.
2. The law of conservation of momentum really is a law of physics, but the expression $p = mu$ is not the correct way to calculate momentum when the particle velocity u becomes a significant fraction of c.

Momentum conservation is such a central and important feature of mechanics that it seems unlikely to fail in relativity.

The classical momentum, for one-dimensional motion, is $p = mu = m(\Delta x/\Delta t)$. Δt is the time to move distance Δx. That seemed clear enough within a Newtonian framework, but now we've learned that experimenters in different reference frames disagree about the amount of time needed. So whose Δt should we use?

One possibility is to use the time measured *by the particle*. This is the proper time $\Delta\tau$ because the particle is at rest in its own reference frame and needs only one clock. With this in mind, let's redefine the momentum of a particle of mass m moving with velocity $u = \Delta x/\Delta t$ to be

$$p = m\frac{\Delta x}{\Delta\tau} \tag{36.32}$$

We can relate this new expression for p to the familiar Newtonian expression by using the time-dilation result $\Delta\tau = (1 - u^2/c^2)^{1/2}\Delta t$ to relate the proper time interval measured by the particle to the more practical time interval Δt measured by experimenters in frame S. With this substitution, Equation 36.32 becomes

$$p = m\frac{\Delta x}{\Delta\tau} = m\frac{\Delta x}{\sqrt{1 - u^2/c^2}\,\Delta t} = \frac{mu}{\sqrt{1 - u^2/c^2}} \tag{36.33}$$

You can see that Equation 36.33 reduces to the classical expression $p = mu$ when the particle's speed $u \ll c$. That is an important requirement, but whether this is the "correct" expression for p depends on whether the total momentum P is conserved when the velocities of a system of particles are transformed with the Lorentz velocity transformation equations. The proof is rather long and tedious, so we will assert, without actual proof, that the momentum defined in Equation 36.33 does, indeed, transform correctly. **The law of conservation of momentum is still valid in all inertial reference frames *if* the momentum of each particle is calculated with Equation 36.33.**

The factor that multiplies mu in Equation 36.33 looks much like the factor γ in the Lorentz transformation equations for x and t, but there's one very important difference. The v in the Lorentz transformation equations is the velocity of a *reference frame*. The u in Equation 36.33 is the velocity of a particle moving *in* a reference frame.

With this distinction in mind, let's define the quantity

$$\gamma_p = \frac{1}{\sqrt{1 - u^2/c^2}} \tag{36.34}$$

where the subscript p indicates that this is γ for a particle, not for a reference frame. In frame S', where the particle moves with velocity u', the corresponding expression would be called γ_p'. With this definition of γ_p, the momentum of a particle is

$$p = \gamma_p mu \tag{36.35}$$

EXAMPLE 36.11 **Momentum of a subatomic particle**

Electrons in a particle accelerator reach a speed of $0.999c$ relative to the laboratory. One collision of an electron with a target produces a muon that moves forward with a speed of $0.95c$ relative to the laboratory. The muon mass is 1.90×10^{-28} kg. What is the muon's momentum in the laboratory frame and in the frame of the electron beam?

MODEL Let the laboratory be reference frame S. The reference frame S' of the electron beam (i.e., a reference frame in which the electrons are at rest) moves in the direction of the electrons at $v = 0.999c$. The muon velocity in frame S is $u = 0.95c$.

SOLVE γ_p for the muon in the laboratory reference frame is

$$\gamma_p = \frac{1}{\sqrt{1 - u^2/c^2}} = \frac{1}{\sqrt{1 - 0.95^2}} = 3.20$$

Thus the muon's momentum in the laboratory is

$$p = \gamma_p mu = (3.20)(1.90 \times 10^{-28}\text{ kg})(0.95 \times 3.00 \times 10^8\text{ m/s})$$
$$= 1.73 \times 10^{-19}\text{ kg m/s}$$

The momentum is a factor of 3.2 larger than the Newtonian momentum mu. To find the momentum in the electron-beam reference frame, we must first use the velocity transformation equation to find the muon's velocity in frame S':

$$u' = \frac{u - v}{1 - uv/c^2} = \frac{0.95c - 0.999c}{1 - (0.95c)(0.999c)/c^2} = -0.962c$$

In the laboratory frame, the faster electrons are overtaking the slower muon. Hence the muon's velocity in the electron-beam frame is negative. γ'_p for the muon in frame S' is

$$\gamma'_p = \frac{1}{\sqrt{1 - u'^2/c^2}} = \frac{1}{\sqrt{1 - 0.962^2}} = 3.66$$

The muon's momentum in the electron-beam reference frame is

$$p' = \gamma'_p\, mu'$$
$$= (3.66)(1.90 \times 10^{-28}\,\text{kg})(-0.962 \times 3.00 \times 10^8\,\text{m/s})$$
$$= -2.01 \times 10^{-19}\,\text{kg m/s}$$

ASSESS From the laboratory perspective, the muon moves only slightly slower than the electron beam. But it turns out that the muon moves faster with respect to the electrons, although in the opposite direction, than it does with respect to the laboratory.

The Cosmic Speed Limit

FIGURE 36.34a is a graph of momentum versus velocity. For a Newtonian particle, with $p = mu$, the momentum is directly proportional to the velocity. The relativistic expression for momentum agrees with the Newtonian value if $u \ll c$, but p approaches ∞ as $u \rightarrow c$.

The implications of this graph become clear when we relate momentum to force. Consider a particle subjected to a constant force, such as a rocket that never runs out of fuel. If F is constant, we can see from $F = dp/dt$ that the momentum is $p = Ft$. If Newtonian physics were correct, a particle would go faster and faster as its velocity $u = p/m = (F/m)t$ increased without limit. But the relativistic result, shown in **FIGURE 36.34b**, is that the particle's velocity asymptotically approaches the speed of light ($u \rightarrow c$) as p approaches ∞. Relativity gives a very different outcome than Newtonian mechanics.

The speed c is a "cosmic speed limit" for material particles. A force cannot accelerate a particle to a speed higher than c because the particle's momentum becomes infinitely large as the speed approaches c. The amount of effort required for each additional increment of velocity becomes larger and larger until no amount of effort can raise the velocity any higher.

Actually, at a more fundamental level, c is a speed limit for *any* kind of **causal influence**. If I throw a rock and break a window, my throw is the *cause* of the breaking window and the rock is the *causal influence*. If I shoot a laser beam at a light detector that is wired to a firecracker, the light wave is the *causal influence* that leads to the explosion. A causal influence can be any kind of particle, wave, or information that travels from A to B and allows A to be the cause of B.

For two unrelated events—a firecracker explodes in Tokyo and a balloon bursts in Paris—the relativity of simultaneity tells us that they may be simultaneous in one reference frame but not in others. Or in one reference frame the firecracker may explode before the balloon bursts but in some other reference frame the balloon may burst first. These possibilities violate our commonsense view of time, but they're not in conflict with the principle of relativity.

For two causally related events—A *causes* B—it would be nonsense for an experimenter in any reference frame to find that B occurs before A. No experimenter in any reference frame, no matter how it is moving, will find that you are born before your mother is born. If A causes B, then it must be the case that $t_A < t_B$ in *all* reference frames.

Suppose there exists some kind of causal influence that *can* travel at speed $u > c$. **FIGURE 36.35** shows a reference frame S in which event A occurs at position $x_A = 0$. The faster-than-light causal influence—perhaps some yet-to-be-discovered "z ray"—leaves A at $t_A = 0$ and travels to the point at which it will cause event B. It arrives at x_B at time $t_B = x_B/u$.

How do events A and B appear in a reference frame S' that travels at an ordinary speed $v < c$ relative to frame S? We can use the Lorentz transformations to find out.

FIGURE 36.34 The speed of a particle cannot reach the speed of light.

(a)

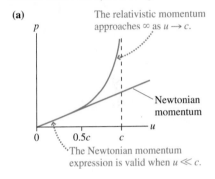

The relativistic momentum approaches ∞ as $u \rightarrow c$.

Newtonian momentum

The Newtonian momentum expression is valid when $u \ll c$.

(b)

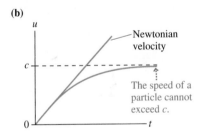

Newtonian velocity

The speed of a particle cannot exceed c.

FIGURE 36.35 Assume that a causal influence can travel from A to B at a speed $u > c$.

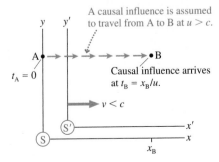

A causal influence is assumed to travel from A to B at $u > c$.

Causal influence arrives at $t_B = x_B/u$.

Because $x_A = 0$ and $t_A = 0$, it's easy to see that $x'_A = 0$ and $t'_A = 0$. That is, the origins of S and S' overlap at the instant the causal influence leaves event A. More interesting is the time at which this influence reaches B in frame S'. The Lorentz time transformation for event B is

$$t'_B = \gamma\left(t_B - \frac{vx_B}{c^2}\right) = \gamma t_B\left(1 - \frac{v(x_B/t_B)}{c^2}\right) = \gamma t_B\left(1 - \frac{vu}{c^2}\right) \quad (36.36)$$

where we first factored out t_B, then made use of the fact that $u = x_B/t_B$ in frame S.

We're assuming $u > c$, so let $u = \alpha c$ where $\alpha > 1$ is a constant. Then $vu/c^2 = \alpha v/c$. Now follow the logic:

1. If $v > c/\alpha$, which is possible because $\alpha > 1$, then $vu/c^2 > 1$.
2. If $vu/c^2 > 1$, then the term $(1 - vu/c^2)$ is negative and $t'_B < 0$.
3. If $t'_B < 0$, then event B happens *before* event A in reference frame S'.

In other words, if a causal influence can travel faster than c, then there exist reference frames in which the effect happens before the cause. We know this can't happen, so our assumption $u > c$ must be wrong. **No causal influence of any kind—particle, wave, or yet-to-be-discovered z rays—can travel faster than c.**

The existence of a cosmic speed limit is one of the most interesting consequences of the theory of relativity. "Warp drive," in which a spaceship suddenly leaps to faster-than-light velocities, is simply incompatible with the theory of relativity. Rapid travel to the stars will remain in the realm of science fiction unless future scientific discoveries find flaws in Einstein's theory and open the doors to yet-undreamed-of theories. While we can't say with certainty that a scientific theory will never be overturned, there is currently not even a hint of evidence that disagrees with the special theory of relativity.

36.10 Relativistic Energy

Energy is our final topic in this chapter on relativity. Space, time, velocity, and momentum are changed by relativity, so it seems inevitable that we'll need a new view of energy.

In Newtonian mechanics, a particle's kinetic energy $K = \frac{1}{2}mu^2$ can be written in terms of its momentum $p = mu$ as $K = p^2/2m$. This suggests that a relativistic expression for energy will likely involve both the square of p and the particle's mass. We also hope that energy will be conserved in relativity, so a reasonable starting point is with the one quantity we've found that is the same in all inertial reference frames: the spacetime interval s.

Let a particle of mass m move through distance Δx during a time interval Δt, as measured in reference frame S. The spacetime interval is

$$s^2 = c^2(\Delta t)^2 - (\Delta x)^2 = \text{invariant}$$

We can turn this into an expression involving momentum if we multiply by $(m/\Delta\tau)^2$, where $\Delta\tau$ is the proper time (i.e., the time measured by the particle). Doing so gives

$$(mc)^2\left(\frac{\Delta t}{\Delta\tau}\right)^2 - \left(\frac{m\Delta x}{\Delta\tau}\right)^2 = (mc)^2\left(\frac{\Delta t}{\Delta\tau}\right)^2 - p^2 = \text{invariant} \quad (36.37)$$

where we used $p = m(\Delta x/\Delta\tau)$ from Equation 36.32.

Now Δt, the time interval in frame S, is related to the proper time by the time-dilation result $\Delta t = \gamma_p \Delta\tau$. With this change, Equation 36.37 becomes

$$(\gamma_p mc)^2 - p^2 = \text{invariant}$$

Finally, for reasons that will be clear in a minute, we multiply by c^2, to get

$$(\gamma_p mc^2)^2 - (pc)^2 = \text{invariant} \quad (36.38)$$

To say that the right side is an *invariant* means it has the same value in all inertial reference frames. We can easily determine the constant by evaluating it in the reference frame in which the particle is at rest. In that frame, where $p = 0$ and $\gamma_p = 1$, we find that

$$(\gamma_p mc^2)^2 - (pc)^2 = (mc^2)^2 \tag{36.39}$$

Let's reflect on what this means before taking the next step. The space-time interval s has the same value in all inertial reference frames. In other words, $c^2(\Delta t)^2 - (\Delta x)^2 = c^2(\Delta t')^2 - (\Delta x')^2$. Equation 36.39 was derived from the definition of the spacetime interval; hence the quantity mc^2 is also an invariant having the same value in all inertial reference frames. In other words, if experimenters in frames S and S′ both make measurements on this particle of mass m, they will find that

$$(\gamma_p mc^2)^2 - (pc)^2 = (\gamma_p' mc^2)^2 - (p'c)^2 \tag{36.40}$$

Experimenters in different reference frames measure different values for the momentum, but experimenters in all reference frames agree that momentum is a conserved quantity. Equations 36.39 and 36.40 suggest that the quantity $\gamma_p mc^2$ is also an important property of the particle, a property that changes along with p in just the right way to satisfy Equation 36.39. But what is this property?

The first clue comes from checking the units. γ_p is dimensionless and c is a velocity, so $\gamma_p mc^2$ has the same units as the classical expression $\frac{1}{2}mv^2$—namely, units of energy. For a second clue, let's examine how $\gamma_p mc^2$ behaves in the low-velocity limit $u \ll c$. We can use the binomial approximation expression for γ_p to find

$$\gamma_p mc^2 = \frac{mc^2}{\sqrt{1 - u^2/c^2}} \approx \left(1 + \frac{1}{2}\frac{u^2}{c^2}\right)mc^2 = mc^2 + \frac{1}{2}mu^2 \tag{36.41}$$

The second term, $\frac{1}{2}mu^2$, is the low-velocity expression for the kinetic energy K. This is an energy associated with motion. But the first term suggests that the concept of energy is more complex than we originally thought. It appears that **there is an inherent energy associated with mass itself.**

With that as a possibility, subject to experimental verification, let's define the **total energy** E of a particle to be

$$E = \gamma_p mc^2 = E_0 + K = \text{rest energy} + \text{kinetic energy} \tag{36.42}$$

This total energy consists of a **rest energy**

$$E_0 = mc^2 \tag{36.43}$$

and a relativistic expression for the *kinetic energy*

$$K = (\gamma_p - 1)mc^2 = (\gamma_p - 1)E_0 \tag{36.44}$$

This expression for the kinetic energy is very nearly $\frac{1}{2}mu^2$ when $u \ll c$ but, as FIGURE 36.36 shows, differs significantly from the classical value for very high velocities.

Equation 36.43 is, of course, Einstein's famous $E = mc^2$, perhaps the most famous equation in all of physics. Before discussing its significance, we need to tie up some loose ends. First, notice that the right-hand side of Equation 36.39 is the square of the rest energy E_0. Thus we can write a final version of that equation:

$$E^2 - (pc)^2 = E_0^2 \tag{36.45}$$

The quantity E_0 is an *invariant* with the same value mc^2 in *all* inertial reference frames.

Second, notice that we can write

$$pc = (\gamma_p mu)c = \frac{u}{c}(\gamma_p mc^2)$$

FIGURE 36.36 The relativistic kinetic energy.

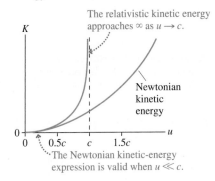

The relativistic kinetic energy approaches ∞ as $u \to c$.

K

Newtonian kinetic energy

0 $0.5c$ c $1.5c$ u

The Newtonian kinetic-energy expression is valid when $u \ll c$.

But $\gamma_p mc^2$ is the total energy E and $u/c = \beta_p$, where the subscript p, as on γ_p, indicates that we're referring to the motion of a particle within a reference frame, not the motion of two reference frames relative to each other. Thus

$$pc = \beta_p E \qquad (36.46)$$

FIGURE 36.37 shows the "velocity-energy-momentum triangle," a convenient way to remember the relationships among the three quantities.

FIGURE 36.37 The velocity-energy-momentum triangle.

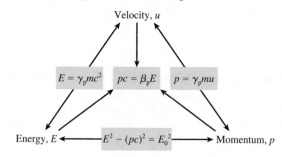

Velocity, u

$E = \gamma_p mc^2$ $pc = \beta_p E$ $p = \gamma_p mu$

Energy, E ←——— $E^2 - (pc)^2 = E_0^{\,2}$ ———→ Momentum, p

EXAMPLE 36.12 Kinetic energy and total energy

Calculate the rest energy and the kinetic energy of (a) a 100 g ball moving with a speed of 100 m/s and (b) an electron with a speed of 0.999c.

MODEL The ball, with $u \ll c$, is a classical particle. We don't need to use the relativistic expression for its kinetic energy. The electron is highly relativistic.

SOLVE a. For the ball, with $m = 0.10$ kg,

$$E_0 = mc^2 = 9.0 \times 10^{15} \text{ J}$$

$$K = \frac{1}{2} mu^2 = 500 \text{ J}$$

b. For the electron, we start by calculating

$$\gamma_p = \frac{1}{(1 - u^2/c^2)^{1/2}} = 22.4$$

Then, using $m_e = 9.11 \times 10^{-31}$ kg, we find

$$E_0 = mc^2 = 8.2 \times 10^{-14} \text{ J}$$

$$K = (\gamma_p - 1)E_0 = 170 \times 10^{-14} \text{ J}$$

ASSESS The ball's kinetic energy is a typical kinetic energy. Its rest energy, by contrast, is a staggeringly large number. For a relativistic electron, on the other hand, the kinetic energy is more important than the rest energy.

STOP TO THINK 36.8 An electron moves through the lab at 99% the speed of light. The lab reference frame is S and the electron's reference frame is S′. In which reference frame is the electron's rest mass larger?

a. In frame S, the lab frame
b. In frame S′, the electron's frame
c. It is the same in both frames.

FIGURE 36.38 An inelastic collision between two balls of clay does not seem to conserve the total energy E.

m m

K K $E_i = 2mc^2 + 2K$

$E_f = 2mc^2$?

Mass-Energy Equivalence

Now we're ready to explore the significance of Einstein's famous equation $E = mc^2$. FIGURE 36.38 shows two balls of clay approaching each other. They have equal masses and equal kinetic energies, and they slam together in a perfectly inelastic collision to form one large ball of clay at rest. In Newtonian mechanics, we would say that the initial energy $2K$ is dissipated by being transformed into an equal amount of thermal energy, raising the temperature of the coalesced ball of clay. But Equation 36.42, $E = E_0 + K$, doesn't say anything about thermal energy. The total energy before the collision is $E_i = 2mc^2 + 2K$, with the factor of 2 appearing because there are two masses. It seems like the total energy after the collision, when the clay is at rest, should be $2mc^2$, but this value doesn't conserve total energy.

There's ample experimental evidence that energy is conserved, so there must be a flaw in our reasoning. The statement of energy conservation is

$$E_f = Mc^2 = E_i = 2mc^2 + 2K \qquad (36.47)$$

where M is the mass of clay after the collision. But, remarkably, this requires

$$M = 2m + \frac{2K}{c^2} \qquad (36.48)$$

In other words, **mass is not conserved.** The mass of clay after the collision is larger than the mass of clay before the collision. Total energy can be conserved only if kinetic energy is transformed into an "equivalent" amount of mass.

The mass increase in a collision between two balls of clay is incredibly small, far beyond any scientist's ability to detect. So how do we know if such a crazy idea is true?

FIGURE 36.39 shows an experiment that has been done countless times in the last 50 years at particle accelerators around the world. An electron that has been accelerated to $u \approx c$ is aimed at a target material. When a high-energy electron collides with an atom in the target, it can easily knock one of the electrons out of the atom. Thus we would expect to see two electrons leaving the target: the incident electron and the ejected electron. Instead, *four* particles emerge from the target: three electrons and a positron. A *positron*, or positive electron, is the antimatter version of an electron, identical to an electron in all respects other than having charge $q = +e$.

In chemical-reaction notation, the collision is

$$e^- \text{ (fast)} + e^- \text{ (at rest)} \rightarrow e^- + e^- + e^- + e^+$$

An electron and a positron have been *created*, apparently out of nothing. Mass $2m_e$ before the collision has become mass $4m_e$ after the collision. (Notice that charge has been conserved in this collision.)

Although the mass has increased, it wasn't created "out of nothing." This is an inelastic collision, just like the collision of the balls of clay, because the kinetic energy after the collision is less than before. In fact, if you measured the energies before and after the collision, you would find that the decrease in kinetic energy is exactly equal to the energy equivalent of the two particles that have been created: $\Delta K = 2m_e c^2$. The new particles have been created *out of energy*!

Particles can be created from energy, and particles can return to energy. FIGURE 36.40 shows an electron colliding with a positron, its antimatter partner. When a particle and its antiparticle meet, they *annihilate* each other. The mass disappears, and the energy equivalent of the mass is transformed into light. In Chapter 38, you'll learn that light is *quantized*, meaning that light is emitted and absorbed in discrete chunks of energy called *photons*. For light with wavelength λ, the energy of a photon is $E_{\text{photon}} = hc/\lambda$, where $h = 6.63 \times 10^{-34}$ J s is called *Planck's constant*. Photons carry momentum as well as energy. Conserving both energy and momentum in the annihilation of an electron and a positron requires the emission in opposite directions of two photons of equal energy.

If the electron and positron are fairly slow, so that $K \ll mc^2$, then $E_i \approx E_0 = mc^2$. In that case, energy conservation requires

$$E_f = 2E_{\text{photon}} = E_i \approx 2m_e c^2 \qquad (36.49)$$

Hence the wavelength of the emitted photons is

$$\lambda = \frac{hc}{m_e c^2} \approx 0.0024 \text{ nm} \qquad (36.50)$$

The tracks of elementary particles in a bubble chamber show the creation of an electron-positron pair. The negative electron and positive positron spiral in opposite directions in the magnetic field.

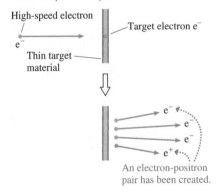

FIGURE 36.39 An inelastic collision between electrons can create an electron-positron pair.

High-speed electron

e^-

Target electron e^-

Thin target material

An electron-positron pair has been created.

e^-
e^-
e^-
e^+

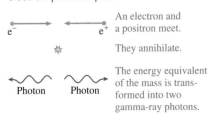

FIGURE 36.40 The annihilation of an electron-positron pair.

e^- e^+ An electron and a positron meet.

They annihilate.

Photon   Photon

The energy equivalent of the mass is transformed into two gamma-ray photons.

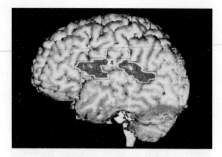

Positron-electron annihilation (a PET scan) provides a noninvasive look into the brain.

This is an extremely short wavelength, even shorter than the wavelengths of x rays. Photons in this wavelength range are called *gamma rays*. And, indeed, the emission of 0.0024 nm gamma rays is observed in many laboratory experiments in which positrons are able to collide with electrons and thus annihilate. In recent years, with the advent of gamma-ray telescopes on satellites, astronomers have found 0.0024 nm photons coming from many places in the universe, especially galactic centers—evidence that positrons are abundant throughout the universe.

Positron-electron annihilation is also the basis of the medical procedure known as a positron-emission tomography, or PET scans. A patient ingests a very small amount of a radioactive substance that decays by the emission of positrons. This substance is taken up by certain tissues in the body, especially those tissues with a high metabolic rate. As the substance decays, the positrons immediately collide with electrons, annihilate, and create two gamma-ray photons that are emitted back to back. The gamma rays, which easily leave the body, are detected, and their trajectories are traced backward into the body. The overlap of many such trajectories shows quite clearly the tissue in which the positron emission is occurring. The results are usually shown as false-color photographs, with redder areas indicating regions of higher positron emission.

Conservation of Energy

The creation and annihilation of particles with mass, processes strictly forbidden in Newtonian mechanics, are vivid proof that neither mass nor the Newtonian definition of energy is conserved. Even so, the *total* energy—the kinetic energy *and* the energy equivalent of mass—remains a conserved quantity.

> **Law of conservation of total energy** The energy $E = \sum E_i$ of an isolated system is conserved, where $E_i = (\gamma_p)_i m_i c^2$ is the total energy of particle i.

Mass and energy are not the same thing, but, as the last few examples have shown, they are *equivalent* in the sense that mass can be transformed into energy and energy can be transformed into mass as long as the total energy is conserved.

Probably the most well-known application of the conservation of total energy is nuclear fission. The uranium isotope ^{236}U, containing 236 protons and neutrons, does not exist in nature. It can be created when a ^{235}U nucleus absorbs a neutron, increasing its atomic mass from 235 to 236. The ^{236}U nucleus quickly fragments into two smaller nuclei and several extra neutrons, a process known as **nuclear fission**. The nucleus can fragment in several ways, but one is

$$n + {}^{235}U \rightarrow {}^{236}U \rightarrow {}^{144}Ba + {}^{89}Kr + 3n$$

Ba and Kr are the atomic symbols for barium and krypton.

This reaction seems like an ordinary chemical reaction—until you check the masses. The masses of atomic isotopes are known with great precision from many decades of measurement in instruments called mass spectrometers. If you add up the masses on both sides, you find that the mass of the products is 0.185 u smaller than the mass of the initial neutron and ^{235}U, where, you will recall, 1 u = 1.66×10^{-27} kg is the atomic mass unit. In kilograms the mass loss is 3.07×10^{-28} kg.

Mass has been lost, but the energy equivalent of the mass has not. As FIGURE 36.41 shows, the mass has been converted to kinetic energy, causing the two product nuclei and three neutrons to be ejected at very high speeds. The kinetic energy is easily calculated: $\Delta K = m_{lost} c^2 = 2.8 \times 10^{-11}$ J.

This is a very tiny amount of energy, but it is the energy released from *one* fission. The number of nuclei in a macroscopic sample of uranium is on the order of N_A, Avogadro's number. Hence the energy available if *all* the nuclei fission is enormous. This energy, of course, is the basis for both nuclear power reactors and nuclear weapons.

FIGURE 36.41 In nuclear fission, the energy equivalent of lost mass is converted into kinetic energy.

The mass of the reactants is 0.185 u more than the mass of the products.

n •→

^{235}U

⇩

^{236}U

⇩

^{144}Ba ^{89}Kr

0.185 u of mass has been converted into kinetic energy.

We started this chapter with an expectation that relativity would challenge our basic notions of space and time. We end by finding that relativity changes our understanding of mass and energy. Most remarkable of all is that each and every one of these new ideas flows from one simple statement: The laws of physics are the same in all inertial reference frames.

CHALLENGE EXAMPLE 36.13 Goths and Huns

The rockets of the Goths and the Huns are each 1000 m long in their rest frame. The rockets pass each other, virtually touching, at a relative speed of 0.8c. The Huns have a laser cannon at the rear of their rocket that fires a deadly laser beam perpendicular to the rocket's motion. The captain of the Huns wants to send a threatening message to the Goths by "firing a shot across their bow." He tells his first mate, "The Goths' rocket is length contracted to 600 m. Fire the laser cannon at the instant the tail of their rocket passes the nose of ours. The laser beam will cross 400 m in front of them."

But things are different in the Goths' reference frame. The Goth captain muses, "The Huns' rocket is length contracted to 600 m, 400 m shorter than our rocket. If they fire as the nose of their ship passes the tail of ours, the lethal laser beam will pass right through our side."

The first mate on the Huns' rocket fires as ordered. Does the laser beam blast the Goths or not?

MODEL Both rockets are inertial reference frames. Let the Huns' rocket be frame S and the Goths' rocket be frame S'. S' moves with velocity $v = 0.8c$ relative to S. We need to describe the situation in terms of events.

VISUALIZE Begin by considering the situation from the Huns' reference frame, as shown in **FIGURE 36.42**.

FIGURE 36.42 The situation seen by the Huns.

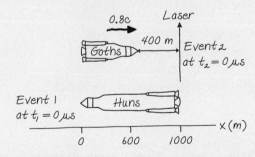

SOLVE The key to resolving the paradox is that two events simultaneous in one reference frame are not simultaneous in a different reference frame. The Huns do, indeed, see the Goths' rocket length contracted to $L_{Goths} = (1 - (0.8)^2)^{1/2} (1000 \text{ m}) = 600 \text{ m}$. Let event 1 be the tail of the Goths' rocket passing the nose of the Huns' rocket. Since we're free to define the origin of our coordinate system, we define this event to be at time $t_1 = 0 \text{ μs}$ and at position $x_1 = 0 \text{ m}$. Then, in the Huns' reference frame, the spacetime coordinates of event 2, the firing of the laser cannon,

are $(x_2, t_2) = (1000 \text{ m}, 0 \text{ μs})$. The nose of the Goths' rocket is at $x = 600 \text{ m}$ at $t = 0 \text{ μs}$; thus the laser cannon misses the Goths by 400 m.

Now we can use the Lorentz transformations to find the spacetime coordinates of the events in the Goths' reference frame. The nose of the Huns' rocket passes the tail of the Goths' rocket at $(x_1', t_1') = (0 \text{ m}, 0 \text{ μs})$. The Huns fire their laser cannon at

$$x_2' = \gamma(x_2 - vt_2) = \frac{5}{3}(1000 \text{ m} - 0 \text{ m}) = 1667 \text{ m}$$

$$t_2' = \gamma\left(t_2 - \frac{vx_2}{c^2}\right) = \frac{5}{3}\left(0 \text{ μs} - (0.8)\frac{1000 \text{ m}}{300 \text{ m/μs}}\right) = -4.444 \text{ μs}$$

where we calculated $\gamma = 5/3$ for $v = 0.8c$. Events 1 and 2 are *not* simultaneous in S'. The Huns fire the laser cannon 4.444 μs *before* the nose of their rocket reaches the tail of the Goths' rocket. The laser is fired at $x_2' = 1667 \text{ m}$, missing the nose of the Goths' rocket by 667 m. **FIGURE 36.43** shows how the Goths see things.

FIGURE 36.43 The situation seen by the Goths.

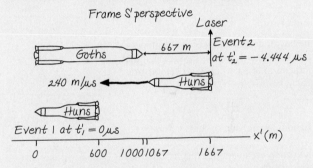

In fact, since the Huns' rocket is length contracted to 600 m, the nose of the Huns' rocket is at $x' = 1667 \text{ m} - 600 \text{ m} = 1067 \text{ m}$ at the instant they fire the laser cannon. At a speed of $v = 0.8c = 240 \text{ m/μs}$, in 4.444 μs the nose of the Huns' rocket travels $\Delta x' = (240 \text{ m/μs})(4.444 \text{ μs}) = 1067 \text{ m}$—exactly the right distance to be at the tail of the Goths' rocket at $t_1' = 0 \text{ μs}$. We could also note that the 667 m "miss distance" in the Goths' frame is length contracted to $(1 - (0.8)^2)^{1/2} (667 \text{ m}) = 400 \text{ m}$ in the Huns' frame—exactly the amount by which the Huns think they miss the Goths' rocket.

ASSESS Thus we end up with a consistent explanation. The Huns miss the Goths' rocket because, to them, the Goths' rocket is length contracted. The Goths find that the Huns miss because event 2 (the firing of the laser cannon) occurs before event 1 (the nose of one rocket passing the tail of the other). The 400 m distance of the miss in the Huns' reference frame is the length-contracted miss distance of 667 m in the Goths' reference frame.

SUMMARY

The goal of Chapter 36 has been to understand how Einstein's theory of relativity changes our concepts of space and time.

General Principles

Principle of Relativity **All the laws of physics are the same in all inertial reference frames.**

• The speed of light c is the same in all inertial reference frames.

• No particle or causal influence can travel at a speed greater than c.

Important Concepts

Space

Spatial measurements depend on the motion of the experimenter relative to the events. An object's length is the difference between *simultaneous* measurements of the positions of both ends.

Proper length ℓ is the length of an object measured in a reference frame in which the object is at rest. The object's length in a frame in which the object moves with velocity v is

$$L = \sqrt{1 - \beta^2}\,\ell \leq \ell$$

This is called **length contraction.**

Time

Time measurements depend on the motion of the experimenter relative to the events. Events that are simultaneous in reference frame S are not simultaneous in frame S′ moving relative to S.

Proper time $\Delta\tau$ is the time interval between two events measured in a reference frame in which the events occur at the same position. The time interval between the events in a frame moving with relative velocity v is

$$\Delta t = \Delta\tau/\sqrt{1 - \beta^2} \geq \Delta\tau$$

This is called **time dilation.**

Momentum

The law of conservation of momentum is valid in all inertial reference frames if the momentum of a particle with velocity u is $p = \gamma_p m u$, where

$$\gamma_p = 1/\sqrt{1 - u^2/c^2}$$

The momentum approaches ∞ as $u \rightarrow c$.

Energy

The law of conservation of energy is valid in all inertial reference frames if the energy of a particle with velocity u is $E = \gamma_p m c^2 = E_0 + K$.

Rest energy $E_0 = mc^2$

Kinetic energy $K = (\gamma_p - 1)mc^2$

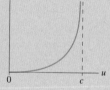

Invariants

Invariants are quantities that have the same value in all inertial reference frames.

Spacetime interval: $s^2 = (c\Delta t)^2 - (\Delta x)^2$

Particle rest energy: $E_0^2 = (mc^2)^2 = E^2 - (pc)^2$

Mass-energy equivalence

Mass m can be transformed into energy $E = mc^2$.

Energy can be transformed into mass $m = \Delta E/c^2$.

Applications

An **event** happens at a specific place in space and time. Spacetime coordinates are (x, t) in frame S and (x', t') in frame S′.

A **reference frame** is a coordinate system with meter sticks and clocks for measuring events. Experimenters at rest relative to each other share the same reference frame.

The **Lorentz transformations** transform spacetime coordinates and velocities between reference frames S and S′.

$$x' = \gamma(x - vt) \qquad x = \gamma(x' + vt')$$
$$y' = y \qquad y = y'$$
$$z' = z \qquad z = z'$$
$$t' = \gamma(t - vx/c^2) \qquad t = \gamma(t' + vx'/c^2)$$
$$u' = \frac{u - v}{1 - uv/c^2} \qquad u = \frac{u' + v}{1 + u'v/c^2}$$

where u and u' are the x- and x'-components of an object's velocity.

$$\beta = \frac{v}{c} \quad \text{and} \quad \gamma = 1/\sqrt{1 - v^2/c^2} = 1/\sqrt{1 - \beta^2}$$

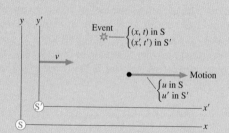

Terms and Notation

special relativity	spacetime coordinates,	time dilation	causal influence
reference frame	(x, y, z, t)	light year, ly	total energy, E
inertial reference frame	synchronized	proper length, ℓ	rest energy, E_0
Galilean principle of relativity	simultaneous	length contraction	law of conservation of total
ether	relativity of simultaneity	invariant	energy
principle of relativity	rest frame	spacetime interval, s	nuclear fission
event	proper time, $\Delta\tau$	Lorentz transformations	

CONCEPTUAL QUESTIONS

1. **FIGURE Q36.1** shows two balls. What are the speed and direction of each (a) in a reference frame that moves with ball 1 and (b) in a reference frame that moves with ball 2?

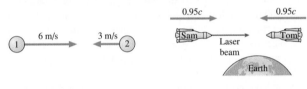

FIGURE Q36.1

FIGURE Q36.2

2. Teenagers Sam and Tom are playing chicken in their rockets. As **FIGURE Q36.2** shows, an experimenter on earth sees that each is traveling at $0.95c$ as he approaches the other. Sam fires a laser beam toward Tom.
   a. What is the speed of the laser beam relative to Sam?
   b. What is the speed of the laser beam relative to Tom?

3. Firecracker A is 300 m from you. Firecracker B is 600 m from you in the same direction. You see both explode at the same time. Define event 1 to be "firecracker A explodes" and event 2 to be "firecracker B explodes." Does event 1 occur before, after, or at the same time as event 2? Explain.

4. Firecrackers A and B are 600 m apart. You are standing exactly halfway between them. Your lab partner is 300 m on the other side of firecracker A. You see two flashes of light, from the two explosions, at exactly the same instant of time. Define event 1 to be "firecracker A explodes" and event 2 to be "firecracker B explodes." According to your lab partner, based on measurements he or she makes, does event 1 occur before, after, or at the same time as event 2? Explain.

5. **FIGURE Q36.5** shows Peggy standing at the center of her railroad car as it passes Ryan on the ground. Firecrackers attached to the ends of the car explode. A short time later, the flashes from the two explosions arrive at Peggy at the same time.

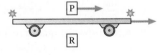

FIGURE Q36.5

   a. Were the explosions simultaneous in Peggy's reference frame? If not, which exploded first? Explain.
   b. Were the explosions simultaneous in Ryan's reference frame? If not, which exploded first? Explain.

6. **FIGURE Q36.6** shows a rocket traveling from left to right. At the instant it is halfway between two trees, lightning simultaneously (in the rocket's frame) hits both trees.

   a. Do the light flashes reach the rocket pilot simultaneously? If not, which reaches her first? Explain.
   b. A student was sitting on the ground halfway between the trees as the rocket passed overhead. According to the student, were the lightning strikes simultaneous? If not, which tree was hit first? Explain.

FIGURE Q36.6

7. Your friend flies from Los Angeles to New York. She carries an accurate stopwatch with her to measure the flight time. You and your assistants on the ground also measure the flight time.
   a. Identify the two events associated with this measurement.
   b. Who, if anyone, measures the proper time?
   c. Who, if anyone, measures the shorter flight time?

8. As the meter stick in **FIGURE Q36.8** flies past you, you simultaneously measure the positions of both ends and determine that $L < 1$ m.

FIGURE Q36.8

   a. To an experimenter in frame S′, the meter stick's frame, did you make your two measurements simultaneously? If not, which end did you measure first? Explain.
   b. Can experimenters in frame S′ give an explanation for why your measurement is less than 1 m?

9. A 100-m-long train is heading for an 80-m-long tunnel. If the train moves sufficiently fast, is it possible, according to experimenters on the ground, for the entire train to be inside the tunnel at one instant of time? Explain.

10. Particle A has half the mass and twice the speed of particle B. Is the momentum p_A less than, greater than, or equal to p_B? Explain.

11. Event A occurs at spacetime coordinates (300 m, 2 μs).
    a. Event B occurs at spacetime coordinates (1200 m, 6 μs). Could A possibly be the cause of B? Explain.
    b. Event C occurs at spacetime coordinates (2400 m, 8 μs). Could A possibly be the cause of C? Explain.

EXERCISES AND PROBLEMS

Problems labeled ▨ integrate material from earlier chapters.

Exercises

Section 36.2 Galilean Relativity

1. ‖ At $t = 1.0$ s, a firecracker explodes at $x = 10$ m in reference frame S. Four seconds later, a second firecracker explodes at $x = 20$ m. Reference frame S′ moves in the x-direction at a speed of 5.0 m/s. What are the positions and times of these two events in frame S′?

2. ‖ A firecracker explodes in reference frame S at $t = 1.0$ s. A second firecracker explodes at the same position at $t = 3.0$ s. In reference frame S′, which moves in the x-direction at speed v, the first explosion is detected at $x′ = 4.0$ m and the second at $x′ = -4.0$ m.
   a. What is the speed of frame S′ relative to frame S?
   b. What is the position of the two explosions in frame S?

3. | A sprinter crosses the finish line of a race. The roar of the crowd in front approaches her at a speed of 360 m/s. The roar from the crowd behind her approaches at 330 m/s. What are the speed of sound and the speed of the sprinter?

4. | A baseball pitcher can throw a ball with a speed of 40 m/s. He is in the back of a pickup truck that is driving away from you. He throws the ball in your direction, and it floats toward you at a lazy 10 m/s. What is the speed of the truck?

5. | A newspaper delivery boy is riding his bicycle down the street at 5.0 m/s. He can throw a paper at a speed of 8.0 m/s. What is the paper's speed relative to the ground if he throws the paper (a) forward, (b) backward, and (c) to the side?

Section 36.3 Einstein's Principle of Relativity

6. | An out-of-control alien spacecraft is diving into a star at a speed of 1.0×10^8 m/s. At what speed, relative to the spacecraft, is the starlight approaching?

7. | A starship blasts past the earth at 2.0×10^8 m/s. Just after passing the earth, it fires a laser beam out the back of the starship. With what speed does the laser beam approach the earth?

8. | A positron moving in the positive x-direction at 2.0×10^8 m/s collides with an electron at rest. The positron and electron annihilate, producing two gamma-ray photons. Photon 1 travels in the positive x-direction and photon 2 travels in the negative x-direction. What is the speed of each photon?

Section 36.4 Events and Measurements

Section 36.5 The Relativity of Simultaneity

9. ‖ Your job is to synchronize the clocks in a reference frame. You are going to do so by flashing a light at the origin at $t = 0$ s. To what time should the clock at $(x, y, z) = (30$ m, 40 m, 0 m) be preset?

10. | Bjorn is standing at $x = 600$ m. Firecracker 1 explodes at the origin and firecracker 2 explodes at $x = 900$ m. The flashes from both explosions reach Bjorn's eye at $t = 3.0$ μs. At what time did each firecracker explode?

11. ‖ Bianca is standing at $x = 600$ m. Firecracker 1, at the origin, and firecracker 2, at $x = 900$ m, explode simultaneously. The flash from firecracker 1 reaches Bianca's eye at $t = 3.0$ μs. At what time does she see the flash from firecracker 2?

12. ‖ You are standing at $x = 9.0$ km. Lightning bolt 1 strikes at $x = 0$ km and lightning bolt 2 strikes at $x = 12.0$ km. Both flashes reach your eye at the same time. Your assistant is standing at $x = 3.0$ km. Does your assistant see the flashes at the same time? If not, which does she see first, and what is the time difference between the two?

13. ‖ You are standing at $x = 9.0$ km and your assistant is standing at $x = 3.0$ km. Lightning bolt 1 strikes at $x = 0$ km and lightning bolt 2 strikes at $x = 12.0$ km. You see the flash from bolt 2 at $t = 10$ μs and the flash from bolt 1 at $t = 50$ μs. According to your assistant, were the lightning strikes simultaneous? If not, which occurred first, and what was the time difference between the two?

14. ‖ Jose is looking to the east. Lightning bolt 1 strikes a tree 300 m from him. Lightning bolt 2 strikes a barn 900 m from him in the same direction. Jose sees the tree strike 1.0 μs before he sees the barn strike. According to Jose, were the lightning strikes simultaneous? If not, which occurred first, and what was the time difference between the two?

15. ‖ You are flying your personal rocketcraft at 0.9c from Star A toward Star B. The distance between the stars, in the stars' reference frame, is 1.0 ly. Both stars happen to explode simultaneously in your reference frame at the instant you are exactly halfway between them. Do you see the flashes simultaneously? If not, which do you see first, and what is the time difference between the two?

Section 36.6 Time Dilation

16. ‖ A cosmic ray travels 60 km through the earth's atmosphere in 400 μs, as measured by experimenters on the ground. How long does the journey take according to the cosmic ray?

17. | At what speed, as a fraction of c, does a moving clock tick at half the rate of an identical clock at rest?

18. | An astronaut travels to a star system 4.5 ly away at a speed of 0.9c. Assume that the time needed to accelerate and decelerate is negligible.
   a. How long does the journey take according to Mission Control on earth?
   b. How long does the journey take according to the astronaut?
   c. How much time elapses between the launch and the arrival of the first radio message from the astronaut saying that she has arrived?

19. ‖ a. How fast must a rocket travel on a journey to and from a distant star so that the astronauts age 10 years while the Mission Control workers on earth age 120 years?
   b. As measured by Mission Control, how far away is the distant star?

20. ‖ You fly 5000 km across the United States on an airliner at 250 m/s. You return two days later at the same speed.
   a. Have you aged more or less than your friends at home?
   b. By how much?
   Hint: Use the binomial approximation.

21. ‖ At what speed, in m/s, would a moving clock lose 1.0 ns in 1.0 day according to experimenters on the ground?
   Hint: Use the binomial approximation.

Section 36.7 Length Contraction

22. | At what speed, as a fraction of c, will a moving rod have a length 60% that of an identical rod at rest?
23. | Jill claims that her new rocket is 100 m long. As she flies past your house, you measure the rocket's length and find that it is only 80 m. Should Jill be cited for exceeding the 0.5c speed limit?
24. ‖ A muon travels 60 km through the atmosphere at a speed of 0.9997c. According to the muon, how thick is the atmosphere?
25. ‖ A cube has a density of 2000 kg/m^3 while at rest in the laboratory. What is the cube's density as measured by an experimenter in the laboratory as the cube moves through the laboratory at 90% of the speed of light in a direction perpendicular to one of its faces?
26. | Our Milky Way galaxy is 100,000 ly in diameter. A spaceship crossing the galaxy measures the galaxy's diameter to be a mere 1.0 ly.
    a. What is the speed of the spaceship relative to the galaxy?
    b. How long is the crossing time as measured in the galaxy's reference frame?
27. ‖ A human hair is about 50 μm in diameter. At what speed, in m/s, would a meter stick "shrink by a hair"?
    Hint: Use the binomial approximation.

Section 36.8 The Lorentz Transformations

28. | An event has spacetime coordinates $(x, t) = (1200$ m, 2.0 μs$)$ in reference frame S. What are the event's spacetime coordinates (a) in reference frame S′ that moves in the positive x-direction at 0.8c and (b) in reference frame S″ that moves in the negative x-direction at 0.8c?
29. ‖ A rocket travels in the x-direction at speed 0.6c with respect to the earth. An experimenter on the rocket observes a collision between two comets and determines that the spacetime coordinates of the collision are $(x', t') = (3.0 \times 10^{10}$ m, 200 s$)$. What are the spacetime coordinates of the collision in earth's reference frame?
30. ‖ In the earth's reference frame, a tree is at the origin and a pole is at $x = 30$ km. Lightning strikes both the tree and the pole at $t = 10$ μs. The lightning strikes are observed by a rocket traveling in the x-direction at 0.5c.
    a. What are the spacetime coordinates for these two events in the rocket's reference frame?
    b. Are the events simultaneous in the rocket's frame? If not, which occurs first?
31. ‖ A rocket cruising past earth at 0.8c shoots a bullet out the back door, opposite the rocket's motion, at 0.9c relative to the rocket. What is the bullet's speed relative to the earth?
32. ‖ A laboratory experiment shoots an electron to the left at 0.9c. What is the electron's speed relative to a proton moving to the right at 0.9c?
33. ‖ A distant quasar is found to be moving away from the earth at 0.8c. A galaxy closer to the earth and along the same line of sight is moving away from us at 0.2c. What is the recessional speed of the quasar as measured by astronomers in the other galaxy?

Section 36.9 Relativistic Momentum

34. | A proton is accelerated to 0.999c.
    a. What is the proton's momentum?
    b. By what factor does the proton's momentum exceed its Newtonian momentum?

35. ‖ At what speed is a particle's momentum twice its Newtonian value?
36. ‖‖ A 1.0 g particle has momentum 400,000 kg m/s. What is the particle's speed?
37. ‖ What is the speed of a particle whose momentum is mc?

Section 36.10 Relativistic Energy

38. | What are the kinetic energy, the rest energy, and the total energy of a 1.0 g particle with a speed of 0.8c?
39. | A quarter-pound hamburger with all the fixings has a mass of 200 g. The food energy of the hamburger (480 food calories) is 2 MJ.
    a. What is the energy equivalent of the mass of the hamburger?
    b. By what factor does the energy equivalent exceed the food energy?
40. | How fast must an electron move so that its total energy is 10% more than its rest mass energy?
41. | At what speed is a particle's kinetic energy twice its rest energy?
42. ‖ At what speed is a particle's total energy twice its rest energy?

Problems

43. | A 50 g ball moving to the right at 4.0 m/s overtakes and collides with a 100 g ball moving to the right at 2.0 m/s. The collision is perfectly elastic. Use reference frames and the Chapter 10 result for perfectly elastic collisions to find the speed and direction of each ball after the collision.
44. | A billiard ball has a perfectly elastic collision with a second billiard ball of equal mass. Afterward, the first ball moves to the left at 2.0 m/s and the second to the right at 4.0 m/s. Use reference frames and the Chapter 10 result for perfectly elastic collisions to find the speed and direction of each ball before the collision.
45. ‖ The diameter of the solar system is 10 light hours. A spaceship crosses the solar system in 15 hours, as measured on earth. How long, in hours, does the passage take according to passengers on the spaceship?
    Hint: $c = 1$ light hour per hour.
46. | A 30-m-long rocket train car is traveling from Los Angeles to New York at 0.5c when a light at the center of the car flashes. When the light reaches the front of the car, it immediately rings a bell. Light reaching the back of the car immediately sounds a siren.
    a. Are the bell and siren simultaneous events for a passenger seated in the car? If not, which occurs first and by how much time?
    b. Are the bell and siren simultaneous events for a bicyclist waiting to cross the tracks? If not, which occurs first and by how much time?
47. ‖‖ The star Alpha goes supernova. Ten years later and 100 ly away, as measured by astronomers in the galaxy, star Beta explodes.
    a. Is it possible that the explosion of Alpha is in any way responsible for the explosion of Beta? Explain.
    b. An alien spacecraft passing through the galaxy finds that the distance between the two explosions is 120 ly. According to the aliens, what is the time between the explosions?
48. ‖ Two events in reference frame S occur 10 μs apart at the same point in space. The distance between the two events is 2400 m in reference frame S′.
    a. What is the time interval between the events in reference frame S′?
    b. What is the velocity of S′ relative to S?

49. ||| A starship voyages to a distant planet 10 ly away. The explorers stay 1 yr, return at the same speed, and arrive back on earth 26 yr after they left. Assume that the time needed to accelerate and decelerate is negligible.
    a. What is the speed of the starship?
    b. How much time has elapsed on the astronauts' chronometers?

50. || In Section 36.6 we saw that muons can reach the ground because of time dilation. But how do things appear in the muon's reference frame, where the muon's half-life is only 1.5 μs? How can a muon travel the 60 km to reach the earth's surface before decaying? Resolve this apparent paradox. Be as quantitative as you can in your answer.

51. || The Stanford Linear Accelerator (SLAC) accelerates electrons to $c = 0.99999997c$ in a 3.2-km-long tube. If they travel the length of the tube at full speed (they don't, because they are accelerating), how long is the tube in the electrons' reference frame?

52. || In an attempt to reduce the extraordinarily long travel times for voyaging to distant stars, some people have suggested traveling at close to the speed of light. Suppose you wish to visit the red giant star Betelgeuse, which is 430 ly away, and that you want your 20,000 kg rocket to move so fast that you age only 20 years during the round trip.
    a. How fast must the rocket travel relative to earth?
    b. How much energy is needed to accelerate the rocket to this speed?
    c. Compare this amount of energy to the total energy used by the United States in the year 2010, which was roughly 1.0×10^{20} J.

53. | A rocket traveling at 0.5c sets out for the nearest star, Alpha Centauri, which is 4.25 ly away from earth. It will return to earth immediately after reaching Alpha Centauri. What distance will the rocket travel and how long will the journey last according to (a) stay-at-home earthlings and (b) the rocket crew? (c) Which answers are the correct ones, those in part a or those in part b?

54. || The star Delta goes supernova. One year later and 2 ly away, as measured by astronomers in the galaxy, star Epsilon explodes. Let the explosion of Delta be at $x_D = 0$ and $t_D = 0$. The explosions are observed by three spaceships cruising through the galaxy in the direction from Delta to Epsilon at velocities $v_1 = 0.3c$, $v_2 = 0.5c$, and $v_3 = 0.7c$.
    a. What are the times of the two explosions as measured by scientists on each of the three spaceships?
    b. Does one spaceship find that the explosions are simultaneous? If so, which one?
    c. Does one spaceship find that Epsilon explodes before Delta? If so, which one?
    d. Do your answers to parts b and c violate the idea of causality? Explain.

55. || Two rockets approach each other. Each is traveling at 0.75c in the earth's reference frame. What is the speed of one rocket relative to the other?

56. || A rocket fires a projectile at a speed of 0.95c while traveling past the earth. An earthbound scientist measures the projectile's speed to be 0.90c. What was the rocket's speed?

57. || Through what potential difference must an electron be accelerated, starting from rest, to acquire a speed of 0.99c?

58. || What is the speed of a proton after being accelerated from rest through a 50×10^6 V potential difference?

59. || The half-life of a muon at rest is 1.5 μs. Muons that have been accelerated to a very high speed and are then held in a circular storage ring have a half-life of 7.5 μs.

a. What is the speed of the muons in the storage ring?
b. What is the total energy of a muon in the storage ring? The mass of a muon is 207 times the mass of an electron.

60. || A solar flare blowing out from the sun at 0.9c is overtaking a rocket as it flies away from the sun at 0.8c. According to the crew on board, with what speed is the flare gaining on the rocket?

61. || This chapter has assumed that lengths perpendicular to the direction of motion are not affected by the motion. That is, motion in the x-direction does not cause length contraction along the y- or z-axes. To find out if this is really true, consider two spray-paint nozzles attached to rods perpendicular to the x-axis. It has been confirmed that, when both rods are at rest, both nozzles are exactly 1 m above the base of the rod. One rod is placed in the S reference frame with its base on the x-axis; the other is placed in the S' reference frame with its base on the x'-axis. The rods then swoop past each other and, as FIGURE P36.61 shows, each paints a stripe across the other rod.

We will use proof by contradiction. Assume that objects perpendicular to the motion *are* contracted. An experimenter in frame S finds that the S' nozzle, as it goes past, is less than 1 m above the x-axis. The principle of relativity says that an experiment carried out in two different inertial reference frames will have the same outcome in both.
    a. Pursue this line of reasoning and show that you end up with a logical contradiction, two mutually incompatible situations.
    b. What can you conclude from this contradiction?

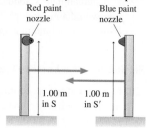

FIGURE P36.61

62. || Derive the Lorentz transformations for t and t'.
    Hint: See the comment following Equation 36.22.

63. || a. Derive a velocity transformation equation for u_y and u'_y. Assume that the reference frames are in the standard orientation with motion parallel to the x- and x'-axes.
    b. A rocket passes the earth at 0.8c. As it goes by, it launches a projectile at 0.6c perpendicular to the direction of motion. What is the projectile's speed in the earth's reference frame?

64. || What is the momentum of a particle whose total energy is four times its rest energy? Give your answer as a multiple of mc.

65. || a. What are the momentum and total energy of a proton with speed 0.99c?
    b. What is the proton's momentum in a different reference frame in which $E' = 5.0 \times 10^{-10}$ J?

66. ||| At what speed is the kinetic energy of a particle twice its Newtonian value?

67. | A typical nuclear power plant generates electricity at the rate of 1000 MW. The efficiency of transforming thermal energy into electrical energy is $\frac{1}{3}$ and the plant runs at full capacity for 80% of the year. (Nuclear power plants are down about 20% of the time for maintenance and refueling.)
    a. How much thermal energy does the plant generate in one year?
    b. What mass of uranium is transformed into energy in one year?

68. | The sun radiates energy at the rate 3.8×10^{26} W. The source of this energy is fusion, a nuclear reaction in which mass is transformed into energy. The mass of the sun is 2.0×10^{30} kg.
    a. How much mass does the sun lose each year?
    b. What percent is this of the sun's total mass?
    c. Estimate the lifetime of the sun.

69. ‖ The radioactive element radium (Ra) decays by a process known as *alpha decay,* in which the nucleus emits a helium nucleus. (These high-speed helium nuclei were named alpha particles when radioactivity was first discovered, long before the identity of the particles was established.) The reaction is $^{226}\text{Ra} \rightarrow {}^{222}\text{Rn} + {}^{4}\text{He}$, where Rn is the element radon. The accurately measured atomic masses of the three atoms are 226.025, 222.017, and 4.003. How much energy is released in each decay? (The energy released in radioactive decay is what makes nuclear waste "hot.")

70. ‖ The nuclear reaction that powers the sun is the fusion of four protons into a helium nucleus. The process involves several steps, but the net reaction is simply $4\text{p} \rightarrow {}^{4}\text{He} + \text{energy}$. The mass of a helium nucleus is known to be 6.64×10^{-27} kg.
   a. How much energy is released in each fusion?
   b. What fraction of the initial rest mass energy is this energy?

71. ‖‖ An electron moving to the right at 0.9c collides with a positron moving to the left at 0.9c. The two particles annihilate and produce two gamma-ray photons. What is the wavelength of the photons?

72. ‖ Consider the inelastic collision $e^- + e^- \rightarrow e^- + e^- + e^- + e^+$ in which an electron-positron pair is produced in a head-on collision between two electrons moving in opposite directions at the same speed. This is similar to Figure 36.39, but both of the initial electrons are moving.
   a. What is the threshold kinetic energy? That is, what minimum kinetic energy must each electron have to allow this process to occur?
   b. What is the speed of an electron with this kinetic energy?

Challenge Problems

73. Two rockets, A and B, approach the earth from opposite directions at speed 0.8c. The length of each rocket measured in its rest frame is 100 m. What is the length of rocket A as measured by the crew of rocket B?

74. Two rockets are each 1000 m long in their rest frame. Rocket Orion, traveling at 0.8c relative to the earth, is overtaking rocket Sirius, which is poking along at a mere 0.6c. According to the crew on Sirius, how long does Orion take to completely pass? That is, how long is it from the instant the nose of Orion is at the tail of Sirius until the tail of Orion is at the nose of Sirius?

75. Some particle accelerators allow protons (p^+) and antiprotons (p^-) to circulate at equal speeds in opposite directions in a device called a *storage ring.* The particle beams cross each other at various points to cause $\text{p}^+ + \text{p}^-$ collisions. In one collision, the outcome is $\text{p}^+ + \text{p}^- \rightarrow e^+ + e^- + \gamma + \gamma$, where γ represents a high-energy gamma-ray photon. The electron and positron are ejected from the collision at 0.9999995c and the gamma-ray photon wavelengths are found to be 1.0×10^{-6} nm. What were the proton and antiproton speeds prior to the collision?

76. A very fast pole vaulter lives in the country. One day, while practicing, he notices a 10.0-m-long barn with the doors open at both ends. He decides to run through the barn at 0.866c while carrying his 16.0-m-long pole. The farmer, who sees him coming, says, "Aha! This guy's pole is length contracted to 8.0 m. There will be a short interval of time when the pole is entirely inside the barn. If I'm quick, I can simultaneously close both barn doors while the pole vaulter and his pole are inside." The pole vaulter, who sees the farmer beside the barn, thinks to himself, "That farmer is crazy. The barn is length contracted and is only 5.0 m long. My 16.0-m-long pole cannot fit into a 5.0-m-long barn. If the farmer closes the doors just as the tip of my pole reaches the back door, the front door will break off the last 11.0 m of my pole."

   Can the farmer close the doors without breaking the pole? Show that, when properly analyzed, the farmer and the pole vaulter agree on the outcome. Your analysis should contain both quantitative calculations and written explanation.

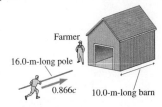

FIGURE CP36.76

Stop to Think 36.1: a, c, and f. These move at constant velocity, or very nearly so. The others are accelerating.

Stop to Think 36.2: a. $u' = u - v = -10$ m/s $- 6$ m/s $= -16$ m/s. The *speed* is 16 m/s.

Stop to Think 36.3: c. Even the light has a slight travel time. The event is the hammer hitting the nail, not your seeing the hammer hit the nail.

Stop to Think 36.4: At the same time. Mark is halfway between the tree and the pole, so the fact that he *sees* the lightning bolts at the same time means they *happened* at the same time. It's true that Nancy *sees* event 1 before event 2, but the events actually occurred before she sees them. Mark and Nancy share a reference frame, because they are at rest relative to each other, and all experimenters in a reference frame, after correcting for any signal delays, *agree* on the spacetime coordinates of an event.

Stop to Think 36.5: After. This is the same as the case of Peggy and Ryan. In Mark's reference frame, as in Ryan's, the events are simultaneous. Nancy *sees* event 1 first, but the time when an event

is seen is not when the event actually happens. Because all experimenters in a reference frame agree on the spacetime coordinates of an event, Nancy's position in her reference frame cannot affect the order of the events. If Nancy had been passing Mark at the instant the lightning strikes occur in Mark's frame, then Nancy would be equivalent to Peggy. Event 2, like the firecracker at the front of Peggy's railroad car, occurs first in Nancy's reference frame.

Stop to Think 36.6: c. Nick measures proper time because Nick's clock is present at both the "nose passes Nick" event and the "tail passes Nick" event. Proper time is the smallest measured time interval between two events.

Stop to Think 36.7: $L_A > L_B = L_C$. Anjay measures the pole's proper length because it is at rest in his reference frame. Proper length is the longest measured length. Beth and Charles may *see* the pole differently, but they share the same reference frame and their *measurements* of the length agree.

Stop to Think 36.8: c. The rest energy E_0 is an invariant, the same in all inertial reference frames. Thus $m = E_0/c^2$ is independent of speed.

Mathematics Review

Algebra

Using exponents:

$$a^{-x} = \frac{1}{a^x} \qquad a^x a^y = a^{(x+y)} \qquad \frac{a^x}{a^y} = a^{(x-y)} \qquad (a^x)^y = a^{xy}$$

$$a^0 = 1 \qquad a^1 = a \qquad a^{1/n} = \sqrt[n]{a}$$

Fractions:

$$\left(\frac{a}{b}\right)\left(\frac{c}{d}\right) = \frac{ac}{bd} \qquad \frac{a/b}{c/d} = \frac{ad}{bc} \qquad \frac{1}{1/a} = a$$

Logarithms:

If $a = e^x$, then $\ln(a) = x$ $\qquad \ln(e^x) = x \qquad e^{\ln(x)} = x$

$$\ln(ab) = \ln(a) + \ln(b) \qquad \ln\left(\frac{a}{b}\right) = \ln(a) - \ln(b) \qquad \ln(a^n) = n\ln(a)$$

The expression $\ln(a + b)$ cannot be simplified.

Linear equations: The graph of the equation $y = ax + b$ is a straight line. a is the slope of the graph. b is the y-intercept.

Proportionality: To say that y is proportional to x, written $y \propto x$, means that $y = ax$, where a is a constant. Proportionality is a special case of linearity. A graph of a proportional relationship is a straight line that passes through the origin. If $y \propto x$, then

$$\frac{y_1}{y_2} = \frac{x_1}{x_2}$$

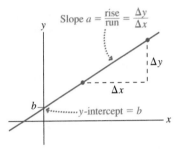

Quadratic equation: The quadratic equation $ax^2 + bx + c = 0$ has the two solutions $x = \dfrac{-b \pm \sqrt{b^2 - 4ac}}{2a}$.

Geometry and Trigonometry

Area and volume:

Rectangle

$A = ab$

Triangle

$A = \frac{1}{2}ab$

Circle

$C = 2\pi r$
$A = \pi r^2$

Rectangular box

$V = abc$

Right circular cylinder

$V = \pi r^2 l$

Sphere

$A = 4\pi r^2$
$V = \frac{4}{3}\pi r^3$

Arc length and angle: The angle θ in radians is defined as $\theta = s/r$.

The arc length that spans angle θ is $s = r\theta$.

2π rad $= 360°$

Right triangle: Pythagorean theorem $\quad c = \sqrt{a^2 + b^2}$ or $a^2 + b^2 = c^2$

$$\sin\theta = \frac{b}{c} = \frac{\text{far side}}{\text{hypotenuse}} \qquad \theta = \sin^{-1}\left(\frac{b}{c}\right)$$

$$\cos\theta = \frac{a}{c} = \frac{\text{adjacent side}}{\text{hypotenuse}} \qquad \theta = \cos^{-1}\left(\frac{a}{c}\right)$$

$$\tan\theta = \frac{b}{a} = \frac{\text{far side}}{\text{adjacent side}} \qquad \theta = \tan^{-1}\left(\frac{b}{a}\right)$$

General triangle: $\alpha + \beta + \gamma = 180° = \pi$ rad

Law of cosines $c^2 = a^2 + b^2 - 2ab\cos\gamma$

Identities:

$$\tan\alpha = \frac{\sin\alpha}{\cos\alpha} \qquad\qquad \sin^2\alpha + \cos^2\alpha = 1$$

$$\sin(-\alpha) = -\sin\alpha \qquad\qquad \cos(-\alpha) = \cos\alpha$$

$$\sin(\alpha \pm \beta) = \sin\alpha\cos\beta \pm \cos\alpha\sin\beta \qquad \cos(\alpha \pm \beta) = \cos\alpha\cos\beta \mp \sin\alpha\sin\beta$$

$$\sin(2\alpha) = 2\sin\alpha\cos\alpha \qquad\qquad \cos(2\alpha) = \cos^2\alpha - \sin^2\alpha$$

$$\sin(\alpha \pm \pi/2) = \pm\cos\alpha \qquad\qquad \cos(\alpha \pm \pi/2) = \mp\sin\alpha$$

$$\sin(\alpha \pm \pi) = -\sin\alpha \qquad\qquad \cos(\alpha \pm \pi) = -\cos\alpha$$

Expansions and Approximations

Binomial expansion:
$$(1 + x)^n = 1 + nx + \frac{n(n-1)}{2}x^2 + \cdots$$

Binomial approximation:
$$(1 + x)^n \approx 1 + nx \quad \text{if} \quad x \ll 1$$

Trigonometric expansions:
$$\sin\alpha = \alpha - \frac{\alpha^3}{3!} + \frac{\alpha^5}{5!} - \frac{\alpha^7}{7!} + \cdots \quad \text{for } \alpha \text{ in rad}$$

$$\cos\alpha = 1 - \frac{\alpha^2}{2!} + \frac{\alpha^4}{4!} - \frac{\alpha^6}{6!} + \cdots \quad \text{for } \alpha \text{ in rad}$$

Small-angle approximation: If $\alpha \ll 1$ rad, then $\sin\alpha \approx \tan\alpha \approx \alpha$ and $\cos\alpha \approx 1$.

The small-angle approximation is excellent for $\alpha < 5°$ (≈ 0.1 rad) and generally acceptable up to $\alpha \approx 10°$.

Calculus

The letters a and n represent constants in the following derivatives and integrals.

Derivatives

$$\frac{d}{dx}(a) = 0$$

$$\frac{d}{dx}(ax) = a$$

$$\frac{d}{dx}\left(\frac{a}{x}\right) = -\frac{a}{x^2}$$

$$\frac{d}{dx}(ax^n) = anx^{n-1}$$

$$\frac{d}{dx}\left(\ln(ax)\right) = \frac{1}{x}$$

$$\frac{d}{dx}(e^{ax}) = ae^{ax}$$

$$\frac{d}{dx}\left(\sin(ax)\right) = a\cos(ax)$$

$$\frac{d}{dx}\left(\cos(ax)\right) = -a\sin(ax)$$

Integrals

$$\int x\,dx = \frac{1}{2}x^2$$

$$\int x^2\,dx = \frac{1}{3}x^3$$

$$\int \frac{1}{x^2}\,dx = -\frac{1}{x}$$

$$\int x^n\,dx = \frac{x^{n+1}}{n+1} \qquad n \neq -1$$

$$\int \frac{dx}{x} = \ln x$$

$$\int \frac{dx}{a+x} = \ln(a+x)$$

$$\int \frac{x\,dx}{a+x} = x - a\ln(a+x)$$

$$\int \frac{dx}{\sqrt{x^2 \pm a^2}} = \ln\left(x + \sqrt{x^2 \pm a^2}\right)$$

$$\int \frac{x\,dx}{\sqrt{x^2 \pm a^2}} = \sqrt{x^2 \pm a^2}$$

$$\int \frac{dx}{x^2 + a^2} = \frac{1}{a}\tan^{-1}\left(\frac{x}{a}\right)$$

$$\int \frac{dx}{(x^2 + a^2)^2} = \frac{1}{2a^3}\tan^{-1}\left(\frac{x}{a}\right) + \frac{x}{2a^2(x^2 + a^2)}$$

$$\int \frac{dx}{(x^2 \pm a^2)^{3/2}} = \frac{\pm x}{a^2\sqrt{x^2 \pm a^2}}$$

$$\int \frac{x\,dx}{(x^2 \pm a^2)^{3/2}} = -\frac{1}{\sqrt{x^2 \pm a^2}}$$

$$\int e^{ax}\,dx = \frac{1}{a}e^{ax}$$

$$\int xe^{-x}\,dx = -(x+1)e^{-x}$$

$$\int x^2 e^{-x}\,dx = -(x^2 + 2x + 2)e^{-x}$$

$$\int \sin(ax)\,dx = -\frac{1}{a}\cos(ax)$$

$$\int \cos(ax)\,dx = \frac{1}{a}\sin(ax)$$

$$\int \sin^2(ax)\,dx = \frac{x}{2} - \frac{\sin(2ax)}{4a}$$

$$\int \cos^2(ax)\,dx = \frac{x}{2} + \frac{\sin(2ax)}{4a}$$

$$\int_0^\infty x^n e^{-ax}\,dx = \frac{n!}{a^{n+1}}$$

$$\int_0^\infty e^{-ax^2}\,dx = \frac{1}{2}\sqrt{\frac{\pi}{a}}$$

Periodic Table of Elements

27	
Co	—Symbol
58.9	

Atomic number —

Atomic mass —

——— Transition elements ———

Period																	
1	1 H 1.0																2 He 4.0
2	3 Li 6.9	4 Be 9.0										5 B 10.8	6 C 12.0	7 N 14.0	8 O 16.0	9 F 19.0	10 Ne 20.2
3	11 Na 23.0	12 Mg 24.3										13 Al 27.0	14 Si 28.1	15 P 31.0	16 S 32.1	17 Cl 35.5	18 Ar 39.9
4	19 K 39.1	20 Ca 40.1	21 Sc 45.0	22 Ti 47.9	23 V 50.9	24 Cr 52.0	25 Mn 54.9	26 Fe 55.8	27 Co 58.9	28 Ni 58.7	29 Cu 63.5	30 Zn 65.4	31 Ga 69.7	32 Ge 72.6	33 As 74.9	34 Se 79.0	35 Br 79.9	36 Kr 83.8
5	37 Rb 85.5	38 Sr 87.6	39 Y 88.9	40 Zr 91.2	41 Nb 92.9	42 Mo 95.9	43 Tc [98]	44 Ru 101.1	45 Rh 102.9	46 Pd 106.4	47 Ag 107.9	48 Cd 112.4	49 In 114.8	50 Sn 118.7	51 Sb 121.8	52 Te 127.6	53 I 126.9	54 Xe 131.3
6	55 Cs 132.9	56 Ba 137.3	71 Lu 175.0	72 Hf 178.5	73 Ta 180.9	74 W 183.9	75 Re 186.2	76 Os 190.2	77 Ir 192.2	78 Pt 195.1	79 Au 197.0	80 Hg 200.6	81 Tl 204.4	82 Pb 207.2	83 Bi 209.0	84 Po [209]	85 At [210]	86 Rn [222]
7	87 Fr [223]	88 Ra [226]	103 Lr [262]	104 Rf [265]	105 Db [268]	106 Sg [271]	107 Bh [272]	108 Hs [270]	109 Mt [276]	110 Ds [281]	111 Rg [280]	112 Cn [285]	113	114	115	116	117	118

——— Inner transition elements ———

Lanthanides 6	57 La 138.9	58 Ce 140.1	59 Pr 140.9	60 Nd 144.2	61 Pm 144.9	62 Sm 150.4	63 Eu 152.0	64 Gd 157.3	65 Tb 158.9	66 Dy 162.5	67 Ho 164.9	68 Er 167.3	69 Tm 168.9	70 Yb 173.0
Actinides 7	89 Ac [227]	90 Th 232.0	91 Pa 231.0	92 U 238.0	93 Np [237]	94 Pu [244]	95 Am [243]	96 Cm [247]	97 Bk [247]	98 Cf [251]	99 Es [252]	100 Fm [257]	101 Md [258]	102 No [259]

An atomic mass in brackets is that of the longest-lived isotope of an element with no stable isotopes.

ActivPhysics OnLine™ Activities (MP)® www.masteringphysics.com

The following list gives the activity numbers and titles of the ActivPhysics activities available in the Pearson eText and the Study Area of MasteringPhysics, followed by the corresponding textbook page references.

1.1 Analyzing Motion Using Diagrams 16
1.2 Analyzing Motion Using Graphs 46
1.3 Predicting Motion from Graphs 46
1.4 Predicting Motion from Equations 49
1.5 Problem-Solving Strategies for Kinematics 49
1.6 Skier Races Downhill 49
1.7 Balloonist Drops Lemonade 52
1.8 Seat Belts Save Lives 49
1.9 Screeching to a Halt 49
1.10 Pole-Vaulter Lands 52
1.11 Car Starts, Then Stops 49
1.12 Solving Two-Vehicle Problems 49
1.13 Car Catches Truck 49
1.14 Avoiding a Rear-End Collision 49
2.1.1 Force Magnitudes 141
2.1.2 Skydiver 141
2.1.3 Tension Change 141
2.1.4 Sliding on an Incline 141
2.1.5 Car Race 141
2.2 Lifting a Crate 141
2.3 Lowering a Crate 141
2.4 Rocket Blasts Off 141
2.5 Truck Pulls Crate 149
2.6 Pushing a Crate Up a Wall 149
2.7 Skier Goes Down a Slope 155
2.8 Skier and Rope Tow 155
2.9 Pole-Vaulter Vaults 155
2.10 Truck Pulls Two Crates 181
2.11 Modified Atwood Machine 181
3.1 Solving Projectile Motion Problems 93
3.2 Two Balls Falling 93
3.3 Changing the x-Velocity 93
3.4 Projectile x- and y-Accelerations 93
3.5 Initial Velocity Components 93
3.6 Target Practice I 93
3.7 Target Practice II 93
4.1 Magnitude of Centripetal Acceleration 102, 194
4.2 Circular Motion Problem Solving 207
4.3 Cart Goes Over Circular Path 207
4.4 Ball Swings on a String 207
4.5 Car Circles a Track 197
4.6 Satellites Orbit 366
5.1 Work Calculations 282
5.2 Upward-Moving Elevator Stops 297
5.3 Stopping a Downward-Moving Elevator 297
5.4 Inverse Bungee Jumper 297
5.5 Spring-Launched Bowler 297
5.6 Skier Speed 297
5.7 Modified Atwood Machine 297
6.1 Momentum and Energy Change 248
6.2 Collisions and Elasticity 265
6.3 Momentum Conservation and Collisions 230
6.4 Collision Problems 230
6.5 Car Collision: Two Dimensions 297
6.6 Saving an Astronaut 230
6.7 Explosion Problems 230
6.8 Skier and Cart 297
6.9 Pendulum Bashes Box 297
6.10 Pendulum Person-Projectile Bowling 230
7.1 Calculating Torques 322
7.2 A Tilted Beam: Torques and Equilibrium 330
7.3 Arm Levers 330
7.4 Two Painters on a Beam 330
7.5 Lecturing from a Beam 330

7.6 Rotational Inertia 319
7.7 Rotational Kinematics 105
7.8 Rotoride: Dynamics Approach 327
7.9 Falling Ladder 327
7.10 Woman and Flywheel Elevator: Dynamics Approach 327
7.11 Race Between a Block and a Disk 335
7.12 Woman and Flywheel Elevator: Energy Approach 318
7.13 Rotoride: Energy Approach 318
7.14 Ball Hits Bat 342
8.1 Characteristics of a Gas 509
8.2 Maxwell-Boltzmann Distribution: Conceptual Analysis 509
8.3 Maxwell-Boltzmann Distribution: Quantitative Analysis 509
8.4 State Variables and Ideal Gas Law 460
8.5 Work Done by a Gas 475
8.6 Heat, Internal Energy, and First Law of Thermodynamics 489
8.7 Heat Capacity 487
8.8 Isochoric Process 478
8.9 Isobaric Process 478
8.10 Isothermal Process 478
8.11 Adiabatic Process 489
8.12 Cyclic Process: Strategies 535
8.13 Cyclic Process: Problems 535
8.14 Carnot Cycle 542
9.1 Position Graphs and Equations 386
9.2 Describing Vibrational Motion 386
9.3 Vibrational Energy 384
9.4 Two Ways to Weigh Young Tarzan 390
9.5 Ape Drops Tarzan 390
9.6 Releasing a Vibrating Skier I 393
9.7 Releasing a Vibrating Skier II 393
9.8 One- and Two-Spring Vibrating Systems 393
9.9 Vibro-Ride 393
9.10 Pendulum Frequency 392
9.11 Risky Pendulum Walk 392
9.12 Physical Pendulum 392
10.1 Properties of Mechanical Waves 566
10.2 Speed of Waves on a String 563
10.3 Speed of Sound in a Gas 574
10.4 Standing Waves on Strings 595
10.5 Tuning a Stringed Instrument: Standing Waves 602
10.6 String Mass and Standing Waves 595
10.7 Beats and Beat Frequency 615
10.8 Doppler Effect: Conceptual Introduction 580
10.9 Doppler Effect: Problems 580
10.10 Complex Waves: Fourier Analysis
11.1 Electric Force: Coulomb's Law 733
11.2 Electric Force: Superposition Principle 733
11.3 Electric Force Superposition Principle (Quantitative) 733
11.4 Electric Field: Point Charge 740
11.5 Electric Field Due to a Dipole 755
11.6 Electric Field: Problems 755
11.7 Electric Flux 785
11.8 Gauss's Law 791
11.9 Motion of a Charge in an Electric Field: Introduction 767
11.10 Motion in an Electric Field: Problems 767
11.11 Electric Potential: Qualitative Introduction 819

11.12 Electric Potential, Field, and Force 844
11.13 Electrical Potential Energy and Potential 844
12.1 DC Series Circuits (Qualitative) 898
12.2 DC Parallel Circuits 904
12.3 DC Circuit Puzzles 906
12.4 Using Ammeters and Voltmeters 906
12.5 Using Kirchhoff's Laws 906
12.6 Capacitance 910
12.7 Series and Parallel Capacitors 910
12.8 *RC* Circuit Time Constants 910
13.1 Magnetic Field of a Wire 928
13.2 Magnetic Field of a Loop 930
13.3 Magnetic Field of a Solenoid 938
13.4 Magnetic Force on a Particle 940
13.5 Magnetic Force on a Wire 947
13.6 Magnetic Torque on a Loop 948
13.7 Mass Spectrometer 942
13.8 Velocity Selector 942
13.9 Electromagnetic Induction 975
13.10 Motional emf 975
14.1 The *RL* Circuit 991
14.2 The *RLC* Oscillator 1043
14.3 The Driven Oscillator 1043
15.1 Reflection and Refraction 661
15.2 Total Internal Reflection 661
15.3 Refraction Applications 661
15.4 Plane Mirrors 659
15.5 Spherical Mirrors: Ray Diagrams 682
15.6 Spherical Mirror: The Mirror Equation 682
15.7 Spherical Mirror: Linear Magnification 682
15.8 Spherical Mirror: Problems 682
15.9 Thin-Lens Ray Diagrams 673
15.10 Converging Lens Problems 679
15.11 Diverging Lens Problems 679
15.12 Two-Lens Optical Systems 679
16.1 Two-Source Interference: Introduction 631
16.2 Two-Source Interference: Qualitative Questions 631
16.3 Two-Source Interference: Problems 631
16.4 The Grating: Introduction and Qualitative Questions 636
16.5 The Grating: Problems 636
16.6 Single-Slit Diffraction 637
16.7 Circular Hole Diffraction 640
16.8 Resolving Power 709
16.9 Polarization 1024
17.1 Relativity of Time 1074
17.2 Relativity of Length 1078
17.3 Photoelectric Effect 1128
17.4 Compton Scattering
17.5 Electron Interference 1135
17.6 Uncertainty Principle 1169
17.7 Wave Packets 1169
18.1 The Bohr Model 1138
18.2 Spectroscopy 1104, 1232
18.3 The Laser 1238
19.1 Particle Scattering 1114
19.2 Nuclear Binding Energy 1253
19.3 Fusion 836
19.4 Radioactivity 1263
19.5 Particle Physics 1266
20.1 Potential Energy Diagrams 1182
20.2 Particle in a Box 1185
20.3 Potential Wells 1194
20.4 Potential Barriers 1206

PhET Simulations (MP) www.masteringphysics.com

The following list gives the titles of the PhET simulations available in the Pearson eText and in the Study Area of MasteringPhysics, with the corresponding textbook section and page references.

1.3	Vector Addition 7, 71, 75		**23.5**	Color Vision 667
1.8	Estimation 27		**23.6**	*Geometric Optics 672, 674
2.1	Equation Grapher 36		**24.4**	*Geometric Optics 703
2.2	Calculus Grapher 40, 44		**25.3**	Conductivity 727
2.3	*The Moving Man 42		**25.3**	Balloons and Static Electricity, John Travoltage 728, 731
4.1	Motion in 2D 86, 102		**25.5**	*Charges and Fields 740
4.1	Maze Game 86, 88		**26.2**	*Charges and Fields 752, 754, 759, 760, 764
4.1	Ladybug Motion in 2D 86		**26.6**	Electric Field of Dreams, Electric Field Hockey 767
4.3	*Projectile Motion 93		**26.6**	Microwaves, Optical Tweezers and Applications 767
4.5	Ladybug Revolution 99		**28.4**	*Charges and Fields 819, 828
5.5	*Forces in 1 Dimension 126, 127		**29.2**	Battery Voltage 843
5.7	*The Ramp 130		**30.2**	Signal Circuit 871
6.1	Molecular Motors 139		**30.5**	Resistance in a Wire, Ohm's Law 880
6.2	Lunar Lander, *The Ramp 142		**30.5**	Battery-Resistor Circuit 883
6.4	*Forces in 1 Dimension, Friction 149		**31.2**	*Circuit Construction (DC Only) 893, 898, 904, 906, 910
6.6	*The Ramp 155		**31.3**	Battery-Resistor Circuit 896
8.2	Ladybug Revolution, Motion in 2D 194		**32.1**	Magnet and Compass 922
8.3	My Solar System 199		**32.5**	Magnets and Electromagnets 931, 939
8.5	Ladybug Motion in 2D 205		**33.1**	*Faraday's Electromagnetic Lab, Faraday's Law 963
9.5	Lunar Lander 236		**33.3**	*Faraday's Electromagnetic Lab 969
10.2	*Energy Skate Park 248, 261		**33.4**	Faraday's Law 972
10.5	*Masses & Springs 257		**33.7**	Generator, *Faraday's Electromagnetic Lab 983
10.6	Stretching DNA 261		**33.8**	*Circuit Construction (AC + DC) 986
11.2	*The Ramp 280		**34.5**	Radio Waves & Electromagnetic Fields 1016, 1023
11.7	States of Matter 292		**35.1**	*Circuit Construction (AC + DC) 1034, 1036, 1041, 1043
12.1	Ladybug Revolution 313		**36.10**	Nuclear Fission 1094
12.3	Torque 317, 322, 340		**37.1**	Blackbody Spectrum 1104
13.3	Gravity Force Lab 357		**37.4**	Neon Lights and Other Discharge Lamps 1108
13.6	My Solar System 366		**37.6**	Rutherford Scattering 1114
14.1	Motion in 2D 379		**38.1**	Photoelectric Effect 1128
14.5	*Masses & Springs 390		**38.6**	Models of the Hydrogen Atom 1141
14.6	*Pendulum Lab 392, 395		**38.7**	Neon Lights and Other Discharge Lamps 1146
15.1	States of Matter 408		**39.1**	Wave Interference 1157
15.2	Gas Properties 411		**39.1**	Plinko Probability 1158
15.4	Balloons & Buoyancy 421		**39.1**	Quantum Wave Interference 1159
15.5	*Energy Skate Park 427		**40.6**	Quantum Bound States 1194
16.1	States of Matter 445, 451		**40.9**	Double Wells and Covalent Bonds 1206
16.5	Gas Properties 453, 457		**40.10**	Quantum Tunneling and Wave Packets 1206
17.2	States of Matter 473, 476		**41.1**	Models of the Hydrogen Atom 1217
17.8	Blackbody Spectrum 493		**41.3**	Stern-Gerlach Experiment 1223
17.8	The Greenhouse Effect 494		**41.6**	Neon Lights and Other Discharge Lamps 1234
18.1	Gas Properties 503		**41.8**	Lasers 1238
18.6	Reversible Reactions 518		**42.5**	Radioactive Dating Game 1260
20.1	Waves on a String, *Wave Interference 561		**42.6**	Alpha Decay, Beta Decay 1264
20.2	*Wave Interference 565, 574		**42.7**	Simplified MRI 1270
20.3	Waves on a String 570, 574			
20.5	Sound, Radio Waves and EM Fields 574			
21.1	Fourier: Making Waves, Waves on a String 592			
21.3	Waves on a String 595			
21.4	*Wave Interference 599			
22.2	*Wave Interference 630, 637			

*Indicates an associated tutorial is available in the MasteringPhysics Item Library.

Answers

Answers to Odd-Numbered Exercises and Problems

Chapter 25

1. a. Electrons added   b. 7.5×10^{10}
3. 2.5×10^{10}
5. 1.9×10^5
9. Right negatively charged, left positively charged
13. a. 0.056 N   b. 2.9
15. a. 58 N   b. 4.7×10^{-35} N   c. 1.2×10^{36}
17. $-(4.1 \times 10^{-4} \text{ N})\hat{\jmath}$
19. a. 1.3×10^{14} m/s^2 toward bead   b. 2.4×10^{17} m/s^2 away from bead
21. a. $(6.4\hat{\imath} + 1.6\hat{\jmath}) \times 10^{-17}$ N
    b. $-(6.4\hat{\imath} + 1.6\hat{\jmath}) \times 10^{-17}$ N   c. 4.0×10^{10} m/s^2 d. 7.3×10^{13} m/s^2
23. $-4.5 \times 10^4\,\hat{r}$ N/C (i.e., toward the bead)
25. 3.3×10^6 N/C, downward
27. $-6.8 \times 10^4\hat{\imath}$ N/C, $3.0 \times 10^4\hat{\imath}$ N/C, $(8.1 \times 10^3\,\hat{\imath} - 3.9 \times 10^4\hat{\jmath})$ N/C
29. a. 0.36 m/s^2 toward glass bead   b. 0.18 m/s^2 toward plastic bead
31. 82 nC
33. 3.1×10^{-4} N, upward
35. 4.3×10^{-3} N, 253° ccw
37. 2.0×10^{-4} N, 45° cw
39. $-1.0 \times 10^{-3}\hat{\imath}$ N
41. $(1.02 \times 10^{-5}\hat{\imath} + 2.2 \times 10^{-5}\hat{\jmath})$ N
43. 0.68 nC
45. $(F_\text{net})_x = \dfrac{-2KQqa}{(a^2 + y^2)^{3/2}}$
47. $(2 - \sqrt{2})\dfrac{KQq}{L^2}$
49. $-\frac{4}{9}q$, $x = \frac{1}{3}L$
51. 6.6×10^{15} rev/s
53. a. 2.3×10^{-6} b. 4.3×10^7 N/C, upward
55. 33 nC
57. a. 1.1×10^{18} m/s^2 b. 1.0×10^{-12} N   c. 6.3×10^6 N/C   d. 69 nC
59. 0.75 μC
61. 1.8×10^5 N/C, 60° ccw from the +x-axis; 1.8×10^5 N/C, 60° cw from the −x-axis
63. a. (4.0 cm, 1.0 cm)   b. (0.0 cm, 2.0 cm)   c. (−2.0 cm, −2.0 cm)
65. a. $\vec{E}_1 = (8.5\hat{\imath} - 2.8\hat{\jmath})$ kN/C, $\vec{E}_2 = 10\,\hat{\imath}$ kN/C,
    $\vec{E}_3 = (8.5\hat{\imath} + 2.8\hat{\jmath})$ kN/C   c. $27\hat{\imath}$ kN/C
67. 14°
69. b. 22 nC
71. b. 5.1 nC
73. 0.11 μC
75. 1.7×10^{-4} N

Chapter 26

1. 7.6×10^3 N/C along the +x-axis
3. 1.0×10^4 N/C at 11° below the +x-axis
5. a. 36 N/C   b. 18 N/C
7. 4000 N/C
9. 1.3×10^5 N/C, 0.0 N/C, 1.3×10^5 N/C
11. a. 2.6×10^4 N/C, left   b. 2.6×10^{-5} N, right
13. a. 7.6×10^4 N/C, left   b. 7.6×10^{-5} N, right
15. 27 nC

17. 1.9 cm
19. 2.7×10^{11}
21. a. 3.6×10^6 N/C   b. 8.3×10^5 m/s
23. 18 cm
25. 3.1×10^{-21} N m
27. 9.0×10^{-13} N$\vec{p}$
29. a. $(-9.7 \times 10^4\hat{\imath} + 9.2 \times 10^4\hat{\jmath})$ N/C
    b. 1.34×10^5 N/C, 136°ccw from the +x-axis
31. $\dfrac{1}{4\pi\epsilon_0}\dfrac{Q}{L^2}(\sqrt{2} - 1)(\hat{\imath} + \hat{\jmath})$
33. a. $\dfrac{2qx}{4\pi\epsilon_0(x^2 + s^2/4)^{3/2}}$
    b. 0 N/C, 768,000 N/C, 576,000 N/C, 358,000 N/C, 158,000 N/C
35. a. $\dfrac{2q}{4\pi\epsilon_0}\left[\dfrac{1}{x^2} - \dfrac{x}{(x^2 + d^2)^{3/2}}\right]\hat{\imath}$
37. $\dfrac{1}{4\pi\epsilon_0}\dfrac{8\lambda d}{4y^2 + d^2}$
39. -0.056 nC
41. $\dfrac{Q}{4\pi\epsilon_0}\dfrac{1}{x\sqrt{x^2 + L^2}}\hat{\imath} - \dfrac{Q}{4\pi\epsilon_0 Lx}\left(1 - \dfrac{x}{\sqrt{x^2 + L^2}}\right)\hat{\jmath}$
43. a. $\dfrac{R}{\sqrt{2}}$ b. $\dfrac{2}{3\sqrt{3}}\dfrac{Q}{4\pi\epsilon_0 R^2}$
45. c. $\dfrac{1}{4\pi\epsilon_0}\dfrac{2Q}{\pi R^2}(\hat{\imath} + \hat{\jmath})$
47. 1.41×10^5 N/C
49. 2.2 mm
51. 1.19×10^7 m/s
53. a. $\dfrac{\frac{4}{3}\pi r^3\rho g + qE}{6\pi\eta r}$ b. 0.067 mm/s   c. 0.049 mm/s
55. 6.56×10^{15} Hz
57. a. $\dfrac{C^2 s^2}{\text{kg}}$ b. $\left(\dfrac{1}{4\pi\epsilon_0}\right)^2\dfrac{2q^2\alpha}{r^5}$, toward ion
59. b. 1.0 mm
61. b. $\dfrac{R}{\sqrt{3}}$
63. 4.2×10^{-4} N

65. a.

    b. $\dfrac{4Q}{L^2}$ c. $\dfrac{8Q}{4\pi\epsilon_0 L^2}\left[1 - \dfrac{x}{\sqrt{x^2 + L^2/4}}\right]$
67. a. $\dfrac{2\eta}{4\pi\epsilon_0}\ln\left(\dfrac{2x + L}{2x - L}\right)\hat{\imath}$

c.

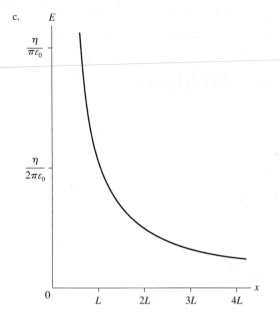

69. −2.3 nC/m

71. a. $k = \dfrac{qQ}{4\pi\epsilon_0 R^3}$ c. 2.0×10^{12} Hz

Chapter 27

1.

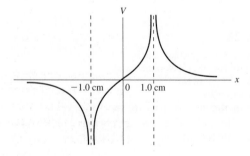

3.

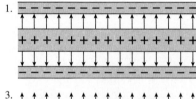

$$\vec{E} = \vec{0}\ \text{N/C}$$

5. No charge
7. Into the front face of the cube; field strength must exceed 5 N/C
9. $1.0\ \text{N}\,\text{m}^2/\text{C}$
11. 1.4×10^3 N/C
13. a. $0.0\ \text{N}\,\text{m}^2/\text{C}$ b. $3.0 \times 10^{-2}\ \text{N}\,\text{m}^2/\text{C}$
15. $3.5 \times 10^{-4}\ \text{N}\,\text{m}^2/\text{C}$
19. $+2q, +q, -3q$
21. $0.11\ \text{kN}\,\text{m}^2/\text{C}$
23. $-1.00\ \text{N}\,\text{m}^2/\text{C}$
25. $2.7 \times 10^{-5}\ \text{C/m}^2$
27. a. $\vec{E} = (25\hat{k})$ kN/C, upward from the plate
   b. 0.0 N/C c. 2.5 kN/C, downward from the plate
29. a. $-0.39\ \text{N}\,\text{m}^2/\text{C}, 0.23\ \text{N}\,\text{m}^2/\text{C}, 0.39\ \text{N}\,\text{m}^2/\text{C},$
   $-0.23\ \text{N}\,\text{m}^2/\text{C}$ b. $0\ \text{N}\,\text{m}^2/\text{C}$
31. a. $-3.5\ \text{N}\,\text{m}^2/\text{C}$ b. $1.2\ \text{N}\,\text{m}^2/\text{C}$
33. $0.19\ \text{kN}\,\text{m}^2/\text{C}$
35. a. 2.0 kN/C b. $0.25\ \text{kN}\,\text{m}^2/\text{C}$ c. 2.2 nC

37. a. −100 nC b. +50 nC
39. a. $2.4 \times 10^{-6}\ \text{C/m}^3$
   b. 1 nC, 10 nC, 80 nC c. 5 kN/C, 9.0 kN/C, 1.8×10^4 N/C
41. -4.51×10^5 C
43. 2.5×10^4 N/C, outward; 0 N/C; 7.9×10^3 N/C, outward
45. $\vec{0}$ N/C, $\dfrac{1}{4\pi\epsilon_0}\dfrac{Q}{r^2}\hat{r}$
47. $\vec{0}$ N/C, $(\eta/2\epsilon_0)\hat{j}$, $-(\eta/2\epsilon_0)\hat{j}$, $\vec{0}$ N/C
49. $(\eta/2\epsilon_0)\hat{j}$, $\vec{0}$ N/C, $(\eta/2\epsilon_0)\hat{j}$, $-(\eta/2\epsilon_0)\hat{j}$
51. a. $\dfrac{\lambda}{2\pi\epsilon_0 r}\dfrac{\vec{r}}{r}$ b. $\dfrac{3\lambda}{2\pi\epsilon_0}\dfrac{\vec{r}}{r}$
53. a. $\dfrac{1}{4\pi\epsilon_0}\dfrac{Q}{r^2}\hat{r}$ b. $\vec{E} = \vec{0}$ c. $\dfrac{1}{4\pi\epsilon_0}\dfrac{Q}{r^2}\left(\dfrac{r^3 - R_{in}^3}{R_{out}^3 - R_{in}^3}\right)\hat{r}$
55. a. $\dfrac{\lambda L^2 dy}{4\pi\epsilon_0\left[y^2 + (L/2^2)\right]}$ b. $\lambda L/(4\epsilon_0)Q_{in}/\epsilon_0$
57. a. $C = \dfrac{Q}{4\pi R}$ b. $\dfrac{1}{4\pi\epsilon_0}\dfrac{Q}{Rr}\hat{r}$ c. Yes
59. a. $\dfrac{Q}{4\pi\epsilon_0 R^2}$ b. $\dfrac{3Qr^3}{2\pi R^6}$

Chapter 28

1. 1.4×10^5 m/s
3. 2.1×10^6 m/s
5. -2.2×10^{-19} J
7. 4.8×10^{-6} J
9. a. $-1.0\ \mu$J b. $1.0\ \mu$J
11. 1.87×10^7 m/s
13. -8.4×10^4 V
15. a. Lower b. −0.712 V
17. a. 1.5 V b. 2.1×10^{-11} C
19. a. 200 V b. 6.3×10^{-9} C
21. a. 1800 V, 1800 V, 900 V b. 0 V, −900 V
23. a. 27 V b. 4.3×10^{-18} J
25. −1600 V
27. a. $\pm\infty$ b. 0, $\pm\infty$
   c.

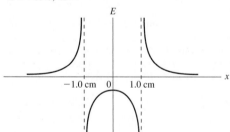

29. a. Positive, positive   b. 1
   c.
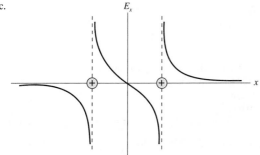

31. 1.4×10^{-3} N
33. a. +103 V   b. 5.40×10^4 V/m
35. ± 12 cm
37. 0.49 m/s
39. a. 1.1×10^{-20} J   b. 2×10^{21} ions
41. 54 kHz
43. a. 2.1×10^6 V/m   b. 9.4×10^7 m/s
45. a. 0.85 m   b. 2.6 m
47. 8.0×10^7 m/s
49. -5.1×10^{-19} J
51. 310 nC
53. 6.8 fm
55. a. Yes   c. 8.21×10^8 m/s
57. a. 2.1×10^{-10} C, 3.0 kV/m, 15 V   b. 2.1×10^{-10} C, 3.0 kV/m, 30 V
   c. 2.1×10^{-10} C, 0.75 kV/m, 3.8 V
59. a. $\dfrac{V_0}{R}$ b. 100 kV/m
61. a. $8.3\,\mu$C   b. 3.3×10^6 V/m
63. 2.1 kV, b is higher
65. a. $\dfrac{2q}{4\pi\epsilon_0 x}\dfrac{1}{\sqrt{1+s^2/4x^2}}$
   b.
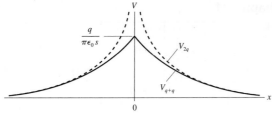

67. $(Q/4\pi\epsilon_0 L)\ln\left[(x+L/2)/(x-L/2)\right]$
69. $Q/4\pi\epsilon_0 R$
71. b. q_1 and q_2 are 10 nC and 30 nC
73. 6.0 cm
75. $v_A = 0.018$ m/s, $v_B = 0.011$ m/s
79. a. $\dfrac{1}{4\pi\epsilon_0}\dfrac{q}{R}dq$ b. $\dfrac{1}{4\pi\epsilon_0}\dfrac{Q^2}{2R}$ c. 2.3×10^{-13} J
81. $\dfrac{3Q}{8\pi\epsilon_0 R^3}\left(R\sqrt{R^2+z^2}+\ln\left(\dfrac{|z|}{R+\sqrt{R^2+z^2}}\right)\right)$

Chapter 29
1. -200 V
3. -0.30 kV

5. 1.5×10^{-6} J
7. 3.0 C
9. $-(20\hat{j})$ kV/m
11.
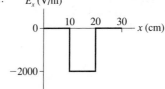

13. -1.0 kV/m
15. a. 27 V/m   b. 3.7 V/m
17. a. 13 pF   b. 1.3 nC
19. 3.0 V
21. 32 μF
23. 150 μF, in series
25. 1.4 kV
27. 1/2
29. a. 1.1×10^{-7} J   b. 0.71 J/m^3
31. a. 62 pC, 9.0 V, 29 kV/m   b. 20 pC, 9.0 V, 90 kV/m
33. a. A   b. -70 V
35. a.
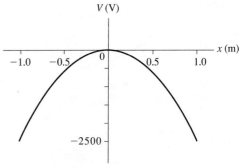

   b. $V(x) = -(2500\,x^2)$ V
37. a. $\vec{E} = -(1.4 \times 10^7 \hat{i})$ V/m, $V = 7 \times 10^4$ V
   b. $E = 0.0$ V/m, $V = 1.4 \times 10^5$ V
   c. $\vec{E} = 1.4 \times 10^7 \hat{i}$ V/m, $V = 7 \times 10^4$ V
39. $\vec{E}_{disk}(z) = \dfrac{Q}{2\pi\epsilon_0 R^2}\left[1-\dfrac{z}{\sqrt{R^2+z^2}}\right]\hat{k}$
41. Point 1: 3750 V/m, downward; point 2: 7500 V/m, upward
43. 1000 V/m, 127° ccw from the +x-axis
45. $Q_{1f} = 2$ nC, $Q_{2f} = 4$ nC
47. 1.1 nC
49. a. ± 32pC, 9.0 V   b. ± 16pC, 9.0 V
51. 7.5 μF
53. 5.0 V, 15 V, 10 V
55. $Q_1 = 45\,\mu$C, $V_1 = 9$ V; $Q_2 = 22\,\mu$C $V_2 = 5.4$ V; and $Q_3 = 22\,\mu$C, $V_3 = 3.6$ V

57. a. $\frac{3}{2}C$ b. 0 V

59. $Q_1 = 0.83$ mC, $Q_2 = Q_3 = 0.67$ mC,
$\Delta V_1 = 55$ V, $\Delta V_2 = 34$ V, $\Delta V_3 = 22$ V

61. $Q_1 = 33$ μC, $Q_2 = 67$ μC, $\Delta V'_1 = \Delta V'_2 = 3.3$ V

63. a. 5.7×10^{-7} J   b. 11.4×10^{-7} J
    c. Work was done on the capacitor.

65. 0.85 kV

67. 0.13 F

69. 2.4×10^{-14} J

73. b. $(10 - z^2)$ V, with z in meters

75. b. 2 μF

77. a. $V = \frac{q}{4\pi\epsilon_0}\left[\frac{1}{\sqrt{x^2 + (y - s/2)^2}} - \frac{1}{\sqrt{x^2 + (y + s/2)^2}}\right]$

    b. $V = \frac{qsy}{4\pi\epsilon_0(x^2 + y^2)^{3/2}}$

    c. $E_x = \frac{qs(3xy)}{4\pi\epsilon_0(x^2 + y^2)^{5/2}}$, $E_y = \frac{qs(2y^2 - x^2)}{4\pi\epsilon_0(x^2 + y^2)^{5/2}}$

    d. $\vec{E}_{\text{on-axis}} = \frac{2p}{4\pi\epsilon_0 r^3}\hat{j}$, yes

    e. $E_{\text{bisecting axis}} = -\frac{p}{4\pi\epsilon_0 r^3}\hat{j}$, yes

79. a. $V_r = \frac{1}{4\pi\epsilon_0}\frac{Q}{R}\left[\frac{3}{2} - \frac{r^2}{2R^2}\right]$ b. 3/2

    c.

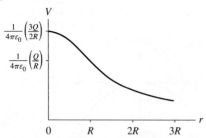

81. a. $\frac{2\pi\epsilon_0}{\ln(R_2/R_1)}$ b. 31 pF/m

Chapter 30

1. 3.0 d
3. 7.6×10^{26} electrons
5. 0.023 V/m
7. 1.0×10^{19} s^{-1}
9. a. 0.80 A   b. 7.0×10^7 A/m^2
11. 130 C
13. 1.88×10^{22}
15. 2.6 mA
17. a. 6.3×10^5 A/m^2 b. 6.5×10^{-5} m/s
19. 1.68 A
21. 5.0×10^{-8} Ω m
23. a. 1.64×10^{-3} V/m   b. 1.10×10^{-5} m/s
25. Tungsten
27. $\frac{1}{2}$
29. Tungsten

31. a. 30 m   b. 1.0 A
33. 4100 Ω
35. 380
37. 0.64 mm
39. Yes, 2.2×10^5 Ω^{-1}m^{-1}
41. a. 75 nA   b. 130 s
43. a. 6.6×10^{15} Hz   b. 1.05×10^{-3} A
45. a. 120 C   b. 0.45 mm
47. 1.4 Ω m
49. 0.50 mm
51. a. $E = \frac{I}{4\pi\sigma r^2}$ b. $E_{\text{inner}} = 3.3 \times 10^{-4}$ V/m, $E_{\text{outer}} = 5.3 \times 10^{-5}$ N/C
53. a. $I(t) = (10$ A$)e^{-t/2.0\text{ s}}$ b. 10 A

    c.

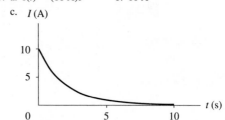

55. 2.0 A, 5.0×10^{-5} m/s
57. 7.2 mm
59. 0.16 V/m
61. 2R
63. a. 4.2×10^5 A   b. Decrease   c. 1.1×10^{-5} J
65. 1.8×10^8 A/m^2
67. a. 2.5 C   b. 1.8 cm
69. 1.01×10^{23}
71. a. 9.4×10^{15} b. 115 A/m^2
73. a. $\eta = \frac{\epsilon_0 I}{A}\left(\frac{1}{\sigma_2} - \frac{1}{\sigma_1}\right)$ b. 3.7×10^{-18} C

Chapter 31

1.

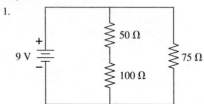

3. 1 A to left
5. a. 0.9 A ccw

    b.

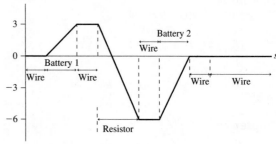

7. 9.60 Ω, 12.5 A
9. 60 W bulb is brighter
11. a. 11.6 A   b. 10.4 Ω
13. 75 Ω
15. a. 0.65 Ω   b. 3.5 W
17. 3.2%
19. 240 Ω
21. 40 Ω
23. 183 Ω
25. 9 V, 1 V
29. 2 ms
31. a. 36 μC, 0.36 A   b. 22 μC, 0.22 A   c. 4.9 μC, 49 mA
33. 18 μF
35. D
37. 93 W
39.

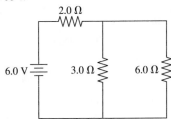

41. 7 Ω
43. 60 V, 10 Ω
45. 9.0 V, 0.50 Ω
47. 1.8 V
49. 1.0 A, 2.0 A, 15 V
51. $65 for the incandescent bulb, $20 for the fluorescent tube
53. a. 0.231 A   b. 0.214 S   c. 7.4%
55. 900 Ω
57. a. 0.505 Ω   b. 0.500 Ω
59.

Resistor	Potential difference (V)	Current (A)
3 Ω	6.0	2.0
4 Ω	6.0	1.5
48 Ω	6.0	0.125
16 Ω	6.0	0.375

61.

Resistor	Potential difference (V)	Current (A)
24 Ω	6.00	0.25
3 Ω	3.00	1.00
5 Ω	3.75	0.75
4 Ω	2.25	0.56
12 Ω	2.25	0.19

63. 9/25 A, left to right
65. 150 V, bottom
67. 0.41 A, left to right
69. a. 65 kΩ   b. 87 V
71. 73 Ω

73. a. $\mathcal{E}$ b. $C\mathcal{E}$ c. $+dQ/dt$

d. $\left(\dfrac{\mathcal{E}}{R}\right)e^{-t/\tau}$

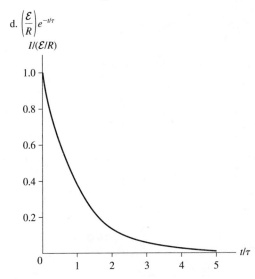

75. 2.0 m, 0.49 mm
77. 20 V
79. a. $\mathcal{E}^2 C$ b. $\mathcal{E}^2 C/2$ c. $\mathcal{E}^2 C/2$ d. Yes
81. 0.60 A

Chapter 32

1. $\vec{B}_1 = (2.0$ mT, into the page), $\vec{B}_2 = (4.0$ mT, into the page)
3. a. 0 T   b. $1.60 \times 10^{-15}\hat{k}$ T   c. $-4.0 \times 10^{-16}\hat{k}$ T
5. $-1.13 \times 10^{-15}\hat{k}$ T
7. 6.3×10^6 m/s in the $+z$-direction
9. 4.0 cm, 0.40 mm, 20 μm to 2.0 μm, 0.20 μm
11. a. 20 A   b. 1.6×10^{-3} m
13. $2.0 \times 10^{-4}\hat{i}$ T, $4.0 \times 10^{-4}\hat{i}$ T, $2.0 \times 10^{-4}\hat{i}$ T
15. a. 0.025 A m^2 b. 1.5 μT
17. 1.4 cm
19. 0.071 T m
21. 7.00 A
23. 1.26×10^{-6} T m
25. 1.0 mm
27. a. $8.0 \times 10^{-13}\hat{j}$ N   b. $5.7 \times 10^{-13}(-\hat{j} - \hat{k})$ N
29. 1.6×10^{-3} T
31. 81 mT
33. 0.131 T, out of page
35. 3.0 Ω
37. 7.5×10^{-4} N m
39. a. 1.26×10^{-11} N m   b. Rotated by $\pm 90°$
41. 0.040 μA
43. $(5.2 \times 10^{-5}$ T, out of page), $\vec{0}$ T
45. 0.77R
47. $(7.9 \times 10^{-5}$ T, into page)
49. #18, 4.1 A
51. a. 1.13×10^{10} A   b. 0.014 A/m^2 c. 1.3×10^6 A/m^2
53. a. 5.7×10^{-6} A   b. 2.9×10^{-8} A m^2
55. $\dfrac{\mu_0 I}{4R}$

57. $0;\ \dfrac{\mu_0 I}{2\pi r}\left(\dfrac{r^2 - R_1^2}{R_2^2 - R_1^2}\right);\ \dfrac{\mu_0 I}{2\pi r}$

59. 1.50 mT, 30° ccw from the $+x$-axis

61. 2.9×10^{-3} T

63. 2.4×10^{10} m/s^2, up

65.

	Ion	Accelerating voltage (V)
a.	O_2^+	96.793
b.	N_2^+	110.25
c.	CO^+	110.29

67. 0.12 T

69. 87 mT

71. a. $\dfrac{\mu g \tan\theta}{I}$, down   b. 11 mT, down

73. 13 T

75. a. $2\pi RIB \sin\theta$ b. 4.3×10^{-3} N

77. a. $\dfrac{\mu_0 I L}{4\pi d\sqrt{(L/2)^2 + d^2}}$ b. $\dfrac{\sqrt{2}\mu_0 I}{\pi R}$ c. 0.900

79. a. $\dfrac{3I}{2\pi R^2}$ b. $\dfrac{\mu_0}{2\pi}\dfrac{Ir^2}{R^3}$ c. Yes

81. a. Horizontal and to the left above the sheet; horizontal and to the right below the sheet   b. $\frac{1}{2}\mu_0 J_s$

Chapter 33

1. 2.0×10^4 m/s
3. a. 1.0 N   b. 2.2 T
5. 6.3×10^{-5} Wb in both cases
7. Decreasing
9. Clockwise current
11. a. 3.9 mV, 20 mA, ccw   b. 3.9 mV, 20 mA, ccw   c. No current
13. 1.6 V
15. $E\ (\times 10^{-4}$ V/m)

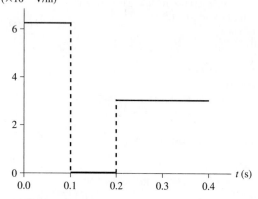

17. a. 4.8×10^4 m/s^2, up   b. 0   c. 4.8×10^4 m/s^2, down
    d. 9.6×10^4 m/s^2, down
19. 1.0 ms
21. 9.5×10^{-5} J
23. 250 kHz to 360 kHz
25. 750 Ω
27. 3.5×10^{-4} Wb
29. 1.6 A, 0.0 A, -1.6 A
31. 8.7 T/s
33. a. -0.0050 V   b. 0.0100 V

35. 44 μA
37. a. 0 μA   b. 160 μA   c. 0 μA
39. a. 0.0 A   b. 79 μA
41. a. 0.93 V   b. 0 V
43. a. 12 500   b. 2.0 A
45. a. $I\ (A)$

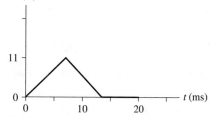

   b. 11 A when halfway in
47. a. 4.0 V   b. 100 A   c. 3.0 V
49. a. $(4.9 \times 10^{-3})f\sin(2\pi ft)$ A   b. 4.1×10^2 Hz, not feasible
51. 0.28 T
53. a. $(vlB\cos\theta)/R$ b. $(mgR\tan\theta)/l^2B^2\cos\theta$
55. 2.5×10^{-4} V
57. 12 V
59. $(R^2/2r)/(dB/dt)$
61. a. 3.9×10^{-4} J/m^3 b. 3.1 A
63. 3.0 s
65. $I\ (A)$

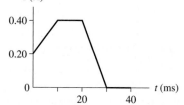

67. a. $\Delta V_L = \left(\dfrac{LI_0}{\tau}\right)e^{-t/\tau}$ b. 0.37 V
69. 1.0 μF
71. 0.50 m
73. a. 76 mA   b. 0.50 ms
75. a. 0.50 A   b. 1.0 A
77. a. $\Delta V_{bat}/R$ b. $I = I_0(1 - e^{-t/(L/R)})$
79. $(\mu v_0 I/2\pi)\ln[(d + l)/d]$
81. a. 0.10 s   b. $2.93\left(\dfrac{(0.10)^2 - 2[0.0707 + (0.293)t]^2}{\sqrt{(0.10)^2 - [0.0707 + (0.293)t]^2}}\right)$ A

   c. I

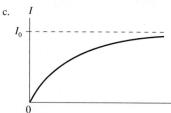

83. a. 32 A   b. 1.3 m/s
85. a. $(\mu_0/2\pi)\ln(r_2/r_1)$ b. 0.36 μH/m

Chapter 34

1. a. $(2.0 \times 10^6$ m/s, 45° from the y-axis) 45°
   b. $(1.47 \times 10^6$ m/s, 16.2° from the y'-axis) 16.2°
3. $-1.0 \times 10^6 \hat{k}$ V/m, $-1.11 \times 10^{-5}\hat{j}$ T

5. 16.3° above the +*x*-axis

9. 1.0 μF

11. 17 μA

13. 3.3×10^{-8} T

15. a. 10.0 nm   b. 3.00×10^{16} Hz   c. 6.67×10^{-8} T

17. a. 3.33×10^{-7} T   b. 13.3 W/m²

19. 980 V/m, 3.3 μT

21. a. 2.2×10^{-6} W/m²   b. 0.041 V/m

23. 3.3×10^{-6} N

25. 60°

27. 30°

29. $(1.73 \times 10^{6}$ V/m, left)

31. a. (0.10 T, into page)   b. 0 V/m, (0.10 T, into page)

33. 1.0×10^{7} m/s parallel to the current

35. a. 0.94 V/m   b. 10 T

37. b. 1.5×10^{-13} A

39.

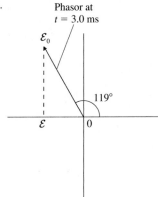

41. 20 V

43. b. 6.67×10^{-6} J/m³

45. a. 3.85×10^{26} W   b. 589 W/m²

47. a. (1/2)*f*   b. (3/4)*f*

49. Yes

51. 1.8×10^{7} V/m

53. 1.3 m

55. 4.9×10^{7} W/m²

57. 8.8 h

59. $(-6.0 \times 10^{5}\,\hat{\imath} + 1.0 \times 10^{5}\,\hat{\jmath})$ V/m

61. 5.2 μV/m

63. a. $E = IR/L$, $B = \dfrac{\mu_0 I}{2\pi r} IR/L$ b. $(I^2 R/2\pi r L$, radially inward)

Chapter 35

1. a. 22×10^{2} rad/s   b. -10 V

3.

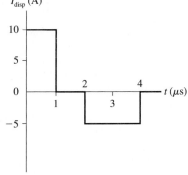

5. a. 50 mA   b. 50 mA

7. a. 1.9 mA   b. 1.9 A

9. a. 80 Hz   b. 0 V

11. a. 95 pF   b. 660 μA

13. 1.6 μF

15. $V_R = 6.0$ V, $V_C = 8.0$ V

17. a. 1000 Hz   b. 2.24 V, 3.53 V, 4.47 V

19. a. 0.80 A   b. 0.80 mA

21. a. 3.2×10^{4} Hz   b. 0 V

23. a. 200 kHz   b. 141 kHz

25. 1.3 μF

27. a. 70 Ω, 72 mA, $-44°$ b. 50 Ω, 0.10 A, 0°   c. 62 Ω, 80 mA, 37°

29. 9.6 Ω

31. 30°

33. 44 Ω

35. a. $(\sqrt{3}RC)^{-1}$ b. $\sqrt{3}\mathcal{E}_0/2$

37. a. 9.95 V, 9.57 V, 7.05 V, 3.15 V, 0.990 V

b.

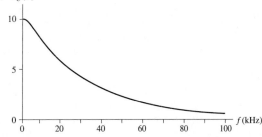

43. 44 Hz

45. a. 50 Hz   b. 4.8 μF

47. a. $\mathcal{E}_0/\sqrt{R^2 + \omega^2 L^2}$, $\mathcal{E}_0 R/\sqrt{R^2 + \omega^2 L^2}$, $\mathcal{E}_0 \omega L/\sqrt{R^2 + \omega^2 L^2}$

b. $V_R \to \mathcal{E}_0$, $V_R \to 0$ c. Low pass   d. R/L

49. a. 69 V   b. 24°   c. 0.17 kW

51. a. 5.0×10^{3} Hz   b. 10 V, 32 V

53. 0.17 A

55. a. 3.6 V   b. 3.5 V   c. -3.6 V

59. a. 11.6 pF   b. 1.49×10^{-3} Ω

61. 14 W in 40 W bulb, 9.6 W in 60 W bulb, 100 W in 100 W bulb

65.

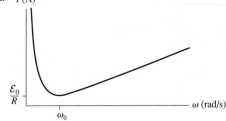

67. a. 0.44 kA   b. 1.8×10^{-4} F   c. 7.4 MW

69. b. 10 V, 12 V

71. b. ∞, ∞ c. $1/\sqrt{LC}$

d.

Chapter 36

1. $x'_1 = 5.0$ m at $t = 1.0$ s, $x'_2 = -5.0$ m at $t = 5.0$ s
3. $v_{sound} = 345$ m/s, $v_{sprinter} = 15$ m/s
5. a. 13 m/s   b. 3.0 m/s   c. 9.4 m/s
7. 3.0×10^8 m/s
9. 167 ns
11. $2.0 \ \mu$s
13. No, bolt 2 hits 20 μs before bolt 1.
15. Yes
17. $0.866c$
19. a. $0.9965c$ b. 59.8 ly
21. 46 m/s
23. Yes
25. 4600 kg/m^3
27. 3.0×10^6 m/s
29. $x = 8.3 \times 10^{10}$ m, $t = 330$ s
31. $0.36c$
33. $0.71c$
35. $0.80c$

37. $0.707c$
39. a. 1.8×10^{16} J   b. 9.0×10^9
41. $0.943c$
43. $u_{50 \ final} = 1.33$ m/s to the right, $u_{100 \ final} = 3.33$ m/s to the right
45. 11.2 h
47. a. No   b. 67.1 y
49. a. $0.80c$ b. 16 y
51. 0.78 m
53. a. 17 y   b. 15 y   c. Both
55. $0.96c$
57. 3.1×10^6 V
59. a. $0.98c$ b. 8.5×10^{-11} J
61. b. Lengths perpendicular to the motion are not affected.
63. a. $u'_y = u_y / \gamma \left(1 - u_x v / c^2 \right)$ b. $0.877c$
65. a. 3.5×10^{-18} kg m/s, 1.1×10^{-9} J   b. 1.6×10^{-18} kg m/s
67. a. 7.6×10^{16} J   b. 0.84 kg
69. 7.5×10^{13} J
71. 1 pm
73. 22 m
75. $0.85c$

Credits

Index

A

Absolute temperature, 450
Absolute zero, 449, 450, 452
Absorbed dose, 1268
Absorption
    excitation by, 1232–33
    in hydrogen, 1233
    of light, 1234, 1235
    in sodium, 1233
Absorption spectra, 1105, 1184, 1190, 1203
Acceleration, 1, 86–87
    angular, 103–05, 205, 313–14
    average, 13, 86, 88, 89
    centripetal, 102–03, 194, 200, 205
    constant, 90–91, 124
    force and, 124–25, 133, 141
    free-fall, 52
    Galilean transformation of, 1005
    instantaneous, 58–60, 89–90
    nonuniform, 58–59
    sign of, 16–17, 61
    tangential, 105–06, 205
    in two dimensions, 192–93
    in uniform circular motion, 102–03
Acceleration constraints, 174–75, 184
Acceleration vectors, 13–14, 86–87
AC circuits, 1031–55
    AC sources, 1034
    capacitor circuits, 1036–38
    inductor circuits, 1041–42
    phasors, 1024
    power in, 1046–49
    RC filter circuits, 1038–41
    resistor circuits, 1034–36
    series RLC circuits, 1042–46
Accommodation, 701, 702
Action/reaction pair, 168. See also Newton's
    third law of motion
    identifying, 169–72
    massless spring approximation and, 179
    propulsion, 171
Activity of radioactive sample, 1261
Adiabatic processes, 479, 488–91
Adiabats, 489
Agent of force, 117
Air resistance, 51, 121, 128, 153
Airplanes, lift in, 429

Allowed transitions, 1232
Alpha decay, 1264
Alpha particles, 836, 1113–14, 1116, 1258, 1263–64
Alpha rays, 1113, 1249
Alternating current (AC), 983. See also
    AC circuits
Ammeters, 900
Amorphous solids, 445
Ampere (A), 875
Ampère-Maxwell law, 1012–14, 1015, 1019–20
Ampère's law, 934–40, 1011–12
Amplitude, 378, 380, 396, 398–99, 566, 570, 573, 579, 593–94
Amplitude function, 595
Angle of incidence, 658, 661, 662
Angle of reflection, 658
Angle of refraction, 661, 662
Angular acceleration, 103–05, 108, 205, 313–14
Angular displacement, 99
Angular frequency, 379, 381
Angular magnification, 704, 705, 706
Angular momentum, 312, 340–45, 1223–25
    conservation of, 342, 368
    hydrogen atom, 1218–19
    quantization of, 1145
    of a rigid body, 341
Angular momentum vector, 340
Angular position, 98–99, 631
Angular resolution, 709
Angular size, 703
Angular velocity, 99–101, 108, 205, 313–14, 342–45
Angular velocity vector, 337–40
Antennas, 1023
Antibonding molecular orbital, 1206
Antimatter, 1093
Antinodal lines, 612, 630
Antinodes, 594, 599, 600, 601
Antireflection coatings, 608–10
Apertures, 657, 698
Archimedes' principle, 420
Arc length, 99
Area vector, 787, 970
Atmospheric pressure, 411–13, 416–17

Atomic clocks, 23, 1076–77
Atomic magnets, 950
Atomic mass, 292, 447, 1250–51
Atomic mass number (A), 447, 1198
Atomic mass unit (u), 447, 1250, 1251, 1254
Atomic model, 120, 264
Atomic number, 1117, 1249
Atomic physics, 1216–47
    electron spin, 1218, 1223–25
    emission spectra, 1231–35
    excited states, 1236–38
    hydrogen atom, 1217–23
    lasers, 1238–42
    multielectron atoms, 1225–28
    periodic table of the elements, 1228–31
Atoms, 217, 453, 1103, 1108. See also
    Electrons; Hydrogen; Nuclear
    Physics; Nucleus; Protons
    Bohr model of, 1138, 1141–46
    electricity and, 725–26
    hard-sphere model of, 453
    nuclear model of, 1114, 1116–17
    plum-pudding or raisin-cake model of, 1112–13
    shell model of, 1222, 1228–31, 1256–57
    structure of, 725, 729, 1118, 1138, 1249
Avogadro's number, 292, 447–48

B

Ballistic pendulum, 253, 344–45
Balmer formula, 1105
Balmer series, 1105
Bandwidth, 1168
Bar charts, 223–24, 249–50, 295–96
Barometers, 416
Basic energy model, 245, 246–47, 279–80
Batteries, 849–50, 881–82, 892–900.
    See also Circuits
    charge escalator model of, 843–44
    emf of, 843
    ideal, 843
    real, 901–03
    short circuited, 902
    as source of potential, 823, 882
Beats, 615–18
    beat frequency, 617

Before-and-after pictorial representation, 223–24

Bernoulli's equation, 426–30

Beta decay, 1258–59, 1264–66

Beta particles, 1258–59

Binding energy, 1145, 1253–55

Binoculars, 675

Binomial approximation, 762, 1086

Biot-Savart law, 925, 927–28, 1006–07

Blackbody radiation, 493, 1104

Blood pressure, 417–18

Bohr radius, 1143, 1221

Bohr's analysis of hydrogen atom, 1141–46, 1147

Bohr's model of atomic quantization, 1138–41, 1217, 1222

Boiling point, 449, 451, 482–83

Boltzmann's constant, 455, 1197

Bonding molecular orbital, 1206

Boundary conditions
    for standing sound waves, 599–601
    for standing waves on a string, 596–99
    for wave functions, 1183–87

Bound states, 1194, 1206

Bound system, 369, 1253

Brayton cycle, 536–38

Breakdown field strength, 748

Bulk modulus, 432–33

Bulk properties, 444

Buoyancy, 419–23

Buoyant force, 419

C

Calories, 477

Calorimetry, 483–85

Camera obscura, 657

Cameras, 695–99
    controlling exposure, 698–99
    focusing, 697
    zoom lenses, 697

Capacitance, 850, 851, 899

Capacitative resistance, 1037

Capacitor circuits, 1036–38

Capacitors
    and capacitance, 849–50
    charging, 764, 912
    current and voltage in AC circuits, 1036
    dielectrics in, 855–58
    discharging, 868, 870, 912
    energy stored in, 854–55
    parallel and series capacitors, 851–85
    parallel-plate capacitors, 764–66
    RC circuits, 909–12

Carbon dating, 1250, 1264

Carnot cycle, 542–46

Carnot engine, 542–46

Cathode rays, 1106–08

Cathode-ray tube (CRT) devices, 768, 1107, 1108

Causal influence, 1089–90

CCD (charge-coupled device), 696, 699

Celsius scale, 449, 452

Center of mass, 312, 394
    in balance and stability, 333
    rotation around, 314–17

Central maximum, 630, 631, 636–37, 639, 640–41

Centrifugal force, 202–03

Centripetal acceleration, 102–03, 194, 200, 205

Charge carriers, 727, 868–69

Charge density, 757

Charge diagrams, 726

Charge distribution. *See also* Electric fields
    continuous, 756–760, 828–29
    symmetric, 781–83

Charge model, 721–25

Charge polarization, 729–31

Charge quantization, 726

Charges, 719, 722, 725–26, 1056
    atoms and electricity, 725
    on capacitors, 850
    conservation of, 726, 892, 896
    discharging, 724
    fundamental unit of, 725, 1111–12
    like charges, 722
    neutral, 723, 726
    opposite, 722
    positive and negative, 725
    units of, 732–33

Charge-to-mass ratio, 767, 942

Charging, 719, 721
    frictional, 721–23, 726
    by induction, 731
    insulators and conductors, 727–28
    parallel-plate capacitors, 849–50

Chromatic aberration, 707

Circuits, 891–920. *See also* AC circuits; DC circuits
    diagrams, 892
    elements, 892
    energy and power in, 896–98
    grounded circuits, 908–09
    Kirchhoff's laws and, 892–96
    LC, 988–90
    LR, 391–93
    oscillator, 920
    parallel resistors, 903–06
    RC, 909–12
    resistor circuits, 906–08
    series resistors, 898–901

Circular-aperture diffraction, 640–42, 708

Circular motion, 3, 98–107, 108, 216. *See also* Rotational motion
    angular acceleration, 103–07
    dynamics of, 193–99
    fictitious forces, 201–04
    nonuniform, 103–107, 205–08
    orbits, 199–201
    period of, 98, 100–01
    problem-solving strategy, 207
    simple harmonic motion (SHM) and, 381–84
    uniform. *See* Uniform circular motion

Circular waves, 572, 610–15

Classically forbidden region, 1195–96

Classical physics, 1015, 1127–29
    end of, 1118–1129

Clocks
    atomic, 1076
    synchronizing, 1070
    time dilation and, 1075

Closed-cycle devices, 529

Coaxial cable, 961, 1002

Coefficients of friction, 149
    kinetic friction, 149, 206
    rolling friction, 149
    static friction, 148

Coefficients of performance, 532, 544

Cold reservoir, 528, 531

Collisional excitation, 1233–34

Collisions, 219, 221–26, 870, 873–74
    elastic, 232, 265–69, 1065
    inelastic, 232–34, 253–54
    mean free path between, 503–04
    mean time between, 874
    molecular, 503–07, 514–20
    pressure and, 503

Color, 667–70
    chromatic aberration, 707
    human color vision, 700
    in solids, 1234–35

Compasses, 923

Component vectors, 75–77

Compression, 431
    adiabatic, 479, 488–89
    isobaric, 458
    in sound waves, 574, 599

Concave mirrors, 682

Condensation point, 450, 451

Conduction
    electrical, 868–70, 1127
    heat, 491–92
    model of, 873–74

Conductivity
    electrical, 878–80
    thermal, 492

Conductors, 724, 727–31
    charge carriers, 868–69

in electrostatic equilibrium, 799–802, 848–49
Conservation of angular momentum, 342–43
Conservation of charge, 726, 877, 892, 896
Conservation of current, 876–78, 892, 896, 903, 920
Conservation of energy, 219, 294–97, 301, 813–14, 897
  in charge interactions, 820
  in double-slit interference, 633
  in fluid flow, 419, 426
  in relativity, 1093–95
  in simple harmonic motion (SHM), 385–86, 440
Conservation laws, 219–20
Conservation of mass, 219, 235, 1093
Conservation of mechanical energy, 254–55, 270
Conservation of momentum, 219, 220, 226–32, 1065
Conservative forces, 288–89, 294, 811
Constant-pressure (isobaric) process, 458–59
Constant-temperature (isothermal) process, 459–60
Constant-volume gas thermometer, 449
Constant-volume (isochoric) process, 457–59
Constructive interference, 594, 605–07, 609–15, 630–39
Contact forces, 117–22
Continuity, equation of, 424–25
Continuous spectra, 1103–04
Contour maps, 614, 823–25
Convection, 491, 492
Converging lenses, 671–73, 702
Convex mirrors, 682
Coordinate systems, 5, 74–77
  displacement and, 8
  right-handed, 339
  rtz, for circular dynamics, 193–94, 209
  tilted axes, 79–80
Correspondence principle, 1191–93
Coulomb (C), 732
Coulomb electric fields, 979
Coulomb's law, 731–32, 733–36, 739, 753, 814, 1006, 1007
Covalent bonds, 1205–06
Critical angle, 333, 665
Critical point, 452
Crookes tubes, 1107
Crossed-field experiment, 1109–10
Crossed polarizers, 1025
Crossover frequency, 1039
Cross product, 338–39, 927

Current, 719, 727, 867–90, 892–94. *See also* Circuits; Electron current; Induced currents
  batteries and, 849–50
  conservation of, 876–78, 892, 896, 903, 920
  creating, 870–73
  displacement, 1011–14
  eddy currents, 968
  magnetic field of, 925–31
  magnetic forces on wires carrying, 946–50
  and magnetism, 923–25
  root-mean-square (rms) current, 1046–47
Current density, 876, 878
Current loops, 930
  forces and torques on, 948–50
  as magnetic dipoles, 932–34
  magnetic field of, 930–31
Curve of binding energy, 1254
Cyclotron, 943–44
Cyclotron frequency, 942
Cyclotron motion, 941–43

D

Damped oscillations, 395–99
Damping constant, 395
Daughter nucleus, 1261, 1263
DC circuits, 1034
de Broglie wavelength, 1135, 1194, 1197, 1204
Decay equation, 1236–38
Decay, exponential, 396, 992
Decay, nuclear, 1263–67
Decay rate, 1237
Decay series, 1267
Decibels (dB), 579–80
Defibrillators, 854, 858
Degrees of freedom, 511–13
Density, 408–09, 421, 756
  mass density, 408, 445
  nuclear density, 1251–52
  number density, 446
  surface charge, 757
Destructive interference, 605, 609–11, 612, 630–39
Deuterium, 1251
Diatomic gases, 447, 453, 487
Diatomic molecules, 264–65, 512–14
Dielectric constant, 857
Dielectric strength, 858
Dielectrics, 855–58
Diesel cycle, 536
Diffraction, 628, 657, 707–09
  circular-aperture, 640–42

  of electrons, 1108, 1135–36
  single-slit, 636–40
Diffraction grating, 634–36
  resolution of, 653
Diffuse reflection, 659
Diodes, 883
Diopter, 701
Dipole moment, 754
Dipoles. *See* Electric dipoles; Magnetic dipoles
Direct current (DC) circuits, 1034. *See also* Circuits
Discharging, 724, 728–29, 870, 909, 910
Discrete spectra, 1105
Disk of charge, electric field of, 761–63
Disordered systems, 518–19
Dispersion, 668–69, 707
Displaced fluid, 420
Displacement, 6–9, 43, 61
  angular, 99
  work and, 281
Displacement current, 1011–14
Displacement vectors, 70–71
Dissipative forces, 289–90, 293–94
Diverging lenses, 671, 675–76, 702
Doppler effect, 580–83
Dose, absorbed, 1268
Dose equivalent, 1268
Dot product of vectors, 284–85, 298
Double-slit experiment, 629–34, 668
Double-slit interference, 629–34, 637–38, 668, 1157
  intensity of, 633–34
  interference fringes, 629–630, 1159
Drag coefficient, 153
Drag force, 121, 152–53
  terminal speed and, 154
Drift speed, electron, 868, 869, 873
Driven oscillations, 398–99
Driving frequency, 398–99
Dynamic equilibrium, 128, 141
Dynamics, 1, 216
  fluid, 423–30
  of nonuniform circular motion, 205–08
  in one dimension, 138–66
  problems, strategy for solving, 142, 159
  of rotational motion, 325–27
  of simple harmonic motion (SHM), 386–89
  in two dimensions, 192–93
  of uniform circular motion, 193–99

E

Earthquakes, 257
Eddy currents, 968

Effective focal length, 697. *See also* Focal length
Effective gravitational force, 146, 202
Efficiency, thermal, 529–30
Einstein, Albert
  explanation of photoelectric effect, 1129–32
  photons and photon model of light, 1132–34
  principle of relativity, 1066–68, 1278
Elasticity, 255, 430–33
  tensile stress, 430–32
  volume stress, 432–33
Elastic collisions, 232, 265–69, 1065
Elastic potential energy, 257–61
Electric charge. *See* Charges
Electric dipoles, 730
  electric field of a dipole, 754–55
  motion in an electric field, 770–72
  potential energy of, 817–18
Electric field lines, 755–56
Electric field strength, 738, 878
Electric fields, 737–42, 750–79, 818–19, 1004, 1012
  changing, producing induced magnetic fields, 981–82
  of charged wires, 924
  of conductors in electrostatic equilibrium, 799–801, 848–49
  of continuous charge distribution, 756–60
  Coulomb and non-Coulomb, 979
  crossed-field experiment, 1109–10
  of disk of charge, 761–63
  of electric dipole, 754–55, 933
  electric potential and, 823, 840
  energy in, 855, 988
  establishing in a wire, 871–72
  finding electric potential from, 840–41
  finding with Gauss's law, 795–99
  Gauss's law, 791–94
  induced, 856, 978–81, 1010, 1013
  insulators in, 856–57
  models of, 751
  motion of charged particle in, 767–69
  motion of dipole in, 770–71
  of plane of charge, 763–64
  of a point charge, 739–40
  of multiple point charges, 752–55
  of parallel-plate capacitor, 764–66, 821–22, 850, 851
  picturing, 755–56
  of ring of charge, 60–61
  of sphere of charge, 764–66
  symmetry of, 781–83, 791
  transformations of, 1005–10

uniform, 766, 811–14
  units, volts per meter, 822
Electric flux, 785–94, 1010, 1012, 1013, 1017, 1019
  of nonuniform electric field, 787–88
Electric force, 121–22, 264–65, 816
Electric potential, 818–838
  batteries and, 843
  in closed circuits or loops, 893
  of conductor in electrostatic equilibrium, 848–49
  electric potential energy vs., 819
  finding from electric field, 840–41
  finding electric field from, 844–48
  inside parallel-plate capacitor, 821–25
  of many charges, 828–30
  of a point charge, 826–27
  sources of, 842–44
Electric potential energy, 811–14
  of a dipole, 817–18
  electric potential vs., 819
  of point charges, 814–17
Electrical oscillators, 1034
Electricity, 719–49, 1056
  charge, 723–26, 1056
  charge model, 721–25
  field model, 736–42
  insulators and conductors, 727–31
  phenomena and theories, 719
Electrodes, 763–64, 765
  capacitance and, 850, 851
  equipotential surface close to, 848
Electromagnetic fields, 1003–16
  forces and, 1004
  Maxwell's equations, 1014–16
  transformation of, 1005–10
Electromagnetic induction, 962–1002
  Faraday's law, 975–78
  induced currents, 963
  induced fields, 978–81
  Lenz's law, 971–74
  motional emf, 964–68
  problem-solving strategy for, 976
Electromagnetic radiation, quantized, 1129
Electromagnetic spectrum, 576
Electromagnetic waves, 561, 575–77, 1016–20, 1066
  Ampère-Maxwell law, 1019–20
  energy and intensity, 1021–22
  generated by antennas, 1023
  light, 628
  Maxwell's theory of, 981–82
  polarization, 1024–26
  radiation pressure, 1022
  speed of, 1020
Electromagnets, 932

Electron capture, 1265
Electron cloud, 725, 1220
Electron configuration, 1227
Electron current, 868–870
Electron spin, 950–51, 952
Electronic ignition, 987
Electrons
  beams, 769
  charge of, 725–26
  charge carriers in metals, 868–69
  diffraction of, 1135
  discovery of, 1108–11
  sea of, 727
  spin of, 1218, 1223–25
Electron volt, 1115
Electroscope, 728–29, 731
Electrostatic constant, 732, 733
Electrostatic equilibrium, 727, 870–71
  conductors in, 799–802, 848–50
Electrostatic forces, and Coulomb's law, 733–36
Electroweak force, 217
Elements, 1117, 1139, 1228–31
emf
  of batteries, 843
  induced emf, 975–78, 983
  motional emf, 964–68
Emission spectrum, 1103–04, 1105, 1234
  hydrogen atom, 1106, 1146–47
Emissivity, 493
Endoscope, 685–86
Energy, 219, 470. *See also* Conservation of energy; Kinetic energy; Mechanical energy; Potential energy; Thermal energy; Work
  allowed energy for quantum mechanics problems, 1187–88
  angular and linear momentum and, 341
  bar charts, 249–50, 295–96, 301
  basic energy model, 245, 246–47, 279–80, 301
  binding energy, 1145, 1253–55
  in circuits, 896–98
  in damped systems, 396–97
  diagrams, 261–65, 813
  in electric fields, 855
  of electromagnetic waves, 1021
  forms of, 245, 246
  ionization energy, 1145
  in magnetic fields, 988
  of photons, 1133–34
  of photoelectrons, 1127–28
  problem-solving strategy, 297, 301
  quantization of, 1136–38, 1184
  relativistic, 1090–95
  rotational, 317–19, 440
  thermodynamic model, 478

Energy bar charts, 249–50, 295–96, 301
Energy density, 855, 858, 988
Energy diagrams, 261–65, 813, 1140
   equilibrium positions, 263
   molecular bonds and, 264–65
Energy equation, 295, 301, 471
Energy-level diagrams, 1140–41
   for hydrogen atom, 1146–47, 1219
   for multielectron atoms, 1226, 1232
Energy levels, 1137
   allowed, 1184, 1187, 1189
   of hydrogen atom, 1143–45, 1219–20
Energy reservoirs, 527–28
Energy transfer, 246, 279–80, 470–71
   heat and energy, 471
   rate of (power), 297–98
Energy transformations, 246, 279–80, 470
Engines. See also Heat engines
   Carnot engine, 542–46
   diesel, 536
   gas turbine engines, 536
   gasoline internal combustion, 536
English units, 24–25
Entropy, 518–520, 557
Environment, 169, 279
Equation of continuity, 425, 429–30
Equation of motion, 387–89, 393, 395
Equilibrium, 133, 138
   dynamic, 128, 141
   electrostatic, 799–802, 848–50, 870–71
   hydrostatic, 414, 416
   mechanical, 128
   phase, 452
   problem-solving strategies, 139, 159
   static, 128, 140
   thermal, 514–20
Equipartition theorem, 510–11
Equipotential surfaces, 823, 825, 826,
   828, 840–41, 845–47
Equivalence, principle of, 358–59
Equivalent capacitance, 851, 852
Equivalent resistance, 899, 903, 904, 905
Escape speed, 363–64, 815
Estimates, order-of-magnitude, 27
Evaporation, 557
Events, 1069–70, 1073–76, 1081–83
   and observations, 1070–71
   time of, 1071
Excitation, 1232
   collisional, 1233–34
Excited states, 1231–35
   lifetime of, 1236–38
Explosions, 220, 234–36
Exponential decay, 396, 992
External forces, 169, 228, 294
Eyepiece, 704, 706
Eyes. See Vision

F

Fahrenheit scale, 449
Far point (FP) of eye, 701
Farad (F), 850
Faraday's law, 963, 975–81, 1015
   electromagnetic fields and, 1009–11
   and electromagnetic waves, 982,
     1017–19
   for inductors, 985–86
Farsightedness, 701
Fermat's principle, 693
Ferromagnetism, 951
Fiber optics, 665–66, 685–86
Fictitious forces, 201–04
Field diagrams, 739–40
Field model, 736–42
Fields, 737. See also Electric fields;
   Electromagnetic fields; Magnetic
   fields
Finite potential wells, 1193–98
First law of thermodynamics, 478–80,
   490–91, 527, 556, 1015
Fission, 1094
Flat-earth approximation, 145
Flow tube, 424
Fluids, 311, 407–40
   Archimedes' principle, 420
   Bernoulli's equation, 426–30
   buoyancy, 419–23
   density, 408–09
   displaced, 420
   dynamics, 423
   equation of continuity, 424–26
Fluorescence, 1106–07
Flux. See Electric flux; Magnetic flux
f-number, 698–99
Focal length, 671, 682, 684
   angular magnification and, 704
   effective focal length of combination
     lenses, 697
Focal point, 671, 674, 676, 682
Force, 1, 116–137
   acceleration and, 124–25, 141, 157–58,
     205
   action/reaction pair, 170, 172–73, 175
   buoyant, 419–23
   combining, 118
   conservative and nonconservative,
     288–89, 811
   contact, 117
   dissipative, 293–94
   drag, 121, 152–54
   electric, 121–22, 733–36
   external, 169
   fictitious, 201–04
   friction, 120–21, 148–52

   gravitational, 119, 357–58
   identifying, 122–23, 130, 133
   impulsive, 221
   as interactions, 127, 173
   internal, 234
   long-range, 117, 736–37
   magnetic, 121–22
   net force, 80, 118
   normal, 120
   restoring forces, 255–57
   SI unit, 126
   spring, 119
   superposition of, 118
   tension, 119–20, 156, 177–81
   thrust, 121
Free-body diagrams, 130–32, 133
Free fall, 51–54, 61
   weightlessness and, 147–48
Free-fall acceleration, 52
Freezing point, 449, 451
Frequency, 378–79, 382
   angular, 379, 381, 388, 594
   beat, 617
   crossover, 1039
   cyclotron, 942
   fundamental, 597, 600–02
   of mass on spring, 385
   natural, 398
   of pendulum, 392, 394
   resonance, 398
   of sinusoidal waves, 566
   of sound waves, 574–75
Friction, 120–21, 128, 148–52
   causes of, 152
   coefficients of, 148–49, 206
   kinetic, 148, 176
   model of, 149
   rolling, 149
   static, 148–49, 176
Fringe field, 765
Fringe spacing, 631–32
Fundamental frequency, 597, 600–02
Fundamental unit of charge, 725, 1111–12
Fusion, 836

G

Galilean field transformation equations,
   1007–09
Galilean relativity, 1008–09, 1061–65
   principle of relativity, 1064
Galilean transformations
   of acceleration, 1005
   of electromagnetic fields, 1007–10
   of position, 1063
   of velocity, 97, 267–68, 1004, 1063,
    1068

Gamma decay, 1266–67
Gamma rays, 1093–94, 1198, 1249,
    1258–59, 1266–67
  medical uses of, 1269
Gases, 408, 445, 446
  ideal gases, 452–56
  ideal-gas processes, 456–61, 478–80
  monatomic and diatomic, 447
  pressure, 411, 505–07
  specific heats of, 485–91
Gas turbine engines, 536–38
Gauge pressure, 415
Gaussian surfaces, 784
  calculating electric flux, 785–87
  electric field and, 785
  symmetry of, 784, 791
Gauss's law, 791–809, 934, 1010, 1015,
    1017
  and conductors in electrostatic
    equilibrium, 799–802
  Coulomb's law vs. , 791, 794
  for magnetic fields, 936
Geiger counter, 859,1259
Generators, 842, 982–83, 1034
Geomagnetism, 923, 931
Geosynchronous orbits, 367
Global positioning systems (GPS), 212
Global warming, 494
Grand unified theory, 217
Gravitational constant (G), 357, 359–61
Gravitational field, 737, 811
Gravitational force, 119, 145–46, 357,
    359–61
  and weight, 357–58
Gravitational mass, 358
Gravitational potential energy, 246,
    248–54, 362–65
  flat-earth approximation, 364
  zero of, 363
Gravitational torque, 324–25, 394
Gravity, 145–46, 311, 736–37, 811–12.
    See also Newton's law of gravity
  little g (gravitational force) and big G
    (gravitational constant), 359–61
  moon's orbit and, 201
  Newton's law of, 145
  on rotating earth, 202–03
  universal force, 356
Gray (Gy), 1268
Greenhouse effect, 494
Grounded circuits, 729, 908–09
Ground state, 1218, 1219, 1220, 1231

H

Half-life, 397, 1077, 1260
Hall effect, 944

Harmonics, 597, 602, 603
Hearing, threshold of, 579
Heat, 279, 293, 471, 475–77
  defined, 476
  in ideal-gas processes, 488
  specific heat and temperature change,
    480–81
  temperature and thermal energy vs., 477
  thermal interactions, 476
  transfer mechanisms, 491–94
  units of, 477
  work and, 471, 476, 490, 527–29
Heat engines, 526–555
  Brayton cycle, 536–38
  Carnot cycle, 542–46
  ideal-gas, 534–36
  perfect, 530, 533
  perfectly reversible, 541–42
  problem-solving strategy for, 535–36
  thermal efficiency of, 529–30
Heat exchanger, 537
Heat of fusion, 482
Heat pump, 551
Heat-transfer mechanisms, 491–94
Heat of transformation, 482–83
Heat of vaporization, 482
Heisenberg uncertainty principle,
    1169–72, 1190
Helium-neon laser, 1241–42
Henry (H), 984
Hertz (Hz), 378, 382
History graphs, 564–65, 566
Holography, 645–46
Hooke's law, 256–57, 289, 361, 387
Hot reservoir, 527
Huygens' principle, 637–39
Hydraulic lifts, 418–19
Hydrogen atom
  angular momentum, 1217–19
  Bohr's analysis of, 1141–46
  energy levels of, 1219
  spectrum, 1146–49
  wave functions and probabilities,
    1220–23
Hydrogen-like ions, 1147–48
Hydrostatics, 413–15
Hyperopia, 701, 702

I

Ideal battery, 843, 902
Ideal-fluid model, 423
Ideal gases, 452–56
ideal-gas heat engines, 534–38
Ideal-gas processes, 456–61, 478–80, 534
  adiabatic process, 479, 488–91, 536–39,
    543–44

constant-pressure process, 458–58
constant-temperature process, 459–60
constant-volume process, 457–59
pV diagram, 456
quasi-static processes, 456–57
work in, 471–75
Ideal wire, 883
Image distance, 660, 667, 676
Image formation
  by refraction, 666–67
  with spherical mirrors, 682–85
  with thin-lenses, 680–81
Image plane, 672
Impedance, 1043
Impulse, 220–26
Impulse approximation, 225
Impulse-momentum theorem, 222–23,
    226
  similarity to work-kinetic energy
    theorem, 281–82
Impulsive force, 221
Inclined plane, motion on, 54–58
Independent particle approximation (IPA),
    1226, 1227
Index of refraction, 576–77, 609, 646,
    662, 663, 664
Induced current, 963–78, 985, 986
  in a circuit, 966–67
  eddy currents, 968
  Faraday's law, 975–78, 1009
  Lenz's law, 971–74
  magnetic flux and, 968–71
  motional emf, 964–65
Induced electric dipole, 754–55
Induced electric fields, 856, 978–981,
    1010, 1013
Induced emf, 975–78, 985, 986
Induced magnetic dipoles, 951–53
Induced magnetic fields, 981–82, 985,
    1013–14
Inductance, 984, 985
Inductive reactance, 1041–42
Inductor circuits, 1041–42
Inductors, 984–88
Inelastic collisions, 232–34
Inertia, 126
  law of, 128
Inertial mass, 126, 318, 358–59
Inertial reference frames, 129, 201, 202,
    1064
Insulators, 724, 727–31, 883
  dielectrics, 856–58
Intensity, 578–80, 1025
  of double-slit interference pattern,
    633–34
  of electromagnetic waves, 1021–22
  of standing waves, 594

Interacting systems, 168–72
  analyzing, 169–70
  revised problem-solving strategy for, 175–77
Interaction diagrams, 169, 170
Interference, 594, 604–610. *See also* Constructive interference; Destructive interference
  of light, 596, 629–34
  in one dimensional waves, 604–07
  mathematics of, 607–10
  and phase difference, 605–07
  photon analysis of, 1159–60
  problem-solving strategy for, 613
  in two- and three-dimensional waves, 610–15
  wave analysis of, 1157–58
Interference fringes, 630, 631–32, 644
Interferometers, 642–46, 1135–36
  acoustical, 643
  atom, 1135–36
  Michelson, 644–45
Internal energy, 470
Internal resistance, 901–02, 905
Inverse-square law, 354, 356, 357
Inverted images, 672
Ion cores, 727
Ionization, 726, 1108
Ionization energy, 1145, 1220, 1231
Ionization limit, 1146, 1147
Ionizing radiation, 1259
Ions, 726
  hydrogen-like ions, 1147–48
Irreversible processes, 516–17
Isobaric (constant-pressure) processes, 458–59, 474, 488
Isobars, 1250
Isochoric (constant-volume) processes, 457, 474, 479, 488
Isolated systems, 220, 228, 231–32, 234
  conservation of energy, 295
  conservation of mechanical energy, 254–55
  second law of thermodynamics, 519–20
Isothermal (constant-temperature) processes, 459–61, 474–75, 479
Isotherms, 460, 485
Isotopes, 1118, 1250

J

Joules (J), 248, 281, 454

K

Kelvin scale, 450, 452
Kelvins (K), 451

Kepler's laws of planetary motion, 354, 355, 365–68
Kinematics, 1, 33–68
  circular motion, 98–107
  with constant acceleration, 45–51
  with instantaneous acceleration, 58–60
  free fall, 51–54
  in two dimensions, 87–91, 108
  uniform motion, 34–38
Kinetic energy, 246, 247–51
  in elastic collisions, 265–69
  in relativity, 1090–91
  of rolling object, 335–36
  rotational, 312, 317–19
  temperature and, 508–09
  work and, 280–82
Kinetic friction, 120, 148–49, 155, 206
Kirchhoff's laws,
  junction law, 878, 892–93
  loop law, 847, 893–95, 989, 991, 1035

L

Laminar flow, 423
Lasers
  creating using stimulated emission, 1238–42
  helium-neon laser, 1241–42
  quantum-well laser, 1197
  ruby laser, 1240–41
Lateral magnification, 674, 684, 698, 704
LC circuits, 988–91
Length contraction, 1078–82, 1085
Lenses, 670. *See also* Cameras; Thin lenses
  aberrations, 707–10
  achromatic doublet, 715
  angular resolution, 709
  in combination, 695–97
  converging, 671
  diffraction limited, 708–09
  diverging, 695, 697, 701
  f-number of, 698–99
  focal length, 695–98
  ray tracing, 670–73, 695–96
Lens maker's equation, 680
Lens plane, 671
Lenz's law, 971–74, 985
Lever arm, 323–25, 394
Lifetime (of excited states), 1236–38
Lift, 429–30
Light, 627–654, 982. *See also* Electromagnetic waves
  absorption or reflection by objects, 1235
  color and dispersion, 667–70
  early theories of, 1103

emission and absorption of, 1103–06
  interference of, 629–34
  models of, 628–29
  photon model of, 629, 1133–34
  properties of, 1278
  ray model of, 629, 641–42, 656–58
  wave model of, 629, 641–42
Light clock, 1074, 1075
Light rays, 656
Light waves, 575–77. *See also* Electromagnetic waves
  Doppler effect for, 582–83
  interference of, 604, 608–10
  polarization of, 1024–26
Light years, 1077
Line of action, 323
Linear acceleration, 12–13
Linear charge density, 757, 759
Linear density, 562, 570
Linear restoring force, 393
Line of charge, 758–60
Line integrals, 934–935, 936, 937, 939
Line of nuclear stability, 1253
Liquid-drop model, 1252
Liquids, 408, 445. *See also* Fluids
  pressure in, 413
Longitudinal waves, 561, 599
Long-range forces, 117, 721, 736
Lorentz force law, 1015
Lorentz transformations, 1082–87, 1090
Loschmidt number, 522
LC circuits, 988–91
LR circuits, 991–93
Lyman series, 1147

M

Macrophysics, 292
Macroscopic systems, 444–46, 470
Magnetic dipole moment, 933, 949
Magnetic dipoles, 922, 931–34
  induced, 951–53
Magnetic domains, 951, 952
Magnetic field lines, 924
Magnetic field strength, 925, 988
Magnetic fields, 923–39, 1009
  Ampère's law, 934–37
  Biot-Savart law, 925, 926, 927, 928
  of current, 927–31
  of current loop, 930–31
  of cyclotron, 944
  energy in, 984, 988
  Gauss's law for, 1010
  induced, 981–82, 1013–14
  of moving charge, 925–27
  properties of, 924
  of solenoids, 938–39

Magnetic fields (*continued*)
  transformations of, 1005–10
  uniform, 038
Magnetic flux, 962, 968–970, 982, 983,
    984, 985, 994
  Faraday's law, 975–78
  Lenz's law, 972
  in nonuniform field, 970–71
Magnetic force, 121–22
  on current-carrying wires, 924–25, 932
  on moving charge, 925–26
Magnetic poles, 922, 932, 948
Magnetic quantum number, 1218
Magnetic resonance imaging (MRI), 939,
    952, 1269–70
Magnetism, 719, 922, 1056
  ferromagnetism, 951
  magnetic properties of matter, 950–53
Magnification, 704–07
  angular, 704
  lateral, 674, 684, 698
Magnifier/magnifying glasses, 675, 703
Malus's law, 1024–25
Manometers, 416
Mass, 126, 133, 138, 144, 1057
  atomic, 447
  conservation of, 219, 1093
  equivalence to energy in relativity,
    1092–94, 1250
  gravitational, 358–59
  inertial, 126, 318
  measurement of, 24
  molar, 448
  molecular, 447
  weight, gravitational force, and, 119,
    144–147
Mass density, 408–09, 445
Mass-energy equivalence, 1092–94, 1250
Massless string approximation, 178–80
Mass number, 1118, 1249
Mass spectrometers, 960, 1117
Matter waves, 1134–36
Maxwell's equations, 1014–16, 1016,
    1066
Mean free path, 503–04
Mean time between collisions, 874
Mechanical energy, 254
  conservation of, 254–55, 318
  conservative forces and, 289–90
Mechanical equilibrium, 128, 471
Mechanical interaction, 471
Mechanical waves, 561
Medium, 561
  displacement of particles by wave, 566
  electromagnetic waves and, 661
  speed of sound in, 574
  wave speed in, 563

Melting point, 445, 449, 482–83
Metal detectors, 984
Metals, 868–69, 879, 882–83, 1056
Meter (m), 23, 1208
Michelson interferometer, 644–45
Micro/macro connection, 502–525
  equilibrium, 517–18
  gas pressure, 505–07
  irreversible processes, 516–17
  molecular collisions, 503–04
  order, disorder, and entropy, 518–19
  second law of thermodynamics, 519–20
  temperature and, 508–09
  thermal interactions and heat, 514–16
Microphysics, 292
Microscopes, 675, 704–05, 710, 711
Millikan-oil-drop experiment, 1111–12
Minimum spot size, 708–09
Mirror equation, 684–85
Mirrors
  plane mirrors, 659–60
  spherical mirrors, 682–85
Models, 1, 21
  atomic model, 120, 221, 264
  basic energy model, 245, 246–47,
    279–80, 301
  Bohr's model of the atom, 1138,
    1141–46
  charge escalator model, 843
  charge model, 721–25
  of electrical conduction, 873–74
  electric field models, 751
  field model, 736–42
  of friction, 149
  ideal-fluid model, 423
  ideal-gas model, 452–56
  nuclear model of atoms, 1114, 1116–17
  particle model, 4–5
  photon model of light, 629, 1133–34
  quantum-mechanical models, 1138–41,
    1148
  raisin-cake model of atoms, 1112–13
  ray model of light, 629, 641–42,
    656–58
  rigid-body model, 311, 313
  shell model of atoms, 1222, 1228–31,
    1256–58
  thermodynamic energy model, 478
  wave model of light, 561–63, 629,
    641–42
Modes, 597, 619
Molar mass, 447–48
Molar specific heats, 481, 510–13
  at constant pressure, 486
  at constant volume, 486
Molecular bonds, 264–65
  covalent, 1205–06

Molecular mass, 447
Molecular vibrations, 265, 1202–03
Moles, 447–48
Moment arm, 323–25, 394
Moment of inertia, 311, 317–21, 335
Momentum, 222
  angular, 340–45
  changes in kinetic energy and, 282
  conservation of, 226–32, 1067
  and impulse, 221–23
  problem-solving strategy for, 223–26
  quantization of angular momentum,
    1145–46
  relativistic, 1087–90
  in two dimensions, 236–37
  velocity-energy-momentum triangle,
    1092
Momentum bar charts, 223
Monatomic gases, 447, 453, 487
Motion, 1–32, 216. *See also* Acceleration;
    Circular motion; Kinematics; Linear
    motion; Newton's laws of motion;
    Oscillations; Projectile motion;
    Relative motion; Rotational motion;
    Simple harmonic motion (SHM);
    Uniform circular motion; Velocity
  of charged particle in electric field,
    767–69
  with constant acceleration, 90–91, 124
  cyclotron, 941–43
  graphical representations of, 1, 2,
    17–19, 45, 56–58, 61
  on inclined plane, 54–58
  in one dimension, 15–19
  types of, 3
  uniform, 34–37
  vectors, 2, 6
Motional emf, 964–68, 1009
Motion diagrams, 3–6, 38
  acceleration vectors, 13–14
  displacement vectors, 9
  examples, 14–16
  velocity vectors, 11–12
Motion graphs, 56–58, 61
Motors, 949, 1048–49
MRI (magnetic resonance imaging), 939,
    952, 1269–70
Myopia, 702–03

N

Natural frequency, 398–99
Near point (NP) of eye, 701
Nearsightedness, 702–03
Neutral buoyancy, 421
Neutrino, 1266
Neutron number, 1250

Neutrons, 1117–18, 1197, 1249
Newtons (N), 126, 281
Newton's first law of motion, 127–29,
    133, 139, 216, 390
Newton's law of gravity, 145–46, 354–76,
    440
Newton's second law of motion, 126–27,
    133, 141–44, 205, 207, 216
    examples of, 155–58
    for oscillations, 387, 570, 571
    for rotational motion, 326, 327, 330
    in terms of momentum, 222
Newton's third law of motion, 172–77,
    216, 472, 505
    conservation of momentum and,
        226–29
    problem-solving strategy for interacting
        objects, 175–77
    reasoning within, 173–74
Newton's zeroth law, 127
Nodal lines, 612
Nodes, 594, 595
Nonconservative forces, 289, 294
Normal force, 120, 203–04
Normalization, 1164–66, 1188
Normal modes, 597
Nuclear decay, 1263–67
Nuclear fission, 1094
Nuclear force, 1254, 1255–56
Nuclear fusion, 836
Nuclear magnetic resonance (nmr), 1270
Nuclear model of the atom, 1114,
    1116–17
Nuclear physics, 1197–98, 1248–79
    biological applications, 1268–71
    decay mechanisms, 1263–67
    nuclear size and density, 1251–52
    nuclear stability, 1252–1255
    nucleons, 1197, 1249
    properties of nuclei, 1278
    shell model, 1256–57
    strong force, 1255–56
Nucleons, 1197, 1249
Nucleus, 1114. *See also* Nuclear physics
    discovery of, 1112–17
    nuclear size and density, 1251–52
Number density, 446

O

Object distance, 667, 676, 679, 682
Objective, 704
Object plane, 672
Ohm (Ω), 880
Ohmic materials, 882, 883
Ohm's law, 881, 882–84, 1035
One-dimensional waves, 564–66

Optical axis, 667
Optical cavity, 1240
Optical instruments, 694–715
    for magnification, 703–07
    resolution of, 707–10
Optics, 628, 716. *See also* Light; Optical
    instruments; Ray optics; Wave optics
Orbital angular momentum, 1218
Orbital quantum number, 1217
Orbits, 354
    circular, 199–201, 355, 371
    elliptical, 355
    energetics of, 369–70
    geosynchronous, 367
    Kepler's laws, 355, 365–68
Order (of diffraction), 635
Oscillations, 257, 262–63, 311, 377–406,
    440. *See also* Simple harmonic
    motion (SHM)
    amplitude of, 378
    angular frequency, 379
    damped, 395–99
    driven, resonance and, 398–99
    frequency of, 378
    initial conditions, 381–84
    period of, 378, 562
    phase of, 382
    turning points in, 363, 379
Oscillators, 378, 388–89
    quantum harmonic oscillator, 1200–05
Otto cycle, 536

P

Parallel-axis theorem, 321, 346
Parallel-plate capacitors, 764–66, 787,
    849–59
    electric field of, 765–66, 841
    electric flux inside, 787–88
    electric potential of, 821–25, 841
Paraxial rays, 667, 676–77
Parent nucleus, 1263–65
Particle accelerators, 943
"Particle in a box"
    energies and wave function, 1183–88
    interpreting the solution, 1188–91
    potential energy function for, 1185,
        1189
Particle model, 4–5, 117
Particles, 4
Pascal (Pa), 410, 417
Pascal's principle, 414
Path-length difference, 605–06, 611, 612,
    630–31, 643
Pauli exclusion principle 1227–28, 1257
Pendulums, 268–69, 311, 391–94
    ballistic, 253–54, 344–45

damped, 397
physical, 394
Penetration distance, 1196
Perfect destructive interference, 605, 606,
    608, 611, 612, 619
Perfectly elastic collisions, 265
Perfectly inelastic collisions, 232
Perfectly reversible engine, 540–42
Period, 98, 366
    of oscillation, 378, 392, 562
    of planetary bodies, 367
    of sinusoidal waves, 566
Periodic table of the elements, 1228–31,
    A-4
Permanent magnets, 932, 952
Permeability constant, 925
Permittivity constant, 733
Phase (oscillation), 382
Phase (wave), 573–74
Phase angle, 1044
Phase changes of matter, 445, 450–52,
    482–83
Phase constant, 382–84, 569
Phase diagram, 451
Phase difference, 573, 605–07, 611
Phase equilibrium, 451
Phasors, 1024
Photodissociation, 265
Photoelectric effect, 1126, 1209
    classical interpretation of, 1127–29
    Einstein's explanation of, 1129–32
Photoelectrons, 1126
Photolithography, 708
Photon model of light, 629, 1133–34
Photons, 628, 1067, 1093, 1132–34.
    See also Light
    absorption and emission, 1103–06
    connecting wave and photon views of
        interference, 1160–61
    energy of, 1133–34
    photon emission rate, 1134
    photon model of light, 629, 1133–34
Photosynthesis, 669–70
Physical pendulum, 394
Pictorial representations, 19–21
Pinhole cameras, 657
Pivot point, 322, 330, 394
Planck's constant, 1093, 1129, 1135
Plane mirror, 659–60
Plane of polarization, 1024
Plane waves, 572, 1016, 1017, 1021
Planets. *See also* Orbits
    extrasolar, 368
    Kepler's laws of planetary orbits, 355,
        365–68
Plasma ball, 827
Point charges, 732, 733

Point charges (*continued*)
  electric field of, 739–40, 751
  electric field of multiple point charges, 752–56
  magnetic field of, 927
  potential energy of, 814–17
Point source of light rays, 657
Polarization
  charge polarization, 729–31
  of electromagnetic wave, 1024–26
  Malus's law, 1025
Polarization force, 265, 729–30, 770
Polarizing filters, 1024
Polaroid, 1024
Poles, magnetic, 922, 932, 948
Population inversion, 1240
Position vectors, 6–7
Position-versus-time graphs, 17–19, 38–39, 45
Positron-emission tomography (PET scans), 1094
Positrons, 1093, 1264
Potassium-argon dating, 1262
Potential differences, 820, 822, 944–45, 964–65
  across batteries, 843–44
  across capacitors, 849–53
  across inductors, 985–87
  across resistors, 894, 895
Potential energy, 246, 295
  conservative force and, 288–89, 294, 811
  elastic, 257–61
  electric, 811–18
  finding force from, 290–92
  gravitational, 246, 248–54, 362–65
  in mechanical energy, 254–55
  at microscopic level, 292–93
  inside the nucleus, 1198
  work and, 288–90
  zero of, 824, 908
Potential-energy curve, 261–63
Potential-energy function, 1182, 1183, 1184, 1189
Potential wells, 1193–1198, 1205
  classically forbidden region, 1195–96
  nuclear physics, 1197–98
Power, 297–300
  in AC circuits, 1046–49
  in DC circuits, 896–98
  of lenses, 701
  of waves, 578–80
Power factor, 1048–49
Poynting vector, 1021
Prefixes, denoting powers of ten, 24
Presbyopia, 701
Pressure, 409–15
  atmospheric, 411–13

blood pressure, 417–18
causes of, 410–11
constant-pressure process, ideal gases, 458–59
in gases, 411, 453–55, 505–07
in liquids, 413–15
measuring, 415
units of, 417
Pressure gauge, 407, 415
Principal quantum number, 1217
Prisms, 668
Probabilities, 1158
  of detecting particle, 1162–64
  of detecting photon, 1160, 1162–63, 1190–91
Probability density, 1161, 1164
  corresponding classical quantity, 1192–93
  radial probability density, 1221
Projectile motion, 3, 91–95, 193
  launch angle, 92
  problem-solving strategy for, 94–95
  reasoning about, 93
Proper length, 1079–80
Proper time, 1075–76
Proportionality, 125
Proportionality constant, 124, 125
Propulsion, 171–72
Protons, 1116, 1197, 1249
Pulleys, 175, 180
pV diagrams, 456

Q

Quadrants of coordinate system, 74
Quanta of light, 1130–32
Quantization, 1129, 1184
  of angular momentum, 1145–46, 1218–19
  of charge, 726
  Bohr's model of atomic quantization, 1138–41
  of energy, 1136–38
Quantum computers, 1279
Quantum harmonic oscillator, 1200–05
Quantum jumps, 1184, 1189, 1198
Quantum-mechanical models, 1182
Quantum mechanics, 1137,
  correspondence principle, 1191–93
  drawing wave functions, 1199
  law of, Schrödinger equation, 1180–1184
  particle in a box, energies and wave function, 1183–88
  particle in a box, interpreting solution, 1188–91
  potential wells, 1193–1198

problem-solving strategy, 1184, 1217
wave functions, 1162
Quantum numbers, 1137, 1225
  in hydrogen atoms, 1217–18
  and Pauli exclusion principle 1227–28
  in protons and neutrons, 1257
Quantum-well laser, 1197
Quasars, 583
Quasi-static processes, ideal gases, 456–57, 473

R

Radial acceleration, 105
Radial axis, 194, 195
Radial probability density, 1221
Radial wave functions, 1221–23
Radians, 99, 379
Radiated power, 493
Radiation, 1258–63
  blackbody radiation, 493, 1104
  medical uses of radiation, 1269
  radiation dose, 1268–69
  radioactivity, 1113, 1249
  thermal radiation, 493–94
Radiation pressure, 1022–23
Radio waves, 982
Rainbows, 669
Rarefactions, 574, 599
Rate equations, 1237–38
Ray diagrams, 657
Ray optics, 655–93
  color and dispersion, 667–70
  ray model of light, 656–58
  reflection, 658–60
  refraction, 661–70
Ray tracing, 670–76, 682–83
Rayleigh scattering, 670
Rayleigh's criterion, 709
RC circuits, 909–12
RC filter circuits, 1038–41
Real images, 672, 675, 677, 680, 682
Red shift, 582
Reference frames, 96–98, 1061–62, 1064
  accelerating, 129
  in Einstein's principle of relativity, 1066, 1078
  in Galilean relativity, 1061–62
  inertial, 129
Reflection, 658–60
  diffuse, 659
  law of, 658, 659
  specular, 658
  total internal reflection (TIR), 664–65
Reflection gratings, 636
Refraction, 661–66
  image formation by, 666–67

index of refraction, 576–77, 609, 646, 662
sign conventions for refractive surfaces, 677
Snell's law of, 661
total internal reflection (TIR), 664–65
Refrigerators, 526, 532–33
coefficient of performance, 532–33
ideal-gas, 538–39
perfect, 533
Relative biological effectiveness (RBE), 1268
Relative motion, 95–98, 108
Relativity, 441, 1008–09, 1278. *See also* Galilean relativity
causal influence, 1089–90
clock synchronization, 1070
Einstein's principle of, 1066–68
energy and, 1090–95
events, 1068–69, 1070–71
Galilean, 1008–09, 1061–65
general, 1061
length contraction, 1078–82
Lorentz transformations, 1082–87
measurements, 1069–70
momentum and, 1087–90
proper time, 1075
simultaneity and, 1071–75
special, 1061
time dilation, 1074–78
Resistance, 880–84
equivalent, 899, 903
internal, 901, 905
Resistivity, 879
Resistor circuits, 906–08, 1034–36
Resistors, 882, 883
Ohm's law and, 893, 898
parallel resistors circuit, 903–05
power dissipated by, 897
series resistors circuit, 1042–46
Resolution,
angular, 709
of diffraction grating, 653
of optical instruments, 707–09
Resonance, 398
LC circuits, 1044–45
mechanical resonance, 398–99
standing-wave resonance, 603
Resonance frequency, 398, 1044
Rest energy, 1091
Rest frame, 1074
Restoring forces, 255–57, 361, 387
Resultant vector, 71
Right-hand rule, 338, 923, 927, 946
Rigid bodies, 312. *See also* Rotational motion
Rigid-body model, 313

Ring of charge
electric field of, 760–61, 844, 872
electric potential of, 829–30
RLC circuits, series, 1042–46
Rocket propulsion, 171, 236
Rolling constraint, 334
Rolling friction, 149, 153
Rolling motion, 334–37
Root-mean-square (rms) current, 1046–47
Root-mean-square speed (rms speed), 506–07
Ropes and pulleys, 177–81
acceleration constraints, 175
massless string approximation, 178–80
tension, 177–78
Rotational kinematics, 98–107, 313–14, 346
Rotational kinetic energy, 312, 317, 508, 511
Rotational motion, 3, 103–05, 311, 313–14, 440
angular momentum, 340–45
about the center of mass, 314–17
about a fixed axis, 327–29
dynamics, 325–27
rolling motion, 334–37
torque, 312, 321–25
vector description of, 337–40
rtz coordinate system, 193–94, 209
Rutherford model of atom, 1114

S

Satellites, 365
orbital energetics, 369–70
orbits, 365–68
s-axis, 35–36
Scalar product, 337–40
Scalars, 6, 70
Scanning tunneling microscope (STM), 1209
Schrödinger equation, 1180–82, 1228
solving, 1183–84
Screening, 800
Sea of electrons, 727
Second law of thermodynamics, 519–20, 527, 533, 541, 545, 556, 1017
Selection rules, 1232
Self-inductance, 984
Series *RLC* circuits 1042–46
Shell model of atom, 1222, 1228–31, 1256–57
Short-range forces, 217, 1197
Sieverts (Sv), 1268
Sign convention
for electric and magnetic flux, 1018
for motion in one dimension, 16–17

for refracting surfaces, 677
for rotational motion, 314
for spherical mirrors, 684
for thin lenses, 680
Significant figures, 2, 25–27
SI units, 2, 23–25
Simple harmonic motion (SHM), 378–81, 569, 595, 1034
and circular motion, 381–84
dynamics of, 386–89
energy in, 384–86
kinematics of, 379–81
Simultaneity, 1071–75, 1089
Single-slit diffraction, 636–40
Sinusoidal waves, 366–72
fundamental relationship for, 567–68
mathematics of, 568–70
standing waves, 593
wave motion on a string, 570–72
Small-angle approximation, 391–93
Snapshot graphs, 563–64, 566
Snell's law, 661, 662, 663, 667, 677
Sodium
emission spectra, 1234
excited states of, 1232
Solenoids, 938–39, 952. *See also* Inductors
Solids, 445–46
color in, 1234–35
induced electric field in, 980
phase changes, 450–52, 482–83
specific heat of, 511–12
Sound intensity levels, 579–80
Sound waves, 561, 574–75
beats, 615–18
Doppler effect, 580–83
standing sound waves and musical acoustics, 599–604
Source charges, 737, 738, 740, 821
Spacetime coordinates, 1069, 1081–1083
Spacetime interval, 1081–82
Special relativity, 1061
Specific heat, 480–81
of gases, 485–91
thermal energy and, 510–13
Specific heat ratio, 489
Spectrometer, 1103
Spectroscopy, 636
Spectrum, 1103. *See also* Absorption spectrum; Emission spectrum
excited states and spectra, 1231–35
hydrogen atom spectrum, 1146–49
Specular reflection, 658
Speed, 35
escape, 363–64, 815
of light, 575, 1020, 1066–67
molecular, 503–04

Speed (*continued*)
root-mean-square (rms), 1046–47
of sound, 574
terminal, 154
velocity vs., 11
wave, 561–63
Sphere of charge, 764, 795–97
Spherical aberration, 707
Spherical mirrors, 682–86
Spherical symmetry, 783, 784
Spherical waves, 572, 610–15
Spin, of electrons, 950–51, 952, 1219
Spin quantum number, 1224
Spontaneous emission, 1238
Spring constant, 255–56
Spring force, 119, 124
Springs. *See also* Oscillations; Simple
harmonic motion (SHM)
elastic potential energy, 257–60
restoring forces and Hooke's law,
255–57, 289
work-kinetic theorem for, 287
Spring scale, 272
Stability and balance, 332–34
Stable equilibrium, 263
Stable isotopes, 1250
Standard atmosphere (atm), 413, 417
Standing waves, 591, 593–603, 1136,
1142, 1190. *See also* Superpositon
electromagnetic waves, 598–99
mathematics of, 594–95
nodes and antinodes, 594
sound waves and musical acoustics,
599–604
on a string, 595–98
State variables, 445–46, 453–54
Static equilibrium, 128, 140. 330–34
Static friction, 121, 148, 157
Stationary states, 1139, 1140, 1141, 1184
allowed energies for, 1187–88
hydrogen atom, 1142, 1143, 1144,
1217–18
Stern-Gerlach experiment, 1223–23
Stick-slip motion, 256–57
Stimulated emission, 1238–42
Stopping potential, 1126, 1127–28, 1131–32
STP (standard temperature and pressure),
455
Strain, 431–32
Streamlines, 424
Stress, 417, 430–32
Strong force, 217, 1197, 1255–56
Subatomic particles, 1111
Sublimation, 451
Superconductivity, 879–80, 933
Superposition, 591–626, 751, 828
beats, 615–18

creating a wave packet, 1167
of electric fields, 751–52, 757, 759
of forces, 118
of magnetic fields, 926–27
principle of, 592–93
of two or more quantum states, 1279
Surface charge density, 757, 821
Surface integrals, 785–86, 788–89
Symmetry
of electric fields, 781–83, 791
of magnetic fields, 930, 931, 934
Systems, 169, 219, 227
disordered, 518–19
energy of, 279–80, 293, 295
isolated, 220, 228
ordered, 518–19
self-organizing, 557
total momentum of, 227–30, 260

T

Tangential acceleration, 105–06, 205
Tangential axis, 194
Telescopes, 706–07, 711
resolution of, 709
Temperature, 449–54, 508–09
absolute, 450
change in, and specific heat, 480
heat and thermal energy vs., 477
Tensile strength, 431–32
Tensile stress, 430–32
Tension force, 119–20, 156
Terminal speed, 154
Tesla (T), 925
Thermal conductivity, 492
Thermal efficiency, 529–30, 537–38
limits of, 540–42
Thermal energy, 246, 247, 254, 279, 292–94
heat and temperature vs., 449, 477
in inelastic collisions, 265
properties of matter, 480–83
and specific heat, 510–13
Thermal equilibrium, 446, 453–54, 476,
514–20
Thermal interactions, 471, 476
Thermal properties of matter, 480–83
Thermal radiation, 1104
Thermodynamic energy model, 478
Thermodynamics, 443–468, 527, 556
first law of, 480–82, 490–91, 556, 1015
nonequilibrium, 557
second law of, 519–20, 527, 533, 541,
545, 556, 1015
Thermometers, 497
Thin-film optical coatings, 608–10
Thin lenses, 670–76
ray tracing, 670–76, 682–83

refraction theory, 676–81
sign conventions for, 680
Threshold frequency, 1126, 1131
Thrust, 121, 171
Time
direction or arrow of, 519–20
measurement of, 10, 23
spacetime coordinates, 1069, 1081
Time constant
in *LR* circuits, 991, 992
in oscillations, 396, 397, 399
in *RC* circuits, 910
Time dilation, 1074–78
Torque, 312, 321–25, 346
angular acceleration and, 313–14
on current loops, 948–50
gravitational, 324–25
net torque, 324, 341
torque vector, 339–40
Total internal reflection (TIR), 664–66
Total momentum, 227–28
Trajectory, 3, 87–89, 141. *See also*
Projectile motion
parabolic, 199, 204
in projectile motion, 193, 216
Transformers, 983–84
Transitions, 1220, 1232
nonradiative, 1235
radiative, 1239
Translational kinetic energy, 508–15
Translational motion, 3, 313
Transmission grating, 636
Transverse waves, 561
Traveling waves, 558–590
amplitude of, 566
displacement, 566
Doppler effect, 580–83
electromagnetic waves, 575–76
frequency of, 566
power, intensity, and decibels, 578–80
sinusoidal waves, 566–72
spherical waves, 572
types of, 561
Triple point, 452
Tsunami, 717
Tunneling current, 1209
Turbulent flow, 423, 424
Turning points, 41, 43, 261, 262, 263, 379
Twin paradox, 1077–78

U

Ultrasonic frequencies, 575
Uncertainty, 1168–69
Uncertainty principle, 1169–72
Uniform circular motion, 79, 98–103,
108, 193–99, 216

acceleration in, 86–87, 101–03
dynamics of, 195–99
simple harmonic motion (SHM) and, 381–84
velocity in, 101–03, 194–95
Uniform electric fields, 766, 770, 811–14
electric potential energy of charge in, 812–14
Uniform magnetic fields, 938
Uniform motion, 34–38
Unit vectors, 69, 77, 740, 927
Units, 2, 23–27
Universal constant, 361
Universal gas constant, 454
Unstable equilibrium, 263
Upright images, 675

V

Vacuum, 411–13
Van Allen radiation belt, 943
Van de Graaf generators, 842
Vapor pressure, 411, 416
Vector algebra, 77–80
addition, 7, 71–73, 77–78
multiplication, 73, 79, 927
subtraction, 8, 73, 79
Vector product, 338–39
Vectors, 2, 6, 69–84
area vector, 787, 970
components, 74–77
displacement, 6–9
magnitude and direction, 70
notation, 6
properties of, 70–74
unit vectors, 69, 77, 740, 927
zero, 8
Velocity, 10–12, 1067
angular, 99–101, 108, 205, 313–14
finding from acceleration, 58, 59
finding position from, 42–45
Galilean transformation of, 97, 267–68, 1004, 1063, 1068
instantaneous, 38–42
Lorentz transformation of, 1086–87
magnetic fields and force dependent on, 1004
momentum and, 222
relative, 95–96
sign of, 16–17, 61
speed vs., 11
in uniform circular motion, 101–03, 194
Velocity vectors, 11
Velocity-versus-time graphs, 39, 40, 42, 45

Venturi tube, 429
Vibrational energy levels, 1202
Virial theorem, 1123
Virtual images, 660, 674–75, 677, 680, 683
Viscosity, 423
Visible spectrum, 576, 668
Vision, 700–03
Visualizing physics problems, 22–23
Volt (V), 819
Voltage, 820, 983
of a battery, 820
of capacitors, 849, 856–59, 1036
Hall, 944–45
of inductors, 967, 986–87
peak, 1034–35
of resistors, 1035
rms, 1046–47
terminal, 843, 901
Voltmeters, 905
Volume, 408–09, 445–46, 453–56
flow rate, 425
ideal gas processes, 449, 456–460
unit, 409
Volume strain, 432
Volume stress, 432–33. *See also* Pressure

W

Waste heat, 530
Water molecules, 770
Watts (W), 297–98
Wave fronts, 572
Wave functions, 1162–63
drawing, 1199
finding, 1186
normalizing, 1166–68, 1188
radial wave function, 1221–23
Wavelengths, 567
de Broglie wavelength, 1135
and index of refraction, 577
of light waves, 576, 632, 668
measuring, 643
of sinusoidal waves, 567–68
of sound waves, 574–75
Wave model of light, 629, 641–42, 662, 707–09
Wave number, 569
Wave optics, 627–654
circular-aperture diffraction, 640–42
diffraction grating, 634–36
interference of light, 629–34
single-slit diffraction, 636–40

Wave packets, 1166–69
photons as, 1134
uncertainty about, 1168–71
Waves, 559–590, 716. *See also* Electromagnetic waves; Light waves; Sinusoidal waves; Sound waves; Standing waves; Traveling waves
amplitude of, 566
circular, 572
displacement, 566
Doppler effect, 580–83
frequency of, 566
longitudinal, 565
matter waves, 1134–36
medium of, 561
phase of, 573–74
plane, 572–73
power, intensity, and decibels, 578–80
sinusoidal, 566–72
spherical, 572
traverse, 561
Weber (Wb), 970
Weight, 119, 138, 146–47
gravitational force and, 357–58
mass vs., 146
Weightlessness, 147–48, 200
Wien's law, 1104
Work, 278–309, 470, 811, 818
basic energy model, 279–80
calculating and using, 282–86
heat and, 471, 476, 490, 527–29
in ideal-gas processes, 471–75
and kinetic energy, 280–82
and potential energy, 288–90, 362
Work function, 1127
Work-kinetic energy theorem, 281–82, 289, 294

X

X-rays, 1108, 1269

Y

Young's double-slit experiment, 629–34, 668
Young's modulus, 431–32

Z

Zero-point motion, 1190
Zero vector, 8, 73
Zoom lenses, 697–98